High-resolution Seafloor Survey and Applications

Ziyin Wu • Fanlin Yang • Yong Tang et al.

High-resolution Seafloor Survey and Applications

Science Press
Beijing

Springer

Ziyin Wu
Key Laboratory of Submarine Geosciences
Second Institute of Oceanography
Ministry of Natural Resources
Hangzhou, Zhejiang, China

Yong Tang
Key Laboratory of Submarine Geosciences
Second Institute of Oceanography
Ministry of Natural Resources
Hangzhou, Zhejiang, China

Fanlin Yang
College of Geodesy and Geomatics
Shandong University of Science and Technology
Qingdao, Shandong, China

ISBN 978-981-15-9749-7 ISBN 978-981-15-9750-3 (eBook)
https://doi.org/10.1007/978-981-15-9750-3

Jointly published with Science Press.
The print edition is not for sale in China (Mainland). Customers from China (Mainland) please order the print book from: Science Press
ISBN of the Co-Publisher's edition: 978-7-03-066031-2

This Springer imprint is published by the registered company Springer Nature Singapore Pte Ltd.
The registered company address is: 152 Beach Road, #21-01/04 Gateway East, Singapore 189721, Singapore

Preface

Oceans cover 71% of the earth's surface area and are rich in natural resources. Historically, difficult physical conditions have largely prevented humans from large-scale development of the ocean as they have done on land. However, the first decade of the twenty-first century has seen humans accelerate the pace of exploring, developing and utilizing marine resources, motivated by increased economic development, population expansion and shortages of on-land resources. The most basic parameter for understanding the ocean and making use of the ocean's resources is seafloor topography. Nearly all ocean-related activities depend on a knowledge of seafloor topography, especially in terms of navigational safety. However, accurate determination of seafloor topography remains an ongoing challenge. This is especially true in the deep sea where typical mapping resolution is a quarter less than that of the moon. To date, 80% of the world's seafloor topography is yet to be mapped using the modern multi-beam sounding technology.

Because the thick layer of intervening ocean inhibits the use of conventional optical and electromagnetic methods, acoustical techniques remain the principal means for detecting and developing understandings of seafloor topography. Commonly used acoustical techniques include multi-beam echo sounders, side-scan sonars and sub-bottom profilers. Multi-beam echo sounders are typically used for determining ocean depths and developing maps of seafloor topography, while side-scan sonars are most commonly used to develop images of the seafloor. Sub-bottom profilers make use of low-frequency acoustic signals to penetrate strata tens or even hundreds of meters below the seafloor, to understand the distribution of sediments and shallow strata. A comprehensive summary and discussion, such as that contained in this monograph, of the basic principles and methods as well as recent technological developments related to these techniques can provide practical assistance for marine exploration practitioners.

Multi-beam echo sounders, side-scan sonars, sub-bottom profilers and other seafloor detection equipment consist of highly integrated arrays of sensors. With the rapid development and integration of navigational, acoustical, electronic and computing techniques, seafloor topographic surveys are characterized by high precision, high density, high resolution data that allow consideration of multiple error sources. Improvements in technical hardware are accompanied by rapid improvements in data processing techniques. Research into data processing and related applications of these techniques is necessary because raw survey data contain errors emanating from multiple sources including self-noises, as well as different parameter settings and marine environments. This monograph contains a summary of relevant data processing techniques and methods that will contribute to the continued development of marine information technology.

The data obtained by multi-beam echo sounders, side-scan sonars and sub-bottom profilers contain abundant marine information. Multi-beam echo sounding includes a variety of information such as water depth, water column properties and bottom acoustic scattering. Mining of these data can be used for the seafloor geomorphology research, the acoustic sediment classification and the target identification in the water column. Side-scan sonar images reveal the fine-scale features of seafloor micro-geomorphology. Sub-bottom profiles reveal sedimentary strata within the top tens of meters of the seafloor, which can be used to detect and identify targets buried below the seafloor including applications in underwater archaeology. Developments in computer science and information technology, including artificial intelligence and machine learning methods, can also be applied for deep data mining and visualization of seafloor topographic data.

Because they play a fundamental role in revealing the basic characteristics, evolution and dynamics of seafloor geology, seafloor topography are the most basic starting points for the submarine geosciences and they provide basic information for exploring and researching submarine resources. Since the 1980s, China has conducted international surveys of submarine polymetallic nodules and has successfully applied it to a seafloor mining area of 75 000 km^2. In the 1990s, China additionally conducted investigations and submitted mining applications for cobalt-rich crusts and seafloor massive sulfide deposits located in international waters. In the past 10 years, gas hydrate resources have been discovered in the South China Sea. Studies demonstrate that the formation and occurrence of these submarine resources are closely related to seafloor topography. The delimitation of the continental shelf in accordance with the United Nations Convention on the Law of the Sea is an additional matter of great concern to coastal states. Various demarcation points, such as the foot of the continental slope, are closely related to seafloor topography. Therefore, an adequate accounting of marine geological resources requires increased study of seafloor topography, resources and evolution.

Based on the above considerations, this monograph presents detailed research on four aspects of seafloor topography: detection technology, data processing technology, data mining and geomorphological research. Professors Wu Ziyin and Yang Fanlin present a detailed outline of this monograph, with a total of 12 chapters. The Chap. 1 was written by Dr. Wang Shenping et al., the Chap. 2 was written by Dr. Zhao Dineng et al., the Chap. 3 was written by Dr. Su Dianpeng et al., the Chap. 4 was written by Prof. Ding Weifeng et al., the Chap. 5 was written by Dr. Luo Xiaowen et al., the Chap. 6 was written by Dr. Ying Jianyun et al., the Chap. 7 was written by Prof. Wu Ziyin et al., the Chap. 8 was written by Dr. Jin Shaohua et al., the Chap. 9 was written by Prof. Huo Guanying, the Chap. 10 was written by Dr. Shang Jihong et al., the Chap. 11 was written by Prof. Tang Yong et al., and the Chap. 12 was written by Dr. Li Huaiming et al. In the process of writing the manuscript, we have benefited greatly from the input and contributions of Academician Jin Xianglong, Academician Li Jiabiao, Prof. Wang Xiaobo and Prof. Zheng Yulong. The preface was written by Prof. Wu Ziyin and Prof. Yang Fanlin, and the monograph was drafted by Prof. Wu Ziyin and Prof. Yang Fanlin.

This monograph has been funded by a series of Chinese research projects supported by several institutions including the National Natural Science Foundation of China (41830540, 41906069, 41930535, 42006073, 52001189), the Scientific Research Fund of the Second Institute of Oceanography, MNR (JG2005, SZ2002), the Zhejiang Provincial Natural Science Foundation (LY21D060002), and the National Key R&D Program of China (2018YFF0212203). English-version publishing is supported by the China International Book Promotion Program. We would additionally like to express our sincere gratitude to Wu Dongqiang, Chen Jianbing, Tu Lunze, Liu Hui, Wang Yubin, Wang Jiachong, Fu Guihe, Tu Zejie, Zhu Zhengren, Wang Min, Xu Qiyao, Anokey Michael, Feng Chenkai, Xin Mingzhen, Han Bing, Li Donghui, Qu Meng, Liu Yaming and Sun Yanfei, who have participated in the editing and translation work.

The surveying of seafloor topography is a multi-disciplinary exercise that involves the integration of information from a variety of technologies. Its theory and methodology are closely related to the means of acquiring these data, and the relevant technologies continue to develop in concert with the relevant scientific disciplines. Due to the limitations of the authors, some omissions and deficiencies in the book are unavoidable. Experts and readers are invited to criticize and correct these works as they see fit.

Hangzhou, China
Qingdao, China
September 2019

Ziyin Wu
Fanlin Yang

Author List

Wu Ziyin, Yang Fanlin, Tang Yong, Luo Xiaowen, Zhao Dineng, Huo Guanying, Zhang Kai, Zhou Jieqiong, Jin Shaohua, Su Dianpeng, Li Xiaohu, Li Huaiming, Wang Shengping, Cao Zhenyi, Ying Jianyun, Shang Jihong, Ding Weifeng, Xu Xuefeng, Zhu Xinke, Zhang Weiyan, Gao Jinyao, Yang Kehong, Liang Yuyang, Ma Weilin, Wang Mingwei, Li Shoujun, Liu Yang, Liu Hongxia, Qi Chao, Liu Zhihao, Zhu Chao, Li He.

Technical Issue Group

Zhao Dineng, Zhang Kai, Qi Chao, Zhou Jieqiong, Su Dianpeng, Wu Dongqiang, Wang Mingwei, Chen Jianbing, Liu Zhihao, Liu Yang, Tu Lunze, Zhu Chao, Liu Hui, Wang Yubin, Wang Jiachong, Fu Guihe, Tu Zejie, Zhu Zhengren, Wang Min, Xu Qiyao, Anokey Michael, Feng Chengkai, Xin Mingzhen, Han Bing, Li Donghui, Qu Meng, Liu Yaming, Sun Yanfei.

Contents

Contributors List

1. Overview of Bathymetric Surveying Technology
   Wang Shengping, Yang Fanlin (E-mail: flyang@sdust.edu.cn), Zhu Xinke and Wang Mingwei
2. Multi-beam Bathymetric Technology
   Zhao Dineng, Yang Fanlin (E-mail: flyang@sdust.edu.cn), Wu Ziyin, Zhu Chao and Liu Hongxia
3. Airborne Laser Bathymetric Technology
   Su Dianpeng, Yang Fanlin (E-mail: flyang@sdust.edu.cn) and Qi Chao
4. Side-scan Sonar and Sub-bottom Profiler Surveying
   Ding Weifeng, Zhao Dineng (E-mail: zhaodineng@sio.org.cn), Wang Mingwei and Liu Zhihao
5. Navigation and Positioning Technology
   Luo Xiaowen (E-mail: cdslxw@163.com) and Wang Shengping
6. Tide Level Measurements and Vertical Datum Conversion
   Ying Jianyun, Cao Zhenyi (E-mail: zycao@sio.org.cn), Xu Xuefeng, Zhou Jieqiong and Yang Fanlin
7. Constructing and Mapping a Seabed DTM
   Wu Ziyin, Yang Fanlin, Zhou Jieqiong (E-mail: janechou@sio.org.cn), Gao Jinyao, Wang Mingwei and Liu Yang
8. Acoustic Seafloor Characterization
   Jin Shaohua, Zhang Kai (E-mail: zk0773@163.com), Wang Mingwei and Wu Ziyin
9. Intelligent Detection and Recognition of Seabed Targets in Side-scan Sonar Images
   Huo Guanying (E-mail: huoguanying@hhu.edu.cn)
10. Applications of Submarine Geomorphology
    Shang Jihong, Liang Yuyang, Li Shoujun, Zhang Kai and Wu Ziyin
    (E-mail: zywu@sio.org.cn)
11. Technology and Practice of Continental Shelf Delimitation
    Tang Yong, Wu Ziyin (E-mail: zywu@sio.org.cn), Zhou Jieqiong and Li He
12. Mineral Resources Assessment of the International Seabed
    Li Huaiming, Li Xiaohu (E-mail: xhli@sio.org.cn), Zhang Weiyan, Wu Ziyin, Yang Kehong and Ma Weilin

About the Authors

Ziyin Wu, Ph.D. second-grade professor of Marine Geology and Geophysics. He is the director of the Key Laboratory of Submarine Geosciences (KLSG) at the Second Institute of Oceanography, Ministry of Natural Resources, Hangzhou, China. He has been engaged in high-resolution submarine geomorphology since 1995. His research focuses on the processing theories and applications of multi-beam echo sounder (MBES), and the studies of high-resolution geomorphic features and evolution based on MBES's data. So far, Prof. Wu has made a series of scientific achievements, including 2 monographs as editor-in-chief, 10 co-authored books, 107 journal papers, 37 authorized China and US patents, and 14 science and technology awards. He has also supervised 4 postdoctoral fellows, 6 Ph.D. students, and more than 10 master students. E-mail: zywu@sio.org.cn.

Fanlin Yang, Ph.D. professor of Marine Geodesy. He is the dean of College of Geodesy and Geomatics, Shandong University of Science and Technology, Qingdao, China. His research focuses on the marine geodesy and navigation. More specifically, his research topics include data processing and applications of multi-beam echo sounder system and airborne LiDAR bathymetry system, and technologies of marine navigation and positioning. So far, Prof. Yang is the (co-)author of 50 papers in refereed scientific journals (Web of Science), (co-) inventor of 22 China and US patents, (co-)winner of 7 science and technology awards in China, and 3 standards as editor-in-chief. He has also supervised 2 postdoctoral fellows, 55 Ph.D. and master students. E-mail: flyang@sdust.edu.cn.

Yong Tang, Ph.D. professor of Marine Geology and Geophysics. He works in the Key Laboratory of Submarine Geosciences (KLSG) at the Second Institute of Oceanography, Ministry of Natural Resources, Hangzhou, China. He is also a member of the Commission on the Limits of the Continental Shelf, United Nations. Professor Tang has been involved in the work on the delimitation of continental shelf over the past 15 years. As the chief scientist and project leader in China-Nigeria, China-Mozambique, China-Seychelles joint cruise and research project, he has contributed himself to the international cooperation on marine geophysics and delimitation of continental shelf. He is also the project leader in the National Key R&D Program, National Natural Science Foundation and National High Technology R&D Program of China. So far, Prof. Tang has published more than 30 scientific papers, 3 monographs and 3 patents. He also has supervised 2 Ph.D. students, 8 master students, 4 bachelor students and 2 international scholars. E-mail: tangyong@sio.org.cn.

1 Overview of Bathymetric Surveying Technology

Bathymetric surveying technology can be divided into four types according to the measurement platform: shipborne measurement (conventional vessels and unmanned vehicles), airborne and spaceborne measurement, underwater autonomous navigation measurement and seafloor in situ observations. This chapter presents an overview of underwater geomorphology detection and imaging instruments and technologies to give readers a comprehensive understanding of the relevant developments in this field. In the following chapters, several widely-used detection technologies are described in detail, including the multi-beam echo sounder (MBES) system, airborne light detection and ranging (LiDAR) bathymetry (ALB) system, side-scan sonar (SSS) system and sub-bottom profile (SBP) system. This information can provide an important reference for researchers and workers in the relevant areas.

1.1 Shipborne Bathymetric Surveying Technology

Bathymetry is the core work of shipborne seafloor mapping and it is the most direct way of bathymetric surveying. Light and electromagnetic waves decay rapidly in water, while sound waves can travel long distances. Shipborne acoustic detection is still one of the main methods of bathymetric surveying. The Global Navigation Satellite System (GNSS) is an accurate and efficient positioning technology used for navigation at sea. The widely-used "GNSS plus detector" bathymetry method is based on the principle of acquiring the ship's coordinates via GNSS navigation and simultaneously obtaining depth data, reflection intensity or submarine images using detection equipment.

1.1.1 Shipborne Detection Technology

Early bathymetry was done by the lead line and was inefficient. In 1913, the Canadian scientist Reginald Aubrey Fessenden invented the echo sounder which had a detection range of 3.7 km. In 1918, French physicist Paul Langevin created modern hydroacoustics and invented the sonar which utilized a sandwich-type transducer to detect underwater long-distance targets, receiving the echo of a submarine for the first time. In the 1960s, the MBES emerged with the introduction of computer technology; modern MBESs with high-precision, high-efficiency, automation and digitization were developed. The data collection mode changed from a single point to a continuous line and finally to a 3D surface (Li 1999).

The main detection instruments used in the bathymetric surveying are the single-beam echo sounder (SBES), MBES, phase-measuring bathymetric side-scan (PMBS) sonar, 3D laser scanning system, dual-frequency identification sonar, synthetic aperture sonar (SAS) and 3D panoramic sonar.

1.1.1.1 Early Bathymetry Methods

The bamboo pole was the first bathymetry tool, and later it developed into a rope thrown into the water with a heavy object at one end. In the mid-15th century, Nikola Coussa invented a simple hydraulic sounder that estimated the water depth based on the amount of water pressure. The Bruker-type sounder was developed around 1851, and later the Siegsby-type sounder and the popular Kelvin sounder appeared. In 1891, British Telecom launched the Lucas-type sounder which was inefficient and could only work in one point or line.

1.1.1.2 Single-Beam Echo Sounder

A significant improvement in marine survey work was the invention of the SBES in the 1920s. Its emergence was a leap in ocean bathymetry technology, with its high speed and continuous recording. Single-beam echoes bathymetry records continuously along a line, thus producing a terrain

Z. Wu et al., *High-resolution Seafloor Survey and Applications*,
https://doi.org/10.1007/978-981-15-9750-3_1

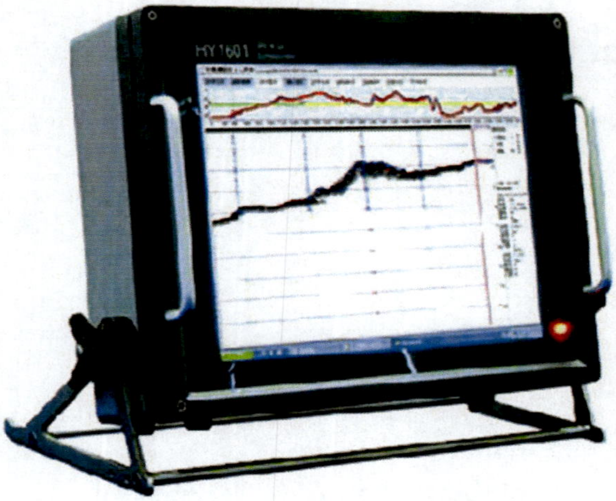

Fig. 1.1 HY1601 single-beam echo sounder

section line. SBES can operate in either single-frequency or dual-frequency mode. The single-frequency sounder is light and emits only one frequency band (low or high), while the dual-frequency sounder emits both high- and low-frequency signals. Bying utilizing these characteristics, the thickness of the sludge layer on the ocean bottom can be acquired by measuring the difference between the distances from the water surface to the bottom surface and the hard stratum surface. The traditional SBES has two shortcomings: firstly, only the points along the survey line are sampled, which represents only a small part of the survey area; secondly, the beam is wide, which can lead to a large depth error when measuring the complex terrain. The SBES is widely used in rivers and shallow sea surveys for its low cost and convenient operation. Figure 1.1 shows an HY1601 SBES.

1.1.1.3 Multi-beam Echo Sounder

The MBES that emerged in the 1970s was a new concept that profoundly changed ocean surveying methods and the quality of the final results. A MBES can produce a 3D map of the seafloor. It can measure the size, shape and height of the underwater target accurately and efficiently within a certain width (3–12 times the water depth). The MBES records thousands of depth points in a plane perpendicular to the track of the surveying vessel at one time (Zhao et al. 2007). Unlike the single-beam echo sounder, the MBES needs to operate with auxiliary equipment, including an attitude indicator, gyro, surface sonic velocimeter, sound velocity profiler and GNSS receiver. This combination of instruments can provide instantaneous position, altitude, heading, speed of sound and other information.

The development of the MBES originated in military research projects funded by the US Naval Research Agency in the 1960s (Li 1999). In 1976, digital computer processing and control hardware was applied to the MBES, resulting in the first MBES system-SeaBeam, which had 16 beams; the lateral measurement range of SeaBeam was about 0.8 times the water depth. During the 1980s and 1990s, various shallow, medium-depth and deep-water MBESs were produced. Figure 1.2 shows the German dual-frequency MBES —the Elac Bottom Chart 1180/1050D, for shallow and medium-depth seafloor mapping. With the widespread use of electronics, computers, new materials and new processes, MBES technology has improved in accuracy, resolution, integration and modular technology, resulting in smaller instruments and more accurate results. See Chap. 2 for a detailed description of multi-beam echo sounding technology.

Fig. 1.2 Dual-frequency MBES, Elac Bottom Chart 1180/1050D

1.1.1.4 Side-Scan Sonar

The appearance of SSS technology can be traced back to near the end of World War II, but it was not used until the late 1950s. In the early 1960s, SSS was developed for commercial use and was adopted for use around the world in the late 1960s. It is based on the echo-sounding principle for underwater target detection where an array of transducers mounted on the bottom of the ship's hull emits the scanning beams which are wide in the downward direction and narrow in the horizontal direction at a certain tilt angle and transmission frequency. After the sound wave propagates to the seafloor or the target, it is reflected and scattered back to the receiving array of the transducer, then processed and stored in the processing unit.

The working frequency of the SSS determines the maximum scanning range and the pulse width directly affects the resolution (Table 1.1). In general, the smaller the width, the higher the resolution. The horizontal beam angle directly affects the horizontal resolution. The angle of the downward beam affects the width of the area covered by the side-scan sonar. The larger the angle, the larger the coverage and the smaller the blind area directly below the sonar.

1.1.1.5 Phase-Measuring Bathymetric Side-Scan Sonar

In 1960, the British Marine Science Institute developed the first SSS for submarine geological surveys. In the mid-1960s, the SSS technology was improved in resolution and image quality. The transducers were mounted on a "tow fish" and towed at depth behind the ship on a cable. In the 1970s, SSSs adapted for specific uses were developed. In the 1990s, the British company Submetrix launched the ISI100 PMBS sonar system for high-density, high-precision measurement of submarine geomorphology. The PMBS sonar measured the seabed depth and produced a 3D map of the surveyed area, an isobath map and a side-scan sonar image (Liu 2001).

This type of device has three characteristics, as follows.

(1) High-density data is collected, generally 2 000–6 000 echo bands (echo angles) on each side of the emission pulse, which is equivalent to a water depth of 7.55 mm on the seafloor.
(2) The transducer array covers a wide area, and the effective width of the surveyed area can reach 10–15 times the water depth.

Table 1.1 Commonly-used SSSs and their performance parameters

Model		Frequency/kHz	Horizontal wave beam angle/(°)	The longest cable length/m	Unilateral range/m	Tow fish size diameter length/cm	Tow fish weight/kg	Operating depth/m
Sonar Beam S-150D		100	1.2	/	1 000	11.2 × 1 360	32	<300
		400	0.3		300			
Klein 3 900		445	0.21	/	150	8.9 × 122	29	25–1 000
		900			50			
Klien 3 000		100	0.70	/	600	8.9 × 122	29	25–1 000
		500	0.21		150			
EdgeTech4100P		100	1.2	1 500	500	11.4 × 140	12	1 000
		500	0.5		200			
Edge Tech 4200-MP	HDM model	100	0.64	6 000	500	/	/	2 000
		400	0.30		150			
	HSM model	100	1.26		500			
		400	0.40		150			
SeaProbe D-S		100/400	0.80	6 000	50-400	26 × 29 × 140	24	Standard 2 500
CS-1		100	1.00	500	500	/	/	3 00
		500	0.4		100			
Benthos C3D		200	5 concurrent angles	10 000	25–300	/	/	2 000
Jwfishers		100	1.50	/	450	10.2 × 140	27	152.4
		600			60			

(3) Swath bathymetry and SSS can collect more seafloor geomorphic information. Compared with the traditional SSS, it can provide the precise coordinates of each pixel, which improves the processing and analysis of the data.

Interferometric sonar is suitable for bathymetry scanning of ports and waterways, so it can replace the sounder and SSS. The main models are the Klein 3500 and GeoSwath Plus (Fig. 1.3, Table 1.2). However, in water depths beyond 200 m, the accuracy of the interferometric sonar is not as good as that of traditional MBES, especially in turbid estuaries and in channels.

1.1.1.6 Dual-Frequency Identification Sonar

The high-resolution dual-frequency identification sonar uses an audio "lens" to produce high-quality images in dark and turbid water, where optical cameras are unusable. The high-resolution dual-frequency identification sonar emits sound waves of two frequencies underwater. When the sound wave encounters an object, it is reflected back and received by the system. After processing of the acoustic image signal, the image of the object is displayed on the display screen. The acoustic imaging system consists of three acoustic lenses and an array transducer. The array transducer has a spatial resolution which can recombine the image based on the strength and time-delay of the reflected signals. Common high-resolution dual-frequency identification sonars include the ARIS and DIDSON models developed by the University of Washington. Figure 1.4 shows the underwater image produced by the DIDSON sonar, and Table 1.3 compares the basic performance of the two models of the DIDSON dual frequency identification sonar.

1.1.1.7 Underwater 3D Panoramic Scanning Sonar

The 3D imaging sonar is similar to a 3D laser scanner. The sonar head emits a sound wave of a fixed frequency. The beam propagates through the water to reach the surface of the object and is reflected back to the sonar head, which converts the signal into an electrical signal. This electric signal is then transmitted to the sonar control unit. The information of the scanned surface is displayed in the form of an image using the operation software of the sonar. Each time the sonar probe emits an acoustic pulse, a

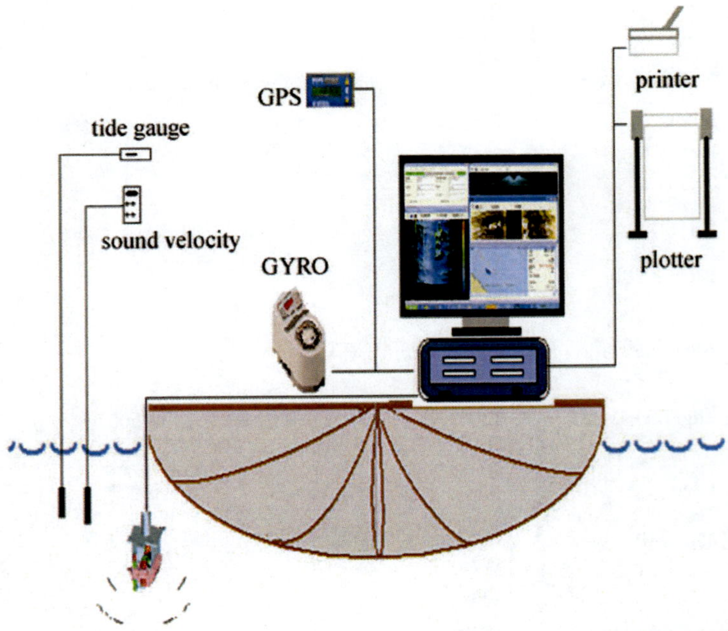

Fig. 1.3 The GeoSwath Plus bathymetry SSS system

Table 1.2 Commonly-used bathymetry SSSs and their performance parameters

Model	Weight in air/kg	Operating frequency/kHz	Wave length/μm	Horizontal opening angle/(°)	Vertical opening angle/(°)	Operating depth/m
Klein 3500	29	100/500	25–400	0.7	40	<2 000
GeoSwath Plus	24	125	128–896	0.85	120	<200
ATLAS FANSWEEP20	30	200	60–350	1.3	161	250–600

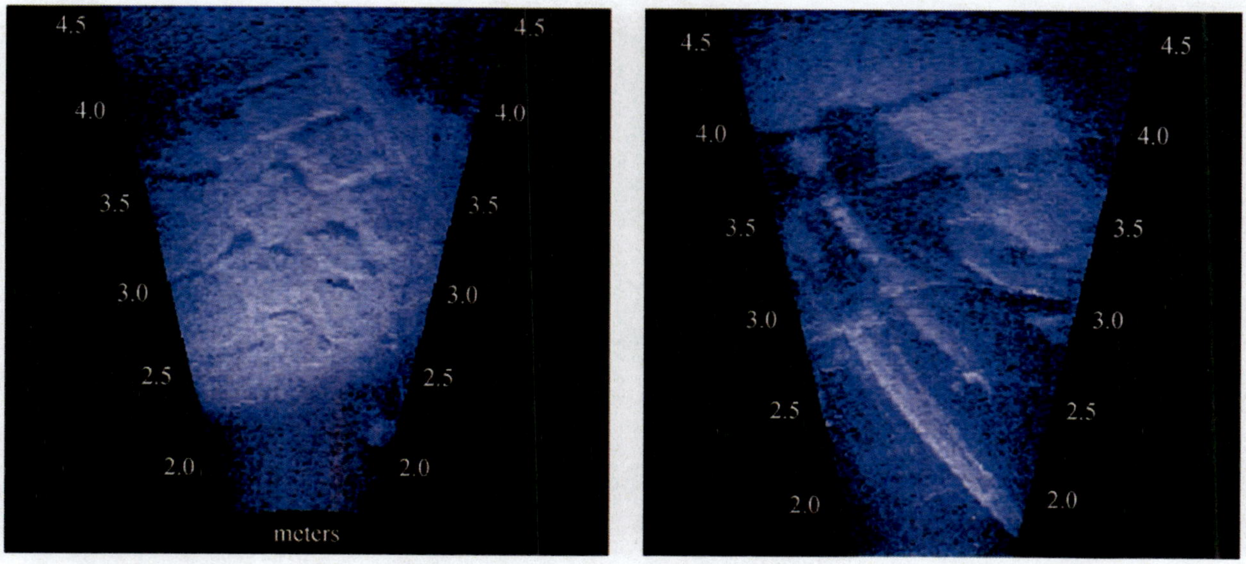

Fig. 1.4 Underwater imaging recorded by the dual-frequency identification sonar DIDSON

Table 1.3 Performance of the DIDSON dual frequency identification sonar

Item	Standard dual frequency recognition sonar		Remote dual frequency identification sonar	
	Low frequency	High frequency	Low frequency	High frequency
Operating frequency/MHz	1.1	1.8	0.7	1.2
Beam width (two-way)/(°)	Horizontal 0.4 Vertical 14	Horizontal 0.3 Vertical 14	Horizontal 0.8 Vertical 14	Horizontal 0.5 Vertical 14
Wave beam	48	96	48	48
Starting distance/m	0.75–23.25	0.38–11.63	0.75–23.25	0.38–11.63
Interval/m	0.75	0.38	0.75	0.38
Detection distance/m	4.5, 9, 18, 36	1.13, 2.25, 4.5, 9	9, 18, 36, 72	2.25, 4.5, 9, 18
Sampling interval in distance/mm	8, 17, 35, 70	2.2, 4.4, 9, 18	17, 35, 70, 140	4.4, 9, 18, 36
Pulse width/μs	16, 32, 64, 128	4, 8, 16, 32	23, 46, 92, 184	7, 13, 27, 54
Frame rate/s^{-1}	4–21		2–10	
Horizontal viewing angle/(°)	29		29	
Power consumption/W	30 (AC115V or DC14–18V)			
Control	Ethernet			
Output form	Ethernet and NTSC			
Size/cm	length 30.7; height 20.6; width 17.1			
Air weight/kg	7			
Water weight/kg	0.6			

scanning sector is formed, and the coordinates of the measuring points can be obtained. The pan and tilt mechanism can scan in a wide range: 360° in the horizontal plane and 130° in the vertical plane. The scanning depth can reach 300 m, with an error of less than 4 cm at a 30-cm range, and an angle error of 1°. The most common 3D scanning sonars currently in use are the BV5000-1350 and BV5000-2250, developed by BlueView Corporation, US (Fig. 1.5, Table 1.4).

1.1.1.8 Synthetic Aperture Sonar

The synthetic aperture line array sonar performs time-sampling of signals reflected from the target as the receiving array is towed above the target, and then processes the data to form a spatial image of the target. The key technologies of SAS imaging include attitude control of the towed array, real-time time-delay correction, inverse scattering image reconstruction, and processing techniques for

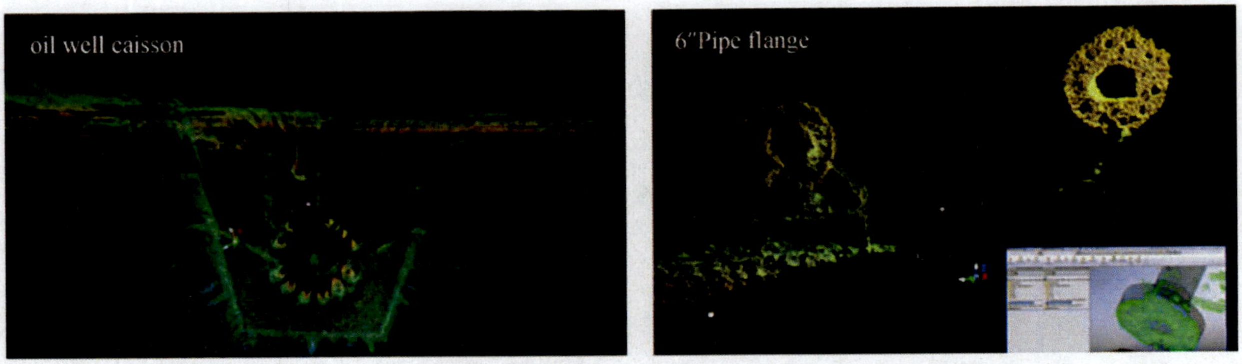

Fig. 1.5 Images of deep sea drilling platform and flange recorded by 3D scanning sonar BV5000-1350; insert shows flange diagram

Table 1.4 3D scanning sonar BV5000 performance parameters

Model	BV5000-1350		BV5000-2250	
Sonar				
Maximum range/m	30		10	
Optimum range/m	1–20		0.5–7	
Beam width	1° × 1°		1° × 1°	
Beam spacing	0.18°		0.18°	
Working frequency/MHz	1.35		2.25	
Interface				
Supply voltage	120–240 VAC	24 VDC	120–240 VAC	24 VDC
Power consumption	Maximum power consumption 45 W			
Data interface	Ethernet/USB	Ethernet/RS485	Ethernet/USB	Cthernet/RS485
Mechanical properties				
Weight in air/kg	22.2	12	20.7	10.5
Weight in water/kg	9.6	4.5	8.4	3.3
Maximum depth/m	300	3 000	300	3 000

acoustic signal fluctuation caused by temporal and spatial changes of the ocean medium. SAS achieves the resolution of a large aperture array with a small aperture array. It has the typical advantage of "constant resolution", which means the resolution of the SAS image does not diminish as the distance increases. Representative products include the SASs developed by the United States DARPA and CEROS, the European Union's Synthetic Aperture Mapping and Imaging (SAMI) sonar, and France's IMBAT3000 SAS. With the support of the National High Technology Research and Development Program (863 Program) of China, the Institute of Acoustics of the Chinese Academy of Sciences successfully developed the first SAS with independent intellectual property rights in China in 2000. At present, SAS can achieve underwater image resolutions of 10-30 cm when the water depth is about 400 m (Fig. 1.6, Table 1.5).

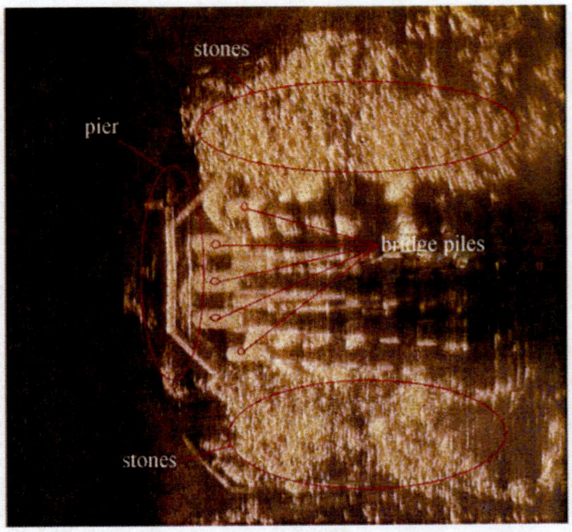

Fig. 1.6 Submarine small artificial target detected by SAS

Table 1.5 Common SASs and their performance parameters

Parameter	UCSB	SAMI	KIWI	DARPA	CEROS	SharkSAS-MF
Center frequency/kHz	600	8	30	50	12.5	15–25
Bandwidth/kHz	4	6	20	10	8	10
Heading resolution/cm	5	100	15	10	4	10
Distance resolution/cm	19	13	5	7.5	10	8
Transmit beam width/(°)	1.5	5	10	8	84	Horizontal 7, vertical 47
Typical working distance/m	70	100–2 500	50	50–1 000	75	Up to 300

1.1.1.9 Mobile 3D Laser Scanning System

A 3D laser scanning system is formed by integrating 3D laser scanning (LS) equipment with a GNSS module, inertial measurement unit, odometer, 360° panoramic camera, assembly control module and high-performance board computer, and attaching them to a rigid platform. Long-distance non-contact measurement can be realized by 3D laser scanning to achieve inter-tidal zone measurements (Fig. 1.7). The plane position of the tidal flat measurement angle using a shipborne mobile 3D laser scanning can reach 10–15 cm, and the elevation position can reach 15–20 cm. The main 3D laser scanners for shipborne mobile measurement are the MDL Dynascan, RIEGL VZ400, Leica HDS8800, FARO Focus3D and Zhonghaida iScan (Table 1.6).

1.1.2 Shipborne Positioning Technology

Seafloor detection in bathymetric surveying requires accurate plane positioning of the survey point, usually obtained from the ship's position. Therefore, accurate ship positioning is an important part of seafloor mapping.

The early positioning methods were mainly based on optical positioning and land-based radio positioning, which were imprecise and cumbersome. Since the end of the 20th century, with the rapid advancement of GNSS technology, marine positioning technology has made breakthrough progress. The widely-used GNSS high-precision positioning technology includes differential point positioning and precise point positioning. The principle of optical positioning at sea is based on the intersection method, similar to that of land positioning systems. Examples are the forward intersection method and the resection method, which were widely used in the 1960s and 1970s.

Radio positioning includes both land-based radio and space-based radio. Land-based radio positioning systems were developed at the beginning of the 20th century. The main part of the system is the ground navigation platform. This method can provide continuous positioning in all weather conditions at long distances, from tens of kilometers to thousands of kilometers. The long-range positioning systems use various radio navigation systems that work on different principles. Different navigation systems (for example, Roland C and Omega) have different data accuracies (Liang 1995).

Fig. 1.7 Shipborne 3D laser operation (left) and scan results (right)

Table 1.6 Commonly-used 3D laser scanners and their performance parameters

Parameter	MDL Dynascan	RIEGL VZ400	Leica HDS8800	FARO Focus3D	Zhonghaida iScan
Measuring distance/m	500	600	2000	153.49	500
Scanning accuracy	±5 cm	5 mm (within 100 m)	10 mm (within 200 m)	2 mm (within 25 m)	±10 cm
Laser divergence		0.3 mrad	0.25 mrad	0.19 mrad	
Scanning view	360°	Vertical 100° × horizontal 360°	Vertical 80° × horizontal 360°	Vertical 305° × horizontal 360°	360°
Acquisition rate/points·s^{-1}	500 000	300 000	8 800	976 000	
Operating temperature/°C	−20 to 60	0 to 40	0 to 50		
Storage temperature/°C	−20 to 70	−10 to 50	−20 to 50	5 to 40	
Protection level	IP65	IP64	IP65	IP65	

The global navigation and positioning technology are one of the most prevalent space-based radio positioning technologies. It emerged in the 1970s and is currently the most widely used and effective method for maritime positioning. GPS single-point positioning is influenced by many factors, such as satellite ephemeris error, ionospheric refraction error and the multipath effect; and its positioning accuracy is 5–20 m. Therefore, it is not suitable for high-precision positioning and navigation. Early GPS differential methods used in China's coastal areas include beacon differential and GPS real-time kinematic (RTK) technology. The Radio Beacon-Differential Global Position System (RBN-DGPS) was established by the China Maritime Safety Administration during 1995–2000 and covers 20 coastal areas of China. GPS RTK is a carrier phase real-time dynamic differential positioning system with an accuracy of a few centimeters. At present, the rapidly developing global navigation technology is available through the United States' GPS, China's BeiDou Navigation Satellite System (BDS), Russia's Global Navigation Satellite System (GLONASS) and the European Union's Galileo. Differential technologies have developed from single-base station differential technology to network RTK technology, and precision single-point positioning technology has also emerged. The accuracy of precise point positioning can reach decimeter or even centimeter scale (Li and Huang 2005). Because it is not limited by the distance to the base station, it has great potential in marine mapping.

GPS positioning technology is detailed in Chap. 5.

1.1.3 Unmanned Shipborne Measurement Platforms

In addition to traditional surveying ships, the current unmanned vehicle-based autonomous measuring platform is also an important means of bathymetric surveying, especially in areas that are difficult to access such as shallow water, reef areas or areas containing hazardous chemicals. An unmanned vessel can be equipped with conventional measuring instruments such as a shallow-water MBES, SSS, SBES and Acoustic Doppler Current Profiler (ADCP). It can work efficiently in shallow water and work collaboratively with the mother ship (Fig. 1.8). In recent years, there has been substantial development in the field of automatic acquisition both in China and worldwide. Intelligent unmanned vehicles are being produced with high-precision GPS-guided, ultrasonic automatic obstacle-avoidance technology, real-time remote data link technology, and composite material technology. In addition, they are equipped with different kinds of detection loads to acquire data automatically.

The unmanned vehicle platform includes many functions such as unmanned remote control, GPS automatic navigation, autonomous navigation and automatic obstacle avoidance control, which can work outside the line of sight. The recorded data, including video footage taken by the unmanned vehicle camera, can be transmitted back in real-time through the system's broadband private network

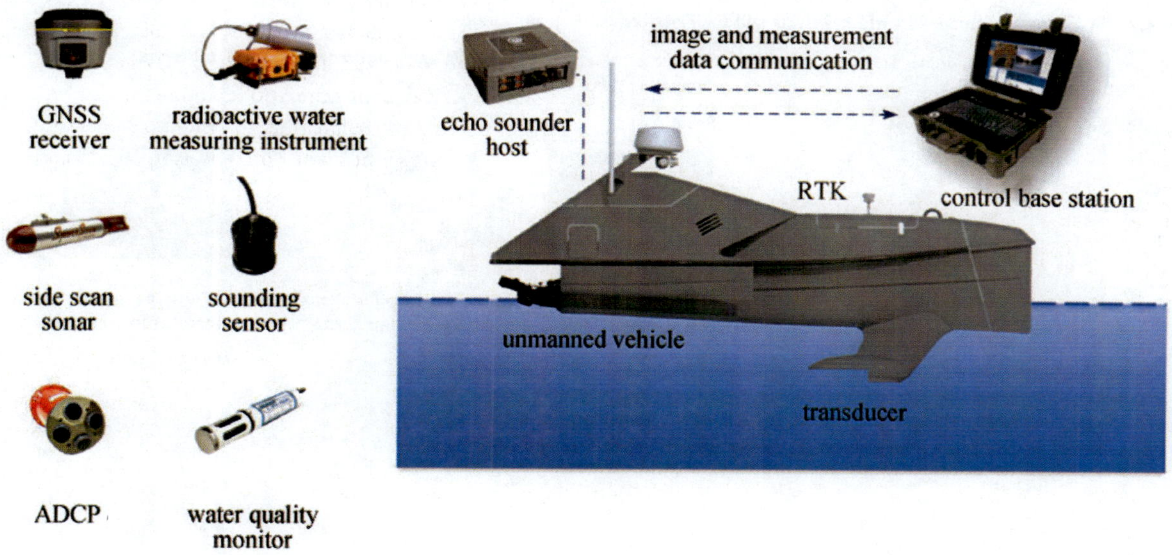

Fig. 1.8 Unmanned vehicle system setup and components (provided by Zhuhai Yunzhou Company)

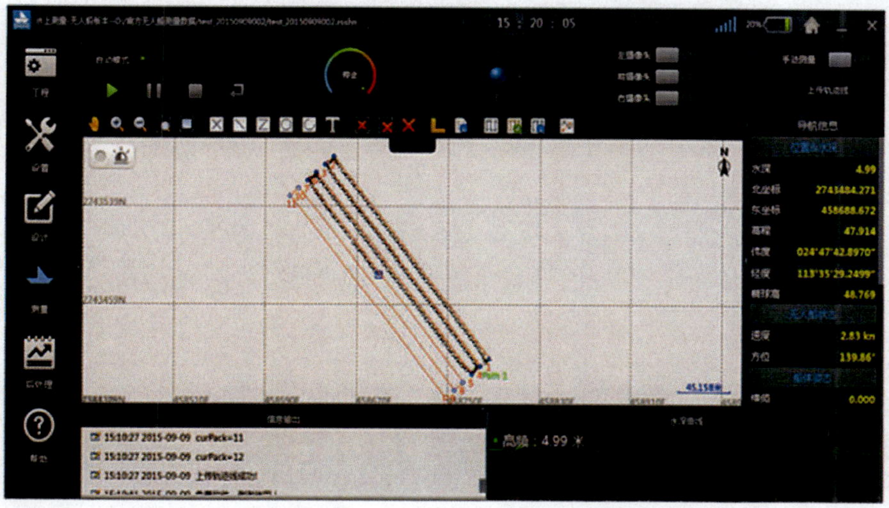

Fig. 1.9 Monitoring interface of unmanned vehicle showing planned survey track (red line) and actual track (black line) (provided by Zhuhai Yunzhou Company)

which can reach 2 MB/s within 15 km. Figure 1.9 shows a plot of an unmanned vehicle's survey plan (red line) and actual track (black line). Generally, under level 4 sea condition with a surface flow of 1 m/s, the navigation accuracy of the unmanned vehicle is ±0.5 m.

The main Chinese companies that develop unmanned surface vessels are Zhuhai Yunzhou, Wuhan Chuhang and Wuhan Laolei Green Bay Company. Internationally, the US, the UK, Germany and other countries also produce and operate a small number of unmanned surface vessels (USVs). Some examples of unmanned vehicles and their performance specifications are shown in Table 1.7.

1.2 Spaceborne and Airborne Geomorphology Detection Technology

1.2.1 Spaceborne Ocean Monitoring Technology

1.2.1.1 Satellite Altimetry Technology

Satellite altimetry technology uses a microwave radar altimeter on a satellite to determine the timing of radar pulses traveling to and from the sea surface, and calculates the distance from the satellite to the measurement point on the sea surface. In addition, it can

Table 1.7 Examples of unmanned surface vessels and their performance specifications

Comparison item	Yunzhou unmanned boat	Water quality monitoring robot fish	C-CAT4 unmanned boat	Sea Serpent
Place of origin	China	Spain	UK	US
Equipment pictures				
Features	Water sampling/monitoring	Water quality monitoring	Customized modifications according to your needs	Customized modifications according to your needs
Data processing	Data backhaul/storage/processing/drawing	Data storage	Data storage	Data storage
Size/m	1–1.5	1.5	4.3	2.5
Weight/kg	20–50	100	650–1 000	>1 000
Life time/h	6–12	2	Custom made	Custom made
Autonomous navigation	Yes	Yes	Yes	Yes
Communication distance/km	10	1	10	
Ultrasonic obstacle avoidance	Yes	No	No	No
Live video	Yes	No	Yes	Yes

determine changes in sea level relative to a reference ellipsoid based on known satellite orbital heights and various error corrections. The basic observations include signal round-trip time, echo signal waveform and echo gain of the automatic gain control value. Figure 1.10 is a schematic diagram of satellite altimetry operation. The sea surface height can be expressed as

$$h = r_s - r_p - h_{\text{alt}} = h_{\text{ass}} + h_s \tag{1.1}$$

where h is the ellipsoidal height of the instantaneous sea surface; r_s is the geocentric moment of the normal projection point of the satellite on the reference ellipsoid; r_p is the geocentric distance of the sub-satellite point P (i.e., the

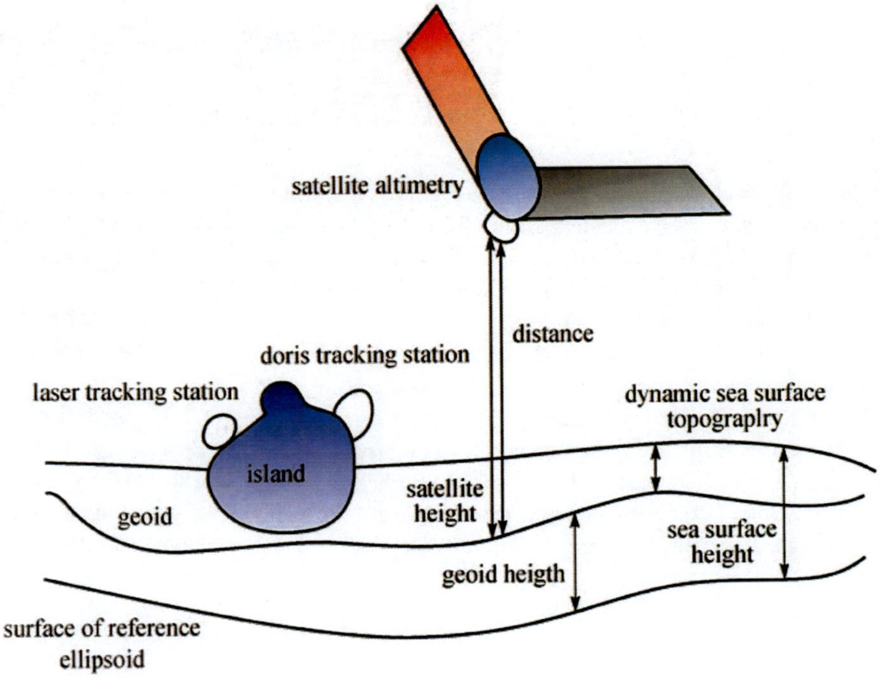

Fig. 1.10 Schematic diagram of satellite altimetry operation and parameters

projection point of the satellite on the mean earth ellipsoid); h_{alt} is the direct measurement of the microwave radar altimeter; h_{ass} is the average sea surface height consisting of the geoid level and the steady sea surface topography; and h_s is the dynamic sea surface topography.

Satellite altimetry provides high-precision and high-resolution geoidal data over large ocean areas, which can be inverted to produce the submarine topography using analytical and statistical methods. When the frequency domain is used to indicate changes in the submarine topography, the satellite altimetry inversion data of the submarine topography can only reflect the low and medium-frequency parts (mild and medium variations) of the submarine topography but cannot detect extreme variations. Satellite altimetry is greatly affected by topographic objects in the vicinity of the ocean or island. Such objects significantly reduce the observation accuracy; therefore, the use of satellite altimetry to invert for bathymetry mapping in the ocean will generally produce satisfactory results only in deep waters. At present, this method is generally not suitable for large-scale mapping.

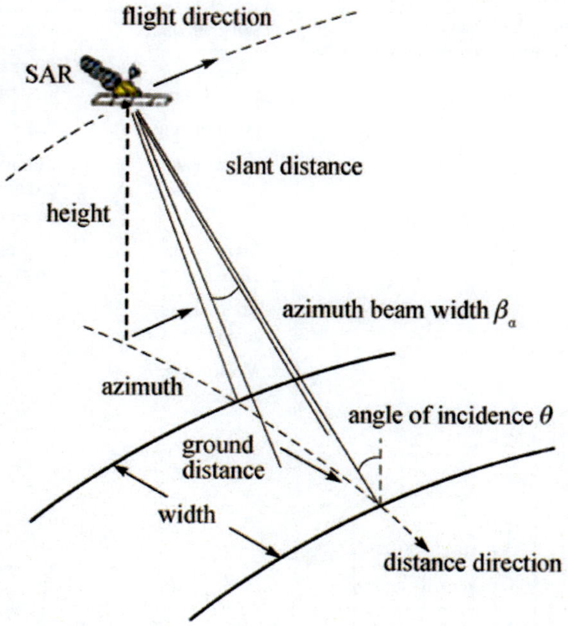

Fig. 1.11 Schematic diagram of remote sensing imaging of spaceborne SAR

1.2.1.2 Ocean Remote Sensing Survey Technology

Spaceborne remote sensing survey technology is an important part of the ocean survey technology system. It has evolved along with high-tech technologies such as satellite technology, electronic technology, opto-electronic technology and microwave technology. Since 1972, a number of satellites with space detectors have been launched, including Landsat, IKONOS, MOS-l and Pleiades, providing continuous quantitative global ocean information in all weather conditions. Remote sensing satellites can collect the following data: sea surface wind speed and direction, wave height, length and spectrum, sea surface temperature, atmospheric water content, sea ice, sea surface topography, ocean level and high-resolution radar images.

At present, satellite sensors for ocean observations acquire information based on the principle of electromagnetic radiation. Remote sensing technology uses electromagnetic waves in the visible light, infrared and microwave bands. Sensors are divided into active and passive types according to their working mode. Active sensors include microwave altimeters, scatterometers and synthetic aperture radar (SAR). Passive sensors operate in the visible and infrared bands and include scanning radiometers and microwave radiometers. Figure 1.11 is a schematic diagram of remote sensing imaging of spaceborne SAR. The parameters of the sensors and main measurements currently used in marine research are shown in Table 1.8.

Spaceborne ocean survey technology has greatly impacted traditional survey technology and broadened our understanding of the oceans and our planet. It provides a macro background and basis for conducting a variety of marine research. Improvements in the performance and resolution of various sensors mean that, spaceborne technology has broad applications in many aspects, such as evaluating the integrity and accuracy of charts, correcting and updating charts, issuing general notices, verifying suspected hazards and determining their location. Remote sensing surveys can also be used to assess and verify water-depth data, to assist in determining areas for priority measurement, and to carry out surveys to determine the geoid, sea topography and ocean currents.

Underwater terrain survey and inversion can be performed using microwave remote sensing technology. The SAR remote sensing imaging of the seabed is based on a three-part process (Fig. 1.12).

(a) The interaction between the underwater topography and tidal currents causes sea surface velocity modulation, forming surface convergence and divergence.
(b) Sea surface flow field modulation causes changes in the short-wave spectrum of the wind-induced ocean waves, leading to changes in the roughness of the sea surface.
(c) Sea surface roughness variations appear as intensity modulation on radar images. The underwater terrain SAR imaging model is based on three transfer equations: the Navier-Stokes equation, the wave action spectrum balance equation and radar backscatter equation.

Table 1.8 Satellite sensors and their measured ocean parameters

Sensor name	Measured ocean parameters
SAR	Wave direction spectrum, mesoscale eddies, ocean internal waves, shallow sea topography, sea surface pollution, sea surface characteristics information, etc.
Infrared sensor	Sea surface temperature
Aqua sensor	Chlorophyll concentration, suspended sediment concentration, marine primary productivity, diffuse attenuation coefficient and other oceanographic optical parameters
Microwave altimeter	Average sea level, geoid, effective wave height, sea surface wind speed, surface flow, gravity anomaly, etc.
Microwave scatterometer	Sea surface wind farm
Microwave radiometer	Sea surface temperature, sea surface wind speed, sea ice water vapor content, rainfall, etc.

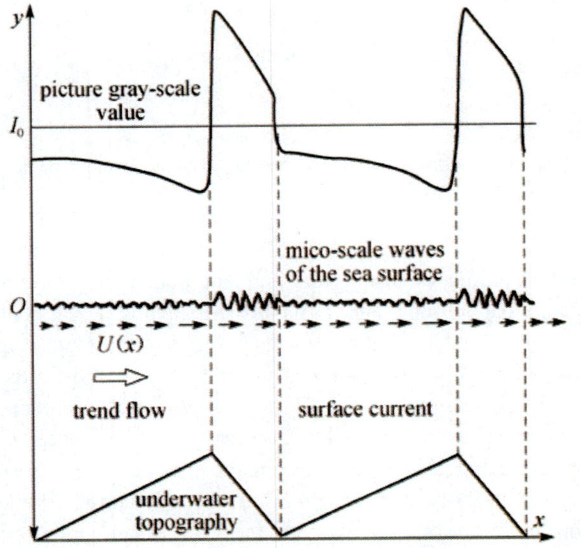

Fig. 1.12 SAR imaging mechanism for shallow underwater terrain

1.2.2 Airborne Ocean Exploration Technology

1.2.2.1 Airborne LiDAR Bathymetry Technology

ALB is an aircraft-based technology where two laser beams are emitted from the aircraft to measure the altitude of the aircraft relative to the sea surface and the seafloor, thereby obtaining the water depth. Generally, depths of 30–50 m can be measured, and the accuracy is about 0.1–0.2 m. ALB can be used for mapping large areas of shallow water (especially water with good visibility). The airborne laser sensor is large and heavy, hence it should be mounted on a manned fixed-wing aircraft; therefore, the data acquisition cost is high. The current mainstream models are the Larse500 in Canada and LADS in Australia (Table 1.9).

1.2.2.2 UAV Aerial Photogrammetry

Spaceborne satellite altimetry and remote sensing methods have low accuracy, whereas airborne LiDAR is costly to run and requires a specific type of aircraft flying within a limited altitude range. The strict timing of the tidal levels and limited maneuverability also restrict the use of these methods. Therefore, low-altitude drones can be effective tools for topographic surveys of tidal flats and island reefs in addition to these measures.

Unmanned aerial photogrammetry uses a light-duty drone equipped with a high- resolution digital color aerial camera to acquire data such as survey area images and measurement status. The data is then transmitted to the ground control system and image processing analysis is performed on a digital photogrammetric workstation. After obtaining the GPS/POS data of the drone and the camera parameters, the elevation information of the object points in the stereo image pair can be extracted through geometric calculation; the elevation information is acquired much faster than via conventional methods. Based on the least-squares theory, the shoals surveying and mapping system calibration technology uses a high-precision aerial triangulation encryption method to collect elevation information with relatively high precision, and uses a secondary internal orientation method to adjust the digital elevation model according to the monitoring and analysis data of the ocean surface level variation. This method can obtain a topographic map of tidal flats and island reef areas on a scale of 1:2 000, but the water depth inversion performance is poor. Currently, under ideal conditions, drones can reach good measurement accuracy in shoal and reef areas at a mapping scale of 1:1 000.

Unmanned aerial photogrammetry has the advantages of a relatively simple aircraft-based operation which offers strong maneuverability, simple maintenance procedures, aerial photography below cloud level, low-risk operation and

Table 1.9 Airborne LiDAR devices and their parameters

Model	Wave length/μm	Pulse repetition frequency/Hz	Sweeping range/m	Aircraft height/m	Flight speed/m · s^{-1}	Maximum depth of measurement/m	Bathymetry accuracy/m
Larse500			268	500	70	40	0.3
LADS	1.06–0.53	168	250	500	70	50	
LADS MKII		900	240	500	90	70	0.3
Hawk eye	1.06–0.53	200		300	15–60	70	0.3
FLASH	1.06–0.53	200				34	

Fig. 1.13 Unmanned aerial photogrammetry using drones to detect oil spills

low altitude resolution. It is effective in obtaining both small-range and large-scale data. At present, unmanned aerial photogrammetry is widely used in many aspects such as sea-state emergency monitoring, unreachable sea area map data acquisition, large-scale mapping and map data acquisition, local updates of marine charts and small-scale 3D model establishment. Figure 1.13 shows a drone operation to detect oil spills.

1.3 Underwater Robots and Technology for the Detection of Submarine Observation Networks

In addition to surface survey vessels and aerial vessels, underwater robots are also an important submarine topography survey vehicle. Underwater surveying equipment has the advantage of near-bottom detection at high resolution. In recent years, in situ long-term submarine monitoring based on underwater observation networks has developed rapidly, enabling real-time observations with online monitoring of submarine topography, submarine disasters and marine ecology (Jin et al. 2016).

1.3.1 Technology of Underwater Robots

The underwater robot is a submarine observation platform that can operate underwater and can accomplish specific tasks independently. Underwater vehicles can be divided into autonomous underwater vehicle (AUV), human-occupied vehicle (HOV) and remotely operated vehicle (ROV). These vehicles are a mobile platform that can carry a variety of sensors to survey the submarine topography.

1.3.1.1 Autonomous Underwater Vehicles

Autonomous underwater vehicles (AUVs) have greatly improved the scope and quality of marine surveying because they can obtain high spatial and temporal resolution data below the water, thus eliminating atmospheric interference. In 1992, Woods Hole Oceanographic Institution (WHOI) developed a deep-ranging AUV autonomous benthic explorer (ABE). On its expeditions ABE found more than 12 new deep submarine hydrothermal vents using the “three-step” method of submarine hydrothermal vent detection. To investigate the submarine hydrothermal activity in the mid-ocean ridge, Mitsui Shipbuilding and the University of Tokyo jointly developed the r2D4 AUV that can explore the ocean to depths of 4 000 m. The AUV weighs approximately 1 600 kg and has a range of 60 km. In 2006, deep-sea lava plains and hydrothermal plumes were discovered in the Indian Ocean. With the development of AUV technology, Kongsberg Maritime of Norway, developed the HUGIN AUV series which can operate at depths of 1 000 to 4 500 m. The vehicles can be used for high-quality marine surveying, hydrographic surveys, rapid environmental assessment and more.

The CR-01 and CR-02 AUVs with a diving depth of 6 000 m were jointly developed by China and Russia (Fig. 1.14). These AUVs detected submarine manganese nodules in the eastern Pacific Ocean at a depth of more than 5 000 m, making China one of the few countries in the world with a 6 000-m level autonomous submersible. During the “Twelfth Five-Year Plan” period, China developed

Fig. 1.14 The AUV CR-02, China

Fig. 1.15 The AUV Qianlong II, China

the "Qianlong II" AUV (Fig. 1.15) which has a maximum working depth of 4 500 m and can be operated independently for 30 h underwater. It was used to detect hydrothermal sulfide resources in the Southwest Indian Ocean. At present, China is at the forefront globally in terms of submersible integration technology.

1.3.1.2 Human Occupied Vehicles

The human occupied vehicle (HOV) provides a platform for marine scientists to explore the deep sea, to observe ocean phenomena directly and to conduct scientific experiments. It can bring people to the depths of the ocean for observation and operation, and realize the dream of humans to explore the ocean.

The US' Alvin submersible is the most frequently used HOV in the world. It is considered to be the most efficient and successful HOV and has made great contributions to deep-sea research. In 1966, the Alvin caused a sensation when it participated in the recovery of a hydrogen bomb dropped by the US off the Mediterranean coast of Spain. This led to the development of HOVs in Russia, France and Japan.

The submersible Jiaolong was a breakthrough in China's HOV technology. In July 2012, it successfully completed a

Fig. 1.16 HOV Jiaolong, China

sea trial to 7 000 m depth. In 2013, it successfully carried out experimental applications for submarine resource exploration and deep-sea scientific research. It is now fully operational (Fig. 1.16).

1.3.1.3 Remote Operated Vehicles

The remotely operated vehicle (ROV) is a submersible vehicle supported by a mother ship. A variety of equipment and sensors can be installed on the ROV for long-term and continuous submarine investigation and sampling operations. ROVs have contributed to marine geological surveys in the world oceans. ROVs were the first submersible vehicles to be developed and applied for work and exploration in the maritime environment. In 1966, the CURV-1 ROV of the US Navy successfully salvaged the hydrogen bomb that fell in the Mediterranean Sea in cooperation with the manned submersible Alvin. The CURV-2 and CURV-3 ROVs were developed shortly afterwards (Fig. 1.17). ROVs for commercial use were introduced in 1975. Hydro Production developed RCV-125 and RCV-150 between 1975 and 1980. Their main task was to connect underwater pipelines and assist in underwater drilling.

ROVs have been widely used in ocean exploration. The US Monterey Bay Aquarium Research Institute (MBARI)'s ROV Ventanta was developed in 1987. It can dive down to 1 800 m and it is the most utilized ROV in the world, having performed more than 3 000 dives. A carbonate sample was obtained from Ventanta's first dive in 1988 and identified as a chimney fragment from an underwater cold spring. WHOI's ROV Jason II produced video footage of the deepest undersea volcanic eruption ever recorded during a dive off the coast of Fiji in 2009, recording the spectacular eruption of magma from the ocean floor for the first time.

ROV CURV-1 ROV CURV-2 ROV CURV-3

Fig. 1.17 US Navy's CURV series ROVs

After nearly 30 years of development, China has independently produced various types of ROVs including floating, crawling and dragging vehicles (Jin et al. 2014). For undersea cable laying, the first self-propelled cable-laying robot CISTAR was developed by Shenyang Automation Research Institute in 2002, for the laying, testing and maintenance of optical cables. Hailong, a sampling robot developed by Shanghai Jiao Tong University, was launched in 2004 and can operate down to a depth of 3 500 m. Hailong is equipped with five cameras of various capabilities for taking still-life monitoring images. It is also equipped with special underwater lighting equipment which can illuminate up to several hundred meters underwater. Its sonar can work in turbid water, and its lifting arms can carry several hundred kilograms. In 2018, the Hailong III reached a depth of 6 000 m.

1.3.2 Technology of Submarine Observation Networks

Observation networks comprise monitoring sensors placed on the seabed and connected through submarine cables and feeder boxes to form a network of submarine observation points. Compared with the traditional shipborne detection method, observation networks provide real-time and online long-term monitoring of the submarine topography and can monitor changes in the marine environment. Observation networks have a wide range of applications in both civilian and military operations. This section only introduces several representative submarine observation networks in China and around the world.

1.3.2.1 Underwater Acoustic Monitoring Network

In the 1950s, the US Navy began building a cable-based underwater observation system for monitoring submarines, the sound surveillance system (SOSUS). The hydrophone network was placed at the bottom of the Atlantic Ocean and expanded to the Pacific coast in the mid-1960 s. The system consists of underwater arrays of hydrophones placed on the seabed, which record and transmit signals to shore processing stations through cables.

In the 21st century, the US Naval Research Bureau launched the Persistent Littoral Undersea Surveillance Network (PLUSNet) Project (Fig. 1.18). It consists of a sensitive hydrophone fixed to the seabed, an electromagnetic sensor and a mobile sensor platform. The fixed observation device and the mobile observation platform can communicate bi-directionally to form a semi-autonomously controlled submarine observation system. The system is designed to use the mobile platform to adaptively process and enhance the reconnaissance, classification, location and tracking of low-noise diesel-electric submarines in shallow waters, especially in the western Pacific. The illustration in Fig. 1.18 shows the large-scale PLUSNet trial conducted in 2006 in the Monterey Bay.

Fig. 1.18 The PLUSNet operation (https://www3.mbari.org/MB2006/UPS/mb2006-ups-links.htm)

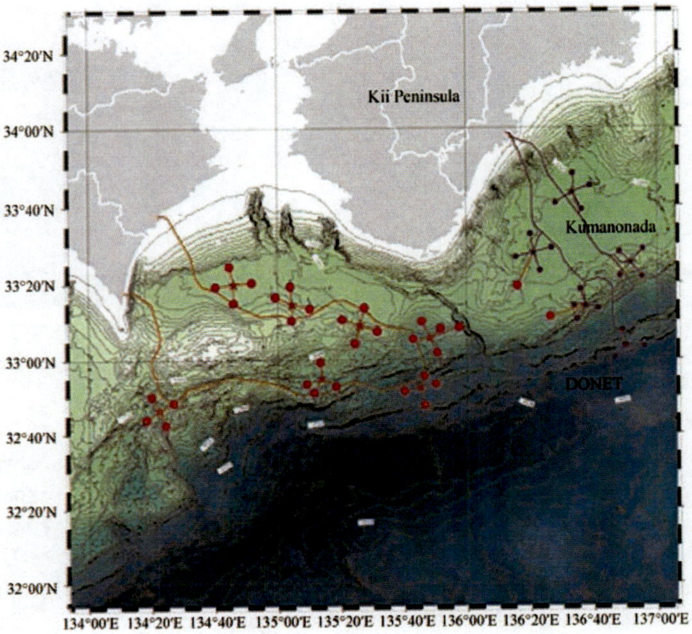

Fig. 1.19 DONET location map

1.3.2.2 Monitoring Networks for Earthquakes and Tsunamis

As an earthquake-prone country, Japan has been at the forefront of the world in seismic surveys. The construction of the dense ocean floor network for earthquakes and tsunamis (DONET) began in 2006 and was completed in July 2011. The main cable length is 300 km. Twenty observatories with five science nodes were installed on the seabed around the South China Sea Trough (Fig. 1.19). Each observatory detects earthquakes and tsunamis accurately and transmits data in real-time to the Yokohama Institute of Geosciences at the Japan Marine Science and Technology Center. Since August 2011, DONET has provided seismic data to the Japan Meteorological Administration and the National Geoscience and Disaster Prevention Institute of Japan for earthquake warning. In the second phase of the project (DONET2) in 2010, the surveillance area was extended westward and an additional 29 observatories with 7 nodes were set up on the Kii Peninsula and offshore (Fig. 1.19).

1.3.2.3 Monitoring Network Systems for Marine Ecological Observations

The New Jersey Shelf Observing System (NJSOS) is a typical example of marine ecological monitoring system. NJSOS originated from the Long-term Ecosystem Observatory at 15 m (LEO-15) in the early and mid-1990s. LEO-15 is located 16 km offshore and it is the first cable-connected observatory for marine ecology with two nodes, one of which serves as the base station of the AUV, as shown in Fig. 1.20. LEO-15 covered an area of only 3 km × 3 km in the 1990s, then it was extended and renamed the Coastal Predictive Skill Experiments (CPSE) in 2012, with an observation area of 30 km × 30 km. In the future, it will be expanded to cover an area of 300 km 300 km with a variety of observation platforms, including satellites, shore-based radars, shipborne towing bodies and gliders. Leo-15 has been continuously monitoring the coastal and transcontinental movements of seawater and sediments for more than 15 years, and recording a variety of biogeochemical processes. CPSE has shown how coastal upwelling zones evolve in three dimensions, interact with coastal topography, and influence plankton distribution and dissolved oxygen. NJSOS plans to promote CPSE's validated

Fig. 1.20 New Jersey Shelf Observing System's LEO-15

approach to the continental shelf for year-round observations, drawing up a list of 10 scientific goals that include applied goals alongside basic research, such as the flow of pollutants, heavy metals, the fate of fish larvae and sediments, and prediction mechanisms for red tides and low oxygen levels on the seafloor. This strategy is applied first to a small piece of ocean, and then successfully expanded to cover a larger region.

1.3.2.4 Underwater Mobile Observation Network Systems

Since 1997, the autonomous ocean sampling network (AOSN) funded by the US Ocean Research Agency has adopted a variety of different types of observation platforms equipped with different sensors to measure ocean parameters in different regions and depths at the same time (Fig. 1.21). In August 2003, the AOSN-II experiment was carried out in Monterey Bay, California. In addition to the traditional observation ship, anchored buoy, and bottom-sitting observation platform, the observation platform also included underwater gliders, AUVs equipped with a CTD, fluorometer and sensors for measuring chlorophyll and other seawater parameters. During a 40-d investigation AOSN-II tested the rising surge of seawater in Monterey Bay and the mobile observation network composed of underwater gliders was able to dynamically track the seawater isotherm based on real-time changes of the marine environment (Zhu 2011).

On the basis of AOSN, the US navy carried out the Adaptive Sampling and Prediction Research Project (Fig. 1.22). An important goal of this project was to investigate the use of multiple underwater gliders to carry out the efficient marine parametric sampling. In August 2006, four Spray underwater gliders and six Slocum underwater gliders were used in the Monterey Bay experiment to investigate the upwelling of the cold current cycle in the northwest part of the bay. The data obtained by the gliders are of high quality, improving researchers' understanding and forecasting ability of ocean phenomena. The gliders show the strong advantages of a distributed, mobile and reconfigurable autonomous sampling network in marine environmental parameter sampling (Zhu 2011).

1.3.2.5 Regional Multi-target Underwater Observation Networks

The North-East Pacific time-series undersea network experiments (NEPTUNE) (Fig. 1.23) is the first regional undersea observation and test system in the world connected by optical cables. NEPTUNE is an undersea network project launched by the US in 1998 with the goal of covering the entire Juan de Fuca Plate with a cable-linked observation system, a ground-breaking innovation in earth science. The entire observation network, constructed by the United States and Canada, is planned to share 2 000 km of fiber optic cable and cover an area of 200 000 km^2. The observation network consists of six nodes (currently five), Folger Passage, Barkley Canyon, ODP1027, ODP889, Middle Valley and Endeavour. It has a network of thousands of undersea observation devices designed to make long-term, continuous and real-time observations of the water layers, seabed and the earth's crust.

NEPTUNE was launched in Canada, with a maximum depth of about 3 000 m, and a system design life of 25 years. Located on the west coast of Vancouver Island, it became operational in late 2009 and is the first regional undersea observation cable network in the world. It has five seabed nodes, an 800-km long ring-shaped cable network

Fig. 1.21 The autonomous ocean sampling network

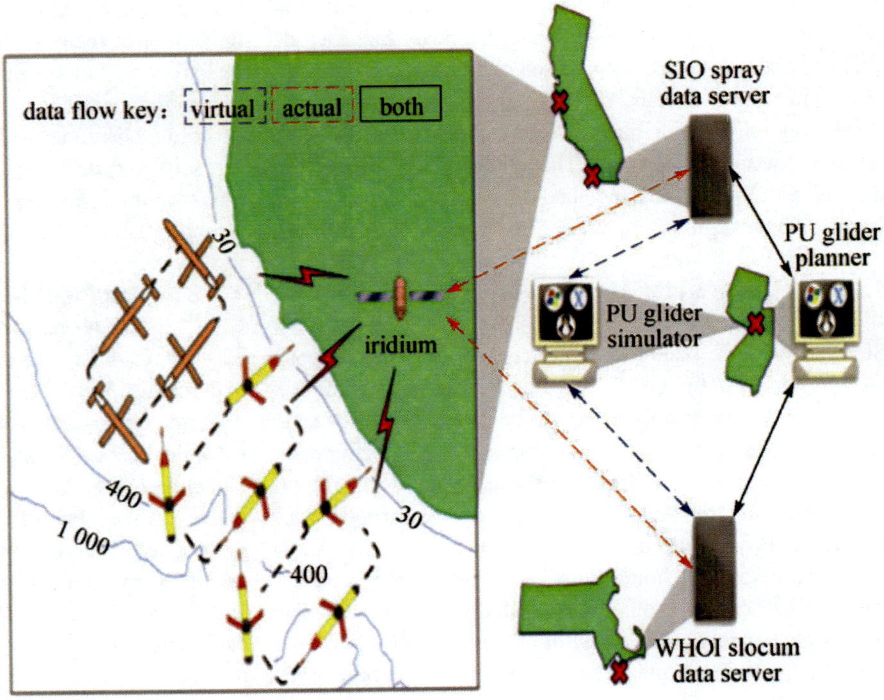

Fig. 1.22 Adaptive sampling and prediction system

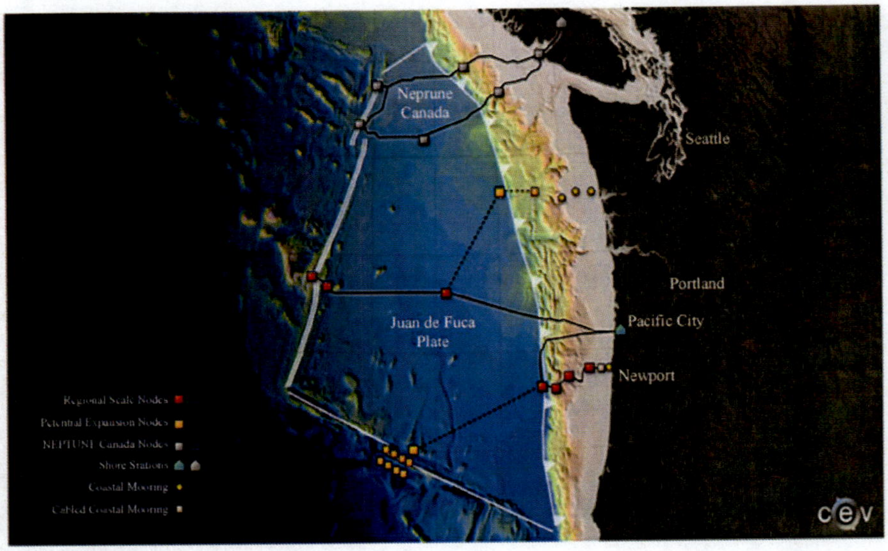

Fig. 1.23 The North-East Pacific time-series undersea network experiments setup in Canada

equipped with seismographs, current meters, cameras, gas sensors, submarine pressure recorders, temperature, salt and depth meters, acoustic imaging systems and other marine instruments. In addition, it has a variety of instruments for physical, chemical, biological and geological measurements (including hydrology, methane hydrate formation and seismic activity detectors) and an observation AUV. NEPTURE can also observe ocean phenomena and processes in the ocean interior within a range of 500 km × 1 000 km of the seabed by using a combination of communications/power cable networks and full-depth mooring systems.

1.3.2.6 Wide-Area Comprehensive Ocean Observation Networks

The Integrated Ocean Observing System (IOOS) of NOAA's management of intergovernmental departments integrates 11 US regional observation networks and forms an overall observation system. The IOOS observations cover the entire

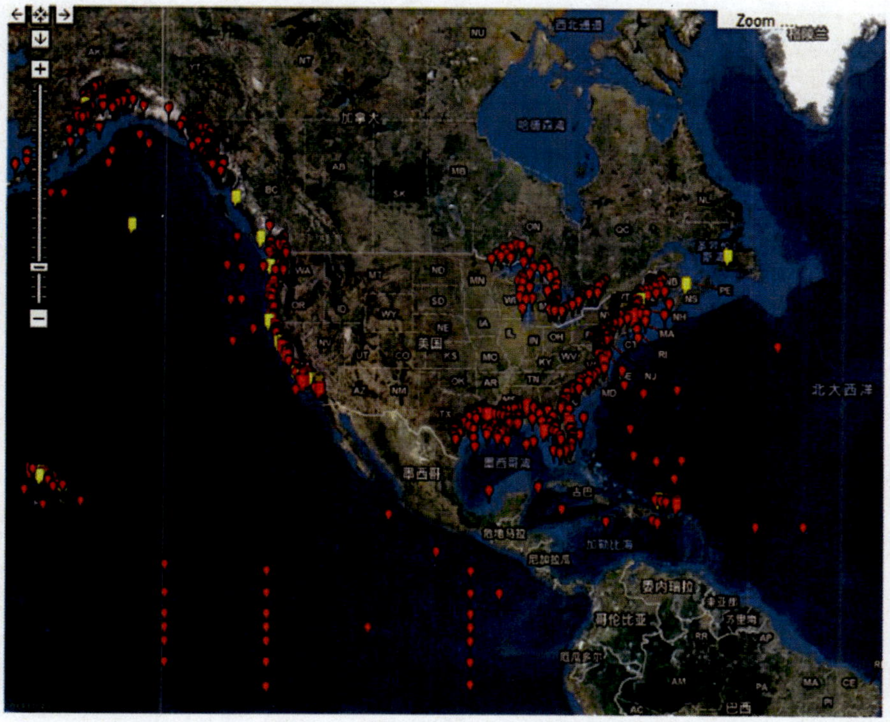

Fig. 1.24 The integrated ocean observing system

offshore waters of the United States (Fig. 1.24). In addition to 535 shore-based stations and more than 130 high-frequency ground-wave radars, the system operates widely distributed underwater observation equipment such as underwater gliders, wave buoys and tsunami warning buoys.

Since 2005, the IOOS operation has incorporated underwater gliders that performed more than 25 722 gliding dives to record temperature and salinity. At present, Slocum and Spray are the most commonly used underwater gliders. The glider technology has great potential in marine disaster prevention and mitigation research. The underwater glider of the Mid-Atlantic Regional Association Coastal Ocean Observing System (MARACOOS) provided accurate forecasts of hurricanes Irene and Sandy that hit the US East Coast in 2011 and 2012, respectively, helping to reduce the damage caused by the weather systems. In addition to helping with hurricane forecasting, gliders can track red tides as they travel, to better understand how to track and predict harmful algal blooms.

1.3.2.7 Development Status of Ocean Underwater Observation Networks in China

In recent years, with the increasing urgency of relevant national needs and the rapid development of marine observation technology, underwater communication technology and general marine technology worldwide, China has carried out experimental underwater observation work in the East China Sea and the South China Sea. Currently, however, there are no operational underwater observation networks in China.

In 2007, Zhejiang University began the construction of the Demonstration System for the Zhai Ruoshan Island submarine observation network (ZERO), and in August 2013 deployed equipment such as chlorophyll meter, turbidity meter, and a color soluble organic matter detector. In October 2014, the ZERO system was launched in the waters around Zhai Ruoshan Island, including the nodes and junction box (Fig. 1.25). In 2009, Tongji University built the Xiao Qushan Experimental Station for the East China Sea Submarine Observation. It consists of an optoelectronic composite cable about 1.1 km long, a junction box for the shore base stations, and observation instruments such as CTDs, ocean bottom seismometers, and acoustic Doppler current profilers. In addition, the ocean elements are observed continuously and in real time. In 2011, Tongji University, Zhejiang University and other universities jointly launched the Key Technology Verification Project of the experimental node in the long-term seabed observation network. This project verified a series of key technologies used in submarine observation networks, such as submarine energy transport and communication technology, underwater ion chromatography and in situ analysis technology, and equipment connections to the Monterey Accelerated Research System in the US. In May 2013, the submarine

(a) (b)

Fig. 1.25 Underwater junction technology. **a** Pool experiment of main junction box; **b** placement of secondary junction box

observation demonstration system led by the South China Sea Ocean Institute of the Chinese Academy of Sciences was completed in Sanya. In 2014, with the support of the Institute of Acoustics of the Chinese Academy of Sciences and the National High Technology Research and Development Program (863 Program) of China, a fiber optic hydrophone array was laid in the waters of Hainan to conduct fiber-optic hydrophone array engineering experiments and related technical research, and to test underwater target detection technology within a radius of 50-80 km. At present, China is establishing an underwater scientific observation network in the East China Sea and the South China Sea to realize all-weather, real-time and high-resolution multi-interface stereoscopic observations from the seabed to the sea surface in the East and South China Seas. The aim is to provide long-term continuous data and an in situ scientific experimental platform for understanding the marine environment of the East China Sea and the South China Sea.

References

Jin XL, Tao CH, Zhu XK et al (2014) China Ocean Engineering and Technology Development Strategy Research Ocean Exploration and Equipment Volume (in Chinese). China Ocean Press, Beijing

Jin XL, Tao CH, Zhu XK et al (2016) Research on the development strategy of China's marine underwater observation system (in Chinese). Chinese Academy of Engineering, Beijing

Li JB (1999) Multi-beam Survey Principle Technology and Method (in Chinese). China Ocean Press, Beijing

Li ZH, Huang JS (2005) GPS Surveying and Data Processing (in Chinese). Wuhan University of Technology Press, Wuhan

Liang KL (1995) GPS Kinematic positioning techniques and their applications (in Chinese with English abstracts). Engineering of Surveying and Mapping 4(4):22–27

Liu YC (2001) Space structure and data processing in marine sounding (in Chinese). Acta Geod Cartogr Sin 30(2):186–187

Zhao JH, Shen WZ, Wu YT et al (2007) Modern Marine Surveying and Mapping (in Chinese). Wuhan University of Technology Press, Wuhan

Zhu XK (2011) Research on ocean sampling method of glider (in Chinese). Dissertation, Graduate School of Chinese Academy of Sciences

2 Multi-beam Bathymetric Technology

Multi-beam bathymetric surveying technology has become ubiquitous with underway marine geological operations, and has now been available to the scientific community for almost four decades (Hughes Clarke 2018). It originated from military research projects funded by the US Naval Research Laboratory in 1964 (Glenn 1970) and by 1977 the first declassified version was available for scientific submarine geomorphological studies (Farr 1980). The technology developed rapidly in the 1980s (e.g., Renard and Allenou 1979; Wölfl et al. 2019), and entered the commercial stage in the 1990s. At present, the multi-beam echo sounding technique is the most commonly used method to perform seafloor topographic mapping (Zhao et al. 2015).

The multi-beam echo sounder (MBES) system is a high-tech product for application to modern marine surveys and is a complex system composed of multiple sensors. Application of acoustic reflection, scattering and acoustic interferometric principles to form the strip-type bathymetric data is combined with the use of hundreds or even thousands of high-density soundings, narrow measurement beams, advanced bottom detection techniques and accurate sound velocity correction methods to ensure high system detection accuracy and high point coordinate reduction accuracy. MBESs have now been developed into a full ocean depth, high-precision and can highly-integrated seafloor mapping technology and can provide a means to promote the use of three-dimensional stereo display editing, virtual reality mapping, acoustic seafloor sediment classification and other deep-seated research fields to form a comprehensive technical system. It has opened up new applications in submarine geomorphology.

This chapter outlines the classification and the principles of an MBES and presents representative instruments and equipment along with marine survey workflows. In the installation, calibration and investigation of MBES systems in particular, the chapter provides a summary of practical work experience that can meet the cognitive needs of practitioners in the field.

2.1 Principle of Multi-beam Sounding Technology

At present, MBES systems are mainly classified into two categories: one that uses beam forming control technology, which is called the beamforming MBES (and is also called traditional MBES) and a second type that uses interferometry technology, which is called the interferometric MBES (and is also called the phase-differencing MBES or bathymetric side-scan). The main difference between the two system types is largely due to beam formers measuring range for each of a set of angles, and interferometers measuring angle for each of a set of ranges (Table 2.1). Depending on the operating frequency of their sonars, these systems can also be classified into high-frequency, intermediate-frequency and low-frequency MBESs, which are used in shallow water, medium water and deep water measurements, respectively.

These MBESs, which are also known as swath echo-sounders or sweep systems, are typically used to determine the depth of the water and the nature of the seafloor. When operating, the projector array emits a broad acoustic fan-shaped pulse at a specified frequency that has a narrow opening along-track angle ($\sim 1°$) and a wide across-track angle ($\sim 150°$), known as a swath. The hydrophone array then listens for their reflections (or echoes) and forms multiple received beams (beamforming) that are much narrower in the across-track direction ($\sim 1°$). The projector and hydrophone arrays are oriented perpendicular to each other, so the strip of the seafloor that is ensonified by the projectors will intersect with the strip of the seafloor that is observed using the hydrophones; one after another, hundreds of narrow beams can then be obtained by across-track. Using the incident angle and the two-way travel time of each narrow beam, the positions and depths of the measurement points can then be calculated. When the vessel is moving, a strip of water depth measurements with a specific width can then be obtained.

Z. Wu et al., *High-resolution Seafloor Survey and Applications*,
https://doi.org/10.1007/978-981-15-9750-3_2

Table 2.1 Comparison between beamforming and interferometric MBES (reference: Bathyswath.com)

Parameter/Function	Beamforming MBES	Interferometric MBES	Notes
Number of depth measurements	60–512	6 000+	Depends on range
Range versus water depth	3–5	10–20	Beam former footprint becomes unacceptably large at far range
Amplification/processing channels	60+	4–5	In a harsh environment, simplicity is important
Horizontal resolution at range	Poor	Good	Beam former footprint becomes unacceptably large at far range
Angular coverage	90°–180° (or beyond 180° using a dual-head system)	260° (including 20° overlap)	
Co-incident side-scan	Partial	True	An interferometer collects amplitude in the same way as its bathymetry: as a time-series
Profile data density	Decreases with reducing grazing angle	Increases with reducing grazing angle	Higher complete profile data confidence with an interferometer
Ability to resolve several targets at the same range	Yes	No	
Ability to resolve several targets at the same angle	No	Yes	
Profile data density	Decreases with reducing grazing angle	Increases with reducing grazing angle	In the first 5 m of horizontal range, a beam former collects slightly more depth samples. Beyond that, an interferometer collects many more
Does it have the capacity to acquire water column information?	Yes	Yes	Interferometric systems can identify targets in the water column but are unable to characterize them accurately due to lack of beam forming angles to locate the target

This section focuses on the basic working principles and the acoustic fundamentals of beamforming MBESs (e.g., Teledyne Reson SeaBat 7125) and interferometric MBESs (e.g., Kongsberg GeoSwath Plus) to provide a reference work for both experienced practitioners and novices in the field.

2.1.1 Beam Directivity

When a nondirectional acoustic pulse (ping) is emitted in water, it travels away from the source with uniform spherical amplitude, which means that its acoustic energy is equal in all directions. This symmetrical spreading is called an isotropic expansion and the projector that produces it is called an isotropic source.

When two neighboring projectors are emitting identical isotropically expanding signals, their acoustic wave patterns will overlap or interfere with each other. The phenomenon by which superposition of two crests or two troughs enhances the energy of the waves is called constructive interference; in contrast, the phenomenon by which superposition of peaks and troughs causes them to cancel each other out, with the net energy being zero, is called destructive interference (Fig. 2.1a). In general, constructive interference occurs at points where the distances to each projector are equal, or where the difference between the two distances is equal to an integer number, of wavelengths of the sound emitted ($n\lambda$, $n = 1, 2, 3, \ldots$). In contrast, destructive interference occurs at points at which the difference between the distances to the projectors is a half-wavelength ($\lambda/2$) or a half-wavelength plus an integer number of wavelengths

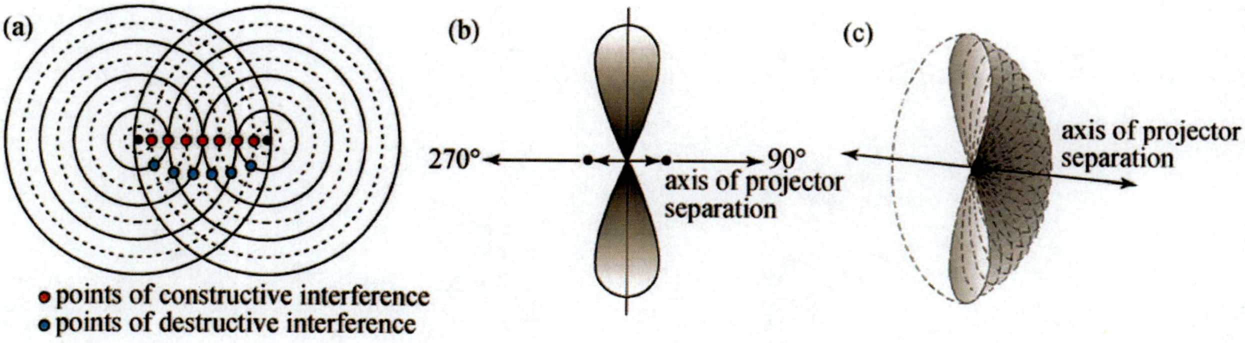

Fig. 2.1 Constructive and destructive interference of two neighboring projectors (**a**), 2D (**b**) and 3D (**c**) beam pattern for two projectors with spacing of $\lambda/2$

($\lambda/2 + n\lambda$, $n = 1, 2, 3, \ldots$) away from the transmitter (L-3 Communications SeaBeam Instruments 2000).

Figure 2.1b and c show a beam pattern when the distance between the two projectors is $\lambda/2$. The distribution of the acoustic energy is clearly visible and the energy varies at different angles; this is called the directivity of the energy. If the energy of the projector array is distributed within a narrow angle, the system's directivity is high. A projector array is composed of multiple projector elements, including linear arrays and circular arrays, and the basic principles of these arrays are similar. As shown in Fig. 2.2, based on an array of two projector elements, the beam pattern of a linear array composed of projector elements can be deduced.

In Fig. 2.2, the part of the beam with the maximum energy is called the main lobe, and the smaller broadside lobes are called the side lobes; the constructive interference occurs among these smaller lobes. Side lobes can cause energy leakage and the resulting echo can interfere with the echo of the main lobe. The central axis of the main lobe is called the maximum response axis (MRA) of the beam pattern, and the beam width at half-power of the main lobe (i.e., at -3 dB relative to the energy of the main lobe) is the beam angle. As more projector elements are used and the array becomes longer, the beam angle will decrease and the directivity will increase in tandem. Reducing the acoustic wavelength or increasing the array length can improve the beam directivity. However, the array length is finite and as the wavelength becomes smaller, which also means the frequency is higher, the beam suffers greater attenuation in water; therefore, the beam directivity cannot be increased indefinitely.

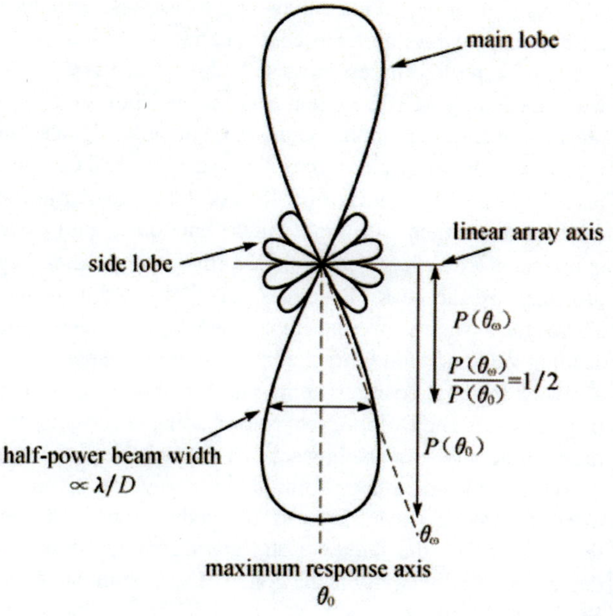

Fig. 2.2 Beam pattern of multi-element linear array

2.1.2 Principle of Beamforming Multi-beam Echo Sounder System

2.1.2.1 Beam Forming

The perpendicular arrangement of the projector and hydrophone line arrays is called a Mills Cross and is named after a pioneering radio astronomy instrument built in New South Wales, Australia (L-3 Communications SeaBeam Instruments 2000). Mills Cross arrays are widely used in MBES transducer arrays and, using this type of array as an example, the principle of beamforming will be introduced here.

While the transducer is running, the projector or hydrophone array produces a transmitted or received beam, respectively, which is wide along the vertical axis direction of the array and narrow along the horizontal axis direction of the array. The strip of the ocean floor that is ensonified by the projectors will intersect with the strip of the ocean floor that is observed by the hydrophones. This only occurs over a small area with dimensions that correspond approximately to the projector and hydrophone array beamwidths, a single narrow beam can be obtained (Fig. 2.3). Therefore, the

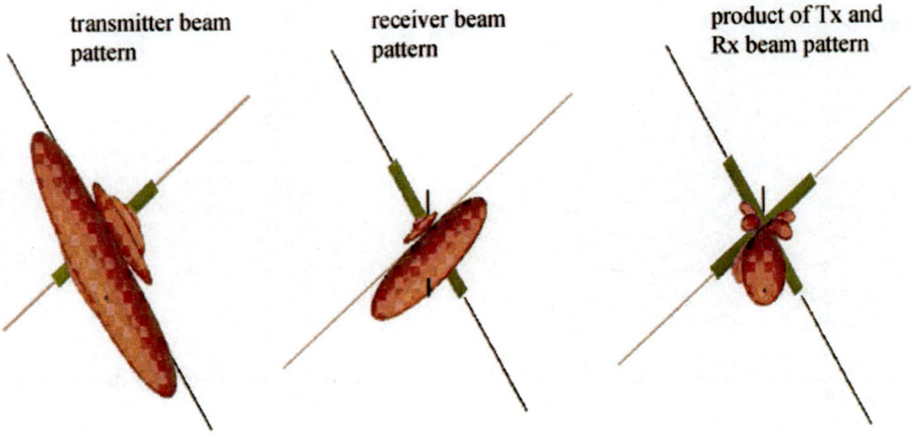

Fig. 2.3 The product of the two orthogonal beam patterns provides a narrow beam when using the Mills Cross array (Hughes Clarke 2010)

beamwidth of this narrow beam, in both the along-track and across-track directions, is affected directly by the beam control results of the corresponding transmitted and received beams.

In a complete transmission and reception period (ping) of an MBES System, the projector array runs only once to generate an acoustic pulse, while the hydrophone arrays acquire multiple received beams by introducing appropriate delays to the hydrophone array elements multiple times. The transmitted and received beams intersect to produce multiple narrow beams with small time intervals, as shown in Fig. 2.4.

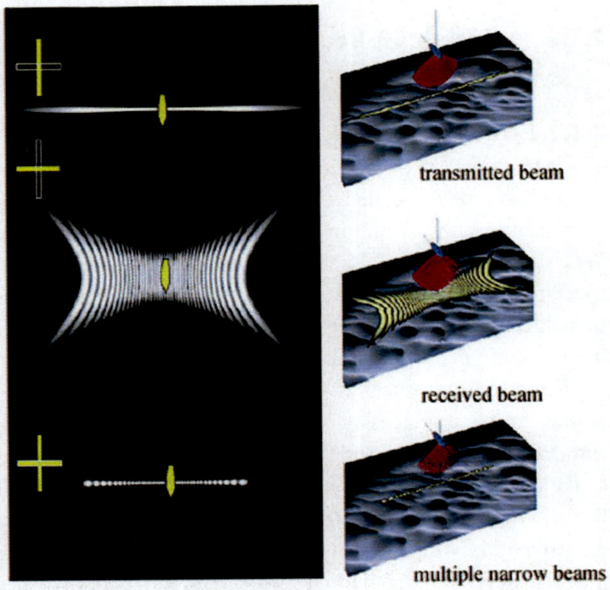

Fig. 2.4 Multiple narrow beams produced by intersection of the transmitted and received beams (Marques 2012)

2.1.2.2 Side Lobe Suppression

The acoustic wave signals that are transmitted or received by the transducer array include the main lobe in the middle, and a series of side lobes on either side of the main lobe (Fig. 2.5). In general, the purpose of the array is to ensonify targets in the direction of the main lobe. However, the existence of side lobes not only reduces the signal-to-noise ratio by leaking acoustic energy in these directions, but also may be echoes from them as well, resulting in erroneous seafloor topography measurements (L-3 Communications SeaBeam Instruments 2000). The strength of side lobes is measured as a fraction of the power projected into them divided by the power projected into the main lobe, and is called the side lobe level. For an array in which all projectors emit the same power level, the first side lobe level is roughly −13 dB (Fig. 2.5a).

To obtain the actual measurement information and reduce the existence of this interference information, it is necessary to apply measures to suppress any side lobes as much as possible when designing MBES system transducers. Side lobe suppression is critical in multi-beam since side lobe signals can fool bottom detection algorithms if it is stronger than the main lobe signals. The technique of the side lobe suppression is called array element "shading", and the fraction of energy that is projected by each projector element divided by the energy projected by the peak element is called its shading value. There are a variety of algorithms used to determine what shading values should be applied to each array element, such as linear shading, cosine shading and binomial shading etc. (Fig. 2.5b–f). The Dolph-Chebyshev shading is currently the most commonly used shading scheme, it can be used to bring all side lobe levels to a uniform value (Fig. 2.5f). Although these shading schemes can be used to reduce the side lobes, they also cause the width of the main lobe to be larger, decreasing the directivity of the system (L-3 Communications SeaBeam Instruments 2000).

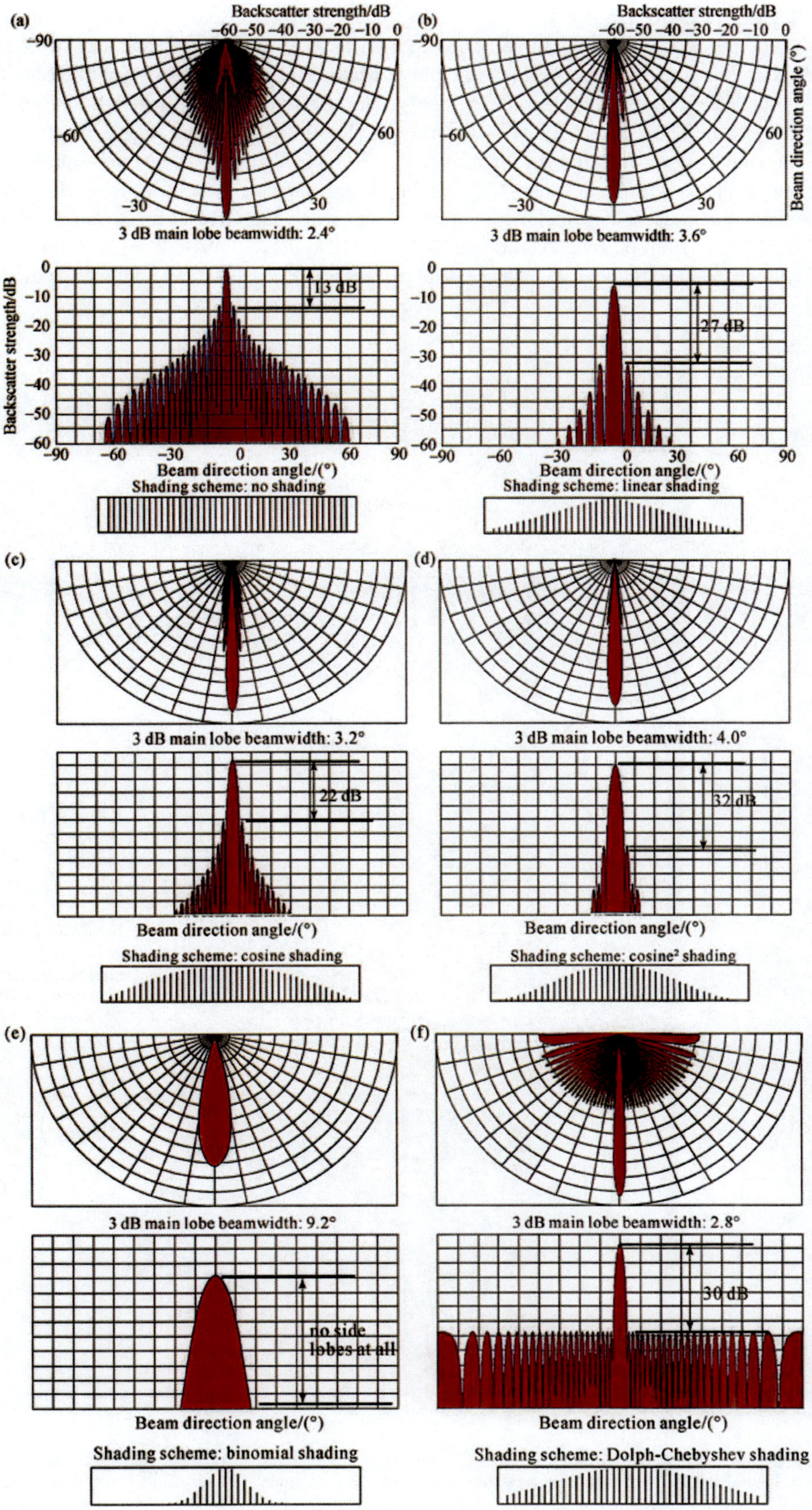

Fig. 2.5 The beam patterns of line array (42 elements 22λ long) after we apply different shading schemes. **a** No shading, **b** linear shading, **c** cosine shading, **d** cosine^2 shading, **e** binomial shading, **f** Dolph-Chebyshev shading (Hughes Clarke 2010)

2.1.2.3 Beam Steering

Using the formation of multiple beams in a linear array of hydrophones as an example, the principle of beam steering in an MBES is now discussed. According to the characteristics of the beam formed by the array, when the linear array direction is at $\theta = 0°$, the signals received by each element are in phase with each other, thus leading to the largest output response. When incident acoustic waves reach the linear array from other directions and the appropriate delay has not been introduced to each element at this stage, the maximum output response cannot be obtained. Therefore, to form beams in other directions, the appropriate delays must be introduced to ensure that each element of the hydrophone array will still be co-directionally stacked in the output signal (Fig. 2.6).

There are three methods for beam steering: the time-delay, the phase-delay and the fast fourier transform (FFT) methods (Hughes Clarke 2010). The time-delay and phase-delay methods steer beams away from the array normal by introduction of appropriate time or phase delaying for

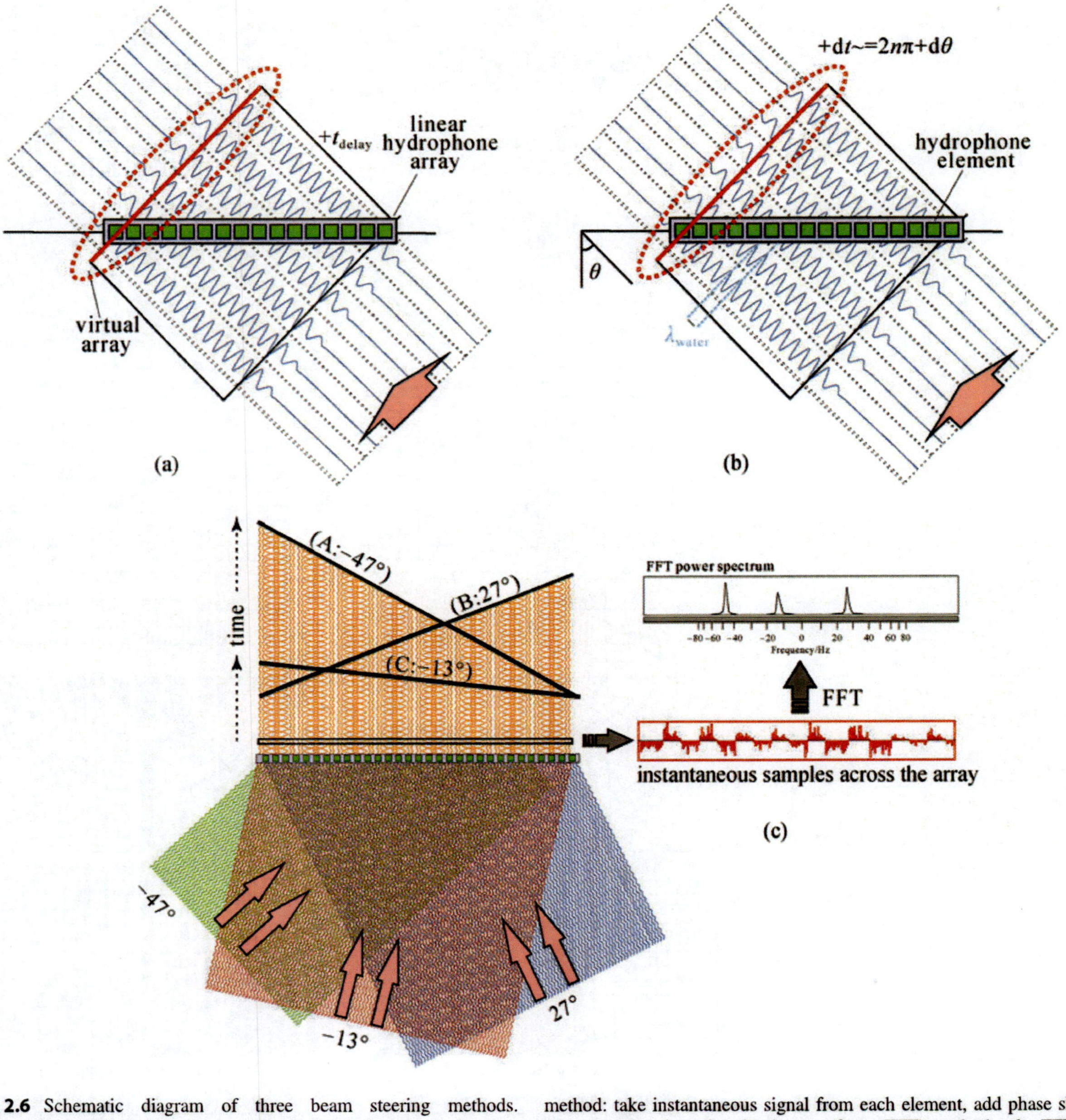

Fig. 2.6 Schematic diagram of three beam steering methods. **a** Time-delay method: take time series of signals from each element, shift them by appropriate amounts, then sum signals; **b** phase-delay method: take instantaneous signal from each element, add phase shift, then sum signals; **c** fast fourier transform (FFT) method: take FFT of instantaneous signals across the array face (Hughes Clarke 2010)

each element, respectively (Fig. 2.6a, b)). However, these two methods require a lot of calculation time and storage, so they are not suitable for real-time calculation with a large number of beams.

The FFT method produces a time series of amplitudes for beams with axes at the beam steering angles (Fig. 2.6c). Echoes recorded from these time series will correspond to locations on the seafloor to which each of these angles is pointing (L-3 Communications SeaBeam Instruments 2000). The FFT method can perform on high-speed digital signal processors (DSPs), allows the beam steering to be accomplished in real time, and is currently the most commonly used method for beam steering.

All the three beam steering methods require knowledge of the sound velocity at the transducer face, errors in sound velocity result in errors in beam steering angle. The beam-width of main lobe increases with steering angle, and the beam becomes a cone (instead of a plane), so we lose the angular resolution (Hughes Clarke 2010).

2.1.2.4 Bottom Detection

The MBES generally uses the amplitude detection method, the phase detection method, and a method that combines amplitude and phase detection to capture the bottom echoes accurately. When the incident angle is small (usually around nadir), the projection area of the beams on the seafloor is small, the amplitude detection method is used to detect the bottom echoes based on finding the center of the echo envelope (Fig. 2.7a). However, for the amplitude detection, the resolvable dimension cannot be smaller than this footprint as the echo is integrated over that dimension (Hughes Clarke 2012).

When the incident angle is large (away from nadir), the projection area of the beams on the seafloor increases accordingly, with the energy of the echo being decentralized, and the echo, which is mainly composed of scattered waves, lasts for a long time, resulting in poor amplitude detection results. However, the phase-detection method uses a split-aperture technique (received array is split into two separate apertures). The two apertures generate slightly different time-series, from which differential phase can be measured (Hughes Clarke 2012). As the incident angle increases, the phase change between the beams also becomes more pronounced. In this manner, discrimination across-track can be achieved based on phase (Fig. 2.7b). For the long, lower grazing angle echoes, phase (and thus feature definition) can be discerned at a scale significantly finer than the beam footprint dimension (Hughes Clarke et al. 1998).

The latest generation of MBES systems uses both amplitude detection and phase detection in the bottom detection process, which not only improves the beam detection accuracy but also reduces the effects of uneven measurement accuracy in the ping section. A detailed review of bottom detection algorithms is provided in De Moustier (1993).

2.1.2.5 Real-Time Motion Compensation

Because of the wind, waves, and tides, the attitude of the vessel can change momentarily during the survey (Fig. 2.8a). Real-time motion compensation refers to decomposition of the rocking motion of the vessel, which compensates for rotation of the acoustic array caused by the rocking of the vessel by controlling the reverse rotation of the transmitted and received beams to ensure that the transmitted or received beam surface remains stable relative to the geographic coordinate system (Hughes Clarke 2003b). Most previous MBES systems have used post-processing methods, while the latest generation of MBES systems is currently adopting real-time motion compensation technology, which provides a better solution to the problem of the uneven soundings caused by attitude changes of the vessel during the survey.

Real-time motion compensation of MBES includes roll, pitch and yaw stabilization. Roll stabilization requires dynamically adjusted receive steering, and is essential if the full swath is to be used. The roll signal from the motion sensor is fed to the beamformer electronics. The beamformer then updates the steering angles for all received beams such that they are stabilized with respect to the local level (Fig. 2.8b). This technique can only be used with time-delay and phase-delay beam steering methods (not for FFT method).

Pitch stabilization requires steered transmit sector, and is more important in deeper water (Hughes Clarke 2012). While electronic pitch stabilization can maintain the central beam close to vertical (thereby ensuring uniform along track density for the central beams), the same steering angle (w. r. t. the line of the transmit array) will over-compensate for the fore-aft displacement of the outermost beams (Fig. 2.8c).

Yaw stabilization requires multiple transmit sectors using different steering, and each sector (including received beams examining that sector) uses discrete frequency bands (Hughes Clarke 2010). A succession of individual transmissions is generated, closely spaced in time (separated in time only by the length of each pulse). Each sector/transmission addresses only a specific subset of the total swath and can thus have a unique steering angle that best aligns subset of the swath (Fig. 2.8d).

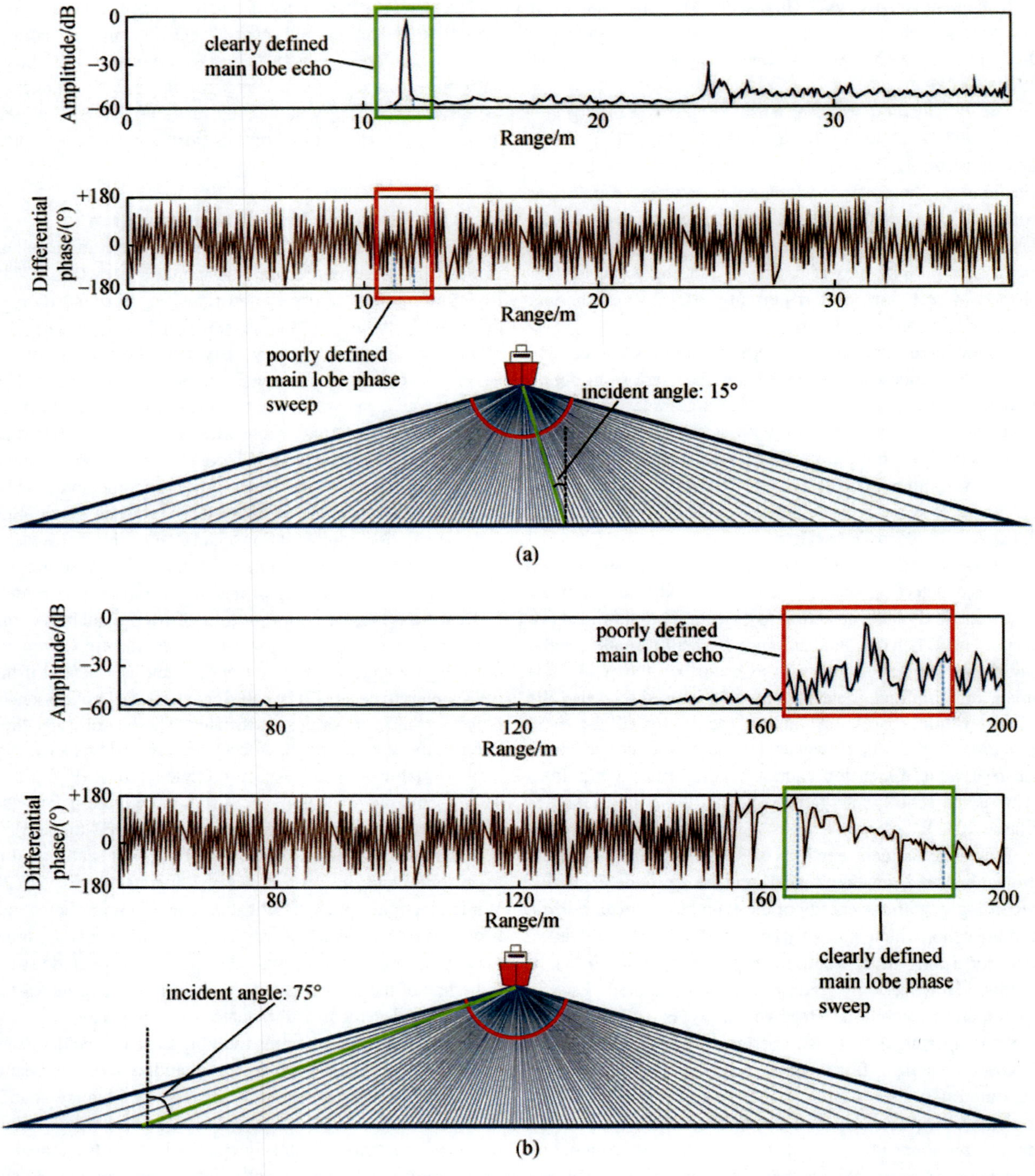

Fig. 2.7 Two bottom detection methods for MBES. **a** The amplitude detection method uses the intensity (amplitude) information for each received beam and works best with nearly vertical incidence (usually around nadir), and **b** the phase detection method uses the phase information for each received beam and works best with high incident angles (away from nadir)

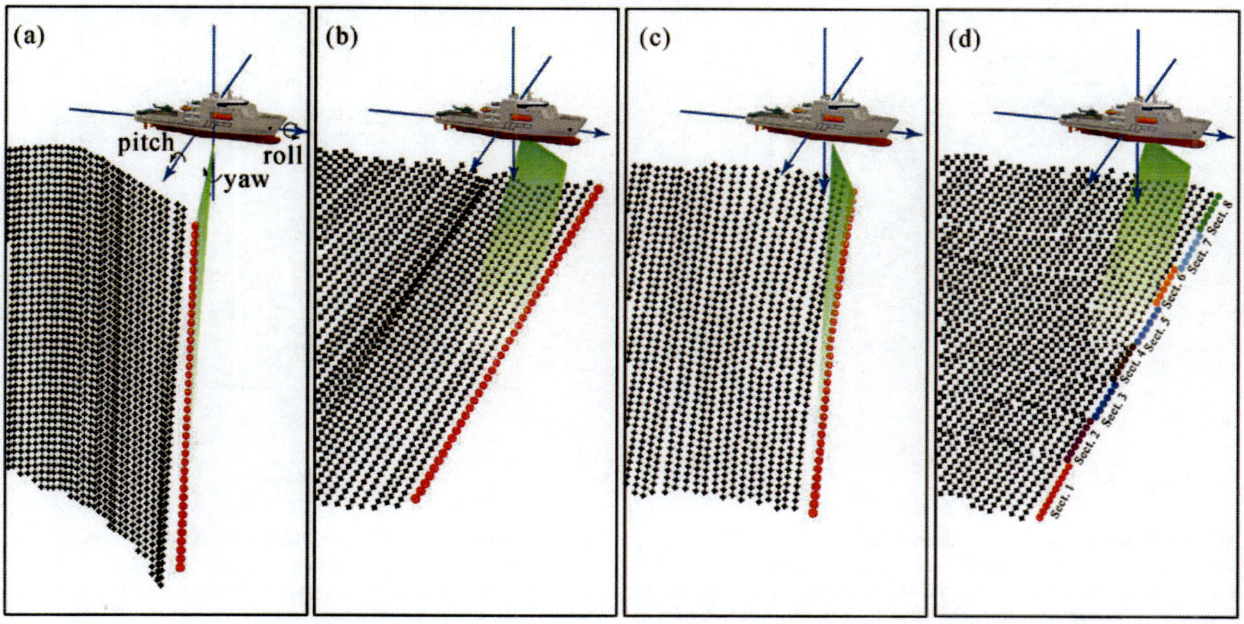

Fig. 2.8 Real-time motion compensation of MBES. **a** No stabilization, **b** roll stabilization, **c** roll and pitch stabilization, and **d** roll, pitch and yaw stabilization

2.1.3 Principle of Interferometric Multi-beam Echo Sounder System

While beamforming MBES is known to provide very accurate bathymetric information, their data acquisition capability is typically limited to 3-5 times water depth. This does not become a major limiting factor until working in waters shoaler than 10-15 m where it can be difficult to efficiently attain full bottom coverage. In many of these areas water turbidity and resolution requirements preclude the use of LiDAR and there are few alternatives for obtaining bathymetry in an efficient manner (Gostnell et al. 2006).

The interferometric MBES uses the phase content of the sonar signal to measure the angle of a wave front returned from a sonar target. It provides high-resolution wide-swath bathymetry in shallow water with swaths of 10-15 times of the instrument altitude; a significant improvement over the typical 3-5 times of the water depth capability of beamforming MBES. What's more, the acquisition process of the interferometric MBES is fast, and a large mass of data can be processed within a short time. The number of sampling points of the interferometric MBES is higher than that used in the beamforming MBES. The interferometric MBES integrates depth detection technology with imaging technology and can obtain depth and high-resolution seafloor backscatter patterns simultaneously.

Figure 2.9 shows the principle of the interferometric MBES. All of these use similar transducer geometry: two or more horizontal arrays arranged one above the other. Each array produces a beam that is narrow in azimuth, and wide in elevation. One of these arrays (e.g., transmitter, Tx) is supplied with a pulse of electrical energy at the sonar frequency, producing a narrow "shell" of sound that moves outwards from the transducer. Where this shell meets the seafloor there is a small "ensonified" patch, which moves across the seafloor as the sound travels outwards. The ensonified patch scatters sound energy in all directions. When this scattered sound is detected back at the receivers (Rxs), the angle that makes with the transducer is precisely measured, the differences between the phase at each of the Rxs are used to calculate the angle (θ) from which the return originated. This angle of origin, in combination with the range calculated from the two-way travel time, provides a discrete location on the seafloor. This sampling is done thousands of times per acoustic ping, providing a cross-track profile of bathymetry. A number of consecutive profiles are combined to build a three-dimensional model of the seafloor.

Because the interferometric MBES uses a phase detection method, it has the disadvantage that the water depth under the keel is inaccurate, so the system must be equipped with an altimeter or a single-beam echo sounder synchronously.

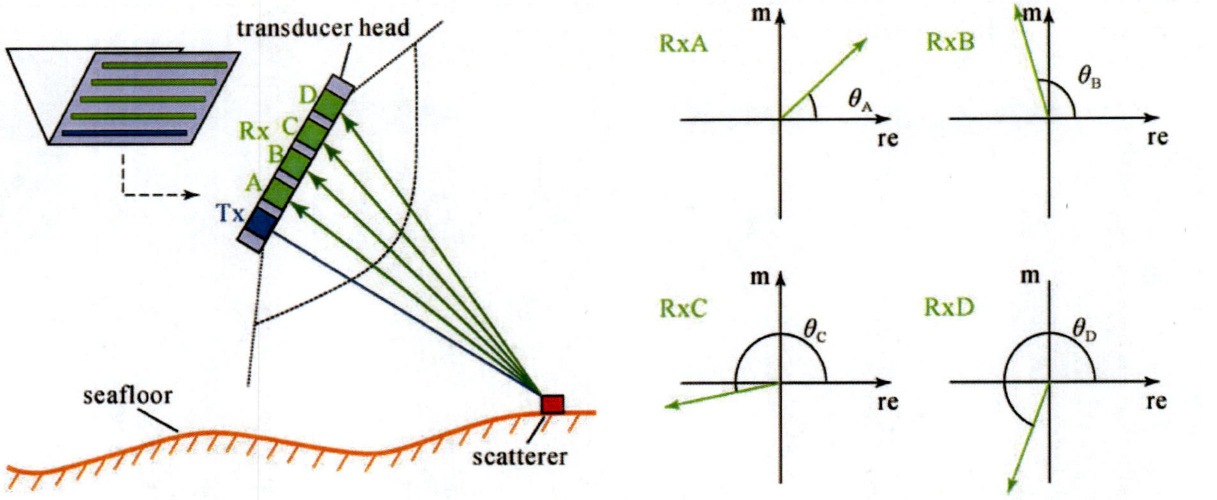

Fig. 2.9 The principle of the interferometric MBES. The primary array of the interferometric MBES consists of two transducers mounted to a "V" plate, each transducer contains multiple ceramic staves, and bottom stave is transmitter (Tx), multiple receivers (Rxs). Phase is measured at each of several Rxs which are spaced at precisely known distances. The angle from which the return originated is calculated using differences in the phase as measured at each Rx (Hiller and Lewis 2004)

2.2 Basic Technical Requirements for Multi-beam Survey

2.2.1 Installation of Multi-beam Echo Sounder System

2.2.1.1 Transducer Installation Method

Shipborne is currently the most commonly used MBES transducer installation approach, and these systems are generally installed in two possible ways: bottom fixed installation and shipside portable installation.

Bottom fixed installation, in which the transducer is installed in a fixed manner on the bottom of the vessel, offers the advantages of improving the safety of the transducer and providing the long-term posture stability, and can also improve the efficiency of the investigation work. However, the disadvantage of this approach is that the installation and disassembly of the transducer must be completed in a dock, and the cost is thus high. The installation method can also be divided into three types: gondola, flush mounted and blister (Fig. 2.10a–c).

The gondola installation is mainly used for deep water MBES systems with large transducers. The transducer, which is positioned below the hull, is generally hidden in a "T" shaped or trapezoidal fairing and is supported using a streamlined bracket (Fig. 2.10a). The advantage of this approach is that the transducer is located far away from the hull, thus allowing it to effectively avoid the hull noise and bubbles that are generated during navigation, and other acoustic equipment (such as an Acoustic Doppler Current Profiler (ADCP) or a sub-bottom profiler) can be installed in the fairing. The disadvantage of this installation is that the attachment resistance will increase strongly (up to or more than 20% of the bare hull resistance), which in turn affects the speed of the vessel and increases the installation cost. In addition, a protruding fairing is more likely to cause the vessel to be stranded at the bottom, to tow fishing net cables and to have difficulty in docking.

Flush mounted installation is primarily used for medium and shallow water MBES systems with moderate or small transducers, or for icebreakers sailing in polar regions. During installation, the transducer is installed in an instrument cabin arranged inside the hull with a structural opening at the bottom of the hull (Fig. 2.10b). When the hull at the installation position is relatively flat, the instrument can be installed directly into the cabin and the opening can be closed using a sound-permeable material; however, when the installation position of the transducer is inclined, it is necessary to design a fairing that can be integrated with the hull. The advantage of this installation method is that it will not lead to increased appendage drag and has little effect on the speed of the vessel. The disadvantage is that bubbles on the hull surface will attach to the surface of the transducer along with the water flow, and it is also easily affected by hull noise, thus reducing the signal-to-noise ratio of the transducer.

Blister installation is mainly used for medium and deep water MBES systems with large transducers. The transducer is installed under the hull and the fairing protrudes by approximately 0.5 m from the hull (Fig. 2.10c). The advantage of this approach is that the bubbles on the surface

Fig. 2.10 Transducer installation method. **a** Gondola installation, **b** flush mounted installation, **c** blister installation, and **d** shipside portable installation

of the hull are not easily attached to the surface of the transducer because of the distance of the transducer from the hull. The disadvantage is that it will add some resistance to the hull (10–20% of the bare hull resistance), and the installation thus requires the design of a special steel structure.

Shipside portable installation is mainly used for shallow-water MBES systems with small transducers. The instrument is mounted on the port and starboard side or on the bow using a rotatable strut (Fig. 2.10d). The advantage of this approach is that the installation and disassembly processes are convenient, but the disadvantage is that the safety of the transducer is poor; the system is susceptible to both noise and bracket jitter, and the installation angle of the transducer must be recalibrated for each installation and recovery process.

2.2.1.2 Transducer Installation Precautions

The transducer installation operations must consider the effects of both noise and jitter on the MBES. The noise types include vessel self-generated noise and environmental noise. Vessel self-generated noise includes: mechanical noise caused by the vessel's diesel engines, gearboxes, drive shafts, propellers and other auxiliary machinery; flow noise caused by the laminar flow associated with speed; the electronic noise of the sonar system itself; cavitation caused by the propellers as a result of the extremely low pressure; and interference from other acoustic instruments in close proximity or with frequencies and harmonics that are close to those of the sonar. The corresponding solutions are described as follows: when installing the transducer, the installation location must be carefully selected to be away from the main engine, the auxiliary machines, the pump and the propeller; the addition of a suitable shroud for the transducers is also essential. The environmental noise includes the noise caused by waves, tides, currents, the weather, ocean earthquakes and other vessels, and the biological noise caused by marine life. However, no good solutions are available to combat the environmental noise at present. Vibration is the main technical problem that is encountered in portable installation on the vessel side and is caused by weak fixation of the transducer mounting rod and insufficient material rigidity.

The following considerations thus apply to MBES installation.

(1) The transducers should be installed in positions where noise is low and bubbles are unlikely to occur.
(2) An attitude sensor should be installed at a position that can accurately reflect the motion of the transducer, and its direction must be parallel to the axis of the vessel.
(3) A gyrocompass must be installed on the headline of the vessel with its direction pointing toward the bow.
(4) The global navigation satellite system (GNSS) antenna should be installed in a relatively open area on the top of the vessel.
(5) The accuracy of spatial relative relationship measurement for each component of the MBES must be better than 0.05 m.
(6) When multiple acoustic instruments are installed on the vessel and need to work synchronously, acoustic synchronizers must be installed to prevent interference from acoustic frequencies similar to those used in the MBES.

2.2.1.3 Installation Offsets

The spatial relationship between all of the sensors in an MBES system (GPS, transducer, motion reference unit etc.) with the vessel's frame of reference is paramount to obtaining high resolution and accurate data. The bathymetric error comes not only from the MBES itself, but also includes the offset errors of each peripheral sensor. Therefore, after installation of the MBES, it is necessary to establish a vessel reference frame (VRF) that is fixed within the vessel and centered on a reference point (Hughes Clarke 2018). To relate the ancillary position and orientation measurements to the sonar, all the position, orientation and sonar sensors must then be located and oriented within this VRF (Fig. 2.11).

The procedure required to establish a VRF includes definition of the offset configurations between the transducers (transmitter and receiver), the motion sensors, and the positioning system antennas, and measurement of the vessel draught (Fig. 2.11). The vessel's central reference point (CRP) is determined upon each new installation of a MBES system, and to best suit the vessels balance and installation criteria (if the MBES is hull or pole mounted) (Lucieer et al. 2018). After the CRP is defined, the next step is to define the installation offsets for the Global Navigation Satellite System (GNSS) antenna, the motion sensor and the transducers. To ensure that the depth charted is the true depth and not the depth under keel, the vessel draught must be taken into consideration during data acquisition (likely at the start and end of a survey) to account for changes in draft due to, for example, fuel usage, although this will depend on the model of the vessel used for survey. All installation offsets are required to be recorded and detailed within the survey report and processing log of the supplied raw data files. A detailed description of the installation offsets is provided in Hughes Clarke (2003a).

2.2.2 Calibration of the Multi-beam Echo Sounder System

Misalignment or mistiming of the sensors relative to each other can create both static biases (e.g., a roll bias) and dynamic residuals (known as wobbles) (Hughes Clarke 2012). Therefore, before the survey can be started, the calibration of the MBES system parameters must be performed via the patch test.

The patch test requires collecting bathymetric dataset over two distinct types of seafloor topography; a flat bottom is used for the roll offset correction whereas a steep slope or a well-defined feature (such as a rock or small wreck) is used for the latency, pitch and yaw offset correction. The dataset for latency, pitch and yaw should be done in water as deep as possible. This is particularly true for the pitch offset correction due to the fact that in shallow water the angle of pitch may not be easily determined due to the lack of resolution (Brennan 2017).

2.2.2.1 Latency Correction

The latency is a time offset between the clocks of the MBES sonar and the positioning system. The latest generation of MBES will incorporate GPS time and the pulse per second (PPS) signal for time synchronisation, as such, no latency is expected in the GPS position. However, it is necessary to complete at least one or two latency tests to prove that the latency is zero for all practical purposes.

As shown in Fig. 2.12a, the dataset for latency correction is collected on a pre-defined line up a steep slope or over a feature. The vessel sweeps the target with a central beam along a pair of overlapping routes at different speeds (e.g., one line should be twice the speed of the other) and in the same direction, the angle of the transmit sector of the system should be as small as possible to increase the transmission frequency. If there is latency, the target obtained by these two measurements is separated; the formula for calculation of the latency is as shown in Eq. (2.1).

$$e_{\text{time}} = h/(v_{\text{h}} - v_{\text{l}}) \tag{2.1}$$

where e_{time} is the positioning time delay (s); h is the distance between the targets (m); v_{h} is the higher vessel speed (m/s); and v_{l} is the lower vessel speed (m/s).

2.2.2.2 Roll Offset Correction

The roll offset value is perhaps the most critical value in the patch test as an error in roll will result in enormous error in sounding depths. As shown in Fig. 2.12b, the dataset for roll offset correction has to be over a flat seafloor. One line is surveyed twice, in reciprocal directions and at the same survey speed. If there is a roll offset, there will be two seafloors sloped in opposite directions and increase linearly from the central beam toward both sides when the soundings are looked at in

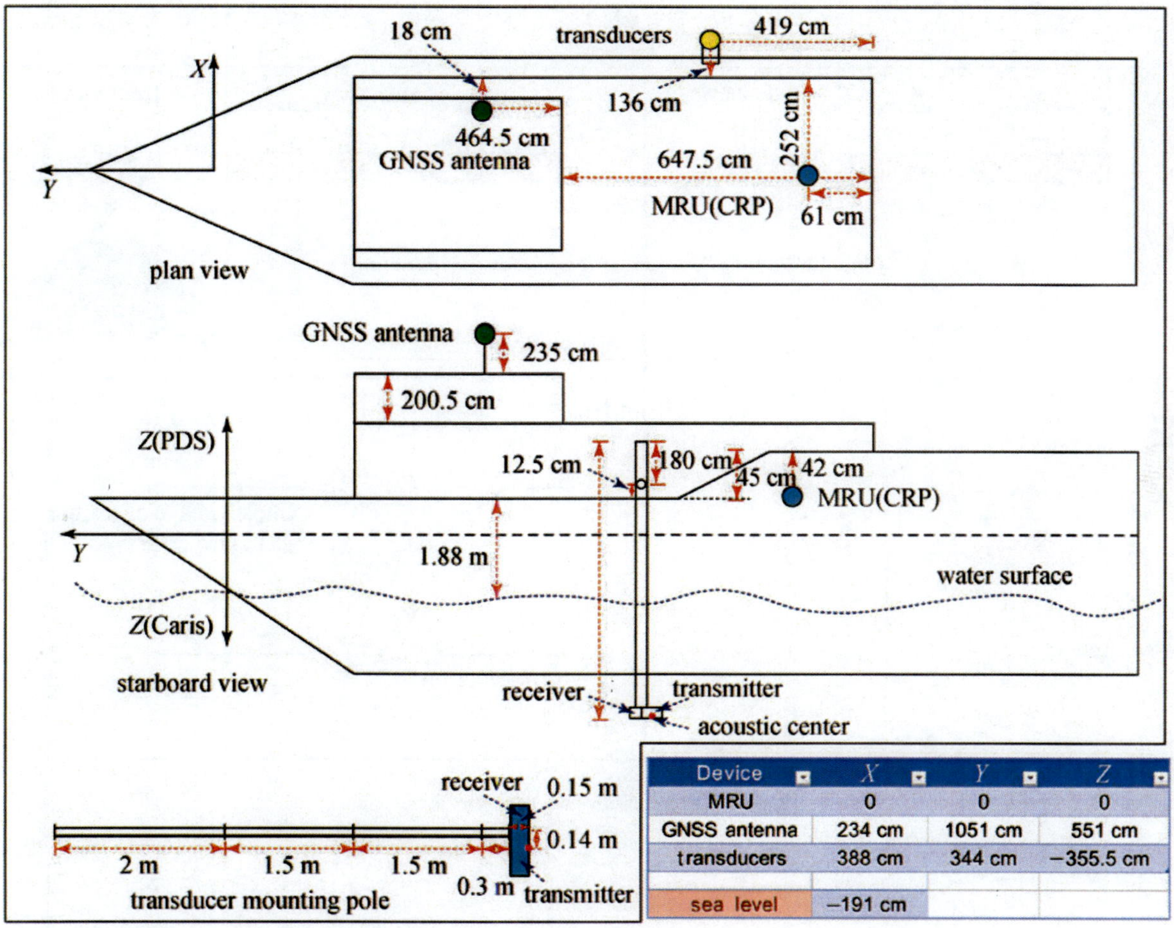

Device	X	Y	Z
MRU	0	0	0
GNSS antenna	234 cm	1051 cm	551 cm
transducers	388 cm	344 cm	−355.5 cm
sea level	−191 cm		

Fig. 2.11 Diagram of the vessel reference frame (VRF) for MBES

profile view. Correction of the roll offset (less than 3°) can be calculated using Eq. (2.2).

$$e_{\text{roll}} = 0.5 \times \arctan(S/X) \tag{2.2}$$

where e_{roll} is the roll offset; S is the depth difference (m); and X is the lateral tracking distance (m).

2.2.2.3 Pitch Offset Correction

A pitch error will result in an along-track position error, which increases greatly with depth. The dataset for pitch offset correction is over the same type of seafloor as the latency dataset, i.e., steep slope or feature on the seafloor. As shown in Fig. 2.12c, after selecting a steep slope on the seafloor and setting a sufficiently long round-trip line, the vessel sweeps the slope at low speed with a stable central beam. The survey lines in reciprocal directions are superimposed, and the same exact parts of the slope that are measured in the different directions are marked. If there is a pitch offset, then two separate slopes are displayed. The formula for calculation of the pitch offset value e_{pitch} is as shown in Eq. (2.3).

$$e_{\text{pitch}} = \arctan(L/2D) \tag{2.3}$$

where L is the lateral separation distance (m) of the slopes; and D is the water depth (m).

2.2.2.4 Yaw Offset Correction

A yaw error will result in a depth position error, which increases with the distance away from nadir. The yaw offset correction is usually the most difficult of the 4 corrections that comprise a patch test. This is especially true if a slope is used for the yaw computation; a well-defined feature generally works much better. The reason for this is that the area that is used for the computation is not directly under the vessel, but in the outer beams and the slope may not be perfectly perpendicular in relation to the course of the vessel.

As shown in Fig. 2.12d, the dataset for yaw offset correction is collected on a pre-defined two parallel survey lines. The lines are to be on either side of a seafloor feature, and should be approximately 2–3 times water depth in separation. The vessel sweeps the feature using edge beams, and should be sufficient overlap between the two survey lines. The bathymetric dataset from the overlapping part are processed and if there is a yaw offset, the measured features

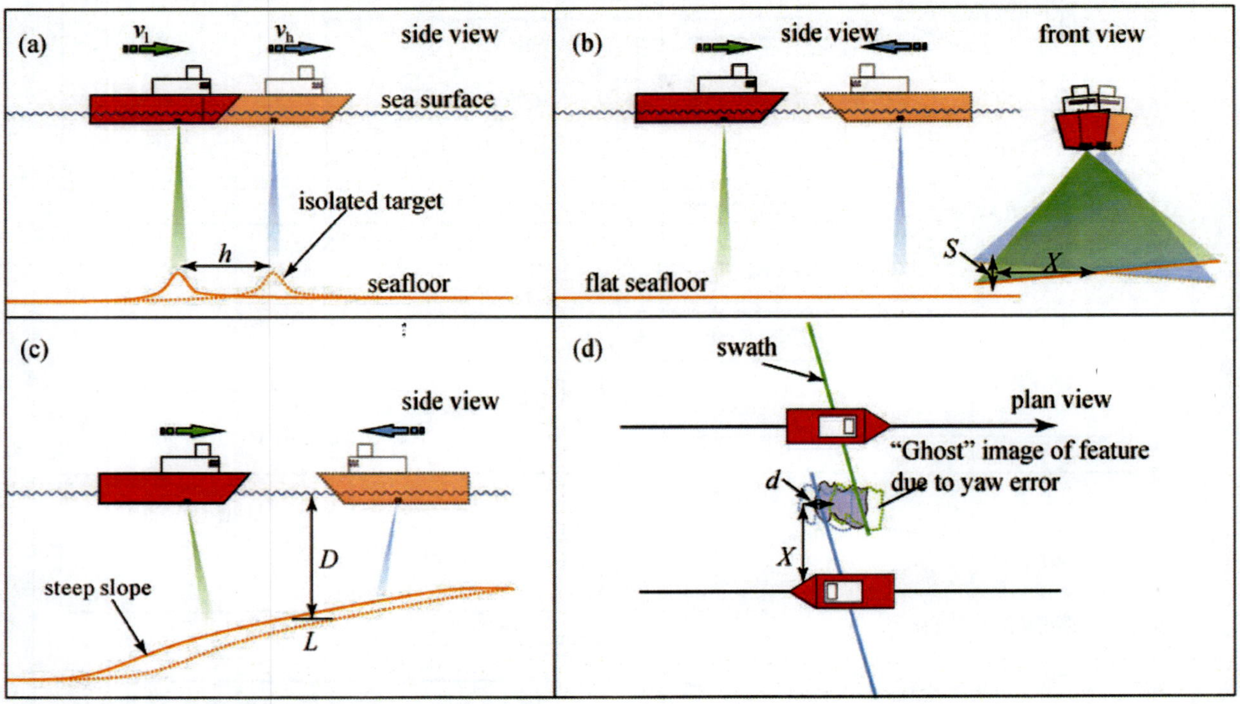

Fig. 2.12 Bathymetric dataset for latency **a** roll offset **b** pitch offset **c** and yaw offset **d** correction

are separated and the offset value is calculated using Eq. (2.4).

$$e_{\text{yaw}} = \arcsin(d/2X) \quad (2.4)$$

where e_{yaw} is the yaw offset; d is the distance that separates the features (m); and X is the relative lateral tracking distance of the beam (m).

2.2.2.5 Order of the Patch Test

The four corrections above must be performed multiple times during the patch test. In general, latency will be solved before pitch; roll will be solved before yaw. The most feasible correction sequences are roll, positioning time delay, pitch and yaw correction, or positioning time delay, pitch, roll and yaw correction. It is common that a larger error than expected in one of the tests will make it necessary to go back and resolve for all previous values.

Each time that a set of correction parameters is obtained and is input into the acquisition system of the MBES to participate in the bathymetric calculations, the calibration line must be repeated until the topographic profile and the water depth contour lines are well matched. A detailed description of the calibration (the Patch Test) is provided in Godin (1998) and NOAA (2014).

2.2.3 Requirements for Laying Out Multi-beam Planned Lines

Unlike single-beam sounding, the survey in multi-beam sounding is generally carried out in full coverage mode. Therefore, to improve the efficiency of the survey, the main track line should be oriented parallel to the depth contour direction of the survey area. To ensure the accuracy of the edge beams, the overlapping of the adjacent swaths should not be less than 10% of the width of a single swath. The direction of check lines should be oriented perpendicular to the main track lines as far as possible, and the check lines should also be evenly distributed to ensure that they are capable of general inspection of the main track lines. The length of the check lines must not be less than 5% of the total length of all track lines, and at least one check line that spans the entire survey area must be arranged.

Bathymetric testing and splicing must also be carried out for the joint parts of adjacent survey blocks between the different types of instruments, different operating periods and different operating units. At the time of survey when using the surveying net method, there must be at least one repeating check line; when performing full coverage bathymetric surveys, the overlap area width should be no less than three to five times the depth of the center beam point. When using the MBES for surveying operations in a survey area with a complex seafloor feature or large terrain fluctuations, the main track line spacing should be narrowed to encrypt the detection process, and the line density must be perfect to reflect the changes in the seabed terrain. Particular shallow points (which are also called "golden" soundings) and obstacles affect the safe navigation of ships and submarines. They are the focus of seafloor mapping. If obstacles are found during the survey, they must be detected in different directions to measure their range and their shallowest depth.

2.2.4 Basic Requirements for Sound Velocity Profile Measurement

One of the main factors that can affect the accuracy of multi-beam sounding is the sound velocity profile (SVP) in water. The SVP varies greatly over different seasons, in different regions and at different depths. Correction of the SVP is the most important and difficult factor to control among all the types of corrections. Generally speaking, the hydrological environment in both offshore and estuarine areas is complex and the SVPs in these areas can vary rapidly, so it is essential to measure the SVP frequently. However, the hydrological environment in deep sea is more stable, which means that the temporal and spatial variations of the sound velocity are small; therefore, rather than measuring the SVP frequently in this case, it is usually sufficient to perform a single acquisition of the SVP each day. The CTD (standing for "Conductivity, Temperature, and Depth"), the sound velocity profiler and the eXpendable BathyThermograph (XBT) are the main instruments used to collect SVPs.

During the multi-beam bathymetric survey, the basic requirements for the SVP measurements are described as follows.

(1) Each time the survey started, the SVP should be collected at least once to gain a preliminary understanding of the sound velocity characteristics of the survey area.
(2) During the survey, the value of the surface sound velocity should be monitored and evaluated in real-time to ensure that the data are accurate and reliable.
(3) In shallow waters, vertical continuous full-depth measurements of SVP must be acquired throughout the journey, and there should be at least one SVP in the 50 km × 50 km range.
(4) In deep seas, vertical continuous full-depth SVPs are again required. The sound velocity at water depths in excess of 2 000 m can be supplemented using the measured SVPs and the historical SVPs. The maximum water depth in this SVP should be greater than that in the current voyage survey block. There should be at least one fully measured SVP in the 100 km × 100 km range.

(5) The SVP should be replaced over time when the bathymetric profile of the seafloor is distorted or shows the long-term symmetrical bend in the monitoring interface.
(6) When the hydrological conditions (i.e., temperature, salinity and turbidity) that affect the SVP change greatly within the survey area, the number of SVP measurements should be increased.
(7) The sound velocity measurement accuracy for each SVP should be better than 1 m/s.
(8) When the SVP is arranged specifically in time and space, the water depth accuracy at the edge beam position of the multi-beam stripe sounding must be in accordance with the relevant specifications.

2.3 Sounding Data Filtering Method Based on CUBE Algorithm

Multi-beam bathymetric surveying technology has greatly advanced the development of marine geosciences, particularly in terms of the leaping evolution of submarine geomorphology from qualitative to quantitative studies (Passalacqua et al. 2015). In addition, remarkable progress has been made in high-resolution submarine geomorphology research in China (Wu et al. 2017; Zhou et al. 2018, 2020).

When compared with single-beam techniques, multi-beam echo sounding has significant advantages that include high precision, high resolution, and full range coverage. However, the massive quantities of MBES data also significantly increase the difficulty and cost of post-processing. The first-generation commercial MBES had only 16 beams (Farr 1980), while the newest generations have more than a thousand beams (Kongsberg 2018). When surveying in shallow water in particular, MBES can generate tens of millions or even hundreds of millions of soundings per hour. Therefore, how to process the bathymetric data from the new generation MBES efficiently has become an important and complex technical issue.

Many studies have addressed automatic filtering of MBES bathymetric data. For example, filtering has been performed by establishing depth and beam angle thresholds (Rezvani et al. 2015), by comparing the angles and gradients between soundings (Bourillet et al. 1996; Pereda García et al. 2016), by establishing local bathymetric models (Debese 2001), and by detecting outliers using statistics (Ladner et al. 2017). However, these automatic processing methods mainly classify the soundings using specific criteria, and as a result, the processing effect is sometimes insufficient.

CUBE is an algorithm that implements computer-assisted hydrographic processing, which is an acronym for "Combined Uncertainty and Bathymetry Estimator". The primary aim of the algorithm is to use as much information as possible from the data to determine the true depth at any point in the survey area from the noisy estimates of that depth, and to quantify an uncertainty on the depth estimate (Calder and Wells 2007). CUBE attempts to answer the fundamental question of "What is the best estimate of depth (and depth uncertainty) at this location?" rather than the more traditional multi-beam bathymetric processing question of "How good is this sounding?".

The algorithm was first developed by Brian Calder and David Wells at the Center for Coastal and Ocean Mapping at the University of New Hampshire in 2000; the Hare-Godin-Mayer (HGM) MBES uncertainty model and multiple hypothesis tracking and model interventions were added to CUBE in late 2001 and early 2002, respectively. The core algorithm has been greatly improved from its initial form (Calder and Mayer 2001, 2003). In recent years, the CUBE has been continually improved to handle the large quantities of bathymetric data. A version of CUBE with CHRT (CUBE with Hierarchical Resolution Techniques) (Calder and Rice 2011; Calder 2019) and network-coupled parallel cluster technology (Venugopal and Calder 2012) are proposed to further improve the processing accuracy and efficiency.

The original CUBE algorithm was first presented in 2001 at the US Hydrographic Conference (Calder and Mayer 2001) and gradually attracted the attention of hydrographic departments worldwide. The Office of Coast Survey of the National Oceanic and Atmospheric Administration adopted the algorithm into its Field Procedure Manual (Calder and Smith 2004; NOAA 2014). The United Kingdom Hydrographic Office (UKHO) Data Archiving Centre found that the bathymetric surfaces that were processed using this algorithm were suitable for hydrographic surveys (Howlett 2010). Netsurvey Ltd studied the hypothesis filtering of CUBE (Mallace and Gee 2005; Mallace and Robertson, 2007). The Chilean Hydrographic Office (SHOA) optimized four of the parameters of CUBE and applied them to CUBE data processing (Vasquez 2007). The Maritime Security Research Center of South Korea compared bathymetric datasets that were processed manually with those processed automatically by CUBE and concluded that the algorithm is both efficient and accurate (Park et al. 2013).

Although not yet universally accepted as data cleaning method by the hydrographic survey industry and hydrographic agencies, who have a particular need to preserve shoal soundings and comply with strict quality control procedures to ensure safety of navigation, CUBE is widely used and of special interest to more applied bathymetric data users and the related scientific community (Lecours et al. 2016). To date, CUBE has been adopted in a variety of commercial and open-source hydrographic software packages, such as CARIS HIPS & SIPS, QPS Fledermaus,

Qimera, HyPack, Geocap Seafloor, MB-Syetem and IFREMER CARAIBES.

2.3.1 Basic Principle of CUBE

Figure 2.13 shows the workflow of CUBE, which can be summarized in the following eight steps.

(1) Two fundamental inputs to the CUBE are horizontal and vertical components of total propagated uncertainty (TPU) (Calder and Wells 2007). The sounding's vertical component of TPU is obtained using the HGM model (Hare et al. 1995, 2011; Hare 2001). The basic idea of this step is to establish a forward propagation prediction model using the relationship between the depth and the instrumental observations (including their uncertainty) to obtain the vertical uncertainty estimates of the sounding. The horizontal component of TPU can be obtained in a similar manner. The algorithm requires that TPU values closely match the performance of the sonar system. Mismatch between TPU and the actual sonar performance can cause inappropriate weighting of soundings and may result in a calculated surface that is not the best estimate of depth (Battilana and Lawes 2010).

(2) The radius of influence around each sounding is calculated using Eq. (2.5):

$$r_{\text{influ}} = \left[G_{\text{res}} \times \left(\frac{\sigma^2_{\text{vertmax}}}{\sigma^2_{\text{vert}}} - 1 \right)^{\frac{1}{\mu_{\text{distexp}}}} \right] - 2.925 \times \sigma_{\text{hz}}, \ \sigma_{\text{res}} < r_{\text{influ}} < 2.925 \times \sigma_{\text{hz}} \quad (2.5)$$

where r_{influ} is the radius of influence around each sounding (m); G_{res} is the resolution of the surface being created (i.e., the grid spacing) (m); $\sigma^2_{\text{vertmax}}$ is the maximum allowable variance according to the user-defined International Hydrographic Organization (IHO) S-44 order (at 1σ) (m^2) (International Hydrographic Organization 2008); σ^2_{vert} is the square of the sounding's vertical uncertainty (variance) at 1σ (m^2); σ_{hz} is the sounding's horizontal uncertainty at 1σ (m); and μ_{distexp} is the scaling factor of the uncertainty, which increases with distance.

(3) The radius of influence around each grid node is calculated using Eq. (2.6):

$$r_{\text{captdist}} = \max\left(\frac{\mu_{\text{captdistsca}}}{100} \times G_{\text{endepth}}, \ \mu_{\text{captdistmin}} \right) \quad (2.6)$$

where r_{captdist} is the radius of influence around each grid node (m); $\mu_{\text{captdistsca}}$ is the scale based on the predicted or estimated depth used to determine the distance from which to accept data, where the value is a percentage of the depth used to limit the radius of influence that one sounding may have on the grid; G_{endepth} is the desired grid node depth (m); $\mu_{\text{captdistmin}}$ is the minimum value of the distance from which the soundings can be accepted. This value is used in conjunction with $\mu_{\text{captdistsca}}$ to limit the radius of influence of an individual sounding. If the sounding's distance to a grid node exceeds the computed capture distance r_{captdist}, the sounding is then not applied to that grid node, even if it was within the initially computed radius of influence r_{influ}.

(4) The uncertainty of each sounding in three dimensions (i.e., the two-dimensional horizontal uncertainty and the one-dimensional vertical uncertainty) is fused and propagated to the one-dimensional vertical uncertainty at each grid node using Eq. (2.7).

$$\sigma^2_{\text{p}} = \sigma^2_{\text{vert}} \left[1 + \left(\frac{S_{\text{dist}} + \mu_{\text{hes}} \sigma_{\text{hz}}}{G_{\text{res}}} \right)^{\mu_{\text{distexp}}} \right] \quad (2.7)$$

where σ^2_{p} is the propagated uncertainty of a sounding after it has been translated to a node (m^2); S_{dist} is the distance from the sounding to the node (m); and μ_{hes} is the value used to scale the horizontal error of each sounding for use in the radius of influence computation.

(5) The soundings are reordered using a median filter to change the sequence in which the input soundings are presented to the node estimation procedure. A queue containing an odd number of input soundings is first constructed. These soundings are inserted into the queue and reordered according to their depth values (with shallowest at the top and deepest at the bottom). When the queue is full, the median sounding (the middle sounding on the queue) is exported to the next process. The exported sounding is then replaced with a new sounding so that the queue is full again, unless there are no more input soundings.

The queue played a vital role in the initial versions of CUBE, which did not support multiple hypotheses. At that time, the queue's purpose was to delay the input of potential outliers into CUBE so that the system could obtain a better depth estimate by the time that the potential outliers had to be processed (Calder 2003). The queue is not quite so important in the current multiple hypothesis version of CUBE because the early outliers would have their own hypotheses and disambiguation would then determine which hypothesis should be picked as the best representation of the seafloor. However, the queue is still used in the current version of CUBE because, as the CUBE manual states, "it does no harm" (Calder and Wells 2007).

(6) When the current sounding has been exported from the median filter to the nodal estimation process, there are

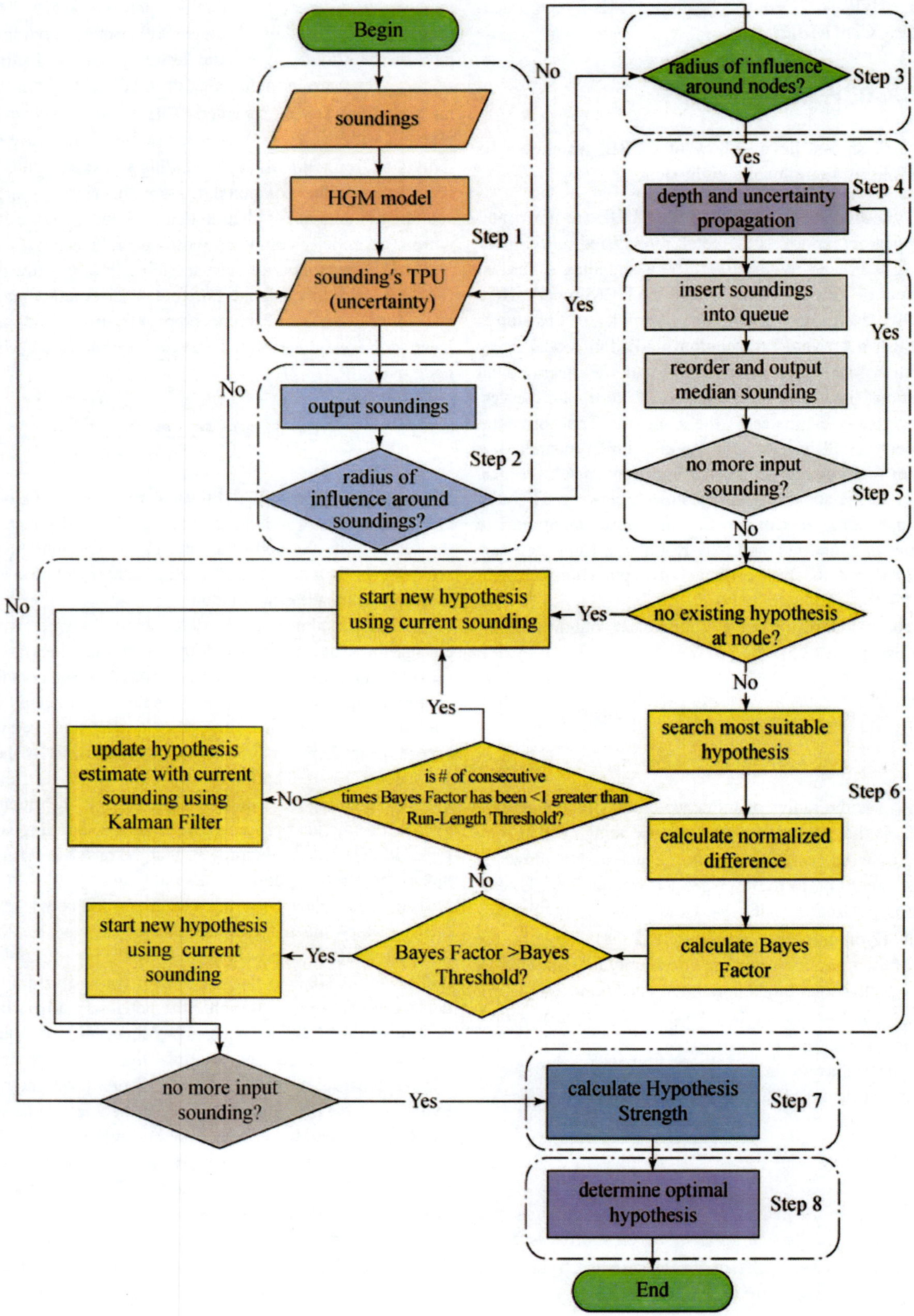

Fig. 2.13 The workflow of CUBE

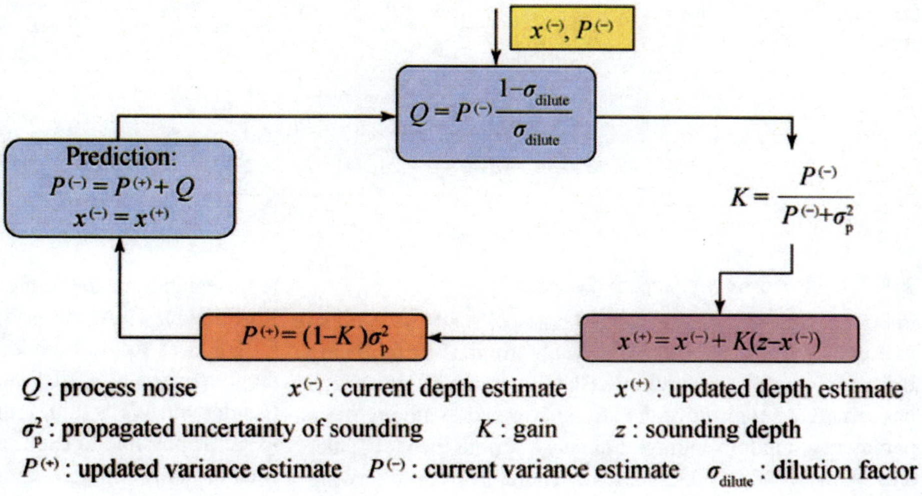

Fig. 2.14 Flow chart of Kalman filtering

two possible situations: (a) If no hypothesis is present at the node, the current sounding will be used as the first depth estimate for a new hypothesis. (b) If hypotheses are present at the node, the most suitable hypothesis for the sounding is determined by computing the normalized difference between each hypothesis and the sounding using Eq. (2.8) (Calder and Mayer 2003; Calder and Wells 2007). If there is more than one suitable hypothesis (i.e., two hypotheses that give identical smallest difference in depth), the last one in the list will then be chosen.

$$e_n = \frac{s_Z - \hat{S}}{\sqrt{\hat{\sigma}^2 + \sigma_{vert}^2}} \tag{2.8}$$

where e_n is the normalized difference; s_Z is the sounding depth (m); $\hat{S}$ is the hypothesis depth estimate (m); $\hat{\sigma}^2$ is the hypothesis variance estimate at 1σ (m^2); and σ_{vert}^2 is the square of the sounding's vertical uncertainty (variance) at 1σ (m^2).

The Bayes Factor must then be calculated using Eq. (2.9) and the results must be compared with the Bayes Factor Threshold and the Run-Length Threshold. If both thresholds pass, the Kalman filter must be run to update to the most suitable hypothesis (Fig. 2.14). If not, a new hypothesis is created as shown:

$$B_n = \begin{cases} \exp\left\{\frac{h_{estimoff}^2 - 2h_{estimoff}e_n}{2}\right\} e_n \geq 0 \\ \exp\left\{\frac{h_{estimoff}^2 + 2h_{estimoff}e_n}{2}\right\} e_n < 0 \end{cases} \tag{2.9}$$

where B_n is the Bayes Factor, and $h_{estimoff}$ is the estimated offset value used to describe the vertical step size that is considered to be significant. Because normalized differences (e_n) are used, the Bayes Factor is computed in terms of the standard normal distribution. Increasing the value of $h_{estimoff}$ will then cause CUBE to create fewer hypotheses, while reducing this value will cause CUBE to create more hypotheses (Calder and Wells 2007).

(7) The algorithm is then applied to find the soundings that will be used for any given node, and to determine how to use these soundings in a Kalman filter to update the hypothesis estimates. When all hypotheses have been found, their qualities will be quantified using the hypothesis strength computation (Eq. (2.10)).

$$\sigma_{stren} = 5 - \frac{\varepsilon_{current}}{\varepsilon_{next}} \tag{2.10}$$

where σ_{stren} is the hypothesis strength, $\varepsilon_{current}$ is the number of soundings in the current hypothesis, and ε_{next} is the number of soundings to be used in the next hypothesis with a decreasing sounding count. The hypothesis strength is used in both data quality control and the disambiguation process. A smaller value of σ_{stren} gives a higher hypothesis strength, which indicates that the hypothesis is then more reliable.

(8) Finally, a disambiguation method is used to determine the optimal hypothesis for each grid node (Calder and Wells 2007).

2.3.2 Parameter Group Optimization Combining CUBE with Surface Filtering

Combining CUBE with surface filtering is actually a mainstream method for MBES data processing (Zhao et al. 2020). However, the selection of the parameter group is the most important step for successful implementation of this method.

Table 2.2 Four parameter groups of CUBE

CUBE parameter groups	$\mu_{captdistsca}$	$\mu_{captdistmin}$	μ_{hes}	$h_{estimoff}$
CUBE default	5	0.5	2.95	4
CUBE deep	20	2	2.95	3
CUBE shallow	4	0.4	0.5	2
CUBE NOAA	0.5	$\frac{G_{res}}{\sqrt{2}}$	1.96	4

2.3.2.1 Parameter Group Selection

$\mu_{captdistsca}$, $\mu_{captdistmin}$, μ_{hes} and $h_{estimoff}$ are the four most important parameters of the CUBE algorithm. By combining these four parameters, multiple CUBE parameter groups can be formed. As a result of the strong correlation among these parameters, random settings can cause serious errors (Calder and Wells 2007; NOAA 2014). Therefore, the following relatively ubiquitous four parameter groups for CUBE are selected (Table 2.2). CUBE Default configuration is the group of default settings for CUBE (Calder and Wells 2007; NOAA 2014); CUBE Deep configuration is intended for use in areas in which small features are not likely (e.g., shifting sand shoals), unimportant (e.g., steep grades and deep water), or can be located separately using side-scan, and this parameter set corresponds with the NOAA's 2006 Complete Gridding standard (CARIS 2015); CUBE Shallow configuration is generally intended for use in areas of critical underkeel clearance, in areas with numerous small features, and in MBESs that can stand on their own without feature-by-feature side-scan correlations; this parameter set corresponds to NOAA's 2006 Object Detection gridding requirements (CARIS 2015).

However, the CUBE NOAA configuration disables the "capture distance scale" function and forces $\mu_{captdistmin}$ to be used by setting the value of $\mu_{captdistsca}$ to 0.5, which was determined to be low enough to be suitable for all grid resolutions. In addition, this configuration sets $\mu_{captdistmin} = \frac{G_{res}}{\sqrt{2}}$ to ensure that no sounding is ever "lost" to the algorithm, but there is no oversampling of data from areas located significantly further away than the grid resolution. Because all the soundings are taken in close proximity to the node, the grid depicts the seafloor in that area most accurately without losing any soundings (NOAA 2014).

For grid-based bathymetric estimation techniques, determining the right resolution at which to work is essential. Appropriate grid resolution can be related, roughly, to data density and thence to sonar characteristics, survey methodology and depth. The seafloor in any particular area has some, generally unknown, level of detail in the topography and a grid-based representation must be capable of achieving the Nyquist sampling rate for the area if an appropriate representation of the seafloor is to be preserved (Calder and Rice 2011). Thus, the grid resolution selection principle of CUBE is as follows: under the premise that each grid node recommended by CUBE absorbs no less than 11 soundings (Calder and Wells 2007), the resolution should be improved as far as possible to capture small topographic features. The filter window (μ_{filter}) automatically eliminates soundings outside by setting a specific window size (which is usually a multiple of the standard deviation).

2.3.2.2 Parameter Group Optimization

As shown in Fig. 2.15, parameter group optimization of the CUBE and surface filtering procedure can be summarized into the following four main steps.

(1) Select a test area that can characterize the typical topographic features for the entire survey area. Analyze the densities of the original soundings to obtain the optimal grid resolution. As shown in Table 2.2, the four parameters are set to form multiple CUBE parameter groups and then the corresponding sets of CUBE surfaces are generated.
(2) Anomalous depth areas based on the various attributes of the CUBE surface are identified and located, including the water depth, the number of hypotheses, the hypothesis strength, and the depth uncertainty. An iterative process should then be used to determine which of the multiple hypotheses for the depth values best represent the seafloor by nominating an alternative hypothesis to act as the depth. When these types of changes have been made to the CUBE surface, the surface must then be updated.
(3) Apply multiple filtering parameters to clean up the outliers automatically based on the updated CUBE surface. Consequently, a series of bathymetric points and grids can then be obtained using the different parameter groups of CUBE and the surface filter.
(4) Use both the sounding and grid analysis methods to compare the bathymetric dataset obtained in Step 3 with the bathymetric dataset that was edited via traditional manual processing. Finally, the optimal parameter group that combines the CUBE with the surface filter for the entire survey area is obtained.

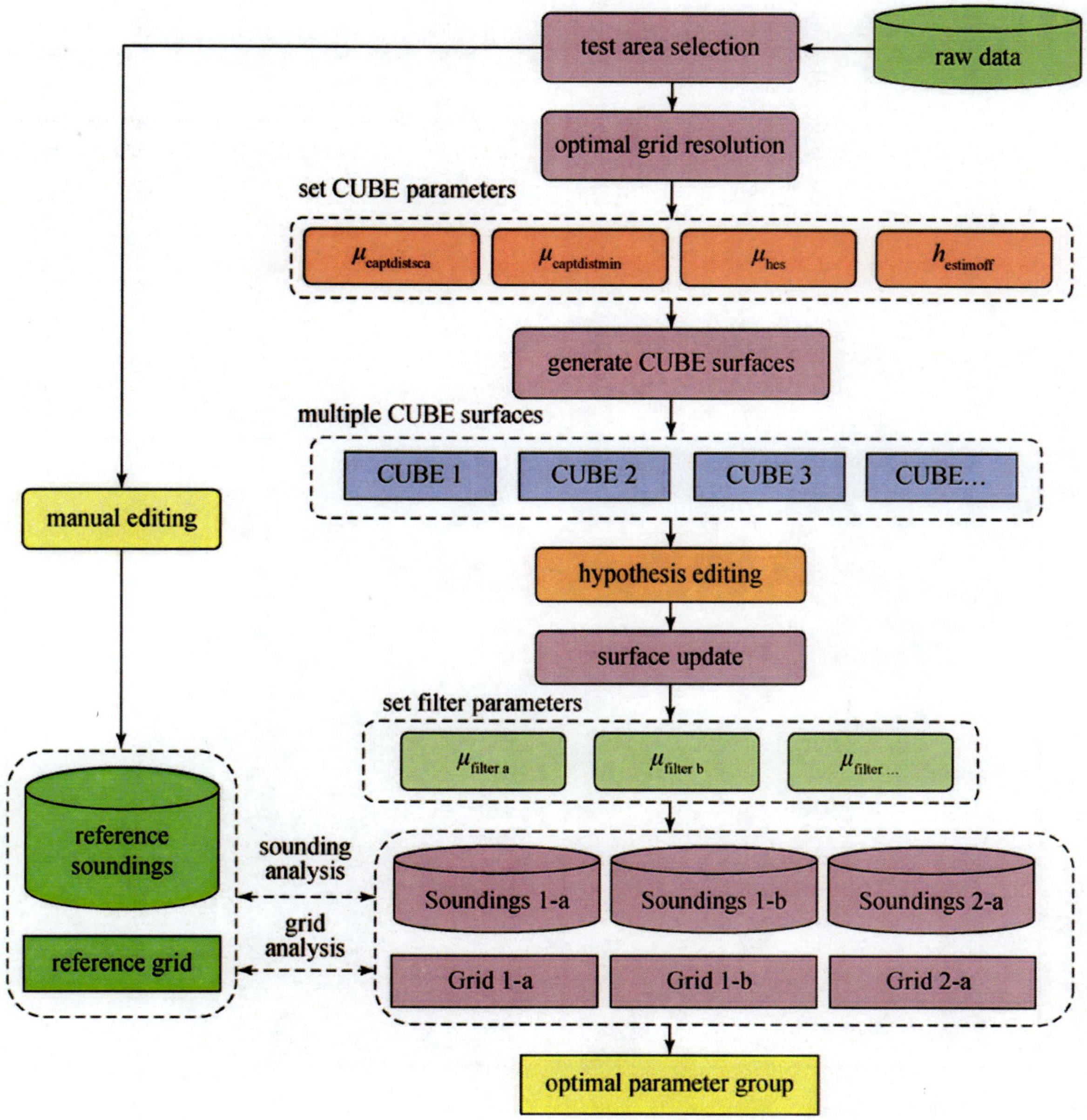

Fig. 2.15 Flow chart of parameter group optimization by combining CUBE with surface filtering

2.3.3 Experiments and Analysis

2.3.3.1 Experimental Data

The dataset was collected in a hydrographic survey of Second Institute of Oceanography, China by the survey vessel *Zhejiayuke 1* in July 2016. The cruise mapped the sand-wave field of the Taiwan Banks, northern South China Sea. The components of the MBES and their accuracies are shown in Table 2.3.

The MBES sonar installation error and the sound velocity profile were compensated during the survey, and the tide was applied in the later stages. As shown in Fig. 2.16, the water depth of the survey area fell in the range of 15–40 m, with an average depth of 28 m. Giant sand waves (heights > 10 m) were well developed in the western part of the area (Zhou et al. 2018, 2020), resulting in great changes in water depth. According to the MBES sonar characteristics, local depth and morphology of the seafloor, the track line spacing was set to ~110 m. Numerous bathymetric holes were found at sand wave crests because of the insufficient coverage of the MBES swath. A typical area (see the red boxed area in Fig. 2.16) is selected for use as an experimental area to validate the proposed method.

Table 2.3 Components of the MBES and their accuracy

Instrument	Model	Accuracy
Multi-beam sonar	Reson Seabat 7125	depth accuracy: 6 mm
Positioning	NavCom SF-3050	<0.1 m
Heading	IXBlue Octans III	±0.1° seclat RMS
Heave		±5 cm or 5% RMS
Roll, pitch		±0.01° RMS
Tide level	RBRduo T.D|tide	±0.05% full scale
Surface sound velocity	Reson SVP 70	±0.05 m/s
Sound velocity profile	RBR concerto C.T.D	±0.2 m/s

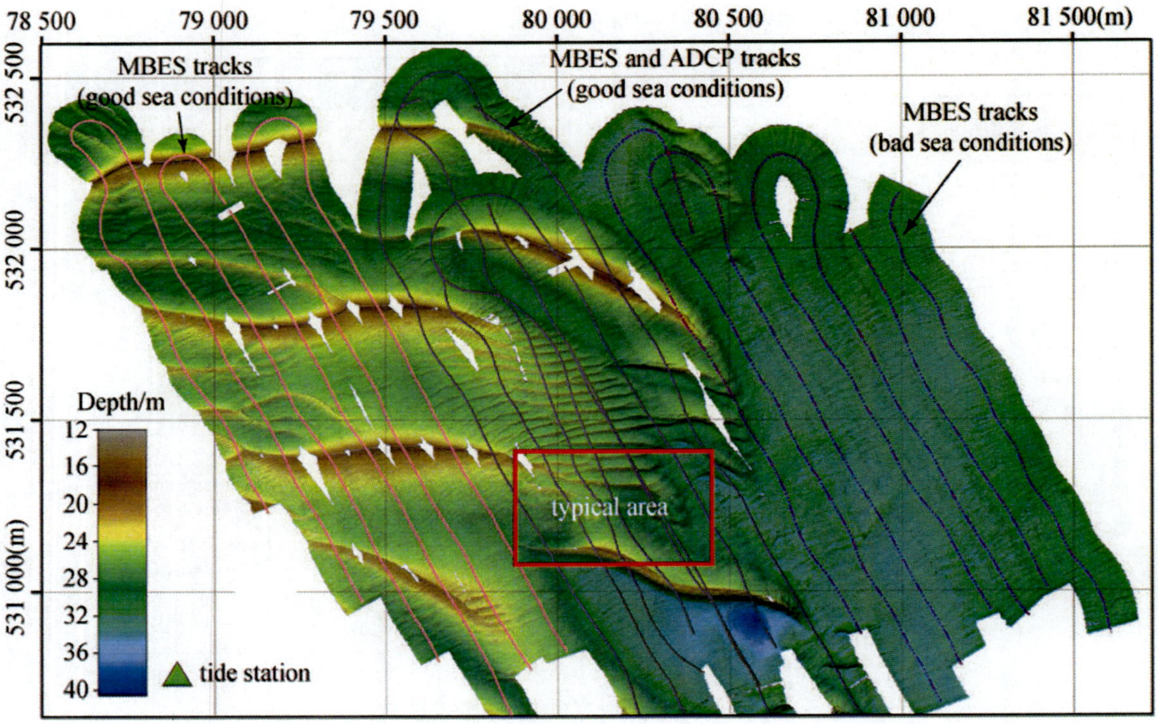

Fig. 2.16 Survey tracks and submarine topography of the survey area

2.3.3.2 Results of Parameter Group Optimization

A sounding density analysis showed that 99.5% of the grid nodes fit the minimum number of the soundings when the grid resolution is 1 m (Fig. 2.17). Therefore, the optimal grid resolution is set to 1 m.

Table 2.4 summarizes the effects of the different parameter groups of CUBE and the surface filter on data processing for the typical area. The soundings analysis (SA) and the grid analysis (GA) show relatively good correspondence. When the MACA + MRCR value is large, and both the MRCA + MRCR value and the grid node difference at the 95% confidence level (ND) value are small (the meanings of these abbreviations are given in the footnote to Table 2.4),

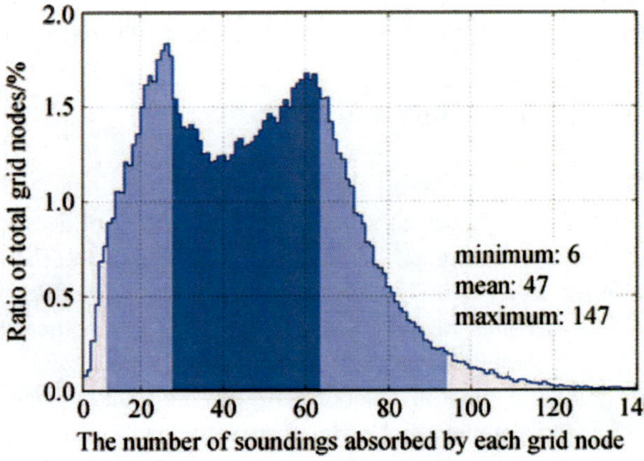

Fig. 2.17 Sounding density analysis of the grid

Table 2.4 Processing effect on the typical area using different parameter groups combining CUBE and surface filter

Line No.	CUBE Parameter groups	μ_{filter}	SA/%						GA/m
			MACA	MACR	MRCA	MRCR	MACA + MRCR	MRCA + MRCR	ND
1	Default	1.5 × Std	94.72	0.831	4.063	0.383	95.10	4.894	0.08
2	Default	2.5 × Std	95.37	0.182	4.071	0.375	95.75	4.253	0.06
3	Default	3.5 × Std	94.51	0.195	4.915	0.381	94.89	5.110	0.12
4	Deep	2.5 × Std	94.93	0.742	3.939	0.385	95.32	4.681	0.09
5	Shallow	2.5 × Std	95.85	0.114	3.537	0.502	96.35	3.651	0.03
6	NOAA	2.5 × Std	95.98	0.102	3.385	0.532	96.51	3.487	0.03

Std: Standard deviation of grid depth; SA: soundings analysis; MACA: manual accepted and CUBE accepted; MACR: manual accepted and CUBE rejected; MRCA: manual rejected and CUBE accepted; MRCR: manual rejected and CUBE rejected; GA: grid analysis; ND: the grid node difference at the 95% confidence level

Table 2.5 Range and step setting of parameter sensitivity analysis

Parameters	Optimal parameter	Parameter range	Step
$\mu_{captdistsca}$	0.5	0.1–20	0.25
$\mu_{captdistmin}$	0.71	0–2	0.1
μ_{hes}	1.96	0.05–5	0.25
$h_{estimoff}$	4	0.1–10	0.5
μ_{filter}(× Std)	2.5	0.5–5	0.5

thus indicating that the processing effect of this parameter group is consistent with that of manual editing, i.e., that this parameter group is suitable for the typical area. As shown in Lines 1-3 of Table 2.4, using the same CUBE parameter group in combination with different surface filtering parameters has specific influences on the processing effects for the final result.

When the filtering window is small (1.5 × Std), some of the correct soundings are removed, and the MACR value increases. When the filtering window is large (3.5 × Std), however, some of the outliers cannot be eliminated automatically, and the MRCA value then increases. By a process of trial and error, the optimal filtering parameter for this area is found to be 2.5 × Std. As shown in Lines 2 and 4-6 of Table 2.4, when the different CUBE parameter groups are combined with the same filtering parameters, it has a significant impact on the processing effect for the final result. From the values of MACA + MRCR, MRCA + MRCR and ND, the optimal parameter group for the typical area is CUBE NOAA and 2.5 × Std.

2.3.3.3 Parameter Sensitivity Analysis

The influence of each parameter variation on the data processing effect is investigated by fixing any four of the parameters and then adjusting the value of the remaining parameter using a specific step size (Table 2.5). As shown in Fig. 2.18, with the exception of $\mu_{captdistmin}$, the changes in the other four parameters can have a major impact on the final processing results. Additionally, the influences of the different parameters are quite different. For example, the MACA + MRCR variation that corresponds to μ_{hes}, $h_{estimoff}$, and μ_{filter} presents a parabolic shape with two sides that are nearly symmetrical. Around the optimal value, the MACA + MRCR value reaches a maximum, thus indicating that the best processing effect has been achieved. As the value of $\mu_{captdistmin}$ increases, the MACA + MRCR value remains almost unchanged at first, then decreases, and finally returns to a constant value again.

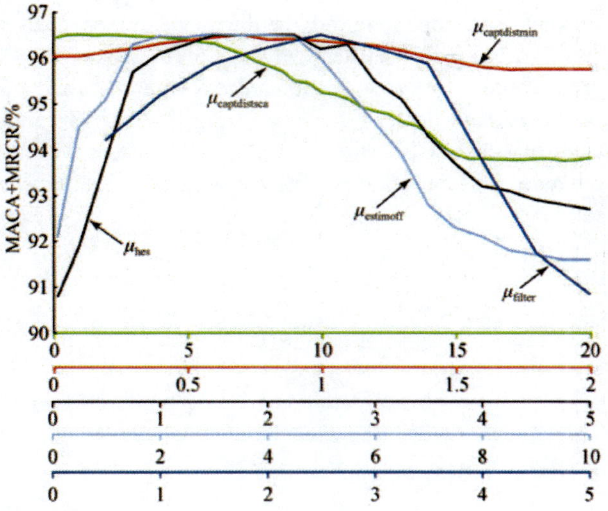

Fig. 2.18 Sensitivity analysis result of the parameter group optimization

2.3.3.4 Application of the Optimized Parameter Group

As shown in Fig. 2.19, the optimal CUBE parameters are used to generate the CUBE surface for the entire survey area. While the raw dataset contains a large number of outliers, the submarine topography that was established using CUBE is almost accurate. However, obvious outliers still exist in individual areas. This is a result of the low signal-to-noise ratio of the data derived by the continuous pulse outliers and manual editing is then required (Fig. 2.19a).

Figure 2.19b shows that the flat areas have only one hypothesis, while in the sand wave fields, and especially at the sand wave crests, multiple hypotheses exist and the hypothesis strength is low (Fig. 2.19c). The reason for the above is that, during the process of uncertainty propagation, the depth near a sounding is assumed to be constant and zero-order prediction is then used to calculate the water depth and the uncertainty propagation (Calder and Mayer 2001). However, the water depth varies greatly in the sand wave fields and these soundings at different depths propagate to the same grid node, thus leading to multiple hypotheses.

In the middle of the area, outliers appear with a perpendicular orientation relative to the tracks at specific intervals, which is caused by interference from the ADCP (Fig. 2.19). Because the number of outliers is relatively small, they can be suppressed effectively using the disambiguation method (Fig. 2.19a). Additionally, for the undulating false topography that is distributed symmetrically along the tracks, and which is caused by shaking the mounting rod during bad weather conditions, no good solution is available at present.

The accuracy of the final CUBE surface is evaluated based on the surface that was generated via traditional manual editing. As shown in Fig. 2.20a, c, with the exception of some larger differences in individual areas due to shaking of the mounting rod, the difference in most areas is less than 3 cm. The statistics show that for 97.2% and 97.8% of the entire survey area, the difference is within 1‰ and 2‰ of the average depth, respectively (Fig. 2.20b). This indicates that the CUBE surface is almost consistent with the reference surface, which meets the accuracy requirements fully.

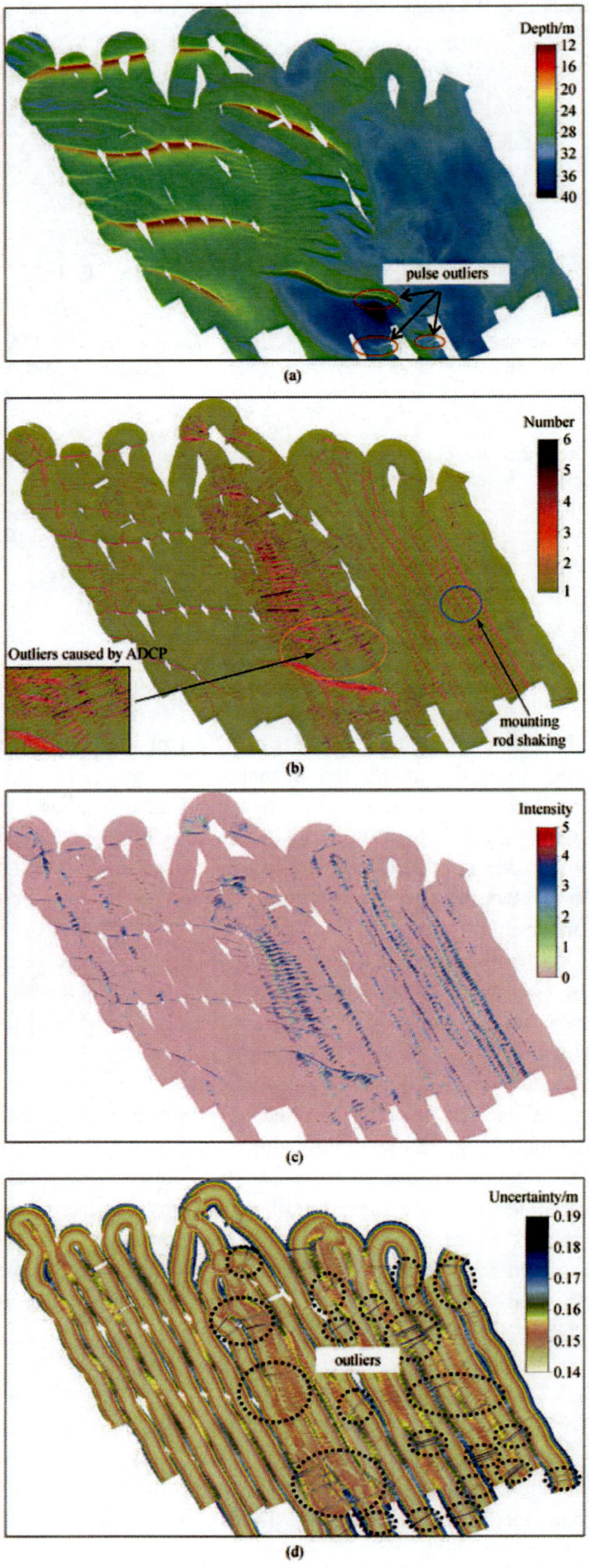

Fig. 2.19 Optimal CUBE surface and auxiliary information. **a** Depth; **b** number of hypotheses; **c** hypothesis strength; and **d** depth uncertainty

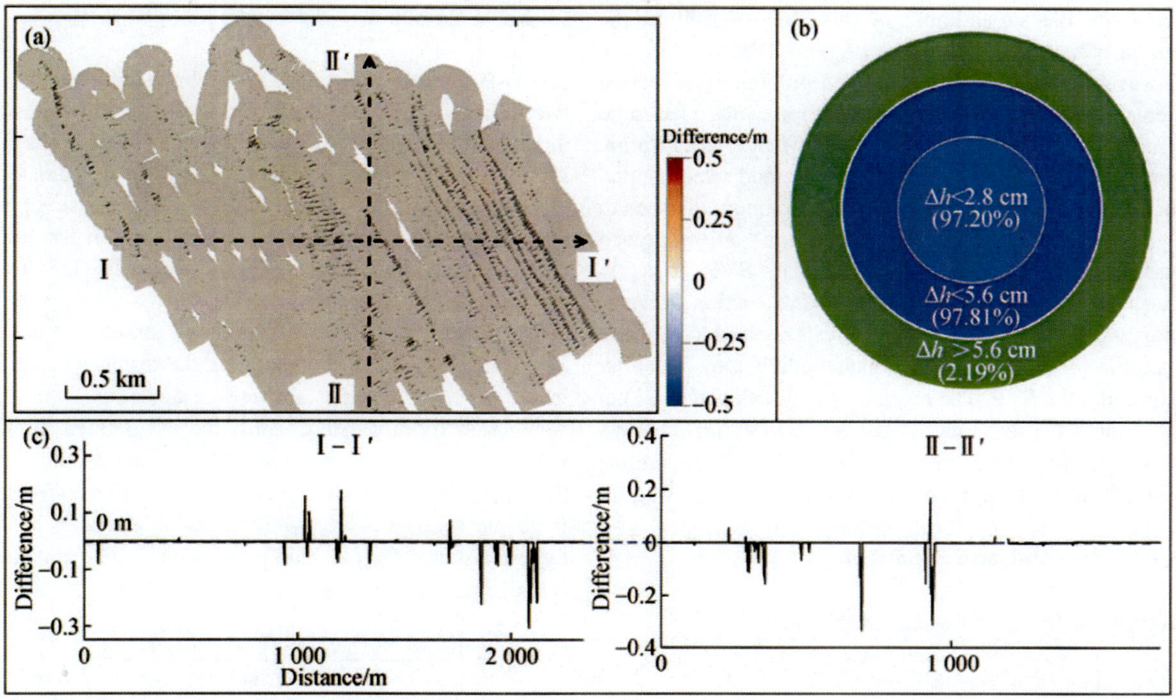

Fig. 2.20 Differences between CUBE and the manually edited surface

2.4 Sound Velocity Correction for Multi-beam Bathymetry Sounding

2.4.1 Automatic SVP Optimization Based on MOV Algorithm

Sound velocity profile (SVP) is the essential parameter necessary for accurate multi-beam bathymetric survey. A MBES system transmits sound waves and receives the round-trip time of their reflection or scattering, and thus it is possible to determine the depth and coordinates of the detected targets using the SVP based on Snell's Law. Generally, the SVP is determined by a device or a sound velocity empirical model. Currently, the sampling frequency of a device for measuring the SVP can achieve 20 Hz. Therefore, for a sink rate of 1 m/s, the number of SVP points will reach 2 000 for a water depth of 100 m, and the quantity of data will be substantial for a survey over a depth of 1 000 m. Because of the high sampling rate, the operational time of ray tracing and beam footprint reduction will increase, lowering the overall efficiency, and preventing the MBES system from working properly. Consequently, many MBES systems limit the number of SVP data points; for example, the deep-water MBES system SeaBeam 2112 (L-3 ELAC Nautik Company, Germany) limits an SVP data file to 30 lines. Another example is the Kongsberg EM series MBES system, the file used by the PU processing unit must be a maximum of 30 kB and it is limited to a maximum number of depth points—1 000 for the EM 2040, EM 710, EM 302 and EM 122 sensors and a maximum of 570 points for older MBESs (Kongsberg 2013). To promote the timeliness of multi-beam surveys and data processing, redundant points in the original SVP must be screened out and at the same time, errors following the streamlining of the SVP must be evaluated and controlled.

In early multi-beam bathymetric surveys, the SVP feature points are selected manually. This is inefficient, and it is difficult to evaluate the SVP accuracy. It is also easy to miss the SVP feature points. Maul and Bishop (1970) applied mean sounding velocity (MSV) to replace the SVP to determine the depth of beams. Typical existing procedures involve subsampling the profile by calculating the mean in vertical bins of fixed size, which could miss the feature points of the original SVP (Beaudoin et al. 2011). Recently, methods of correction based on the equivalent sound speed (ESS) method have been developed, and some models have been applied in fake terrain data correction (Geng and Zielinski 1999; Kammerer 2000; Yang et al. 2007). Because the ESS Method does not use measured SVPs, the correction effects will not be ideal in areas with dramatically changing water depths, and errors caused by non-SVP reasons will be masked because of the complexity of multi-beam echo-sounding errors. Therefore, detailed studies are required on the streamlining of the SVP as well as the retention of feature points of the original SVP.

To overcome the above-mentioned problem, this section presents a new streamlining and evaluation method based on the maximum offset of sound velocity (MOV) algorithm. Based on the measured SVP data, this method selects sound velocity data points by calculating the maximum distance to the sound-velocity-dimension based on an improved Douglas-Peucker algorithm to streamline the SVP. To assess the accuracy of the streamlined SVP, this section uses ray tracing and the percentage error analysis method to evaluate the accuracy of the sounding data both before and after streamlining the SVP. The results show that this method not only ensures the accuracy of the beam footprint after ray tracing but also reduces the computing time, improving the efficiency of the multi-beam survey and data processing.

2.4.1.1 Method and Modules

1. **Improved Douglas-Peucker Algorithm**

(1) Douglas-Peucker algorithm.

The Douglas-Peucker (D-P) algorithm is a method commonly used for line simplification in computer display applications of highly complex multi-segment graphical line objects (Douglas and Peucker 1973). The basic idea of this method (Fig. 2.21a) is to first define the threshold Q of the algorithm and then to join lines from the endpoint P_1 to the endpoint P_n, calculating the vertical distance S_i from the remaining points P_i to this line. Then the largest vertical distance $S_{i\max}$ is compared with Q. The equation to calculate S_i can be written as

$$S_i = \left| \frac{(P_{nx} - P_{1x}) \times (P_{1y} - P_{iy}) - (P_{1x} - P_{ix}) \times (P_{ny} - P_{1y})}{\sqrt{(P_{nx} - P_{1x})^2 + (P_{ny} - P_{1y})^2}} \right| \tag{2.11}$$

where P_{1x}, P_{1y} are the coordinates of the first endpoint P_1, respectively; P_{nx}, P_{ny} are the coordinates of the last endpoint P_n; and (P_{ix}, P_{iy}) are the coordinates of ith point P_i. S_i is the vertical distance between P_i and the lines joined from P_1 to P_n.

If $S_{i\max} \le Q$, all the points are deleted except the two endpoints. Otherwise, point P_i is held back and the line is divided into two parts from P_i. Then lines are joined between P_1, P_i and P_n, respectively, and the calculation of S_i by Eq. (2.11) is performed again. If the largest vertical distance is still greater than Q, the above process is repeated until all the vertical distances S_i are less than Q.

(2) MOV algorithm.

The D-P algorithm is applied to the line simplification for a two-dimensional linear object with two axes in the same units. However, for SVP data, the x-coordinate represents sound velocity with unit of m/s, and the y-coordinate represents depth with unit of m. In this case, the basic physical and mathematical model of the D-P algorithm has already changed, and thus it cannot be used to calculate the distance for streamlining the SVP.

The MOV algorithm is improved based on the D-P algorithm (Fig. 2.21b) by changing the distance calculation into a maximum offset calculation in the sound-velocity-dimension. Thus, for the defined threshold value T, the profile data are joined from the endpoint P_1 to the endpoint P_n. The distance D_i from the remaining points P_i to the sound-velocity- dimension is calculated, and the largest distance $D_{i\max}$ is compared with T. The equation to calculate D_i can be written as

$$D_i = \left| \frac{(P_{i\text{dep}} - P_{1\text{dep}}) \times (P_{nv_s} - P_{1v_s})}{P_{n\text{dep}} - P_{1\text{dep}}} + P_{1v_s} - P_{iv_s} \right| \tag{2.12}$$

where $P_{1\text{dep}}$, P_{1v_s} are the depth value and the sound velocity of the first endpoint P_1 of the SVP; $P_{n\text{dep}}$, P_{nv_s} are the depth value and the sound velocity of the last endpoint P_n of the SVP; $P_{i\text{dep}}$, P_{iv_s} are the depth value and the sound velocity of ith point P_i; and D_i is the vertical distance between the point P_i and the lines joined from P_1 to P_n.

The remaining steps are the same as for the D-P algorithm above; i.e., iterate until all the distances D_i are less than T. Remarkably, the automatic selection of the threshold value T is the key step.

Therefore, the improved D-P algorithm is applicable to the line simplification of the two-dimensional system with different units. This method for SVP streamlining is called the maximum offset of sound velocity method, but it could be extended to other similar two-dimensional line simplification problems.

2. **Basic Modules of the Method**

To evaluate whether the streamlined SVP meets the desired accuracy requirements, this section is divided into two parts: SVP streamlining, and accuracy analysis of the multi-beam sounding data processing using the streamlined SVP. Therefore, the method is divided into two modules: the streamlining module and the evaluation module (Fig. 2.22).

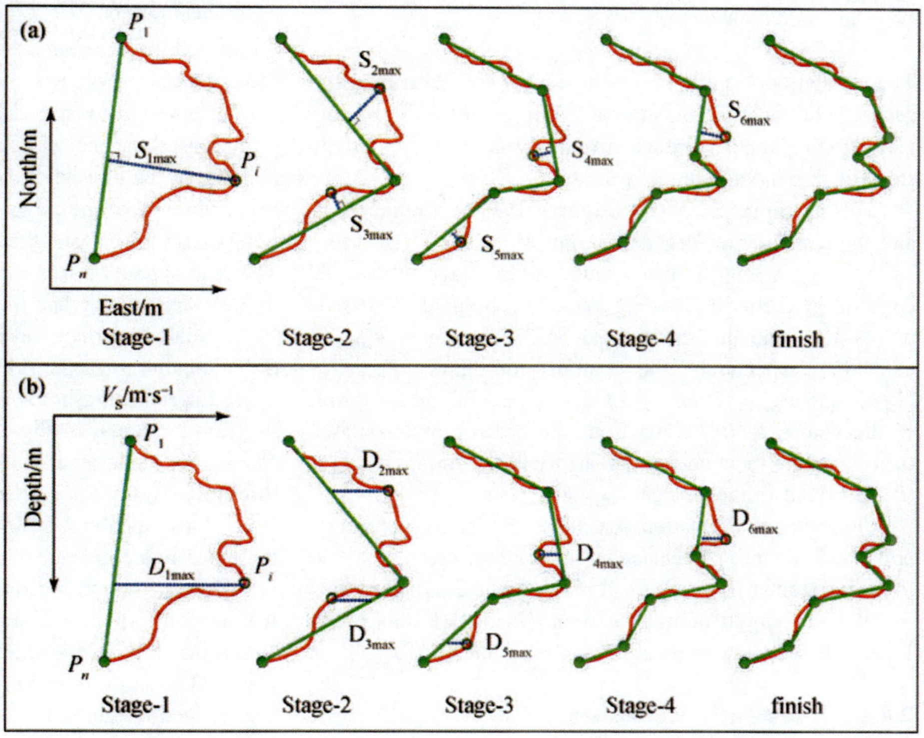

Fig. 2.21 Comparison of D-P algorithm before and after the improvement. The solid red line represents the original line (**a**) and the original SVP (**b**) the solid green line represents the streamlined line (**a**) and the streamlined SVP (**b**)

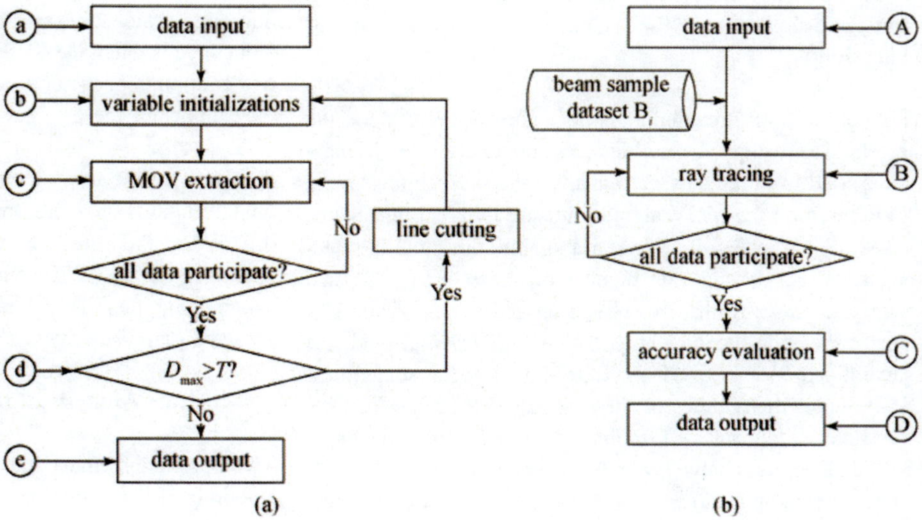

Fig. 2.22 Flow diagrams of the streamlining module (**a**) and the evaluation module (**b**)

(1) The streamlining module.

The streamlining module is used for streamlining the SVP. Its core is the MOV algorithm described above, and the main implementation steps are as follows:

(a) Data input: input original SVP data points V_{orig} and the threshold value T;

(b) Variable initializations: initialize and store streamlined SVP data points V_{simp} and other variables;

(c) MOV extraction: use round-robin algorithm and Eq. (2.12) to obtain the maximum offset D_{max} and the corresponding sound velocity data points P_k;

(d) Line cutting of SVP: judge the numeric size of D_{max} and T. If $D_{max} \leq T$, go to Step (e); if $D_{max} > T$, store P_k into V_{simp}, divide the line into two parts from P_k, then return to Step (c);

(e) Data output: store the two endpoints of V_{orig} into V_{simp}, and then output V_{orig} and V_{simp}.

(2) The evaluation module.

The evaluation module is used to assess the beam footprint accuracy both before and after streamlining the SVP. Its core is ray tracing and percentage error analysis and the four main steps of this module are as follows:

(a) Data input: input the original SVP data points V_{orig} and the streamlined SVP data points V_{simp}.

(b) Ray tracing: use round-robin algorithm and constant-gradient ray tracing with the original SVP data points V_{orig} and the streamlined SVP data points V_{simp} to calculate, respectively, the position coordinates F_{iorig} ($i\in[1, n]$) and F_{isimp} ($i\in[1, n]$) of the corresponding footprints of the beams B_i ($i\in[1, n]$) from the beam sample dataset; then calculate their percentage errors in the horizontal ε_{ihori} ($i\in[1, n]$)and in the vertical ε_{idepth} ($i\in[1, n]$).

(c) Accuracy evaluation: calculate the mean percentage errors and variance percentage errors of all the beam footprints in the horizontal μ_{hori}, σ_{hori} and in the vertical μ_{depth}, $\sigma_{depth.}$

(d) Data output: output the streamlined SVP data points V_{simp} and accuracy evaluation data σ_{hori} and $\sigma_{dept.}$

2.4.1.2 Key Technical Issues

1. **Impact of Threshold on SVP Streamlining, and Automatic Search for Computation Interval of Threshold**

The core of SVP streamlining is the automatic selection of the threshold. To study the problem, we randomly select a measured SVP from the dataset to analyze the streamlining of the SVP based on the MOV algorithm using different thresholds.

As shown in Fig. 2.23, when a smaller threshold is selected, such as 0.05 m/s, the number of streamlined SVP data points decreases from 214 to 21; a reduction rate of 90.19%. The correlation coefficient between the original and the streamlined SVP is 0.999 99, indicating that the streamlined SVP retains the features of the original SVP best. When the threshold is doubled, i.e., 0.1 m/s, the number of streamlined SVP data points is halved, which is a reduction rate of 94.86%. The correlation coefficient is 0.999 94, indicating that the streamlined SVP retains the features of the original SVP better. When the threshold is 0.25 m/s, the streamlined SVP data points reduce another three points, the reduction speed slows markedly, the reduction rate is 96.26%, and the correlation coefficient is 0.999 92. When the threshold is 0.5 m/s, the reduction rate decreases to 97.66%, the bottom of the profile is streamlined into a straight line, the detailed features of the original SVP rapidly disappear, and the correlation coefficient falls to 0.999 04. When the threshold increases to 1.0 m/s, the original SVP data points are streamlined into four feature points; the reduction rate is 98.13%, and the correlation coefficient is 0.998 64. When the threshold is 3.0 m/s, the profile reduction rate reaches 98.60%, the correlation coefficient reduces sharply to 0.953 15, the profile is simplified to a two-layer water layer model, and the features of the original profile disappear almost completely (Table 2.6). With further increase of the threshold, the ultimate profile only retains the two endpoints of the original SVP; the corresponding reduction rate is 99.07%, and the correlation coefficient is 0.834 57. Therefore, it can be seen that the threshold selection is crucial to SVP streamlining and to the retention of feature points.

To analyze further the impact of threshold selection on SVP streamlining, the 11 selected SVPs are streamlined using different thresholds. The threshold and reduction rate diagram is shown in Fig. 2.24a. The MOV algorithm performs well in streamlining all types of SVP. When the threshold is 0.02 m/s, the mean reduction rate of SVP streamlining is about 60%. When the threshold is 0.12 m/s, the mean reduction rate is around 82%. When the threshold is 0.22 m/s, the mean reduction rate is approximately 90%, that is, only 10 points are retained from 100 data points. When the threshold is 0.50 m/s, the mean reduction rate is about 93%, and following that the curve tends to be parallel. The reduction rate curve in Fig. 2.24a increases in stages, indicating that streamlining based on this method is discontinuous. When the threshold is larger than a certain numerical value, the rate no longer increases. The critical value here is when the SVP data points are streamlined to the two endpoints of the original, the rate at this time is $2/n$ (n is the number of the original SVP data points).

The solid red line in Fig. 2.24b indicates the second derivative of the mean reduction rate of the 11 SVPs. When the threshold is 0.04 m/s, the value decreases sharply to about 0.1. When the threshold is 0.18 m/s, the value gradually decreases to zero. It follows that for a particular SVP streamlining, there is an interval [T_1, T_2] of the threshold, and T_1 and T_2 are the values at which the reduction rate mutates.

2. **Accuracy Analysis of the Streamlined SVP**

Two problems must be made clear before using the streamlined SVP: first, it should be established whether the streamlined SVP influences the ultimate velocity refraction correction; and second, the relationship between the threshold and the sounding data accuracy should be determined. As the multi-beam system is a complex system comprising many subsystems with a variety of possible error sources, the use of the method of contrast, between terrains built before and after streamlining the SVP, to estimate the accuracy may not be able to avoid the introduction of other errors. To overcome this problem, the percentage error analysis method is introduced to estimate the impact on sounding data accuracy using the streamlined SVP.

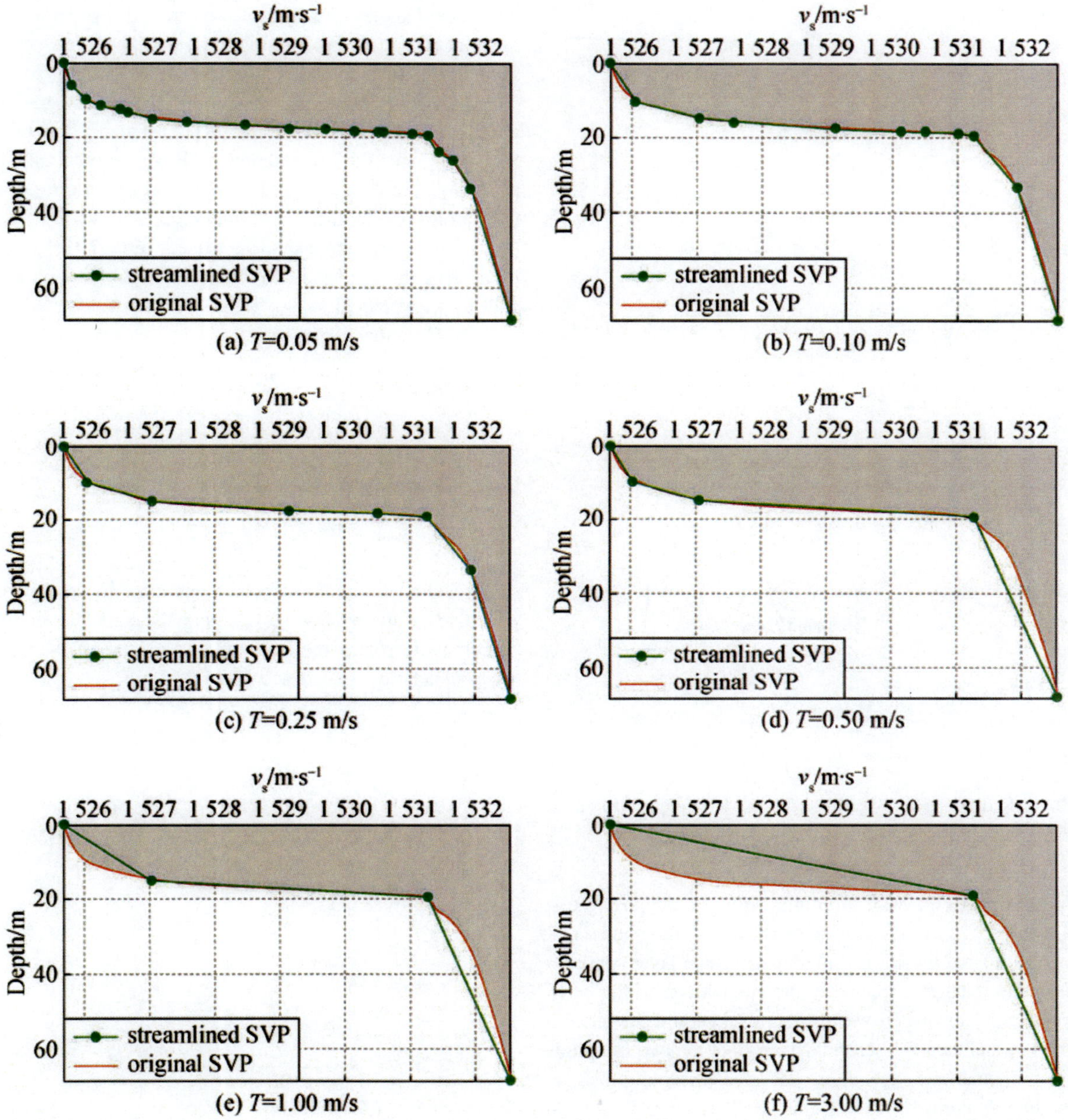

Fig. 2.23 Cases of simplified SVP under different thresholds. The solid red line represents the original SVP; the solid green line represents the streamlined SVP; v_s represents sound velocity

Table 2.6 Cases of simplified SVP under different thresholds

Threshold T/m · s^{-1}	n^{a}_{orig}	n^{b}_{simp}	Reduction rate/%	Correlation coefficient r^{c}
0.05	214	21	90.19	0.999 99
0.1	214	11	94.86	0.999 94
0.25	214	8	96.26	0.999 92
0.5	214	5	97.66	0.999 04
1	214	4	98.13	0.998 64
3	214	3	98.60	0.953 15

Note [a] The number of original SVP data points; [b] the number of streamlined SVP data points; [c] the correlation coefficient between the original and the streamlined SVP data

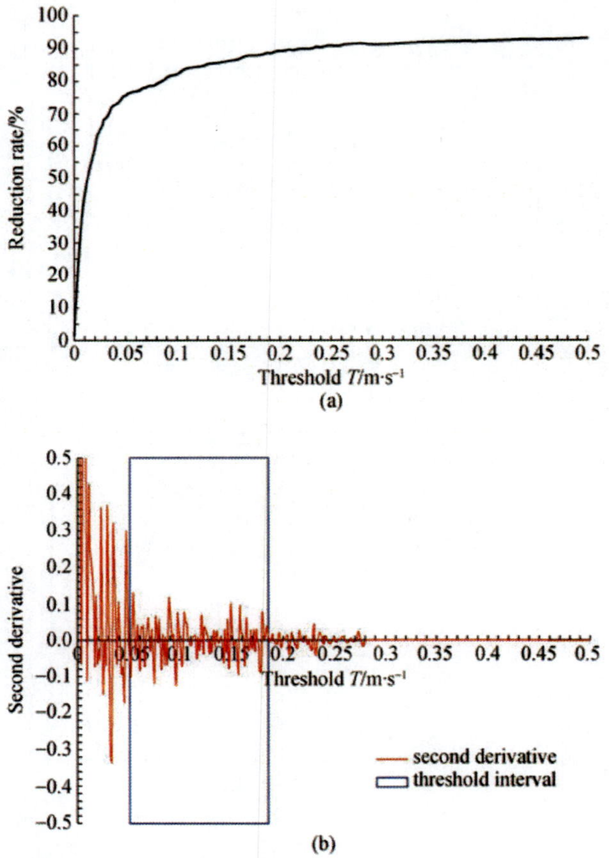

Fig. 2.24 Reduction rate **(a)** and second derivative plotted **(b)** versus threshold T

As shown in Fig. 2.25, to estimate quickly, supposing that the sounding data consist of values in the horizontal (across-track) and in the vertical (depth), inside-layer constant-gradient ray tracing is applied to the comparison of the beam footprint reduction when we use the original and the streamlined SVP, and then the percentage errors of the two footprints in the horizontal and vertical are analyzed (assuming the corresponding footprint of the original SVP is correct). As the accuracy of the SVP has a significant impact on the edge beam, to eliminate the impact of incident angle on water depth error, the incident angles of the selected beam in this paper are in the range of 60° ± 1°. For a certain threshold, the implementation steps of this method are as follows:

(1) Starting from the coordinate origin, ray tracing is applied to the same beam B_i ($i \in [1, n]$) using the original SVP V_{orig} and the streamlined SVP V_{simp}, obtaining footprints $F_{i\text{orig}}$ and $F_{i\text{simp}}$ ($i \in [1, n]$). The absolute errors of the two footprints in the horizontal $\delta_{i\text{hori}}$ and in the vertical $\delta_{i\text{depth}}$ ($i \in [1, n]$) are calculated using Eq. (2.13), and the percentage errors of the two footprints in the horizontal $\varepsilon_{i\text{hori}}$ and in the vertical $\varepsilon_{i\text{depth}}$ ($i \in [1, n]$) are calculated using Eq. (2.14), respectively:

$$\begin{cases} \delta_{i\text{hori}} = \left| F_{i\text{orig}_{\text{hori}}} - F_{i\text{simp}_{\text{hori}}} \right| \\ \delta_{i\text{depth}} = \left| F_{i\text{orig}_{\text{depth}}} - F_{i\text{simp}_{\text{depth}}} \right| \end{cases} \quad (i \in [1, n]) \tag{2.13}$$

$$\begin{cases} \varepsilon_{i\text{hori}} = 100\% \times \frac{(\delta_{i\text{hori}})}{F_{i\text{orig}_{\text{hori}}}} \\ \varepsilon_{i\text{depth}} = 100\% \times \frac{(\delta_{i\text{depth}})}{F_{i\text{orig}_{\text{depth}}}} \end{cases} \quad (i \in [1, n]) \tag{2.14}$$

(2) Calculate the mean percentage error in the horizontal μ_{hori} and in the vertical μ_{depth}, and the standard deviation percentage error in the horizontal σ_{hori} and in the vertical σ_{depth} using Eqs. (2.15) and (2.16).

$$\begin{cases} \Delta_{\text{hori}} = \frac{\sum_{i=1}^{n} \delta_{i\text{hori}}}{n} \\ \mu_{\text{hori}} = \frac{\sum_{i=1}^{n} \varepsilon_{i\text{hori}}}{n} \\ \sigma_{\text{hori}} = \sqrt{\frac{1}{n} \sum_{i=1}^{n} (\varepsilon_{i\text{hori}} - \mu_{\text{hori}})^2} \end{cases} \quad (i \in [1, n]) \tag{2.15}$$

$$\begin{cases} \Delta_{\text{depth}} = \frac{\sum_{i=1}^{n} \delta_{i\text{depth}}}{n} \\ \mu_{\text{depth}} = \frac{\sum_{i=1}^{n} \varepsilon_{i\text{depth}}}{n} \\ \sigma_{\text{depth}} = \sqrt{\frac{1}{n} \sum_{i=1}^{n} (\varepsilon_{i\text{depth}} - \mu_{\text{depth}})^2} \end{cases} \quad (i \in [1, n]) \tag{2.16}$$

Fig. 2.25 Sketch of the percentage error analysis method

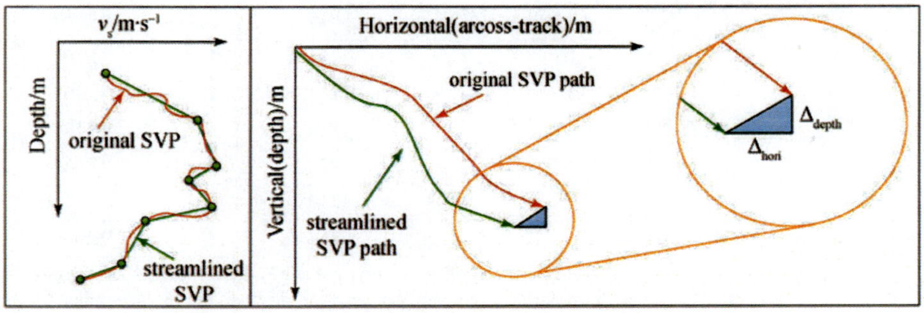

Table 2.7 Relation of thresholds and water depth percentage errors

Threshold T/m · s^{-1}	Reduction rate avg.[a]/ %	In the horizontal			In the vertical		
		Δ_{hori}[b]/ m	μ_{hori}[c]/ %	σ_{hori}[d]/ %	Δ_{depth}[e]/ m	μ_{depth}[f]/ %	σ_{depth}[g]/ %
0.01	44.2	0.00	0.00	0.00	0.00	0.00	0.00
0.05	76.1	0.01	0.00	0.01	0.01	0.00	0.01
0.1	82.1	0.01	0.00	0.02	0.02	0.00	0.02
0.25	90.0	0.03	0.00	0.05	0.03	0.00	0.05
0.5	93.4	0.06	0.00	0.10	0.05	0.00	0.10
0.75	94.1	0.09	0.00	0.15	0.10	0.00	0.15
1	94.3	0.13	-0.01	0.20	0.12	-0.01	0.19
2	95.3	0.26	0.01	0. 41	0.27	0.01	0.43
3	96.1	0.32	0.01	0.51	0.31	0.01	0.49
4	96.2	0.37	0.02	0.6	0.38	0.02	0.61
5	96.3	0.41	0.02	0.68	0.42	0.03	0.70
6	96.4	0.45	0.03	0.75	0.47	0.03	0.76
7	96.5	0.46	0.02	0.77	0.48	0.02	0.78
8	96.5	0.46	0.02	0.77	0.47	0.02	0.78

Note [a]The average of reduction rate under different thresholds; [b]the absolute error in the horizontal; [c]the mean percentage error in the horizontal; [d]the standard deviation percentage error in the horizontal; [e]the absolute error in the vertical; [f]the mean percentage error in the vertical; [g]the standard deviation percentage error in the vertical.

The results of the mean percentage errors measured after streamlining the SVP under different thresholds are shown in Table 2.7. Continuously streamlining the SVP, because of the increase of the threshold, leads to the increase in the standard deviation of the percentage errors of the sounding data (in the horizontal and the vertical). However, the change of the mean percentage errors in the horizontal μ_{hori} and the vertical μ_{depth} is insubstantial. This is because the mean percentage errors of plus or minus with the same absolute value offset each other when the number of statistics is large. When the threshold is 0.01 m/s or smaller, the beam footprint reduction using the corresponding streamlined SVP almost does not produce any standard deviation percentage error in the vertical.

As shown in Table 2.7 and Fig. 2.26, with the threshold $T \in [0, 1]$ increasing, there is a nonlinear relation between the threshold and the sounding data. When the threshold is greater than 7 m/s, the slope of the curve is zero and the SVP is streamlined to the simplified two-endpoints model.

Table 2.8 shows the relation of thresholds, reduction rate, and water depth percentage errors of deep sea SVP which are collected in deep sea. The results show the algorithm has the same streamlined effects on both shallow sea SVP and deep sea SVP, and it has no special requirements for the number of SVP points.

The causes of inaccurate sound speed in a multi-beam echo-sounding system include inaccurate values of surface sound speed and sound traveling in the water column (Beaudoin et al. 2009). Although the International Hydrographic Organization standard for the accuracy of charted depths is 1% for depths greater than 30 m (International Hydrographic Bureau 2008), in a realistic scenario only 0.3% to 0.5% can be assigned to inaccurate sound speed (Peyton et al. 2009). With the threshold increases, the SVP is continuously streamlined (Fig. 2.26), but this also leads to the increase in percentage error in the vertical; thus, for this reason, the SVP streamlining must meet proper standards. Therefore, according to the test

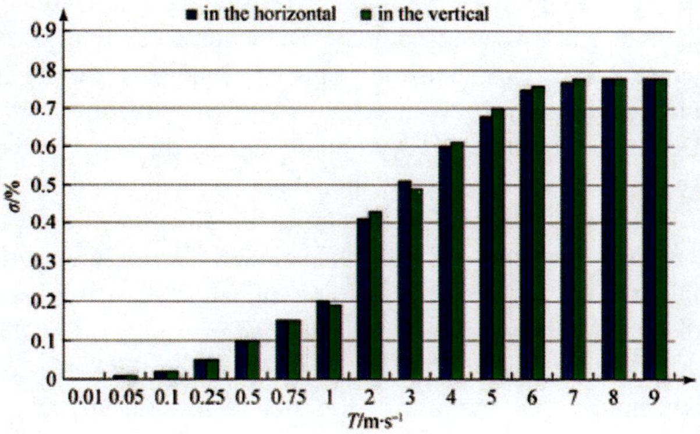

Fig. 2.26 Standard deviation percentage error σ versus threshold T

Table 2.8 Relation of thresholds and water depth percentage errors

Threshold T/m · s^{-1}	n^{a}_{orig}	n^{b}_{simp}	Reduction rate avg. ^{c}/%	In the horizontal		In the vertical	
				μ^{d}_{hori}/%	σ^{e}_{hori}/%	μ^{f}_{depth}/%	σ^{g}_{depth}/%
0.001	66 827	29 666	55.6	0.00	0.00	0.00	0.00
0.005	66 827	27 261	59.2	0.00	0.00	0.00	0.00
0.01	66 827	14 707	78.0	0.00	0.00	0.00	0.00
0.05	66 827	3 712	94.4	0.01	0.03	0.01	0.04
0.1	66 827	1 820	97.3	0.02	0.06	0.02	0.07
0.25	66 827	523	99.2	0.03	0.10	0.03	0.11
0.5	66 827	168	99.7	0.03	0.23	0.04	0.24
1	66 827	45	99.9	-0.04	0.57	-0.05	0.55

Note [a]The number of original SVP data points; [b]the number of streamlined SVP data points; [c]the average reduction rate under different thresholds; [d]the mean percentage error in the horizontal; [e]the standard deviation percentage error in the horizontal; [f]the mean percentage error in the vertical; [g]the standard deviation percentage error in the vertical

results, it is suggested that the percentage error introduced by the SVP should be controlled to within 0.1%, which can serve as a standard to streamlining original SVPs.

3. **SVP Automatic Streamlining and Optimization**

Through the above analysis, it can be seen that there is a close relationship between SVP streamlining and evaluation. A streamlined SVP is produced from the streamlining module, and it is assessed in the evaluation module to judge whether it is satisfactory in terms of accuracy, or whether further streamlining is needed to meet the requirements. Defining and automatically optimizing the threshold is the key to reducing the computational load.

From the above consideration, the flow of the SVP automatic streamlining and evaluation can be described as follows (Fig. 2.27): (1) Input the original SVP$_{in}$; (2) import the streamlining module, obtain the interval [T_1, T_2] of the threshold T based on MOV, and set T_1 as the initial value of threshold T (the purpose of this step is to limit the calculating width of the threshold to reduce computational burden); (3) import the streamlining module again, and calculate an intermediate result SVP$_T$ based on the threshold T; (4) import the evaluation module, input SVP$_T$, and judge whether the standard deviation percentage σ_{depth} is less than 0.1%; and (5) output the result as SVP$_{out}$ if it meets the requirements, otherwise automatically adjust T by the defined step, and return to Step (3) and repeat until SVP$_T$ meets the requirements. The value of the steps can be defined by external input or from the automatic calculation based on the threshold interval.

This flow uses the automatic threshold selection as a link between the streamlining module and the evaluation module, to realize the automatic optimization process of the survey SVP.

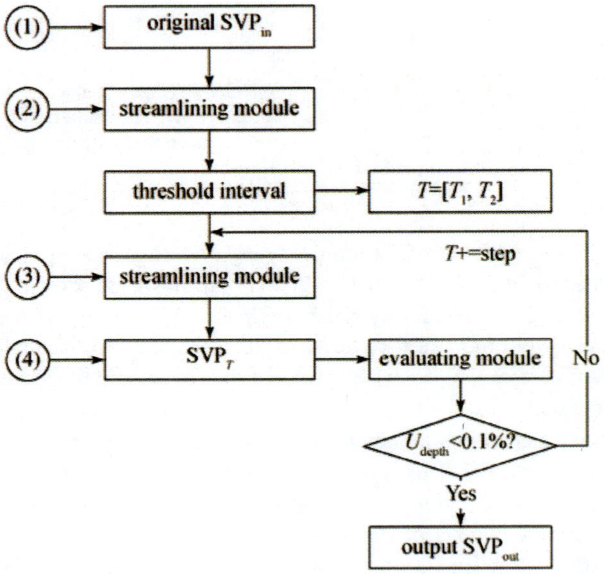

Fig. 2.27 Flow diagram of the automatic optimization of SVP

2.4.1.3 Examples of Time-Dependent Analysis

To test the impact of streamlining the SVP on the efficiency of data processing, a comparison is made between the processing of multi-beam sounding data both before and after the streamlining process. The multi-beam processing software used in the study is Caris HIPS 7.1. The sound velocity correction module of this software is used for sound ray tracing of the SVP data, and to compute the operation time statistically.

1. **Shallow Sea SVP**

The SVP acquisition instrument used in the shallow sea (<100 m) is XR-420-CTD of Canadian RBR Corporation.

Its sampling frequency is 1 Hz, the operational temperature range is from −5 °C to 35 °C, the precision is 1 °C, the conductivity range is from 0 to 8.5 S/m, the precision is ±0.000 3 S/m, the depth range is 500 m, and the precision is 0.05% of full scale, using the Wilson empirical formula to calculate sound velocity.

The XR-420-CTD was used to obtain 11 SVPs during a multi-beam survey in the East China Sea in July and August 2012. For further study, the 11 SVPs are divided into three types (Fig. 2.28a): (1) positive gradient type - sound velocity increases with the increase of depth; (2) negative gradient type - sound velocity decreases with the increase of depth; and (3) small gradient type - sound velocity stays almost the same with the increase of depth. The fundamental features of sound velocity in this area are: (1) sound velocity ranges from 1 523 to 1 541 m/s, and the change of sound velocity in the vertical can be 18 m/s in less than 100 m depth, indicating the complex hydrological environment in this area; (2) two obvious spring layers appear at about 3.5 and 20 m depth; the first may be caused by the sun or surface currents and the second may be caused by the mixing of two different currents; (3) the main SVP type is the negative gradient type, which is different from the SVP in most waters; and (4) the small gradient type may be caused by water mixing after the passage of a typhoon, which causes the layered characteristics of the SVP to disappear in the vertical.

The MBES system used in shallow sea SVP study is the SeaBeam 1180. The water depth is from 40 to 50 m, 40 data files are selected, the survey line length is 498 km, the data size is 390 MB, and the number of original data points is 5.83×10^6.

The SVP data in the study are synchronous with the multi-beam sounding data (Fig. 2.28a). One SVP is selected from each of the three types of profile and the SVP is streamlined on the condition that the standard deviation percentage in the vertical σ depth is less than 0.1% under the different thresholds.

When the threshold is 0 (Fig. 2.29a), that is, using the original SVP, the operation time of sound ray tracing is 58 s. As the threshold is increased, the SVP streamlining rate continues to rise and the corresponding time needed for sound ray tracing tends to decrease nonlinearly; when the streamlining rate is 90%, the time needed is 17 s.

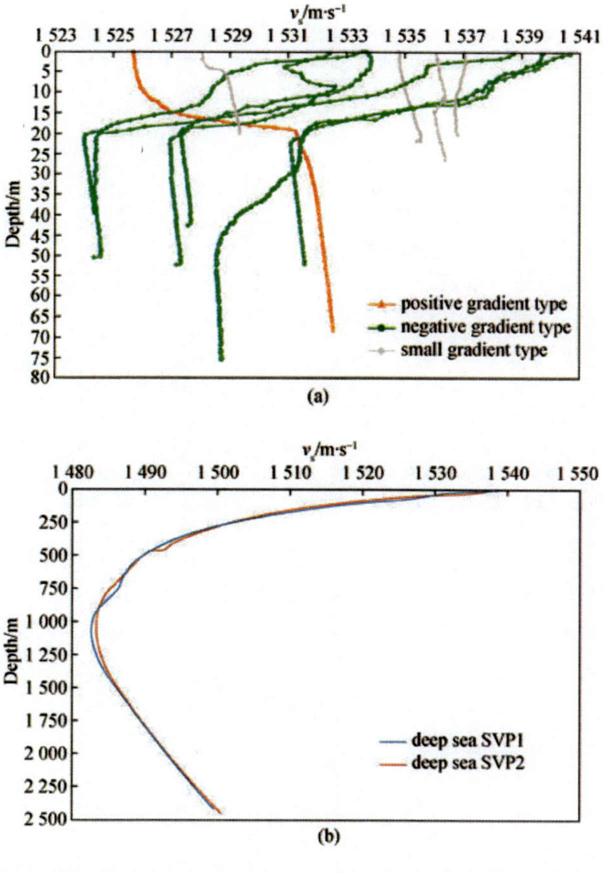

Fig. 2.28 Three types of shallow sea SVPs (**a**) and deep sea SVPs (**b**)

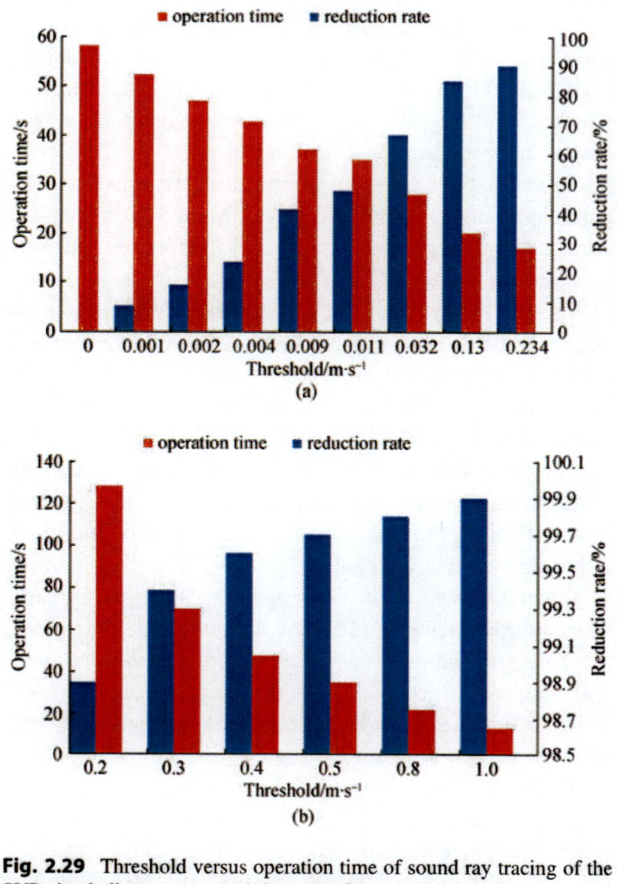

Fig. 2.29 Threshold versus operation time of sound ray tracing of the SVPs in shallow sea (**a**) and deep sea (**b**)

2. **Deep Sea SVP**

The SVP acquisition instrument used in deep sea (>2 000 m) is SBE-911 plus CTD of USA Sea-Bird Electronics Inc. Its sampling frequency is 24 Hz, the operational temperature range is from −5 to 35 °C, the precision is ±0.001 °C, the conductivity range is 0 to 7 S/m, the precision is ± 0.0003 S/m, the depth range is 6 800 m, and the precision is 0.015% of full scale, using the Wilson empirical formula to calculate sound velocity.

The SBE-911 plus CTD was used to obtain 2 SVPs during a multi-beam survey in the South China Sea in April and June 2014 (Fig. 2.28b). Each profile has more than 60 000 sampling points. We can see that the sound velocity characteristics of deep sea are more stable than shallow sea (Fig. 2.28a). The typical deep sea profile may be divided into several layers having different characteristics and occurrence (Urick 1967). Just below the sea surface (0-200 m) is the surface layer, in which the velocity of sound is susceptible to daily and local change of heating, cooling, and wind action. Below the surface layer lies the seasonal thermocline (200–400 m), in which the temperature changes with depth. Underlying the seasonal thermocline is the main thermocline (400–1 100 m), which is affected only slightly by seasonal changes. Below the main thermocline and extending to the sea bottom is the deep isothermal layer (1 100–2 500 m) having a nearly constant temperature, in which the velocity of sound increases with depth because of the effect of pressure on sound velocity.

To verify the validity of the method, the synchronously measured multi-beam sounding data are used to compare the multi-beam data processing timeliness before and after the streamlining of the SVP.

The MBES system used in deep sea SVP study is the SeaBeam 3012. The water depth is from 2 500 to 2 700 m, 2 data files are selected, the survey line length is 98 km, the raw data size is 377 MB, and the number of original data points is 3.73×10^6.

The SVP data in the study are synchronous with the multi-beam sounding data (Fig. 2.28b). The standard deviation percentage in the vertical σ depth is less than 0.5% under the different thresholds.

When the threshold is 0.2 (Fig. 2.29b), the operation time of sound ray tracing is 128 s. As the threshold is increased, the SVP streamlining rate continues to rise and the corresponding time needed for sound ray tracing tends to decrease nonlinearly; when the streamlining rate is 99.9%, the time needed is 13 s.

Therefore, it can be seen, on the precondition of accuracy, that streamlining the SVP based on the MOV algorithm can reduce the post-processing time dramatically, and operational efficiency improves significantly, indicating that this method has practical value for engineering applications.

2.4.2 Area Difference Method

The theory of the area difference method involves obtaining the correction value for accurate beam positioning by calculating the difference between the integral areas of the sections formed by the original SVP and the accurate SVP, and then completing the sound velocity correction. First, it is necessary to understand the concept of the equivalent SVP.

2.4.2.1 Equivalent SVP

The position at which a sound pulse impinges on the seafloor is almost the same for a family of SVPs with the same travel time, the same surface sound velocity and the same area underneath them (Geng and Zielinski 1999). Therefore, when we calculate the beam footprint position, a simple constant-gradient equivalent SVP can be used to replace the actual complex SVP for ray tracing purposes (Fig. 2.30).

Assume that the sound velocity $c(t, z)$ is a function of the water depth z and the time t. Two SVPs, $c_1(t, z)$ and $c_2(t, z)$ have the same surface value c_0 and the same initial depth z_0, as shown in Fig. 2.31. The integral area of the section formed by these SVPs is

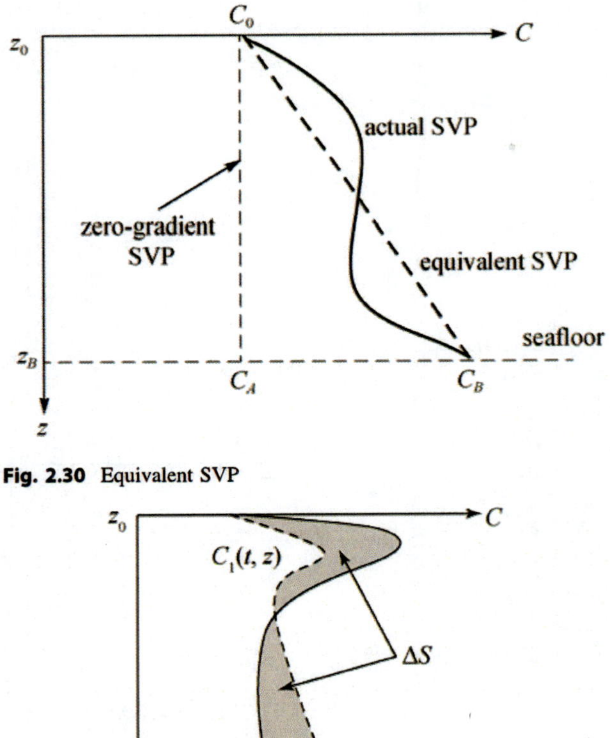

Fig. 2.30 Equivalent SVP

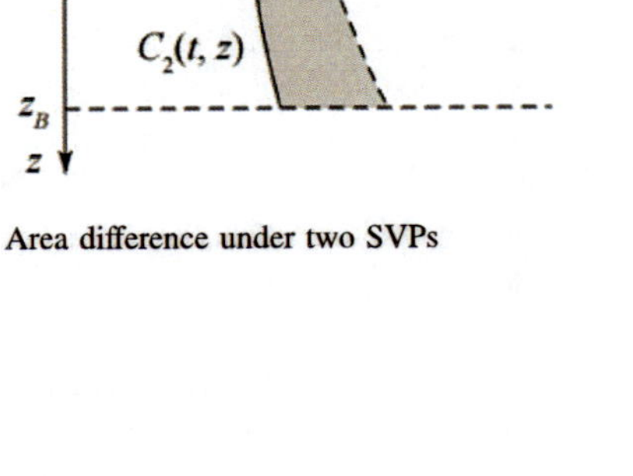

Fig. 2.31 Area difference under two SVPs

$$\Delta S = \int \mathrm{d}S = \int_{Z_0}^{Z_B} [C_1(t,\ z) - C_2(t,\ z)]\,\mathrm{d}z \tag{2.17}$$

The relative area difference is defined as

$$\varepsilon_S = \frac{\Delta S}{\int_{Z_0}^{Z_B} C(t,\ z)\mathrm{d}z} = \frac{\Delta S}{S} \tag{2.18}$$

where $C_1(t, z)$ and $C_2(t, z)$ are the two SVPs, and z_B is the beam projection point at the bottom of the water.

Equations (2.17) and (2.18) show that the depth estimate difference related to these two SVPs is almost a function of the relative area difference ε_S alone. If a sound ray is traced within two different SVPs using the same travel time and the same initial incident angle, and these two SVPs also have the same surface sound speed and the same area between the ray and the depth axis, the calculated positions of their footprints will also be the same. Therefore, a simple SVP can be obtained to replace the actual complex SVP, as illustrated in Fig. 2.30. An actual SVP in the ocean can then be represented approximately using one of these SVP types. The SVP is a function that varies with space and time and if a simple equivalent SVP can be found for ray tracing, the computational efficiency will be greatly improved.

2.4.2.2 Relative Area Difference Method

The theory of the relative area difference method is described as follows. First, the beam footprint position is obtained by calculating a simple reference SVP (such as the zero-gradients SVP); then, the correction value for the beam position is obtained from the difference in the integral area of the section formed by the reference SVP and the actual SVP. Finally, the correct beam footprint position is obtained. As shown in Fig. 2.32, it is assumed that the difference in the integral area of the section formed by the actual SVP and the reference SVP is ΔS_i at layer i; here, we discuss the horizontal position and depth characteristics of each beam caused by ΔS_i.

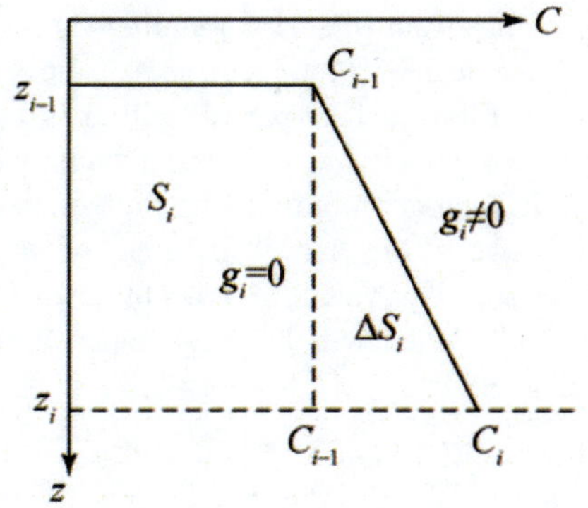

Fig. 2.32 Difference in the integral area of the section formed by the actual SVP and the reference SVP

Consider the actual SVP (where $g_i \neq 0$), where the horizontal range y_i and the depth z_i of the beam passing through layer i can be calculated as follows:

$$\begin{cases} y_i = y_{i-1} + \frac{\cos\theta_{i-1} - \cos\theta_i}{pg_i} \\ z_i = z_{i-1} + \frac{\sin\theta_i - \sin\theta_{i-1}}{pg_i} \end{cases} \tag{2.19}$$

Assuming that the sound velocity within the layer remains constant (i.e., $g_i = 0$), the horizontal range y_i' and the depth z_i' of the beam are defined as follows:

$$\begin{cases} y_i' = y_{i-1} + C_{i-1} t_i \sin\theta_{i-1} \\ z_i' = z_{i-1} + C_{i-1} t_i \cos\theta_{i-1} \end{cases} \tag{2.20}$$

The beam footprint position accuracy cannot meet the accuracy requirements if the result above is used. However, ray tracing will be time consuming if the actual SVP is used. Therefore, to simplify the calculation process and improve the precision, the area difference method is used to obtain the beam footprint position correction value. Additionally, there is a representative SVP error in the multi-beam survey, and the original SVP is corrected based on this error to obtain the actual SVP; the area difference method can thereby be used to obtain the beam position more accurately.

As shown in Fig. 2.32, the relative area difference ε_{s_i} is

$$\varepsilon_{s_i} = \frac{\Delta S_i}{S_i} = \frac{C_i - C_{i-1}}{2C_{i-1}} \tag{2.21}$$

When $g_i = 0$, the actual depth and the horizontal relative error can be obtained as follows:

$$\begin{cases} \varepsilon_{z_i} = f_{z_i} = \frac{z_i' - z_i}{z_i - z_{i-1}} \\ \varepsilon_{y_i} = f_{y_i} = \frac{y_i' - y_i}{y_i - y_{i-1}} \end{cases} \tag{2.22}$$

Using Snell's law and Eq. (2.21), we obtain

$$\sin\theta_i = (1 + 2\varepsilon_{S_i}) \sin\theta_{i-1} \tag{2.23}$$

By substituting Eqs. (2.13), (2.19), (2.20), (2.21), and (2.23) into (2.22), we obtain

$$\begin{cases} f_{z_i}(\varepsilon_{s_i},\ \theta_{i-1}) = \frac{\cos\theta_{i-1}}{2\varepsilon_{s_i}} \ln\left\{(1 + 2\varepsilon_{s_i}) \frac{1 + \cos\theta_{i-1}}{1 + \sqrt{1 - [(1 + 2\varepsilon_{s_i})\sin\theta_{i-1}]^2}}\right\} - 1 \\ f_{y_i}(\varepsilon_{s_i},\ \theta_{i-1}) = \frac{[\sin\theta_{i-1}/(2\varepsilon_{s_i})] \ln[(1 + 2\varepsilon_{s_i})(1 + \cos\theta_{i-1})/(1 + \sqrt{1 - [(1 + 2\varepsilon_{s_i})\sin\theta_{i-1}]^2})]}{\sqrt{1 - [(1 + 2\varepsilon_{s_i})\sin\theta_{i-1}]^2} - \cos\theta_{i-1}/\sin\theta_{i-1} - (1 + 2\varepsilon_{s_i})\sin\theta_{i-1}} - 1 \end{cases} \tag{2.24}$$

As shown in Eq. (2.24), the horizontal and depth relative errors in layer i are related to the incident angle and the integral area difference of this layer only.

Using Eq. (2.22), the error-correction functions for a constant sound velocity in the layer can be obtained as

$$\begin{cases} y_i = y_{i-1} + \frac{y_i' - y_{i-1}}{1 + f_{y_i}(\varepsilon_{S_i}, \theta_{i-1})} \\ z_i = z_{i-1} + \frac{z_i' - z_{i-1}}{1 + f_{z_i}(\varepsilon_{S_i}, \theta_{i-1})} \end{cases} \tag{2.25}$$

If the entire water column is regarded as a single layer, then the incident angle of the sound ray, sound velocity of sound, the initial transducer position and the one-way travel time of sound are given by θ_0, C_0, (y_0, z_0) and T, respectively. First, the initial position of the beam is calculated when the sound velocity gradient is 0 and the correction value for the beam position is then calculated using the relative area difference method to obtain the final position of the beam (z_B, y_B):

$$\begin{cases} z_B' = z_0 + C_0 T \cos\theta_0 \\ y_B' = y_0 + C_0 T \sin\theta_0 \\ z_B = z_0 + \frac{z_B' - z_0}{1 + f_z(\varepsilon_S, \theta_0)} \\ y_B = y_0 + \frac{y_B' - y_0}{1 + f_y(\varepsilon_S, \theta_0)} \end{cases} \tag{2.26}$$

The relative area difference method first selects a relatively simple SVP; then, it establishes a relationship with the actual SVP based on the area difference principle and corrects the calculation results for the reference SVP, which simplifies both the ray tracing process and the calculation complexity. In this multi-beam data processing process, errors from outdated SVPs can be corrected using area differences between outdated and current SVPs, and an equivalent constant-gradient SVP can be found to give an accurate SVP; the relative area difference method is used to simplify the calculations.

2.4.3 The Equivalent SVP Model with Three Layers to Removal of Refraction Artifacts

2.4.3.1 Basic Principle of the Model

As shown in Fig. 2.33, the model has three water layers, includes five parameters: surface sound velocity c_0, the sound velocity c_1 and c_2 in the bottom of the first and second layers, and depths z_1 and z_2 of the first and second layers. The sound velocities of the upper two layers have a constant gradient. The sound velocity c_2, depth z_2 and sound velocity gradient of the third layer are the same as one of the most bottom layers of measured SVP. Therefore, the sound velocity c_2 and the depth z_2 of the second layer are known. But the parameters of first layer seem to be unknown which include surface sound velocity c_0, sound velocity c_1 and depth z_1. Surface sound velocity has much influence on sound ray tracing for linear array (Beaudoin et al. 2004).

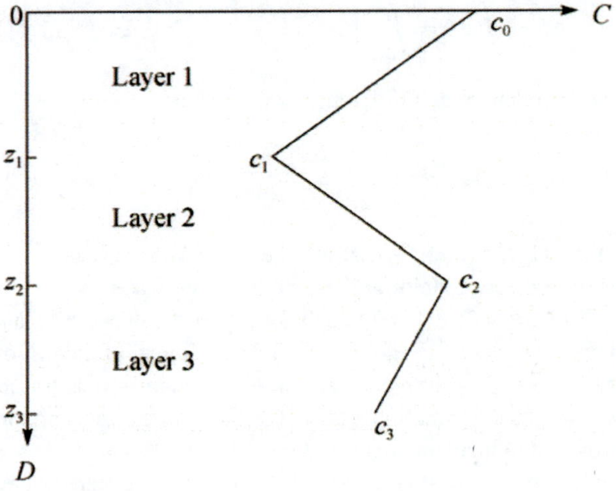

Fig. 2.33 The model of SVP

At first, we use Eq. (2.27) to correct surface sound velocity errors according to the interpolation of measured SVPs. Then new positions of beam footprints can be recalculated after the correction of surface sound velocity errors. Next, we assume that the original water column only has one layer and the sound velocity is 1 500 m/s. Then the travel time and launch angle of each beam can be recalculated using the across-track distance and depth, instead of using raw records. The surface sound velocity c_0 in our model can be also considered as 1 500 m/s. If we assume that the two upper layers have the same thickness, the depth z_1 will be half of the depth z_2 of the second layer. There will be only one unknown parameter left, i.e., sound velocity c_1. The variance of c_1 will continuously change the area between SVP and depth axis, causing the 3D positions of beam footprints to continuously move. Because we only want to seek an equivalent SVP to displace unknown real SVP, we can make above suppositions. Then a Fibonacci search algorithm is adopted to obtain the sound velocity c_1. The sound velocity c_1 will be acquired if it becomes minimal for the squared sum of difference between the same positions of overlapping beams. Finally, an equivalent SVP is obtained. The spatial position of each beam will be recalculated according to the equivalent SVP.

$$\Delta\theta = \frac{\Delta c_{a_0}}{c_{a_0}} \tan(\theta - \beta - \rho) \tag{2.27}$$

In order to limit the impact of terrain and to avoid seeking an equivalent SVP when a single SVP cannot control the real scope, multi-beam data must be divided into many sub-regions. Each ping in sub-regions is projected along-track direction. However, a discontinuity will occur at the edges of sub-regions. To solve this problem, the model parameters of each sub-region will be only represented by the central ping. The specific parameters of all other pings will be calculated through the interpolation of the c_1 estimates between sub-regions.

The stop criterion of the search can be expressed as

$$d_{\min}^2 = \min \sum \left(z_i^1 - z_i^2\right)^2 \tag{2.28}$$

where subscript i denotes the number of points, and superscripts denote the number of the swaths.

To avoid being trapped in a local minimum, we need to limit an initial search scope. The initial average value c_{in} of velocity c_1 can be calculated based on the assumed lunch angle and travel time in each sub-region. The search scope of sound velocity c_1 is $[c_{\text{in}}, 1\,600]$ or $[c_{\text{in}}, 1\,400]$ according to whether the refraction artifacts are like "smile" or "frown". It is so large a scope that can cover any actual area under the profile.

2.4.3.2 Experiments and Analysis

At first, a flat seafloor with depth of 100 m is simulated. There are eight layers with constant sound velocity gradient (Fig. 2.34). The sound velocities of those layers are respectively 1 534.5, 1 545.5, 1 536.0, 1 523.1, 1 509.0, 1 517.6, 1 535.8 and 1 543.5 m/s from the surface layer to bottom layer. The bottom depth of those layers are respectively 6.3, 26.0, 43.7, 45.5, 50.0, 52.6, 64.2 and 100.0 m. It is assumed that the measured sound velocity is 1 534.5 m/s and only one layer exists. There are 127 beams for the equidistant MBES. The distance between neighboring beam footprints is about 4 m on the seafloor. Each swath has 15 beams in the overlap. The inaccurate sound velocity causes a concave artifact.

Our algorithm obtained good results after the removal of the sound refraction artifacts according to the simulated data (Table 2.9). Although the refraction artifacts are serious, the maximal error of depth is only 0.02% after the data is corrected. The depth standard of IHO is $\sqrt{(0.5^2 + (0.013 \times 100)^2} = 1.4$ m with a probability of 95% (International Hydrographic Bureau 2008). The simulated result easily meets the requirement. The maximal error of horizontal position is only $\pm$ 0.06 m (Table 2.9). It also easily meets the requirement of IHO in positional precision. The similar oblique seafloor is also simulated (Fig. 2.35a). A good corrected result is also obtained (Fig. 2.35b). Therefore, this method can be applied in both flat and oblique seafloor.

During July 2 to August 21, 2005, we carried out a multi-beam survey project in the South China Sea. The MBES system used was SeaBeam 2100. The data presented in this paper are a part of investigating data. The scope of our example is from (13.00°N, 112.50°E) to (15.24°N, 115.31°E). Data are shown as a sun-illuminated terrain image (Fig. 2.36 a). It includes 42 swaths, with a few swaths are complementary and irregular. The maximal depth is about 6 500 m and the minimal depth is about 350 m. To explain the problem more clearly, we extracted 7 swaths to expatiate the algorithm (the polygon region in the northeast corner of Fig. 2.36a). The extracted data are also illustrated in a sun-illuminated terrain image (Fig. 2.37a). Then we transform all beam footprints from geographical coordinates to Cartesian coordinates. A narrow and flat seafloor (thick line positions in Fig. 2.36a) is selected to check roll mount angle bias. Figure 2.38a is the profile of the selected seafloor. The scale of the depth axis is bigger than that of the across-track axis. An obvious roll mount angle bias can be found in Fig. 2.38a. According to the manual adjustment, a bias of 23′ is obtained. Next, all data are corrected according to this bias (Fig. 2.38b).

Because many swaths are longer than 100 km, a single SVP cannot perfectly represent such large areas. Furthermore, to avoid the influence of terrain, every two swaths should be divided into several sub-regions according to the scope of depth. Otherwise, the equivalent SVP model will not represent the real SVP. Figure 2.39 shows how to subdivide the left two swaths in Fig. 2.37 into seven

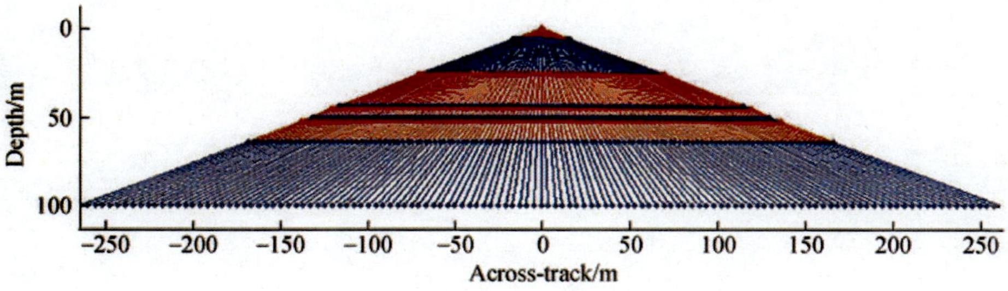

Fig. 2.34 Refraction artifacts in flat seafloor caused by inaccurate SVPs

Table 2.9 The differences in across-track distances and depths before and after the removal of the refraction artifacts

Beam	1	5	10	15	20	25	30	35	40	45	50	55	60	64
dx1/m	−1.43	−1.33	−1.20	−1.08	−0.96	−0.84	−0.73	−0.61	−0.51	−0.40	−0.29	−0.19	−0.08	0
dx2/m	0.06	0.05	0.05	0.05	0.05	0.04	0.04	0.04	0.03	0.03	0.02	0.02	0.01	0
dz1/%	−1.62	−1.37	−1.08	−0.83	−0.60	−0.41	−0.24	−0.10	0.01	0.11	0.18	0.22	0.25	0.26
dz2/%	0.02	0.02	0.02	0.02	0.02	0.01	0.01	0	−0.01	−0.01	−0.02	−0.02	−0.02	−0.02

Note dx1: the differences in across−track distances before removing artifacts; dx2: the differences in across−track distances after removing artifacts; dz1: the differences in depths before removing artifacts; dz2: the differences in depths after removing artifacts

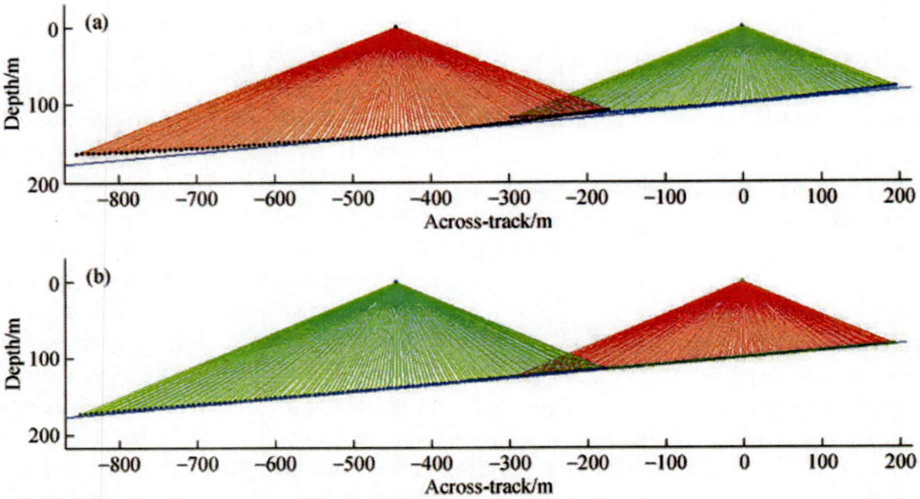

Fig. 2.35 Refraction artifacts in oblique seafloor caused by inaccurate SVPs (**a**) and the result after removing artifacts (**b**)

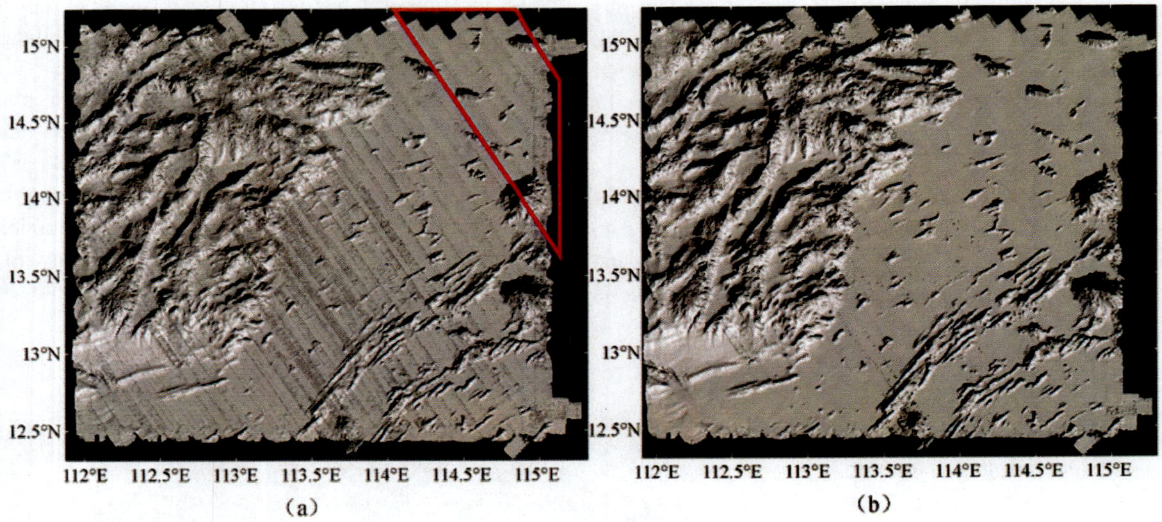

Fig. 2.36 The raw sun-illuminated swath terrain (**a**) and the corrected result (**b**)

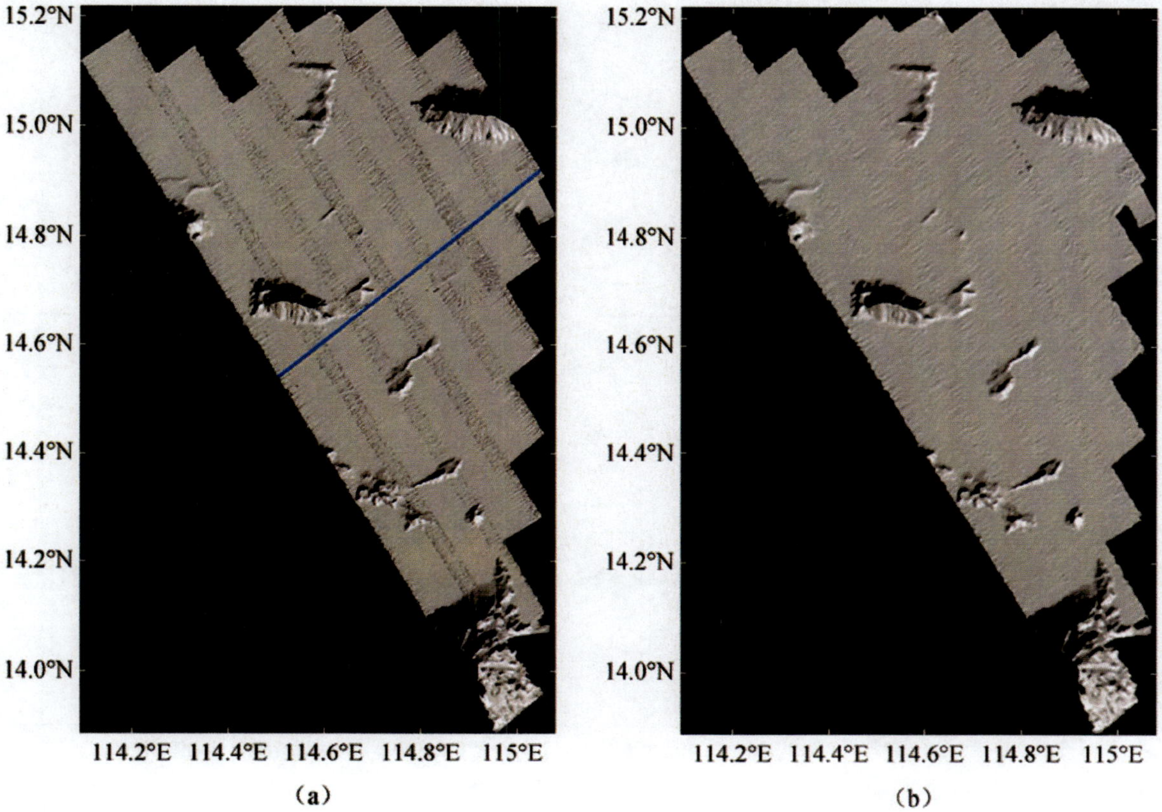

Fig. 2.37 The raw sun-illuminated swath terrain (**a**) and the corrected result (**b**) of extracted 7 swaths from Fig. 2.36a

sub-regions. Although the way of dividing may be a little random, we should follow the principles. First, each sub-region should not be too long and should have almost depth; second, flat and slope seafloors should be divided into different sub-regions. Then the parameter c_1 of each sub-region can be computed using our algorithm. If we suppose that the equivalent SVP model parameters correspond to the center of the sub-region, each ping will have different parameters through interpolating. From Fig. 2.38b we can see that there is a minor "smile" in each swath. It can be inferred that there are refraction errors in the survey data. A result can be obtained after the artifact is removed according to our algorithm (Fig. 2.38c). After that, some redundant edge beams are automatically deleted according to their positions in the overlap, because they usually have big errors due to the flaws of the device (Fig. 2.38d). Finally, a mosaic terrain image is obtained (Figs. 2.36b and 2.37b).

By comparing Fig. 2.36a, b, it can be seen that a good result is obtained for real data after roll mount angle bias and the refraction artifact are removed. Otherwise, the neighboring swaths can not be mosaiced. The division into sub-regions improves the accuracy of equivalent SVP model parameters. Interpolating on model parameters eliminates the discontinuity among the edges of sub-regions. After the model parameters are obtained, i.e., an equivalent SVP is formed, each beam footprint is recalculated by the equivalent SVP. We believe the equivalent SVP should be the replacement of unknown real SVP.

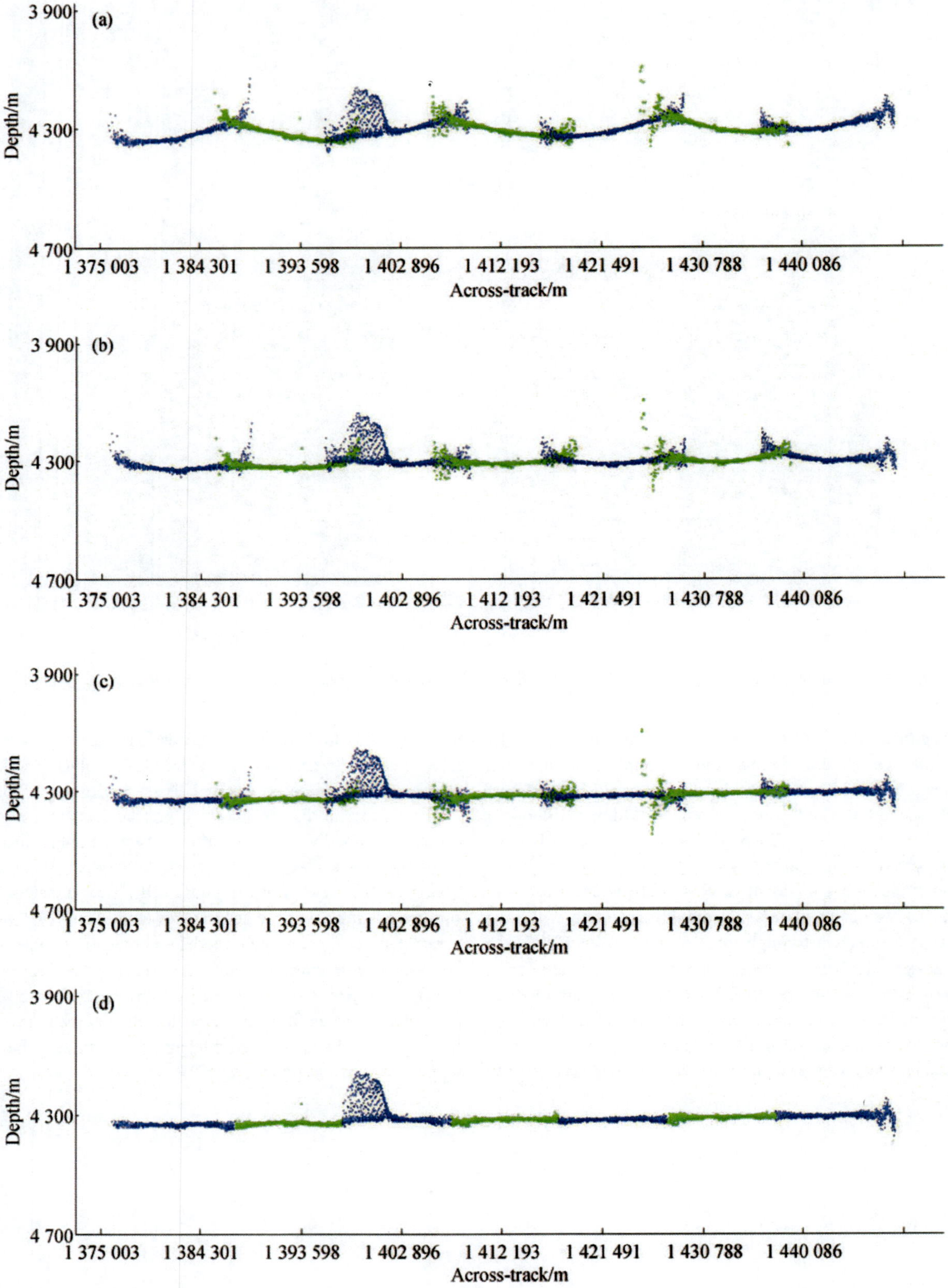

Fig. 2.38 The correcting processes: raw 7 swaths (**a**), after roll mount angle bias is removed (**b**), after sound refraction artifacts are removed (**c**), and after redundant edge beams are automatically deleted (**d**)

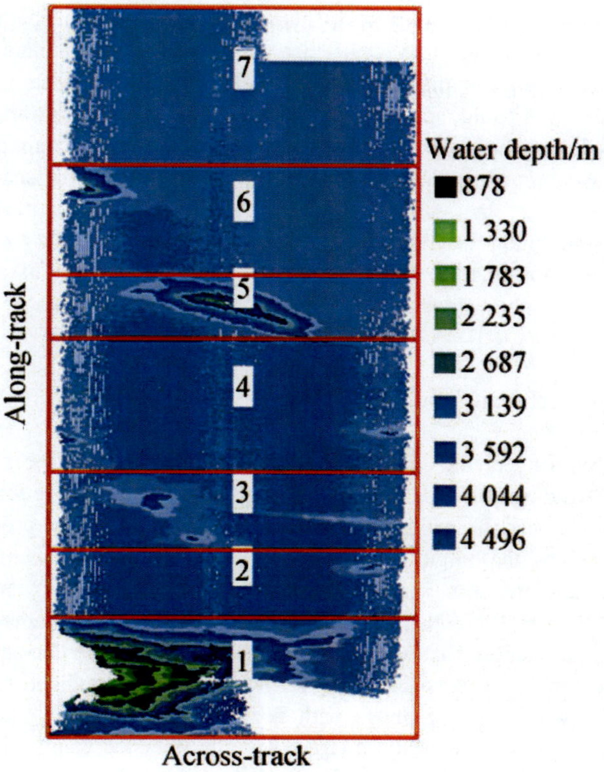

Fig. 2.39 Region segmentation of the left two swaths in Fig. 2.37

2.5 Multi-beam Backscatter and Water Column Data Processing Method

Multi-beam and side-scan sonar systems are the most commonly used tools for seafloor mapping. The former is mainly used for bathymetric measurements, but can also perform geomorphic imaging; the latter is mainly used in geomorphic imaging, and the latest interferometric side-scan sonar systems can also provide bathymetric information along with the geomorphic imaging function. The multi-beam sonar has higher bathymetric accuracy but provides lower image resolution than the side-scan sonar.

2.5.1 Multi-beam Sonar Backscatter Data Processing

2.5.1.1 Principle of Multi-beam Sonar Scattering Imaging

Multi-beam sonars (when used for imaging, multi-beam echo sounders (MBESs) are commonly referred to as multi-beam sonar systems) can not only measure the water depth to perform high-resolution seafloor topographic mapping, but also can use the returned backscatter intensity to map the seafloor sonar image, which plays a very important role in analysis and interpretation of the seafloor. These images can be used to perform inversion of the bottom characteristics, for seafloor target detection and recognition, including qualitative description of fish behavior, and in obstacle avoidance for ships (Hughes Clarke et al. 2006). At present, several methods are available to perform multi-beam seafloor imaging:

(1) Beam average method

Each received narrow beam only takes one acoustic intensity value or average acoustic intensity value, from which the number of average acoustic intensities obtained is the same as the number of sounding points.

(2) Pseudo-side-scan imaging method

This method usually refers to the two additional wide beams that are transmitted by the multi-beam transducer independently of its bathymetric function and involves sampling the amplitude time series in the sector that is covered by the wide beam.

Both multi-beam sonar and side-scan sonar can acquire the backscatter intensity of echoes and use the data to form acoustic images of the seafloor. From this perspective, these two measurement instruments are quite similar, but they are also quite different in terms of their image deformation and resolution characteristics. Side-scan sonar usually has higher resolution, and when the towing measurement method is used, its transducer array is close to the seafloor and thus produces a relatively larger incident angle, which makes objects cast larger areas of shadow. Therefore, it is easier to identify objects using side-scan sonar than using multi-beam sonar. While the multi-beam transducer is usually fixed on the ship, meaning that the transducer is far away from the seafloor and resulting in the sonar images being less distorted, the image resolution is also reduced accordingly.

(3) Snippet time-series method

In this method, amplitude-time series sampling is carried out for each narrow beam received to obtain multiple intensity values with high resolution.

When sampling the echo intensity, the measured object is the beam footprint of the seafloor. For bathymetric survey, only the average travel time or the phase change at the center of the beam footprint can be detected, which also indicates the average slant distance from the transducer center to the seafloor within the sonar propagation area. For seafloor sonar imaging, a time-series observation of the backscatter intensity is detected and each of these time-series observations is much smaller than the beam footprint. The number of time-series samples per unit time is several times or even more than ten times the number of bathymetric samples (depending on the sonar image resolution). Each sequential sample is still the interface between the transmitting beam mode of a spherical surface and the receiving beam mode of an annular surface formed in the time period $[t, t + dt]$ (for sequential waves or continuous waves (CW), where dt is the pulse width). The working principle of this method is shown in Fig. 2.40.

After each ping measurement, a set of time-series echo intensity observations on the line of intersection between the fan and the seafloor is formed. After several pings, the echo intensity can be obtained at different positions in the survey area. To acquire the sonar images, the echo intensity must be converted from time series to transverse distance series. When the echo intensity is measured, the travel time, the azimuth, and the depression angle of the beam are also obtained. The slant distance correction of the beam can also be carried out easily using ray tracing and the horizontal distances; the depths of the time series sampling points in each beam can thus be calculated by interpolation.

When compared with the two previous methods, the snippet time-series method not only avoids the problem of backscatter data and bathymetric data mixing, but also allows sonar images to be acquired with a high signal-to-noise ratio (SNR) and high resolution simultaneously, so this method is more widely used at present. In addition to the principal axis direction, the spatial positions of the other intensity samples in each beam are acquired using an interpolation method while assuming that the seafloor is flat within the beam range. This inaccurate assumption may mean that the backscatter intensity data and the corresponding spatial position data are unable to be fused accurately, which becomes more obvious in the case of complex terrain changes.

(4) Coherent imaging method

Similar to the principle of interferometric multi-beam sounding, the angle of arrival at each seafloor detection point can be estimated by sampling and coherently processing the output signal from each received narrow beam. Therefore, the spatial position and echo intensity are obtained for each point. Then, based on the actual underwater acoustic environment and the angle response characteristics, the echo intensity image data can be modified to obtain a seafloor image with good spatial resolution. The final seafloor acoustic image resolution obtained using this method is necessarily higher than that of the previous methods. Bathymetric side-scan sonar, which combines

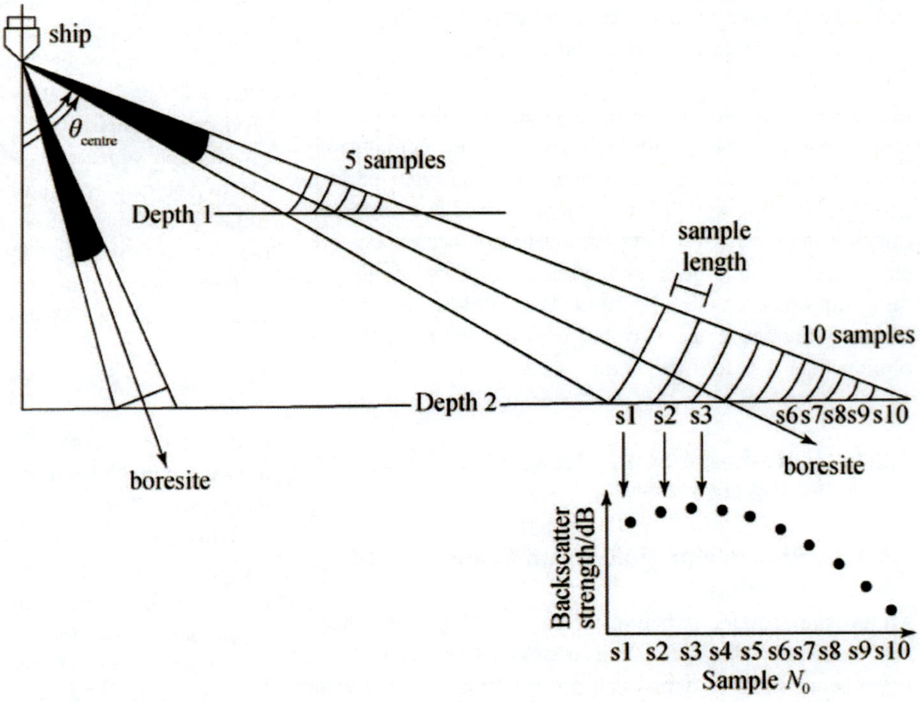

Fig. 2.40 Time-series sampled schematic diagram of sonar image in a single beam footprint (de Oliveira Junior and Hughes Clarke 2007)

bathymetric measurements with high-resolution imaging, was designed based on this method.

(5) Multi-beam SAS imaging method

This method is based on the principles of multi-beam sounding and synthetic aperture sonar (SAS) technology. During measurements, the sonar will send a signal toward the seafloor at each heading position and will receive the processed echo signal in the lateral distance direction. After synthetic aperture processing of the output signal of each beam, a high-resolution beam on the heading and better seafloor images that contain more target information can be obtained (Wei et al. 2019).

After the sonar acquires the echo intensity, it is necessary to quantify this intensity and express it in the form of images. The echo intensity in the acoustic radiation area is quantized using a specified gray level to form a sonar gray image that reflects the echo intensity level while also reflecting the physical properties of the seafloor sediments. In fact, the conversion of the echo intensity into gray levels corresponds to the echo intensity of the gray levels of the image and thus quantifies the echo intensity. There are many quantization methods that can be applied and their use is generally determined according to the specific situation. After quantization, a sonar image is formed in which each pixel can be determined using two sets of parameters, i.e., the pixel position (i, j) and the corresponding pixel grayscale $f(i, j)$.

2.5.1.2 Correlation Between Backscatter Intensity and Sediment Type

The overall sound wave process from transmission to reception constitutes the sonar equation, which is a relation that integrates the sound transmission medium, the target, the background interference, and the sonar equipment parameters. The sonar equation can be used to design the working parameters of sonar systems and estimate the system's detection capability. The equation can also reflect the variety of the seafloor sediments when applied to interpret the characteristics of the seafloor.

It is assumed that the sound wave emission intensity is the source level (SL); the transducer directivity index is represented by DI; the transmission loss (TL) is the energy loss generated during transmission; the sound wave energy loss caused by marine noise is the noise level (NL); the energy level of the backscatter or reflected signal generated by the target is represented by BS; and the energy level of the recovered signal is the excess level (EL), which can be considered to be the sum of the reflected energy from an infinite number of multiple point reflectors in the insonified area (Hellequin 1998). The echo intensity change in the above process can then be described using the sonar equation:

$$\mathrm{EL} = \mathrm{SL} - 2\mathrm{TL} + \mathrm{BS} - \mathrm{NL} + \mathrm{DI} \tag{2.29}$$

According to the sonar equation, the direct interaction between the transmitted beam and the seafloor is reflected in the BS, which can be understood to be an indication of the reflection and scattering abilities of the submarine medium with regard to sound waves. The backscatter intensity BS is dependent on the type of sediment, the topographic conditions and the area on the bottom insonified by the beam (AE) and can be expressed as

$$\mathrm{BS} = \mathrm{BS_B} + 10\lg\mathrm{AE} \tag{2.30}$$

where $\mathrm{BS_B}$ is the seafloor backscatter intensity with angular dependence. According to Eqs. (2.29) and (2.30), $\mathrm{BS_B}$ can be expressed as

$$\mathrm{BS_B} = \mathrm{EL} - \mathrm{SL} + 2\mathrm{TL} + \mathrm{NL} - \mathrm{DI} - 10\lg\mathrm{AE} \tag{2.31}$$

As long as the terms on the right side of Eq. (2.31) can be obtained accurately, $\mathrm{BS_B}$ can also be calculated. The regional distributions of the different types of bottom sediments can be inverted according to the relationship between these sediments and the bottom materials, and this process is called seafloor classification. The $\mathrm{BS_B}$ is related not only to the sediment type but also to the incident angle of the beam. Figure 2.41 shows the change in the measured average backscatter intensity for different sediments with respect to the grazing angle (Lurton and Lamarche 2015).

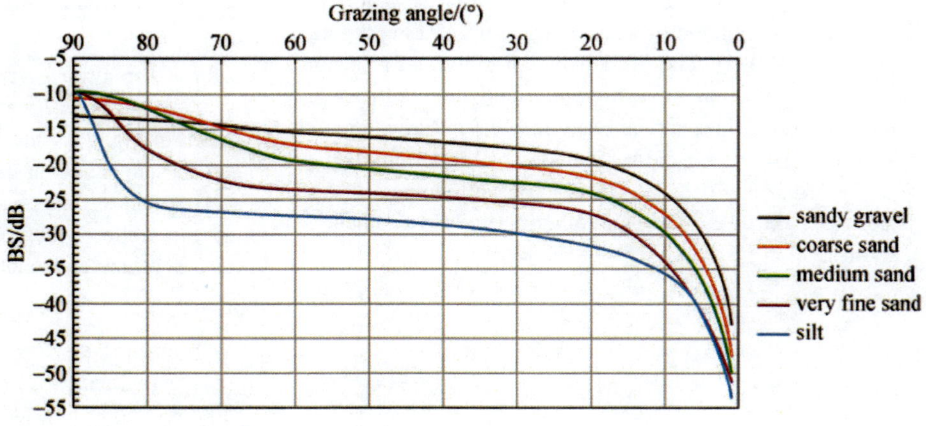

Fig. 2.41 Variation of the mean backscatter intensity versus grazing angle for the different classes of sediment at 100 kHz

By remote sensing of the sediment's acoustic characteristics (e.g., the reflection coefficient, sound velocity, attenuation, and scattering), the acoustic seafloor classification process can be used to recognize the physical properties of the seafloor (such as sediment types and particle sizes); the process offers high efficiency, has continuous and abundant data, and provides a rapid and reliable seafloor classification method. Multi-beam sonar can not only obtain high-precision bathymetric data, but can also obtain high-resolution seafloor backscatter intensity data, so it is highly advantageous to use multi-beam sonar images for seafloor sediment classification.

The echo intensity can be regarded as a function of the target or sediment type, the sound beam frequency and the incident angle of the beam. Because different sediment types (e.g., bedrock, gravel, sand, mud) have various physical properties, including particle size, porosity, and density, acoustic signals that even have the same direction of incidence and intensity can produce echo signals with different backscatter intensities (or amplitudes). The echo signal depends on the incident angle of the sound waves, the roughness of the seafloor, the acoustic parameters of the sediments (e.g., density, sound velocity, attenuation, scattering) and the propagation status of the sound waves in water, thus reflecting the characteristics of different types of seafloor sediments. This shows that there is a certain correspondence between the backscatter intensity and the sediment type. However, the same type of sediment in different regions can produce different backscatter intensities because of physical characteristics such as water content, density, and mechanical intensity, along with the different sedimentary environments of the seafloor. Therefore, it is inaccurate to classify the seafloor sediment solely by establishing the relationship between the backscatter intensity and the sediment type.

2.5.2 Multi-beam Water Column Data Processing

The multi-beam sonar water column image (WCI) reflects the reflection or scattering information of a target object in the entire water column of a sound wave penetrating region (Colbo et al. 2014). The WCI plays an important role not only in the detection of shipwrecks, navigation obstacles, torpedoes, submarines and other civilian and military targets, but also in monitoring the seafloor for hydrothermal vents, gas layer leakage (Elhegzy 2012), ocean internal waves (Hughes Clarke et al. 2006) and other marine environmental activities.

The latest generation of MBES systems can generally record WCI data. At the same time, many scholars have done a lot of research on how to use the WCI data. Auke (2010) used WCI data to analyze the imaging capabilities for objects such as shipwreck masts. Marques (2012) noted that multi-beam WCI technology could recognize and position suspended objects accurately in seawater. Hughes Clarke et al. (2014) proposed the use of multi-beam WCI data to determine the shallowest depth of a wreck accurately. Liu et al. (2019) proposed an adaptive soft threshold algorithm to reduce the WCI noise.

2.5.2.1 WCI Imaging Principle

During multi-beam sonar operation, the transmitter continuously emits sound waves, the echo signals that are backscattered from the water column to the seafloor are finally received by the receiver. Traditional bathymetric measurements only detect the average round trip time or the phase change at the center of the beam footprint, the water column data are used to provide a time series observation of the backscattering intensity along the detection beam direction, where the sampling number is hundreds of times that used in simultaneous bathymetric measurements. Without considering the sound velocity, the water depth environment, the seafloor fluctuations or other factors, each survey line can obtain a triangular cylinder when the vessel is sailing (Fig. 2.42).

Because the water column data are collected in the equal time interval sampling mode, the number of sampling points of each beam sequence in the current water depth environment is determined based on the pulse width and the sampling frequency set by the MBES. The positions of the sampling points in the transducer coordinate system can be calculated by combining the sound velocity in water and the incident angle of the beam.

$$R = \frac{n \cdot C}{f} \tag{2.32}$$

where R is the distance from the current sampling point to the transducer, n is the sampling point sequence number, C is the velocity of sound wave propagation in seawater, and f is the sampling frequency at the sampling points.

In general, there are three ways to display WCI data: a course display, a vertical display and a beam array display.

(1) Course display

As shown in Fig. 2.43, the WCI data along the track line are stacked and are shown in the course graph (Fig. 2.43), where the x-axis represents the course and the z-axis points vertically toward the seafloor, thus forming a two-dimensional plane. When the vessel continuously acquires samples along the

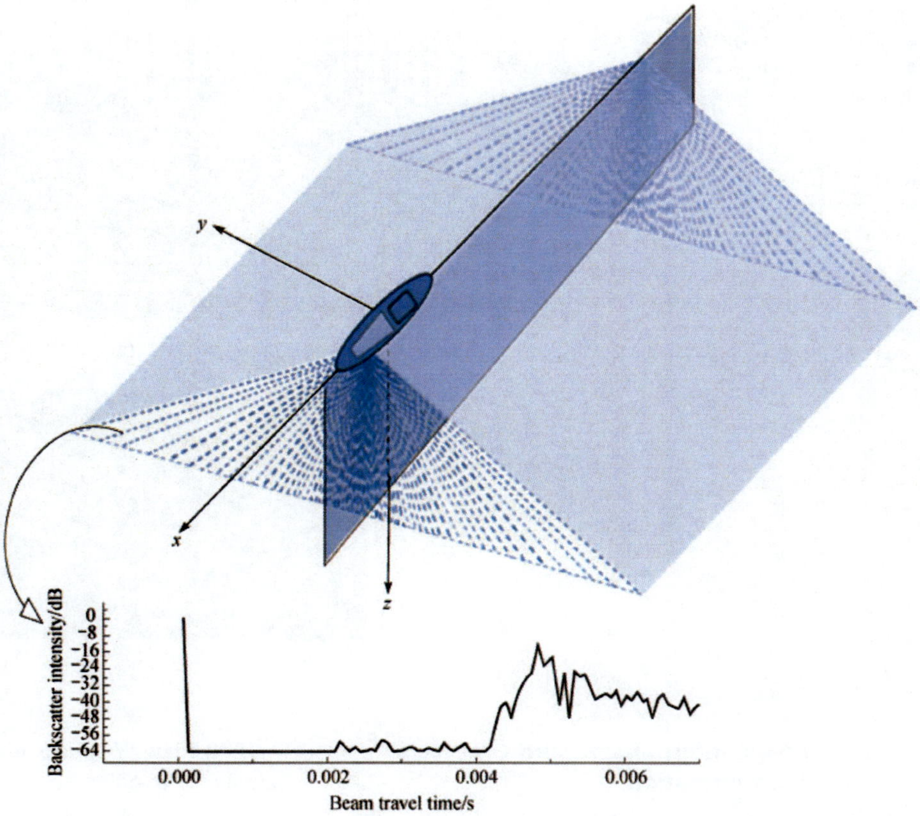

Fig. 2.42 Principle of water column data collection

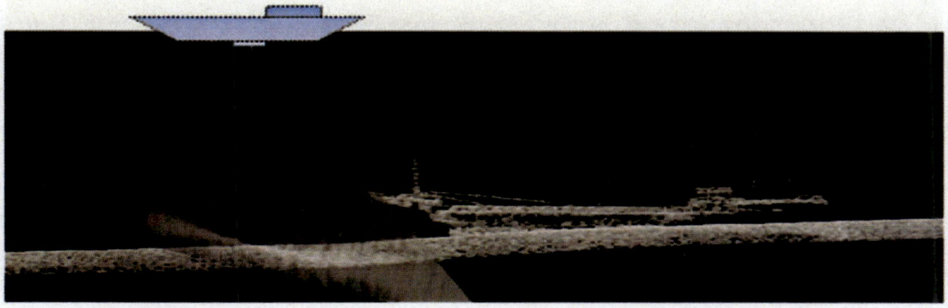

Fig. 2.43 Course stacking graph

survey line direction, all water column sampling points are projected onto the *x*-*z* plane according to their positions and the backscattering intensity; via this process, the course graphs are generated. These graphs can be used to view the changes in the water column at the current measuring line over a period of time.

(2) Vertical and beam array display

As shown in Fig. 2.44, the vertical graph of the WCI data (Fig. 2.44a) is used to view the full WCI data from the sea surface to the seafloor during each transmission and reception period (ping), where the *y*-axis is perpendicular to the course direction and points toward the starboard direction of the ship and the *z*-axis points vertically toward the seafloor, thus forming a two-dimensional plane. At a specified moment, the backscattering intensity data that were received from the ping were plotted on the *y*-*z* plane based on the imaging principle and a vertical graph of the WCI was generated. The WCI beam array graph (Fig. 2.44b) is a two-dimensional image that is generated by orderly arrangement of all the sampling points on all beams based on the size of the beam angle, which is convenient for viewing the characteristics of backscattering intensity change among the beams.

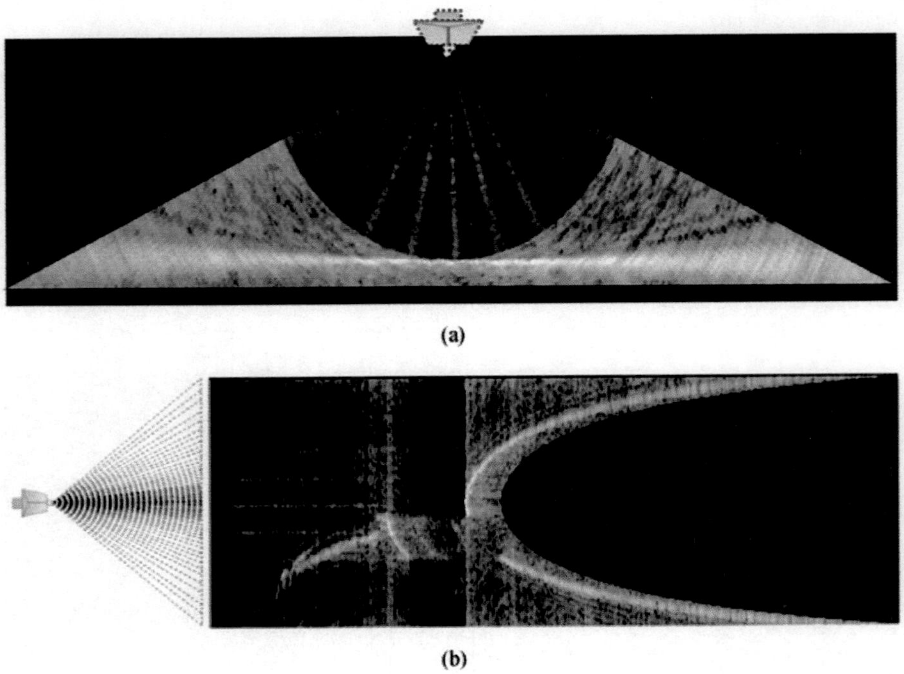

Fig. 2.44 Vertical display and beam array display of WCI. **a** Vertical graph of WCI; **b** beam array graph of WCI

2.5.2.2 Image Interpolation and Grayscale Transformation

The WCI data that are acquired from the surface to the seafloor are uneven, i.e., they are dense around the transducer but sparse far away from the transducer. Therefore, it is necessary to apply image interpolation technology to the image formation process. In addition, because of changes in the equipment parameters, the marine environment and other factors, the backscattering intensity may not reflect the target characteristics accurately. When the backscattering intensity is quantized directly into an image, the uneven image gray level phenomenon will also appear. Image grayscale transformation is then required to ensure that the image reflects the real situation of the water column more reasonably and clearly.

In the raw WCI data, the backscattering intensity data are arranged in order according to the beam angle and the sequence number of each sampling point. When the regression is calculated with respect to the rectangular coordinate system, there are no data between the beam edges, which results in an incomplete image being displayed. Therefore, it is necessary to interpolate the backscattering intensity points with an empty value in the beam angle range. In general, neighborhood interpolation can be used to interpolate the raw WCI data. Figure 2.45 shows the contrast effect in detailed areas before and after eight neighborhood interpolations.

In the WCI data, when the backscattering intensity is quantized into the gray scale, the gray scale of the original

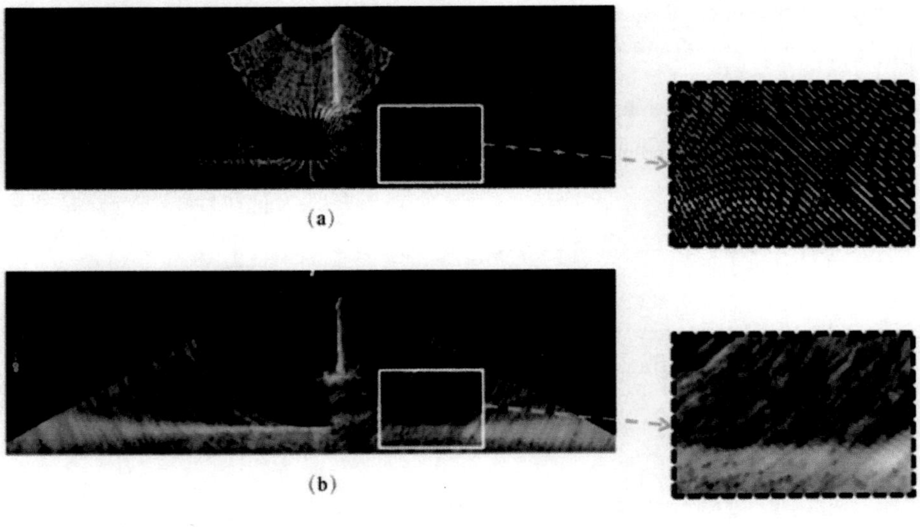

Fig. 2.45 Image interpolation before **a** and after **b** contrast

image can be transformed to achieve an improved image display effect. After equalization, the image texture becomes clearer and the features are more obvious, while its grayscale almost covers the entire grayscale level, thus enhancing the image contrast. Experiments have shown that the histogram equalization gray level transformation method is effective for WCI processing. Figure 2.46 shows a rendering of multiple transformations of the target region shown in Fig. 2.45.

2.5.2.3 Denoising

In the original WCI data generated using the backscatter intensity, there are numerous noise sources in addition to the objects, which affects the target identification accuracy. The transmitting side lobe is generated and is attached to all angles near the main lobe when that lobe is formed, and side lobes at various angles have different energies because of the differences in beam steering. The transmitting side lobe

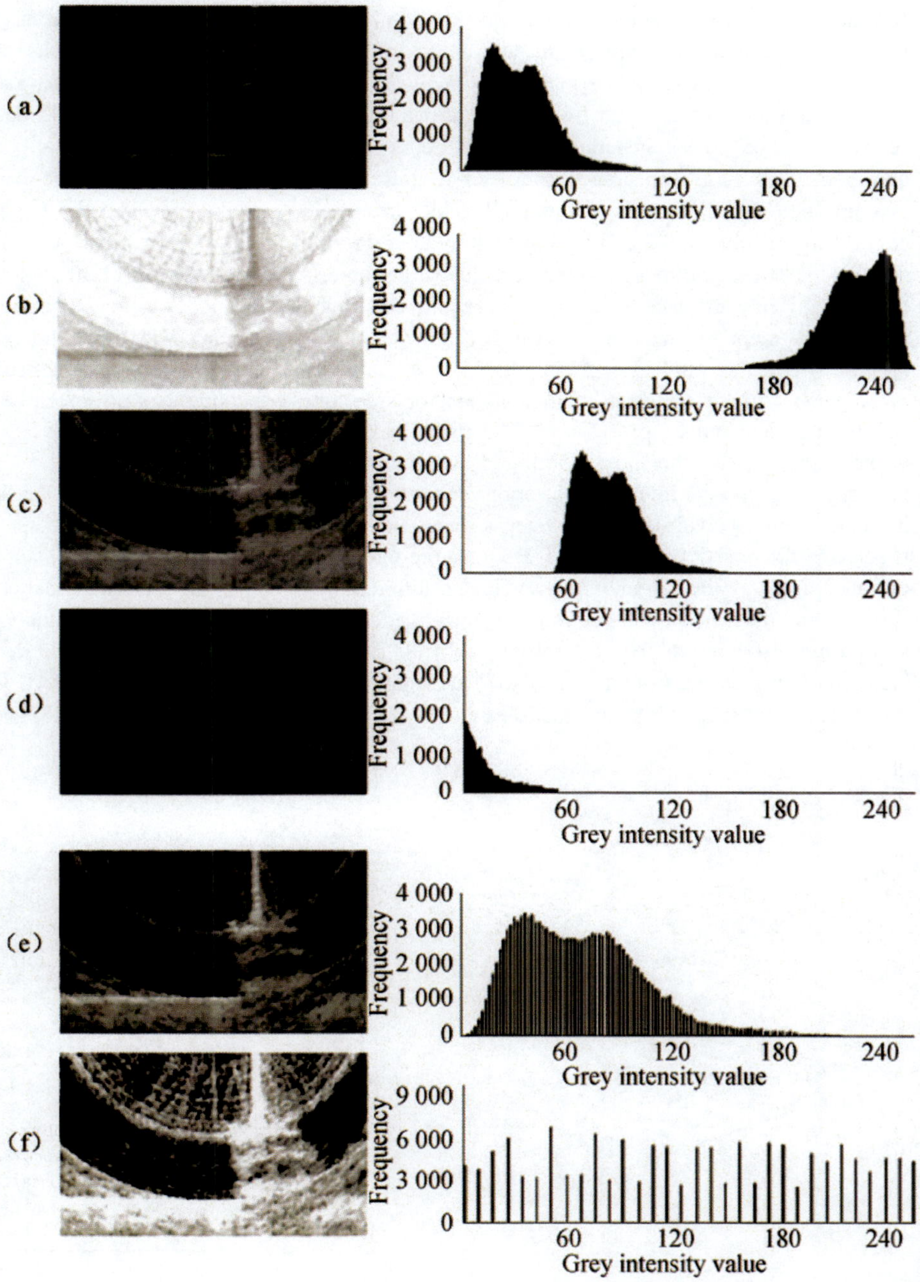

Fig. 2.46 Grayscale transformations and corresponding histograms. **a** Original image; **b** gray scale inversion; **c** brightening; **d** dimming; **e** linear change; **f** histogram equalization

appears as a false image of a "convex" curve in the original WCI. The receiving beam is similar to the transmitting beam when it is formed and also produces a side lobe that is attached to all directions near the main lobe. This lobe appears as a "concave" curve noise, i.e., the same distance noise. These side lobes and the background noise are reflected in the images of various water columns.

The WCI track profile (the yellow section in Fig. 2.47) is selected to observe the "convex" curve noise caused by the transmission of the side lobes at the target of the WCI at the current cross-section location, as shown on the top-right of Fig. 2.47. The vertical-sector graphs that are close to and parallel to the target location are selected for analysis and the false target artefacts that result from the combination of the "convex" curve noises generated at each depth can be observed. This false appearance can lead to false target contours before the target is in the WCI, which results in confusion. The pitch noise that is generated by the receiving side lobe is distributed in a concentric arc that is the same as the transducer's distance characteristic, while the radial noise is the same as the single beam direction and has a specific width.

In general, there are three main methods for side lobe suppression and noise processing. The first method is backscatter intensity threshold filtering. Based on the differences in backscatter intensity for various substances when irradiated using acoustic waves in water, a specific threshold is selected for rejection of the noise. Because the intensities of the noise and the target will vary under different circumstances, this method will not produce an ideal effect. The second method is the miniature slant range (MSR) effective area retention method, in which only the water data within the MSR are reserved, while the other data are not. However, because the effective area of the MSR only accounts for 70% or less of the entire WCI, the data utilization rate when using this method is too low. The third method involves use of a multi-sector mode to collect, transmit and receive beams with different frequencies in each sector. The noise only exists in a single sector in this case and does not affect the other sectors. While its range of influence is reduced, the noise in the single sector is not eliminated and that has an impact on the judgment of the true target shape.

Because of the different distributions of the target and noise intensities in varying situations, it is unlikely that side lobe artefacts and background noise will be removed effectively with single backscatter intensity thresholds. The adaptive soft threshold denoising algorithm can be used to improve the image quality by allowing the angle sequence of the whole image to be divided into two regions—the background area and the target-noise-mixed area—through the adaptive thresholds calculated based on the mathematical characteristics of all angle arrays. Next, the target and noise thresholds in the target-noise-mixed-region angle array are calculated in turn, and the noise is compressed by compression factors, demonstrating that the background, radial and side lobe noise can be suppressed effectively (Liu et al. 2019).

(1) Determination of the threshold parameters

Based on the discrepancies in the degree of discreteness of the angle array backscatter intensity and vibration shown in the beam array chart, the variable region threshold parameter A can be calculated to determine the appropriate range for the angle array in which the side lobe noise is located. By adjusting the variable noise threshold parameter B to

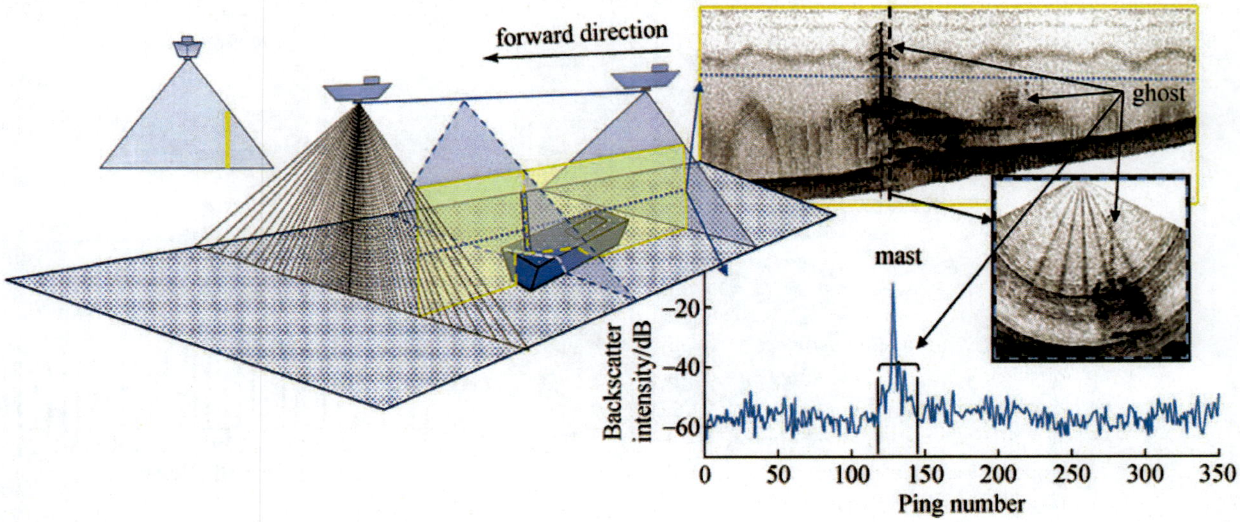

Fig. 2.47 Performance of the transmitting side lobe in the image from the WCI data

determine the noise position accurately within the current angle sequence and by compressing the estimated noise data using the compression factor *C*, the suppression effect on the co-range noise can predictably be achieved. Unsurprisingly, a method to determine the threshold parameters *A*, *B* and *C* forms an essential part of this algorithm.

The upper part of Fig. 2.48a shows the beam array chart, where each column shows the sequence at the current sampling point. Because there is a lobe gap between the main lobe and the side lobe for each receiving beam, the echo signal is not continuous and the angle sequence curve that formed is serrated. The vibration of the data is too significant to indicate the distribution of the backscatter intensity points properly when using the original data to compute the standard deviation. Therefore, a higher-order polynomial is used to fit the valid data (in which the backscatter intensity is consistently greater than −64 dB within the angle sequence), and then the standard deviation and the standard deviation curve of the fitting angle sequence are obtained (Fig. 2.48b). The curve, in which the standard deviation of the fitting sequence of the WCI background area is steadily less than 4 dB, has a more stable standard deviation than the original angle array's standard deviation curve. Nevertheless, when some targets appear in water or at the bottom of the sea, the deviation rises dramatically, mostly in the area where the standard deviation is no less than 4 dB. In addition, the relative fluctuation of the standard deviation increases as the number of sampling points increases, before it peaks at the bottom of the sea.

Given the above attributes, the threshold parameter *A* within a specified region is available in the standard deviation curve (Fig. 2.48b) of the fitting angle sequence as follows:

$$f_i - A\begin{cases} \geq 0, \text{ when } f_i \text{ denotesangle sequence of target and noise area} \\ <0, \text{ when } f_i \text{ denotesangle sequence of background area} \end{cases} \tag{2.33}$$

where f_i denotes the standard deviation of the *i*th fitting angle sequence; f_i is larger than the angle sequence of *A* and is regarded as a mixed target and background area, while conversely, the area where f_i is less than the angle sequence of *A* is regarded as the background area. The acquisition frequency, the pulse parameter configuration and the acquisition mode can differ on the same instrument; hence, the background noise is slightly different. However, when the version of the same instrument and the environment remain fixed, the standard deviation can only be calculated when the optimal area boundary parameter has been obtained.

Dividing the region using the region-characteristic parameter *A* gives angle arrays that are distributed in both the mixed-target-noise area and the background area. For example, when f_i is greater than *A*, the angle sequence is located in the mixed area composed of the target and the co-range noise, and the noise threshold parameter *B* is used as the real target in the angle sequence and the threshold (Fig. 2.48c), which divides the noise backscatter intensity:

$$B = M + b\ *\sigma \tag{2.34}$$

$$f_{ij} - B = \begin{cases} \geq 0, \text{ target backscatter intensity point} \\ <0, \text{ sidelobe noise and background backscatter intensity point} \end{cases} \tag{2.35}$$

where *M* is the average of the original angle sequence, σ is the standard deviation of the original angle sequence, *b* is the multiplying parameter, and f_{ij} denotes the backscatter ntensity value of the *i*th angle sequence at the position of beam *j*. In addition, the parameter *B* is the target and side lobe co-range noise threshold in this sequence, and the change required in the boundary threshold range is achieved by adjusting the multiplying parameter *b*. The backscatter point, which is higher than the threshold, is the true target intensity point. During the processing period, the backscatter intensity and position remain unchanged; conversely, the backscatter intensity point, which is lower than the threshold, is the co-range noise point.

For the angle sequence S_2 that is located in the target-noise-mixed area (Fig. 2.48a, right), the standard deviation σ of the original angle arrays is calculated. For those (S_1) located in the background area, the same method is used to find the threshold parameter *B* and determine the location of the radial noise within the background noise area (Fig. 2.48c).

After both the side lobe noise and the radial noise are located, to highlight the object outline and then obtain a clear WCI, it is necessary to suppress the noise and increase the intensity difference between the noise area and the target area. Use of the compression factor *C* to compress the side lobe noise and the radial noise gives the processed backscatter intensity value g_{ij}:

$$g_{ij} = x_{\min} + c|f_{ij} - x_{\min}|, (0 \leq c \leq 1) \tag{2.36}$$

where $x_{\min}$ denotes the minimum intensity range recorded by the instrument and *B* denotes the compression factor, which indicates the compression rate of the noise intensity point. Through adjustment of the size of the compression factor *C*, the background noise can be preserved to varying degrees.

We select the angle sequence S_4 (Fig. 2.48a, right) passing through the weak-echo targets in the target and noise area and then compress the noise. In Fig. 2.48d, right, the brownish red histogram (above) shows the results of differencing of the noise and the minimum boundary value, while

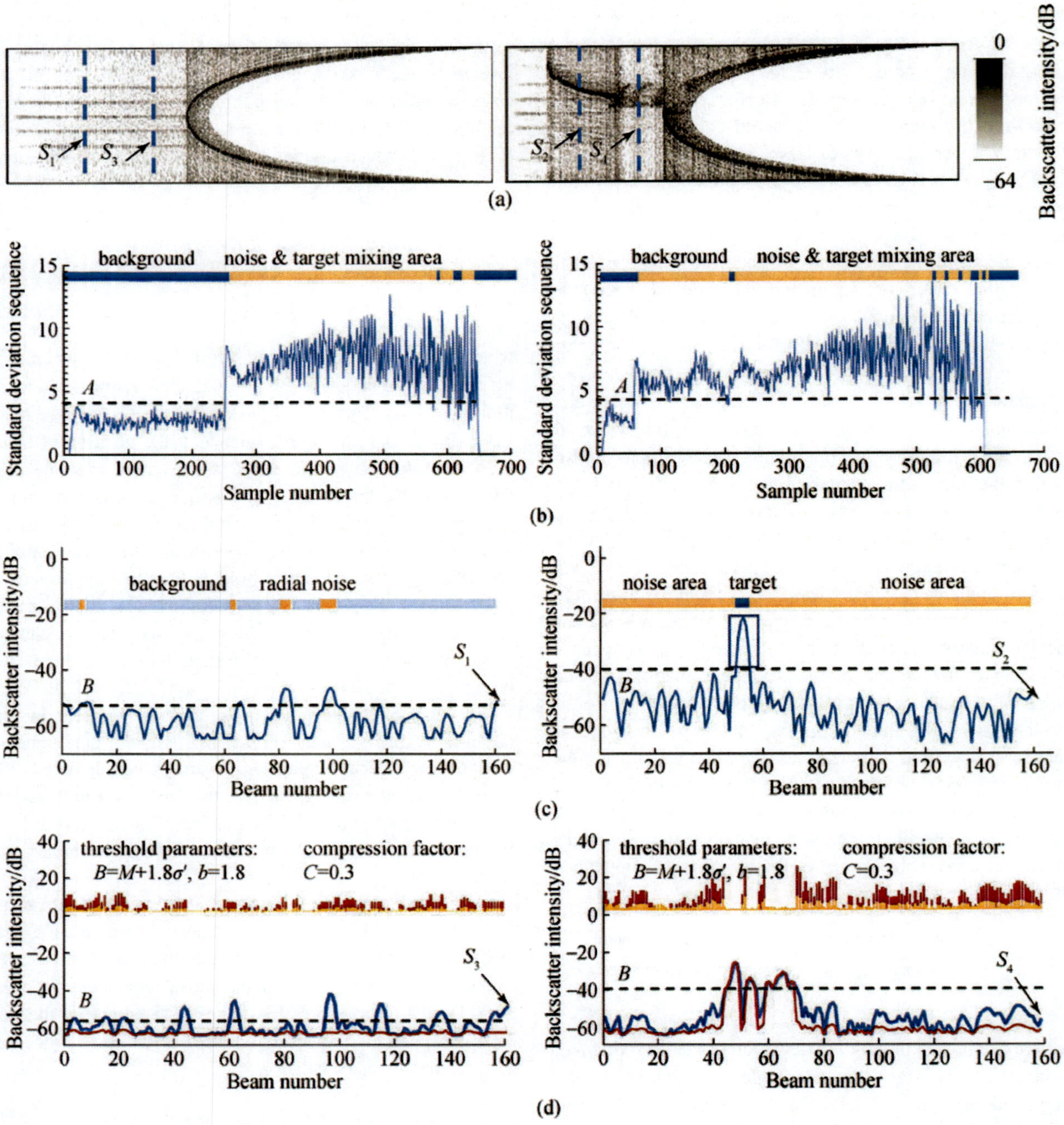

Fig. 2.48 Determination of the optimal threshold parameters A and B and signal suppression. **a** Image without the water target (left), image with the water target (right). **b** Standard deviation of the image without the water target (left), standard deviation of the image with the water target (right); angle sequence curve (blue); threshold line (black). **c** Angle sequence in the background area (left), angle sequence in the target area (right). **d** Process of radial noise compression in the background area (left), process of weak-echo target angle sequence compression (right)

the yellow histogram shows the differencing after compression, the blue dotted line (below) shows the original signal curve, and the red line represents the signal curve after denoising.

The presence of radial noise in the background area data makes radial-channel noise appear in the WCI, where it is displayed as several peaks on the angle sequence curve, and this noise must also be suppressed. Angle sequence S_2

(Fig. 2.48a, left) in the background area is selected; using the noise threshold parameter *B*, the radial noise is then located and is set to be equal to the average of the current angle sequence; then, by compressing the entire sequence using the compression factor *C*, the denoised angle sequence in the background area is finally given (Fig. 2.48d, left).

(2) Sensitivity analysis of threshold parameters

The denoising effect of the algorithm is strongly influenced by three factors: the threshold parameter of the target and the background area (*A*), the target and noise threshold parameter in the target angle sequence (*B*), and the compression factor *C*.

To analyze the effect of the threshold parameter selection on the denoising results, testing of the sensitivities of three threshold parameters and adjustment of these parameters is performed to give the corresponding WCI (Fig. 2.49). Figure 2.49a shows the results of adjusting parameter *A* only, which mainly affects the judgment of the background and the target area: when parameter *A* is too small, the background and the target area cannot be distinguished successfully, but if it is too large, it causes some weak-echo targets to be treated as background noises. Figure 2.49b shows the results of only adjusting the parameter *B*, where the main effect is on the outlines of objects: as *B* increases, fewer backscatter intensity points will be regarded as the target; however, as *B* decreases, more false targets will be obtained and the target would therefore appear to be bulgy. Figure 2.49c shows the results of adjusting the compression factor *C*: the top image is the original WCI; the middle image shows the processing of radial noise in the background area. When combined with what is shown in Fig. 2.49b, the size of the compression factor plays a major role in the WCI display, and the intensities of all types of noises can be adjusted. Consider the effects of the parameters on the processing results: when the region threshold parameter $A = 4$, the noise threshold parameter $B = 1.8$ and the compression factor $C = 0$, the target outline in WCI is both vivid and intact; the side lobe illusion, and the radial and background noises are suppressed; and thus a WCI with a clearer target outline and position can be obtained.

2.5.2.4 Case Analysis

The MV G.B. Church freighter, which was 54 m long and had a basically intact mast and hanger bracket and complete ropes on the deck, sank in 1991 to a depth of approximately 24–27 m. Because photographs were taken before the sinking, the sizes of the bulges, and the features and locations of the protrusions are well documented. The model of the experimental MBES is the EM 3002, which simultaneously collects the bathymetric data and the WCI data.

In the course graph of the WCI (Fig. 2.50), an obvious object that is suspected to be a sunken ship is found. After image filtering and equalization (Fig. 2.51), a clearer WCI is obtained, from which the size and depth information of the target can basically be determined. The course graph shows a stacked view of the backscattering intensity of the entire beam, which may cover up other information contained in the water column. It is necessary to use the vertical display graph (shown in Fig. 2.52) to aid in judgment and analysis.

The course graph within the ping of the mast is displayed vertically and includes the current ping information (beam angle, beam number, sound velocity and frequency). In this vertical graph, the target object can be seen clearly, and the minimum range curve generated by the first echo and the abnormal backscattering intensity on the common range curve of the target object can also be observed. At the same time, the depth at the shallowest point of the mast can be accurately determined to be 4.1 m.

The seafloor location obtained by multi-beam sounding is shown by the recording of the measured points during bottom tracking (Fig. 2.53). At this time, because the mast is small and there are few sampling points, the mast will be ignored in the seafloor topography obtained by sounding, thus resulting in certain hidden danger to navigation safety.

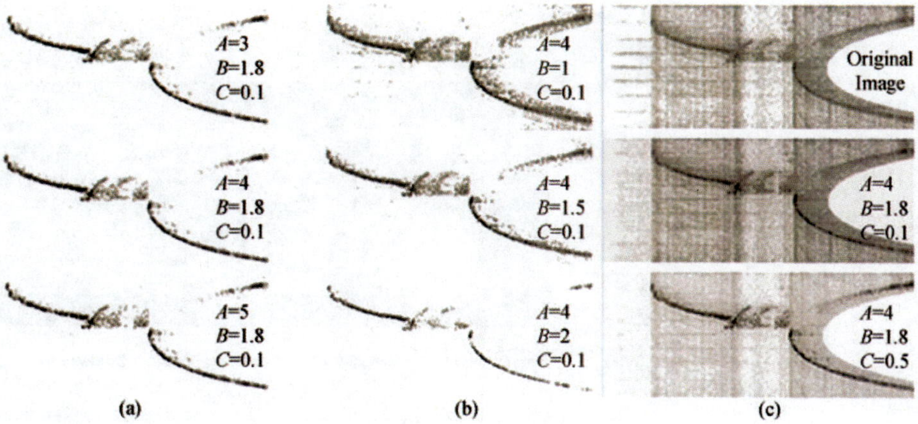

Fig. 2.49 Process of radial noise compression in the background area

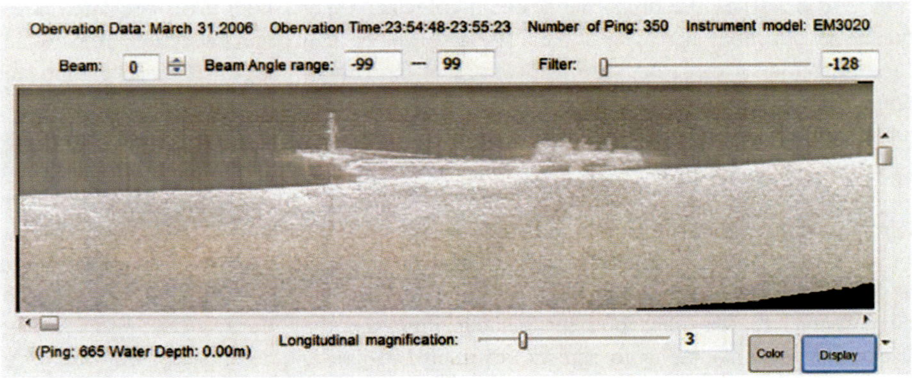

Fig. 2.50 Course display

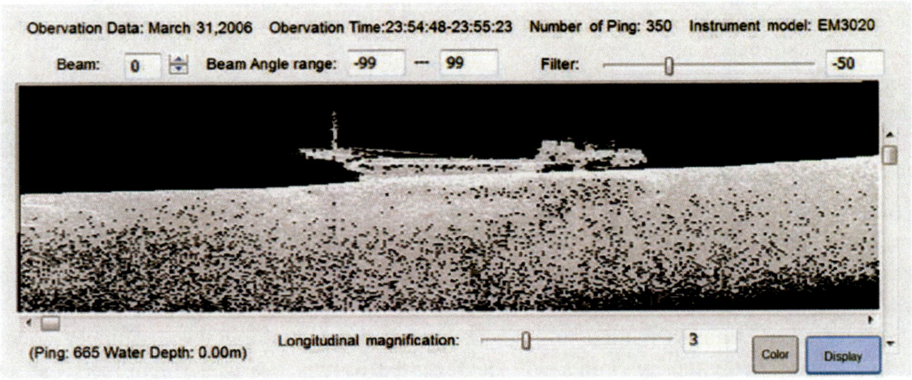

Fig. 2.51 Course display (after equalization and filtering)

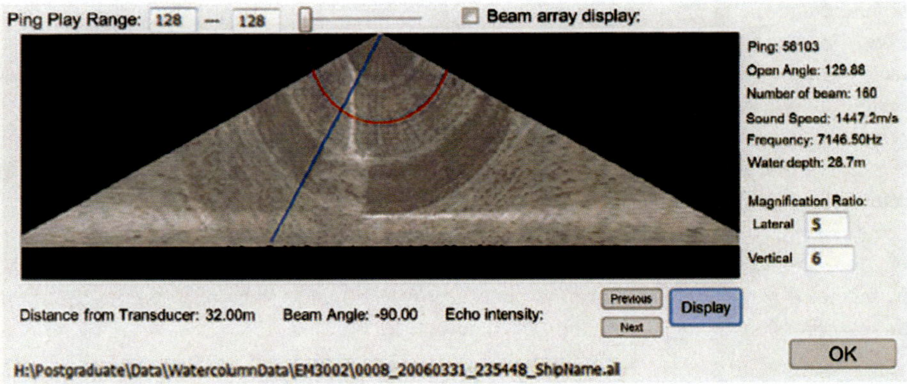

Fig. 2.52 Vertical display of the mast

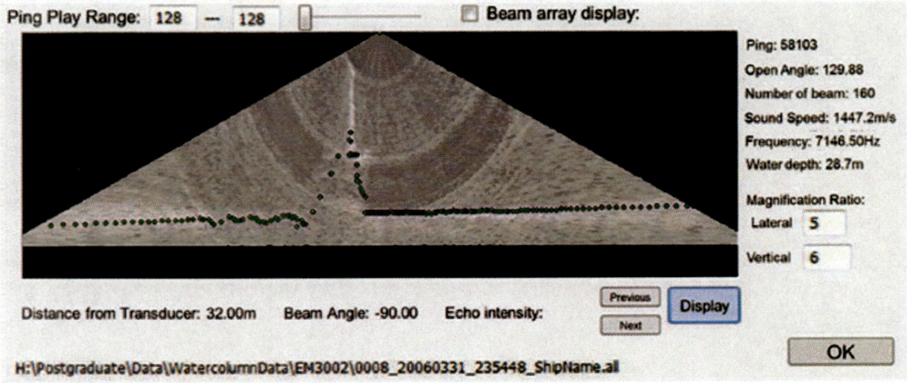

Fig. 2.53 Bottom tracking mode used to obtain the location of the seafloor

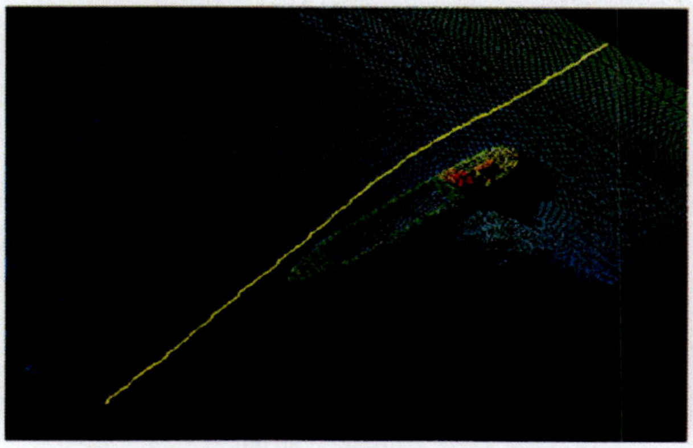

Fig. 2.54 Bathymetric graph

The water depth data in the experimental area were mapped in three dimensions (Fig. 2.54) and the hull profile can be observed clearly in the image. However, because of the inherent defects in the automatic bottom tracking algorithm, which relies on the water depth data alone, the existence of the mast is shown inadequately. The shape and size of the wreck, the mast and even the ropes can be observed clearly in both the course graph (Fig. 2.55) and the vertical graph of the WCI (Fig. 2.56).

The image (Fig. 2.57c_1–c_3) is obtained by denoising using the adaptive soft threshold denoising algorithm. When compared with Fig. 2.57b_1–b_3, the co-range noise and radial noise distribution in the original WCI accounts for up to 80% of the entire image, and that noise is evidently suppressed in the vertical sector graph after processing. This

Fig. 2.55 Course graph of WCI

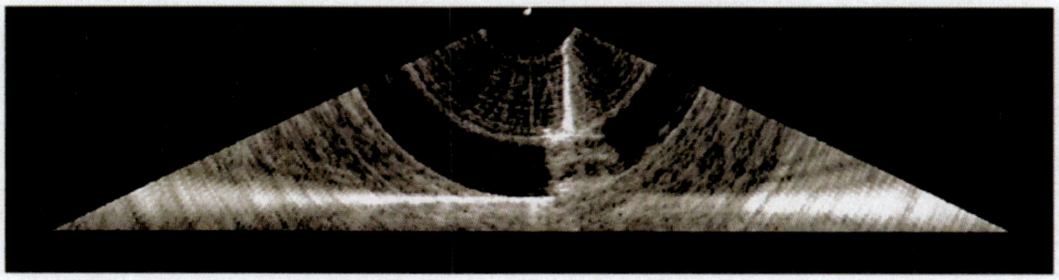

Fig. 2.56 Vertical graph of WCI

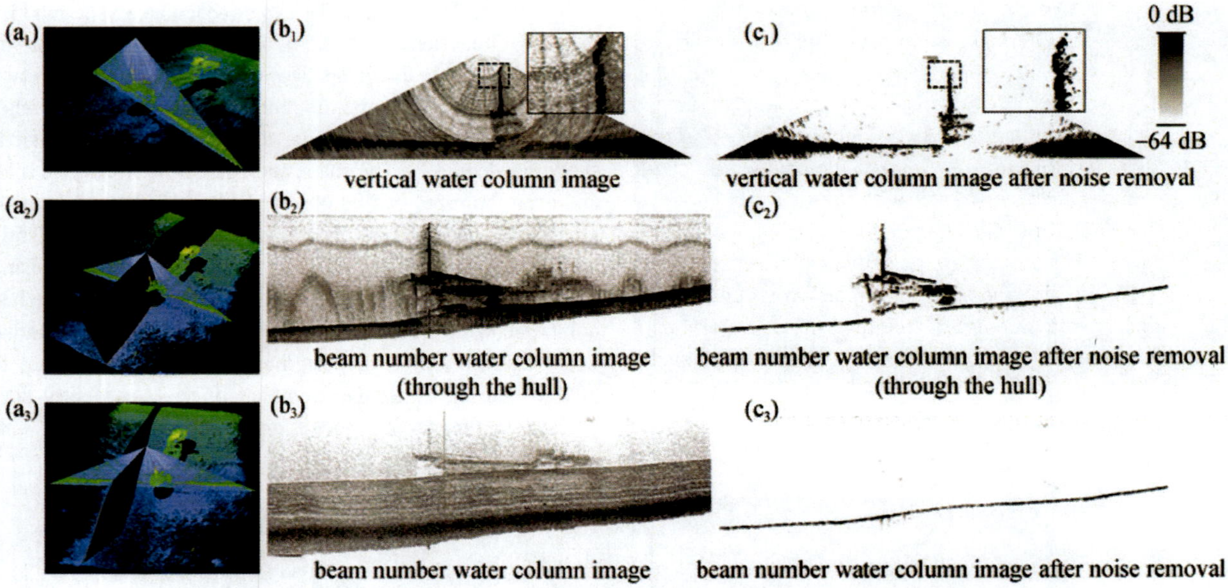

Fig. 2.57 Comparison of multi-beam water images of underwater wreck before and after denoising. $\mathbf{a}_1$–$\mathbf{a}_3$ Composite of wreck data and water image; $\mathbf{b}_1$–$\mathbf{b}_3$ original water image; $\mathbf{c}_1$–$\mathbf{c}_3$ water image after denoising

process obviously improves the display effect of WCI, making the results sharper and more effective, with more details remaining, and the image is thus easier to read. For the convex curve illusion of the side lobe generated in the mast position, as shown in Fig. 2.57b$_2$, good results are also obtained after processing, which means that the mast and other slender objects can be distinguished clearly from the background (as shown in Fig. 2.57c$_2$). In Fig. 2.57c$_3$, the false images that were generated in the other beams are also processed because of the co-range noise. Therefore, this algorithm has a positive effect on removal of the co-range noise and the illusions.

Analysis of the shipwreck images shows that the multi-beam WCI data contain more detailed recordings and have greater reproducibility than the bathymetric data. The time-series changes in each beam or course section of the water column can be viewed in the course display graph of the WCI. The vertical graph shows selected location features in detail and enables analysis and judgment of the angle and distance of the target. The combination of multi-beam bathymetric information with WCI information can enable 3D reconstruction of target details to an improved extent. Multi-beam WCI information thus has great potential for practical application.

References

Auke V (2010) Mast tracking capability of EM3002D using water column imaging. Dissertation, University of New Brunswick

Battilana D, Lawes G (2010) Fansweep 20 total propigated uncertainty, does it reflect reality? Paper presented at the FIG Congress 2010, Sydney, Australia, 11–16 April 2010

Beaudoin JD, Calder B, Hiebert J et al (2009) Estimation of sounding uncertainty from measurements of water mass variability. Int Hydrogr Rev 11:20–38

Beaudoin JD, Hughes-Clarke JE, Bartlett JE (2004) Application of surface sound speed measurements in post-processing for multi-sector multibeam echosounders. Int Hydrogr Rev 5(3):17–32

Beaudoin JD, Smyth S, Furlong A et al (2011) Streamlining sound speed profile pre-processing: case studies and field trials. Paper presented at the US Hydrographic Conference, Tampa, Florida, 25–28 April 2011

Bourillet JF, Edy C, Rambert F et al (1996) Swath mapping system processing: bathymetry and cartography. Mar Geophys Res 18(2–4):487–506. https://doi.org/10.1007/bf00286091

Brennan CW (2017) Multibeam Calibration: The Patch Test. Available via DIALOG. https://www.r2sonic.com/wp-content/uploads/2020/03/The-New-Patch-Test.pdf. Accessed 15 May 2020

Calder B (2003) Automatic statistical processing of multibeam echosounder data. Int Hydrogr Rev 4(1):53–68

Calder B (2019) Parallel variable-resolution bathymetric estimation with static load balancing. Comput Geosci 123:73–82. https://doi.org/10.1016/j.cageo.2018.11.011

Calder B, Mayer L (2001) Robust automatic multi-beam bathymetric processing. Paper presented at the US Hydrographic Conference, Norfolk, Virginia, 21–24 May 2001

Calder B, Mayer L (2003) Automatic processing of high-rate, high-density multibeam echosounder data. Geochem Geophy Geosy 4(6):1048. https://doi.org/10.1029/2002gc000486

Calder B, Rice G (2011) Design and implementation of an extensible variable resolution bathymetric estimator. Paper presented at the US Hydrographic Conference, Tampa, Florida, 25–28 April 2011

Calder B, Smith S (2004) A time comparison of computer-assisted and manual bathymetric Processing. Int Hydrogr Rev 5(1):10–23

Calder B, Wells D (2007) CUBE user manual. Available via DIALOG. http://ccom.unh.edu/sites/default/files/publications/Calder_07_CUBE_User_Manual.pdf. Accessed 25 June 2019

CARIS (2015) Caris hips and sips, Version: 10.2.2. Available via DIALOG. http://www.teledynecaris.com/docs/4.4.11/hips%20and%20sips/index.html#page/CARIS%2520HIPS%2520and%2520SIPS%2520Help%2FA26A69.html%23. Accessed 25 June 2019

Colbo K, Ross T, Brown C et al (2014) A review of oceanographic applications of water column data from multibeam echosounders. Estuar Coast Shelf Sci 145:41–56. https://doi.org/10.1016/j.ecss.2014.04.002

De Moustier C (1993) Signal processing for swath bathymetry and concurrent seafloor acoustic imaging. In: Moura JMF and Louttie IMG (eds) Acoustic signal processing for ocean exploration, pp 329–354

de Oliveira Junior AM, Hughes Clarke JE (2007) Recovering wide angular sector multibeam backscatter to facilitate seafloor classification. Paper presented at the US Hydrographic Conference 2007, Norfolk, Virginia, 15–17 May 2007

Debese N (2001) (2001) Use of a robust estimator for automatic detection of isolated errors appearing in the bathymetry data. Int Hydrogr Rev 2(2):32–44

Douglas DH, Peucker TK (1973) Algorithms for the reduction of the number of points required to represent a digitized line or its caricature. Cartographica: The International Journal for Geographic Information and Geovisualization 10(2):112–122

Elhegzy H (2012) Gas plumes analysis using multi-beam EM710 water column. Dissertation, University of New Brunswick

Farr HK (1980) Multibeam bathymetric sonar: sea beam and hydro chart. Mar Geod 4(2):77–93. https://doi.org/10.1080/15210608009379375

Geng XY, Zielinski A (1999) Precise multibeam acoustic bathymetry. Mar Geod 22:157–167. https://doi.org/10.1080/014904199273434

Glenn MF (1970) Introducing an operational multi-beam array sonar. Int Hydrogr Rev 47:35–39

Godin A (1998) The calibration of shallow water multibeam echo-sounding systems, M Eng. Report, Department of Geodesy and Geomatics Engineering Technical Report No. 190, University of New Brunswick, Fredericton, New Brunswick, Canada, pp 182

Gostnell C, Yoos J, Brodet S (2006) NOAA test and evaluation of phase differencing bathymetric sonar technology. Paper presented at the Canadian Hydrographic Conference, Halifax, Canada, 6–9 June 2006

Hare R (2001) Error budget analysis for US Naval Oceanographic Office (NAVOCEANO) hydrographic survey systems. Available via DIALOG. http://www.academia.edu/10086841/Error_Budget_Analysis_For_NAVO. Accessed 25 June 2019

Hare R, Eakins B, Amante C (2011) Modelling bathymetric uncertainty. Int Hydrogr Rev 6:1–14

Hare R, Godin A, Mayer L (1995) Accuracy estimation of canadian swath (multibeam) and sweep (multitransducer) sounding systems. Available via DIALOG. http://www.dfo-mpo.gc.ca/Library/191216.pdf. Accessed 25 June 2019

Hellequin L (1998) Statistical characterization of multi-beam echo-sounder data. Paper presented at the OCEANS 98, Nice, France, 28 September–1 October 1998

Hiller T M, Lewis K (2004) Getting the Most out of High Resolution Wide Swath Sonar Data. Proceedings of the 14th International Symposium of the Hydrographic Society. The Hydrographic Society Special Publication Number 53, Paper 8

Howlett C (2010) Considerations and advantages of accepting CUBE surfaces as survey deliverables. Paper presented at the NSHC 29th Conference, Brest, 11–16 September 2010

Hughes Clarke JE (2003a) A reassessment of vessel coordinate systems: what is it that we are really aligning? Paper presented at the US Hydrographic Conference, Biloxi, Mississippi, 10–12 February 2003

Hughes Clarke JE (2003b) Dynamic motion residuals in swath sonar data: ironing out the creases. Int Hydrogr Rev 4(1):6–23

Hughes Clarke JE (2006) Applications of multi-beam water column imaging for hydrographic survey. The Hydrographic Journal 4:1–33

Hughes Clarke JE (2010) GGE3353-imaging and mapping II: submarine acoustic imaging methods. Available via DIALOG. http://www.omg.unb.ca/GGE/SE_3353.html. Accessed 1 July 2019

Hughes Clarke JE (2012) Optimal use of multibeam technology in the study of shelf morphodynamics. In Michael ZL, Christopher RS, Philip RH (eds) IAS Special Publication# 44. Sediments, morphology, and sedimentary processes on continental shelves: Advances in Technologies, Research, and Applications. Wiley-Blackwell, Oxford, pp 1–28

Hughes Clarke JE (2018) Multibeam echosounders. In: Micallef A, Krastel S, Savini A (eds) Submarine geomorphology. Springer International Publishing, New York, pp 25–42

Hughes Clarke JE, Brucker S, Czotter K (2014) Improved definition of wreck superstructure using multi-beam water column imaging. Journal of the Canadian Hydrographic Association 68:1–2

Hughes Clarke JE, Gardner JV, Torresan M. Mayer LA (1998) The limits of spatial resolution achievable using a 30 kHz multibeam sonar: model predictions and field results. Paper presented at the OCEANS 98, Nice, France, 28 September -1 October 1998

Hughes Clarke JE, Lamplugh M, Czotter K (2006) Multi-beam water column imaging: Improved wreck least-depth determination. Paper presented at the Canadian Hydrographic Conference, Halifax, Canada, 6–9 June 2006

International Hydrographic Bureau (2008) IHO Standards for Hydrographic Survey (5th edition). Available via DIALOG. https://www.iho.int/iho_pubs/standard/S-44_5E.pdf. Accessed 25 June 2019

Kammerer E (2000) A new method for the removal of refraction artifacts in multi beam echosounder systems. Dissertation, University of New Brunswick

Kongsberg M (2013) EM series multibeam echo sounders datagram formats. Available via DIALOG. https://epic.awi.de/id/eprint/29970/1/Kon2010a.pdf. Accessed 1 July 2019

Kongsberg M (2018) EM 124–12 kHz multibeam echo sounder. Available via DIALOG. https://www.kongsberg.com/globalassets/maritime/km-products/product-documents/em-124-multibeam-echo-sounder. Accessed 01 July 2019

L-3 Communications SeaBeam Instruments (2000) Multibeam sonar theory of operation. Available via DIALOG. https://www3.mbari.org/data/mbsystem/sonarfunction/SeaBeamMultibeamTheoryOperation.pdf. Accessed 01 July 2019

Ladner RW, Elmore P, Perkins AL et al (2017) Automated cleaning and uncertainty attribution of archival bathymetry based on a priori knowledge. Mar Geophys Res 38(3):291–301. https://doi.org/10.1007/s11001-017-9304-9

Lecours V, Dolan MFJ, Micallef A et al (2016) A review of marine geomorphometry, the quantitative study of the seafloor. Hydrol Earth Syst Sc 20(8):3207–3244. https://doi.org/10.5194/hess-20-3207-2016

Liu HX, Yang FL, Zheng SQ et al (2019) A method of sidelobe effect suppression for multi-beam water column images based on an adaptive soft threshold. Appl Acoust 148:467–475. https://doi.org/10.1016/j.apacoust.2019.01.006

Lucieer V, Picard K, Siwabessy J, Jordan A, Tran M, Monk J (2018) Seafloor mapping field manual for multibeam sonar. In: Przeslawski R, Foster S (eds) Field manuals for marine sampling to monitor australian waters. National Environmental Science Programme (NESP), Marine Biodiversity Hub, pp 42–64

Lurton X, Lamarche G (2015) Backscatter measurements by seafloor-mapping sonars, guidelines and recommendations. Available via DIALOG. https://www.niwa.co.nz/static/BWSG_REPORT_MAY2015_web.pdf. Accessed 15 May 2019

Mallace D, Gee L (2005) Multibeam processing-the end to manual editing. Int Hydrogr Rev 6(1):1–11

Mallace D, Robertson P (2007) Alternative use of CUBE: how to fit a square peg in a round hole. Paper presented at the US Hydrographic Conference, Norfolk, Virginia, 15–17 May 2007

Marques C (2012) Automatic mid-water target detection using multi-beam water column. Dissertation, University of New Brunswick

Maul GA, Bishop JC (1970) Mean sounding velocity, a brief review. Int Hydrogr Rev 47(2):87–92

NOAA. (2014) Field procedures manual. Available via DIALOG. https://nauticalcharts.noaa.gov/publications/docs/standards-and-requirements/fpm/2014-fpm-final.pdf. Accessed 25 June 2019

Park Y, Jung ND, Kim JS et al (2013) Performance validation of surface filter based on CUBE algorithm for eliminating outlier in multibeam echo sounding. Available via DIALOG. https://www.hydrographicsociety.org/documents/hydrographicsociety.org/downloads/ifhs_news_no_1_-_yosup_park_et_al.pdf. Accessed 25 June 2019

Passalacqua P, Belmont P, Staley DM et al (2015) Analyzing high resolution topography for advancing the understanding of mass and energy transfer through landscapes: a review. Earth Sci Rev 148:174–193. https://doi.org/10.1016/j.earscirev.2015.05.012

Pereda García R, Piña García F, De Luis R et al (2016) Model for the processing and estimation of dual frequency echo sounder observations in detailed bathymetries. Mar Geod 39(3–4):305–320. https://doi.org/10.1080/01490419.2016.1193577

Peyton DR, Beaudoin JD, Lamplugh M et al (2009) Optimizing sound speed profiling for hydrographic surveys. Int Hydrogr Rev 2:20–38

Renard V, Allenou JP (1979) Sea beam, multi-beam echo-sounding in "Jean Charcot" -description, evaluation and first results. Int Hydrogr Rev 1:35–67

Rezvani MH, Sabbagh A, Ardalan AA (2015) Robust automatic reduction of multibeam bathymetric data based on M-estimators. Mar Geod 38(4):327–344. https://doi.org/10.1080/01490419.2015.1053639

Urick RJ (1967) Principles of underwater sound for engineers. McGraw Hill, New York

Vasquez M (2007) Tuning the CARIS implementation of CUBE for patagonian waters. Dissertation, University of New Brunswick

Venugopal R, Calder B (2012) Hydrographic data processing on a robust, network-coupled parallel cluster. Paper presented at the 6th International Conference on High-Resolution Survey in Shallow Water, Wellington, New Zealand, 21–26 February 2012

Wei B, Zhou T, Li HS et al (2019) Theoretical and experimental study on multibeam synthetic aperture sonar. J Acoust Soc Am 145:3177. https://doi.org/10.1121/1.5109392

Wölfl AC, Snaith H, Amirebrahimi S et al (2019) Seafloor Mapping-the challenge of a truly global ocean bathymetry. Front Mar Sci 6:283. https://doi.org/10.3389/fmars.2019.00283

Wu ZY, Jin XL, Zhou JQ et al (2017) Comparison of buried sand ridges and regressive sand ridges on the outer shelf of the East China Sea. Mar Geophys Res 38(1–2):187–198. https://doi.org/10.1007/s11001-016-9278-z

Yang FL, Li JB, Wu ZY et al (2007) A post-processing method for the removal of refraction artifacts in multibeam bathymetry data. Mar Geod 30:235–247. https://doi.org/10.1080/01490410701438380

Zhao DN, Wu ZY, Zhou JQ et al (2015) A new method of automatic svp optimization based on MOV algorithm. Mar Geod 38(3):225–240. https://doi.org/10.1080/01490419.2015.1006798

Zhao DN, Wu ZY, Zhou JQ et al (2020) Parameter group optimization by combining CUBE with surface filtering and its application. Journal of Geodesy and Geoinformation Science 3(2):81–92. https://doi.org/10.11947/j.JGGS.2020.0209

Zhou JQ, Wu ZY, Jin XL et al (2018) Observations and analysis of giant sand wave fields on the Taiwan Banks, northern South China Sea. Mar Geol 406:132–141. https://doi.org/10.1016/j.margeo.2018.09.015

Zhou JQ, Wu ZY, Zhao DN et al (2020) Giant sand waves on the Taiwan Banks, southern Taiwan Strait: Distribution, morphometric relationships, and hydrologic influence factors in a tide-dominated environment. Mar Geol 427:106238. https://doi.org/10.1016/j.margeo.2020.106238

3 Airborne Laser Bathymetric Technology

The traditional measurement of submarine topography mainly depends on shipborne sonar systems, e.g., the single beam echo sounder and multi-beam echo-sounding system (MBES). However, the measurement rate is low and the measurement cost is high for such measurements owing to the limitations of the acoustic characteristics, and it is thus difficult to carry out rapid and large-area measurements (Kuus 2008; Guenther 1985).

Laser scanning measurement technology obtains highly accurate three-dimensional (3D) coordinates of topography through non-contact active measurement, which overcomes the limitations of traditional acoustic measurement technology. The advantages of the laser scanning measurement technology include a fast scanning speed, high density of measuring points, high measurement efficiency, high measurement accuracy, and strong initiative. Airborne LiDAR bathymetry (ALB), which has been developed in the past 20–30 years, is an effective technique for deriving submarine topography in areas of shallow water. The ALB system integrates the laser, global navigation satellite system (GNSS), automatic control, aviation, computer, and other cutting-edge technologies and is especially suited to the rapid detection of complex terrains, such as that of shallow waters and areas near island reefs. The main advantages of the ALB system are as follows.

(1) The ALB system has a high measurement capability in shallow waters. The minimum detection depth of the ALB system reaches 0.15 m, allowing measurement of the water depth in shallow waters where vessels cannot reach.
(2) The measurement efficiency of the ALB system is independent of the water depth. The wide coverage range of the MBES is generally 4–8 times the water depth, while the wide coverage range of ALB is fixed, being related only to the flight altitude and scanning nadir angle. The wide coverage range of ALB can reach 320 m when the flight altitude is 600 m and the scanning nadir angle is 15°.
(3) The ALB system has high efficiency, high density, and low cost. According to statistics, the cost of ALB in shallow waters is only 6-10% the cost of the MBES. However, the measurement efficiency of ALB is approximately 20 times that of the MBES, and the density of the measuring point can reach the decimeter level (Niemeyer and Kogut 2014).
(4) The ALB system efficiently obtains integrated surface and underwater topography data, which greatly improves the efficiency of obtaining a seamless mosaic of integrated surface and underwater topography (Fig. 3.1).

3.1 Working Principle of the ALB System

The ALB system is a working system that uses a laser transmitting system to transmit laser signals for scanning and measuring the sea surface and detects the laser waveform signals of the sea surface and seafloor through a receiving system; the submarine topography and water depth can thus be determined through photoelectric conversion and digital signal processing (Doneus et al. 2015).

The ALB system has good flexibility because it uses fixed-wing aircraft and helicopters as platforms. Adopting laser scanning, GNSS and inertial navigation technologies, the ALB system makes fast, large-area and high-density scanning measurements. The ALB system has therefore been widely applied to the bathymetric surveying of coastal continental shelves. In addition to making conventional submarine topography measurements, the ALB system can also detect navigation obstacles and underwater moving targets (such as submarines) with high probability owing to its high coverage. For landing sites without bathymetric information, the ALB system can acquire bathymetric

Z. Wu et al., *High-resolution Seafloor Survey and Applications*,
https://doi.org/10.1007/978-981-15-9750-3_3

Fig. 3.1 Seamless mosaic integrating surface and underwater data (Piel et al. 2007)

information quickly and safely, thereby improving the operational capability of rapid-reaction troops. Additionally, the ALB system measures not only turbidity, temperature, and salinity in the sea area but also the siltation of ports in the field of ocean engineering.

The detection depth of the ALB system is limited by the serious absorption and scattering of laser light by water. The ALB system cannot completely replace an acoustics bathymetric system, especially for deep water. However, the ALB system may be indeed an effective direct bathymetric method to be used offshore in that it can supplement the low ocean bathymetric capability of shipborne systems.

3.1.1 System Composition

The ALB system mainly comprises an airborne system and ground processing system (Guenther 1989). The airborne system comprises laser transceiver, scanner, optical receiving, data acquisition, control, real-time display and other subsystems. The ground processing system is mainly responsible for the post-processing of data (e.g., bathymetric information processing and aircraft attitude correction) to produce digital products (e.g., submarine topographic maps, ocean charts, profiles, and digital elevation models). Specifically, the ALB system can be divided into six major components.

3.1.1.1 Measurement System

The measurement system uses a laser transceiver and scanning prism.

3.1.1.2 Positioning and Attitude Measurement System

The GNSS/inertial measurement unit (IMU) integrated navigation system is usually employed for positioning and measuring attitude.

3.1.1.3 Data Processing and Analysis System

The data processing and analysis system comprises a height counter, depth counter, and data controller that record the position, water depth and other data.

3.1.1.4 Control and Monitoring System

The control and monitoring system includes the system monitor, navigation display, control keyboard, and data display. Operators control and monitor the system in real time on the control platform.

3.1.1.5 Ground Processing System

The ground processing system includes a computer, system console, and drawing system and is used to process collected data and draw maps.

3.1.1.6 Aircraft and Maintenance Equipment

The aircraft and maintenance equipment provides flight status parameters and a working power supply.

3.1.2 System Working Principles

3.1.2.1 Scanning Measurement

The working principle of the ALB system is to use laser scanning equipment to realize longitudinal scanning along the flight direction and then use the scanning rotating prism to realize horizontal scanning. At the same time, the point cloud data of ground objects over a wide coverage range of strip areas can be obtained using the precise position information provided by the GNSS and the flight attitude data (i.e., heading, roll, pitch and acceleration) provided by the inertial navigation system (INS) (Optech 2015).

The strip broadening of the bathymetric point can be realized by adopting a scanning method that uses local movements of the scanning mirror (Fig. 3.2). Currently, there are presently two common scanning modes, i.e., conical scanning and linear scanning (Guenther 1985). Conical scanning mainly employs circular or elliptical spiral trajectories (e.g., the ABS system, SHOALS system, and LARSEN 500 system (Fig. 3.3)) while linear scanning uses horizontal parallel trajectories (e.g., the LADS MKII system (Fig. 3.4)). In the flight measurement, to achieve full-coverage bathymetry, the high-frequency laser light is transmitted in the horizontal scanning mode to form a scanning band of several hundred meters that is perpendicular to the flight direction (Pan et al. 2015).

The system structure is relatively simple when taking the dual-color ALB system with elliptical scanning (Fig. 3.5) as an example. A dual laser with wavelengths of 532 and 1064 nm can form an elliptical laser footprint at the sea surface after reflection through a high-speed rotating reflector prism. To measure the angle (Φ) of mirror rotation, an angle encoder and mirror are fixed together on the rotating shaft of the mirror driving motor. The normal line of the mirror is at a certain angle to the rotating shaft of the driving motor, and the rotating shaft of the mirror is inclined at 45°. To achieve a wide coverage range of the scanning measurement, the horizontally incident laser beam, located on (or parallel to) the vertical plane of the rotating shaft of the driving motor, is deflected to the sea surface in different directions. The ALB receiving system then receives the reflection signals of the surface and seafloor to calculate the water depth.

The measuring point density and the strip width can be adjusted by setting system parameters (e.g., the laser emission frequency and scanning angle) when the ALB system

Fig. 3.2 Scanning measurement of the ALB system

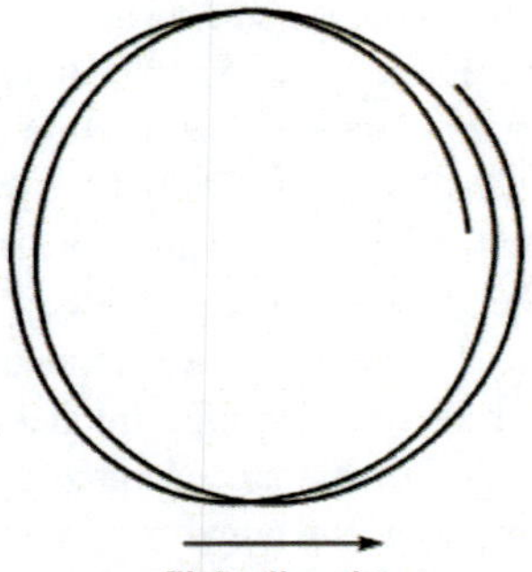

Fig. 3.3 Circular scanning mode

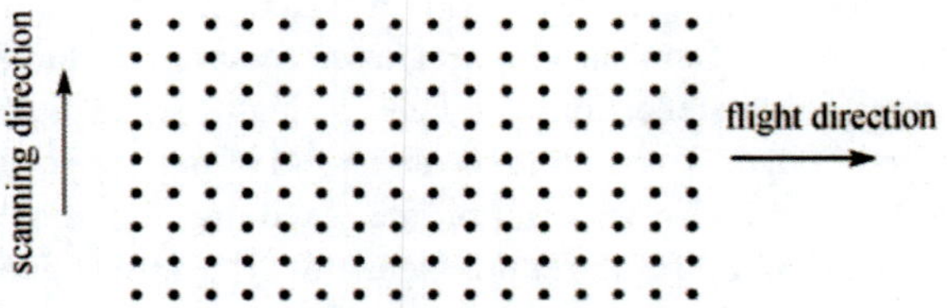

Fig. 3.4 Linear scanning mode

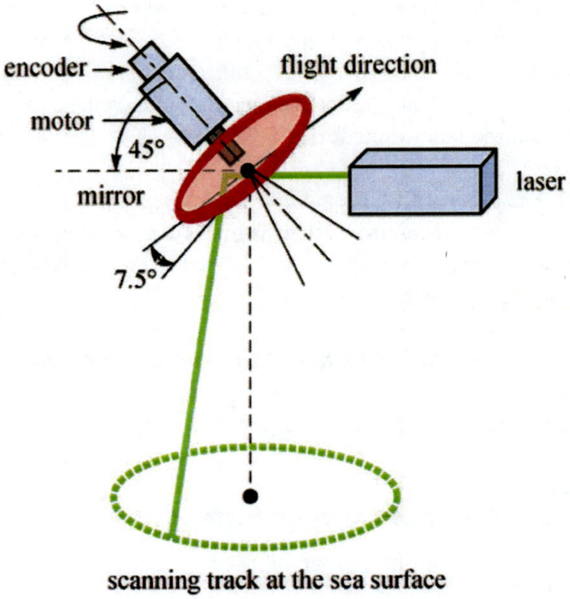

Fig. 3.5 Measurement principle of the elliptical scanning mode

performs a scanning measurement. Combining the flight altitude and speed, a reasonable match between the point density and strip width can be achieved according to the measurement purpose. If the point density needs to be large, the strip width needs to be narrow and the flying speed is limited. In contrast, the wide coverage range greatly increases when the point density is reduced.

3.1.2.2 Attitude Measurement

Attitude data (e.g., heading, roll, pitch and acceleration data) of the aircraft can be obtained through the INS, which comprises an IMU and navigation computer. The IMU comprises three single-axis accelerometers and three single-axis gyroscope accelerometers. The accelerometers are employed to measure the force and thus determine the position, velocity, and attitude of the carrier. The gyroscopes are employed not only to establish the reference coordinate system but also to monitor the angular velocity signals of the carrier, which are relative to the navigation coordinate system. These signals are transmitted to the navigation computer to compensate for systematic errors, and the relative attitude matrix calculation, gravity correction, acceleration integration, and velocity integration are then completed. Ultimately, the positioning, navigation, and attitude information of the carrier in the navigation coordinate system, including three positions, three velocities, and three attitudes respectively, is output (Guenther et al. 1996).

The initial position and attitude need to be provided to the INS for acceleration conversion and integral calculation. The initial position of the carrier can be obtained by the GNSS immediately. However, it takes some time to initially align the attitude. The initial horizontal attitude can be determined from the output of the accelerometer in a completely static mode, and the initial azimuth angle can be calculated by monitoring the rotation speed of the Earth with a gyroscope. After the initialization, the time-stamped attitude data can be obtained by the IMU in real time during the flight and then used to post-process the ALB data.

3.1.2.3 Positioning Measurement

The differential GNSS (DGNSS) receiver records the position information of the aircraft in real time. The receiver has three main functions: (i) providing the precise 3D position of the laser scanner sensor in air; (ii) providing external data for the INS to eliminate the drift of the gyroscope system and participates in the correction calculation of the gyroscope system at the same time; and (iii) providing navigation data for the navigation display.

At present, to realize precise positioning, GNSS and IMU integrated navigation is employed by most ALB systems in what is called the position and orientation system (POS). In fact, both the DGNSS and IMU can obtain the positioning alone, but each system has its advantages and drawbacks. The DGNSS can provide high measurement accuracy and its errors do not accumulate over time. However, the dynamic performance of the DGNSS is poor (e.g., it easily locks) and the output frequency is low. The IMU can realize positioning continuously but its positioning errors accumulate over time. The DGNSS and IMU are therefore complementary in terms of positioning, and the data of the two systems are combined to obtain highly accurate and highly reliable positioning data, which are mainly processed using the Kalman filter.

3.1.2.4 Bathymetry

ALB is an active telemetry technology. Studies have shown that blue-green light with a wavelength of 520–535 nm has the weakest light absorption, and this band is thus called the ocean optical window. The ALB system was developed to obtain water depths according to the propagation characteristics of blue-green light. According to the number of bands, ALB systems can be divided into dual-color and monochrome laser bathymetry systems.

(1) Dual-color laser bathymetry

The dual-color ALB system employs an infrared laser with a wavelength of 1064 nm and a blue-green laser with a wavelength of 532 nm. The system uses a laser transmitter installed in the lower part of the aircraft to transmit laser pulses that are inclined to the sea surface at a certain angle. In the example of the collinear scanning of infrared light and blue-green light, light from both lasers is both transmitted downward. After reaching the sea surface, the infrared laser light is reflected by the sea surface while the blue-green laser light continues to penetrate the water column at a certain refractive angle to the seafloor. Light from both lasers then returns along the incident path and is received by the laser receiving system. The instantaneous water depth can be obtained from the incident angle of the laser light, the refractive index of water, and the round-trip propagation time of blue-green light (Fig. 3.6). The measuring point in the geographic coordinate system and the depth of water based on the depth datum can be obtained by synthetically processing the positioning information measured using the GNSS, flight attitude information, and tide data. Finally, data in *x*-*y*-*z* format can be imported into computer-aided design, geographical information system, or other digital mapping software for mapping.

Owing to the collinear scanning, the round-trip propagation time of blue-green light in water is obtained by deducting the infrared laser travel time from the blue-green laser travel time (Fig. 3.7).

The angle of refraction can be calculated from the incident angle of the laser (θ_i), refractive index of air (n_{air}), and refractive index of water (n_{water}) as

$$\theta_i' = \arcsin\left(\frac{n_{\text{air}}}{n_{\text{water}}} \cdot \sin\theta_i\right) \tag{3.1}$$

c_{water} denotes the velocity of laser light in water and is expressed as

$$c_{\text{water}} = \frac{c}{n_{\text{water}}} \tag{3.2}$$

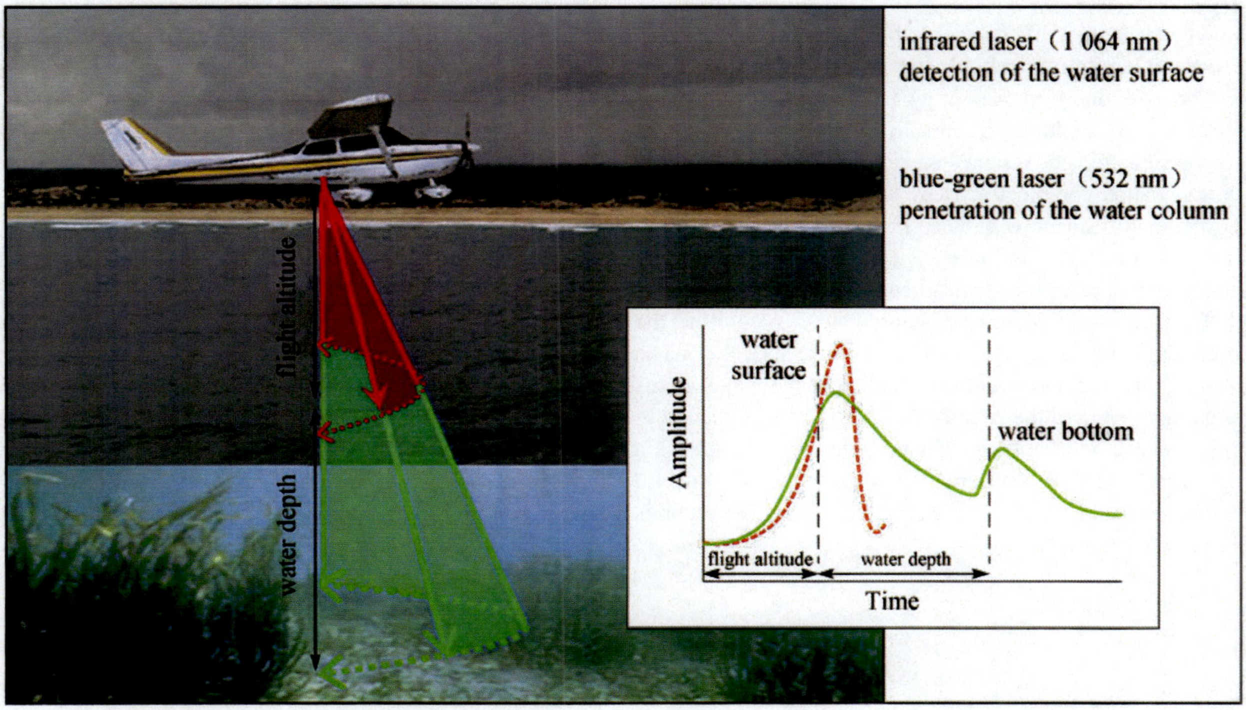

Fig. 3.6 Principles of the ALB system (Kuus et al. 2008)

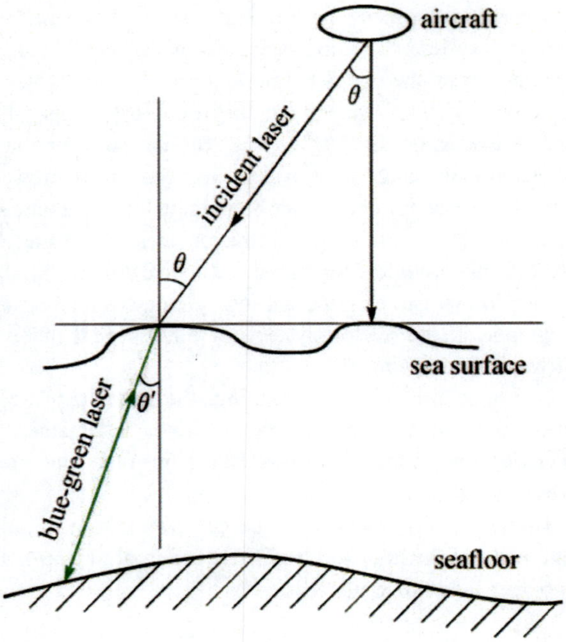

Fig. 3.7 Diagram of the laser propagation path of the ALB system

where the refractive index in water (n_{water}) at 532 nm is 1.334 and c is the velocity of laser light in air. The formula for calculating the instantaneous water depth (D) is

$$D = \frac{1}{2}\frac{c}{n_{\mathrm{water}}} \cdot \Delta t_i \cdot \cos\theta_i' \tag{3.3}$$

where Δt_i is the difference in the round-trip travel time between the infrared and blue-green laser light.

The positioning of measuring points involves the transformation of multiple coordinate systems, including the scanner coordinate system, inertial navigation coordinate system, carrier coordinate system, local horizontal coordinate system, and geodetic coordinate system. Through the transformation of these coordinate systems, measuring points are finally converted to the geodetic coordinate system.

The origin of the scanner coordinate system is at the reference point of laser emission (or receiving), the x-axis points in the flight direction of the aircraft, the y-axis points to the right wing of the aircraft, the z-axis is perpendicular to the x-y plane and points downward, and o-xyz constitutes a right-hand coordinate system. The relative position calculation of measuring points in the scanner coordinate system is expressed as

$$\begin{bmatrix} x_i \\ y_i \\ z_i \end{bmatrix} = \begin{bmatrix} 0 \\ \frac{1}{2}c \cdot \Delta t_i^{\mathrm{IR}} \cdot \sin\theta_i + \frac{1}{2}\frac{c}{n_{\mathrm{water}}} \cdot \Delta t_i \cdot \sin\theta_i' \\ \frac{1}{2}c \cdot \Delta t_i^{\mathrm{IR}} \cdot \cos\theta_i + \frac{1}{2}\frac{c}{n_{\mathrm{water}}} \cdot \Delta t_i \cdot \cos\theta_i' \end{bmatrix} \tag{3.4}$$

where x_i, y_i, z_i are coordinates of the ith measuring point in the scanner coordinate (SC) system and Δt_i^{IR} is the round-trip travel time difference of the ith laser.

(2) Monochrome laser bathymetry

Earlier ALB systems employed a dual-color laser because the blue-green laser light having a wavelength of 532 nm reflected faintly on the sea surface and it was thus difficult to obtain the accurate travel time of the sea surface return. However, taking advantage of the monochrome laser as a source of emission not only simplifies the system structure but also eliminates the synchronization of the dual-color laser. Therefore, employing a monochrome laser has become a goal of the design of the ALB system so as to improve the bathymetric accuracy. A monochrome laser ALB system that employs the blue-green laser with a wavelength of 532 nm as the laser source has been already developed. A semiconductor- pumped (Nd: YAG) laser loaded on an aircraft emits high-power, narrow-pulse blue-green laser light. Some of the laser light reaches the sea surface and is reflected back to the laser receiving system while some continues to penetrate the water column and reaches the seafloor. Part of the laser light is received by the laser receiving system after reflection. The water depth can be calculated according to the time difference of the round-trip travel between the sea surface return and the seafloor return (Wang et al. 2015). Except for there being only one laser color, the principle of the monochrome laser system is the same as that of the dual-color ALB system.

3.1.3 System Calibration

The ALB system mainly comprises a GNSS receiver, IMU, laser scanner, and other sensors. Because the geometric centers of the sensors do not coincide and the main axes are not parallel, systemic misalignment errors that result in the maximum error of the system are generated. The offset of the general geometric center is easily measured. However, the placement angle differences between the laser scanner and the INS (e.g., the heading error, pitch error and roll error) are difficult to measure directly. Therefore, in compensating for systemic misalignment errors, the system must be calibrated to accurately determine the placement angles among the sensors in advance.

Although the ALB system can measure water and land simultaneously, the acquired laser information of the water is not as accurate as that of the land because the laser is restricted by various geophysical environmental factors in the case of ocean. Taking into account the accuracy of the land laser information and the apparent extent of the ground features, the system calibration for land is selected in the ALB system. To ensure the quality of the system calibration, the calibration field should contain flat terrain, regular landmarks, checkpoints (i.e., obviously inclining terrain or features, such as spires), and highly reflective targets. Generally, similar to the calibration of the MBES system, the parameters of the system placement angle are determined by two overlapping survey lines that are of the same speed but in opposite directions in the calibration field.

3.2 Main Technical Parameters of the ALB System

The ALB system generally obtains bathymetric data with a density of 2 m × 2 m. Even denser data can be obtained from a lower flight altitude. The bathymetric accuracy meets the first-level bathymetric standard of IHO S-44.

3.2.1 Maximum Penetration Depth

Because the absorption of laser light is easier than that of a sound wave in water, the ALB system only measures a maximum depth of dozens of meters. The maximum penetration depth is an important index with which to evaluate the performance of the bathymetric system. The bathymetric capability of the system mainly depends on water quality parameters and system parameters (e.g., the flight altitude and receiving field of view (FOV) angle). The theoretical maximum detection depth of the system can be expressed as

$$L_m = \ln(P_m/P_b)/(2\Gamma) \tag{3.5}$$

where P_b is the power of the background light and Γ is the water effective attenuation coefficient. P_m is a system parameter that is expressed as

$$P_m = P_L R A \eta/(\pi H^2) \tag{3.6}$$

where P_L is the peak power of the laser, R is the seafloor reflectivity, A is the receiving area, η is the receiving efficiency, and H is the flight altitude (Guenther 1985). P_b and Γ depend on the natural conditions and water characteristics of the sea area while P_b is related to the sunlight.

In practice, owing to the uncertainty of the background light signal power and the seafloor reflectivity, the laser maximum penetration depth is generally estimated using the Secchi depth. To obtain the Secchi depth, a white disc is gradually submerged into water until the disc is just invisible. For a typical ALB system, it is generally believed that the laser maximum penetration depth is 2–3 times the depth of the Secchi disc in clear water and 3–5 times the depth of the Secchi disc in turbid water (Guenther 1985). At present, the maximum penetration depth is approximately 80 m and the bathymetric accuracy is less than 0.3 m for a water depth of around 50 m. Turbid water restricts the maximum penetration depth of the ALB system. Conversely, the maximum penetration depth of the ALB system can be used to retrieve the water turbidity, which is meaningful to environmental protection departments.

3.2.2 Shallowest Detection Depth

In the case of the ALB system, owing to the limitation of the laser pulse width and the superposition of the water column backscattering signals near the water surface, the sea surface signal is superimposed with the seafloor signal in extremely shallow areas. The ALB shallowest detection depth, therefore, can be calculated by novel method of processing the waveform data. The ALB system must have good depth detection in shallow water to realize a high-precision seamless mosaic survey of land and sea. The ability to obtain a shallower water depth is important in studying changes to coastal zones and beaches. With the development of science and technology, the depth detection capability of the ALB system has improved from an initial 2 m to a current 0.15–0.20 m. For instance, the shallowest detection depth of the CZMIL system developed by Teledyne Optech is 0.15 m.

To improve the depth detection capability of the ALB system in the shallowest water, an important problem to solve is how to accurately separate the sea surface signal and the seafloor signal from the superimposed waveform signals. At present, to separate the superposed waveform signals easily, the mainstream solution is to employ a narrow laser pulse, high-speed detector, small receiving FOV angle, narrow-band interference filter, and orthogonal polarization to receive signals (Harsdorf et al. 1999), thereby reducing the shallowest detection depth of the system.

3.2.3 Density of Measuring Points

The measuring point density is an important factor in evaluating data quality. The density of ALB bathymetric points (i.e., the number of measuring points per square meter) can be expressed as

$$\rho = \frac{r}{2v[\sin 0.5\varphi(\tan\theta \cdot H)]} \quad (3.7)$$

where ρ is the density of ALB bathymetric points, r is the laser repetition frequency, v is the flight speed, H is the flight altitude, φ is the scanning angle, and θ is the nadir angle of the laser beam.

The laser repetition frequency is one of the most important system parameters for the ALB system and is proportional to the measuring point density under certain conditions (i.e., the flight altitude, flight speed, and scanning angle are constant). The development of high-power and high-repetition-frequency lasers is an effective way to increase the density of measuring points and also a great challenge. The ALB laser repetition frequency has now reached 550 kHz while the density of measuring points has reached 0.12 m × 0.12 m (i.e., 69 points/m^2).

3.2.4 Bathymetric Accuracy

The bathymetric accuracy is an important index of the measurement of submarine topography. In August 2009, Airborne Hydrography AB (AHAB) of Sweden conducted a bathymetric comparison experiment of an ALB system and MBES system, providing a valuable data resource for bathymetric accuracy comparison and analysis of these two systems.

The comparison experiment used the Hawk-Eye II ALB system. The bathymetric frequency was 4 kHz, the (root-mean-square) water depth measurement accuracy was 0.25 m, and the maximum penetration depth was 2–3 times the depth of the Secchi disc. The density of measuring points reached 1.8 m × 1.8 m with a flight altitude of around 250 m. In addition, the Simrad EM MBES system was employed for multi-beam bathymetry. To ensure that the effects of environmental factors and weather conditions were equal, the experiment was carried out in the same sea area and in the same time period. The bathymetric data obtained by the two systems were then compared quantitatively. It is concluded that the bathymetric data collected by Hawk-Eye II basically coincide with those collected by the MBES system. The water depth differences between these two systems are mostly less than 10 cm (Fig. 3.8). These results show that the ALB system meets the accuracy requirements of hydrographic surveying and charting.

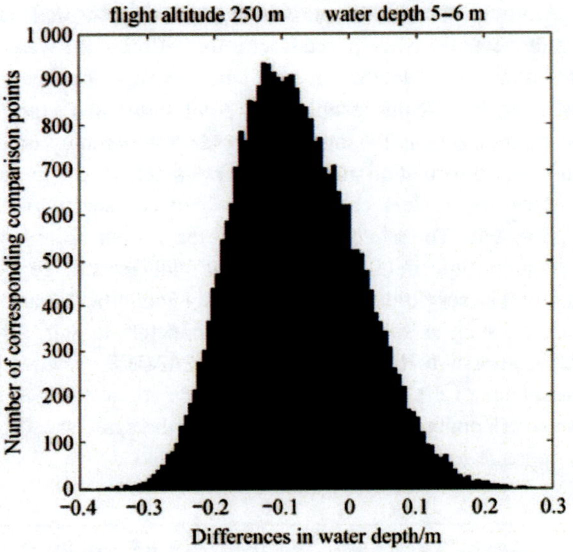

Fig. 3.8 Comparison of the bathymetric accuracy between the ALB system and MBES system for shallow water (Airborne Hydrography AB 2010)

3.3 Wave Correction Technology Based on the ALB Point Cloud

The ultimate goal of the ALB system is to obtain the 3D positions of the measuring points in the geographic coordinate system. The measurement data of each sensor are directly acquired through field acquisition. Therefore, to obtain the laser point cloud and correct for the effects of system error, the point cloud data should be processed accordingly.

The data acquired by the ALB system mainly comprise laser ranging data, GNSS data, attitude data, and tide data. The basic flow of data processing is as follows: (i) the instantaneous slant distance from the sea surface to the seafloor is extracted from laser waveform data, and the relative position of the measuring point in the scanner coordinate system is calculated; (ii) combining GNSS data and attitude data, the absolute geographical positions of the sea surface and the seafloor points are calculated; (iii) the wave effect on the sea surface is corrected; (iv) according to tide data, the instantaneous slant distance from the sea surface to the seafloor is reduced to the depth on the chart.

The processing and correction of the ALB and MBES bathymetric point cloud data are roughly similar. The spatial reduction, attitude correction, and tidal level correction are identical for these two systems. The change in the laser refraction index in water is small and can be negligible. Both systems require wave correction but the correction methods are different. The MBES system employs attitude sensors to directly measure wave fluctuations and realize wave correction. Aircraft cannot make contact with the water surface and other methods are needed for the wave correction of the ALB system.

The wave correction of ALB technology is the main correction of the measurement environment. The accuracy of wave correction directly affects the overall measurement accuracy of the bathymetry system. To obtain the precise seafloor elevation and the corresponding water depth information based on the depth datum, the wave information and the tidal level should be corrected using the measurement data (Tulldahl and Steinvall 2004). Finally, the geoid height of the seafloor point can be calculated directly without correction of the wave and tidal level. The spatial structure of the ALB system is shown in Fig. 3.9.

There are three wave correction methods; i.e., the non-correction method, filtering method, and inertial navigation auxiliary correction method (Guenther et al. 2000). They are introduced separately in the following.

3.3.1 Non-Correction Method

The non-correction method based on a GNSS provides highly precise 3D coordinates of the aircraft platform without the correction of the wave and tidal level. The seafloor is employed as the transition surface to obtain the seafloor elevation and the corresponding water depth directly. The depth reduction process without a synchronous tide can avoid the interference of wave correction and tidal level correction completely. The disadvantage is that model information of the geoid height and the surface topographic height must be known in advance. The detailed calculation process of the non-correction method is as follows.

1. The seafloor point geodetic height (h_d) is calculated from h, h_1 and h_2 as

$$h_d = h - h_1 - h_2 \tag{3.8}$$

2. The height of the mean sea level (i.e., the distance from the mean sea level to the reference ellipsoid surface, h_m) is calculated employing the geoid model and the sea surface topographic model:

$$h_m = N - \zeta \tag{3.9}$$

where N is the geoid height (i.e., the distance from the geoid to the reference ellipsoid surface) and ζ is the difference between the geoid and the multi-year mean sea level (i.e., submarine topography).

In addition to using the above method, h_m can also be obtained directly by satellite altimetry or indirectly by GNSS elevation measurement at the leveling point near the coastal tide station.

3. The distance between the seafloor point and the mean sea level is calculated from h_d and h_m as

$$H_m = h_m - h_d \tag{3.10}$$

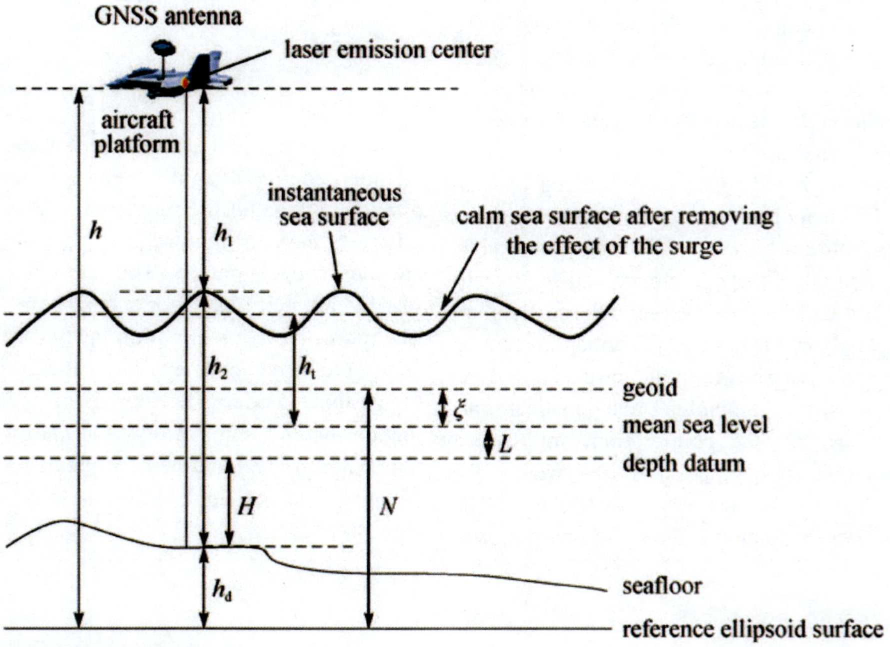

Fig. 3.9 Spatial structure of the ALB system

4. The depth H from the seafloor point to the depth datum is calculated as

$$H = H_m - L \tag{3.11}$$

where L is the difference between the mean sea level and depth datum.

The geoid height of the seafloor point can be obtained directly by calculating the difference between the geoid height N and geodetic height h_d.

3.3.2 Filtering Method

The wave correction is the calculation of an ultra-short-term mean sea level (not affected by the wave but by the tide) in nature. Different types of ALB systems have different methods of determining the mean sea level. For the dual-color laser bathymetry system, there are two cases (non-collinear scanning and collinear scanning) of determining the mean sea level. For non-collinear scanning, the infrared laser shoots vertically to the sea surface and the blue-green laser shoots to the sea surface at a certain angle. The beam angle of the infrared laser is wide and the spot diameter at the sea surface is thus 20–30 m. The mean sea level in this range can therefore be obtained by processing the waveform. For collinear scanning, the accelerometer and attitude sensor on the aircraft provide the attitude and vertical motion in real time, so that the corresponding mean sea level can be determined using the filtering method in a certain period of time. In addition, scanning using the monochrome laser bathymetry system is similar to dual-color laser collinear scanning. If the flight altitude is controlled within 500 m, the diameter of the green laser spot at the sea surface is only 50 cm. The corresponding mean sea level can be determined using the filtering method in a certain period of time.

The filtering method is the process of realizing optimal estimation. DGNSS technology is employed to obtain the precise geodetic height of the laser center. The instantaneous distance from the aircraft to the sea surface and the instantaneous slant distance from the sea surface to the seafloor are measured using the laser. The wave information can be obtained not only by filtering the multiple means around the measuring points but also by employing mathematical tools (e.g., the wavelet analysis, the Fourier transform, and a digital filter) according to the period of the wave. The advantage of this method is that wave correction can be realized using the dense point cloud of the system itself without knowing the geoid height and the sea surface topography height in advance. The disadvantage is that the filtering method can filter out details of the seafloor.

3.3.3 Inertial Navigation Auxiliary Correction Method

The above two methods are based on the high-precision geodetic height h provided by the positioning system. The inertial navigation auxiliary correction method overcomes the limitation of needing a highly precise geodetic height. To correct for a change in flight altitude, the acceleration information is measured by the INS on the aircraft platform. The main steps of converting the instantaneous water depth into the depth on the chart are as follows.

1. On the basis of the starting time t_1 for a certain period of time $\Delta t = [t_1, t_2]$, the height variation Δh of the laser emission center at time t ($t_1 \leq t \leq t_2$) is calculated.
2. The height variation of the instantaneous sea surface fluctuation is calculated according to

$$h_1' = h_1 - \Delta h \tag{3.12}$$

3. The mean of h_1' is calculated ass

$$\overline{h_1'} = \sum h_1'/n \tag{3.13}$$

where n is the number of samples.

4. The wave correction value Δh_b from h_1' and $\overline{h_1'}$ is calculated as

$$\Delta h_b = \overline{h_1'} - h' \tag{3.14}$$

In summary, the non-correction method and filtering method require a highly precise geodetic height that is provided by the positioning system. These two methods are relatively easy to implement in coastal and reef areas and are thus widely applied. The mean sea level obtained by satellite altimetry has poor accuracy and is thus unreliable. Therefore, the non-correction method cannot be employed as the preferred scheme without the geoid and sea surface topographic model. The inertial navigation auxiliary correction method requires only the highly precise flight altitude variation and not the high-precision geodetic height in advance and its application is thus more flexible and extensive.

3.4 Refraction Correction Based on Sea Surface Profile and Ray Tracing

3.4.1 Effect of Refraction on ALB

The effects of external environmental factors, such as the wind, waves, tides, and other random fluctuations on the sea surface, can change the geometry of the air-water interface. In addition, the backscatter effect in water column, MIE scattering effect arising from small particles, and diffuse reflection on the seafloor will also induce ranging errors into ALB data. As a result, the propagation path of the green laser will deviate, significantly affecting the geolocation accuracy of the seafloor points. According to the study of Westfeld (Westfeld et al. 2017) for an ALB system with small beams, the effects of waves on the planimetry and depth coordinates are significant. Planimetric effects may reach several decimeters or even meters, and depth coordinate errors also reach several decimeters. In terms of the International Hydrographic Organization (IHO) Standards for Hydrographic Surveys (S-44), the maximum allowable Total Vertical Uncertainty (TVU) for Special Order is 0.261 m when the water depth is 10 m (IHO 2008). The maximum refraction depth error may exceed the accuracy requirement of IHO S-44, even without considering other errors. Therefore, it is necessary to develop an effective method to reduce the ALB refraction error.

The refraction effect on the bathymetric accuracy of ALB has been previously studied. Guenther (Guenther 1985) proposed a method of reducing the refraction impact by spreading the laser beam to cover multiple wave cycles. This correction method was based on averaging the widened laser beams to smooth the sea surface. However, at present, the laser divergence angle is usually 1 mrad, which is much less than 20 mrad when this classical method was designed. Steinvall (Steinvall et al. 1994) demonstrated that the large laser footprints could average the wave effects. On the basis of Snell's law, he assumed that the laser beam was refracted at a flat sea surface. By evaluation, he found out that the maximum refraction depth error was 1-2% of the depth. The FLASH system was used in his experiments at 300 m operating altitude. The laser was a frequency-doubled Nd: YAG laser having emissions at 532 nm and 1064 nm with a pulse length of approximately 7 ns. The receiver FOV was 5–50 mrad and the nadir angle was 20°. The laser beam divergence was automatically controlled via a beam expander to values between 2 and 10 mrad (Steinvall et al. 1993). However, for the current advanced ALB system with a footprint of less than 0.5 m, only very small part of waves can be covered; and thus, a sea surface slope can lead to significant measurement error. Tulldahl (Tulldahl et al. 2000) estimated the effects of refraction on the planimetric position and water depth of ALB by establishing a two-dimensional sea surface slope model. During the refraction correction, the water depth was corrected by a one-dimensional slope of the water surface, but the planimetric position was not corrected. Westfeld (Westfeld et al. 2017) investigated the effects of waves on the 3D point accuracy in ALB. Based on the intensity distribution within the laser cross section and the inclinations of the corresponding water surface elements, the refraction of the laser beam was modeled differentially, in a rigorous manner. This method is based on simulations of typical wave patterns, not the actual water surface represented from the captured laser scanner points themselves. Therefore, improving the bathymetric accuracy of ALB by reducing the refraction errors requires an effective refraction correction method to track each laser on the sea surface.

3.4.2 Condition of Refraction Correction

For a single-wavelength ALB system, the green laser is emitted to obtain the sea surface and seafloor. By decomposing waveform and extracting the sea surface and seafloor signals, the respective travel times of a green laser reflected by the sea surface and seafloor can be derived. Considering the system calibration parameters, scanning nadir angle, real-time position, and attitude information, the geometric location of laser point on the sea surface can be calculated. After the refraction correction, water depth with high accuracy can be obtained. The light structure of ALB is illustrated in Fig. 3.10.

Fresnel equations describe reflection/refraction of a planar wave at a planar interface. Based on the refraction property of Fresnel's law (Chiu et al. 1999), the incident light is decomposed into the planimetric direction and depth direction, where P is the laser point on the sea surface.

$$\begin{cases} T_H = \dfrac{n_{\mathrm{w}}}{n_{\mathrm{a}}} \cdot \dfrac{4\cos i_{\mathrm{laser}} \cos r_{\mathrm{laser}} \sin^2 r_{\mathrm{laser}}}{\sin^2(i_{\mathrm{laser}} + r_{\mathrm{laser}})} \\ T_V = \dfrac{n_{\mathrm{w}}}{n_{\mathrm{a}}} \cdot \dfrac{4\cos i_{\mathrm{laser}} \cos r_{\mathrm{laser}} \sin^2 r_{\mathrm{laser}}}{\sin^2(i_{\mathrm{laser}} + r_{\mathrm{laser}}) \cos^2(i_{\mathrm{laser}} - r_{\mathrm{laser}})} \end{cases} \tag{3.15}$$

where T_H and T_V refer to the refraction components in the planimetric direction and depth direction, respectively; n_{a} is the refractive index of air ($n_{\mathrm{a}} = 1$); n_{w} is the refractive index of water ($n_{\mathrm{w}} = 1.33$); i_{laser} is the laser incident angle on a sloping sea surface; and r_{laser} is the laser refraction angle on a sloping sea surface.

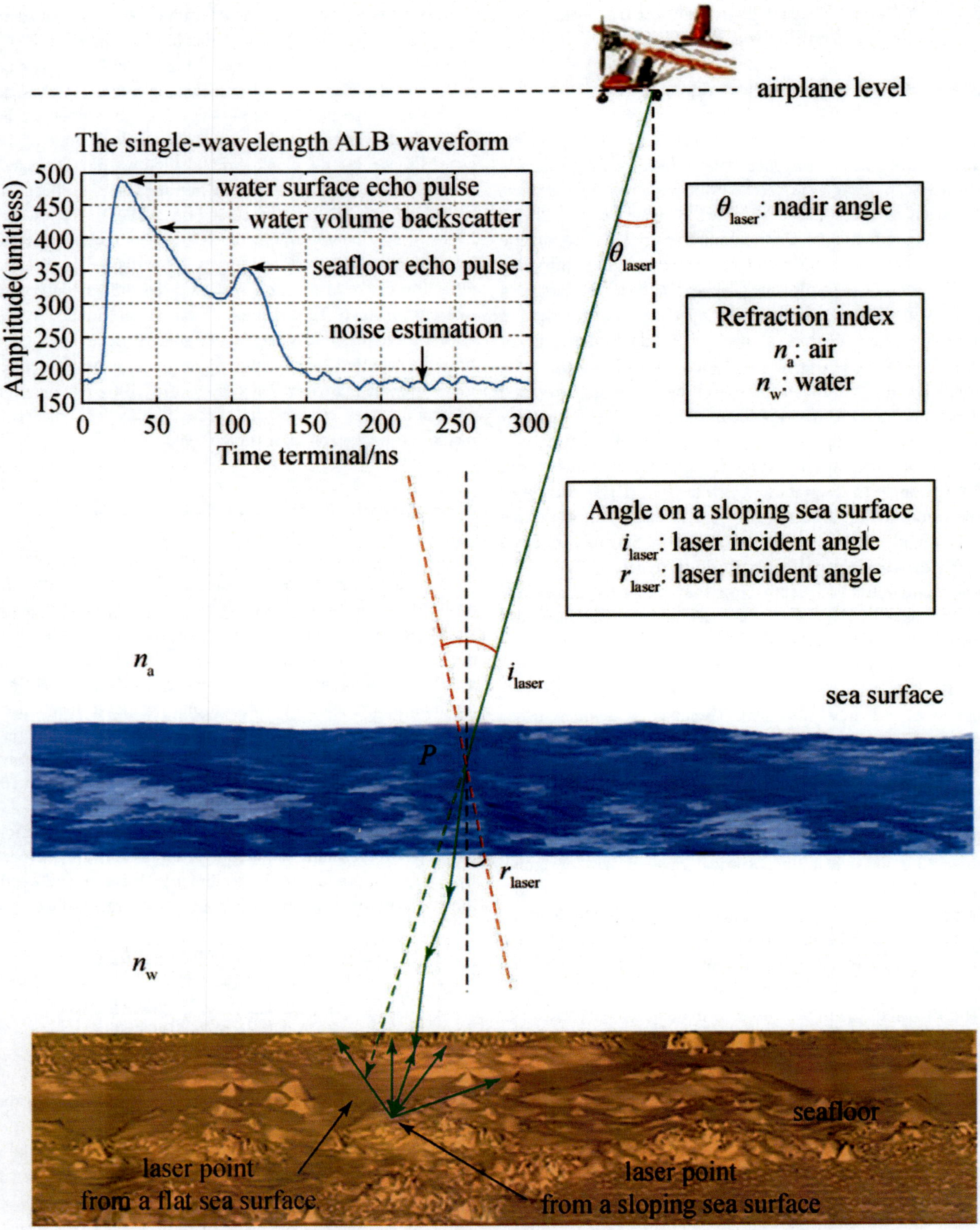

Fig. 3.10 Schematic illustration of ALB (Tessendorf 1999; Mandlburger et al. 2013)

We assume that the green laser is linear polarized light. The transmissivity T_1 through the air–water interface can be represented as shown in Eq. (3.16).

$$T_1 = T_H \cos^2 \delta + T_V \sin^2 \delta \tag{3.16}$$

where δ is the azimuth angle between the vibration plane and incident plane. After averaging the azimuth angle δ, T_1 can be expressed as follows:

$$T_1 = \frac{1}{2}|T_H + T_V| \tag{3.17}$$

The wind speed v (m/s) has a significant effect on the transmissivity T_2, which is influenced by sea surface foam and other external factors. The relationship (Li et al. 2008) between T_2 and v can be expressed as

$$T_2 = \begin{cases} 1 - 1.2 \times 10^{-5} \times v^{3.3} & v<9 \\ 1 - 1.2 \times 10^{-5} \times v^{3.3} \times (0.225v - 0.99) & v \geq 9 \end{cases} \tag{3.18}$$

In summary, the total transmissivity T can be calculated using Eq. (3.19).

$$T = T_1 \times T_2 \tag{3.19}$$

According to Eq. (3.19), the relationships among the total transmissivity, wind speed, and incident angles are calculated, which are presented in Fig. 3.11.

As shown in Fig. 3.11, the total transmissivity is inversely proportional to the wind speed above the sea surface. When the laser incident angle is less than 35°, the total transmissivity is almost constant; when the incident angle is between 35° and 55°, the total transmissivity decreases linearly with increases in the incident angle; and when the incident angle is beyond 55°, the total transmissivity decreases sharply with increases in the incident angle. Improving the efficiency of refraction correction should guarantee sufficient total transmissivity, in other words, better data quality. Guenther (1985) investigated the wave height regime influenced by wind speeds to ensure an ALB system to be fully operational. Beaufort 3 corresponds to the wind of gentle breeze (less than 5.4 m/s). We investigated on the transmissivity of the air-water interface influenced by winds and laser scanning angles to ensure the refraction correction. The results showed that the data should be obtained when the winds and scanning angles are less than 10 m/s and less than 25°, respectively. It indicated that the refraction correction can be well operated as long as the ALB data were obtained obeying the criteria stated by Guenther (1985), and the wind speed is approximately equal to 5 m/s when the ALB data used in this paper were obtained.

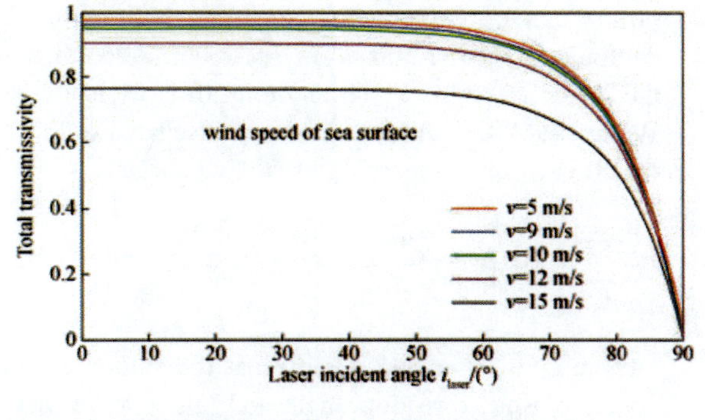

Fig. 3.11 Factors influencing the total transmissivity

3.4.3 Refraction Correction Methodology

In the process of the refraction correction, first, the applicable condition is described to ensure the method can be effectively used. Then, the procedure of the sea surface fitting is derived in detail, which is used to calculate the slope of the laser point on sea surface. Finally, each laser path is successively calculated to get the seafloor coordinate. The detailed procedure of the refraction correction is illustrated in Fig. 3.12.

3.4.3.1 Sea Surface Profile

The sea surface profile is very changeable and can be represented as the superposition of many cosine functions with different wave heights, different periods, different phases and different directions based on wave spectrum theory (Peachey 1986; Boccotti 2004; Martins et al. 2016; Vimala et al. 2014). The parameters of every cosine function can be fitted by the discrete point clouds on the sea surface measured by the 532 nm green laser. With the fitted sea surface profile, the air-water interface can be precisely derived to trace every laser path and eliminate the error caused by refraction at the air-water interface.

(1) The coordinate frame of the laser points on the sea surface. The sensor frame is established for the ALB system with a monochrome laser and circular scanning. As shown in Fig. 3.13, the coordinate origin o is the laser reference center, the z-axis is defined as the zenith direction in the frame, the y-axis is defined as the forward direction of the airplane and is normal to z, and x-axis completes the right-hand system.

In the sensor frame (laser scanner frame), the geometric location (x_P, y_P, z_P) of laser point P on the sea surface is described using Eq. (3.20).

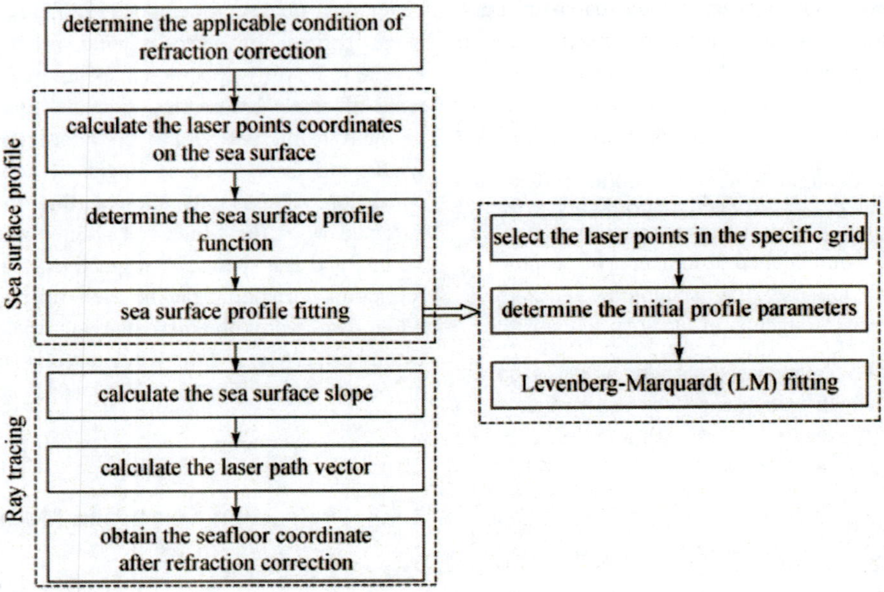

Fig. 3.12 Flowchart of the refraction correction

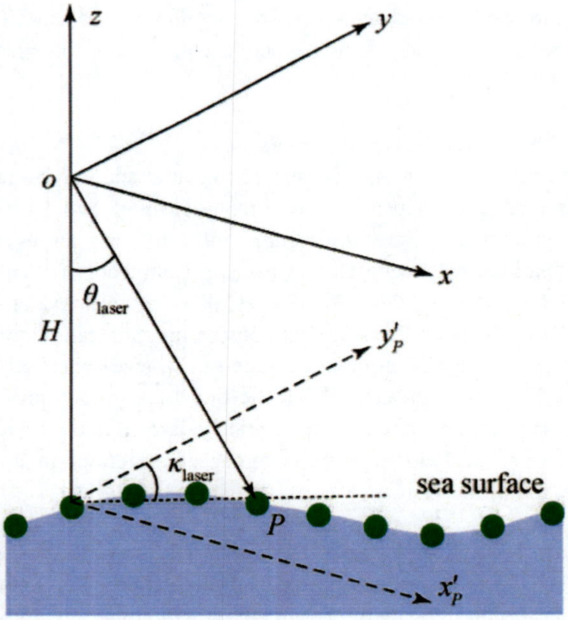

Fig. 3.13 The geometry of laser points on sea surface

$$\begin{cases} x_P = \frac{1}{2}c\Delta t_1 \sin\theta_{laser} \sin\kappa_{laser} \\ y_P = \frac{1}{2}c\Delta t_1 \sin\theta_{laser} \cos\kappa_{laser} \\ z_P = -\frac{1}{2}c\Delta t_1 \cos\theta_{laser} \end{cases} \quad (3.20)$$

where θ_{laser} is the nadir angle; κ_{laser} is the azimuth angle of the incident laser, which is clockwise from the y direction; c is the light speed in air; and Δt_1 is the derived travel time of the green laser from the laser transmitter to the air-water interface, which can be obtained by waveform fitting (Wong and Antoniou 1991; Wang et al. 2015). Actually, the refraction correction at the air-water interface for laser points is carried out in the local vertical reference frame, which usually defines the z-axis as the zenith direction; the x-axis directs to the east direction; y-axis completes the right-hand system. Thus, a series of coordinate transformations should be done frame by frame, which includes the sensor frame, carrier frame, and local vertical reference frame. After the transformations, a sea surface laser point (x_P, y_P, z_P) in the local vertical reference frame can be obtained, which will be used in Eqs. (3.21), (3.22), (3.24) and (3.31).

(2) Sea surface profile based on the wave spectrum. This profile can be expressed as the superposition of the continuous Fourier transform spectrum. According to the linear wave theory (Hamylton 2011; Méhauté and Wang 1996), the profile can be expressed as Eq. (3.21), which is a continuously differentiable surface.

$$z = \sum_{i=1}^{m} \zeta_i \cos\left[w_i t - \frac{w_i^2}{g}(x cos\alpha_i + y \sin\alpha_i) + \varepsilon_i\right] \quad (3.21)$$

where ζ_i, w_i, t, α_i and ε_i represent the amplitude, frequency, time, direction angle and initial wave phase, respectively; g = 9.8 m/s^2; and m is the number of superposition for cosine waves, which is set to 30 in this paper. The data used in this paper was captured at a wind speed of 5 m/s. Based on JONSWAP spectrum (Komen et al. 1994), when the wind is less than 5.5 m/s, the PSD (power spectral density) is mainly concentrated upon the angular frequency range [1.1, 4.0] and to finely illustrate details of the sea surface, the angular frequency interval should be less than 0.2. Therefore, the angular frequency ω_i, one of the fitted parameters, is set from 1.1 to 4.0 with a 0.1 interval and m is equal to 30.

(3) Sea surface profile fitting. The laser points reflected by the sea surface can be captured by the green laser. First, in every strip, the laser points of sea surface captured by the ALB system are searched and selected at different scale grids, e.g., typically 5 m × 5 m, 10 m × 10 m, etc., based on the planimetric coordinate of laser points. In every specific grid, the selected laser points are used to fit the sea surface profile using Eq. (3.21). Second, in the fitting process, the initial parameters (amplitude ζ_i and angular frequency ω_i) are calculated based on JONSWAP spectrum at a given wind speed. The data used in this paper was captured at a wind condition of 5 m/s; therefore, the angular frequency ω_i is set from 1.1 to 4.0 with a 0.1 interval and i is set as i = 1, 2, 3, …, 30. 30 corresponding amplitudes ζ_i are calculated based on JONSWAP spectrum and all the 30 angles α_i and 30 phases ε_i are set to be 0. Then, in every grid, all 120 final parameters are estimated iteratively by substituting the initial parameters and the coordinates of captured laser points using LM (Levenberg- Marquardt) algorithm. Finally, in each grid, the slope of the sea surface of every laser point is consecutively calculated using Eq. (3.22) which is the partial differential of Eq. (3.21). A sample of 100 m × 100 m fitting profiles is illustrated in Fig. 3.14b, whereas the captured laser points of the sea surface are illustrated in Fig. 3.14a, which are used to fit the sea profile.

3.4.3.2 Ray Tracing

To obtain the actual laser point coordinates on the seafloor, ray tracing must be carried out based on the sea surface slope.

(1) Calculating the sea surface slope. As shown in Fig. 3.15, the sea surface slope angle φ is the included angle between the sloping surface C and the sea level B; P is the laser point on the sea surface; the surface A is the sea surface; the surface B is the sea level; and the sloping surface C is the actual reflection surface, which is the tangent plane of the surface A at the point $\boldsymbol{P}$.

As noted previously, the surface A can be presented as the sea surface profile in Eq. (3.21). Therefore, the sea surface slope $\tan\varphi_x$ and $\tan\varphi_y$ at (x_P, y_P, z_P) in the x and y directions can be obtained by calculating the partial differential of Eq. (3.21), and the total sea slope is equal to the root sum square (RSS) of the slopes in the x and y directions, i.e., $\tan\varphi=\sqrt{\tan^2\varphi_x+\tan^2\varphi_y}$.

$$\begin{bmatrix}\tan\varphi_x\\ \tan\varphi_y\end{bmatrix}=\begin{bmatrix}\sum_{i=1}^{m}\zeta_i\frac{w_i^2}{g}\cos\alpha_i\sin\left[w_it-\frac{w_i^2}{g}(x_P\cos\alpha_i+y_P\sin\alpha_i)+\varepsilon_i\right]\\ \sum_{i=1}^{m}\zeta_i\frac{w_i^2}{g}\sin\alpha_i\sin\left[w_it-\frac{w_i^2}{g}(x_P\cos\alpha_i+y_P\sin\alpha_i)+\varepsilon_i\right]\end{bmatrix}\tag{3.22}$$

(2) Calculating the laser path vector for ray tracing. The laser propagation path is illustrated in Fig. 3.16. $\boldsymbol{N}$ is the unit normal vector of the reflection surface C; $\boldsymbol{I}_{\text{laser}}$ is the unit vector of the incident ray above the sea surface; and $\boldsymbol{R}_{\text{laser}}$ is the unit vector of the reflected ray beneath the sea surface. Among the three vectors, $\boldsymbol{N}$ and $\boldsymbol{I}_{\text{laser}}$ can be obtained from the sea surface slope and the geographic coordinate of the laser point $\boldsymbol{P}$, respectively. However, $\boldsymbol{R}_{\text{laser}}$, which has three unknown variables (R_x, R_y, R_z), must be calculated by solving three simultaneous equations.

Based on the sea surface slope $\tan\varphi_x$ and $\tan\varphi_y$ in the x and y directions calculated in Eq. (3.22), the unit normal vector $\boldsymbol{N}$ of this reflection surface can be expressed as

$$N=\left[\frac{-\tan\varphi_x}{\sqrt{1+\tan^2\varphi_y+\tan^2\varphi_x}},\frac{-\tan\varphi_y}{\sqrt{1+\tan^2\varphi_y+\tan^2\varphi_x}},\frac{1}{\sqrt{1+\tan^2\varphi_y+\tan^2\varphi_x}}\right]=(N_x,N_y,N_z)\tag{3.23}$$

where (N_x, N_y, N_z) refers to the three vector components in x, y and z directions, respectively.

Based on the geographic coordinate (x_P, y_P, z_P) of the laser point on the sea surface in the local vertical reference frame, the unit vector $\boldsymbol{I}_{\text{laser}}$ of the incident ray above the sea surface can be expressed as

$$\boldsymbol{I}_{\text{laser}}=\left(\frac{x_P}{\sqrt{x_P^2+y_P^2+z_P^2}},\frac{y_P}{\sqrt{x_P^2+y_P^2+z_P^2}},\frac{z_P}{\sqrt{x_P^2+y_P^2+z_P^2}}\right)=(I_x,I_y,I_z)\tag{3.24}$$

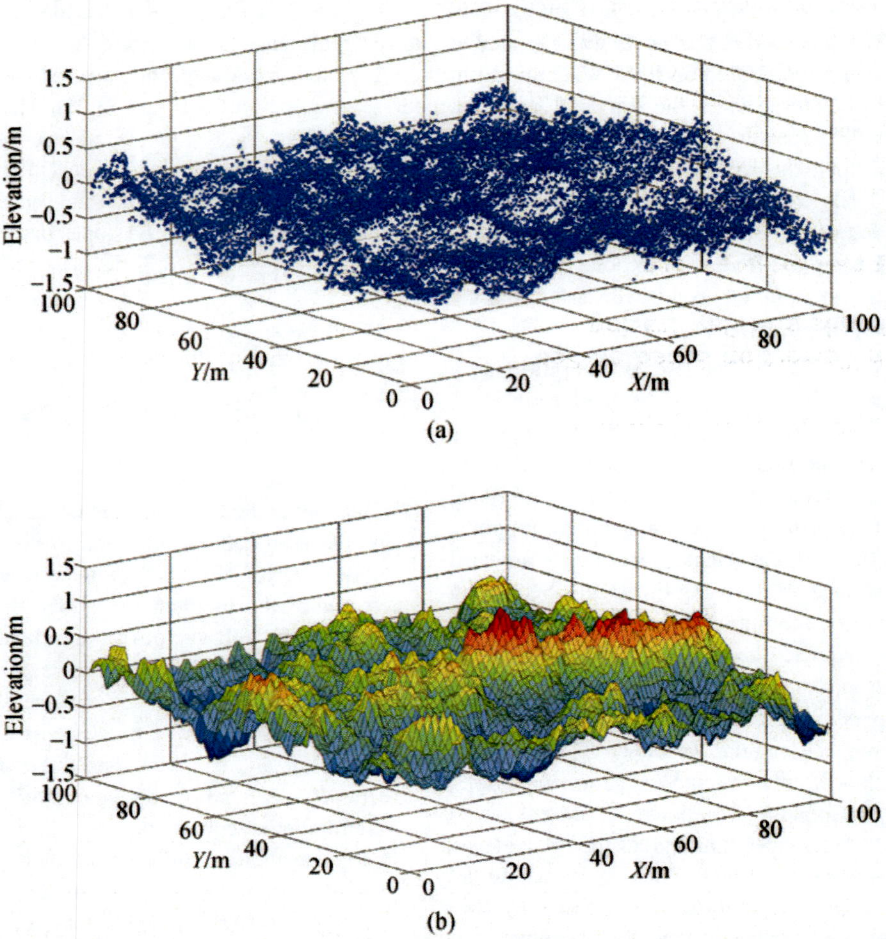

Fig. 3.14 Instantaneous sea surface model. **a** 3D original laser point on the sea surface captured by the ALB system; and **b** fitted sea surface model using laser points captured by the ALB system

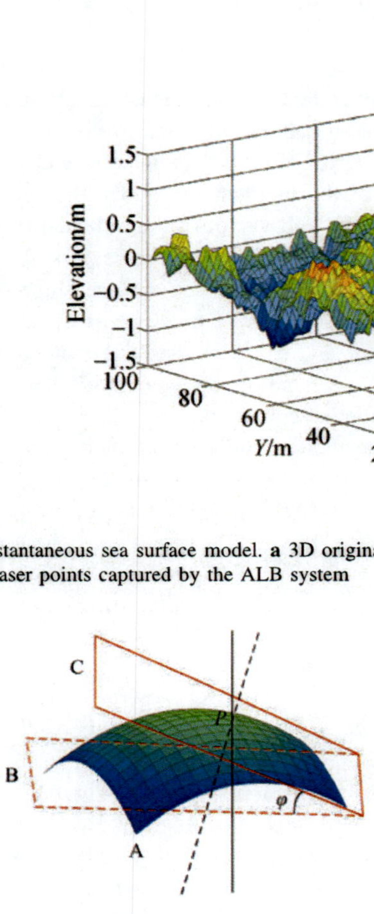

Fig. 3.15 Schematic illustration of sea surface slope

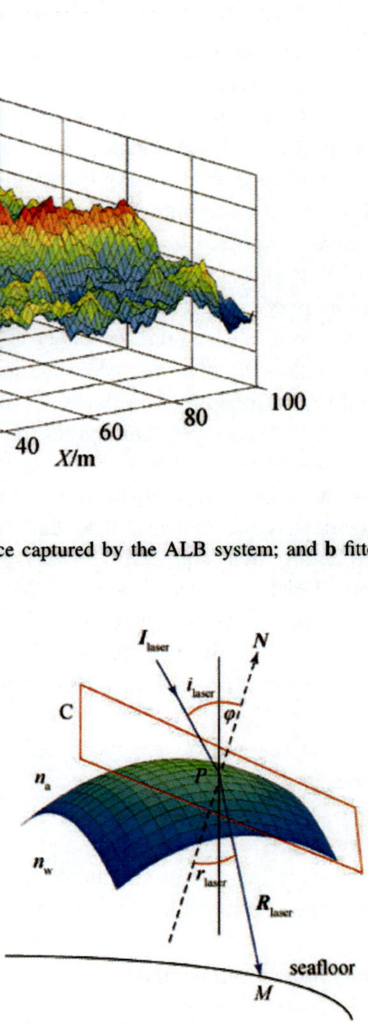

Fig. 3.16 Schematic illustration of ray tracing

By combining Eqs. (3.23) and (3.24), the acute laser incident angle i_{laser} can be obtained by Eq. (3.25) based on the law of cosines, as follows:

$$\cos i_{\text{laser}} = -\frac{\boldsymbol{I}_{\text{laser}} \cdot \boldsymbol{N}}{|\boldsymbol{I}_{\text{laser}}||\boldsymbol{N}|} = -(I_x N_x + I_y N_y + I_z N_z) \quad (3.25)$$

Following Snell's law, the laser refraction angle r_{laser} can be expressed as

$$\sin r_{\text{laser}} = \frac{n_{\text{w}} \sin i_{\text{laser}}}{n_{\text{a}}} \quad (3.26)$$

where n_{a} is the refractive index of air ($n_{\text{a}} = 1$); and n_{w} is the refractive index of water ($n_{\text{w}} = 1.33$). Considering that the laser refraction angle r_{laser} is the acute angle between the unit vector $\boldsymbol{R}_{laser}$ of the reflected ray and the unit normal vector $\boldsymbol{N}$, r_{laser} can also be expressed as

$$\cos r_{laser} = -\frac{\boldsymbol{R}_{laser} \cdot \boldsymbol{N}}{|\boldsymbol{R}_{laser}||\boldsymbol{N}|} = -(R_x N_x + R_y N_y + R_z N_z) \quad (3.27)$$

By substituting Eqs. (3.26) and (3.27) into $\sin^2 r_{laser} + \cos^2 r_{laser} = 1$ and taking Eq. (3.25) into consideration, the resulting equation is as follows:

$$\left(\frac{n_w}{n_a}\right)^2 [1 - (I_x N_x + I_y N_y + I_z N_z)^2] + (R_x N_x + R_y N_y + R_z N_z)^2 = 1 \quad (3.28)$$

Considering that $\boldsymbol{R}_{laser}$ is a unit vector, $\boldsymbol{R}_{laser}$ can be expressed as

$$|\boldsymbol{R}_{laser}|^2 = R_x^2 + R_y^2 + R_z^2 = 1 \quad (3.29)$$

Because the three vectors $\boldsymbol{N}$, $\boldsymbol{I}_{laser}$ and $\boldsymbol{R}_{laser}$ are coplanar, the three vectors should follow the coplanar equation:

$$\boldsymbol{I}_{laser} \times \boldsymbol{N} \times \boldsymbol{R}_{laser} = 0 \quad (3.30)$$

By constructing three simultaneous equations according to Eqs. (3.28), (3.29) and (3.30), (R_x, R_y, R_z) can be calculated. Finally, the geographic coordinate (x_M, y_M, z_M) of the laser point M on the seafloor in the local vertical reference frame can be calculated as

$$\begin{bmatrix} x_M \\ y_M \\ z_M \end{bmatrix} = \begin{bmatrix} x_P \\ y_P \\ z_P \end{bmatrix} + \frac{1}{2}\frac{n_a}{n_w} c\Delta t_2 \cdot \begin{bmatrix} R_x \\ R_y \\ R_z \end{bmatrix} \quad (3.31)$$

where Δt_2 is the travel time from the sea surface to the seafloor. The water depth D after the refraction correction can be presented as

$$D = \frac{1}{2} \cdot \frac{n_a}{n_w} \cdot c\Delta t_2 |R_z| \quad (3.32)$$

References

Airborne Hydrography AB (2010) Article on shallow water surveys LiDAR vs multibeam. Available via DIALOG. https://drive.google.com/file/d/1zXR-BZmxfbR7AinQCxgVqL2kiDM-4JtR/view?tdsourcetag=s_pctim_aiomsg. Accessed 1 July 2019

Boccotti P (2004) A method to obtain the directional wave spectrum. Ocean Eng 31(5–6):539–545. https://doi.org/10.1016/j.oceaneng.2003.10.002

Chiu MH, Lee JY, Su DC (1999) Complex refractive-index measurement based on Fresnel's equations and the uses of heterodyne interferometry. Appl Optics 38(19):4047–4052. https://doi.org/10.1364/AO.38.004047

Doneus M, Miholjek I, Mandlburger G et al (2015) Airborne laser bathymetry for documentation of submerged archaeological sites in shallow water. Paper presented at the ISPRS/CIPA Workshop "Underwater 3D Recording and Modeling", Piano di Sorrento (Napoli), 16–17 April 2015

Guenther GC (1985) Airborne laser hydrography system design and performance factors. Available via DIALOG. https://apps.dtic.mil/dtic/tr/fulltext/u2/a488936.pdf. Accessed 29 June 2019

Guenther GC (1989) Airborne laser hydrography to chart shallow coastal waters. Sea Technol 30(3):55–59

Guenther GC, Brooks MW, Larocque PE (2000) New capability of the "SHOALS" airborne LiDAR bathymeter. Remote Sens Environ 73 (2):247–255. https://doi.org/10.1016/S0034-4257(00)00099-7

Guenther GC, Thomas RWL, Larocque PE (1996) Design considerations for achieving high accuracy with the SHOALS bathymetric Lidar system. In: Proceedings of SPIE 2964, CIS selected papers: laser remote sensing of natural waters: from theory to practice, St. Petersburg. https://doi.org/10.1117/12.258353

Hamylton S (2011) The use of remote sensing and linear wave theory to model local wave energy around Alphonse Atoll Seychelles. Estuar Coast Shelf S 95(4):349–358. https://doi.org/10.1016/j.ecss.2011.08.035

Harsdorf S, Janssen M, Reuter R et al (1999) Submarine Lidar for seafloor inspection. Meas Sci Technol 10(12):1178–1184. https://doi.org/10.1088/0957-0233/10/12/309

IHO (2008) IHO standards for hydrographic surveys. International Hydrographic Bureau. Available via DIALOG. https://www.iho.int/iho_pubs/standard/S-44_5E.pdf. Accessed 29 June 2019

Komen GJ, Cavakeri L, Donelan M et al (1994) Dynamics and modelling of ocean waves. Cambridge University Press, Cambridge. https://doi.org/10.1017/CBO9780511628955

Kuus P (2008) Bottom tracking issues and recognition thereof using Shoals-3000 green laser beam in dense fields of Zostera Marina and Laminaria sp. Dissertation, University of New Brunswick

Kuus P, Clarke JA, Brucker S (2008) SHOALS 3000 surveying above dense fields of aquatic vegetation- quantifying and identifying bottom tracking issues. In: Proceedings of the Canadian hydrographic conference and national surveyors conference, Canada

Li T, Ao F, Ouyang C (2008) Study on the effects of blue-green laser propagation through sea interface (in Chinese). In: Proceedings of national network coding conference, Qingdao, July 2008. National Defense Industry Press, Beijing, pp 1243

Mandlburger G, Pfennigbauer M, Pfeifer N (2013) Analyzing near water surface penetration in laser bathymetry—A case study at the River Pielach. Paper presented at the ISPRS workshop laser scanning, Antalya, 11–13 November 2013

Martins K, Blenkinsopp CE, Zang J (2016) Monitoring individual wave characteristics in the inner surf with a 2-dimensional laser scanner (LiDAR). J Sensors 2016:1–11. https://doi.org/10.1155/2016/7965431

Méhauté B, Wang S (1996) Calibration of the linear wave theory. Water Waves Generated by Underwater Explosion 27:111–151. https://doi.org/10.1142/9789814350884_0003

Niemeyer J, Kogut T (2014) Airborne laser bathymetry for monitoring the German Baltic Sea coast. Paper presented at the annual conference of the German Society for Photogrammetry, HafenCity University Hamburg, Hamburg, 26–28 March 2014

Optech Incorporated (2015) Airborne laser terrain mapping (ALTM) intelligent waveform digitizer operation and processing manual. Available via DIALOG. https://www.environmental-expert.com/companies/teledyne-optech-incorporated-55983/. Accessed 29 June 2019

Pan ZG, Glennie C, Hartzell P et al (2015) Performance assessment of high resolution airborne full waveform Lidar for shallow river bathymetry. Remote Sens 7(5):5133–5159. https://doi.org/10.3390/rs70505133

Peachey D (1986) Modeling waves and surf. ACM Siggraph, Paper presented at the 13th conference on computer graphics and interactive techniques, New York, 18–22 August 1986

Piel S, Populus J, Chust G et al (2007) Recommended operating guidelines (ROG) for LiDAR surveys. Mapping European Seabed Habitats. Available via DIALOG. https://www.researchgate.net/publication/237740709. Accessed 29 June 2019

Steinvall OK, Koppari KR, Karlsson UC (1993) Experimental evaluation of an airborne depth-sounding LiDAR. Opt Eng 32(6):1307–1321. https://doi.org/10.1117/12.135859

Steinvall OK, Koppari KR, Karlsson UC (1994) Airborne laser depth sounding: system aspects and performance. In: Proceedings of SPIE ocean optics XII, Bergen, 13–15 June 1994

Tessendorf J (1999) Simulating ocean water. SIGGRAPH 99 Course Notes. Available via DIALOG. https://citeseerx.ist.psu.edu/viewdoc/download?doi=10.1.1.161.9102&rep=rep1&type=pdf. Accessed 29 June 2019

Tulldahl HM, Steinvall KO (2004) Simulation of sea surface wave influence on small target detection with airborne laser depth sounding. Appl Optics 43(12):2462–2483. https://doi.org/10.1364/AO.43.002462

Tulldahl M, Andersson M, Steinvall OK (2000) Airborne laser depth sounding: improvements in position and depth estimates by local correlations for sea surface slope. In: Proceedings of oceans 2000 MTS/IEEE conference and exhibition, Providence, 11–14 September 2000

Vimala J, Latha G, Venkatesan R (2014) Validation of *in-situ* wave spectrum with Jonswap in the Indian Ocean. Int J Remote Sens 35 (14):5384–5393. https://doi.org/10.1080/01431161.2014.926421

Wang CS, Li QQ, Liu YX et al (2015) A comparison of waveform processing algorithms for single-wavelength LiDAR bathymetry. ISPRS-J Photogramm Remote Sens 101:22–35. https://doi.org/10.1016/j.isprsjprs.2014.11.005

Westfeld P, Maas HG, Richter K et al (2017) Analysis and correction of ocean wave pattern induced systematic coordinate errors in airborne LiDAR bathymetry. ISPRS-J Photogramm Remote Sens 128:314–325. https://doi.org/10.1016/j.isprsjprs.2017.04.008

Wong H, Antoniou A (1991) Characterization and decomposition of waveforms for LARSEN 500 airborne system. IEEE Trans Geosci Remote Sens 29(6):912–921. https://doi.org/10.1109/36.101370

4 Side-scan Sonar and Sub-bottom Profiler Surveying

Side-scan sonar and sub-bottom profiler are two types of equipment commonly used in submarine geomorphological exploration. Side-scan sonar is applied to detect micro-topography and surface features of the seafloor, whereas a sub-bottom profiler is used to detect the stratigraphic structure beneath the seafloor as well as submarine features buried within the stratum. Some manufacturers have integrated these two devices to improve the efficiency of submarine exploration. This chapter introduces these two techniques of exploration of submarine geomorphology and shallow strata.

4.1 Side-scan Sonar Detection Technology

Side-scan sonar emits wide-angle conical or fan-shaped pulses downward to the seafloor perpendicular to the direction of travel of the sensor through the water. The backscattered signal of the incident sound waves from the submarine substrate is used to determine the morphology of the seafloor and to detect specific underwater targets. Thus, a continuous acoustic image that directly reflects the morphology and distribution of submarine micro-topography is generated. The technology of side-scan sonar was originally developed in the late 1950s. Since the first application of such technology to seafloor geological surveying by the British Institute of Marine Sciences in 1960, the resolution and quality of side-scan sonar imagery have been improved considerably. Sonar equipment suitable for different purposes was developed in the 1970s, and various types of side-scan sonar system have been developed since the 1990s (Wu et al. 2005; Liu et al. 2005).

Depending on the deployment of transducer probes, side-scan sonar equipment can be divided into two types: ship-borne and towed sonar systems (Wu et al. 2005). Ship-borne sonar system usually consists of acoustic transducers installed on both sides of the hull. The working frequency of such equipment is generally low (<10 kHz), the sweeping range of the seafloor is wide, and the working efficiency is high. Towed sonar system can be divided into two sub-types depending on the height of the tow body above the seafloor: high-towed sonar system deployed near the sea surface and deep-towed sonar system deployed near the seafloor. A high-towed sonar system can provide side-scan sonar imagery and bathymetric data, and can operate at high speed (up to 8 km/h) when towed approximately 100 m below the sea surface. A deep-towed sonar system is generally deployed only a few tens of meters above the seafloor; consequently, as the tow depth can be reasonably deep, the operational speed is relatively slow compared with a high-towed sonar system. However, the side-scan sonar imagery obtained by a deep-towed sonar system has high resolution, which can be used to distinguish pipelines and other objects with a diameter smaller than 10 cm. Currently, most towed sonars are deep-towed systems. Following technological development, some deep-towed side-scan sonar systems are now available that have high-speed operation capability, obtaining high-resolution side-scan sonar imagery of the seafloor at 10 km/h.

Side-scan sonar technology is widely applied because of the high resolution of the obtained imagery and the low cost of the equipment. The technology is used in surveys of submarine topography and geomorphology because of its ability to detect reefs, sunken ships, pipelines, cables, and various other submarine targets. It is also applied in the fields of marine geodesy and cartography, marine geological surveying, submarine engineering construction, detection of submarine obstacles and sediments, and submarine mineral exploration.

Z. Wu et al., *High-resolution Seafloor Survey and Applications*,
https://doi.org/10.1007/978-981-15-9750-3_4

4.1.1 Principle and Composition of Side-Scan Sonar

4.1.1.1 Basic Principle

The working principle of the side-scan sonar is shown in Fig. 4.1. The left and right linear transducer arrays of the sonar equipment have sector directivity, and the open angle of the sonar equipment is θ_V in the vertical plane and θ_H in the horizontal plane. When the acoustic transducers transmit an acoustic pulse, a narrow trapezoidal area of seafloor (such as the trapezoid *ABCD* on the left) can be illuminated to the left and right sides of the direction of travel, where the near transducer bottom edge (Line *AB*) of the illumination trapezoid is smaller than the far transducer bottom edge (Line *CD*). Acoustic waves traveling in the form of spherical waves scatter when they encounter the seafloor within the illumination trapezoid. Some scattered wave energy will return to the transducer along the original path and be received by the transducer and converted into electrical pulses. Generally, the echoes from the near distance arrive back at the transducer before the echoes from the far distance. Therefore, the echoes from the seafloor immediately below the transducer return first, while the echoes from the seafloor at longer distances that travel along an inclined path arrive later, generating the left and right trapezoidal submarine topography maps illuminated continuously along the heading direction.

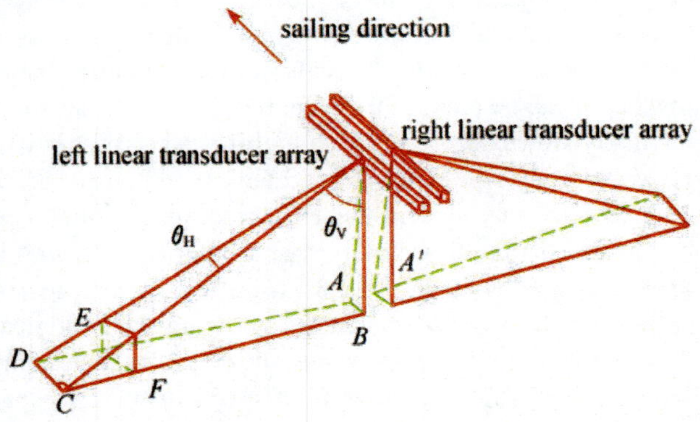

Fig. 4.1 Side-scan sonar working principle

Following the transmission of a narrow pulse signal by the side-scan sonar equipment, a long sequence of pulse echoes is received over a certain period. In general, a hard, rough and prominent seafloor produces strong echoes, whereas a soft, flat, and concave seafloor generates weak echoes. Echoes are not received from areas of the seafloor blocked by protruding targets, i.e., acoustic shadow areas, as shown in Fig. 4.2 (Xu and Wei 2006). Thus, the variation of the amplitude of the received echo pulse sequence contains information on the fluctuation of the seafloor condition. A single acoustic emission can derive information in relation to the seafloor in a narrow band on both sides of the transducer, and the received information is displayed as a line. Side-scan sonar transmits and receives acoustic waves at certain time intervals as the survey ship sails forward. The sonar system displays the line data received after each pulse, producing a two-dimensional image of submarine topography and geomorphology.

4.1.1.2 Basic Composition

Side-scan sonar systems generally include workstations (sonar data acquisition software and transceiver processing units), sonar towfish, tow cables, GNSS receivers and other external equipments.

The workstation forms the core of any side-scan sonar system. It controls the entire system and it controls functions such as data collection, processing, display, storage, image mosaicking, and post-processing. A workstation consists of two elements: hardware and software. The hardware generally comprises a high-performance computer host and a sonar transceiver processing unit. The software includes both system software and application software (i.e., sonar acquisition, display and post-processing software), as shown in Fig. 4.3.

The towfish of a towed sonar system is a streamlined stable tow body that comprises a head and a tail. The head of a towfish comprises the fish head, transducer, and other tow

Fig. 4.2 Principle of side-scan sonar image generation

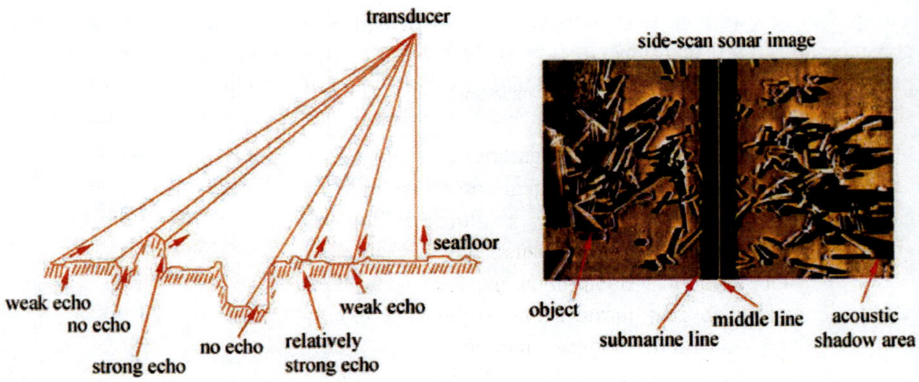

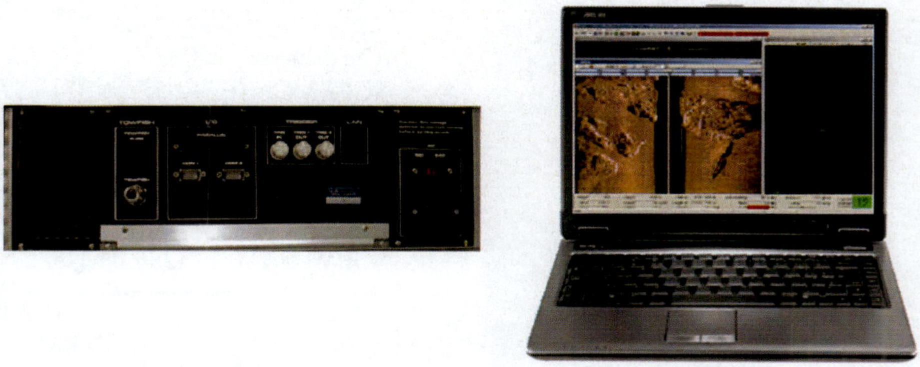

Fig. 4.3 Klein 5000 side-scan sonar transceiver processing unit and workstation computer host (https://kleinmarinesystems.com/)

parts (a tow rod or tow hook). Usually, there is an elongated transducer array on either side of the towfish. The radiation surface of the array is sealed with polyurethane vulcanized rubber to ensure water-tightness and sound transmission performance. The tail of the towfish consists of an electronic warehouse and a tail fin unit. The tail fin is present to keep the towfish balanced when being towed through the water. The tow cable establishes the mechanical and electrical connections between the transceiver processing unit and the towfish, and the towfish is placed at the optimal operating depth depending on the speed and length of the towfish, as shown in Fig. 4.4.

A Global Navigation Satellite System (GNSS) receiver, which is an external component of a side-scan sonar system, provides real-time positioning for side-scan sonar data. Users can configure different types and functions of GNSS according to their needs. Side-scan sonar systems have a standard interface to realize a real-time connection with the NMEA-0183 standard positioning equipment.

4.1.2 Basic Workflow and Methods

4.1.2.1 Installation of a Side-Scan Sonar System

Figure 4.5 shows the connections between the various components of a side-scan sonar system. The "slip ring" refers to the slip ring on a winch to which a tow cable is connected. The other end of the slip ring is connected to a deck cable, which is then connected to the transceiver processing unit. The GNSS navigation unit is connected to the transceiver processing unit via a serial cable. The transceiver processing unit is connected to the workstation host via a network cable. The workstation host is equipped with sonar data acquisition software. By

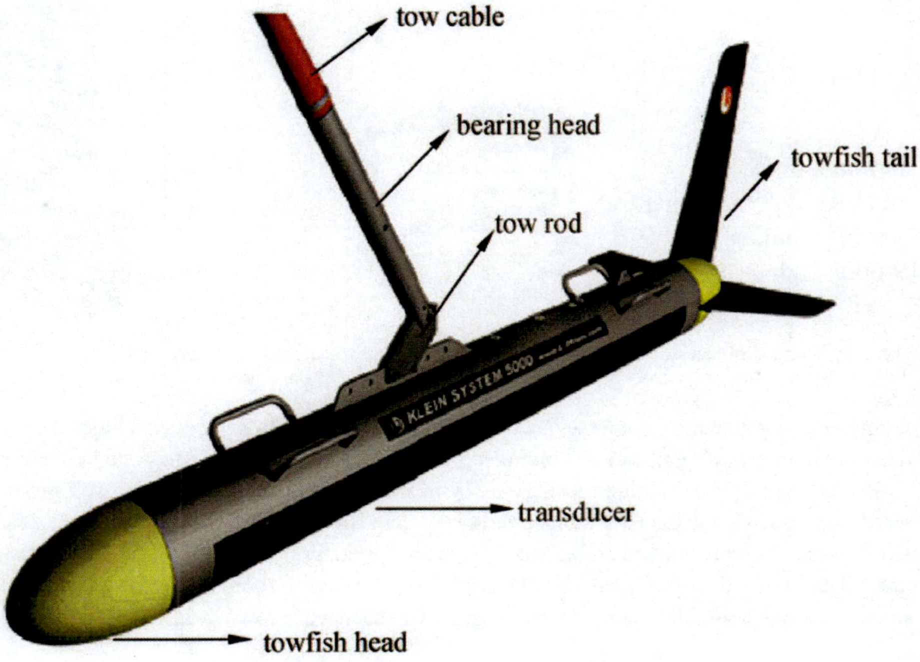

Fig. 4.4 Klein 5000 side-scan sonar towfish (https://kleinmarinesystems.com/)

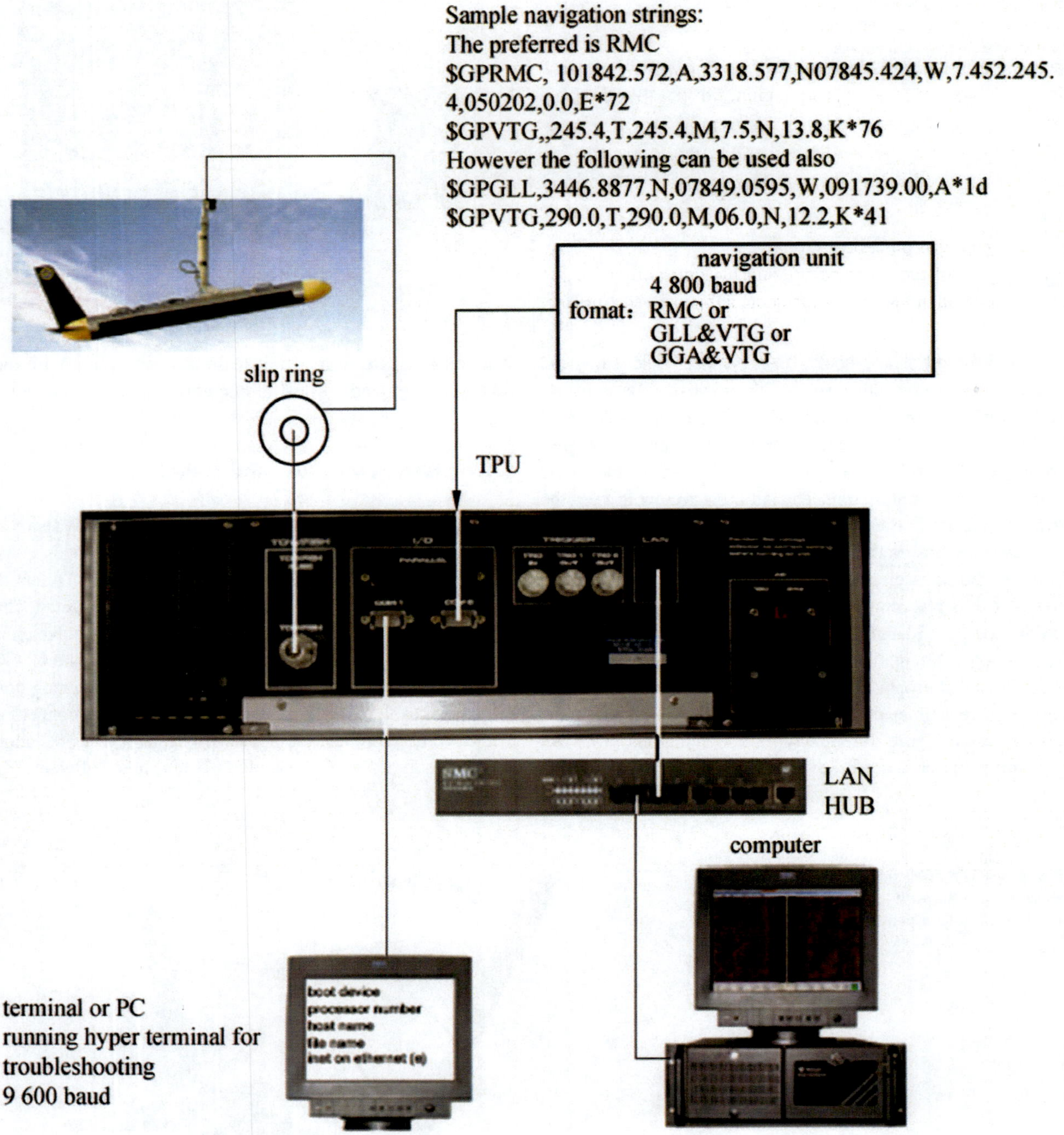

Fig. 4.5 Diagram of side-scan sonar system connections (https://kleinmarinesystems.com/)

setting relevant parameters in the acquisition software, the transceiver processing unit can be controlled.

Because attitude and altitude sensors are installed directly within the towfish, the data are transmitted to the acquisition workstation through the tow cable together with the sonar data. Therefore, the hardware installation of a side-scan sonar system is reasonably simple, and no other signals need to be accessed other than the GNSS signals. When installing instruments onboard, it is necessary to ensure that the workstation is well-grounded to avoid interference from electrical signals such as alternating current. In the process of shallow sea area investigation, the instrument operation room should have good visibility to ensure the safety of the towfish in the water during fieldwork.

4.1.2.2 Layout and Parameter Setting of Side-Scan Sonar

In a field survey using side-scan sonar, the sweeping method, overlap swath width, resolution, speed, towfish height, and streamer length should all be determined according to the survey area environment and sweeping requirements. Based on this, the direction and spacing of the line layout should be designed. An effective tow velocity is a key parameter in side-scan sonar fieldwork because it is related to the detection range and the image resolution. In theory, the tow velocity could be designed according to Eq. (4.1) (Xu and Wei 2006):

$$V = L \times \frac{C}{2} \times \frac{1.94}{R \times H} \tag{4.1}$$

where V is the maximum tow velocity in knots (1 knot $\approx$ 0.5 m/s), L is the target scale, C is the sound velocity (generally, 1 500 m/s), R is the selected port and starboard sweep range (m), and H is the number of measurement points expected on the target. It can be seen from Eq. (4.1) that the maximum tow velocity is inversely proportional to the selected sweep range when the target dimension is fixed; the larger the selected sweep range, the lower the required tow velocity. In addition, when high-resolution images are needed, the target acoustic signal coverage must be dense and thus a lower tow velocity is required. Therefore, it is necessary to consider the relationship between the detection effect and the work efficiency in field detection. Generally, when a target is detected, the ship speed should slow and, whilst ensuring the safety of the towfish, the tow body should be positioned as close to the seafloor as possible.

The entire side-scan sonar system should be tested on land or on the ship deck before launching a survey. It is important to determine that all components can operate normally and that the water inlet part is checked to ensure no water leakage. During the measurement process, each parameter should be debugged according to the actual situation to ensure the sonar image brightness remains moderate, the energy of the far and near profiles remains balanced, and the sonar images of the seafloor are clear.

Usually, there are two methods of side-scan sonar submarine sweeping: coarse sweeping and fine sweeping. For sweeping of a large sea area, coarse sweeping should be undertaken first, and then fine sweeping should be performed once any prospective targets are found. Fine sweeping confirms the existence of a suspected target and it can distinguish the type, position, and height of the target with the sonar image.

Seafloor sweeping must perform full coverage of the survey area. The survey design should produce parallel lines and the effective working distance of adjacent survey lines should have overlapping strips; accordingly, missing areas between adjacent lines can be avoided. When submarine micro-topography is detected, adjacent line spacing of double the effective distance removes the need to design overlapping strips.

Fine sweeping of the seafloor determines the position and height of a target in a sonar image corresponding to the same target in a preliminary sweeping image. It determines the sweep frequency, transmission pulse width, sweeping range, tow velocity, and depth of the towfish before performing a specific sweep. For fine sweeping, the direction of the fine sweep should be aligned parallel to the direction of the target as much as possible, and the effective range should be reflected in the single-sided intermediate position of the sonar image according to the target image.

4.2 Sub-bottom Profiler Detection Technology

4.2.1 Principles of Sub-bottom Profiler Technology

Sub-bottom profiler detection, is a method for fine-scale detection of shallow subsurface formations in a continuous navigation mode based on hydro-acoustic principles. By transmitting high-frequency acoustic signals from active sources and using the acoustic impedance difference between different submarine lithological strata to generate reflected echoes, the method can generate continuous recording profiles that directly reflect the structure and spatial distribution of different lithological strata beneath the seafloor after being collected by professional software. Compared with offshore single-channel and multi-channel seismic detection, sub-bottom profiler detection has the characteristics of high vertical resolution, high signal repetition, and high excitation velocity and it represents an indispensable technical approach for fine detection of submarine shallow strata.

The principle of sub-bottom profiler detection is similar to that of multi-channel reflection seismic exploration. First, a longitudinal acoustic wave is excited artificially, which during propagation can generate a reflection echo at the lithological interface of a formation. The reflected signal is received and stored by an acoustic transducer or a single receiving cable, and then the structure and shape of the submarine formation can be displayed in real time. As shown in Fig. 4.6, acoustic signals excited from transmitter O propagate in seawater and submarine sedimentary strata. As different lithological strata are encountered, in accordance with the principles of the geometry of seismic reflection, the incident wave generates a reflection wave at angle β that is equal to incident angle α at the interface. The transmission angle γ is determined by the sound wave

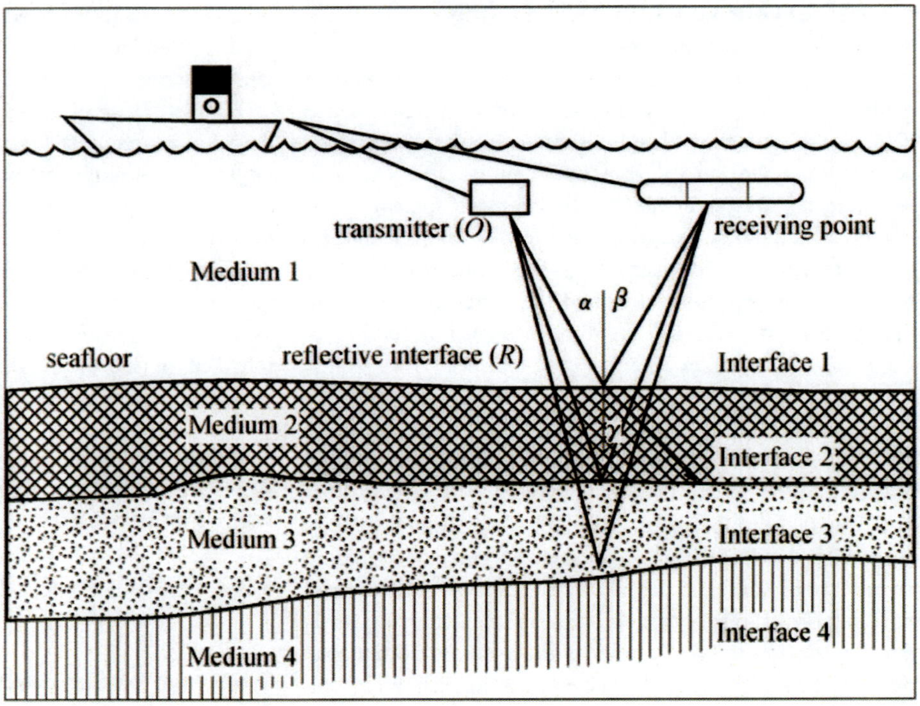

Fig. 4.6 Geometric kinematic model of acoustic signal for sub-bottom profiler detection

propagation velocity of the upper and lower layers at the interface. If the sound wave propagation velocity of Medium 2 below the seafloor interface is greater than that of Medium 1 above the interface, transmission angle γ is greater than incident angle α, and vice versa. As the transmitted wave continues to propagate downward, new reflected and transmitted waves continue to be generated at subsequent acoustic impedance interfaces until the energy of the transmitted wave is too small to generate a reflected echo of effective energy. To form a strong reflective echo at an interface, there must be an obvious difference in the acoustic impedance condition at that interface. Thus, reflection coefficient R at an interface is not equal to zero when the sound wave excited by shallow strata detection is incident at an angle that is nearly perpendicular to the reflective interface. The equation can be expressed as follows (Liu 1978):

$$R = \frac{Z_2 - Z_1}{Z_2 + Z_1} = \frac{\rho_2 v_2 - \rho_1 v_1}{\rho_2 v_2 + \rho_1 v_1} \neq 0 \quad (4.2)$$

where ρ and v are the density of the sedimentary strata and the propagation velocity of sound waves within the strata, respectively, and their product is called wave impedance. Subscripts 1 and 2 denote the strata above and below the interface, respectively. According to the above equation, the condition that means the reflection interface can form a reflection echo is that $\rho_2 v_2$ is not equal to $\rho_1 v_1$; the greater the difference between $\rho_2 v_2$ and $\rho_1 v_1$, the stronger the reflection echo energy. The greater the amplitude of the reflection signal received by the transducer, the greater the gray value of the acoustic reflection in-phase axis displayed by the recording section and the clearer the reflection strata interface. Therefore, a reflection interface is also called a wave impedance interface, which is consistent with the lithological interface of the actual strata and it represents different lithological horizons.

Figure 4.7 shows abundant sedimentary strata revealed by sub-bottom profiler detection in Zhoushan (East China Sea). In this figure, the concave interface reflects a paleo-river valley on which are deposited plentiful oblique bedding strata that are covered by horizontal sedimentary strata. It indicates the existence of a wide range of channels in the sea area during the geological glacial period, following which abundant terrestrial material was deposited in different periods as the sea level rose. According to the dynamic characteristics of acoustic reflection, the sub-bottom profiler data can also be applied to inverse the elastic parameters of submarine substrates and sediments. In marine geological research, based on the amplitude-phase information of acoustic reflection blanks, enhanced reflections, and turbid reflections in shallow stratigraphic sections, it is possible to study the collection of shallow seafloor gas, migration channels of submarine fluids, fluids of seafloor mud volcanoes, migration of cold spring systems, and changes of the physical properties of sediments.

Figure 4.8 shows the typical reflection characteristics of a cold spring system in the northeastern continental slope

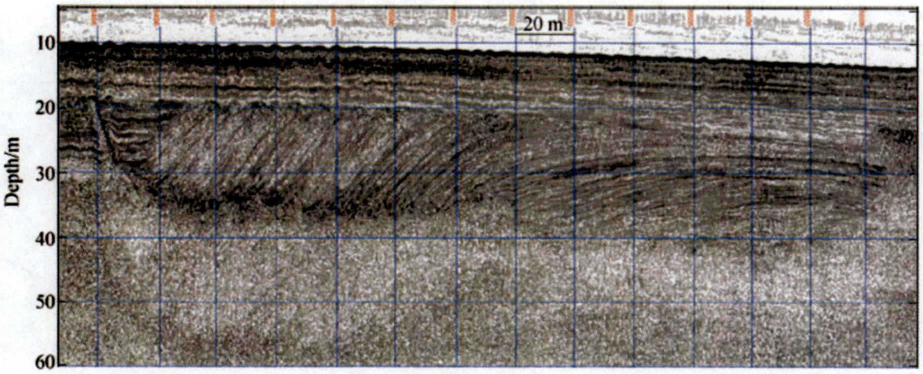

Fig. 4.7 Rich sedimentary strata revealed by sub-bottom profiles in the Zhoushan sea area (East China Sea)

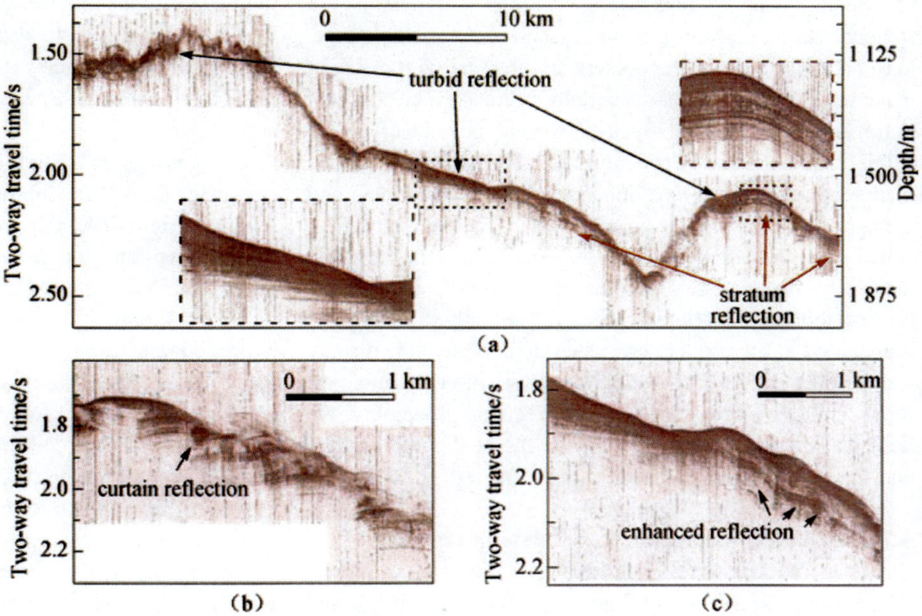

Fig. 4.8 Typical reflection feature related to shallow gas in sub-bottom profile in the northeastern continental slope of the South China Sea. **a** Acoustic turbidity, **b** acoustic curtain, and **c** enhanced reflection (Liu et al. 2015)

of the South China Sea, which reveals the reflection characteristics of submarine shallow gas with different amplitudes in the sub-bottom profile data. Sub-bottom profiling is also very effective for revealing targets within water bodies, e.g., high-pressure shallow gas spillage. As shown in Fig. 4.9, a high-pressure shallow gas puncture spillage within a water body is revealed by high-resolution sub-bottom profiler surveying. In the area of the high-pressure shallow gas spillage in the middle of the section, the interface of the horizontal sedimentary bedding on both sides is shielded by the high-pressure gas layer. With the overflow of the high-pressure shallow gas, the horizontal seafloor undergoes erosion and collapse, and the surface reflection of the overflowing gas appears in the water above the seafloor.

4.2.2 Basic Workflow and Methods

Sub-bottom profiler detection is an efficient and high-resolution reflected acoustic detection method that is applied mainly to detect submarine topography and shallow sedimentary strata as well as bottom conditions. The source device or the transducer array transmits sound pulses vertically downward at a certain time interval. After the sound pulse passes through the seawater and touches the seafloor, part of the sound energy is reflected back to the transducer, while the remainder of the sound energy continues to propagate deeper into the strata. Echoes are returned from successive interfaces until the energy of the sound wave is gradually lost. At this time, the effective echo received is the maximum depth detection range. At present, all-digital sub-bottom profiler detection devices adopt the sonar

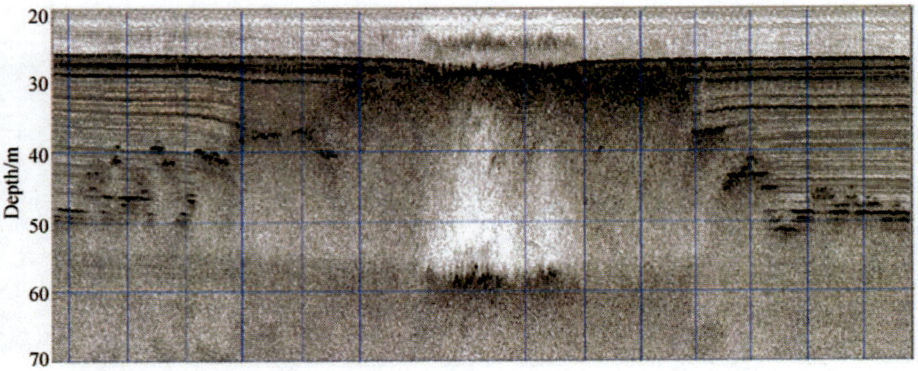

Fig. 4.9 Reflection characteristics of a sub-bottom profile when shallow gas spills from the sea surface (Dongji Island)

principle. The transmitting array transmits a frequency-modulated pulse within a certain frequency range. When the pulse signal encounters an interface with different wave impedance, it generates a reflected pulse. The reflected pulse signal is received by the acoustic transducer or the single-channel streamer and amplified. The received signal is then sent to a control unit onboard where it is sampled and converted into a digital signal by an A/D converter. The signal is amplified by a digitally controlled amplifier and then sent to a digital signal processor (DSP) digital processing module for related processing. Finally, the signal is transmitted to the acquisition system to complete the display and storage processing. After the acquisition system performs the time-depth conversion and related processing, the field can display the distribution of the sedimentary strata and the internal fine structure in real-time.

4.2.2.1 Installation and Commissioning of Offshore Equipment

GNSS is necessary for sub-bottom profiler detection in the field, it provides real-time positioning information for sub-bottom profile data that is stored within the data record. The bathymetric equipment provides a reference for real-time seafloor tracing of sub-bottom profile data. Therefore, sub-bottom profile needs to be connected with navigation, positioning, and bathymetric equipment and to have real-time communication. Figure 4.10 illustrates the field connections of a sub-bottom profile detection system in which the content of the dashed boxes and the dashed arrows indicate that no transceiver-integrated acoustic transducer is required, whereas a single streamer is required for connection.

The connection system of sub-bottom profiler comprises two main parts: the ship-borne system and the underwater towing system. The ship-borne system includes a GNSS, a data acquisition station and an electric energy box. The underwater towing system includes a seismic sound source (such as a boomer plate or an electric spark) and a combined hydrophone floating cable receiving system. In the field, the survey ship sails along survey lines preset by the navigation software. The GNSS provides the position of the ship in real time, which is recorded and stored in the navigation software. The navigation software synchronously outputs the positional information to the acquisition system in real time. The acquisition system automatically collects reflected echoes according to the recording parameters (e.g., source excitation interval, recording length and sampling rate) set by the users and it stores the recorded echoes together with the concurrent positional information on a hard disk. At the same time, the acquired continuous sections are displayed dynamically on the visualization screen. By interpreting the dynamic continuous sections, the operating staff can assess the quality of the collected data in real time, modify the filter parameters as necessary, and understand the structure and

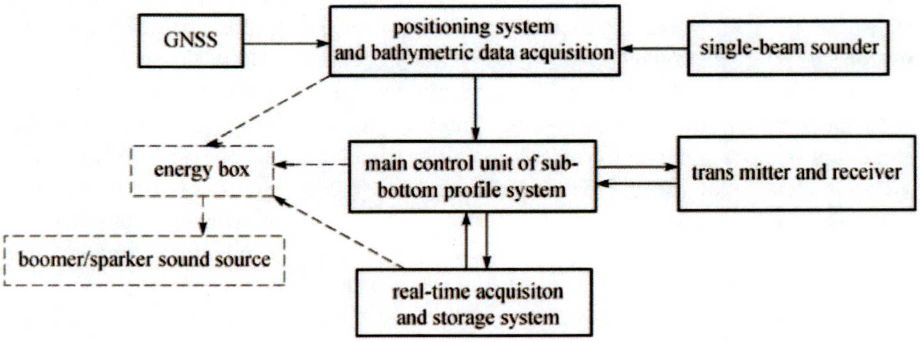

Fig. 4.10 Schematic of working system for in-field sub-bottom profiler detection

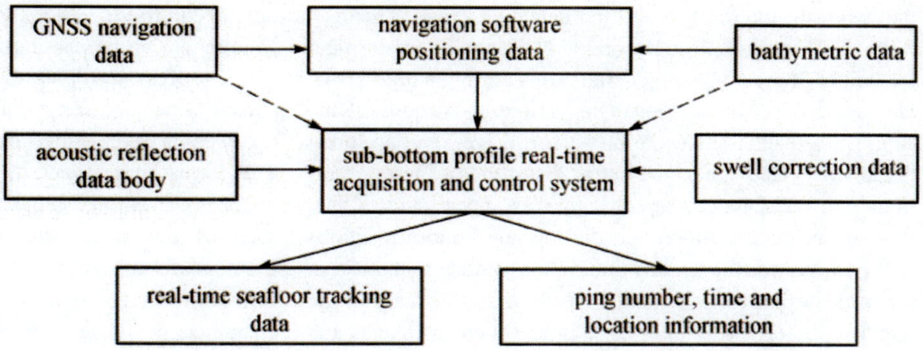

Fig. 4.11 Schematic of input and output data during sub-bottom profiler detection

morphology of the detected strata and other information. There are two modes of sub-bottom profiler detection: active and passive working modes. In the active working mode, the functions of excitation, acquisition, recording, and playback are controlled uniformly by the acquisition system; whilst in the passive working mode, the functions of the energy box and the acquisition system are controlled by the navigation software, and the acquisition, recording, and playback of the echo signals are controlled by the acquisition system. The active mode is usually applied to the equal time interval triggering mode, while the passive mode is usually applied to the equal distance acquisition mode. In field operation, reasonable selection of the appropriate mode of operation should be made according to work requirements.

4.2.2.2 Sound Wave Transmission and Acquisition

Compared with multi-beam data acquisition, the sub-bottom profiler data acquisition is simpler because it does not need access to heading, three-dimensional attitude, sound velocity profile, clock synchronization, and other information; it needs access only to positioning information and one-dimensional vertical attitude data (i.e., swell correction data). The control parameters of field acquisition software are mainly set up for sub-bottom profiler detection, including the communication of the system host computer, working channel, transmitting waveform, transmitting frequency, transmitting interval, transmitting energy, gain and serial port parameter setting for GNSS data access.

In deep-water sub-bottom profiler detection, it is necessary to set seafloor tracing parameters because deep-water detection needs to avoid recording the propagation signals from above the seafloor; otherwise, it will be a time-consuming process and it will require considerable hardware resources to process the reception and storage of large amounts of data. Thus, this affects the source transmission interval, the continuity of horizontal sound wave coverage, and reflection in-phase axis. Deep-water sub-bottom profiler detection generally adopts Multi-ping technology, i.e., under the premise of avoiding recording water body propagation signals, the acoustic transducer continuously transmits multiple acoustic signals to the water body in a short time. At the same time, the transducer continuously receives and monitors the reflected echoes, which are spliced according to the receiving time to generate a continuous reflection profile. The reflection profile also needs to display the actual seafloor that is traced automatically; otherwise, the depth displayed by the profile will not match the actual depth. Therefore, in field deep-water sub-bottom profiler detection, it is very important to accurately trace the reflected signal from the seafloor to avoid incorrect cross-section signals and depth values.

4.2.2.3 Data Transmission and Recording

The raw sub-bottom profiler data includes acoustic datagram, GNSS navigation datagram, GNSS time datagram, swell correction data, and related data output during acquisition (e.g., seafloor tracing data). Figure 4.11 lists the relevant data information that could be accessed and output during sub-bottom profiler detection. All data are transmitted in real time to the designated recording location and recorded on a hard disk in a common format. At present, the usual recording format is the SEG-Y format defined by the American Geophysical Society. This format is used to store the sub-bottom profiler data, although some key parameters are not fixed. In practice, some special modifications and definitions, such as channel number definition and positioning label definition need to be implemented.

4.3 Processing Technology and Method of Side-Scan and Sub-bottom Profile Data

4.3.1 Side-Scan Sonar Data Processing Technology and Method

Side-scan sonar detection is imaged based on range. In underwater detection, there is a certain relationship between the height of the towfish and the detection range. Theoretically, the imaging effect is good if the range is greater than the height of the towfish (generally, it is set in the ratio 10:1). However, this will still result in various distortions of the sonar image such as severe target shrinkage, overlap, positional offset and other geometric distortions. These distortions will make targets in sonar images appear shorter than they are in reality. Moreover, the top and bottom positions will be inverted, the sonar illumination surface compressed, the back surface elongated, and the slope and distance of the target will increase. To correctly interpret sonar images and to eliminate the influence of various factors such as effects linked to instruments, bathymetry and towfish attitude, it is necessary to process the raw sonar data. This involves correcting some geometric distortions and suppressing irrelevant noise interference such that the sonar image can truly reflect the relationship between the actual seafloor geomorphology and its relative position, restore the intensity of acoustic backscatter data, and provide reliable data for correct interpretation and classification of the sonar image.

A sonar image is also a type of digital graphics image. Some conventional image processing techniques are also applied in sonar image processing, e.g., image denoising, image enhancement, image detection and image mosaic processing. In addition, sonar backscatter data represent a digital signal, which can be manipulated using special signal processing methods such as gain processing, digital filter processing and convolution processing that improve the resolution of the signal-to-noise ratio. Depending on the intended purpose, various methods can be applied to achieve comprehensive processing of side-scan sonar data.

To process sweeping data, the data format must first be parsed. Side-scan sonar data are generally recorded in XTF (extended Triton format) format (Triton Imaging Inc. 2014). This format was created by the Triton Company of the USA to meet the storage requirements of different data sources such as raw sonar datagram, navigation datagram, and bathymetric datagram. The advantages of the XTF data format are that it is easy to expand and that the user can customize new packets according to the data characteristics. Since its creation, the format has undergone continuous updating and it was developed into the X37 version in 2014.

Data stored in the XTF format can be understood as a data pool that stores different data sources received synchronously in the field. For side-scan sonar data, the commonly applied data sources include sonar reflection data, GNSS data, and three-dimensional attitude data. These data sources, which are mutually independent in the data pool, can be freely written and accessed, which improves the data writing and access rate.

The XTF record format consists of a record file header and various data packets. These data packets correspond to different data sources and they have corresponding packet definition headers, e.g., a side-scan sonar data definition header XTFPINGHEADER and bathymetric data definition header XTFBATHYHEADER. These data file headers are stored in binary format, which makes it convenient for users to access and extract parameters quickly. The packet header is followed immediately by the sonar datagram that is also stored in binary format. The specific XTF format structure is

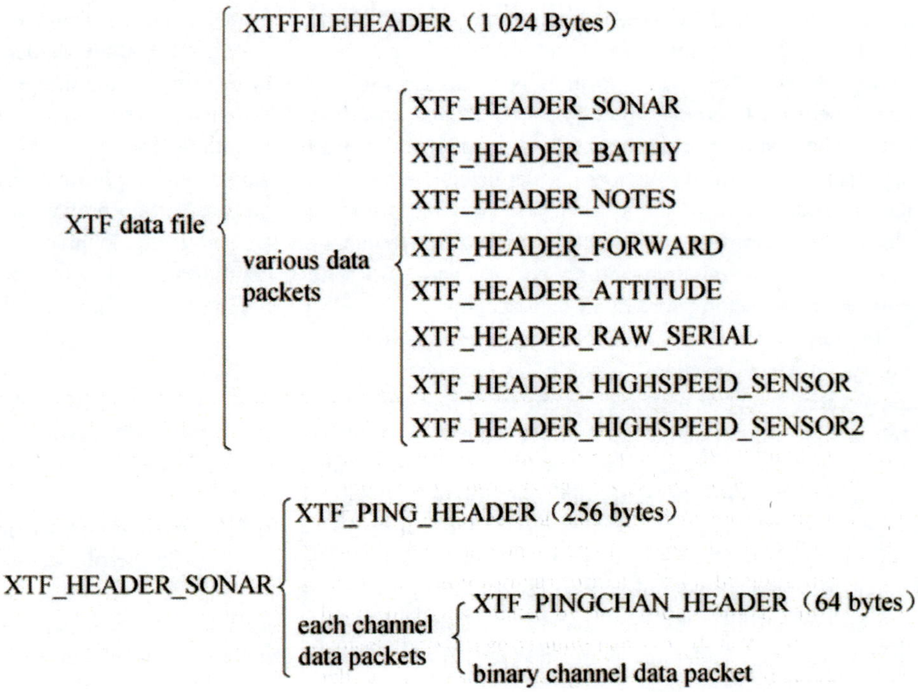

Fig. 4.12 Data structure diagram of XTF format file and side-scan sonar (modified from Ma and Xiong 2011)

shown in Fig. 4.12. For a side-scan sonar data, commonly available commercial software can effectively identify the sonar datagram and it can call the side-scan data and auxiliary information of multiple channels. Moreover, the user can select the valid channel data according to the requirements and perform subsequent related processing.

Some sonar acquisition systems also record raw data in Q-MIPS format. This format was created by the Isis Company of the USA and it also stores different data sources in binary data stream format. Compared with the XTF format, this format is relatively simple. It consists of a file header followed immediately by a Ping record. The file header is also defined as 1 024 bytes and it includes various field setting parameters and navigation parameters. The Ping record includes echo signals and footnotes. The Q-MIPS format is reasonably old and it is unsuitable for multi-sensor, large-capacity, and multi-source data storage; therefore, XTF is the format preferred for recording data in most system software.

Figure 4.13 shows the conventional processing workflow of side-scan sonar data and the corresponding resulting data, among which the conventional processing of sonar images includes a data energy balance, bottom tracing, water body correction, slant range correction, speed correction, trace smoothing, mosaic mapping, and result output. At present, the processing results of side-scan sonar data are usually output in GeoTiff image format with plane coordinates, which is convenient for subsequent interpretation of the results.

4.3.2 Side-Scan Sonar Image Denoising

Noise generally appears as a high-frequency random characteristic in an acoustic image. Its energy often has substantial deviation from the gray level of the surrounding pixels and it is often distributed as small mottling. According to this feature, suppression of sonar data noise is generally filtered using a certain data model, which can be implemented successively in the neighborhood, space and frequency domains. The main methods include neighborhood averaging, spatial domain smoothing, frequency domain smoothing, medium value filtering and wavelet domain denoising.

The main characteristic of side-scan sonar imaging is to reflect the shape and property of seafloor target objects through variation of the gray level in a sonar image. Therefore, the emphasis of sonar image enhancement processing is to adjust the gray level of the scattered image. The main factor contributing to an uneven gray level distribution in a sonar image is the intensity of the echo signals received at the proximal and distal ends of the sonar image. A sonar image often shows strong gray levels at the proximal end and weak gray levels at the distal end, which reduces the resolution of the image to a certain extent. Therefore, before mosaic processing, the gray level of a sonar image in a full-coverage test area must be adjusted uniformly. Time-varying gain (TVG) can be applied to adjust the gray

Fig. 4.13 Processing workflow of side-scan sonar data (Feng et al. 2014)

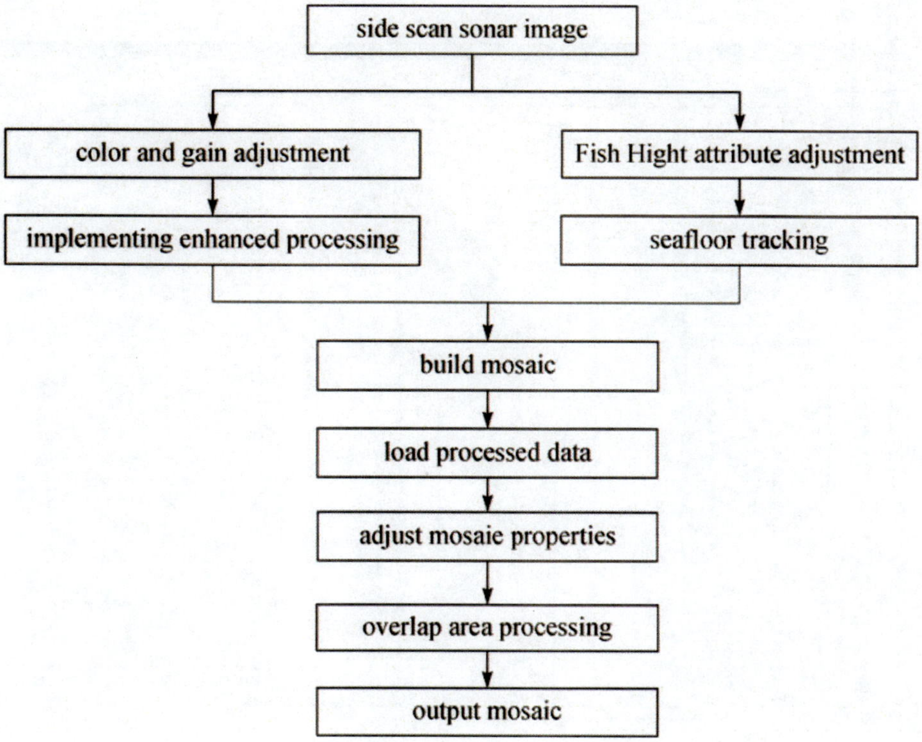

level difference and to eliminate differences in echo signal strength caused by different receiving times, which can eliminate the problem of gray level imbalance in a sonar image. By adjusting the TVG curve manually, sonar gray-scale image enhancement processing can be realized. Meanwhile, by adjusting the color distribution and the contrast of an image, the visual effect of the sonar image can be improved and effective display of sonar image information can be enhanced. In addition, automatic gain control (AGC) can equalize the difference between adjacent signals, which makes the energy distribution between signals more uniform, the image appear smoother, and highlights the weak signal. Figure 4.14 shows side-scan sonar images before and after TVG and pseudo color adjustment processing. It can be seen that after TVG processing, the gray level of the proximal end of the side-scan acoustic profile is reduced and that of the distal end is obviously enhanced. Clearly, the gray level of the entire profile is better balanced, and the continuity of the sand waves and sand ridges in the profile is enhanced. At the same time, the stereoscopic display effect of the image is enhanced and the resolution of the sonar signal is improved by adjustment of the color distribution. This enhances the overall visual effect of the sonar profile and it provides favorable conditions for tracing the spatial distribution characteristics of sand waves and sand ridges. Figure 4.15 shows a comparison of the effects of TVG and AGC processing. It is evident that TVG processing can lift the edge signal at the distal end of the profile, and equalize the distal end and proximal end energy levels. The gray level of the profile remains similar, but the gray level difference between adjacent signals remains very large at points of abrupt change of geomorphology and sediment. Image shadows are obvious and part of the characteristic geomorphology is concealed.

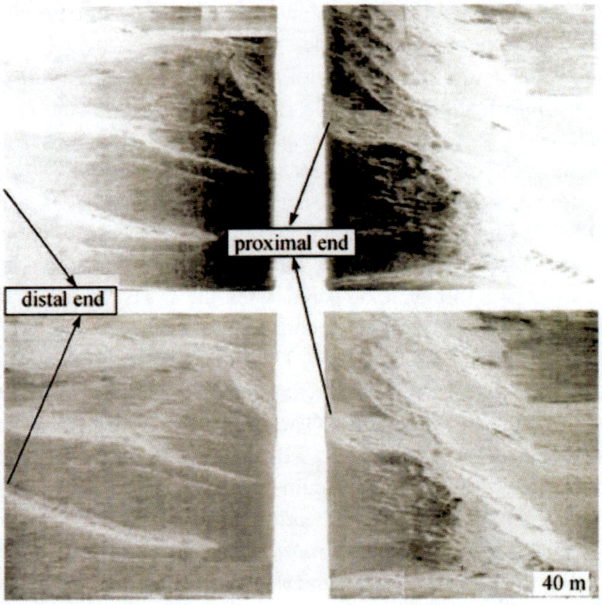

Fig. 4.14 Improvement of side-scan image display after TVG processing

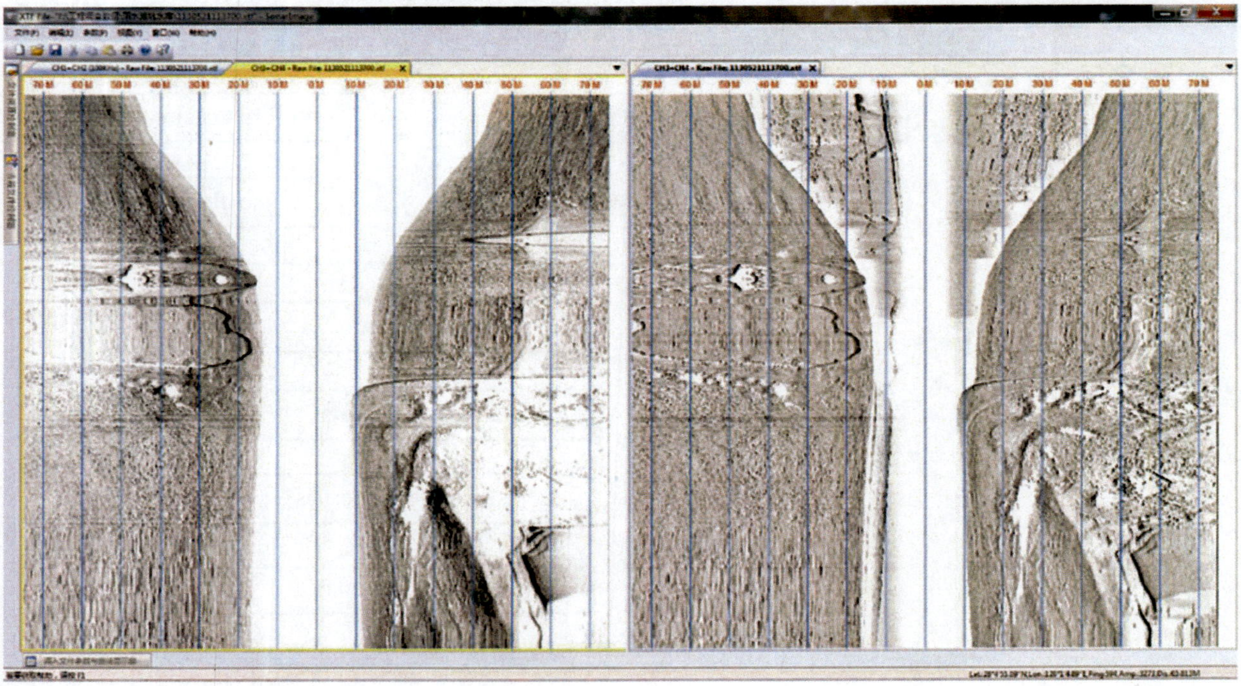

Fig. 4.15 Comparison of results of TVG and AGC processing

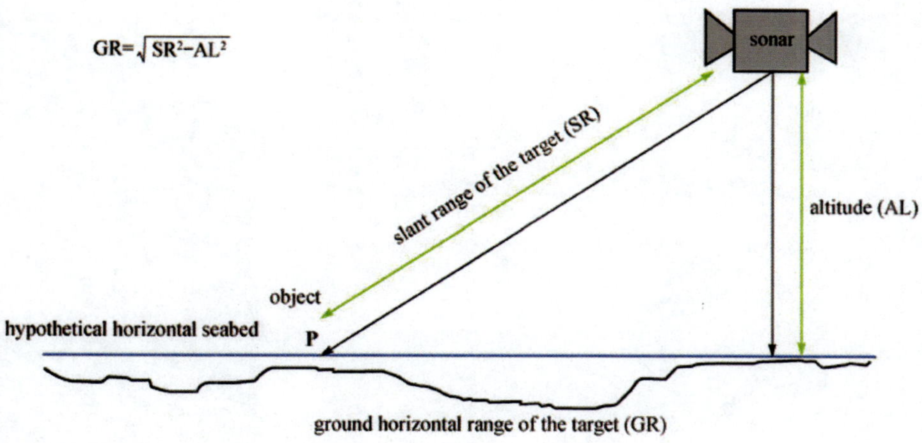

Fig. 4.16 Front and rear of side-scan sonar seafloor line tracing

After AGC processing, such differences no longer exist and shadowing on the profile is not obvious. The overall energy characteristics of the profile are the same and the geomorphological information is displayed completely.

Before mosaic processing of a side-scan sonar image, effective tracing of the seafloor line must be completed. Figure 4.16 shows the spatial positional relationship of a side-scan sonar probe in underwater detection. It can be seen that the sonar probe is positioned at a certain height from the seafloor, which is called its altitude. The effect of altitude extends the propagation time of the signal transmitted and received via sound waves, which distorts the echo signal image and must be eliminated during processing. Tracing the altitude from the seafloor involves tracing the height of the towfish above the seafloor in real time, removing the display of the water body in the sonar image, and then splicing the left and right sonograms of the side-scan sonar image. Figure 4.17 shows effective tracing of the signal scattered by the seafloor in an original echo waveform. There is a certain distance between the effectively traced seafloor signal and the initial arrival signal, which is the actual altitude. In the process of actual surveying, the altitude from the seafloor is affected by the towing mode, changes of water depth, sea conditions, and other factors. Therefore, it is necessary to use software to trace the seafloor automatically. If such tracing is inaccurate, a large gap will remain when the left and right channels of sonar images are assembled, which leads to low accuracy and affects the quality of the mosaic acoustic images. The gap between the left and right channels of sonar images can be eliminated by picking via human–computer interaction. As shown in Fig. 4.18, the combination of automatic tracing and manual interactive picking not only traces the seafloor line of the sonar profile with high accuracy, but it also improves the connection accuracy of the

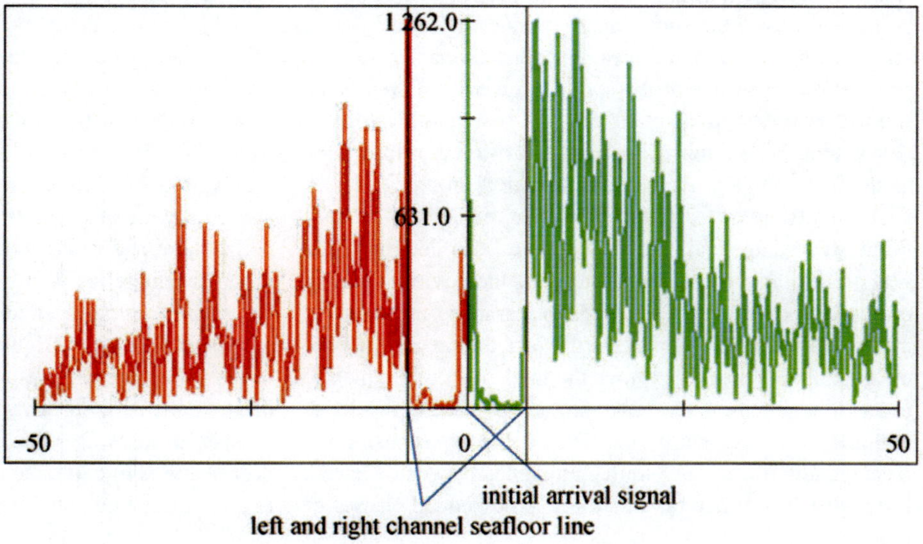

Fig. 4.17 Submarine line tracing position of received echo waveform

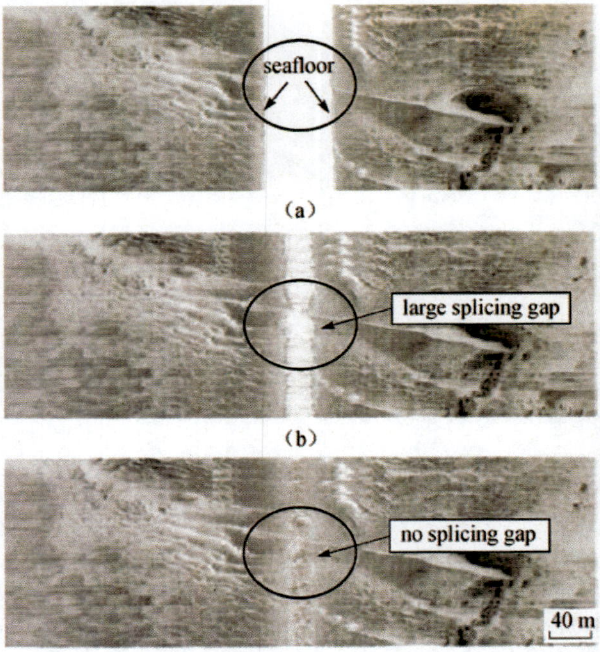

Fig. 4.18 Comparison of sonar images before and after seafloor ray tracing with side-scan sonar. **a** Before seafloor tracing; **b** after automatic seafloor tracing; and **c** after manual seafloor picking

Fig. 4.19 Results of using the average algorithm for mosaicking multiple side-scan sonar survey lines in a certain sea area (the splicing traces in the overlapping areas of adjacent survey lines are largely non-existent, and the scattering energy distribution of the entire mosaic section is uniform)

left and right channels of sonar images, laying the foundation for subsequent side-scan sonar image mosaicking.

To ensure full-coverage surveying of side-scan sonar, the sweeping range of adjacent survey lines usually has an overlapping area of approximately 20%, and the quality of the overlapping area splicing trace processing will directly affect the sonar image mosaic processing; therefore, the splicing trace processing of the overlapping area must be performed before the sonar image mosaicking. The splicing trace of the overlapping area is adjusted mainly for the penetration properties of the sound map, and different software adopts different methods for penetration property adjustment. For example, SonarWiz software develops three methods: Cover Up, Average, and Shine Through, whereas CODA software develops three different methods: Threshold Furthest, Average and Threshold Closest. To determine the best processing method and to suppress the obvious splicing trace of the overlapping area, it is necessary to use different methods to explore and process the penetration attributes of the overlapping area to improve the final effect of the mosaic processing. Figure 4.19 shows profiles of mosaic results of multiple lines obtained by the SonarWiz software using the Average algorithm. The overlapping areas between adjacent lines appear very natural, and there is no sudden change of scattering energy or substantial difference between the lines, making the image suitable for interpretation and application of the mosaic results.

4.3.3 Slant Range Correction Processing

The distance recorded by side-scan sonar is the slant range from the sonar probe to each point recorded, not the horizontal range (the actual horizontal range from the projection of the towfish trace on the seafloor to a target point). The offset from the feature point to the ship trace cannot be measured directly because of the existence of the slant range, which must be calculated. In addition, the slant range makes the target appear compressed in the vertical trace direction, and the degree of compression varies with the change of the horizontal range to the target. The closer the target is to the sonar, the more serious the compression phenomenon. The longer the slant range is, the closer the actual horizontal range; therefore, the compression effect of the target is reduced.

To eliminate the influence of the slant range, appropriate correction processing must be performed before mosaic processing. The key to slant range correction processing is to calculate the horizontal range of the target to be measured on the sonar profile, and then the

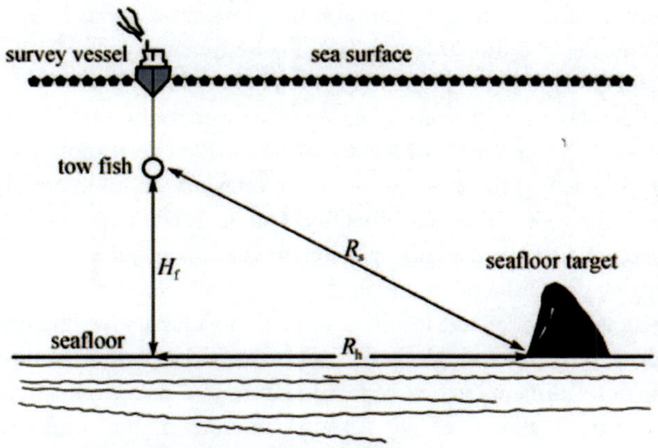

Fig. 4.20 Schematic of the method for calculation of the horizontal range of a target object on an original side-scan profile (Cervenka and De Moustier 1993)

horizontal offset can be calculated using the Pythagorean theorem (Fig. 4.20). The calculation for slant range correction can be based on the following equation (Teng 1998; Deng et al. 2002):

$$D = \sqrt{R^2 - H^2} \tag{4.3}$$

where D is the horizontal range, R is the slant range that is the value shown on the raw profile, and H is the vertical range of the sonar probe from the seafloor, which can be obtained by accurate tracing of the seafloor. This equation assumes a flat seafloor, but can induce distortion if this assumption is significantly invalid (Ingo 2018).

Figure 4.21 shows a comparison of the effects before and after the slant range correction. There is a distortion of the bottom object in the section before slant range correction, and the relative range relation is not accurate, which would lead to misunderstanding in subsequent interpretation. After slant range correction, image distortion is minimal. Moreover, the object is displayed more clearly, it is displayed intact, the size of the object is reasonable, and the relative positional relationship is correct, which provides a reliable basis for subsequent target recognition and extraction.

4.3.4 Seafloor Target Extraction and Sediment Identification

Side-scan sonar transmits acoustic pulses in the form of fan-shaped waves that travel through the water on either side of the towfish. These sound waves are both backscattered and reflected when they encounter objects within the seafloor or the water column. Blocking objects prevent the sound waves from reaching certain parts of the seafloor, and the degree of blocking depends on the height of the towfish relative to the seafloor and the height of the target. Blocking creates an acoustic shadow area that is generally dark in color on the recorded profile. However, sonar acoustic shadows can provide more detail than direct reflections of a target, as illustrated in Fig. 4.22, which shows the acoustic shadows cast by a ship on the seafloor, clearly revealing the acoustic shadowing effects of deck booms, upper bridges and chimneys.

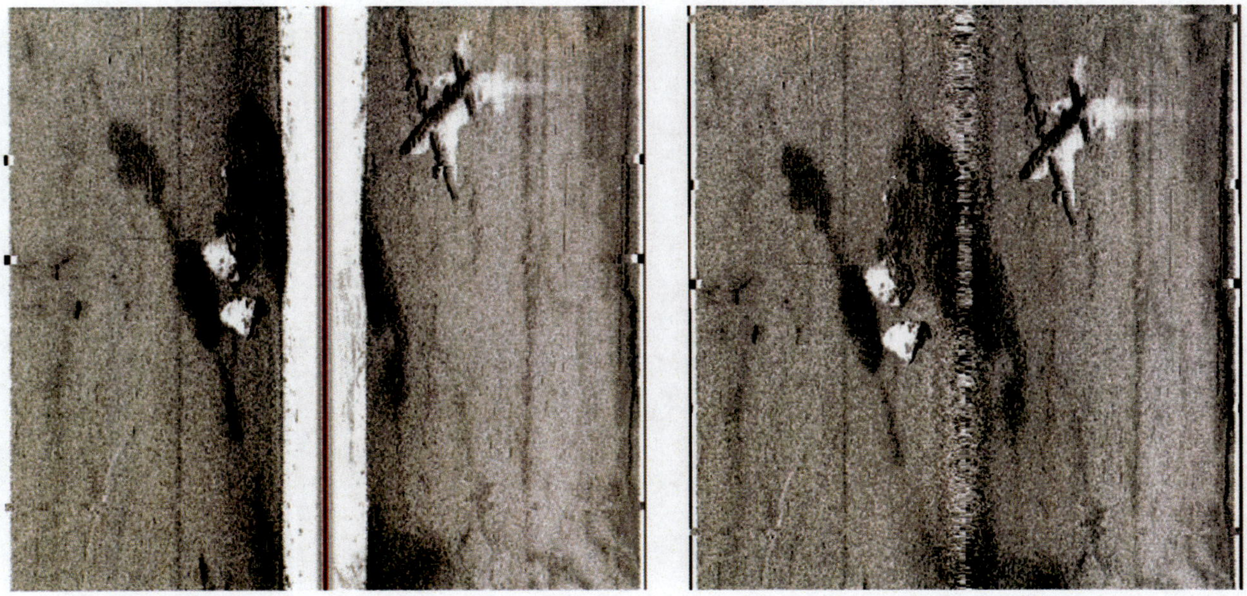

Fig. 4.21 Raw sonar image before slant range correction (left) and actual submarine geomorphology map after slant range correction (right)

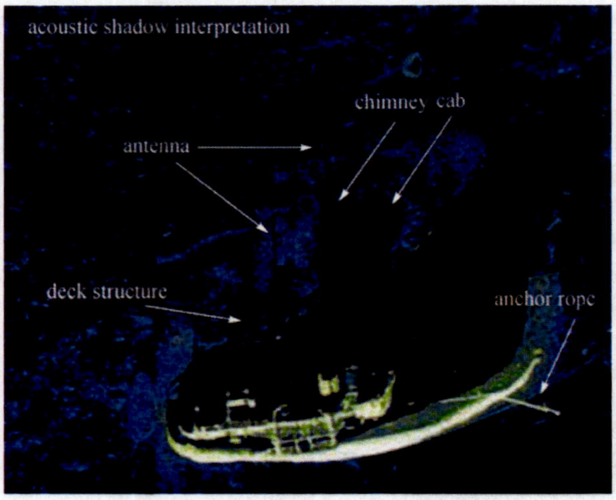

Fig. 4.22 Acoustic shadowing of a wrecked ship on seafloor

The combination of acoustic shadowing and amplitude-intensified reflection of side-scan sonar is the main basis for identification of seafloor targets. Using the illumination technology in computer graphics, various geomorphological features and special targets on the seafloor can be identified effectively, such as the bedrock shown in Fig. 4.23, the wrecked ship in Fig. 4.24, and the fish school in Fig. 4.25. With advanced computer graphics technology, the sonar reflection characteristics of targets are obvious, making identification easy.

Side-scan sonar is an important tool for improving our understanding of the submarine world, particularly in terms of seafloor development and research. This technique is applicable not only to underwater high-resolution imaging but also to continuous monitoring of seafloor changes and acoustic classification of seafloor sediment types (Yang 2003).

Seafloor sediment classification generally uses the backscatter intensity of the sonar data, which corresponds to a gray-scale image of seafloor echo intensity. Echo intensity is a complex physical quantity that is related to various factors such as the transmission frequency and the type of substrate. Different sediment types might have similar echo intensity, which means it is difficult to classify sediments directly using a single gray-scale value. The texture of different sediments varies because texture is a direct reflection of the roughness of the surface structure of the seafloor; therefore, it can be applied to the classification of substrates (Xiong et al. 2014). There are many feature quantities to describe texture, e.g., texture density, texture direction and texture thickness. Generally, the gray level co-occurrence matrix (GLCM) method is applied to calculate texture, and a series of texture feature statistics can be obtained by weighting them with different weight matrices. Feature statistics that can be applied to extract texture information from images include contrast, entropy, inverse difference, gray correlation, energy, angular second-order range and variance. To avoid relying too heavily on a single form of texture, the fractal dimension of submarine topography can also be extracted, which reflects the roughness of the topography. Ultimately, methods for the classification of side-scan sonar images are rich and varied and each method has its own advantages and disadvantages. Practical use demands a multi-parameter combination, supplemented by actual sample controls and verification, to obtain accurate classification results.

Fig. 4.23 Side-scan sonar image of the submarine bedrock

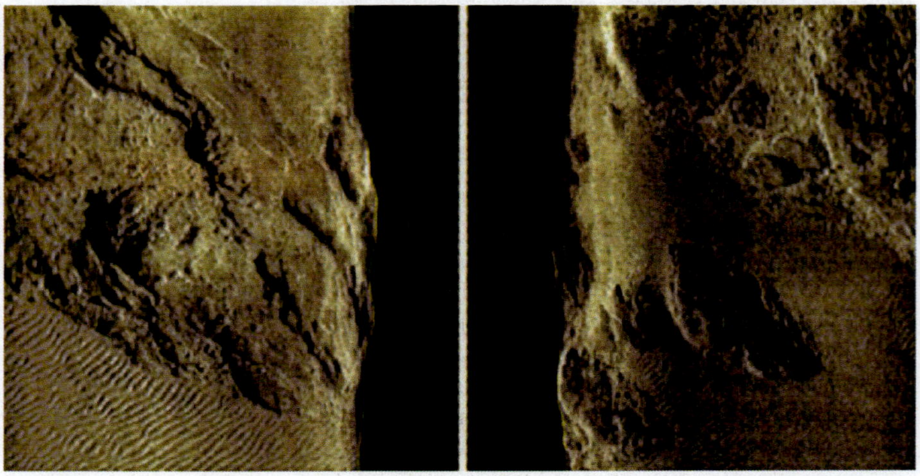

Fig. 4.24 Side-scan sonar image of a wrecked ship

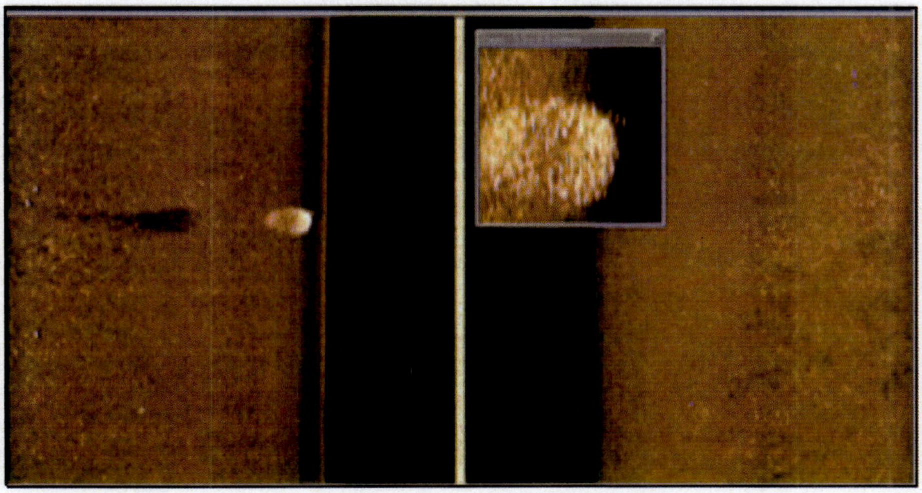

Fig. 4.25 Side-scan sonar image of fish school

4.3.5 Sub-bottom Profile Data Processing Technology and Method

The sub-bottom profile of the Chirp signal has very high stratigraphic resolution for detecting submarine shallow strata and it has good penetrating power in unconsolidated weak strata. Based on the amplitude attenuation features of the Chirp signal, the sediment properties of sedimentary strata can also be studied (Panda et al. 1994; Stevenson et al. 2002). The Chirp signal has the advantages of good repeatability of wavelet, wide frequency band, and adjustable range. After the modulated wideband signal is matched by filtering or cross-correlation and envelope processing, the effects of random noise and seafloor siren-like interference suppression are obvious. The signal-to-noise ratio and resolution of the reflection profile are effectively improved by this process. Therefore, the sub-bottom profile device requires a DSP hardware processing module in the host hardware to complete matched filtering and amplitude envelope processing of the reflected echo. However, cross-correlation and amplitude envelope processing will lose some effective information on reflected echo loss, such as phase information, and the envelope profile will reduce the resolution of the signal, affecting the application of many digital processing methods, e.g., predictive deconvolution for suppressing multiple waves, frequency distance (FX) domain deconvolution for improving resolution, and migration processing. Therefore, the original reflection sequence signal needs to be obtained before the fine sub-bottom profile of the Chirp signal is processed. According to research (Baradello 2014), the Chirp original reflection data should be processed by a professional process, following which a reflection profile with high resolution and a high signal-to-noise ratio can be obtained.

For the reflections from Boomer and Sparker sources received by a single-channel towline, conventional processing generally requires a frequency filtering process, e.g., the commonly applied bandpass filtering, high-pass filtering, and matched filtering. If the random interference noise energy is strong and the effective reflection signal is submerged and difficult to identify, then a multi-channel seismic processing method using two-dimensional FK filter processing technology can be considered. If the detection area has multiple concurrent seismic reflection data, then a superimposed velocity spectrum of the multi-channel seismic data can be applied to a post-stack migration of the sub-bottom profile reflection data to suppress diffraction waves and rotary waves on the profile. Thus, the reflection interface can be located correctly and the imaging accuracy of the sub-bottom profile reflection data could be improved. Similar to the reflection signal received by a two-in-one acoustic transducer array, the data of a single-channel towline must be normalized and the energy balanced before imaging. If the wind and waves are reasonably high during the survey, which can result in obvious undulation on the profile, it is necessary to correct for this effect. For deep-water sub-bottom profile data, delay correction is also required. For example, Fig. 4.26 lists the relevant processing workflow that must be performed for sub-bottom profile data. The processing modules indicated by the dashed arrows and dashed boxes in the figure are specific to data received by a single-channel towline, whereas the solid arrows and the processing modules in the solid boxes apply to both a single-channel towline and a transducer array.

Usually, many single-channel towline data are applied to simple predictive deconvolution processing to suppress the multi-wave interference of strong energy existing in the sub-bottom profile section of a shallow-water area. For the acoustic signal received by a transducer array, if the original amplitude data contain positive and negative waveforms

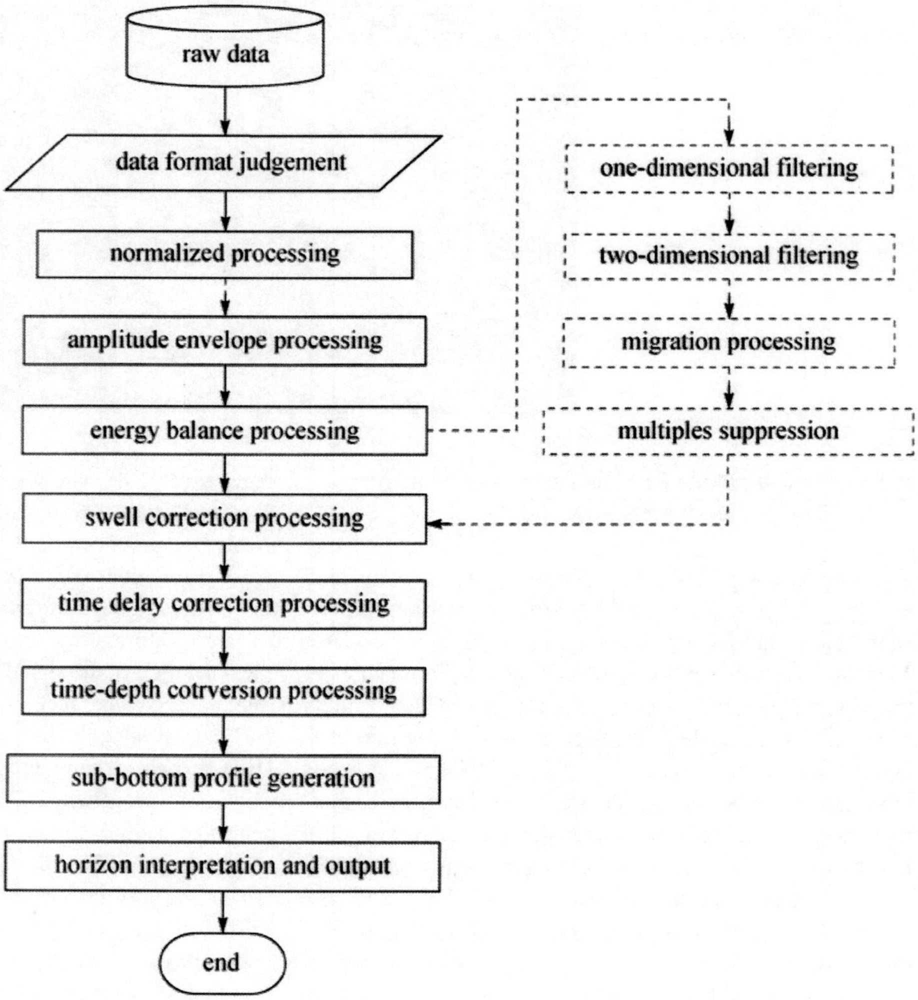

Fig. 4.26 General processing workflow of sub-bottom profile data (dotted lines/boxes indicate the selection is based on characteristics of single-channel towline data)

(such as the sub-bottom profiler produced by EdgeTech), processing can use all the method modules applicable to single-channel towline data. However, some sub-bottom profilers (such as the Chirp III system produced by Teledyne Benthos) have matched filtering and amplitude envelope processing in the host DSP module. After envelope processing, the reflected data do not contain a negative amplitude signal and the relevant phase information is lost. Moreover, many processing methods are not applicable, e.g., frequency filtering, two-dimensional filtering, and deconvolution processing. Currently, it is difficult to suppress interference such as multiple waves on the data profile. To suppress such interference effectively, it is necessary to obtain correlation data both before and after envelope processing to perform multiple wave processing on these data.

For offshore engineering exploration, the interpretation of sub-bottom profile data is also very important for tasks such as the acquisition of reflective horizons (Ding et al. 2008) and the delineation of feature targets. The technology of efficient-accurate acquisition and interpretation represents a necessary means to obtain and apply sub-bottom profile data; however, interpretation technology and methods are also needed (Ding et al. 2012a, b).

4.3.6 Sub-bottom Data Post-Processing Methods

There are many post-processing methods for sub-bottom profile detection data and there are multiple methods for processing amplitude signals and profile images. For image processing of a reflection profile, some software has been developed that incorporates professional image processing methods such as gray histogram linear and nonlinear transformation, gray level equalization processing, Gaussian smoothing, median filtering, and edge refinement processing. The purpose of post-processing is to suppress interference noise on the profile to highlight the effective reflection energy and to improve the stratigraphic resolution of sub-bottom profile detection of the reflected image.

The purpose of data processing is to improve the resolution and signal-to-noise ratio of the sub-bottom profile, improve the quality of the data profiles, and provide a basis for interpreting the mapping of subsequent data profiles. For the features of sub-bottom profile data, the appropriate processing modules and method parameters should be selected during the actual process to achieve the corresponding processing objectives.

4.3.6.1 Processing Methods of Wind and Wave Correction

For ship-borne sub-bottom profilers, when sea waves are large, the survey ship could sway violently, which could cause the profiler to move up and down in tandem, creating an undulating effect on the reflection profile. In Fig. 4.27, the left panel shows how a ship affected by wind and waves at different heights can cause the position of the acoustic waveform transmitted and received by the sub-bottom profiler to be misaligned (see the right panel). To correct this effect, swell filtering is required. Figure 4.28 shows the results of an actual sub-bottom profile affected by wind and waves. In the upper panel, the wind and waves have serious impact on the acquisition profile, reducing the resolution of the in-phase axis of the profile reflection, affecting the interpretation and tracing of the reflection horizon, and making interpretation of the profile difficult. Wind and wave correction processing generally adopts model channel cross-correlation technology and a method of combining the average filter with a median filter (Ding et al. 2012a, b). The lower panel in Fig. 4.28 shows how the undulating appearance attributable to the wind and wave effect is suppressed by processing.

4.3.6.2 Processing Methods of Sub-bottom Profile Delay Correction in Deep Water

For sub-bottom profile data in deep water, delay correction is important, especially in areas where the seafloor is rugged and steep (Ding et al. 2015). A sub-bottom profile of deep water is generally stored in the internationally adopted SEG-Y format. However, this format can only store up to

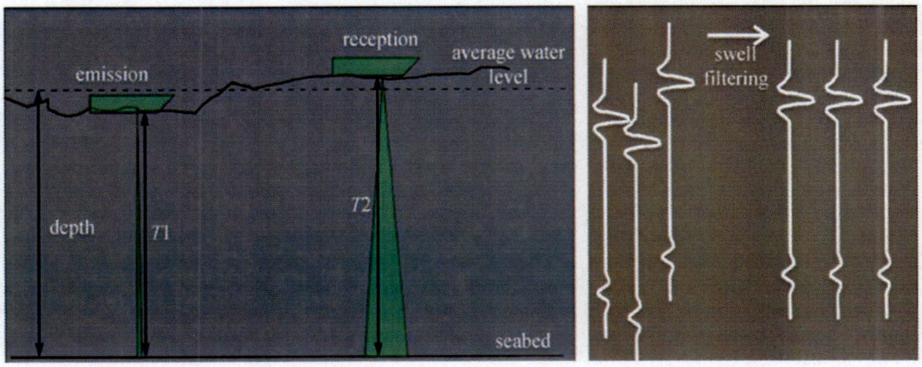

Fig. 4.27 Schematic of the impact of sea wind and waves on sub-bottom profile detection

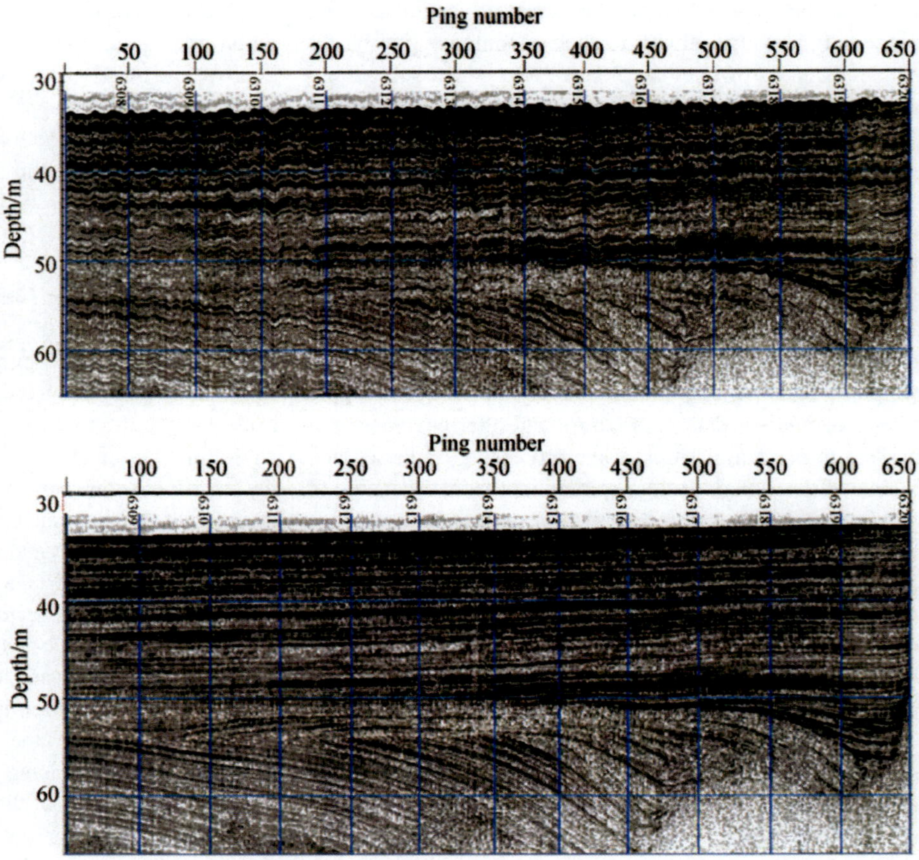

Fig. 4.28 Effect of wind and waves on sub-bottom detection data: raw profile (upper) and corrected profile (lower)

$2^{15} - 1 = 32\ 767$ sample data. For reflection data with a sampling rate of 20 kHz, only a two-way travel time data body of $32\ 767 \times 0.05 = 1\ 638$ ms can be stored. At an average speed of 1 500 m/s, only a data body in the depth range of 1 228 m can be recorded. If the data exceeds this depth range, the recording system will automatically enable a recording delay function according to the traced seafloor position, thereby avoiding most of the water travel time in deep-water acquisition, reducing the actual data acquisition, and meeting the storage requirements of the SEG-Y data format.

When recording sub-bottom profile data in deep water after the application of the delay function, if delay correction is not performed, the shape and the stratigraphic structure of reflection profile will be misaligned with the actual stratum. As shown in Fig. 4.29, illustrating an raw SEG-Y data profile, the seafloor appears very flat and the depth of the seafloor is shown as <10 m, which is not the case. Therefore, the data body of the graph must be corrected for the corresponding delay. The corrected data profile is shown in Fig. 4.30. It can be seen from the corrected profile that the actual seafloor becomes increasingly deep along the

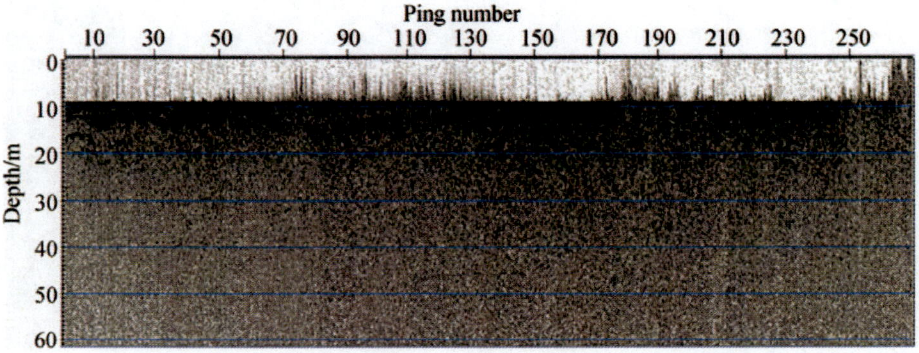

Fig. 4.29 SEG-Y sub-bottom profile without delay correction in deep water

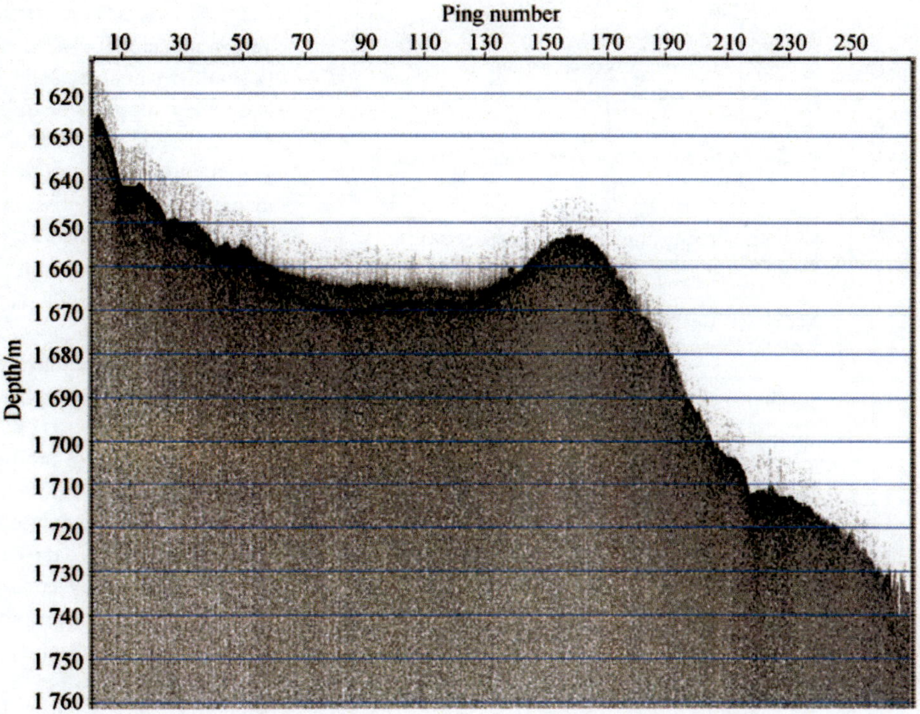

Fig. 4.30 Result of the correction for time delay in deep water of the sub-bottom profile data shown in Fig. 4.29

direction of travel of the exploration ship, and the range of variation is >300 m. The actual depth is in the range 1 425–1 740 m. Clearly, the topographical shape of the seafloor is not flat seafloor, as shown in Fig. 4.29, but violently abrupt. After correction, the depth values of the seafloor and the strata below are corrected and displayed correctly.

4.3.6.3 Raw Data Processing of Chirp Signals

For raw reflection data of the Chirp signal, wavelet cross-correlation and amplitude envelope processing are not performed; consequently, the reflection data contain rich unmodified amplitude and phase information. For raw Chirp signal data, processing methods for single or multi-channel reflection data can be applied to test the efficacy of various noise suppression and processing methods for improving stratigraphic resolution. As the effective frequency band of the raw Chirp reflection data is wide, high and contains rich phase information, conventional frequency filtering can be effective, and when suppressing the field DC interference signal, 100 Hz high-pass filtering processing is especially effective. To reduce the influence of reverberation interference in shallow water, predictive deconvolution processing for suppressing multiple waves can be applied. In addition, FX domain deconvolution processing, which increases the spatial continuity of the reflected in-phase axis and the resolution of the longitudinal stratum, can also be applied. Conventional migration processing can be applied when the detection zone has reliable sedimentary stratigraphic velocity data; however, if there are no multi-channel seismic data in the actual detection, it is generally difficult to obtain reliable multiple speed data. Conventional migration processing is limited, but when combined with the features of sub-bottom profile data, particularly its shallow penetration depth and hyperbolic reflection, the sound wave velocity of diffracted waves can be estimated from the diffraction wave hyperbola present in the profile (using the hyperbolic correction method). By estimating the sound wave velocity of diffracted waves, an migration processing profile of the effective convergence diffracted wave can be obtained by applying the Stolt migration method with only a single

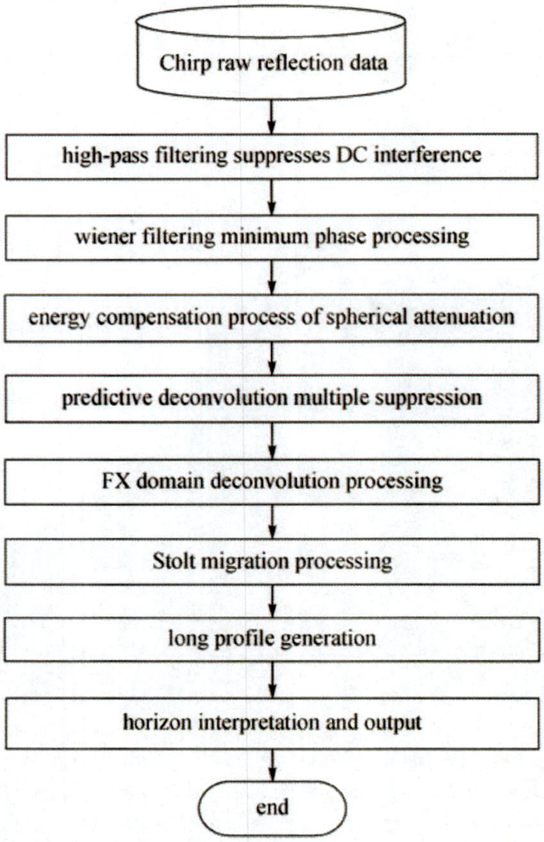

Fig. 4.31 Processing workflow of raw sub-bottom profile data of Chirp signal

velocity data. Figure 4.31 shows the effective processing flow for the raw data profile of the Chirp signal (Baradello 2014), and Fig. 4.32 shows a comparison of the results of a raw profile with the envelope profile after application of the above process. It can be seen that after the raw data are processed by the special module, the profile information is very large, the reflection is more abundant in the same phase axis, and the stratigraphic resolution is much higher in the envelope profile. The processed profile is more effective in detecting submarine microlayers and buried targets, and it can be used more effectively to mine the information in the profile.

4.3.6.4 Two-Dimensional Filtering Method

Two-dimensional F-K filtering technology is widely applied in multi-channel seismic data processing because it can effectively suppress random noise and coherent interference. The sub-bottom profile data received by a single-channel towline do not exist in the common shot gather record. This means there is no concept of wave number K, which prevents application of two-dimensional F-K filtering in the processing of sub-bottom profile data. If the Ping number of the sub-bottom profile data is assumed as the offset in a multi-channel earthquake, and a hypothetical wave number K parameter is generated, the two-dimensional F-K filtering technique could be applied to sub-bottom profile data processing and its effect is obvious. Figure 4.33a shows the raw data of a sub-bottom profile received using a Sparker source in a sea area. The random interference noise that is evident in the profile submerges the effective reflection signal, which means the profile data cannot be interpreted. Using conventional one-dimensional frequency analysis as shown in Fig. 4.33b, the corresponding one-dimensional frequency curve, shows that high-frequency interference noise in the range 1 900–2 600 Hz is dominant and that the effective reflection signal frequency below 1 000 Hz is not prominent. After application of one-dimensional frequency band-pass filtering in the range 100–1 200 Hz on the raw profile data, the profile shown in Fig. 4.34d is obtained. In this figure, the random noise is reasonably obvious and the effective reflection remains mixed with the random noise. If two-dimensional F-K transform analysis is applied, the corresponding two-dimensional F-K spectrum energy group shown in Fig. 4.34b is obtained. It can be seen that the strong energy cluster is concentrated mainly near the 0 Hz/km wave number and that the frequency is in the range 200–800 Hz. According to the distribution features of the effective energy mass on the two-dimensional spectrum, if the band of energy outside the 0 Hz/km range in the profile data of Fig. 4.34 is stripped and the energy mass above 800 Hz is removed, then the two-dimensional F-K inverse transformation can be obtained, as shown in Fig. 4.35. It can be seen from Fig. 4.35 that the random noise is suppressed, the effective reflection is continuous with the same phase axis, and the energy is prominent, indicating the efficacy of the application of F-K filtering to sub-bottom profile data.

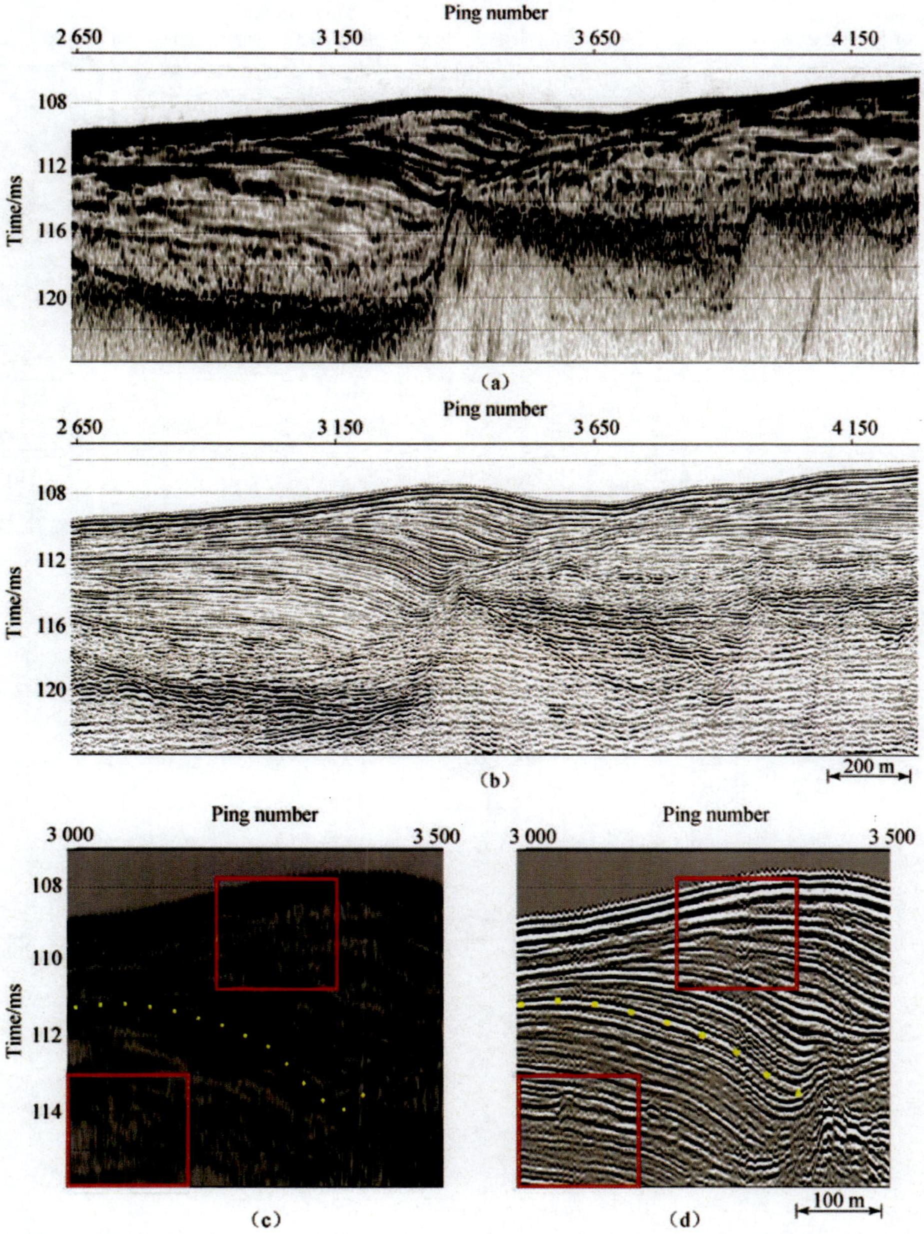

Fig. 4.32 Comparison between the traditional envelope section **a** and the minimum-phase seismic processed section (gain, predictive deconvolution, band-pass filter, FX-deconvolution and Stolt migration) (**b**). In the envelope section **c** a most information has been lost (red boxes) compared to the minimum-phase processing (**d**). The yellow dot line shows the same seismic reflection, which is irregular in the envelope section, and well resolved by using the minimum-phase method (Baradello 2014)

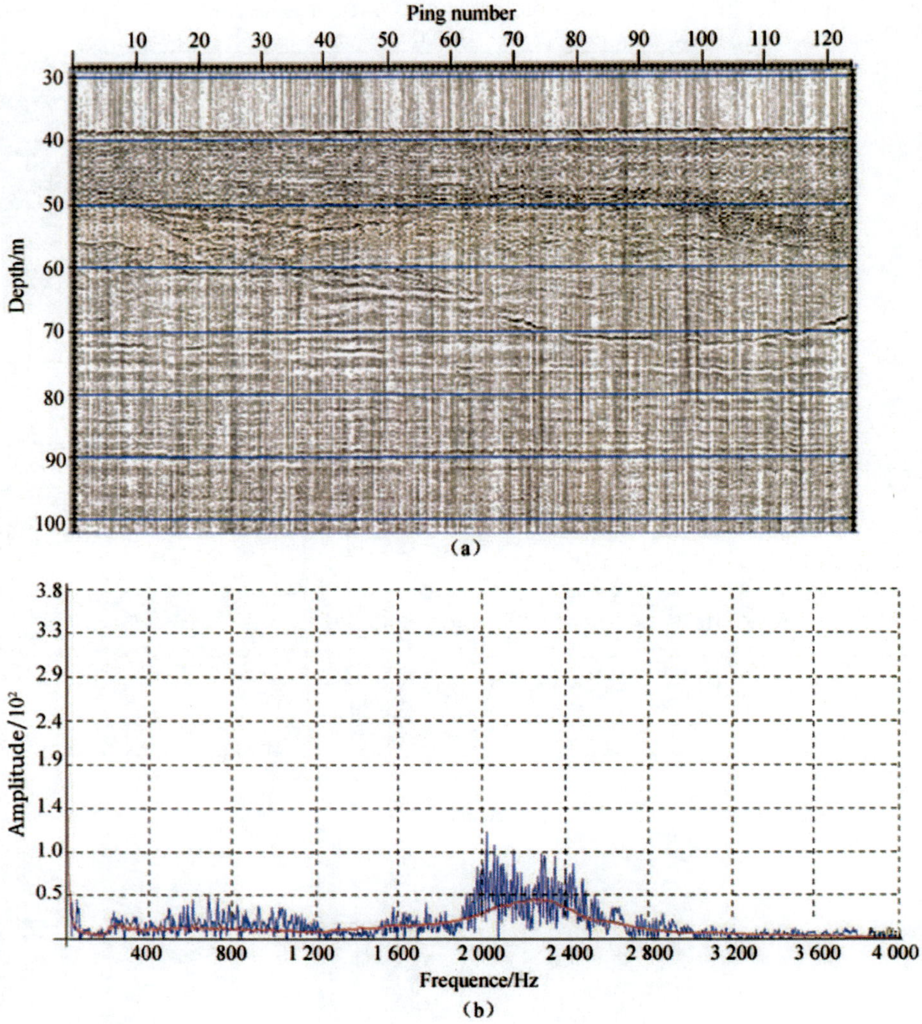

Fig. 4.33 Raw sub-bottom profile data (**a**) and corresponding one-dimensional frequency curve (**b**) received by Sparker source in a certain sea area

Fig. 4.34 Profile of the original data in Fig. 4.33 when processed by one-dimensional bandpass filtering (**a**) and the corresponding two-dimensional F-K spectrum (**b**)

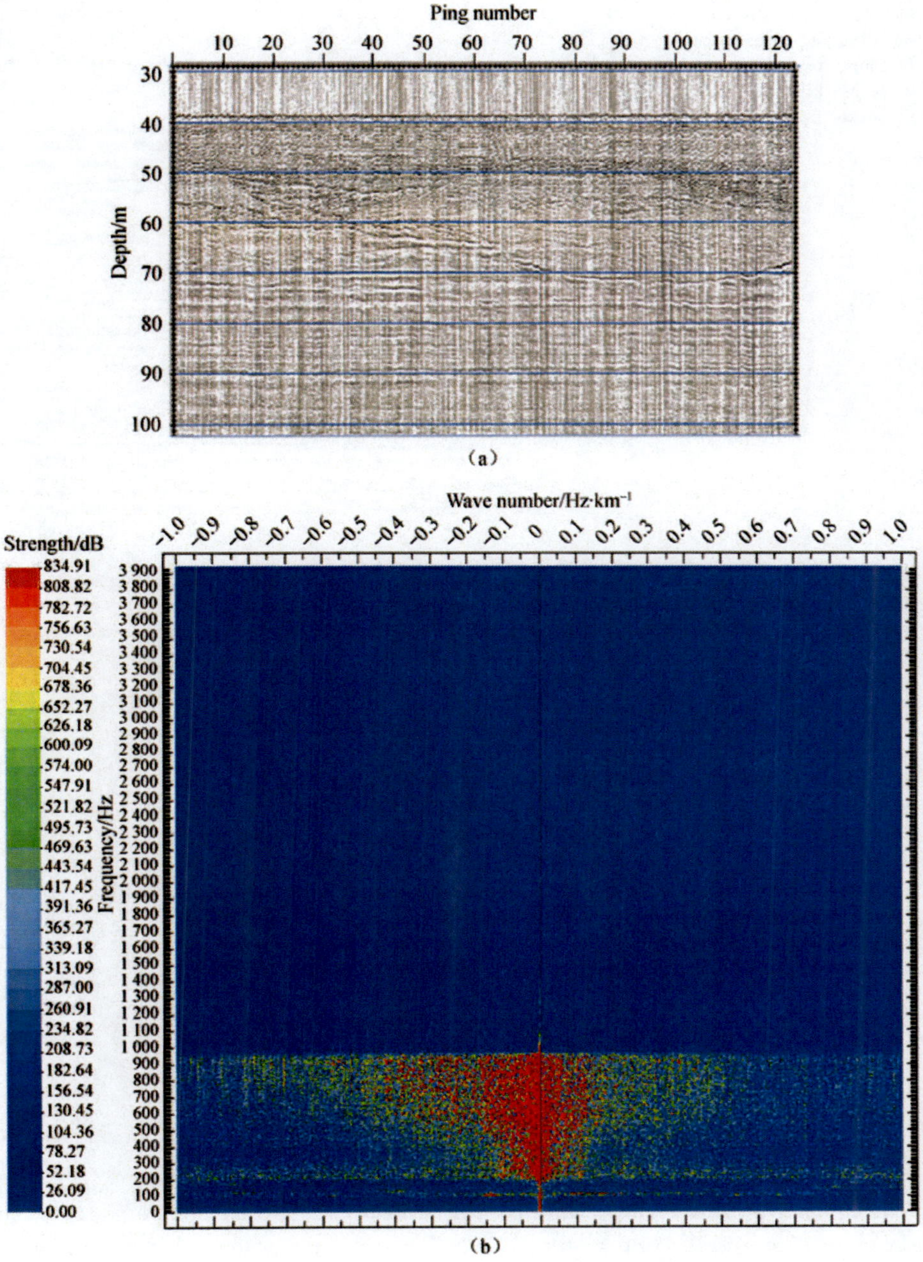

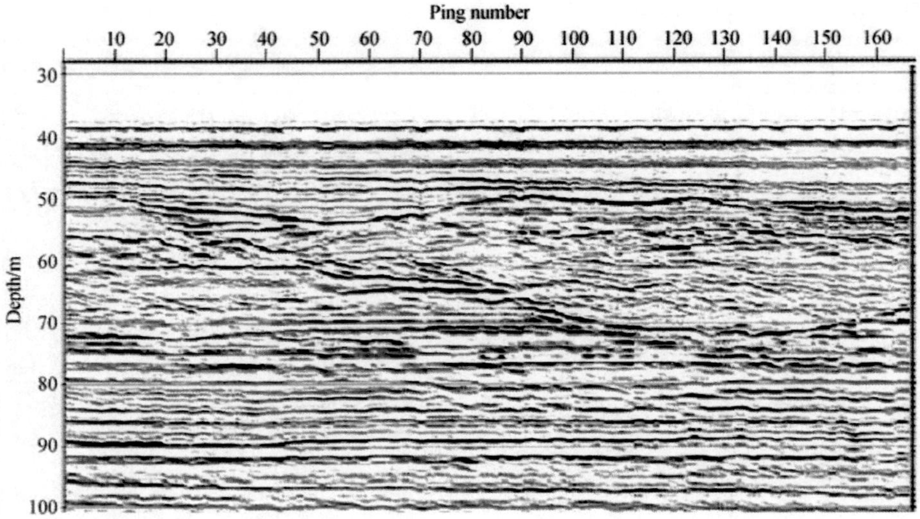

Fig. 4.35 Result of the one-dimensional frequency filtering process in Fig. 4.34 when processed by two-dimensional F-K filtering

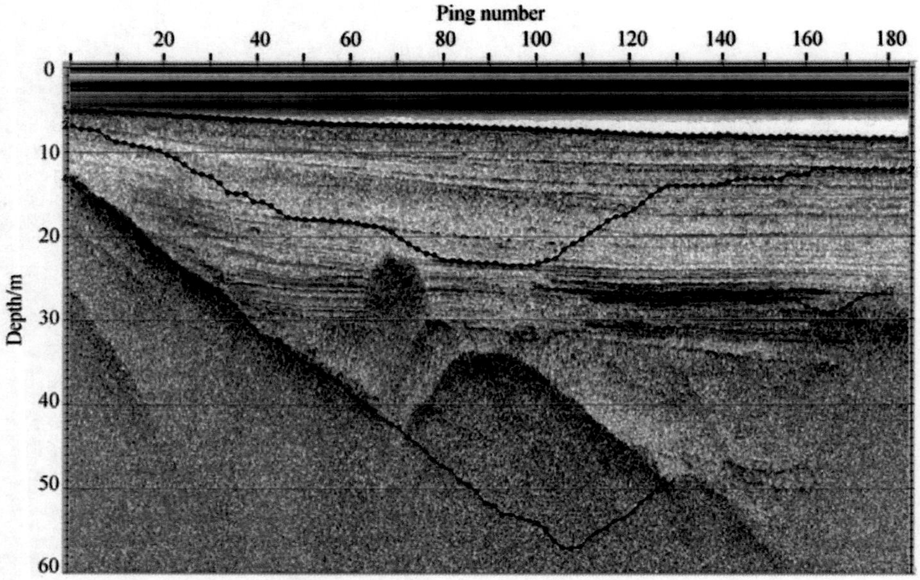

Fig. 4.36 Constrained sliding window energy ratio method for automatic picking of the reflected in-phase axis. Except for the first layer of seafloor reflection in-phase axis that is correctly picked up, the results for the automatic picking of the following three layers of reflection in-phase axis all have problems

4.3.6.5 Automatic Picking Method for Stratum in a Sub-bottom Profile

Reflection in-phase axis picking technology is important for sub-bottom profile detection and it is a vital step in the interpretation of sub-bottom profile data. Efficient picking technology can obtain effective digital results in profile interpretation, which can accelerate the interpretation of reflection profiles and improve application of the results (Ding et al. 2012a, b). However, the actual detection and reflection data are complex and variable, and it is difficult to obtain reliable interpretation results using a single method. The automatic pick-up method of the reflected in-phase axis needs to comprehensively consider the features of the energy and waveform phase of the reflected data. The energy ratio method uses the amplitude information of the reflection data to accurately pick up the seafloor reflection in-phase axis from under the background noise; however, the results reflecting the in-phase axis between layers with similar pick-up energy are generally poor. The cross-correlation method utilizes the waveform phase features of the reflected data to accurately pick out reflected in-phase axes with similar energy and uniform phase between layers. In automatic picking of the reflection phase in a reflection interface with gentle changes, the combination of the two can achieve better interpretation results. However, for a reflective interface with large fluctuations, it is easy to pick up the deviation using an automatic picking method. Nevertheless, it is generally difficult to obtain accurate results and therefore human–computer interaction is required to modify the results. Figure 4.36

Fig. 4.37 Result of using the correlation analysis method and human–computer interaction picking technology to re-pick the error tracing and layer interpretation layer of Fig. 4.36

shows the automatic tracing result of the energy ratio method. It can be seen that the in-phase axis with large changes is not accurate and error tracing occurs. Figure 4.37 effectively corrects the results of the energy ratio error tracing using cross-correlation and human–computer interaction pick-up methods.

The reflection amplitude energy ratio method is automatically traced by sliding time window segmentation, the formula for which is as follows:

$$A_1 = \frac{\sum_{t=T_2}^{T_3} x^2(t)}{\sum_{t=T_1}^{T_2} x^2(t)} \quad (4.4)$$

where $x(t)$ is the recorded sample value, T_1 is the starting point of the previous time window, T_2 is the end point of the previous time window (and also the starting point of the latter time window), and T_3 is the end point of the latter time window.

The cross-correlation automatic picking method adopts a cross-correlation matching algorithm between the model trace and the to-be-picked trace, which obtains the position to be picked up by comparing the maximum value of the cross-correlation coefficient in a fixed time window. The maximum correlation coefficient can be calculated as follows:

$$r_{xy}(t_0) = \sum_{t=t_1}^{t_2} x_t y_{t-t_0} \quad (4.5)$$

where x_t is the model reference trace that generally averages the amplitude of the previously picked trace, and y_{t-t_0} is the data trace that needs to be picked up for which the automatically related time range is actively controlled by the interpreter.

References

Baradello L (2014) An improved processing sequence for uncorrelated chirp sonar data. Mar Geophys Res 35:337–344. https://doi.org/10.1007/s11001-014-9220-1

Cervenka P, De Moustier C (1993) Sidescan sonar image processing techniques. IEEE J Ocean Eng 18(2):108–122. https://doi.org/10.1109/48.219531

Deng XQ, Gong DC, Luo R (2002) Research on geocoding technology of Side-scan sonar image (in Chinese with English abstracts). Ocean Mapping 22(4):14–17

Ding WF, Luo JH, Lai XH et al (2008) The research of interactive interpretation picking for sub-bottom profile (in Chinese with English abstracts). Mar Sci 32(9):1–6

Ding WF, Feng X, Fu XM et al (2012) Marine wave correction research on single channel seismic data and a sub-bottom profile (in Chinese with English abstracts). Acta Oceanol Sin 34(4):91–98

Ding WF, Pan GF, Gou ZK et al (2012) The research of interactive auto pickup of seismic events based on energy ratio and cross-correlation (in Chinese with English abstracts). Acta Oceanol Sin 34(3):87–91

Ding WF, Li JB, Su XH et al (2015) Technology research and development for acoustic sub-bottom survey in deep sea (in Chinese with English abstracts). Acta Oceanol Sin 37(6):70–77. https://doi.org/10.3969/j.issn.0253-4193,2015.06.007

Feng J, Zhao TH, Yang Y et al (2014) Analysis of mosaic techniques for side-scan sonar image (in Chinese with English abstracts). Bull Surv Mapp 9:66–69. https://doi.org/10.13474/j.cnki.11-2246.2014.0296

Ingo K (2018) Sidescan Sonar. In: Micallef A, Krastel S, Savini A (eds) Submarine geomorphology. Springer International Publishing, New York, pp 13-24

Liu GD (1978) Marine geophysical exploration (in Chinese). Geological Publishing House, Beijing

Liu BH, Ding JS, Pei YL et al (2005) Marine geophysical survey techniques and their applications to offshore engineering (in Chinese with English abstracts). Adv Mar Sci 23(3):374–384. https://doi.org/10.1360/gs050303

Liu BR, Song HB, Guan YX et al (2015) Characteristics and formation mechanism of cold seep system in the northeastern continental slope of south china sea from sub-bottom profiler data (in Chinese with

English abstracts). Chin J Geophys 58(1):247–256. https://doi.org/10.6038/cjg20150122

Ma WD, Xiong XM (2011) Visualization and preprocessing of side-scan sonar images based on IDL language (in Chinese with English abstracts). Microcomput Appl 30(2):41–44

Panda S, LeBlanc LR, Schock SG (1994) Sediment classification based on impedance and attenuation estimation. J Acoust Soc Am 96 (5):3022–3035. https://doi.org/10.1121/1.411266

Stevenson IR, McCann C, Runciman PB (2002) An attenuation-based sediment classification technique using chirp sub-bottom profiler data and laboratory acoustic analysis. Mar Geophys Res 23(4):277–298. https://doi.org/10.1023/a:1025708024518

Teng HZ (1998) Research on image processing method of seafloor geomorphological sonar (in Chinese with English abstracts). Dissertation, PLA Institute of Surveying and Mapping

Triton Imaging Inc. (2014) eXtended Triton Format (XTF). Triton. Available via DIALOG. https://www3.mbari.org/products/mbsystem/formatdoc/XtfFileFormat_X26.pdf. Accessed 28 June 2019

Wu ZY, Zheng YL, Chu FY et al (2005) Research status and prospect of sonar-detecting techniques near submarine (in Chinese with English abstracts). Adv Earth Sci 20(11):1210–1217. https://doi.org/10.3321/j.issn:1001-8166.2005.11.007

Xiong MK, Wu ZY, Li SJ et al (2014) Wavelet neural network identification and classification of sediment seafloor sonar images based on genetic algorithms (in Chinese with English abstracts). Acta Oceanol Sin 36(5):90–97. https://doi.org/10.3969/j.issn.0253-4193.2014.05.010

Xu F, Wei JJ (2006) Sonar technology and its application topic: side-scan sonar (in Chinese). Physics 35(12):1034–1037

Yang FL (2003) Fusing and classifying multi-beam sonar and side-scan sonar data (in Chinese with English abstracts). Dissertation, Wuhan University

5 Navigation and Positioning Technology

The Global Navigation Satellite System (GNSS) is not a single constellation system but a comprehensive constellation system including the GPS, GLONASS, BDS and Galileo. The GPS was approved in 1973 and it has been 41 years since its first satellite was launched in 1978. GLONASS was created and operated by the Soviet Union and then Russia in 1976 and 1993.

Back in the mid-1990s, to break the monopoly of the US in the market of satellite positioning, navigation and timing, the European Union worked on an ambitious civil global navigation project, called the Global Navigation Satellite System, which provided huge market benefits and increased the employment opportunities of Europeans. The plan was implemented in two steps. The first step was the establishment of a first-generation global navigation satellite system (known then as GNSS-1 and later as EGNOS), which integrated the US GPS with the Russian GLONASS system. The second step was the establishment of a second-generation global navigation satellite system completely independent of the US GPS and Russian GLONASS system, namely the Galileo satellite navigation and positioning system, which is under construction. A GNSS, also known as a space-based positioning, navigation and timing system, plays an important role in providing temporal/spatial benchmarks and all location-related real-time dynamic information. It is important for the space and information infrastructure of a country as well as an important symbol of a country's status and national strength. Because the GNSS system plays an indispensable role in national security and economic and social development, world powers are competing to develop independent satellite navigation systems. It is expected that by 2020, there will be four GNSS, namely the existing GPS of the US, the GLONASS of Russia, the Galileo of the European Union and the BeiDou Satellite Navigation System (Compass) of China.

The research and application of marine science uses not only satellite navigation systems but also acoustics, seabed topography, gravity and magnetic navigation. The different features of different navigation systems are discussed below.

5.1 Development of GNSS

5.1.1 GPS

The GPS was created by the US Department of Defense with an investment of more than $1 billion for the Star Wars program. The role of the GPS was to assist the navigation of ships and planes of the US military and the command of the US Army. The GPS was developed in three phases.

The first stage (1973–1978) was the project demonstration stage. The second stage (1979–1985) was the engineering development and system testing phase, in which test results met predetermined design objectives. At that time, seven experimental satellites were in orbit, providing limited navigation capability. The third stage improved the performance of the system and put an entire system into operation. According to the original plan, the operational satellites were to be put into orbit by the space shuttle fleet in batches beginning in 1986 and the system was to be complete and operational, providing three-dimensional positioning, by 1989. However, this phase was postponed until February, 1989 because of the Space Shuttle *Challenger* disaster. By around 1990, 12 satellites had been put into orbit to achieve two-dimensional positioning. The GPS satellite network was completed in 1993.

The GPS is a global positioning, navigation, and timing system comprising 24 satellites. At any point on Earth, signals from more than four satellites can be received at any time, meaning that the information transmitted by the GPS

Z. Wu et al., *High-resolution Seafloor Survey and Applications*,
https://doi.org/10.1007/978-981-15-9750-3_5

satellites covers the entire surface of the Earth (Dana 1997; Gabbay and Konotov 2019; Hofmann-Wellenhof et al. 2012). The space constellation of the GPS comprises 21+3 satellites, of which 21 are working satellites and the other three are backup satellites. The 24 satellites are evenly distributed over six orbital planes, each of which contains four satellites. The orbital plane has an inclination angle of 55° relative to the equatorial plane, the orbital planes are 60° from each other, and the orbital right ascensions of ascending nodes are 60° apart from each other. The arguments of latitude between satellites in orbital planes are 90° apart from each other, and the satellites in one orbital plane are 30° ahead of the corresponding satellites in adjacent orbital planes to the west. Each satellite orbit is elliptical, with an average altitude of about 20 200 km and an operating period of approximately 11 h and 58 min. The ground control segment of the GPS system mainly comprises six earth stations distributed around the world, including satellite monitoring stations, a master station and standby master stations, and information injection stations, located in Colorado, Gaithersburg and Hawaii and on the Ascension Islands in the South Atlantic, and on Kwajalein in the South Pacific. The user equipment part of the GPS comprises GPS receiver hardware and corresponding data processing software as well as a microprocessor and terminal equipment. The hardware of the GPS receiver includes the host computer, antenna, and power supply and is mainly used to receive signals sent by a satellite, obtain the observed values, and extract the broadcast ephemeris and the corrected satellite clock from the navigation message, before the navigation and positioning work is completed through data processing. GPS software refers to a variety of post-processing software packages, which are usually provided by manufacturers. The main role of the software is to process the observation data precision and thus obtain precise positioning results.

The main technical measures to be used in the GPS modernization plan are as follows. (i) Turn off the selective availability (SA) signal to improve the precision of signals in space and eliminate the pseudo-range error generated by SA. (ii) Update the structure of the GPS signal and add a third civil signal, L5, which has the advantages of a greater bandwidth, higher symbol rate, greater transmission power and improved data parity checking compared with the L1 signal. (iii) Initiate the GPS Block III satellite program, focusing on abandoning the current medium Earth-orbiting (MEO)-orbiting GPS constellation in favor of the high Earth orbit + geostationary orbit (GEO) constellation, while the GPS Block III satellite will adopt M-code beam technology. (iv) Renew GPS ground facilities, improve the ability to monitor satellite signals, make the control network more powerful and improve the accuracy and security of the GPS in civilian and military fields.

5.1.2 BeiDou Navigation Satellite System

In 1983, Academician Chen Fangyun, who won the Two Bombs and One Satellite Medal for Meritorious Service, and his collaborators proposed the idea of using two synchronous fixed-point satellites for positioning and navigation, known as the double-star positioning system. The system comprises two synchronous fixed-point satellites at a certain distance or angle of longitude from each other, a master ground station for operating and control, and several ground-based subscriber stations. The positioning procedure is as follows. The master ground station sends signals to the subscriber station via the two synchronous fixed-point satellites. Upon receiving the signal, the subscriber station responds, and the response signal is returned to the master ground station via the two satellites. The range between the subscriber stations and the two satellites can be determined by the signal round trips among the master station, two satellites and subscriber stations. Assuming that the approximate distance from the location of the subscriber station to the center of the Earth is known, the position of the subscriber station can be calculated using a mainframe computer located at the master ground station, given the position of the three points (i.e., two satellites and the center of the Earth) in space and the distances from the known subscriber station to the three points. The master ground station then notifies the subscriber station of position information via satellite. The subscriber station has transmitting and receiving equipment, and short telegrams can thus be exchanged between the master ground station and the subscriber station. The BeiDou Navigation Satellite System was developed on this basis. Two BeiDou-1 working satellites were launched in October and December 2000 while a backup satellite was launched in May 2003, marking the establishment of the first generation of the BeiDou navigation system. The double-star positioning system formed by BeiDou-1 satellites provides active positioning services to China's mainland and the surrounding areas of the Taiwan Strait. The BeiDou-1 satellite navigation and positioning system built by China has broad application prospects for the national economy besides its important role in national defense.

The BeiDou Navigation Satellite System comprises three parts: the space segment, ground segment and user segment. The system provides high-precision, high-reliability positioning, navigation and timing services for all types of users in all weather throughout the day. It allows short-message communication and has initial capabilities of regional navigation, positioning and timing, with a positioning accuracy of 10 m, velocity measurement accuracy of 0.2 m/s and timing accuracy of 10 ns. The second-generation BeiDou system, namely the BeiDou Regional Navigation Satellite System, comprises 14 satellites, including five GEO satellites, five inclined geosynchronous orbit (IGSO) satellites, and four MEO satellites. On December 27, 2012, the BeiDou Regional Satellite Navigation System was officially put into operation and Version 1.0 of the BeiDou System Spatial Signal Interface Control Document was released. The system then officially served Asian and Pacific regions. On December 27, 2013, a press conference on the first anniversary of the BeiDou Navigation Satellite System providing regional services was held in the Press Release Hall of the State Council Information Office. Two system documents, "BeiDou System Open Service Performance Specification (Version 1.0)" and "BeiDou System Spatial Signal Interface Control Document (Version 2.0)", were officially released (Yang et al. 2014; Li et al. 2014; Tang et al. 2014).

On November 23, 2014, the Maritime Safety Committee of the International Maritime Organization considered and adopted the Navigation Safety Circular of the BeiDou Navigation Satellite System, marking the BeiDou Navigation Satellite System formally becoming an integral part of the Global Radio Navigation System and gaining international legitimacy for maritime applications.

In March 2015, the 17th BeiDou navigation satellite was launched, marking the start of the shift of China's BeiDou Navigation Satellite System from regional operation to global expansion. The BeiDou Global Navigation Satellite System will comprise five GEO satellites and 30 non-geostationary orbit (non-GEO) satellites. The GEO satellites are located at 58.75°E, 80°E, 110.5°E, 140°E and 160°E, respectively. The non-GEO satellites are 27 MEO satellites and three IGSO satellites. The MEO satellites have an orbital altitude of 21 500 km and an orbital inclination of 55° and are distributed evenly on three orbital planes. The IGSO satellites have an orbital altitude of 36 000 km, are distributed evenly on three inclined synchronous orbits, and have an orbital inclination of 55°. The trajectories of the subsatellite points of the three IGSO satellites coincide with each other, having a longitude of 118° and a phase difference of 120°.

5.1.3 Galileo System

In December 2005, the first Galileo test satellite, GIOVE-A, entered its intended orbit and began transmitting signals to the ground in January 2006, marking the opening of the Galileo system. The Galileo system is a European autonomous, independent global multi- mode satellite positioning and navigation system, providing high-precision, high-reliability positioning services fully under non-military control and management. Galileo, which was initiated by the European Space Agency and European Union with major financial support, will not only free Europe from dependence on the US and Russia for traffic management and telemetry facilities, but also bring economic benefits to European equipment manufacturing and application services, while creating many new jobs.

The Galileo system is compatible with the GPS of the US and the GLONASS of Russia. In the future, users can use a receiver to collect data of an individual system or a combination of systems to meet positioning and navigation requirements. The Galileo system can distribute positioning information in real-time with meter-level accuracy, which is not possible for existing satellite navigation systems. The Galileo system also guarantees that services are provided in many cases requiring certain safety, such as the running of trains, the use of navigational vehicles, and aircraft control, and that users are notified within seconds of failure of the system (Benedicto et al. 2000; Paziewski and Wielgosz 2015; Prasad and Ruggieri 2005).

In March 2013, the European Space Agency first used satellites in orbit for ground positioning. In September 2015, the ninth and tenth Galileo Full Operational Capability satellites were launched on board a Soyuz rocket. The completed Galileo system will comprise 30 navigation satellites, of which 27 will be working satellites and three will be backup satellites. The satellites will have an altitude of 23 616 km and be located in three orbital planes with an inclination of 56°.

Galileo will be the world's first global satellite navigation and positioning system for civilian purposes. When the system becomes operational, users around the world will use multi-format receivers to obtain more navigation and positioning satellite signals, which will greatly improve the precision of navigation and positioning behind the scenes. This is the convenience brought by the Galileo project. The Galileo project is an autonomous, independent, global, multi-modal satellite positioning and navigation system for Europe, providing high-precision, high-reliability positioning services, having complete non-military control and management, and providing navigation and positioning functions with global coverage.

5.1.4 GLONASS System

GLONASS is a satellite navigation system that has been constructed by the Soviet Union and Russia since the early 1980s. The system comprises a satellite constellation, ground monitoring and control stations, and user equipment. The GLONASS satellite constellation comprises 24 satellites, evenly distributed over three nearly circular orbital planes, each having eight satellites, orbiting at an altitude of 19 100 km with an orbital period of 11 h and 15 min and an orbital inclination of 64.8°.

In contrast with the GPS of the US, the GLONASS system distinguishes different satellites by the carrier frequency applying frequency division multiple access (the GPS is based on code division multiple access, and the satellite is distinguished according to the modulation code). Each GLONASS satellite broadcasts carriers on two frequencies, namely $L_1 = (1\,602 + 0.562\,5k)$ MHz and $L_2 = (1\,246 + 0.437\,5k)$ MHz, where $k = 1 - 24$ is the frequency serial number of each satellite. Two types of pseudorandom noise code, S code and P code, are modulated on the carrier of a GLONASS satellite. Russia has adopted an open policy for the GLONASS system, which is not encrypted but used by both the military and civilians.

GLONASS satellites were launched into orbit three at a time by Proton launch vehicles. Each satellite has a three-axis stabilized system, an overall mass of 1 400 kg, and a designed orbital life of 5 a. All GLONASS satellites use precision clocks as their frequency reference. The first GLONASS satellite was launched on October 12, 1982. The GLONASS system is mainly used for navigation and positioning and, like the GPS, can also be widely used for various levels and types of surveying applications, geographical information system applications, and time–frequency applications (Geng and Shi 2017; Cai and Gao 2013; Shi et al. 2013; Al-Shaery et al.2013; Wanninger 2012).

GLONASS has three parts: a satellite constellation, ground support system, and user equipment. The GLONASS constellation comprises 27 working satellites and three backup satellites. The twenty-seven satellites are evenly distributed over three near-circular orbital planes, which are 120° apart from each other, with each orbital plane having eight satellites. The satellites on the same plane are separated by 45° and orbit at 23 600 km. The ground support system comprises a system control center, central synchronizer, telemetry and telecontrol station (including a laser tracking station), and field navigation control equipment. The system control center and central synchronization processor are located in Moscow while the telemetry and telecontrol stations are located in St. Petersburg, Ternopil, Enichesk and Komsomolskaya. The user equipment, namely receivers, can receive navigation signals transmitted by the satellites and measure their pseudo-range and rate of pseudo-range change and simultaneously extract and process navigation messages from satellite signals. The receiver processor processes the data and calculates location, speed and time information for the user.

Although GLONASS satellites were successfully networked and put into operation in 1996, the system has not been maintained continuously for various reasons and the number of operational satellites reduced to six by the end of 2000. With the improvement of the economic situation in Russia, the Russian government formulated the "save GLONASS"satellite program and began to modernize the system. A satellite constellation of 24 GLONASS-M and GLONASS-K satellites was rebuilt in 2010. The launch of a new GLONASS-KM satellite in 2015 further improved the performance of the system by improving the ground control system and coordinate system such that they were aligned with the ITRF framework and by improving the stability of satellite clocks.

5.2 Marine Navigation and Positioning Technology

Navigation and positioning are the basis of marine investigation and surveying. The accuracy of navigation and positioning directly affects the quality and credibility of survey data, which further affects marine engineering and marine scientific research. Navigation and positioning play an important role in the investigation of oceanic resources, especially in the exploration and research of mining areas. A high level of research and development of underwater navigation and positioning technology guarantees the safe submergence of instruments and equipment for wider application in ocean surveys.

Navigation and positioning technology used in marine surveying can be categorized in terms of its use in the navigation and positioning of surface vessels, navigation and positioning of underwater vehicles and carriers, and integrated navigation using nautical charts.

5.2.1 Navigation and Positioning Technology for Surface Vessels

GPS is widely used for the navigation and positioning of surface vessels. GPS receivers receive signals transmitted by navigation satellites and thus obtain information of the geodetic coordinates, elevation and time of the current position for the purposes of positioning, navigation, and measuring elevation.

Satellite navigation and positioning technology are widely used in marine investigations, marine surveying, marine scientific research, marine engineering, marine development and military operations. The technology is popular for its excellent characteristics, such as high precision, fast and convenient operation, and all-weather use.

5.2.1.1 Real-Time Kinematic Technology/Post Processed Kinematic Technology

Real-time kinematic (RTK) technology is a real-time dynamic relative positioning technology based on carrier-phase observations (Fig. 5.1). In RTK operation, two or more GNSS receivers receive satellite signals simultaneously. One receiver is placed on a point with known coordinates as a reference station and the other acts as a mobile station. The reference station transmits its observation and

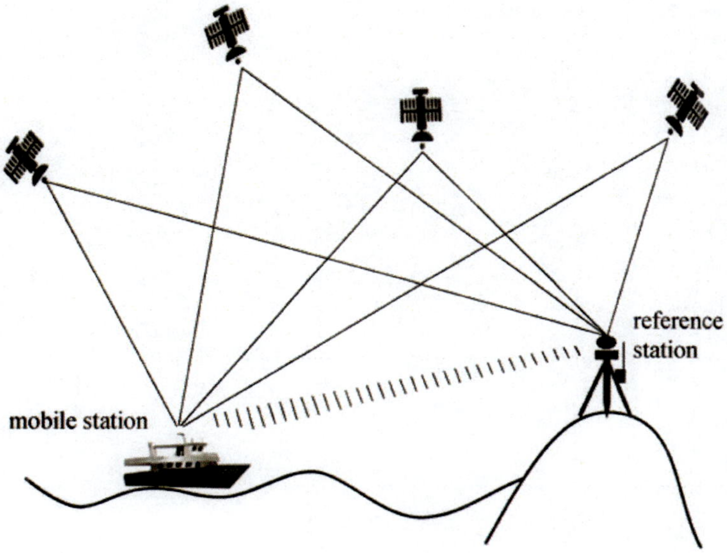

Fig. 5.1 Schematic diagram of RTK technology

coordinate information to the mobile station through the data chain. The mobile station provides a differential observation in the system for real-time processing, and a positioning result is obtained.

RTK technology has been applied rapidly because of its high speed, high precision, no limitation of sight condition, no accumulation of error among measurement results, and centimeter-level positioning results given by the mobile station at any time in the measurement process. However, RTK technology has some shortcomings. With an increase in the distance between the mobile station and reference station, the spatial correlation of various errors decreases rapidly, increasing the observation time. Additionally, the integer cycle ambiguity cannot be fixed and only a floating-point solution is obtained. Therefore, the distance between the mobile station and coordinates of the mobile station are not reliable because they are determined on the basis of only one reference station.

Post processed kinematic (PPK) technology is like RTK technology but uses carrier-phase observations for post-processing. Compared with RTK technology, PPK technology achieves centimeter-level positioning accuracy and users do not need a communication chain. However, PPK technology does not provide real-time positioning results.

Marine geodesy mainly focuses on the accuracy of positioning. The post-processing method not only ensures high positioning accuracy but also reduces the communication limitations of differential signal real-time transmission in the case of long-range positioning applications.

5.2.1.2 Satellite-Based Augmentation System Technology

A satellite-based augmentation system (SBAS) can broadcast the ephemeris error, satellite clock error, ionospheric delay and other correction information to the user to improve the positioning accuracy of the original satellite navigation system, through a satellite navigation augmented signal transponder onboard a GEO satellite. There are presently many SBASs around the world, such as WAAS in the US, SDCM in Russia, EGNOS in Europe, MSAS in Japan and GAGAN in India. The positioning accuracy is generally 1–3 m.

An SBAS comprises GEO satellites, a monitoring station, an upload station and a main control station as shown in Fig. 5.2. The working principles of an SBAS are as follows. (1) The navigation satellites are monitored by widely distributed monitoring stations with known locations to obtain the pseudo-range, carrier-phase observations and other original positioning data, and the positioning data are sent to the master station. (2) Various positioning correction information of each satellite is calculated at the master station and sent to the GEO satellite through the upload station. (3) The GEO satellite broadcasts the correction information to a vast number of users, thereby improving the positioning accuracy.

The main features of an SBAS are as follows. (1) In error processing, the master control station separates the spatial correlation and calculates the ephemeris error, star clock error and atmospheric propagation delay error to improve the positioning accuracy. (2) Not only the correcting data but

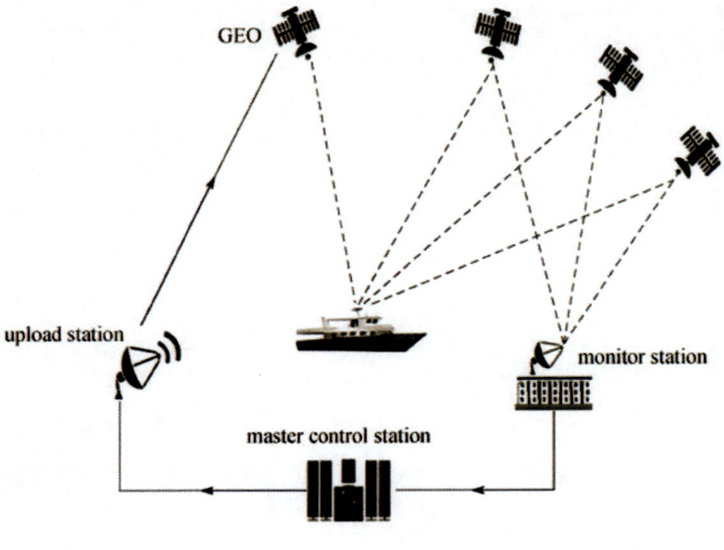

Fig. 5.2 Schematic diagram of an SBAS

also sophisticated information are sent by the master control station, which increases the number of satellites in the constellation and improves the availability and continuous service of the system. (3) It is not necessary for the user equipment to set up the radio frequency receiving part of the data chain separately, and it is only required to leave the receiver out of a receive channel to add a message extraction and processing program.

In the practice of precise positioning at a long distance on oceans, the technology of providing corrective services by satellite communication is essentially a high-precision and real-time positioning technology of the wide-area difference.

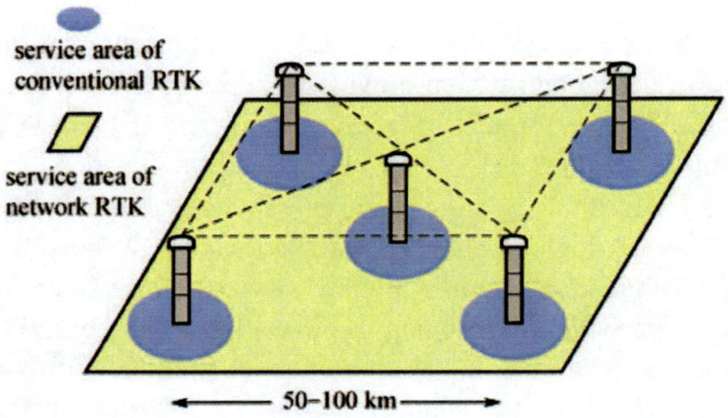

Fig. 5.3 Comparison of the operating range between network RTK and conventional RTK technologies

5.2.1.3 Network RTK Technology

Network RTK technology, also known as multi-base station network technology, is a new real-time dynamic positioning technology developed on the basis of conventional RTK technology, computer technology, and communication network technology. Network RTK technology has positioning accuracy of a centimeter level, and the distance between reference stations can reach 50–100 km (Fig. 5.3). Network RTK technology is thus superior to conventional RTK technology in terms of coverage, positioning accuracy, system reliability, and operation cost (Hu et al. 2003). Network RTK technologies such as the virtual reference station (VRS), Flachen Korrektur parameter (FKP), master-auxiliary concept (MAC), augmentation reference station (ARS) and combined bias interpolation (CBI) are relatively mature.

(1) VRS

The positioning principle of VRS is as follows. The processing center first receives the observation data of each reference station and the approximate coordinates of the mobile station in the reference station network in real time, and a virtual reference station is generated with approximate coordinates. Spatial distance-related errors, such as tropospheric and ionospheric delays at the virtual reference station, are then modeled to generate VRS virtual observations. Observation data in standard format or the corrections at the virtual reference station are finally sent to the mobile station to realize the real-time, high-precision positioning of the mobile station.

(2) FKP

The FKP is based on the state space model. First, the effects of the ionosphere and geometric signal errors in the network are calculated by the data processing center. The effects of errors are then described as north–south and east–west regional parameters and broadcast. Finally, error corrections are calculated by the mobile station according to these parameters and its own position.

Table 5.1 Comparison of various network RTK technologies

Content	VRS	FKP	MAC	ARS	CBI
R & D unit	Trimble	Geo++	Leica	Southwest Jiaotong University	Wuhan University
Calculation precision	High	High	Higher	High	High
Solution stability	High	High	Higher	High	High
Error modeling	Server side	User side	User side	Server side	Server side
Communication mode	Two-way	One-way	One-way and two-way	Two-way	One-way
Mathematical model	Double difference observation model, interpolation model	Un-difference observation model of the whole network, kalman filtering	Double difference observation model, compatibility of all models	Double difference observation model, kalman filtering	Double difference observation model, interpolation model
Reference station	A master reference station, all reference stations participated in the solution	No master reference station, select the nearest 3 reference stations	A master reference station (not necessarily the nearest)	Do not select master reference station	Flexible selection based on the relative position of mobile station and reference station

(3) MAC
The MAC, which is based on a multi-reference station, multi-satellite observation, multiple frequencies, and multi-signal processing algorithms, draws on the advantages of VRS format standardization and the FKP in data processing. Adopting the MAC, all relevant observation data, such as dispersive and non-dispersive difference corrections, representing the ambiguity of the whole cycle level and the whole unknown number level, are transmitted from the reference station network to the mobile station in a highly compressed form as the corrected data of the network.

(4) ARS
VRS technology has the shortcomings of a single baseline difference, not making full use of redundant information, transmitting original observation data, and there being a large amount of data. Meanwhile, MAC technology has the shortcomings of a single epoch needing to transmit correction information of multiple reference stations and a large amount of data. Aiming at overcoming those shortcomings, ARS technology combines the observation data of the reference station network and uses the multi-baseline solution method to conduct the difference positioning of the mobile station. Furthermore, the weighted average of the error correction of the reference station is regarded as the observation value of the ARS.

(5) CBI
The basic idea of CBI is that the ionospheric and tropospheric errors are calculated as a whole when calculating the correction information of the reference station, and the integrated error correction information of all the reference stations is processed in a unified manner, instead of all the correction information of each reference station being transmitted to the user.

Each network RTK algorithm has its advantages and disadvantages but achieves centimeter-level positioning accuracy. VRS and MAC technologies are presently the most popular. Table 5.1 comprehensively compares the solution accuracy and stability of different technical methods.

5.2.1.4 Precise Point Positioning

Traditional standard point positioning uses pseudo-range observations as well as satellite orbit and satellite clock error parameters provided by the broadcast ephemeris for navigation and positioning. Limited by the accuracy of the broadcast ephemeris and pseudo-range observation, standard point positioning has only the accuracy of a few meters to tens of meters. Precise point positioning (PPP) is a technological revolution for satellite positioning technology following the development of RTK technology and network RTK technology. PPP changes the scene such that only the differential positioning mode can be used to achieve higher accuracy (Li and Zhang 2009). Using PPP, a single receiver may achieve high-precision dynamic and static positioning. PPP has the advantages of efficiency and low cost and is suitable for use in various environments, and it also provides a new solution for large-scale control network data

Table 5.2 Product information of precise ephemeris and clock errors

GPS ephemeri	Orbit Accuracy (cm)	Clock error Accuracy (ns)	Dealy	Sampling interval (min)	Release authorities
Broadcast	100	5	real-time	–	CDDIS, SOPAC, IGN
Ultrafast (predicted)	5	3	real-time	15	CDDIS, IGS CB, SOPAC, IGN, KASI
Ultrafast (actual)	3	0.150	3–9 h	15	CDDIS, IGS CB, SOPAC, IGN, KASI
Fast	2.5	0.075	17–41 h	15 5	CDDIS, IGS CB, SOPAC, IGN, KASI
Ultimately	2.5	0.075	12–18 d	15 5	CDDIS, IGS CB, SOPAC, IGN, KASI

processing. PPP has become a research hotspot of GNSS technology.

PPP technology uses carrier-phase and pseudo-range observations collected by a single GNSS receiver to achieve highly precise positioning through the precision satellite orbit and clock products provided by IGS and other organizations (Table 5.2). The satellite positioning error includes orbit error, satellite clock error, ionospheric delay error, tropospheric delay error and the receiver clock offset. The global reference station network is established to obtain high-precision satellite orbit and clock error, and ionospheric delay is eliminated through ionosphere-free combination. The tropospheric delay and receiver clock offset are solved as unknown parameters together with station coordinates and satellite ambiguity parameters to obtain highly precise positioning results.

With the improved precision of precise ephemeris and clock error products and the optimization and perfection of various error correction models, post-processing PPP technology can achieve centimeter-level precision while real-time PPP technology can achieve decimeter- and even centimeter-level precision. However, rapid ambiguity estimation and other important issues needed to solve in real-time large-scale application.

5.2.2 Underwater Navigation and Positioning Technology

With sharp population growth and the depletion of conventional land resources, countries all over the world are paying more and more attention to the search for alternative and potential resources from the land to the oceans and placing greater emphasis on the investigation of oceans. Conventional investigation methods no longer meet research requirements. Deep-sea submersible vehicles, remotely operated vehicles, and autonomous underwater vehicles will therefore be applied in the fine exploration of deep-sea resources. How to ensure the safety of the cruising and operation of underwater vehicles as far as possible is a topic worthy of intense research. It is well known that electromagnetic waves attenuate quickly in water. Therefore, GPS navigation and positioning technology, which is widely used by surface vessels, is not suitable for the navigation and positioning of underwater vehicles.

The navigation and positioning of underwater vehicles can be carried out in a variety of ways, as described briefly below.

5.2.2.1 Direct Navigation

1. **Acoustic navigation**
   Acoustic navigation comprises baseline navigation and Doppler acoustic navigation and involves the transmission of acoustic signals between underwater vehicles and a mother vehicle. Baseline navigation includes the use of long baselines, short baselines, and ultra-short baselines as well as combinations of baselines. Several baselines can also be used for integrated navigation. A baseline navigation system comprises an acoustic emission transducer array and acoustic transponder array. On the basis of the acoustic Doppler effect, the velocity of a vehicle moving underwater relative to the seabed is measured in real-time through acoustic telemetry. Using the direction of motion of the vehicle, the position and trajectory of the vehicle are obtained by integral calculation. Doppler acoustic navigation has been applied to the Alvin submersible.
2. **Inertial navigation**
   On the basis of Newton mechanics, the acceleration and attitude of a vehicle are measured using a gyroscope and accelerometer, and the velocity is obtained by integrating the acceleration in the time domain, for the purpose of navigation. Inertial navigation has been widely applied in the military, such as for submarine navigation, spacecraft

navigation and weapon guidance. Inertial navigation is an autonomous form of navigation and has high accuracy over the short term, but the navigation error accumulates with time and needs to be corrected for by other means.

5.2.2.2 Aided Navigation

1. **Simultaneous localization and mapping**
   Simultaneous localization and mapping (SLAM), also known as concurrent mapping and localization, was first proposed by Smith, Self and Cheeseman, and has been applied in the navigation and positioning of a mobile robot in an unknown environment (Luo and Hong 2004). The mobile robot identifies the signatures in an unknown environment using sensors and then estimates the global coordinates according to the relative positions and odometer readings between the robot and signatures. SLAM is also applied by underwater vehicles but needs further exploration because it still faces many problems in the navigation of mobile land robots; e.g., ultra-multidimensionality, data association and the accumulation of errors.
2. **Aided navigation using submarine topography, gravity, or magnetic fields**
   Submarine topography, gravity, or magnetic field data of an exploration area are acquired to develop a high-resolution database. When a submersible is underwater, the real-time data acquired by the probe and the known databases are used for correlation analysis and abnormality matching. The navigation positions of underwater vehicles are therefore corrected in time for the purpose of aided navigation.

The following is an example of underwater aided navigation and positioning technology based on multi-beam bathymetry.

Accurate navigation guarantees the safety of submarines and underwater vehicles traveling over long distances. The inertial navigation system has become the core navigation equipment of the submarine but it needs to correct periodically with external information. The system is usually calibrated by astronomical, radio and satellite navigation information. However, in the case of underwater autonomous navigation vehicles, such as unwired deep-sea submersible vehicles, GPS and radio signals cannot be used for calibration, while aided navigation can solve the problem effectively.

(1) Basic principle of submarine vehicle aided navigation based on submarine topography
Topographic data of the relevant area is first read from a fine submarine topography database stored in a computer according to the position information of the submarine provided by the inertial navigation system. Then, together with topographic information measured by the multi-beam sonar bathymeter, data are sent to the data processing computer for terrain matching to obtain the best matching point. The position information of the matching point is used to correct the inertial navigation system and effectively improve the positioning accuracy. It is therefore important for submarine vehicle aided navigation based on submarine topography to index and display mass data and select a topography matching algorithm quickly. The real-time topography (i.e., real-time chart) obtained by real-time sensors is compared with reference topography data (i.e., reference chart) stored in the computer to obtain the updated position of the submersible.

(2) B-tree indexing algorithm
The multi-beam bathymetric system differs from the traditional single-beam bathymetric system in that it adopts a strip-type measurement that fully covers the seabed. However, it also generates a large amount of survey data. There are hundreds of gigabytes of original multi-beam data of only the continental shelves, and the data quantity becomes even more astonishing if original multi-beam data obtained by subsequent projects are included. If only used for cartographic purposes, there is generally no need for so much raw data on a small or medium scale, the gridded data are enough. However, the original multi-beam data are needed for submarine aided navigation. The matching accuracy is improved using the raw data because fine topography is matched in real-time. It is therefore important to extract the matching data from the original multi-beam database quickly in real-time. The basic idea is to divide the multi-beam data of the whole sea area regularly and to build an index layer based on the minimum boundary, in which the quadtree, also known as the B-Tree, is built. The smallest node of the quadtree corresponds to the smallest partitioning frame. The query speed of the target frame is greatly accelerated by indexing using the quadtree.

(3) Wavelet-based fast matching algorithm
Submarine terrain matching mainly comprises a feature space, similarity measurement and search strategy. When matching, the similarity criterion is used to measure the similarity between the input image and reference image. Researchers have studied many similarity criteria, including the mean absolute difference correlation, normalized cross-correlation number, statistical correlation, matched filter, local correlation, root-mean-square error, and mask correlation. The multi-beam bathymetric system obtains the submarine topography within a certain range of the seabed in real-time and the coefficient of the normalized cross

correlation of mean removal is therefore regarded as the similarity measure; i.e., the region with the largest correlation coefficient is considered as the matching point.

As a time–frequency localization method, wavelet analysis is an extension of the Fourier transform. A basic wavelet function that satisfies certain conditions is first found to construct a family of wavelet functions through translation and expansion. The family is used to approximate the signal being studied, for ease of analysis and processing. The wavelet transform, which projects signals into a series of nested subspaces, is a powerful tool for multiresolution analysis. The two-dimensional biorthogonal wavelet transform is equivalent to the one-dimensional biorthogonal wavelet transform in the row direction and column direction. Therefore, in the actual transformation, the topographic map is decomposed in one dimension along the row direction using low-pass and high-pass filters and then decomposed in the column direction. The wavelet transform divides the terrain elevation data into sub-band images according to different bandwidths and resolutions. Each layer of the wavelet is thus decomposed into four sub-bands, namely LL with low frequency in vertical and horizontal directions, LH with low frequency in the horizontal direction and high frequency in the vertical direction, HL with low frequency in the vertical direction and high frequency in the horizontal direction, and HH with high frequency in vertical and horizontal directions. LL contains the average information of the image while LH, HL and HH contain the direction information due to the change in terrain elevation. The following algorithm mainly uses information in the sub-band LL. The steps of the stepwise iterative precise matching algorithm are as follows.

(1) The reference image and real-time image are decomposed using the two-layer wavelet transform.
(2) In the low-frequency sub-band image LL, the normalized coefficients of correlation between the two images are calculated point by point.
(3) The correlation coefficients are sorted, the coordinates where the correlation coefficient is a maximum are found, and the coordinates of the point in the original reference image after wavelet inverse transform are calculated.
(4) The correlation coefficients are calculated point by point in the five neighboring cells of the position coordinates in the obtained reference image. The best matching point is where the correlation coefficient is maximal. The coordinate information output of this point is regarded as the adjusting parameter with which to correct the inertial navigation positioning error due to time accumulation.

5.2.3 Navigation Technology Based on Electronic Charts

5.2.3.1 Overview of an Electronic Chart

An electronic chart display and information system (ECDIS), also known as an electronic chart system, is an electronic system that combines all navigation information, such as chart information, navigation and positioning information, ship motion parameters, swath bathymetry, and radar information. Relevant technology emerged in the late 1970s and developed rapidly from the 1980s owing to promotion of the International Hydrographic Organization (IHO) and International Maritime Organization (IMO). The IHO and IMO began cooperating in July 1986, with the IHO setting up a special committee to study ECDIS. In January 1987, the IHO/IMO Coordination Working Group was established under the chairmanship of Dr. Zickwoll G. In June of the same year, Norway and Denmark led a North Sea engineering experiment to develop methods of producing the regional Electronic Chart Database and the modalities for revising electronic charts. In November, COE experts met at The Hague to discuss and draft ECDIS performance standards and technical specifications. In October 1988, the IHO passed technical specifications. Following the success of the test of electronic charts in the North Sea, the Nornews Express, a 4 500-t cruise ship, was used in a practical test (SEATRANS project).

The ECDIS Committee of the IHO has formed six groups to formulate standards, including general standards, revised standards, terminology standards, color and symbol standards, database standards, and quality standards. Configuration requirements for an ECDIS, including a central processing unit, two storage settings and three displays (Fig. 5.4), are given in the general standards.

5.2.3.2 Creating a Special Electronic Navigation Map

An ECDIS is not a mandatory norm and its standards are constantly being revised and perfected. Many countries, including China, are exploring the possibility of the practical application of an ECDIS in navigation. However, so far, an ECDIS is only an auxiliary navigation means for navigation and does not completely replace the traditional paper chart. Figure 5.4 shows that even an ECDIS based on the general standards has a complex composition. Therefore, if an ECDIS is established in strict accordance with relevant international norms, it will require extremely complex engineering and the participation of state authorities. It is thus difficult to apply an ECDIS in marine investigations in the short term, but we can use the basic idea of electronic charts to provide services for marine investigation and establish an electronic map system that meets the needs of marine investigation at this stage.

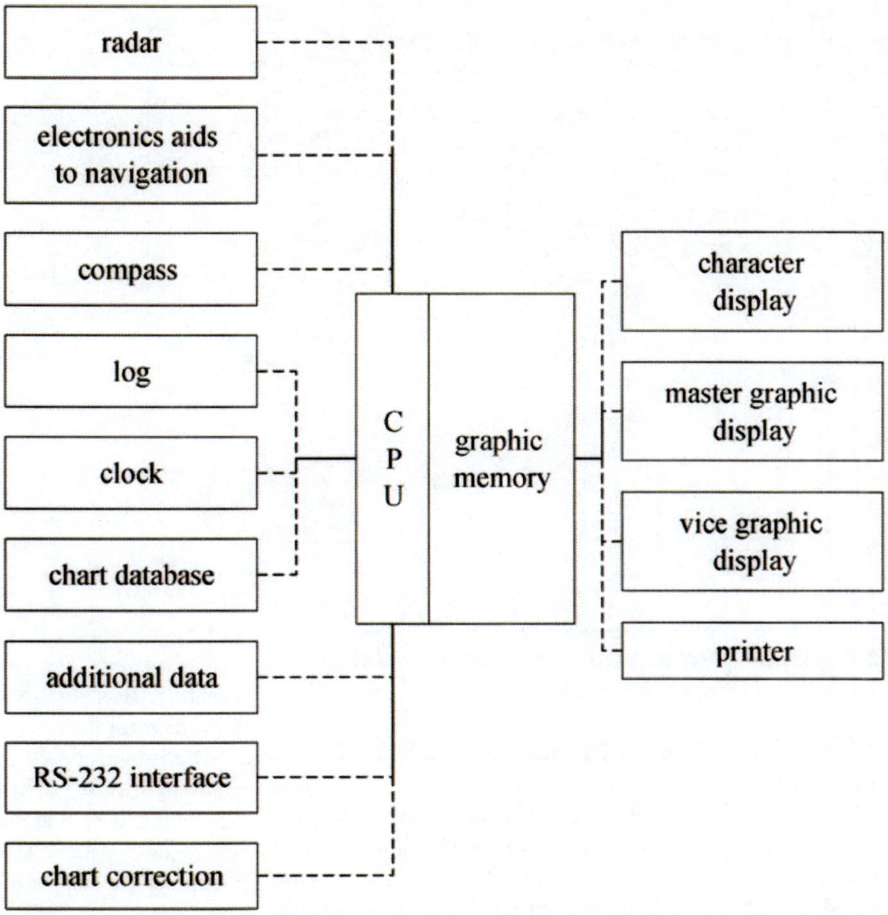

Fig. 5.4 Basic components of an ECDIS

On the basis of the developed mapping system, an electronic map for the navigation and positioning of marine investigation can be constructed; such a map is obviously different from general graphics files. The navigation background to display the terrain and environment elements of the survey area should be only an auxiliary function of the electronic map. The main purpose of the map is to provide reference information for real-time investigation, such as displaying the real-time depth of the mother ship and underwater vehicle to ensure safe navigation, geological information of the operation area for chief decision-making, and information for the design of surveying areas and lines.

Thereby, the electronic map should include a navigation layer, a deep water layer and a professional layer. An index layer (Fig. 5.5) should also be provided to display information of the current operation quickly. The flexibility and scalability of the layered structure make it possible for more practical information to be added to the electronic map in time.

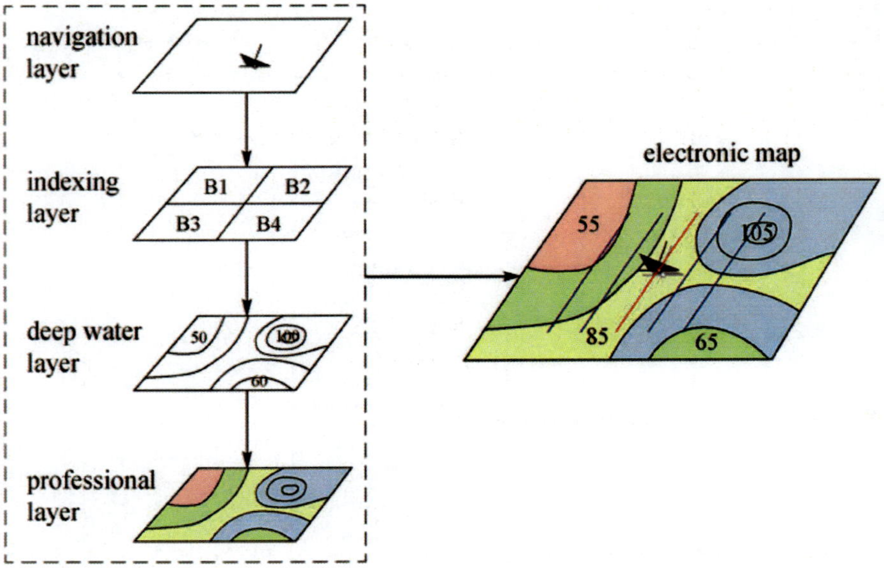

Fig. 5.5 Composition of a special electronic navigation map

5.3 Data Format and Correction Model for Error in a GNSS

5.3.1 RINEX Format Parsing for a GNSS

NMEA-0183, developed by the National Marine Electronics Association, is the most widely used navigation and positioning data format and a protocol for GNSS receivers. Most common GPS receivers, GPS data processing software, and navigation software comply with or are compatible with the protocol. There are many statements defined in the NMEA-0183 protocol, but only $GPGA, $GPGSA, $GPGSV, $GPRMC, $GPVTG and $GPGLL are commonly used. Currently, on the basis of the difference in positioning modes, the output statements of domestic multimode receivers add prefixes, such as $BD and $GN, to the BeiDou positioning system.

The navigation and positioning data obtained from the NMEA-0183 file by differential or single-point positioning only meet the basic requirements of marine navigation and positioning. The original GNSS observation data need to be processed when carrying out new research using GNSS observation data; e.g., acquisition of the centimeter-level tide, highly precise sea waves and marine water vapor from GNSS data.

In GPS data processing, the observation data are obtained from GPS receivers used for field observation. When a receiver observes in the field, the observation data are usually recorded in the internal memory of the receiver or a removable storage medium. After observation, data must be transferred to a computer for processing and analysis. It is therefore necessary to use a unified format for GPS data processing software when processing data from different GPS receivers. The emergence of the receiver independent exchange format (RINEX) solves the problem of the unified GPS data format.

RINEX, proposed by Werner Gurtner at the College of Astronomy, University of Berne, Switzerland in 1989, is a standard data format independent of the receiver. The format stores data with a text file, which is independent of the manufacturer and the model of the receiver. To meet various requirements, RINEX has undergone several changes in format, with RINEX 2 being the most widely applied version. RINEX 3 emerged with the Galileo program in Europe, the modernization of the GPS in the US and GLONASS in Russia, and the use of SBASs and the BeiDou Navigation Satellite System (Gurtner and Estey 2007).

Similar to the observation file, the navigation data file comprises a header portion and data portion. Navigation data, also called ephemeris data, have two types of ephemeris format, namely the GPS broadcast ephemeris and precision ephemeris. The broadcast ephemeris of GPS satellites is a type of forecasting ephemeris that is determined and provided by the ground control part of the GPS and broadcast to all users worldwide by GPS satellites.

The precision of the broadcast ephemeris is artificially reduced to 50–100 m when implementing SA policy.[1] After the abolition of SA policy, the mean square error of the three-dimensional points of the satellite given by the broadcast ephemeris is approximately 5–7 m.

The broadcast ephemeris was initially tracked and observed by five monitoring stations distributed around the

[1]SA Policy: Selective Availability, Choice of Availability, US Policy to Limit GPS Positioning Accuracy.

world, and the observation data were then sent to the master control station. The P-code observations collected at the master control station were used to estimate satellite parameters, such as the position, velocity, solar stress-optical coefficient, clock correction, clock drift and drift velocity based on the Kalman filter. The parameters were then used to estimate the satellite position and clock correction at the following time. At the same time, the corresponding orbit parameters were obtained by fitting these results. Navigation messages were finally broadcast.

Since GPS satellites were first put into operation, the orbital accuracy of the broadcast ephemeris has improved, which is attributed to the launching of new satellites (e.g., Block IIR and Block IIR-M) with better performance and a series of measures taken by relevant agencies; e.g., reducing the data age of navigation messages, strengthening the management of satellite clocks, improving the positioning accuracy of the ground tracking station, and improving the Kalman filter method. The precision ephemeris of GPS satellites is a high-precision ephemeris developed to satisfy the needs of precision application fields, such as geodesy and geodynamics, with precision of 3–5 cm. To meet the needs of high-precision real-time positioning, the International GNSS Service (IGS) began providing precise ephemeris forecasts, which are the most accurate, most widely used, and most convenient precision ephemerides. These ephemerides are freely available on the Internet. The following is a description of the RINEX format of the GPS navigation message.

In the data records of the GPS navigation message file with RINEX format, the clock and orbit data of each satellite are stored according to satellite and reference times. Each satellite has eight rows of data at one reference time. Row 1 gives the pseudo-random code of the satellite, the reference point of the satellite clock, and the corrected model parameters while Rows 2–8 give the broadcast orbit data of the satellite. Navigation messages are usually updated every 2 h and some satellites may therefore have data for multiple different reference times.

5.3.2 Main Error in a GNSS and Model Correction

In GPS single-point positioning, the pseudo-range and carrier-phase observations of the non-differential are used instead of the differential, which can weaken or eliminate the effects of various observation errors. All error terms must therefore be corrected. In general, there are two ways to deal with these errors. The first is to correct the errors through accurate modeling, such as the accurate correction of the satellite antenna phase center deviation and its change, the effect of tidal loads, and relativistic effects using existing models. The second is to estimate the errors and parameters that cannot be accurately modeled or to eliminate them by combining observations; e.g., the tropospheric zenith wet delay is difficult to model but can be accounted for by adding parameters. This section focuses on the first error correction method. There are two parts to the correction of the antenna phase center, one being the correction of the antenna phase center deviation (relative to the antenna reference point) and the other being the correction of the antenna phase center change (relative to the average antenna phase center).

The deviation of the satellite antenna phase center is the deviation between the satellite center of mass and the antenna phase center of the satellite transmission signal. The orbital force model used in satellite orbit determination is based on the centroid of the satellite; i.e., the satellite position calculated using the precise ephemeris provided by the IGS and other organizations is the position of the satellite center of mass. The GPS signal is transmitted from the satellite antenna phase center, and the distance (phase or pseudo-range) between the satellite antenna phase center and the ground GPS receiving antenna can be measured. It is therefore necessary to focus on the deviation correction between the GPS satellite mass center and the satellite antenna phase center in PPP. The deviation of the satellite antenna phase center differs for GPS satellites of different series.

GPS satellites transmit right circularly polarized radio waves and the observed carrier phase is thus related to the relative orientation of the satellite and receiver antennas. The wind-up of phase is such that if the receiver or satellite antenna spins around its central axis (vertical axis) once, the carrier phase observation changes for up to one cycle (one wavelength). The satellite antenna rotates one cycle in less than half an hour. During the rotation, it is necessary to correct or eliminate the relevant phase data (Bar-Sever 1996; Kouba 2003).

The solid tide refers to the periodic deformation of the solid Earth under the action of the tidal forces of the Sun and Moon (Kouba 2009).

Different from the effects of solid Earth tides and ocean loads, polar tides are not equated to zero approximate by averaging over 24 h. The maximum polar tide displacement can reach 25 mm in elevation and 7 mm in the plane direction.

Similar to solid earth tides, ocean loads have diurnal and semi-diurnal cycles, but they are generated by the load of the lower crust in response to ocean tides (Kouba 2009), and the displacement caused by ocean tides is one order of magnitude less than that caused by solid tides. Ocean loads are so regional that they do not have a permanent component, empirically. The effect has to be considered for single epoch positioning with precision of 5 cm or 24-h static positioning precision.

5.4 GNSS PPP Data Processing Method

The emergence of PPP provides a new solution for large-scale and high-precision dynamic positioning. Non-differential PPP technology replaces the traditional differential dynamic positioning technology, drastically reduces the reliance of large-scale and long-distance measurements on ground reference stations, improves work efficiency, and reduces users' costs. In recent years, PPP technology, which is booming and has broad application prospects, has become one of the main development directions in the field of satellite navigation and positioning technology.

The concept of PPP was proposed for Doppler PPP in the age of the Meridian satellite in the US in the 1970s. After the development of the GPS satellite positioning system, because of the low precision of C/A or P code, the use of the original phase observation was explored for PPP in the mid-1980s, which is called non-differential point positioning. However, there are too many parameters to simultaneously estimate in the positioning estimation modeling; e.g., satellite clock error, receiver clock error, tropospheric delay of each epoch, phase ambiguity parameters of satellites, and three-dimensional coordinates of stations. It is impossible to propose a solution where the estimation equation is rank deficient, and the difficulty of the problem meant that the method was temporarily shelved in the 1980s. In the mid-1990s, the IGS began providing globally precise ephemeris and precise satellite clock error products and three types of precise ephemeris with different accuracy levels, namely afterwards, fast and predicted ephemerides, and corresponding precise satellite clock error products with 15- and 5-min intervals. This provided a new solution for non-differential phase PPP. In 1997, Zumberge of the US Jet Propulsion Laboratory (JPL) proposed a method of using the IGS precise ephemeris and data from the GPS tracking network to obtain a satellite clock error in the 5-s interval adopting GIPSY software. In the point positioning equation, only the tropospheric parameters, receiver clock error, and three-dimensional coordinates of the station are estimated. Results show that the precision of 24-h continuous static positioning is 1–2 cm while the precision of single-epoch dynamic positioning is 2.3–3.5 dm. The feasibility of PPP using non-differential phase observations was demonstrated (Zumberge et al. 1997). Heroux et al. of NRCan studied the non-differential PPP method and achieved centimeter-level accuracy in processing long-time static observation data (Kouba and Héroux 2001). The German Research Centre for Geosciences and Geodetic Survey of Canada developed corresponding PPP software systems, which achieved the same accuracy of precision static and dynamic positioning.

In the following years, the mathematical model, error correction, and related algorithms of PPP were studied in depth (Kouba and Héroux 2001; Gao et al. 2001; Han and Ou 2003; Zhang et al. 2005). BERNESE, which is GPS data processing software, introduced the function of PPP processing with non-differential phase observations in Version 4.2 (Hugenobler et al. 2001) and improved the function in Version 5.0. P3 software was developed by the University of Calgary in Canada. TriP software and PANDA software were developed by Wuhan University. Several companies released PPP data processing software around 2007. The main examples were the addition of a PPP algorithm module on the basis of original differential positioning in GrafNav 7.8; the Canadian company Applanix developing POSPac AIR software, which has PPP processing capability; the Swiss company Leica developing its own PPP data processing software IPAS PPP; and the Norwegian company TerraTec developing the dynamic positioning software TerraPOS based on the PPP mode.

PPP technology has been applied in marine mapping. High-precision GPS data processing software applications, such as GIPSY of JPL in the US, EPOS in Germany, and Bernese in Switzerland, have PPP functions. The tropospheric delay data of global IGS stations provided daily by JPL are calculated using PPP technology. Chen et al. (2004) of the Hong Kong Polytechnic University applied PPP technology to GPS buoy positioning to monitor sea level changes. Luo et al. (2015) combined PPP technology with EGM2008 to acquire tide level data.

5.5 Extended Application of a GNSS in Oceanography

The field of marine science investigates natural phenomena, properties, and changing regularities as well as topics related to the exploitation and use of the oceans. The field examines the oceans, which cover 71% of the Earth's surface, including the seawater, substances dissolved and suspended in seawater, organisms in the oceans, bottom deposits, and the submarine lithosphere, as well as the atmospheric boundary layer and estuarine coastal zones. Marine science includes basic research on physical, chemical, biological, and geological processes of the ocean, the exploitation and use of marine resources, and applied research on maritime military activities.

GNSS is widely used on land while some research has been carried out on the oceans. In recent years, with the development of GNSS and the expansion of marine scientific research, studies have begun investigating how to better apply GNSS in oceanographic research and expand the application of GNSS. Including the author's research on

GNSS oceanography in recent years, this section discusses problems affecting the determination and calculation of the GNSS high-precision tidal level, how to determine the current velocity and wave amplitude of buoys using a non-differential GNSS, how to measure the ocean atmosphere and ocean geomagnetic disturbance using an onboard GNSS receiver, and other ocean research topics.

The application of a GNSS is an important method of submarine topographic surveying for underwater stereotactic positioning and non-tide correction. This section also introduces related extended techniques.

5.5.1 Oceanographic Research and Application of a GNSS

5.5.1.1 GNSS Underwater Stereotactic Positioning System

A GNSS realizes the global positioning and navigation of open land, sea and outer space. Radio waves do not propagate through seawater and thus cannot be used directly for positioning in broad lakes and underwater spaces. However, the acoustic wave has good propagation characteristics in water. The integration of GNSS technology and sonar technology expands the application of GNSS technology under water and makes positioning possible in underwater spaces as on land. The underwater GNSS positioning system is based on the variable-surface long baseline of differential GNSS technology. The positioning signals of underwater targets are detected and the travel times from a target to buoys are measured in real-time to realize the positioning of the system through the long baseline network formed by several GNSS intelligent buoys on the water. The system regards the surface buoy as the GNSS satellite, the underwater acoustic signal as the GNSS radio signal, and the underwater target as the GNSS receiver, which moves arbitrarily in the water. Each part of the system is connected wirelessly through an acoustic signal underwater and a radio signal on the water. The system is called an underwater GNSS system. As early as 1989, it was proposed to move the underwater long-baseline system to the surface through the radio chain to establish a floating long-baseline system. However, the GPS was in its construction stage, and the development of GPS technology has not since been able to solve the high-precision real-time positioning of the water surface baseline Dana (1995). studied the surface floating long-baseline system and proposed a technology in 1991 whereby the surface buoy transmits the positioning signal and measures the delay of the signal, the underwater transponder receives the signal from the surface buoy and responds, and the distance from the buoy to the target is then calculated. A patent of the technology was applied for. Thomas (1994) proposed another underwater acoustic measurement method, with the underwater targets transmitting position signals and buoys on the surface receiving the underwater signals. The technology is supported by the French Defense Department for its high flexibility. The first prototype of the GPS intelligent buoy was developed by ACSA in January 1995, and a department called ORCA was established to develop the GPS intelligent buoy and a naval operation scheme suitable for different conditions. A sea trial was successfully performed and test results were published at an ION conference in 1995. In July 1996, ORCA began producing GPS intelligent buoys with a positioning accuracy of 1–10 m, buoy time resolution of 0.5×10^{-5} s, and measurement accuracy of 1×10^{-4} s (https://www.underwater-gps.com/).

5.5.1.2 GNSS Ocean Remote Sensing

GNSS reflectometry (GNSS-R), also known as passive reflectometry and interferometry (PAR1S), was first proposed by Martin-Neira (1993) for altimetry. Subsequently, GNSS-R has received much attention and its applications have been widened and its methodology has been thoroughly researched. The GNSS was originally applied for satellite navigation and positioning (i.e., the GPS), where reflected signals are rejected as harmful multipath signals in conventional measurements. Auber et al. (1994) reported that GPS reflected signals can be received and detected. Subsequently, American scientists have been keenly aware that the GNSS reflection signals may become a new means of microwave remote sensing. The applications of GNSS-R in ocean remote sensing (Larson et al. 2008) include the inversion of the significant wave height of the surface wind, tidal level and seawater salinity (Sabia et al. 2007). Other applications are the monitoring of oil spills (Valencia et al. 2011) and the water levels of lakes and reservoirs.

5.5.1.3 GNSS Tidal Observations

In offshore submarine topographic surveying, both single-beam bathymetry and multi-beam bathymetry need to transmit height data through tide data. Traditional tidal methods include the use of water gauges and automatic tidal gauge observations. Tidal observations using pressure automatic tide gauges are common. The traditional single-station tidal gauge method has effects on a small scale and many inconveniences, such as the specific selection of the tidal gauge location and special care and maintenance for the instrument. The multi-station tide gauge method has the same disadvantages but the complexity of the post-processing of tide gauge data. In recent decades, with the development and popularization of GNSS positioning technology, many organizations have explored how to measure the change in the tidal level with high precision using a GNSS. Results show the feasibility of replacing the traditional tide gauge with the GNSS tide gauge (Zhao 2007; Zhao et al. 2008; Ouyang et al. 2005).

With the development of GPS technology, GPS-RTK or PPP technology is being used at the tidal level. The DGPS tidal level measurements can be classified as static and dynamic methods. The static method is a tidal level measure method in which the receiving antennas of the GPS tide stations are placed on buoys or survey ships near the shore or at fixed points at sea and the dynamic carrier-phase difference measurement is made with the GPS receivers on the shore to obtain the instantaneous sea surface height of GPS tide stations. The dynamic method places a GPS receiving antenna on a ship and measures the dynamic carrier-phase difference with the shore GPS receiver to obtain the instantaneous sea surface height of the ship.

A PPP tidal level measurement involves placing a single dual-frequency GPS receiving antenna on a ship and solving the instantaneous sea surface height of the measurement ship using the precise ephemeris and precise clock error.

5.5.1.4 GNSS Wave Measurement

Waves, as an important movement phenomenon in the ocean, strongly affect human activities offshore and near shore. It is necessary to monitor changes in waves and to study their movement law in real-time. In recent decades, the growth of human activity on the oceans has greatly promoted the research of ocean observation methods and the development of instruments and equipment. Various wave measuring instruments and methods have been developed (Shi et al. 2000). The GNSS signal has a large coverage range and can be used in all weather, with the functions of precision positioning and velocity measurement (He et al. 2003; Xiao et al. 2000; Hebert et al. 1997), and it can thus be used to monitor wave motion in real-time. The three-dimensional instantaneous velocity of the carrier can be measured in real-time using the GNSS Doppler measurement of the ship or buoy to inverse the real-time wave height, cycle, and wave direction at the measurement point. At present, visual and optical methods do not meet the requirements of severe-weather and night observation. Gravity, acoustic and water pressure wave measuring equipment is expensive and does obtain all physical ocean parameters, such as the position, wave height cycle and wave direction. A GNSS has many advantages, such as all-weather operation, high accuracy, simplicity and low cost. The above shortcomings are overcome by inversing the basic information of accurate GNSS measurement.

5.5.1.5 GNSS Water Vapor Measurement

Water vapor is a matrix of transmission between water and heat, directly affects the vertical stability of the atmosphere, and plays an important role in weather changes. The content of water vapor is fickle and plays an important role in the transfer of atmospheric energy and the evolution of the weather system, and the phase change of water vapor is directly related to rainfall. The content of water vapor in the atmosphere is a necessary parameter for forecasting rainfall intensity at the mesoscale or local scale. The change in water vapor with time and space in the atmosphere is important in meteorological forecasting, especially in the monitoring and forecasting of meso- and micro-scale disastrous weather with a horizontal scale of about 100 km and a life history of only a few hours; e.g., rainstorms, hail, thunderstorms, gales and tornados. Water vapor is one of the most difficult meteorological parameters to describe because of its spatiotemporal variability. In addition, water vapor is one of the main gases of the greenhouse effect.

There are many limitations to water vapor detection; e.g., conventional balloon detection has low temporal and spatial resolution and an increasing cost, the water vapor microwave radiometer is expensive, spaceborne radiometers and satellite infrared radiometers are limited because of the existence of clouds, and lidar is expensive and difficult to operate and cannot function in all weather. The poor observation accuracy of water vapor and the insufficient temporal and spatial resolution have been major obstacles to improving the accuracy of weather forecasts.

GPS meteorology has been developed over the past decade via the atmospheric delay noise of GPS signals. GPS detection has many advantages over other methods, such as good timeliness, high temporal resolution, easy maintenance and renewal, low cost, all-weather observations and no effect of aerosols or clouds. GPS meteorology is becoming a hot spot of GPS research. With the rapid development and application of the GPS, more attention is being paid to the application of GPS water vapor data in atmospheric research and weather forecasting (Baker 1998; Bevis et al. 1991). The GPS is mainly used to solve the water vapor content on land owing to its high coordinate precision. However, it does not reflect the actual characteristics of water vapor accurately. Therefore, the topic of how to solve the GPS water vapor in the ocean has attracted much attention. Dodson et al. (Dodson and Baker 1998; Rocken et al. 1995) showed that the accuracy of the water vapor content obtained using the mobile GPS is equivalent to that of the static solution. In addition, the development of GPS PPP technology, especially dynamic PPP technology, provides positioning accuracy at the centimeter level, presenting the opportunity to solve the content of ocean water vapor (Zhang and Andersen 2006; Kouba and Heroux 2001; Zumberge et al. 1997).

5.5.1.6 GNSS and Ocean Geomagnetic Disturbance

The ionosphere is an important part of the solar-terrestrial space environment and lies in the range of 60–2 000 km above the ground. The ionosphere is one of the most important parts of solar-terrestrial space and directly affects human life. The Earth's ionosphere is not only an important part of the Earth's atmosphere but also has the most complex structure. Various physical and chemical changes in the ionosphere are closely related to solar radiation, particulate radiation, magnetospheric disturbances, geomagnetic field changes, and upper atmosphere motion (Yuan and Ou 1999). Ionospheric research has leaped forward with the development of ionospheric GPS technology. The study of the ionosphere using dual-frequency GPS receivers has many advantages over the use of other satellite detection techniques. Results show that the GPS can be used not only to detect the occurrence of ionospheric disturbances, magnetic storms, and other anomalies but also to identify their types, spatial and temporal scales, and propagation characteristics parameters through the design of effective methods or combining with other traditional ground observation networks, such as vertical altimeter networks and high-frequency Doppler networks (Yuan and Ou 1999).

The geomagnetic field is a complex dynamical system that originates from the current system outside the Earth and the induced current within the Earth (Hu et al. 2005; Danilov 2001). Marine magnetic surveying has been widely applied in marine survey units located in China. Technicians have designed a series of processing methods and corrective measures to eliminate the effects of errors on magnetic survey data. In the process of geomagnetic total field measurement, the short-term change in the geomagnetic field is most important and can reach 40–60 nT (Yao et al. 2002). Therefore, how to determine a magnetic disturbance or storm using a GNSS is a new direction of developing a GNSS for marine research.

5.5.2 Further Development of Oceanographic Research and Application of a GNSS

5.5.2.1 GNSS Real-Time PPP

In current research on marine applications, the real-time navigation and positioning is mainly based on the compensated GNSS differential information provided by a global differential reference station network with an accuracy of meters. However, the Global Differential Information Network is expensive and its accuracy does not meet the requirements of specific high-precision marine engineering, such as those of the high- precision GNSS underwater stereo positioning system and high-precision real-time bathymetric data acquisition. There is thus an urgent need to develop a real-time PPP technology suitable for marine applications.

5.5.2.2 GNSS Real-Time Water Vapor Acquisition

GNSS detection has the advantages of good timeliness, high temporal resolution, easy maintenance and renewal, low cost, all-weather observation, and no effect of aerosols or clouds. At the same time, the GPS water vapor in the ocean is not affected by various factors and it thus accurately reflects the actual characteristics of water vapor. In addition, ocean factors strongly affect the terrestrial climate, and the real-time acquisition of ocean water vapor will better guarantee a leap forward in weather forecasting. It is therefore crucial work in the next step to obtain GNSS water vapor in real-time.

5.5.2.3 Refinement of the GNSS Magnetic Disturbance Model

Magnetic disturbances relating to particle flow radiation are mainly caused by solar activity. A short-term change in the geomagnetic field is due to solar activity and geomagnetic activity on the Earth. The total electron content of the ionosphere obtained by a multi-frequency GNSS responds well to the generation of magnetic disturbances.

It is therefore necessary in the marine application of GNSS to solve how to combine ionospheric monitoring based on a GNSS with the correction of disturbance characteristics using the geomagnetic diurnal effectively, explore mechanisms, determine the magnitude, start time, and end time of a geomagnetic disturbance in a geomagnetic survey, and then obtain a highly precise geomagnetic diurnal correction model.

5.5.2.4 GNSS Carrier High-Precision (High-Frequency) Velocity Measurement

The GNSS has a large coverage range, is able to operate in all weather, and provides a precise velocity measurement through the GPS phase measurement of a ship or buoy. The GNSS can measure the three-dimensional instantaneous velocity of the carrier in real-time and obtain various characteristics of waves. However, to obtain the characteristics of waves with sampling intervals ranging from 0.25 to 4 s, it is necessary to develop a high-precision, high-frequency marine GNSS real-time processing technique based on phase to better inverse real-time ocean parameters, such as the wave height, cycle and wave direction at the measurement points.

5.5.2.5 Further Application of Fitting by Selection

Compared with observations and applications of a GNSS on land, the observation and application of a GNSS on the sea differ in that the change in elevation is slight between adjacent or multiple epochs. Meanwhile, the change in tide level is regular, which provides a basis for the application and research of fitting by selection.

References

Al-Shaery A, Zhang S, Rizos C (2013) An enhanced calibration method of GLONASS inter-channel bias for GNSS RTK. GPS Solut 17 (2):165–173. https://doi.org/10.1007/s10291-012-0269-5

Auber JC, Bibaut A, Rigal JM (1994) Characterization of multipath on land and sea at GPS frequencies. Paper presented at the 7th international technical meeting of the satellite division of the institute of navigation (ION GPS 1994), Salt Lake City, Utah, 20–23 September 1994

Benedicto J, Dinwiddy S, Gatti G et al (2000) GALILEO: Satellite system design and technology developments. European Space Agency, Paris

Baker HC (1998) GPS water vapour estimation for meteorological applications. Dissertation, University of Nottingham. https://doi.org/10.1121/1.411266

Bar-Sever Y (1996) A new model for GPS yaw attitude. J Geodesy 70 (11):714–723. https://doi.org/10.1007/BF00867149

Bevis M, Businger S, Herring TA et al (1991) GPS meteorology: remote sensing of atmospheric water vapor using the global positioning system. J Geophys Res 97(D14):15787–15801. https://doi.org/10.1029/92JD01517

Cai CS, Gao Y (2013) Modeling and assessment of combined GPS/GLONASS precise point positioning. GPS Solut 17(2):223–236. https://doi.org/10.1007/s10291-012-0273-9

Chen W, Hu C, Li Z et al (2004) Kinematic GPS precise point positioning for sea level monitoring with GPS buoy. J Glob Position Syst 3(1–2):302–307. https://doi.org/10.5081/jgps.3.1.302

Dana PH (1995) An overview of the global positioning system (GPS). Available via DIALOG. https://www.utexas.edu/depts/grg/gcraft/notes/gps/gps.html. Accessed 10 Nov 2020

Dana PH (1997) Global Positioning System (GPS) time dissemination for real-time applications. Real-Time Syst 12(1):9–40. https://doi.org/10.1023/A:1007906014916

Danilov AD (2001) F2-region response to geomagnetic disturbances. J Atmos Sol-Terr Phys 63(5):441–449. https://doi.org/10.1016/s1364-6826(00)00175-9

Dodson AH, Baker HC (1998) The accuracy of GPS water vapour estimation. Paper presented at the national technical meeting of the institute of navigation, Long Beach, California, 21–23 January 1998

Gao Y, Shen XB (2001) Improving ambiguity convergence in carrier phase-based precise point positioning. Paper presented at the proceedings of ION GPS-2001, Salt Lake City, Utah, 11–14 September 1978

Gabbay R, Konotov TL (2019) Method, system and software for navigation in global positioning system (gps)- denied environments. U.S. Patent US20190234740A1, 1 Aug 2019

Geng J, Shi C (2017) Rapid initialization of real-time PPP by resolving undifferenced GPS and GLONASS ambiguities simultaneously. J Geodesy 91(4):361–374. https://doi.org/10.1007/s00190-016-0969-7

Gurtner W, Estey L (2007) RINEX-The receiver independent exchange format-version 3.0. ftp://igs.org/pub/data/format/rinex301.pdf. Accessed 10 Nov 2020

Han BM, Ou JK (2003) Precise point positioning based on undifferenced GPS data (in Chinese with English abstracts). Geomatics and Information Science of Wuhan University 28(4):409–412

Han CH, Yang YX, Cai ZW (2011) BeiDou navigation satellite system and its time scales. Metrologia 48(4):S213–S218

He HB, Yang YX, Sun ZM et al (2003) Mathematic model and error analyses for velocity determination using gps doppler measurements (in Chinese with English abstracts). J Inst Surv Mapp 20(2):9–82

Hebert CJ, Keith J, Ryan S et al (1997) DGPS kinematic carrier phase signal simulation analysis for precise aircraft velocity determination. Paper presented at the 53rd annual meeting future of navigation: facing the challenges, Albuquerque, Kansas, 30 June-2 July 1997

Hofmann-Wellenhof B, Lichtenegger H, Collins J (2012) Global positioning system: theory and practice. Springer Science & Business Media, Berlin

Hu GR, Khoo HS, Goh PC et al (2003) Development and assessment of GPS virtual reference stations for RTK positioning. J Geodesy 77:292–302. https://doi.org/10.1007/s00190-003-0327-4

Hu HB, Gong SG, Lin CS (2005) Detection of faint magnetic signal in geomagnetic disturbance (in Chinese with Engish abstracts). J Detect Control 27(5):41–44

Hugentobler U, Schaer S, Springer T et al (2001) CODE IGS analysis center technical report 2000. In: IGS central bureau (ed) IGS-2000 technical reports. Jet propulsion laboratory publications, Pasadena, pp 73–82

Kouba J (2003) Testing of the IERS2000 sub-daily earth rotation parameter model. Stud Geophys Geod 47(4):725–739. https://doi.org/10.1023/A:1026338601516

Kouba J (2009) A guide to using international GNSS service (IGS) products. Geodetic Survey Division, Natural Resources Canada. Available via DIALOG. https://igscb.jpl.nasa.gov/. Accessed 10 Nov 2020

Kouba J, Héroux P (2001) Precise point positioning using IGS orbit and clock products. GPS Solut 5(2):12–28. https://doi.org/10.1007/PL00012883

Landau H, Vollath U, Chen X et al (2002) Virtual reference station systems. J Glob Position Syst 01(02):137–143. https://doi.org/10.5081/jgps.1.2.137

Larson KM, Small EE, Gutmann ED et al (2008) Use of GPS receivers as a soil moisture network for water cycle studies. Geophys Res Lett 35(24):L24405. https://doi.org/10.1029/2008GL036013

Li ZH, Zhang XH (2009) New techniques and precise data processing methods of satellite navigation and positioning (in Chinese). Wuhan University Press, Wuhan

Li M, Qu LZ, Zhao QL et al (2014) Precise point positioning with the BeiDou navigation satellite system. Sensors 14(1):927–943. https://doi.org/10.3390/s140100927

Luo RH, Hong BR (2004) Simultaneous localization and mapping based on multisensor fusion (in Chinese with English abstracts). J Harbin Inst Technol 36(5):566–569

Luo XW, Gao JY, Jin XL et al (2015) The latest application to determine tide height in a large-scale sea using shipborne GPS based on precise point positioning and EGM 2008. J Coast Res 73:319–324. https://doi.org/10.2112/SI73-056.1

Martin-Neira M (1993) A passive reflectometry and interferometry system (PARIS)-application to ocean altimetry. ESA J 17(4):331–355

Ouyang YZ, Lu XP, Huang M et al (2005) The application of GPS altimetry in hydrographic surveying without tidal observation (in Chinese with English abstracts). Hydrogr Surv Charting 25(1):6–13

Paziewski J, Wielgosz P (2015) Accounting for Galileo-GPS inter-system biases in precise satellite positioning. J Geodesy 89 (1):81–93. https://doi.org/10.1007/s00190-014-0763-3

Prasad R, Ruggieri M (2005) Applied satellite navigation using GPS, GALILEO, and augmentation systems. Artech House, Boston

Rocken C, Hove TV, Johnson J et al (1995) GPS/STORM-GPS sensing of atmospheric water vapour for meteorology. J Atmos Oceanogr Technol 12:468–478. https://doi.org/10.1175/1520-0426(1995)012%3c0468:GSOAWV%3e2.0.CO;2

Sabia R, Caparrini M, Camps A (2007) Potential synergetic use of GNSS-R signals to improve the sea-state correction in the sea surface salinity estimation: application to the SMOS mission. IEEE Trans Geosci Remote Sens 45(7):2088–2097. https://doi.org/10.1109/TGRS.2007.898257

Shi MC, Gao GP, Bao XW (2000) Marine surveying method (in Chinese). Ocean University of China Press, Qingdao, pp 145–147

Shi C, Yi WT, Song WW et al (2013) GLONASS pseudorange inter-channel biases and their effects on combined GPS/GLONASS precise point positioning. GPS Solut 17(4):439–451. https://doi.org/10.1007/s10291-013-0332-x

Tang WM, Deng CL, Shi C et al (2014) Triple-frequency carrier ambiguity resolution for Beidou navigation satellite system. GPS Solut 18(3):335–344. https://doi.org/10.1007/s10291-013-0333-9

Thomas HG (1994) New advanced underwater navigation techniques based on surface relay buoys. Paper presented at the proceedings of OCEANS'94:oceans engineering for today's technology and tomorrow's preservation, Brest, France, 13–16 September 1994

Valencia E, Camps A, Park H et al (2011) Oil slicks detection using GNSS-R. Paper presented at the IEEE international geoscience and remote sensing symposium, Vancouver, British Columbia, Canada, 24–29 July 2011

Wanninger L (2012) Carrier-phase inter-frequency biases of GLONASS receivers. J Geodesy 86(2):139–148. https://doi.org/10.1007/s00190-011-0502-y

Xiao Y, Sun ZM, Cheng GY (2000) Precise determination of velocity for airborne gravimetry using the GPS doppler observations (in Chinese with Engish abstracts). J Wuhan Tech Univ Surv Mapp 25 (2):113–117

Yao JJ, Sun Y, Zhao HJ (2002) Theoretical analysis of geomagnetic diurnal variation observation data (in Chinese with Engish abstracts). Hydrogr Surv Charting 22(6):8–10

Yang YX, Li JL, Wang AB et al (2014) Preliminary assessment of the navigation and positioning performance of BeiDou regional navigation satellite system. Sci China Earth Sci 57(1):144–152. https://doi.org/10.1007/s11430-013-4769-0

Yuan YB, Ou JK (1999) The effects of instrumental bias in GPS observations on determining ionospheric delays and the methods of its calibration (in Chinese with Engish abstracts). Acta Geod et Cartogr Sin 28(2):110–114. https://doi.org/10.3321/j.issn:1001-1595.1999.02.001

Zhang XH, Andersen OB (2006) Surface ice flow velocity and tide retrieval of the amery ice shelf using precise point positioning. J Geodesy 80(4):171–176. https://doi.org/10.1007/s00190-006-0062-8

Zhang XH, Liu JN, Rene F (2005) Application of dynamic point positioning with sub-decimeter accuracy in aerial surveying (in Chinese with English abstracts). Paper presented at the 8th Annual Meeting of China Association of Global Positioning System Technology Applications, Beijing, 1–3 October 2005

Zhao JH (2007) Modern marine surveying and charting (in Chinese). Wuhan University Press, Wuhan

Zhao JH, Wang SP, Zhang HM et al (2008) Long-distance and on-the-fly GPS tidal level measurement based on GPS PPK/PPP (in Chinese with English abstracts). J Inst Surv Mapp 33(9):34–37

Zumberge JF, Heflin MB, Jefferson DC et al (1997) Precise point positioning for the efficient and robust analysis of GPS data from large networks. J Geophys Res 102(B3):5005–5017. https://doi.org/10.1029/96JB03860

6 Tide Level Measurements and Vertical Datum Conversion

The tide level refers to the distance between the free sea level measured at a fixed point and the datum level in the ocean. The change in tide level is a comprehensive result from various factors, including the periodic change caused by the tide-generating forces of celestial bodies and non-periodic changes caused by wind, air pressure, land runoff and other factors. Through analysis and calculation, tide level data can provide tidal correction parameters for depth measurements and original data for calculations of average sea surface, theoretical sea level datum, tidal harmonic constants and other parameters. These results can be applied to transportation, engineering, military and other industries, as well as basic marine research, such as submarine topography and geomorphology. For example, in submarine topography, it is usually necessary to calibrate the tides to the datum level to eliminate the influence of tidal changes (Baramonov et al. 1983; Shi et al. 2016).

6.1 Technical Methods of Tide Level Measurements

The tidal range is a significant environmental parameter in coastal areas of China. A comprehensive understanding of tidal ranges allowed ancient Chinese coastal people to anticipate tidal variations. Spectacular ocean tides were observed in August of the lunar calendar and then recorded in an ancient book (*Qi fa*, written by Mei Cheng during the Western Han Dynasty of the second century BC) (Fang et al. 1986).

With improvements in science and technology, tide level measurements have developed from the early analog of water gauges and float-type tidal gauges to digital tide level measurements. At present, tidal measurement technology can be divided into two types, conventional tide station and remote sensing, which uses GPS and real-time kinematic (RTK) phase difference technology (Chen 1980; Ruan 2001).

6.1.1 Conventional Tide Level Measurement Techniques

6.1.1.1 Tide Level Measurements Using Water Gauges

A water gauge is used to measure the tide level and the observed values are usually read manually (Fig. 6.1). Water gauges can be made of corrosion resistant, easy-to-clean enamel, polymer or stainless steel plates, and their scale is generally up to 1 cm. They are usually designed as one of three types: vertical, inclined or hanging hammer. In an area with a large tidal range, multiple short piles can also be arranged to observe the tides in relay mode. The equipment for water gauge-assisted tide level measurements is easy to operate, but the accuracy of observed data is often affected by waves. Additionally, in long-term observations, the data cannot be processed automatically. At present, it is only used for tide level calibration with various automated tide gauges and temporary observations.

6.1.1.2 Float-Type Tide Gauge

A float-type tide gauge usually consists of a buoy, balance hammer, steel wire, rope wheel and recording device (Fig. 6.2). It uses a buoy that rises and falls with the water surface, and rotates a steel wire that drives the rope wheel; the rope wheel drives a recording pen to draw lines on paper with a uniform rotation rate to achieve automatic tide level recording. With technological developments, the recording method changed from traditional paper and ink to digital recording and storage. The float-type tide gauge is usually equipped with wave-absorbing tide gauge wells. A tidal gauge well is generally divided into island-type and shore-type wells that are mostly round with an inner diameter of 0.7–1.0 m. There is water intake at the bottom of the shaft, which follows tidal changes and eliminates the influence of waves.

The float-type tide station suffers from the following disadvantages: large construction investment for the tide

Z. Wu et al., *High-resolution Seafloor Survey and Applications*,
https://doi.org/10.1007/978-981-15-9750-3_6

Fig. 6.1 Tide level observations using a water gauge (NOAA 2007)

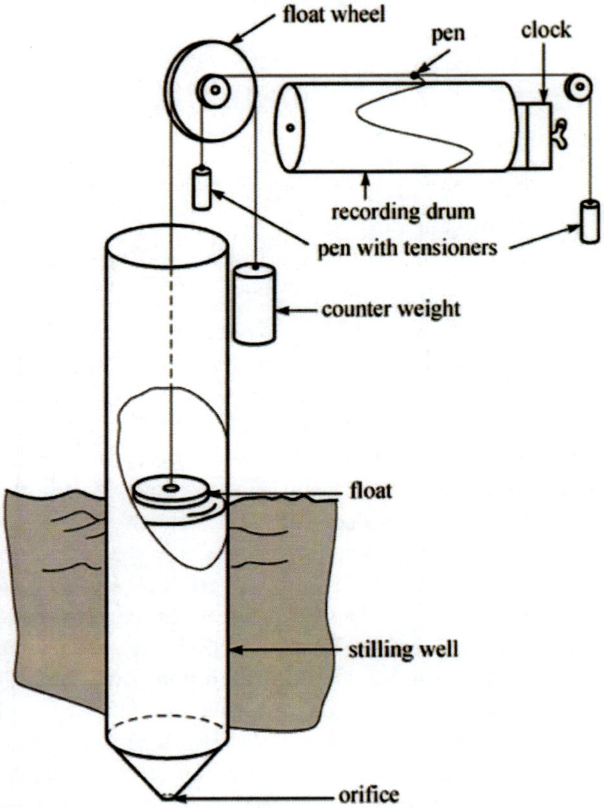

Fig. 6.2 Schematic of a float-type tide gauge (IOC 2006)

well and a heavy subsequent maintenance load. The design and installation of tidal gauge wells have an important influence on tide measurements. The suction effect of waves and the difference between internal and external well velocity and density can cause water level differences between the inside and outside of the well. When the wave damping hole is polluted and blocked by marine organisms, it causes observation errors of tide time and tide level. After long-term use in harsh sea conditions, the mechanical rotation mechanism of a float-type tide gauge may be subjected to wear, resulting in the corrosion of bearings and other transmission devices, and causing measurement errors (Beckers et al. 1981; Zhu and Huo 1990; Kang and Zhang 1993).

6.1.1.3 Pressure Tide Gauge

In the 21st century, tide gauges based on underwater pressure, acoustic and radar principles gradually replaced the float-type tide gauge (IOC 2006). Common underwater pressure gauges include a bubble gauge or a pressure probe gauge. In a bubble gauge, the air source continuously transmits gas underwater through the airway at a constant speed; when the gas fills up the underwater measuring cup, the excess gas flows out intermittently in the form of small bubbles. At this time, the pressure at the mouth of the cup and the current static pressure of seawater are in a state of dynamic equilibrium, which can be converted into water level by measuring the pressure of the gas and the density of the water (Fig. 6.3). This type of tide gauge is more commonly used in the US and the UK. It has many advantages, such as less marine biological pollution, low power consumption, high precision, good stability and strong adaptability, which is especially suitable for applications in cold and frozen environments.

The pressure probe gauge uses an underwater sensor and converts the pressure into a water level through the density of seawater and the acceleration of gravity:

$$H = (P - P_a)/(\rho_w g) \tag{6.1}$$

where H is the water depth of the sensor, P is the pressure recorded by the sensor, P_a is the atmospheric pressure at the

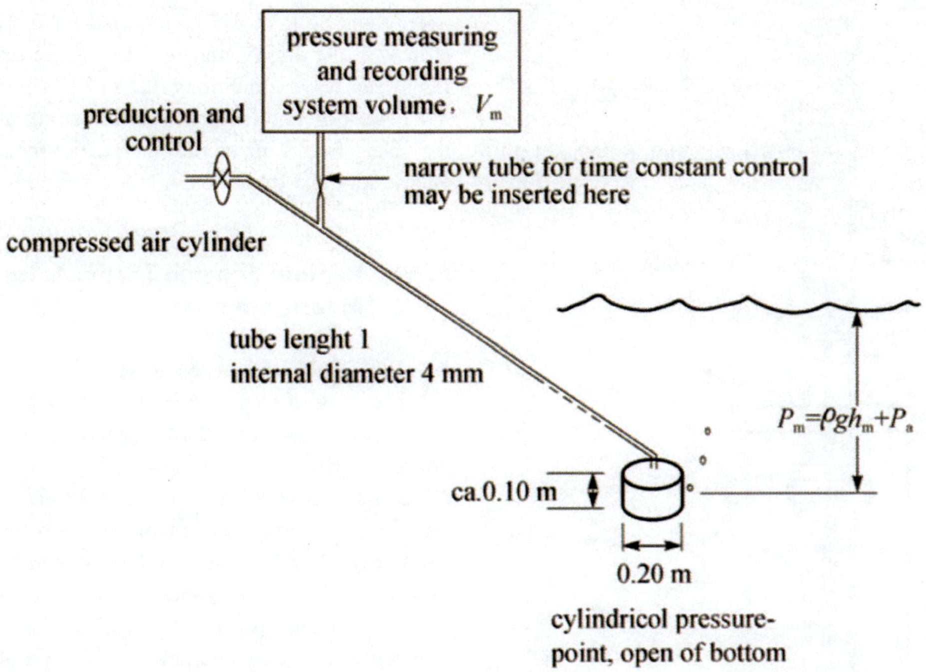

Fig. 6.3 Schematic of a bubble tide gauge (IOC 2006)

sea surface, ρ_w is the average density of the seawater in the water column above the sensor and g is the local gravity acceleration.

Because the pressure recorded by the underwater sensor is equal to the sum of atmospheric and water level pressures, it is necessary to observe atmospheric pressure synchronously to correct the water level. The pressure probe gauge has a convenient arrangement and good environmental adaptability, and is widely used in short-term tide level measurements.

6.1.1.4 Acoustic Tide Gauge

The principle of acoustic tide gauges is to emit sound wave pulses through a sound source, calculate the elapsed time of sound wave traveling from source to water surface and back to source, and then obtain the distance from the sea surface by the sound velocity:

$$H_c = H_d - C \times \Delta T/2 \quad (6.2)$$

where H_c is the tide height, H_d is the height of the probe at a datum level, C is the sound velocity and ΔT is the two-way travel time of the acoustic signal.

Sound speed is affected by temperature, air pressure and humidity, so it is necessary to measure relevant meteorological parameters synchronously for sound velocity corrections. If necessary, temperature sensors are installed at different heights for more accurate sound velocity corrections. An acoustic tide gauge uses tubes to allow sound to pass through and accumulate energy within the tube and avoid the influence of wind, sand and other factors in the air on the transmission of the sound wave. An acoustic tide gauge has advantages, such as non-contact measurements, no mechanical parts, no need for tidal wells and self-calibration of sound velocity with temperature changes. Moreover, the acoustic tube which is not affected with wave attenuation like tidal well is filtered out wave by taking the average value with repeated sampling when installing the acoustic tube, it is suggested to choose the direction with uniform illumination to avoid sunlight on the upper part and shadow on the lower part. If this rule is not observed, the temperature gradient might be too large to guarantee reliable measurements. The acoustic duct vents should be as close as possible to the high tide line installation and the upper exhaust holes should not be blocked. Figure 6.4 shows an acoustic water level gauge, which is composed of an acoustic probe, an acoustic pipe, an external protection well and an instrument host system. The system is structurally simple and capable of filtering out disturbance caused by gravitational wave energy. It can accurately measure the tide changes at sea level and is suitable for coastal hydrological observation (Fan 1997).

6.1.1.5 Radar Tide Gauge

Radar tide gauge (Fig. 6.5) is one of the most widely used tide gauges in the world (Woodworth et al. 2015). There are two types of radar tide gauge. One is similar to an acoustic tide gauge in principle while it calculates the time for a radar pulse to transmit from source to the sea surface and back to a receiver to obtain the distance. The other calculates the phase difference between the emitted and received pulses to obtain

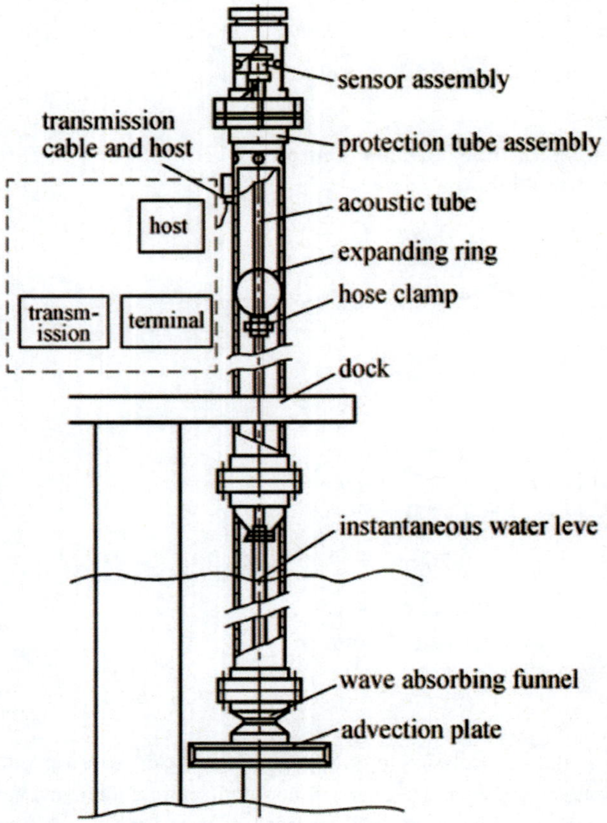

Fig. 6.4 Schematic of an acoustic tide gauge

Fig. 6.5 Radar tide gauge

the sea surface distance. The advantages of a radar tide gauge include simple maintenance, small instrument drift and low power consumption. It can be powered by batteries or solar energy and no tidal well is required for installation and use. However, the disadvantage of this method is that under different sea wave conditions, the radar will cause deviations due to the different footprints on the sea surface. Especially in the case of cold air or hurricanes, the measurement differences radar and traditional methods could be large.

6.1.2 Remote Sensing Methods for Tide Level Measurements

6.1.2.1 GNSS Observations

The accuracy of sounding is affected by a variety of factors, among which the vertical changes caused by draught, fluctuation, water level and attitude are often intertwined, and it is difficult to separate them completely. The employment of traditional sounding modes will inevitably bring certain errors and ultimately affect the sounding accuracy. With the development of Global Navigation Satellite System (GNSS) technology, dynamic positioning reached the centimeter level, indicating a mature technology. The variations in ground height obtained by shipborne GNSS antenna directly reflects the dynamic change in vertical synthesis of the transducer. GNSS high-precision three-dimensional positioning results provide a technical means for correcting the vertical dynamic effect of sounding. GPS-RTK bathymetric survey without tidal observation is a technique that can modify and deduce the geodetic height based on the results of high-precision dynamic GNSS altimetry to obtain the final water depth (Fig. 6.6).

The water depth is calculated by

$$C = (H - h) - S \tag{6.3}$$

$$B = S - (H - h) \tag{6.4}$$

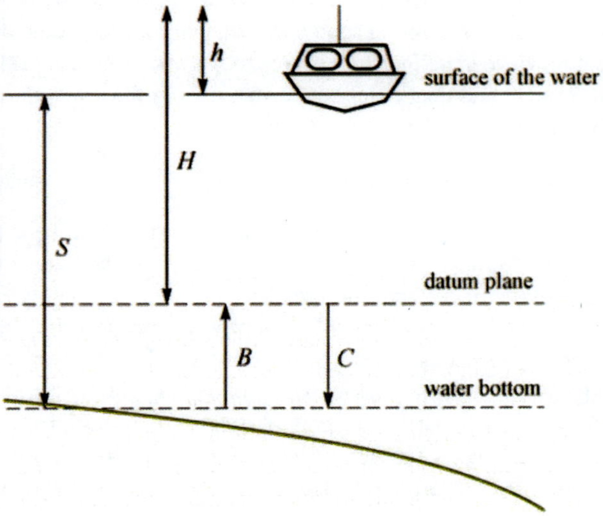

Fig. 6.6 Schematic for determining the final water depth

where the observational data (Fig. 6.6) are h: distance from GPS antenna to water surface; H: geodetic height measured by GPS receiver, which can be converted into the normal height; S: distance from the water surface to the bottom measured by a sounder.

Therefore, the data that need to be calculated are B: distance from water bottom to water level, i.e., water depth; C: distance from the water level to the bottom, i.e., water bottom elevation (Fig. 6.6).

It is worth noting that the data obtained by GPS are geodetic heights. Therefore, after data collection, certain means or models are needed to convert the geodetic height to the datum level of water depth or underwater terrain.

Bathymetric survey without tidal observations can be divided into differential and single point positioning, according to the GPS positioning method. In the application of differential positioning in a bathymetric survey, the method is selected according to the distance between the measuring area and the reference station. For example, the local differential positioning methods of RTK and post-phase difference (PPK) can be adopted for coastal sounding. The satellite difference method can be adopted for pelagic bathymetric surveys, but this method requires expensive service fees (Li et al. 2009).

Precise point positioning (PPP) technology has become a hot topic in global positioning research in recent years. Its principle is that single point positioning is carried out using the orbital and clock correction information of IGS satellites under the assistance of only one GPS receiver. Using this system, a positioning accuracy on the order of centimeters can be achieved (Liu and Ye 2002; Han and Ou 2003).

6.1.2.2 Tide Measurement with GPS

Measuring tides by differential GPS is a new method. High-precision differential GPS technology is used to obtain accurate three-dimensional position changes of a measuring point that is then converted, through coordinate system transformation, into points in the local coordinate system. This represents position changes in the zenith direction, namely the instantaneous elevation changes of the sea surface. It is worth noting that variations in sea level are affected by tides and waves, so wave effects should be filtered out. Normally, fast Fourier transform (FFT) is used to separate waves and tides. If a non-differential technology is adopted, tide measurements with GPS include not only the change in water level but also the movement of the Earth's crust so that the influence of the Earth's tide should be eliminated during usage.

6.2 Analysis and Forecast of Tide Level Data

Once Isaac Newton discovered the law of universal gravitation, he put forward the theory of equilibrium tides, which explained the Earth's tidal phenomena and laid the foundation for tidal theory research. Pierre-Simon Laplace explained tidal phenomena in the ocean from the viewpoint of hydrodynamics. He also put forward the basic equation of tidal dynamics and described tide-generating forces in the first expansion formula, which made a great contribution to the development of tidal analysis. However, tidal theory can only frame basic rules and characteristics of changes in ocean tidal phenomena; actual observations are still necessary to accurately understand the size and variation of tides in specific sea regions.

6.2.1 Harmonic Analysis and Prediction of Tide Level Data

6.2.1.1 Harmonic Analysis of Tide Levels

The tidal harmonic constant can be obtained by harmonic analysis of observed tidal data. Tidal harmonic analysis refers to the analysis and calculation of tidal harmonic constants from tidal statistics and measured tidal data. Tidal harmonic constants can be used to calculate tidal level data at any time.

The tidal observation curve is composed of many components. The tide level at a certain time is calculated as

$$\xi(t) = a_0 + \sum_{j=1}^{m} R_i \cos(\sigma_j t - \theta_j) + \gamma(t) \tag{6.5}$$

$$\text{or } \xi(t) = a_0 + \sum_{j=1}^{m} (a_j \cos \sigma_j t + b_j \sin \sigma_j t) + \gamma(t) \tag{6.6}$$

where a_0 is the average sea level, R_i is the constituent tide amplitude, θ_j represents the constituent of the initial phase, σ_j is the constituent of the angular velocity, $\gamma(t)$ is a non-tide induced remainder, and the positive integer m is the number of constituent tides. $\gamma(t)$ includes irregular perturbations caused by meteorological factors, errors in observation, data and truncation errors and ignored tidal components. It also has random characteristics, which is equivalent to "noise" in physics.

In the analysis of measured tide levels, only a limited number of main tidal components can be selected to obtain a good result. The factors that affect the calculation results are the selection of tidal constituents, the length of the observation period and the time interval between two observation records. To obtain stable tidal harmonic parameters, 355 or 369 days are usually selected for data analysis. In this way, the frequency aliasing between the major constituent tides can be eliminated and almost all of the short-period constituent tides can be separated.

6.2.1.2 Tide Level Forecasting

Prediction of tide levels is the inverse operation of tidal analysis. The harmonic constants of the constituent tides are obtained by analyzing annual tidal data, and then the time-by-time tidal level prediction is carried out according to

$$\xi(t) = a_0 + \sum_{j=1}^{m} f_j h_j \cos\left[\sigma_j t + (V_0 + u) - g_i\right] \tag{6.7}$$

where m is the number of tidal constituents, f is the nodal factor for predicting daily tidal constituents, $(V_0 + u)$ is the initial phase angle of the predicted daily tidal constituent in Greenwich Mean Time (GMT) and a_0 is the height of the annual or perennial average sea surface, h_j g_i are tidal constituent amplitude and phase respectively (Huang and Huang 2005).

6.2.2 Numerical Models for Calculating Tide Levels

For areas lacking tidal observation data, it is generally impossible to predict the tides accurately. Even within the same basin, the number of observation stations is usually limited. To understand the distribution and variation in tide levels from point and surface over a large area, numerical models can be used with basic equations of tidal dynamics (Fang et al. 1986).

Tidal motion is generally described by partial differential equations. In a numerical model, these equations are first discretized and then solved on the computer. Common methods include the finite difference method, finite element method and boundary element method.

6.2.2.1 Finite Difference Method

The finite difference method replaces the differential in the original equation with the difference, which then constitutes the difference equation. The finite difference equation and its solution converge to the differential equation and its solution. For the difference equation, stability conditions must be satisfied and the physical effects caused by numerical calculations must be limited. In the early days, finite difference grids were mostly rectangular, so their use in complex boundary conditions was limited. In recent years, grid generation technology has been used to refinement of the grid where needed. Moreover, grid generation technology adopts adaptive grids, especially unstructured grids developed in recent years, which is closer to the shape of the computing domain and has great advantages.

6.2.2.2 Finite Element Method

The finite element method for solving differential equations divides the computational domain into several units, applies the variational principle to each small unit and forms linear algebraic equations from all the computational areas to solve for each unit. In this method, the size and shape of the element can be determined according to the calculation need. Where the gradient of the variable is large, the computing unit can be smaller and denser; where the gradient is small, the cell may be larger and less dense. Finally, the coefficient matrix of the formed linear algebraic equations is sparse, which can save computing time after processing (Wu and Han 1988).

6.2.2.3 Boundary Element Method

The boundary element method is developed from the finite basic solution method or boundary integral equation method. The boundary of the computational domain is divided into units. The boundary of a two-dimensional problem is a curve and the elements are a series of straight lines; the boundary of a three-dimensional problem is a surface and the elements are a series of planes (triangles or quadrangles) with an unknown number to be solved existing on the unit. For the whole domain boundary integral, it can be reduced to the sum of the element integrals. Finally, linear algebraic equations including each unit are formed to determine the unknown number of each unit (Brebbia and Walker 1980).

6.3 Water Level Corrections for Bathymetric Surveys

The water level can be divided into the astronomical tide level (main part) and residual water level according to the disturbance mechanism. As the main part of the water level, the astronomical tide is obtained by the tidal harmonic constants, while the mean sea level is regarded as a special constituent tidal with an angular velocity of zero. The residual water level is a short-term anomaly caused by weather and other factors, which shows a strong spatial correlation. The essence of water level corrections is to remove the time-varying influence of sea surface tides from the instantaneous sounding data and transform the instantaneous depth into a time-independent steady depth field (Liu 2003). In practice, different vertical datum levels are selected according to requirements. For example, the submarine topographic map usually uses National Vertical Datum 1985 and the charts adopt the theoretical lowest tide level in

China. Because it is impossible to observe the tide at all points in a survey area, the water level correction uses several points as a guide to a whole area. Some tide stations are set up in the survey area and the change in water level at a measuring point is calculated from the measured water level (Bao et al. 2006).

Before a bathymetric survey is carried out, it is usually necessary to collect tidal data in the survey area to understand tidal properties to conduct water level zoning. If no historical data are available, the analysis may also be based on the natural geography of the sea area (e.g., seabed topography, coastal shape) or installation of temporary tide stations for short-term surveys. Water level zoning is mainly divided into three situations:

1. The survey area is small with the same tidal properties. In this case, the water level at each point in the survey area is generally in the same plane, so a single tide station can be set up (water level measurement area is located in the same horizontal plane) and the water level data of this station uses a single station water level correction. If several tide stations are set up (water level measurement area is located in the same line or plane, but not in the same horizontal plane), distance weighted interpolation is used to correct the water level.
2. The survey area is large with the same tidal properties, and the tidal level is not on the same horizontal plane. According to tidal propagation, the water level can be corrected by the zoning method, transit-time differential method, and least squares method.
3. The tidal properties are different within the survey area. If the survey area is large, the tidal properties are different in space. In this case, the survey area is divided into sub-areas according to their tidal properties, to ensure each sub-area has the same tidal properties. Then, the interpolation method, zoning method, transit-time differential method and least squares method are used to correct the water level in a sub-area.

6.3.1 Effective Distance of Tide Stations

Calculating the effective distance of a tide station is significant for the arrangement of tide stations and model determination of water level corrections. According to the tidal harmonic constant of two existing tide stations in a survey area, the difference in maximum instantaneous water level is calculated. The corresponding distance of the sounding accuracy is calculated according to the distance between the two tide stations; that is, the effective distance of the tide stations as required by the sounding accuracy. The formula can be expressed as:

$$d = \frac{\delta_z S}{\Delta h_{\max}} \tag{6.8}$$

where d is the effective distance (km) of a tide station, δ_z is the sounding accuracy (cm), S is the distance (km) between two stations and $\Delta h_{\max}$ is the maximum possible tidal level difference (cm) between two stations at the same time.

Using Eq. (6.8) to estimate the effective distance, the key is to obtain $\Delta h_{\max}$ using one of three following methods (Liu 2003):

1. Synchronous observation comparison. Using synchronous observation data from two stations, the water level change, expressed as a curve, over a few days during the spring tide (from the mean sea level) is drawn to determine $\Delta h_{\max}$.
2. Analytical calculation. The quasi-harmonic tidal model formed by the four main tidal components of two stations is used to calculate $\Delta h_{\max}$.
3. Numerical calculation comparison. This uses the harmonic tidal model, composed of 11 tidal components from two stations, to calculate the water level of the two stations for a period of time and select $\Delta h_{\max}$.

6.3.2 Water Level Correction of a Single Station (Time Interpolation Method)

When the survey area is within the effective range of a tide station, the water level changes at all points in the survey area are the same as the station. Therefore, the water level data of a station can be used for water level corrections. The single station water level correction method is the most common tidal interpolation method in field data processing. For example, with a depth datum (e.g., Fig. 6.7), $Z(t)$ represents the water level correction at the observation time (tide

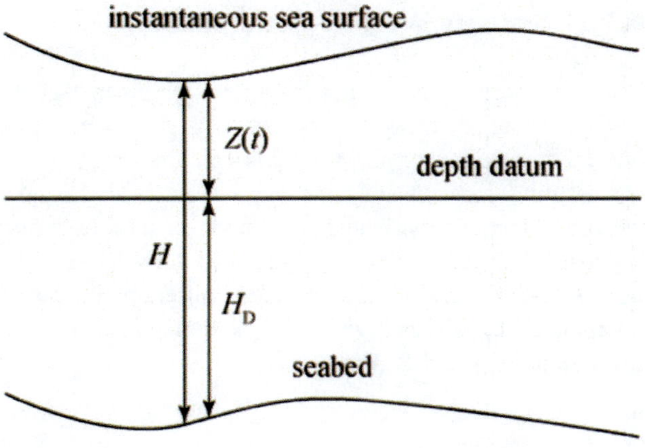

Fig. 6.7 Schematic of single station water level corrections

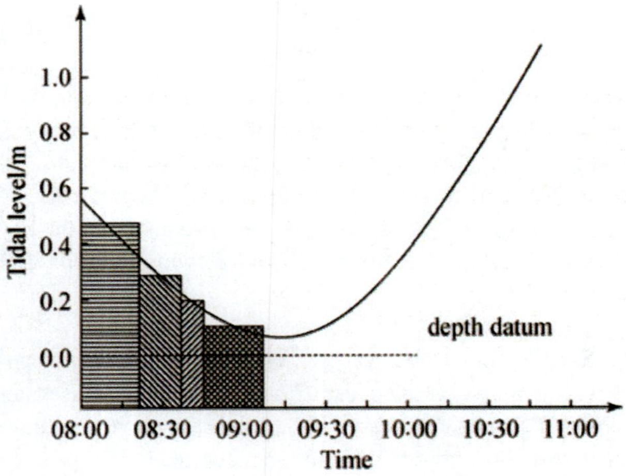

Fig. 6.8 Graphical representation of single station water level corrections

level calculated from the depth datum), H represents the instantaneous water depth so that the water depth H_D is calculated from the depth datum.

$$H_D = H - Z(t) \tag{6.9}$$

The water level observation data from tide stations are discrete values. To determine water level corrections at different times, time interpolation is required. Generally, an analytical method is adopted, and the graphical method is used in the early stage. The graphical method involves drawing the water level curve, with the abscissa representing time and ordinates representing water level corrections (Fig. 6.8). The water level correction at any time can be determined. To determine water level corrections at any time in the measurement period, the analytical method uses observations as sampling points and then performs time interpolation. The common interpolation methods are linear, polynomial and spline.

6.3.3 Distance Weighted Interpolation

When the survey area is not large, it is assumed that the water level of all measuring points is in the same line or plane. After the line or plane is confirmed, the water level of a measuring point at any time can be obtained. Distance weighted interpolation is also a common method to correct water level. If the survey area is located between tide Stations A and B, the water level of all measuring points can be interpolated by distance weighted interpolation using data from Stations A and B (Fig. 6.9).

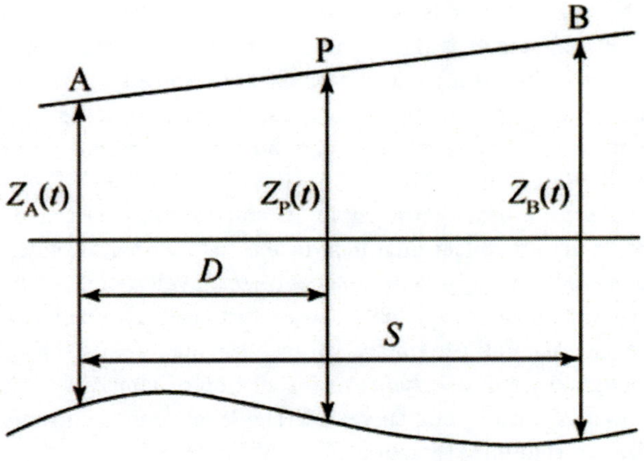

Fig. 6.9 Distance weighted interpolation of two stations

This method is also applicable for three stations (Fig. 6.10). The premise of this method is that the instantaneous sea level among three stations is on the same plane.

The model for two station distance weighted interpolation is

$$Z_P(t) = Z_A(t) + \frac{Z_B(t) - Z_A(t)}{S} D \tag{6.10}$$

and the model for three station distance weighted interpolation is:

$$\begin{vmatrix} X_P - X_A & Y_P - Y_A & Z_P(t) - Z_A(t) \\ X_B - X_A & Y_B - Y_A & Z_B(t) - Z_A(t) \\ X_C - X_A & Y_C - Y_A & Z_C(t) - Z_A(t) \end{vmatrix} = 0 \tag{6.11}$$

where $Z(t)$ is the water level of the corresponding point (station) at a certain time and (X, Y) are the plane coordinates of the corresponding point (station).

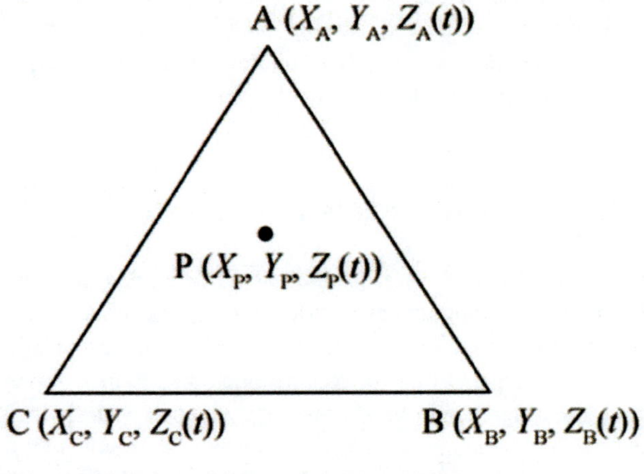

Fig. 6.10 Distance weighted interpolation of three stations

6.3.4 Zoning Method

When the measuring point is beyond the effective distance of a tide station, the zoning method, transit-time differential method and least squares method can be used to correct the water level. The essence of the zoning method is to divide a survey area into several zones according to the position of tide stations and tidal propagation direction and then interpolate the water level curve of each zone. For measuring points within the effective distance of tide stations, the water level can be corrected directly with the observed water level of the tide station. For measuring points beyond the effective distance of tide stations, the water level curve can be interpolated, and then the water level can be corrected using this curve (Fig. 6.11). The basic assumptions of the zoning method include two aspects: (i) the tidal properties are the same in a survey area; (ii) the tidal propagation between two stations is uniform. That is, the change in co-phase time and tidal height between two stations is proportional to their distance.

The co-phase time of a tide refers to the time when the same phase wave points, such as peaks and troughs, between two stations occur in different places; the co-phase tidal height refers to the height of the same phase wave points between two stations. It is assumed that the wave propagation between Stations A and B is uniform, so t_A and t_B are the co-phase tide times and $\tau_{AB} = t_B - t_A$ represents the time difference between the two stations (Fig. 6.12). The distance between Stations C and A is equal to one-third of the distance between Stations A and B, so the co-phase tide time (e.g., high tide) of Station C should be equal to $t_C = t_A + \tau_{AB}/3$, and the co-phase tidal height at Station C should be on the line of the simultaneous heights of A and B.

In the case of uniform tidal wave propagation, the number of water level zones between two tide stations can be determined by the following equation:

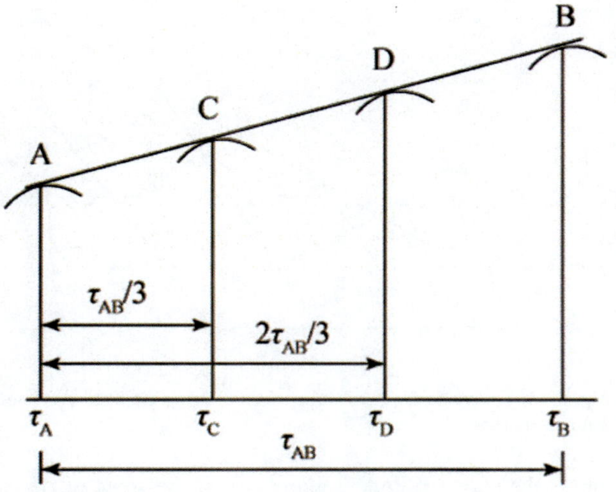

Fig. 6.12 Diagram of co-phase time and tidal height

$$K = \frac{\Delta h_{max}}{\delta_z} \tag{6.12}$$

When zoning is carried out, the maximum difference in water level corrections in adjacent zones should not exceed the depth sounding accuracy δ_z, and the zoning boundary should be basically perpendicular to the propagation direction of waves (Fig. 6.13). After zoning, the water level corrections at a certain time in sub-zones, such as 1, 2 and 3 (Fig. 6.13), can be calculated according to the water level at Stations A or B at the same time.

When a survey area is not narrow and water level changes cannot be described by the same curve after zoning, it is necessary to continue zoning (Fig. 6.14). If there are three tide stations in a survey area, zoning between two stations is first carried out, so that there are water level curve controls at both ends of each zone. For example, in ZoneII (Fig. 6.14), one end is the water level curve of Station C and the other

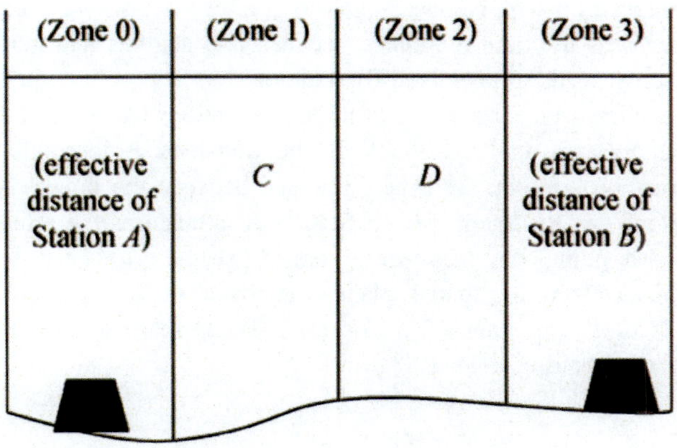

Fig. 6.11 Diagram of the two station zoning method

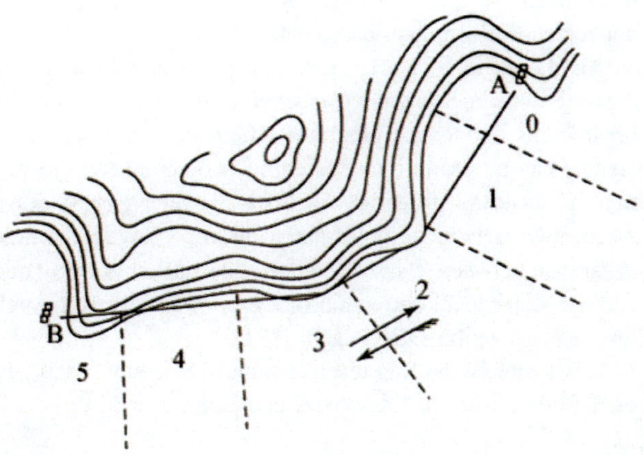

Fig. 6.13 Diagram of water level correction using the zoning method at two stations

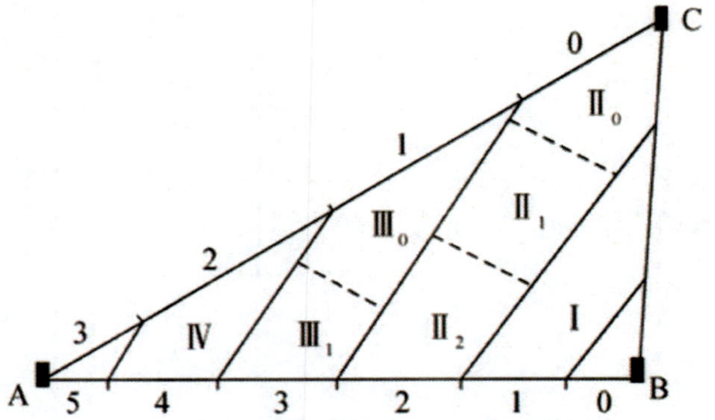

Fig. 6.14 Diagram of water level correction using the zoning method at three stations

end is the curve of Zone 2 along AB. If the value of $\Delta h_{\max}$ from the water level curves at both ends is greater than the bathymetric accuracy δ_z at the same time, the zone should be further divided into sub-zones II_0, II_1 and II_2. The II_1 water level curve is obtained by interpolating the curve of the second zone at Station C and Side AB.

For a larger survey area, the number of tide stations may be more than three. The method is still based on two and three station zoning and water level correction is carried out after the survey area is zoned. In practical application, water level corrections can be completed by computer programming according to the basic zoning method principles.

6.3.5 Transit-Time Differential Method

Water level corrections using the transit-time differential method calculates the water level at a measuring point through characteristics of the correlation function in digital signal processing. First, the time difference in tides between two stations is calculated to obtain the difference between measuring points and tide stations. Then, the water level of a measuring point is determined according to time naturalization. This method is convenient for computer-based calculations.

As shown in Fig. 6.15, the tidal properties in a survey area are the same. If the water level at Stations A and B are regarded as signals, then the time difference between the two signals can be found by researching two signal waveforms; that is, the time difference in tides between two stations. According to the position of the point to be solved, the time difference between it and the datum tide station is calculated and the water level correction of the point is solved through time naturalization (Xie et al. 1988).

It is assumed that the discrete sampling sequences X_n, Y_n of the tide curves of Stations A and B are

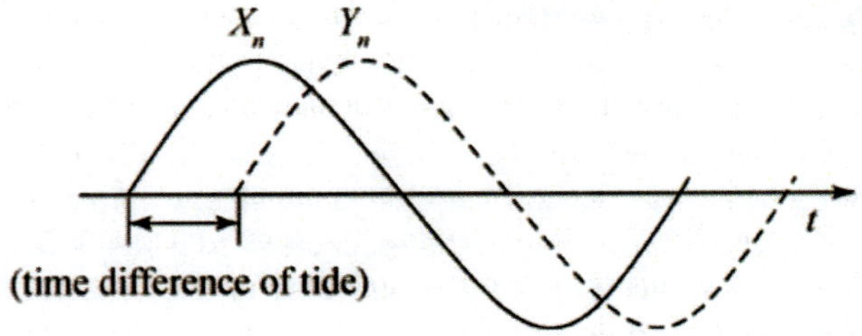

Fig. 6.15 Water level curve of two tide stations

$$\begin{cases} X_n = T_{\text{A}}(t_0 + n\Delta t)\ (n = 0,\ 1,\ \cdots,\ N) \\ Y_n = T_{\text{B}}(t_0 + n\Delta t)\ (n = 0,\ 1,\ \cdots,\ N) \end{cases} \tag{6.13}$$

where t_0 is the initial synchronous moment of the two tide stations, Δt is the sampling interval and N is the total number of samples. The water level curves of two stations are shown in Fig. 6.15.

First, the similarity of water level curves from two stations is analyzed. The degree of similarity of the two curves is determined by the correlation coefficient (R) of a certain sample:

$$R = \frac{\sum_{n=0}^{N} X_n Y_n}{\sqrt{\sum_{n=0}^{N} X_n^2 \sum_{n=0}^{N} Y_n^2}} \tag{6.14}$$

The closer $|R|$ gets to 1, the more similar two curves are; conversely, the closer $|R|$ gets to 0, the less similar two curves are. Because of the time difference between the two stations, the water level curve of one station must be delayed to determine the similarity of the curves. For example, by extending Y_n to τ, which changes into $Y_\tau = T_{\text{B}}(t_0 + n\Delta t - \tau)$, the correlation coefficient of X_n and Y_τ is R_τ:

$$R_\tau = \frac{\sum_{n=0}^{N} X_n Y_\tau}{\sqrt{\sum_{n=0}^{N} X_n^2 \sum_{n=0}^{N} Y_\tau^2}} \tag{6.15}$$

When τ is equal to τ_0 and R_τ reaches the maximum, it indicates that Y_n is most similar to X_i after a delay by τ_0; that is, τ_0 is the time difference between two stations and generally needs to be solved by iteration.

Similarly, for the case of three tide stations (A, B and C), as long as the conditions of the transit-time differential method are met, the time difference between the three stations can be determined. If Station A is taken as the reference point, the measuring point $\text{P}(X_{\text{P}}, Y_{\text{P}}, \tau_{\text{AP}})$ must be located on the spatial plane composed of $\text{A}(X_{\text{A}}, Y_{\text{A}}, 0)$, $\text{B}(X_{\text{B}}, Y_{\text{B}}, \tau_{\text{AB}})$ and $\text{C}(X_{\text{C}}, Y_{\text{C}}, \tau_{\text{AC}})$. Thus, the time delay of the measuring Point P (τ_{AP}) is

$$\tau_{AP} = \frac{(X_P - X_A)[(Y_C - Y_A)\tau_{AB} - (Y_B - Y_A)\tau_{AC}] + (Y_P - Y_A)[(X_B - X_A)\tau_{AC} - (X_C - X_A)\tau_{AB}]}{(X_B - X_A)(Y_C - X_A) - (Y_B - Y_A)(X_C - X_A)} \tag{6.16}$$

Changing the observation time of each tide station into the time corresponding to the moment t of the measuring Point P results in $t_A = t + \tau_{AP}$, $t_B = t + \tau_{AP} - \tau_{AB}$, $t_C = t + \tau_{AP} - \tau_{AC}$. The water level values $Z_A(t_A)$, $Z_B(t_B)$ and $Z_C(t_C)$ at the corresponding time of Stations A, B and C, respectively, are then calculated. Based on these assumptions, the measuring Point $P(X_P, Y_P, \tau_{AP})$ must be located on the spatial plane composed of $A(X_A, Y_A, 0)$, $B(X_B, Y_B, \tau_{AB})$ and $C(X_C, Y_C, \tau_{AC})$, so the water level correction $Z_P(t)$ of the measuring Point P at observation time t is

$$\begin{cases} Z_P(t) = Z_A(t_A) + (M+N)/[(X_B - X_A)(Y_C - X_A) - (Y_B - Y_A)(X_C - X_A)] \\ M = (X_P - X_A)\{(Y_C - Y_A)[Z_B(t_B) - Z_A(t_A)] - (Y_B - Y_A)[Z_C(t_C) - Z_A(t_A)]\} \\ N = (Y_P - Y_A)\{(X_B - X_A)[Z_C(t_C) - Z_A(t_A)] - (X_C - X_A)[Z_B(t_B) - Z_A(t_A)]\} \end{cases} \tag{6.17}$$

6.3.6 Least Squares Fitting Method

The least squares fitting method is similar to the transit-time differential method, but in addition to calculating the time difference in the tides between each point, it also considers the ratio of tidal range and datum level deviation. This method performs least squares fitting for a water level sequence at two known tide stations and determines tidal transfer parameters (ratio of tidal range γ_{AB}, time difference of tide τ_{AB} and datum deviation ε_{AB}) between two stations.

The tidal transfer parameters of the point relative to the reference station are calculated and then the water level of the point is found by interpolation.

As shown in Fig. 6.16, the water level at Station B relative to reference Station A is

$$T_B(t) = \gamma_{AB} \cdot T_A(t + \tau_{AB}) + \varepsilon_{AB} \tag{6.18}$$

Theoretically, there should be

$$\gamma_{AB} = 1/\gamma_{BA}, \tau_{AB} = -\tau_{BA}, \varepsilon_{AB} = -\varepsilon_{BA} \cdot \gamma_{AB} \tag{6.19}$$

If P is the point to be calculated, then:

$$\begin{cases} \gamma_{AP} = 1 + (\gamma_{AB} - 1)R_{AP}/R_{AB} \\ \tau_{AP} = \tau_{AB} \cdot R_{AP}/R_{AB} \\ \varepsilon_{AP} = \varepsilon_{AB} \cdot R_{AP}/R_{AB} \end{cases} \tag{6.20}$$

The instantaneous water level at Point P is

$$T_P(t) = \gamma_{AP} \cdot T_A(t + \tau_{AP}) + \varepsilon_{AP} \tag{6.21}$$

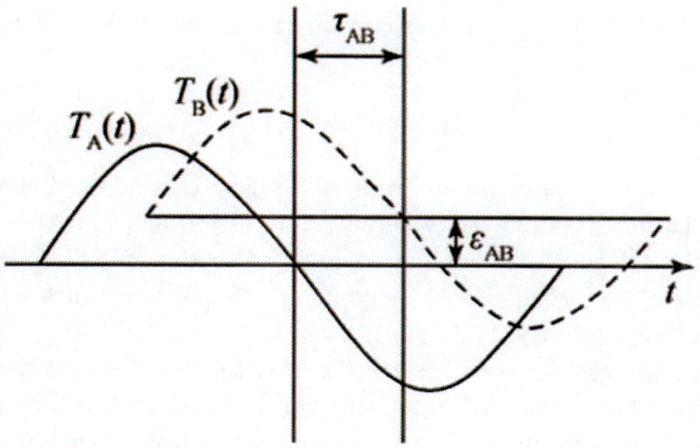

Fig. 6.16 Principle diagram of least squares fitting for a water level curve

where $T_P(t)$ is the water level at Point P deduced from tide Station A.

The least squares method is also suitable for computer programming. It assumes that the discrete sampling sequence of the tide level curves of Stations A and B is

$$\begin{cases} T_A(t_0 + n\Delta t)\ (n = 0,\ 1,\ \cdots,\ N) \\ T_B(t_0 + n\Delta t)\ (n = 0,\ 1,\ \cdots,\ N) \end{cases} \tag{6.22}$$

where t_0 is the initial moment, Δt is the sampling interval, $\Delta t = 5$, 10, 20, 30, 60 min, and N is the total number of samples.

The error in water level compared between two tide stations is expressed as

$$v_n = \gamma_{AB} T_A(t_0 + n\Delta t + \tau_{AB}) + \varepsilon_{AB} - T_B(t_0 + n\Delta t) \quad (n = 0,\ 1,\ \cdots,\ N) \tag{6.23}$$

When Eq. (6.23) is linearized, it can be written as a matrix:

$$\boldsymbol{V} = \boldsymbol{AX} - \boldsymbol{L} \tag{6.24}$$

where $\boldsymbol{V}$ is the correction matrix, $\boldsymbol{A}$ is the coefficient matrix, $\boldsymbol{X} = [\Delta\gamma, \Delta\tau, \Delta\varepsilon]^T$ is the unknown parameter correction vector and $\boldsymbol{L}$ is the constant vector. The elements of matrix $\boldsymbol{A}$ are

$$[T_A(t_0 + n\Delta t + \tau_0) \quad \gamma_0 T'_A(t_0 + n\Delta t + \tau_0) \quad 1], \quad (n = 0,\ 1,\ \cdots,\ N) \tag{6.25}$$

where γ_0, τ_0 and ε_0 are the approximate values of unknown parameters γ_{AB}, τ_{AB}, ε_{AB} and $T'_A(t_0 + n\Delta t + \tau_0)$ is the derivative of T to τ. The elements of matrix $\boldsymbol{L}$ are

$$[T_B(t_0 + n\Delta t) - \gamma_0 T_A(t_0 + n\Delta t + \tau_0) - \varepsilon_0], \quad (n = 0,\ 1,\ \cdots,\ N) \tag{6.26}$$

According to the least squares principle, $[V^TV] = \min$, then:

$$X = (V^TV)^{-1}A^TL \quad (6.27)$$

In these calculations, an iterative method is adopted and the initial values are usually $\gamma_0 = 1$, $\tau_0 = 0$ and $\varepsilon_0 = 0$. In the calculation of coefficient matrix A, continuous discrete data are required and a function interpolation, such as quadratic-spline, is generally used.

In general, when a survey area is small and within the effective distance of a station, water level observations of a tide station can be directly used for corrections. When the range is large and exceeds the control of one station, the observation data of two stations are used for zoning correction. When two stations cannot be controlled effectively, the triangulation band method for three stations is adopted. An interpolation algorithm is required for all correction methods, but the interpolated elements are different, so various methods, such as linear interpolation, distance weighted interpolation, transit-time difference and least squares, need to be used. The specific interpolation algorithm is determined by many factors, such as whether the tide station coordinates are known and whether the tides are spread evenly.

6.4 Vertical Data and Transformation Approaches for Hydrography

6.4.1 Several Common Marine Survey Data

6.4.1.1 Mean Sea Level

1. **Definition and algorithm for mean sea level**

The mean sea level is also known as sea level. The average sea surface position over a certain period of time is the elevation starting point in geodesy. It is obtained from hourly tide level observations over a corresponding period and the height is generally calculated from the zero point of a local tide gauge station.

The water level recording device of a tide gauge station has its own zero record. The depth of the recorded zero point underwater varies by location depending on the device establishment. This is collectively referred to as the water gauge zero point and the most primitive water level observations are relative to the zero point. If water level observations are a continuous curve $y(t)$, the mean sea level in time T can be expressed as:

$$\mathrm{MSL}_T = \frac{1}{T}\int_0^T y\mathrm{d}t \quad (6.28)$$

where MSL_T is the mean sea level.

In general, water level observations from a tide gauge station consist of one hour observation sequences. Therefore, the commonly used method in actual calculations is to directly take the average of the observations over a certain period of time. This is also considered to be a tidal cycle, such as 24 h, one month, one year and many years:

$$\mathrm{MSL}_n = \frac{1}{n}\sum_{i=1}^{n} h_i \quad (6.29)$$

where h is a water level observation and n is the number of observations.

For observations over a day, the value would be 24, and the value is determined according to the actual observation period (e.g., one month, one year or many years). A long-term average value can also be calculated from short-term average sea level data. For example, the monthly mean sea level can be calculated from daily mean sea level values, the annual average can be obtained from the monthly average and multiple annual averages can be used to produce a multi-year record. These mean sea level records are called daily, monthly, annual and multi-year mean sea level.

Using tidal harmonic analysis, the mean sea level can be obtained while calculating the tidal harmonic constant. In fact, mean sea level as the starting point of tidal oscillation can be a zero-frequency tidal constituent. Mean sea level calculated in this way is the equilibrium surface corresponding to tidal vibration and it represents sea level height. When the observation time is long enough, the deviation between results from harmonic analysis and the arithmetic mean is very small. For example, with one year of tidal observation data, the results calculated by the two methods are very consistent. The average method is essentially a filtering algorithm. The elimination of non-tidal components through the calculation of mean sea level depends on the observation time of water levels. Using long-term tidal observation data, the average can be used to eliminate the influence of non-tidal factors and obtain a high-precision sea level record (Zhao et al. 2006).

2. **Stability of mean sea level**

Because the observations cannot be exactly the whole tidal period, mean sea level is affected by the remaining tidal components and the short-term mean sea level also contains the contribution of long-period tidal waves. In addition, non-tidal factors, mainly meteorological, are different at various time periods. On a longer timescale, they can be regarded as noise and, in a shorter time span, it is a signal with certain regularity. This results in variable mean sea level stability for different lengths of time. Because daily mean sea level is significantly influenced by tidal and non-tidal factors, its value can be greatly affected within one month. Due to the influence of annual meteorological factors and tidal waves (S_a and S_{Sa}) caused by half-year variations, the mean sea level shows considerable changes each month. The annual mean sea level is also affected by long-cycle tidal waves. For example, there are two periods (8.6 a and 18.61 a) that result from interactions of the moon and tides, but the magnitude is small and so the impact on cyclical changes in annual mean sea level is not obvious.

The calculated mean sea level for each year can be regarded as equal-precision observations of the ideal undisturbed sea level, as well as annual mean sea level accuracy obtained by the direct adjustment principle. This is an ideal undisturbed sea level estimate over multiple annual mean sea levels (MSL_{my}) with average values and valuations as unit weight variance:

$$MSL_{my} = \frac{1}{n}\sum_{i=1}^{n} MSL_{yi} \quad (6.30)$$

$$\hat{\sigma}_0^2 = \frac{1}{n-1}\sum_{i=1}^{n} (MSL_{yi} - MSL_{my})^2 \quad (6.31)$$

While the variance in multi-year mean sea level ($\hat{\sigma}_{my}^2$) is

$$\hat{\sigma}_{my}^2 = \frac{\hat{\sigma}_0^2}{n} \quad (6.32)$$

With the increase of the number of years n, the multi-year mean sea level has high precision, which can be regarded as an ideal undisturbed sea level and can be used as a benchmark for mean sea level changes on other timescales.

In oceanography, mean sea level stability is usually calculated based on the maximum mutual difference of mean sea level heights at different locations. If the mean sea level sequence shows a specific probability distribution, the relationship between range and medium error can be established. Fang et al. (1986) conducted a statistical analysis for the maximum deviation of mean sea level and multi-year mean sea level offshore China in 1986. The statistical results are shown in Table 6.1.

The accuracy of mean sea level is $\hat{\sigma}_{my}$ and the number of years n required for multi-year mean sea level calculations are:

$$n = \sqrt{\frac{1}{n(n-1)\hat{\sigma}_{my}^2}\sum_{i=1}^{n} (MSL_{yi} - MSL_{my})^2} \quad (6.33)$$

In previous studies, the accuracy of multi-year mean sea level was usually defined by a 95% confidence probability and the error magnitude was taken as 1 cm. At this point, there is an assumption that the annual mean sea level obeys normal distribution:

$$P(|MSL_y - MSL_{my}| < 1\text{ cm}) = 0.95 \quad (6.34)$$

$$\left|\frac{MSL_y - MSL_{my}}{\hat{\sigma}/n}\right| = 1.96 \quad (6.35)$$

$$n = 1.96^2\hat{\sigma}_0^2 \approx 3.84\hat{\sigma}_0^2 \quad (6.36)$$

From this, the number of observation years required for several tidal stations in China's coastal areas to reach a mean sea level of 1 cm is obtained with 95% probability (Table 6.2). It can be seen in Table 6.2 that the stability of the mean sea level is quite different in various sea areas. The Yellow Sea coast maintains a fairly stable mean sea level over an intersection period, while the situation in the Bohai Sea is more complicated.

Long-term trends in mean sea level, especially sea level rise, have caused widespread concern among geometricians, oceanographers and even politicians in recent decades, and has become an important topic of interdisciplinary research. Absolute sea level changes are mainly caused by factors, such as the melting of polar ice caps from global warming, which have certain impacts on the determination of stable mean sea level.

Table 6.1 Variable deviation in mean sea level on different timescales in the China's offshore waters

Observation time	1 month	3 months	6 months	1 year	2 years	5 years
Maximum deviation of mean sea level from multi-year mean sea level/cm	60	40	25	10	8	5

Table 6.2 Number of observation years required for an accurate mean sea level (1 cm) at several important tidal stations in China's coastal areas

Tide station	Weihai	Rushankou	Lianyungang	Yingkou	Qinhuangdao	Tanggu
Time required/a	18	16	16	50	28	118

6.4.1.2 National Height Datum

Currently, countries or regions define elevation benchmarks based on the long-term mean sea level of one or several tidal stations. The US defines their respective elevation data at Portland's tidal station, Japan uses the Tokyo Linguin tidal station and European multi-year mean sea level is derived from Amsterdam's tidal station.

For historical reasons, China's elevation data were rather chaotic, with more than 10 stations, including Daxie Zero, Wujing Zero, Luoxing Tower Zero, Zhujiang Base, Dalian Base, Kanmenji and so on. Some of these represented the mean sea levels of tidal stations, while others used the nautical depth datum of a port. In 1954, the General Survey and Mapping Bureau used part of the coastal level observation data to form 24 loops to adjust the network and define the mean sea level elevation of Qingdao (1953–1954) and Kanmen stations. As a constraint, the "1954 Yellow Sea Mean Sea Level Reference" was established. In 1956, the mean sea level of the Qingdao Dagang tide station from 1950 to 1956 was selected as the national unified elevation datum. This benchmark has lasted for 30 a.

In principle, the long-term mean sea level should be used to define the elevation reference, at least taking into account the average interannual cycle caused by the intersection tide. This stipulates that the mean sea level observation time for the elevation starting plane should be no less than 19 a. This length of time is based on a stability analysis (95% confidence probability at 1 cm accuracy) that contains the influence of the intersection tide in Qingdao. Based on such considerations and long-term observation data, the 1985 National Elevation Benchmark was established. It was used for a total of 28 years from 1952 to 1969. The specific calculation used 10 sets of data with a 19 a moving average, and finally averaged the 10 sets. The national level is 62.26 m. The difference between the 1985 National Elevation Benchmark and the 1956 Qingdao elevation system is only 2.9 cm.

6.4.1.3 Depth Datum

1. **Basic principles for determining the nautical depth datum**

The observed ocean depth measurements can be conveniently reduced to a depth relative to the instantaneous sea level after corrections (e.g., draught, sound velocity). Because instantaneous sea level has obvious time variation characteristics, this reference introduces uncertainty to the bathymetric data. To facilitate bathymetric data representation and management, it is necessary to select a stable datum.

The long-term mean sea level has good stability, so it is the ideal starting point for depth. The purpose of surveying and charting is mainly for navigation. Therefore, determining the depth datum of charts needs to take into account both navigational safety and channel utilization. A depth datum can be described as below the local stable mean sea level, so that the instantaneous sea level is rarely lower than the surface. At a specific time, local tidal range changes need to be considered. The depth datum is defined relative to the local stable (or long-term) mean sea level (Zhao and Liu 2008).

For the determined depth datum, the definition of the depth datum guarantee is a percentage of the ratio of low tide times to the total number of times above the depth datum surface over a certain period of time.

$$\text{Charted depth datum navigation guarantee rate} = \frac{\text{Low tide times above the datum}}{\text{Total number of low tides}} \times 100\% \quad (6.37)$$

The depth datum of a nautical chart in China is the theoretical minimum tide surface with a guarantee of about 95%.

2. **Theoretical depth datum calculation**

Coastal countries use different calculation models according to the tidal nature of their sea regions: (1) mean low water springs use the formula $L{=}H_{M_2}+H_{S_2}$ to calculate the difference between the datum plane and mean sea level. Countries that use it include Italy, former Yugoslavia, Germany, Albania, Greece, Canada (Atlantic coast), Denmark, Belgium, Norway, Indonesia, Argentina and Panama; (2) mean low water where the position of the depth datum at mean sea level is determined by the amplitude of the M_2 tidal wave. The countries that use this model include the US (Atlantic coast), Sweden (North Sea) and the Netherlands; (3) mean lowest water level where a formula is used to calculate the difference with average sea level. The countries that use this model include the US (Pacific coast, Alaska) and the Philippines; (4) lower water, expressed by the formula $H_{M_2}+H_{S_2}+H_{K_1}+H_{O_1}$. The countries using it include those on the Indian Ocean coasts and Japan; (5) lowest observed low water level with a formula 1.2 $(H_{M_2}+H_{S_2}+H_{K_2})$, which is used by France, Portugal and Brazil. China, the Soviet Union, Democratic People's Republic of Korea (DPRK), Vietnam and other countries have adopted theoretical depth data.

The theoretical depth datum is also called the theoretical lowest water level, and its calculation was proposed by Vladimirskiy (Fang et al. 1986). The basic principle is the superposition of eight tidal components, M_2, S_2, N_2, K_2, K_1, O_1, P_1 and Q_1, to calculate the lowest possible water level relative to the average tidal vibration (long-term average sea level); the contributions of neritic tides M_4, MS_4 and M_6 and long-period tides S_a and S_{Sa} are taken into account.

The tidal height relative to the mean sea level after the superposition of the eight main tidal levels can be expressed as

$$h(t) = \sum_{i=1}^{8} f_i H_i \cos(\sigma_i t + V_{0i} + u_i - g_i) \tag{6.38}$$

The lowest tide position represented by the tidal height is taken as the L value of the depth datum, which is defined as

$$L = -\min\left[\sum_{i=1}^{8} f_i H_i \cos(\sigma_i t + V_{0i} + u_i - g_i)\right] \tag{6.39}$$

Because the forecasted tidal height is a trigonometric function of time t, it is obvious that this problem is difficult to solve analytically. Therefore, a solution is sought by simplification and transformation. For simplicity, the following symbols are used:

$$\begin{aligned} f_{M_2}H_{M_2} &= M_2 & \sigma_{M_2}t + V_{0M_2} - g_{M_2} &= \varphi_{M_2} \\ f_{S_2}H_{S_2} &= S_2 & \sigma_{S_2}t + V_{0S_2} - g_{S_2} &= \varphi_{S_2} \\ &\cdots\cdots & &\cdots\cdots \\ f_{Q_1}H_{Q_1} &= Q_1 & \sigma_{Q_1}t + V_{0Q_1} - g_{Q_1} &= \varphi_{Q_1} \end{aligned} \tag{6.40}$$

Therefore, after the intersection of the tidal phase angle is corrected, Eq. (6.39) is rewritten as

$$\begin{aligned} h(t) = {} & M_2 \cos\varphi_{M_2} + S_2 \cos\varphi_{S_2} + N_2 \cos\varphi_{N_2} + K_2 \cos\varphi_{K_2} \\ & + K_1 \cos\varphi_{K_1} + O_1 \cos\varphi_{O_1} + P_1 \cos\varphi_{P_1} + Q_1 \cos\varphi_{Q_1} \end{aligned} \tag{6.41}$$

The tidal phase angle φ is represented by the Doodson number combination of basic astronomical variables, and the relationship is

$$\begin{aligned} \varphi_{M_2} - \varphi_{Q_1} &= \varphi_{K_1} - (g_{K_1} + g_{O_1} - g_{M_2}) = \varphi_{K_1} + a_1 = \tau_1 \\ \varphi_{S_2} - \varphi_{O_1} &= \varphi_{K_1} - (g_{K_1} + g_{P_1} - g_{S_2}) = \varphi_{K_1} + a_2 = \tau_2 \\ \varphi_{N_2} - \varphi_{Q_1} &= \varphi_{K_1} - (g_{K_1} + g_{Q_1} - g_{N_2}) = \varphi_{K_1} + a_3 = \tau_3 \\ \varphi_{k_2} &= 2\varphi_{K_1} + 2g_{K_1} - g_{k_2} = 2\varphi_{K_1} + a_4 \end{aligned} \tag{6.42}$$

Thus, the expression of tidal height with tidal phase angle as a variable (time t is included in the tidal phase angle) is

$$\begin{aligned} h(t) = {} & M_2 \cos\varphi_2 + O_1 \cos(\varphi_{M_2} - \tau_1) + S_2 \cos\varphi_S + P_1 \cos(\varphi_{S_2} - \tau_2) \\ & + N_2 \cos\varphi_N + P_1 \cos(\varphi_{N_2} - \tau_3) + K_2 \cos(\varphi_{K_1} + a_4) + K_1 \cos\varphi_{k_1} \end{aligned} \tag{6.43}$$

The superposition $A\cos\varphi + B\cos(\varphi - \tau)$ of each pair of constituent tides can be transformed as

$$A\cos\varphi + B\cos(\varphi - \tau) = (A + B\cos\tau)\cos\varphi + B\sin\tau\sin\varphi \tag{6.44}$$

and then expressed as

$$A\cos\varphi + B\cos(\varphi - \tau) = R\cos(\varphi - \varepsilon) \tag{6.45}$$

where

$$R = \sqrt{A^2 + B^2 + 2AB\cos\tau}\; \varepsilon = \arctan\frac{B\sin\tau}{A + B\cos\tau} \tag{6.46}$$

Substituting Eqs. (6.44)–(6.46) into Eq. (6.43), the complex function with the phase angle of four constituent tides as the variable is

$$\begin{aligned} h(\varphi_{K_1}, \varphi_{M_2}, \varphi_{S_2}, \varphi_{N_2}) = {} & K_1 \cos(\varphi_{K_1}) + K_2 \cos(2\varphi_{K_1} + a_4) \\ & + R_1 \cos(\varphi_{M_2} - \varepsilon_1) + R_2 \cos(\varphi_{S_2} - \varepsilon_2) + R_3 \cos(\varphi_{N_2} - \varepsilon_3) \end{aligned} \tag{6.47}$$

Note that the amplitude and retardation of the last three terms are also a function of the K_1 tidal phase angle:

$$\begin{aligned} R_1 &= \sqrt{M_2^2 + O_1^2 + 2M_2O_1\cos\tau_1}\; \varepsilon_1 = \arctan\frac{O_1\sin\tau_1}{M_2 + O_1\cos\tau_1} \\ R_2 &= \sqrt{S_2^2 + P_1^2 + 2S_2P_1\cos\tau_2}\; \varepsilon_2 = \arctan\frac{P_1\sin\tau_2}{S_2 + P_1\cos\tau_2} \\ R_3 &= \sqrt{N_2^2 + O_1^2 + 2N_2O_1\cos\tau_3}\; \varepsilon_3 = \arctan\frac{Q_1\sin\tau_3}{N_2 + Q_1\cos\tau_3} \end{aligned} \tag{6.48}$$

It is still very difficult to determine the extremum of Eq. (6.47) directly. Therefore, further approximation is warranted. First, they are simplified into minimum forms:

$$\varphi_{M_2} - \varepsilon_1 = 180^\circ \quad \varphi_{S_2} - \varepsilon_2 = 180^\circ \quad \varphi_{N_2} - \varepsilon_3 = 180^\circ \tag{6.49}$$

Thus, the expression of tidal height becomes a univariate function with the K_1 tidal phase angle as an independent variable:

$$\begin{aligned} h(\varphi_{K_1}) = {} & K_1 \cos(\varphi_{K_1}) + K_2 \cos(2\varphi_{K_1} + a_4) \\ & - (R_1 + R_2 + R_3) \end{aligned} \tag{6.50}$$

The expression is discretized against the independent variable at an appropriate interval between the K_1 tidal phase angles and a set of function values is obtained. The minimum value (the sign is negative and the absolute value is largest) is used to describe the difference L of the reference depth relative to mean sea level and L is usually expressed as an absolute value.

6.4.1.4 Geoid

There is an infinite number of level surfaces, of which mean sea surface is called the geoid. The shape surrounded by the geoid is called the geoidal shape. Because the geoid is one of the levels, it has all the characteristics of a level surface.

Studying the geoidal shape is an important task in geodesy. The irregularity of the geoid is determined by the complexity of material distribution in the Earth and the irregularity of ground fluctuations. To facilitate this study, the Earth is regarded as a regular ellipsoid, which is divided into many layers. Assuming similar material density in the same layer, the sum of the masses of all of the layers is equal to the total mass of the Earth. The gravity plane obtained under these assumptions is called the normal gravity plane. To obtain the geoid, the difference between the geoid and the normal gravity plane is estimated and the normal gravity plane is corrected according to this difference. At present, there is no precise global geodetic model. Therefore, countries around the world use different geoids according to their specific conditions. China set up the tidal station in Qingdao to obtain the average sea surface of the Yellow Sea. The geoid is the level above this mean sea level. In other words, the gravity level at any point on the geodetic level of China is equal to the gravity level of the average Yellow Sea surface at the tide gauge station.

6.4.1.5 Reference Ellipsoid

In the past, due to technical constraints, humans could not survey the entire ellipsoid of the Earth. Only the geodetic data of individual countries and local regions could be used to derive the elements of the ellipsoid (e.g., long axis radius, flattening rate). The sphere is called the reference ellipsoid.

Among these, the WGS84 ellipsoid has become the most commonly used reference within the wide application of GPS. The geometric definition of the WGS84 coordinate system is the origin at the Earth's center, the *Z*-axis pointing in the defined BIH1984.0 Protocol Earth Polar (CTP) direction and the *X*-axis pointing to the intersection of the zero meridian plane of the BIH1984.0 protocol and the CTP equator. The *Y*-axis and the *X* and *Z* axes form the right-handed coordinate system. The reference ellipsoid corresponding to the WGS84 adopts the recommended geodetic constant of the 16th IUCC Congress.

The CGCS2000 is China's new generation geodetic coordinate system. It is a geocentric geodetic coordinate system with the center of the Earth's mass as the origin. In addition to the Earth's mass center as the origin, its geometry is defined as the *Z*-axis pointing to the International Earth Rotation Service (IERS) reference pole and the *X*-axis the intersection of the IERS reference meridian plane with the equatorial plane that passes through the origin and is orthogonal to the *Z*-axis. The axes form a right-handed, solid-centered coordinate system. The CGCS2000 reference ellipsoid is an equipotential ellipsoid. This is defined as an ellipsoid with an equipotential surface. The geometric center of the CGCS2000 reference ellipsoid coincides with the origin of the coordinate system, and the rotation axis coincides with the *Z*-axis. The reference ellipsoid is both a reference surface for geometric applications and a reference surface for the normal gravitational field on the surface of the Earth. Refer to the literature for the definition of ellipsoid constants (Chen 2008).

6.4.2 Typical Technology of Marine Vertical Datum Conversion

At present, the important vertical data for surveying and mapping in China include mean sea level, national elevation datum, theoretical depth datum, geoid and reference ellipsoid. The Specifications for Hydrographic Survey clearly stipulates that the drying heights of the nautical depth, dry beach and dry reef are calculated from the depth reference plane. The elevation of bare rock is the same as island and land elevations, in that the national elevation reference or the island elevation reference close to it is used. For islands and bare rocks far away from the mainland, the average sea surface calculated by longer term local observations is used as the elevation starting point.

The relationship between each vertical datum and chart element is shown in Fig. 6.17.

China has nearly 3×10^6 km^2 of marine area, and there are many neighboring maritime countries. The inconsistency of reference data creates problems between land maps and charts, as well as difficulties in splicing between charts. It is difficult to achieve effective splicing using marine survey maps from neighboring countries and this complicates the demarcation of exclusive economic zones and continental shelves. The issue of maritime delineation between China and neighboring countries does not have a simple solution. At present, competition for maritime rights is increasing. Unifying sea and land benchmarks can improve the division of maritime boundaries, avoid contradictions and even potentially settle conflicts with neighboring countries when developing and utilizing marine resources, and safeguard China's maritime rights and interests (Zhai et al. 2003).

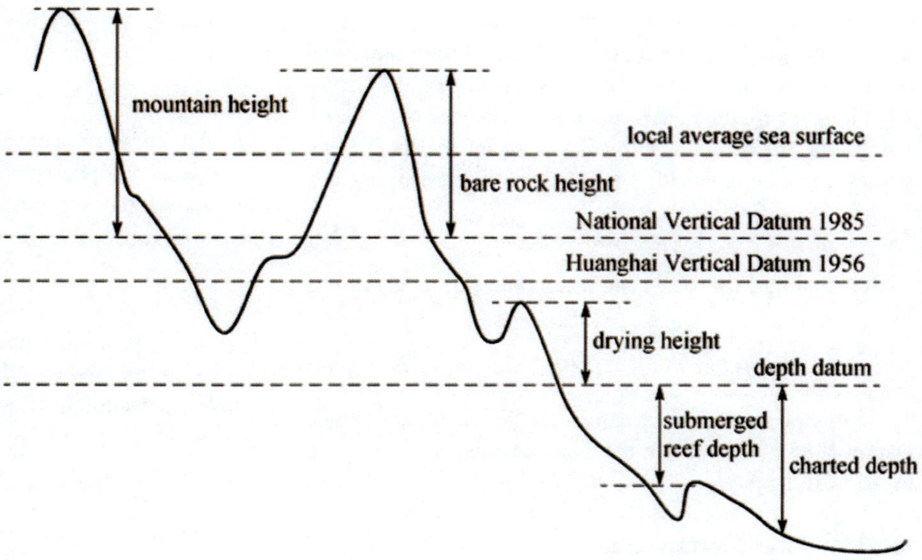

Fig. 6.17 Relationship between each vertical datum and chart element

6.4.3 Conversion Between Depth Datum and Quasi-Geoid

6.4.3.1 Tidal Station Area

Figure 6.18 shows the relationship between several vertical data in a tidal station.

In Fig. 6.18, the local mean sea level MSL_0 is zero tidal observations relative to the water gauge. The depth datum L is a vertical plane based on the local mean sea level. The water depth H_D of the seabed is the vertical distance between the seabed and the depth datum plane. Converting the National Vertical Datum 1985 h at a certain point on the seabed to a more practical nautical depth H_D, the relationship between the various parameters can be obtained:

$$H_D = h - h_L \tag{6.51}$$

where h_L is the National Vertical Datum 1985 for the depth datum. Based on the relationship between the vertical data in Fig. 6.18, h_L can be obtained by the following equation:

$$h_L = L - MSL_{85} \tag{6.52}$$

In other words, based on the relationship between the depth datum and the local mean sea level, the depth datum plane is derived from the National Vertical Datum 1985 for the local mean sea level.

When a long-term tidal station is deployed, the leveling conjunction is carried out at a point in the vicinity and the land elevation of the water gauge at zero tidal observation is

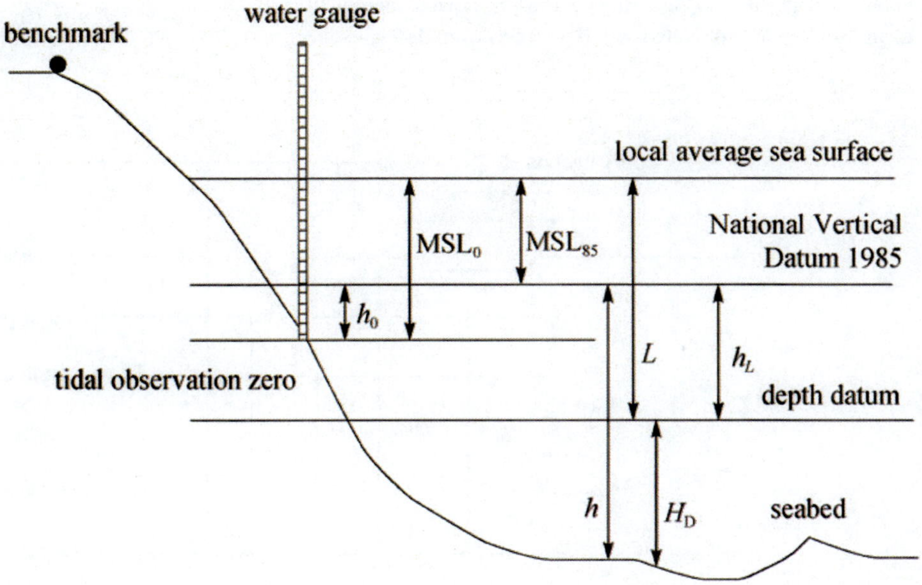

Fig. 6.18 Schematic of the relevant vertical datum relationship in a tide station

obtained. Alternatively, through the regional quasi-geoid model, the geodetic coordinates (*B, L*) of the zero tidal observation are substituted and the National Vertical Datum 1985 for the zero tidal observation h_0 is obtained (Fig. 6.18). Using this, the National Vertical Datum for the local mean sea level can be obtained. The relational expression is

$$\mathrm{MSL}_{85} = \mathrm{MSL}_0 - h_0 \tag{6.53}$$

Substituting Eqs. (6.51) and (6.52) into Eq. (6.53) results in

$$H_\mathrm{D} = h + L - (\mathrm{MSL}_0 - h_0) \tag{6.54}$$

At this point, the depth datum for the National Vertical Datum 1985 h at a point in the tidal station area is converted to nautical depth H_D.

6.4.3.2 Inter-Station Area

Figure 6.19 shows the relationship between several vertical data at a point P between tidal stations (Ke 2012).

In an inter-station area, the tide can be checked with a GPS buoy. The plane position of a point P in Fig. 6.19 is located at (B_P, L_P) by GPS RTK, the geodetic height of the GPS antenna is *H* and the vertical distance of the antenna to the water surface is H_a. The geodetic height of the mean sea level is obtained by

$$H_\mathrm{MSL} = H - H_\mathrm{a} \tag{6.55}$$

Then, the elevation anomaly at this point is obtained by the quasi-geoid model and the normal height of the mean sea level is

$$h_\mathrm{MSL} = H_\mathrm{MSL} - \varsigma \tag{6.56}$$

Using the constructed depth datum model, the depth value *L* at point P is determined and the normal height of mean sea level calculated by Eq. (6.56) can describe the separation h_L between the depth datum at point P and the quasi-geoid:

$$h_L = L - h_\mathrm{MSL} \tag{6.57}$$

At this point, the conversion between the normal height of the seabed topography and the nautical depth can be achieved by the separation between the two bases. The conversion model is

$$H_\mathrm{D} = h - h_L \tag{6.58}$$

where *h* is the normal height of any point and H_D is the nautical depth based on the depth datum, which is generally positive. Substituting Eqs. (6.55)–(6.57) into Eq. (6.58) gives the final conversion model:

$$H_\mathrm{D} = h + H_\mathrm{a} + \varsigma - L - H \tag{6.59}$$

6.4.4 Conversion Between Depth Datum and Reference Ellipsoid

After the relationship between the seamless depth datum and the quasi-geoid is described, the conversion between the seamless depth datum and the reference ellipsoid can be achieved. The separation between the two base surfaces can be expressed as

$$H_L = h_L + \varsigma \tag{6.60}$$

where H_L is the geodetic height of the depth datum, h_L and ς are the normal height and elevation anomaly of the seamless depth datum, respectively.

Solving for H_L, the conversion between the nautical depth and the geodetic height can be realized. The conversion model is

$$H_\mathrm{D} = H - H_L \tag{6.61}$$

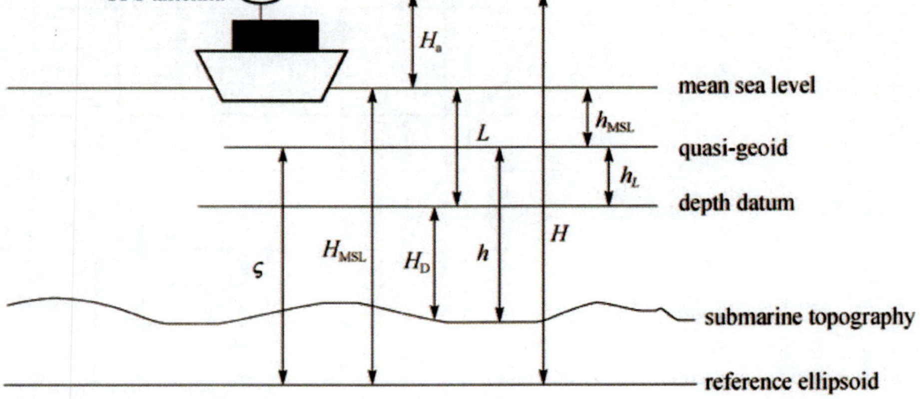

Fig. 6.19 Schematic of vertical reference relationships between tidal stations

In Eq. (6.61), H is the geodetic height of a certain point and the nautical depth H_D is obtained by subtracting the separation between the seamless depth datum plane and the reference ellipsoid datum plane.

References

Bao JY, Liu YC, Chao DB et al (2006) Computations and analyses of chart datum to coastal tide gauges of China (in Chinese with English abstracts). Geomat Inf Sci Wuhan Univ 31(3):224–228. https://doi.org/10.3321/j.issn:1671-8860.2006.03.009

Baramonov AH, Kushner BM, Zabuldayev Вн (1983) Measurement methods and equipment of marine hydrological parameters (in Chinese). Translator: Zhao JM, Peng XL, Lin LM. China Ocean Press, Beijing

Beckers C, Franklin R, Smith T (1981) Sensor subystem for the next generation tide and water level measurement system. Paper presented at the IEEE Oceans 81, Boston, MA, USA, 16–18 September 1981

Brebbia CA, Walker S (1980) Boundary element techniques in engineering. Appl Ocean Res 2(4):185. https://doi.org/10.1016/0141-1187(80)90025-5

Chen JY (2008) China modern geodetic datum—China geodetic coordinate system 2000 (CGCS 2000) and its framework (in Chinese with English abstracts). Acta Geod Cartogr Sin 37(3):269–217. https://doi.org/10.3321/j.issn:1001-1595.2008.03.001

Chen ZY (1980) Tidology (in Chinese). Science Press, Beijing

Fan YM (1997) Discuss about the application of model acoustic water level gauge (in Chinese with English abstracts). Ocean Technol 16 (3):22–28

Fang GH, Zheng WZ, Chen ZY, Wang J (1986) Analysis and prediction of tides and tidal currents (in Chinese). China Ocean Press, Beijing

Han BM, Ou JK (2003) Precise point positioning based on undifferenced GPS data (in Chinese with English abstracts). Geomat Inf Sci Wuhan Univ 28(4):409–412. https://doi.org/10.3321/j.issn:1671-8860,2003.04.008

Huang ZK, Huang L (2005) Principle and calculation of tides (in Chinese). China Ocean University Press, Qingdao

IOC (2006) Manual on sea level: measurement and interpretation, volume 4 : An update to 2006. IOC Manuals and Guides No. 14. Paris, p 80

Kang SL, Zhang Q (1993) Review on technical features of NGWLMS (in Chinese with English abstracts). Ocean Technol 12(1):38–44

Ke H (2012) Research on the theory and implementation method of marine seamless vertical reference surface. Dissertation, Wuhan University

Li KF, Ouyang YZ, Lu XP et al (2009) Sounding base on GPS precise point positioning (in Chinese with English abstracts). Hydrograph Surv Chart 29(6):1–4. https://doi.org/10.3969/j.issn.1671-3044.2009.06.001

Liu JN, Ye SR (2002) GPS precise point positioning using undifferenced phase observation (in Chinese with English abstracts). Geomat Inf Sci Wuhan Univ 27(3):234–240. https://doi.org/10.3321/j.issn:1671-8860.2002.03.002

Liu YC (2003) Spatial structure of marine sounding and its data processing (in Chinese). Surveying and Mapping Press, Beijing

NOAA (2007) NOAA 200th foundations: tides and currents. Available via DIALOG. https://celebrating200years.noaa.gov/foundations/tides/welcome.html. Accessed 7 July 2019

Ruan R (2001) Tide measuring and its technology development (in Chinese with English abstracts). Ocean Technol 20(3):68–71. https://doi.org/10.3969/j.issn.1003-2029.2001.03.018

Shi MC, Gao GP, Bao XW (2016) Methods of marine survey (in Chinese). China Ocean University Press, Qingdao

Woodworth PL, Pugh DT, Plater AJ (2015) Sea-level measurements from tide gauges. In: Shennan I, Long AJ, Horton BP (eds) Handbook of sea-level research, 1st edn. John Wiley & Sons, Chichester, pp 557–574

Wu JH, Han QS (1988) Theory, method and application of computational fluid dynamics (in Chinese). China Science Press, Beijing

Xie XJ, Zhai GJ, Huang MT (1988) Water level correction by time difference method (in Chinese with English abstracts). Hydrograph Surv Chart 8(3):22–26

Zhai GJ, Huang MT, Bao JY (2003) The requirement and status of hydrographic datum (in Chinese with English abstracts). Hydrograph Surv Chart 23(4):54–58. https://doi.org/10.3969/j.issn.1671-3044.2003.04.017

Zhao JH, Liu JN (2008) Multi-beam sounding and image data processing (in Chinese). Wuhan University Press, Wuhan

Zhao JH, Zhang HM, Hughes Clark JE (2006) Establishment of local seamless vertical datum (in Chinese with English abstracts). Geomat Inf Sci Wuhan Univ 31(5):448–450. https://doi.org/10.3321/j.issn:1671-8860.2006.05.018

Zhu GW, Huo SM (1990) The status and future development of tide and water level measurement technology in China (in Chinese with English abstracts). Ocean Technol 9(1):75–79

7 Constructing and Mapping a Seabed DTM

A digital terrain model (DTM) is the basis for submarine visual computing and automatic mapping of the seabed. A map is necessary for displaying the information of marine surveys. This chapter discusses three methods of constructing a seabed DTM, the merge-normalization (MN) method for multi-source bathymetric data fusion, the topological construction method of a submarine topographic map, the methods of constructing submarine geomorphologic maps, and construction of submarine topography three-dimensional (3D) maps based on DTMs.

7.1 Three Methods of Constructing a Seabed DTM

Mathematically, points that are arbitrarily distributed and have no definite relation among them are called discrete points. In fact, discrete data, which have the property of discrete points, often represent a natural object or phenomenon. To facilitate computer processing, grid points with certain structural relations are needed to replace discrete points. Discrete data are gridded by a mathematical model to obtain grid data. Various grid structures are used to meet different needs. A grid structure is suitable for computer image processing, a uniform regular mesh or non-equispaced rectangular mesh is generally suitable for computer graphics representation and analysis, and a curved grid is often used for the computer simulation of fluid dynamics. Grid data representing ground attributes are often referred to as a DTM, which is also known as a digital elevation model (DEM).

Commonly used mesh interpolation methods include inverse distance weighted interpolation, spline function interpolation, and Kriging interpolation.

7.1.1 Inverse Distance Weighting Method

The inverse distance weighted (IDW) interpolation method takes the numerical value at the mesh node that is the result of the influence of its adjacent discrete point values. Its numerical value is a weighted function of the adjacent discrete point values. The weight is calculated according to the distance.

$$f(x,\ y) = \left[\sum_{i=1}^{n} w(d_i) z_i\right] \Big/ \sum_{i=1}^{n} w(d_i) \quad (7.1)$$

where $w(d_i)$ is the weight function, z_i is the value of the ith discrete point, and d_i is the distance between the ith point and the grid point. The surrounding discrete points and grid points have a different amount of influence on $f(x, y)$ owing to their different positions. These influences are expressed by the weight function $w(d_i)$, which is mainly related to distance and sometimes to direction. If the discrete sample points are uniformly distributed in the four directions around the grid point, then the direction factor might not be considered. It has been demonstrated that $w(d_i)= 1/d_i^2$ is a good choice, that is, the size of $w(d_i)$ is determined by the inverse ratio of d_i squared. If the weight is calculated this way, the influence of the adjacent discrete sample points on the grid point decreases rapidly as distance increases. Hence, this interpolation method is called the IDW method.

In the interpolation process, how to select the known discrete sample points for estimating the value of each grid point is significant. If the discrete points are uniformly distributed and each discrete point is of the equal significance, then a number of discrete points can be searched for near each grid point to interpolate the grid point values one by one, thus establishing a mesh DTM. Usually, there are two ways to select the discrete points: one is that the number of

Z. Wu et al., *High-resolution Seafloor Survey and Applications*,
https://doi.org/10.1007/978-981-15-9750-3_7

sample points is determined based on the nearest n known points around the grid point; the other is to determine the sampling radius r, and any points that fall within a circle of radius r centered at the grid point are selected.

In our mapping system MBMap, the distance-weighted interpolation method is improved. We call the new method the azimuth-distance weighted interpolation method. That is, the value of grid node (x, y) is calculated as follows:

$$f(x, y) = \sum_{i=1}^{n} \lambda_i Z_i \quad (7.2)$$

where n is the number of total equipartitioned angles in the four quadrants with grid node (x, y) as the origin. The formula of the weight coefficient λ_i is as follows:

$$\lambda_i = \prod_{\substack{j=1 \\ j\neq 1}}^{N} d_j^2 \Bigg/ \sum_{K=1}^{N} \prod_{\substack{j=1 \\ j\neq 1}}^{N} d_1^2 \quad (7.3)$$

where d_j and d_1 are the distances from the nearest discrete sample point to be estimated in the ith and lth equipartition angles, respectively. The most obvious advantage of this method is that it saves operation time and only considers the N discrete sample points nearest the grid points instead of all N discrete points, which is necessary for the gridding interpolation of multi-beam data. At the same time, the N discrete sample points are selected according to the azimuth, which can reflect some local topographic changes and ensure the interpolation accuracy of multi-beam data.

7.1.2 Spline Function Method

The approximation of bivariate functional surfaces has always been an important mathematical basis of DTM interpolation. According to the vertical displacement between a discrete sample point and a mathematical surface, bivariate function approximation can be divided into two types: pure two-dimensional (2D) interpolation and surface fitting. The former requires that the differences in height between all the discrete sample points in the interpolation range and the corresponding points on the mathematical surface are zero. The latter does not require the mathematical surface to strictly pass through the discrete sample points, but the surface needs to satisfy some conditions; for example, the sum of the squares of the height differences between all the discrete sample points and the corresponding points must be minimized, that is, the least-squares condition must be met. Bivariate function approximation has long been used in the design of shell surfaces and mathematical lofting, but it may not be suitable for DTM interpolation. First, topographies do not have ready-made physical rules similar to those of elastic material surfaces, so there is no expression of a surface function in a real sense. Second, topographies are macroscopically rough, so the commonly used smooth condition of a surface is not applicable, and the convexity preserving condition of a mathematical surface should be considered. For this reason, the necessity of spline interpolation is discussed from the perspective of a pure 2D interpolation, but not as a form of surface fitting.

In general, complex topography needs to be approximated by a complicated mathematical surface, but this often complicates the problem. Hence, the complex topography is usually cut into a certain regular shape according to its characteristics and interpolated in blocks. For multi-beam data, such interpolation in blocks should not only be limited to topography, but also the size of block that can be input determined by the balance of interpolation accuracy and operation efficiency each time, then surface approximation interpolations in blocks are carried out frequently. For similar and equal size blocks, for more complex topography, more discrete sample points should be used to support the mathematical surfaces. Two-dimensional full-term polynomials of mathematical surfaces are described as follows:

$$f(x, y) = \sum_{i=0}^{m} \sum_{j=0}^{m} C_{ij} x^i y^j \quad (7.4)$$

where there are n undetermined coefficients c_{ij} $(i, j = 0, 1, 2, \ldots, m)$. To determine n undetermined coefficients, n linear equations are needed. Further, there needs to be a sample point value $\{x_1, y_1, z_1\}$ at n different plane positions in the interpolation block $(l = 1, 2, \ldots, n)$ to enable the system of linear equations of order n to have a unique solution. The relation between n and m is $n = (m + 1)^2$. In DTM block interpolation, polynomials of $m > 3$ are often called higher-order polynomials. When the missing polynomial is selected, the order of the linear system is n minus the number of missing terms. The interpolation quality of the mathematical surface that approximates the actual terrain is judged by the following three factors:

1. Convexity preserving requirement. If the numbers of fluctuations between the mathematical surface and the actual terrain are equal and the corresponding ridge and valley line positions and orientations are approximately the same, then convexity is well preserved. Otherwise, convexity is poorly preserved.
2. Fidelity requirement. In the region approximated by blocks, if the condition $\max |f(x, y) - z(x, y)| \leq \delta h$, is satisfied for mathematical surface $f(x, y)$ and real terrain $z(x, y)$, the mathematical surface meets the fidelity criterion. The allowable error δh depends on the specific situation. If this value is small, fidelity is good; if this value is large, fidelity is poor.

3. Smoothness requirement. For a curve, smoothness refers to the continuity of its curvature. For a real smooth surface, if its approximated surface contains edges or corners, the smoothness of the approximated surface is considered to be poor.

It is clear that a high-quality approximation of a curved surface should have good convexity and fidelity, and, if the actual terrain is smooth, the approximated curved surface should have good smoothness. The above three requirements are not mutually independent: convexity and fidelity are closely related. However, these three requirements are not compatible with each other; for example, convexity and smoothness are not compatible. Different areas of application differ in their need for each of these three requirements. The approximated curved surface of a seabed DTM primarily requires good fidelity and convexity. Multi-beam data can reflect various local topographic changes thanks to their high resolution, and it is hence especially necessary to emphasize good fidelity and convexity in this case.

It is easy to ensure smoothness because the higher order polynomials of $m > 3$ have $m - 1$ continuous derivatives. As long as these polynomials can be differentiated enough times, they will have a good smoothness. However, if a higher polynomial order is used, more discrete sample points are needed. Although the approximated surface has more points that coincide with the actual topography, in some parts between the discrete sample points, the height gap between the approximated surface and the actual topography is likely to increase sharply. In theory, higher order polynomials mean that more coefficients must be adjusted and additional sample points must be added to the calculation. Moreover, it is more difficult to control the oscillations between the sample points, and it will be more difficult to control the coefficients and sample points that have to be added. As far as practical application is concerned, higher-order linear equations must be solved to determine the coefficients of higher-order polynomials, and smaller disturbances such as rounding error and sample point measurement error can make the solution unstable. To solve the contradiction between smoothness, fidelity, and convexity, we can adopt a further partition in which each small block is approximated by a low-order polynomial. At the same time, according to the specific situation and conditions, the approximated surfaces of each small piece can be spliced together. Taking a one-dimensional case as an example, all adjacent piecewise approximated curves can be required not only to coincide with each other but also to have a continuous slope and curvature at the piecewise junction so that the approximated curve of the whole splice is also a continuous smooth curve. The quadratic derivability of a function is a prerequisite for curvature continuity, and a cubic polynomial is the lowest-order polynomial conforming to this condition. The curve approximated by a piecewise cubic polynomial is just a spline curve drawn by a cartographer with an active curve gauge. The binary spline function is also known as the piecewise polynomial function, and the binary polynomial spreads on each piece, whose order is usually not higher than three, yields

$$Z = \sum_{i=3}^{3} \sum_{j=3}^{3} C_{ij} x^i y^j$$

$$\boldsymbol{C} = \begin{bmatrix} c_{00} & c_{01} & c_{02} & c_{03} \\ c_{10} & c_{11} & c_{12} & c_{13} \\ c_{20} & c_{21} & c_{22} & c_{23} \\ c_{30} & c_{31} & c_{31} & c_{33} \end{bmatrix} \tag{7.5}$$

The decomposed matrix form is $Z = \boldsymbol{X}^{\mathrm{T}}\boldsymbol{C}\boldsymbol{Y}$, where $\boldsymbol{X}^{\mathrm{T}} = (1\ x\ x^2\ x^3)$ and $\boldsymbol{Y}^{\mathrm{T}} = (1\ y\ y^2\ y^3)$.

7.1.3 Kriging Interpolation Method

The Kriging interpolation method is based on the theory of regionalized variables and uses the semi-variant function as a tool. Under the condition that the variance of the estimation is minimal, the Kriging interpolation method considers the randomness of the terrain change as well as the correlation of the discrete samples. It gives the optimal linear unbiased estimation. At the same time, the estimated variance of the interpolation is given, which is not available in other interpolation methods. The estimated variance σ_E^2 is the mathematical expectation of the deviation squared between the actual terrain depth z_v and the estimated value $z_v{*}$, that is, $\sigma_E^2 = E\left[(z_v - z_v{*})^2\right]$.

The Kriging interpolation method is very similar to the distance weighting method in form. It uses the weight of the discrete sample point to estimate the value of the grid point, but the process of calculating the weight of the grid point is more complicated. First, the spatial structure is analyzed, and the difference in the data values and spatial distribution is analyzed using the values of the known points. Then, a variogram is produced to determine the suitability of the interpolation method. Second, the average or sliding window method is used to calculate the value of each grid point. The weight of the sample point is determined by the spatially continuous equation of the variation graph. Actual Kriging methods are divided into two types: one is the ordinary Kriging calculation method, which assumes that the variation of each grid value is not affected by any structural elements, and the other is the general-purpose Kriging algorithm, which assumes that the variation of each grid value is affected by the structural elements. Here only the basic principles of the ordinary Kriging method are discussed:

$$z_v{}^* = \sum_{i=1}^{n} \lambda_i z(x_i) \tag{7.6}$$

where the estimated value $z_v{}^*$ of any point can be obtained by a linear combination of n effective sample points in the influence range of the study region:

$$E\left[\sum_{i=1}^{n} \lambda_i z(x_i)\right] = \sum_{i=1}^{n} \lambda_i E[z(x_i)] = m \tag{7.7}$$

where λ_i is the weighted coefficient related to effective sample point $z(x_i)$, and the sum of weighted coefficients is 1 (Eq. 7.8).

$$\sum_{i=1}^{n} \lambda_i = 1 \tag{7.8}$$

To ensure the estimation is unbiased, the condition $E[z_v - z_v{}^*] = 0$ should be satisfied. Suppose the terrain change in the study area is stable. That is, the expected value of $z_v{}^*$ exists and is equal to a constant m, that is, $E[z_v{}^*] = m$. Because $E[z(x_i)] = m$, the above formula is an unbiased mathematical expression.

Another important equation needs to be established to ensure that the estimated variance is minimized. Using the semi-variance function, the estimated variance can be expressed as covariance, where $C(x_i, x_j)$ denotes the covariance between each sample point in the study area and $C(x_0, x_j)$ represents the covariance between the point to be estimated and another sample point. In addition, $C(x_0, x_0)$ represents the variance of the point to be estimated. To minimize the estimated variance σ_E^2 under unbiased conditions, the Lagrange operator is used to find the conditional extremum.

For n weighted coefficients of $\lambda_1, \lambda_2, \ldots, \lambda_n$, Lagrange operators are used to obtain partial derivatives and make them zero.

Hence, the Kriging equation can be rewritten as follows:

$$F = \sigma_E^2 - 2\mu\left(\sum_{i=1}^{n} \lambda_i - 1\right)$$
$$\sigma_E^2 = \sum_{i=1}^{n}\sum_{j=1}^{n} \lambda_i \lambda_j C(x_i, x_j) + C(x_0, x_0) - 2\sum_{i=1}^{n} \lambda_i c(x_0, x_i) \tag{7.9}$$

$$\begin{cases} \frac{\partial F}{\partial \lambda_i} = 0 \\ \frac{\partial F}{\partial \mu} = 0 \end{cases} \quad i = 1, 2, \cdots, n$$
$$\begin{cases} \sum_{j=1}^{n} \lambda_j C(x_i, x_j) - \mu = C(x_0, x_i) \; i = 1, 2, \cdots, n \\ \sum_{i=1}^{n} \lambda_j = 1 \end{cases} \tag{7.10}$$

This formulation contains $n + 1$ unknown variables and $n + 1$ equations, expressed using compact matrix coefficients as follows:

$$\boldsymbol{K} = \boldsymbol{\lambda M}$$

$$\boldsymbol{\lambda} = \begin{bmatrix} \lambda_1 \\ \lambda_2 \\ \lambda_3 \\ \vdots \\ \lambda_n \\ -\mu \end{bmatrix}$$
$$\boldsymbol{M} = \begin{bmatrix} C(x_0, x_1) \\ C(x_0, x_2) \\ \vdots \\ C(x_0, x_n) \\ 1 \end{bmatrix}$$
$$\boldsymbol{K} = \begin{bmatrix} C(x_1, x_1) & C(x_0, x_2) & \cdots & C(x_1, x_n) & 1 \\ C(x_2, x_1) & C(x_2, x_2) & & C(x_2, x_n) & 1 \\ & \vdots & & \ddots & \vdots \\ C(x_n, x_1) & C(x_n, x_2) & \cdots & C(x_n, x_n) & 1 \\ 1 & 1 & & 1 & 0 \end{bmatrix} \tag{7.11}$$

Usually, $\boldsymbol{K}$ is called the Kriging matrix. By solving the Kriging equation, we can obtain all the weighted coefficients of the points to be estimated ($\lambda_1, \lambda_2, \ldots, \lambda_n$). The value of the grid node is calculated by Eq. (7.6).

7.2 MN Method for Multi-source Bathymetric Data Fusion

Accurate and reliable bathymetric data are the basis of marine activities. A high-resolution digital bathymetric model (DBM) can be used not only for the construction of charts but also for the study of geomorphological features and genesis, seabed tectonic processes, marine resources, hydrodynamic flow models, biological habitats, and ecosystems (Smith et al. 1997; Bolmer et al. 2004; Wang et al. 2018; Zhou et al. 2018). The systematic study of marine bathymetry has lasted for more than a century, from extremely inefficient handlead sounding in the early years to the current multi-beam echo sounding (MBES) with high precision, resolution, and efficiency. However, more than 80% of the global sea has not been measured by MBES because of the vast global ocean area and the high cost of shipborne MBES (Mayer et al. 2018). Therefore, historical bathymetric data are still needed when constructing a large-scale DBM. At present, oceanic bathymetric data are collected independently for different purposes (Wu et al. 2016, 2018), resulting in many differences in the source, format, and resolution.

The construction of a DBM is also called data gridding. Gridding is a combination of 2D sampling, interpolation, and 2D extrapolation (Tobler et al. 1970). In current multi-source bathymetric data fusion research, the fusion method is mainly based on Kriging, bicubic spline, tension spline and other interpolation algorithms. For example, Jakobsson et al. (2012) fused multi-beam, single-beam, and historical bathymetric data to compile the International Bathymetric Chart of the Arctic Ocean and Arndt et al. (2013) obtained the International Bathymetric Chart of the Southern Ocean based on various sources of bathymetric data. Beaman et al. (2011) combined multi-beam, single-beam and coastline data that were collected in different periods to construct the DBM of the edge of George V Land in eastern Antarctica, and then they observed more details about the submarine geomorphology in this area. Although the above methods have fused multi-source bathymetric data, they only interpolate areas without data mechanically. Unfortunately, the spatial resolution of the fused data is low (>500 m), so their methods cannot ensure the interpolation precision of sparse data areas or data gaps.

To solve the above problem, we propose the merge-normalization (MN) method. The method combines multi-beam, single-beam, and electronic navigational charts with Shuttle radar topography mission (SRTM) dataset. This method fills the data gaps while retaining the detail of high-resolution topographic features. It was used to form a high-precision DBM of the Mariana Trench with a resolution of 100 m.

7.2.1 Data Sources

The Mariana Trench is the deepest region in the world and is located on the easternmost edge of the Philippine Sea. Due to its unique geographical location, the Mariana Trench has always been a hotspot of global marine research studied by scientists from all over the world (Nakanishi and Hashimoto 2011; Gardner et al. 2014). In recent years, China has organized several integrated marine surveys in this region. A high-precision DBM of the trench area is necessary for carrying out deep-sea exploration in a submersible, *in-situ* observation of the seabed, and numerical simulation of plate subduction (Zhou et al. 2015). Because of the low resolution and poor precision of the existing data in the Mariana Trench, it is difficult to meet the needs of marine scientific research and engineering applications. Additionally, the data sources for this area are diverse and many data gaps exist (Fig. 7.1), making it an ideal study area for developing bathymetric data fusion techniques.

The main sources of data used in this chapter are multi-beam bathymetric data, single-beam bathymetric data, electronic navigational chart data and SRTM dataset.

The main data source for the MN data fusion method is high-precision multi-beam bathymetric data, which have been collected since 2003. A total of 45 cruises of multi-beam bathymetric data have been collected in the Mariana Trench, of which, 35 cruises were operated by the National Geophysical Data Center (NGDC), National Oceanic and Atmospheric Administration (NOAA). Another 10 cruises were operated by the Japan Agency for Marine-Earth Science and Technology (JAMSTEC).

In addition, a large number of single-beam soundings have been collected by the NGDC since 2000. Because the corresponding information of the collected single-beam bathymetric data such as information about the echo sounders and positioning systems are not provided, it is difficult to evaluate the accuracy of these single-beam soundings. Moreover, the single-beam soundings have a nonuniform spatial distribution, and are distributed densely in survey areas using multi-beam echo sounders, e.g., the region near Guam, US.

Meanwhile, since 2005, four electronic navigational charts in the Mariana Trench were collected by the Office of Coast Survey of the NOAA. No relevant metadata description other than some basic information about electronic navigational charts is provided for them, so their accuracy cannot be fully evaluated.

Although there is a large amount of bathymetric data, there are still data gaps of up to thousands of square kilometers. In this study, the SRTM dataset, which is currently considered to be a mainstream international bathymetric dataset, is used to fill these gaps. The SRTM_15 dataset is the latest version of the SRTM dataset and its predicted bathymetric data were retrieved based on observational data and satellite altimetry.

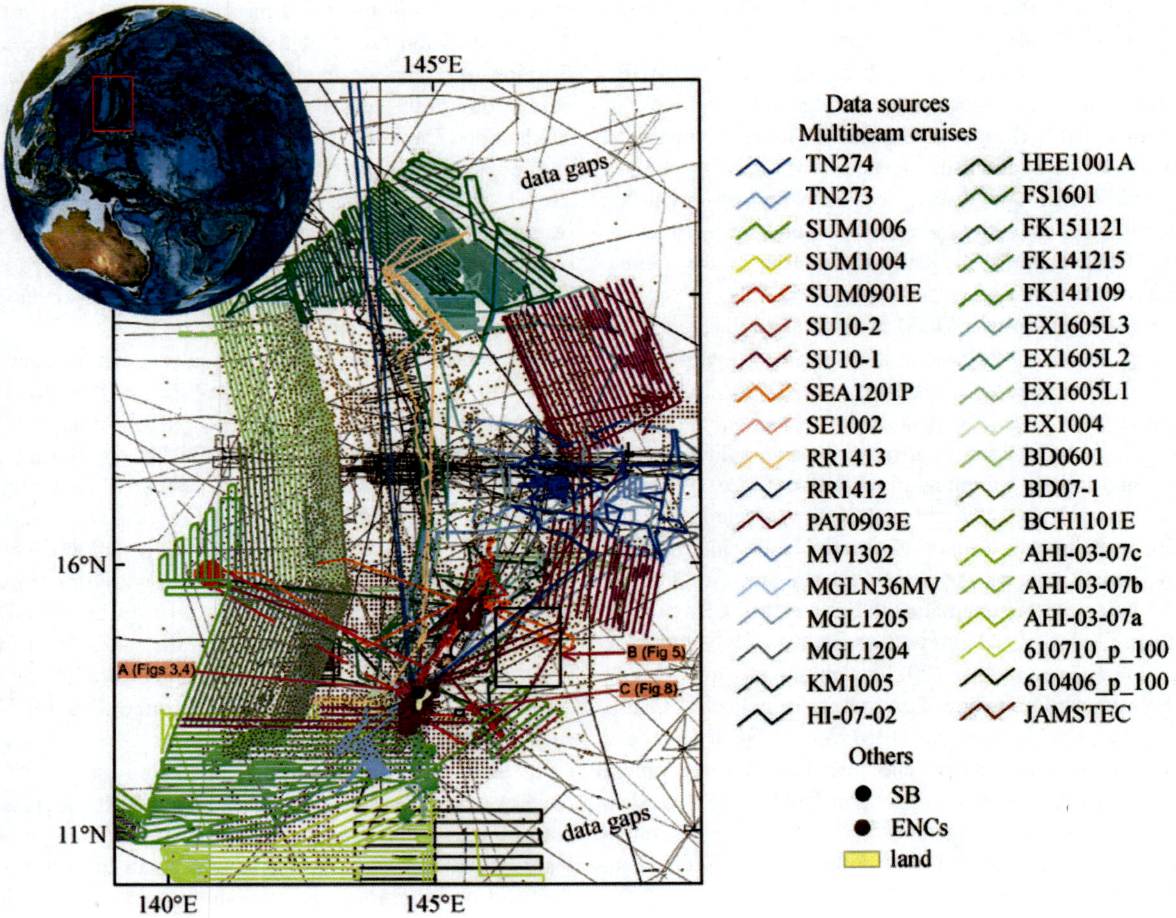

Fig. 7.1 Location and data sources of the Mariana Trench. SB: single-beam data, ENCs: electronic navigational chart data

7.2.2 MN Data Fusion Method

The core aim of the MN method is to merge and normalize bathymetric data. The method consists of the following four main steps, as shown in Fig. 7.2:

(1) Data Pre-processing and Accuracy Assessment

According to the various sources and types of bathymetric data, the multiple sources of raw data are carefully processed. Taking multi-beam echo soundings as an example, system errors are eliminated by processes such as tide correction, sound velocity profile correction, and transducer installation deviation calibration (Wu et al. 2017). Outliers are also identified and eliminated by manual or automatic filtering. However, the values of random errors can only be obtained using the accuracy of the bathymetric data. A bathymetric database is then constructed to manage the processed multi-source bathymetric data.

It is important to assess the accuracy of the multi-source bathymetric data to ensure the quality of the final grid data. The accuracy of multi-source data from the same area is evaluated using the mean difference (Jakobsson et al. 2012) based on the central beam soundings of multi-beam swath-bathymetric data as follows:

$$\delta = \frac{1}{q}\sum_{i=1}^{q}(D_{S_i} - D_{M_i}) \qquad (7.12)$$

where δ is the mean bathymetric difference, q is the number of bathymetric data, D_{M_i} denotes the multi-beam bathymetric data, and D_{S_i} denotes multi-source bathymetric data, which have negative values.

According to the *International Hydrographic Organization Standard for Hydrographic Surveys*, the bathymetric deviation should be within 2% of the bathymetry. If the deviation is less than 2% of the depth, no further processing is required. Otherwise, they need to be processed using a least-squares adjustment based on the central beam of the

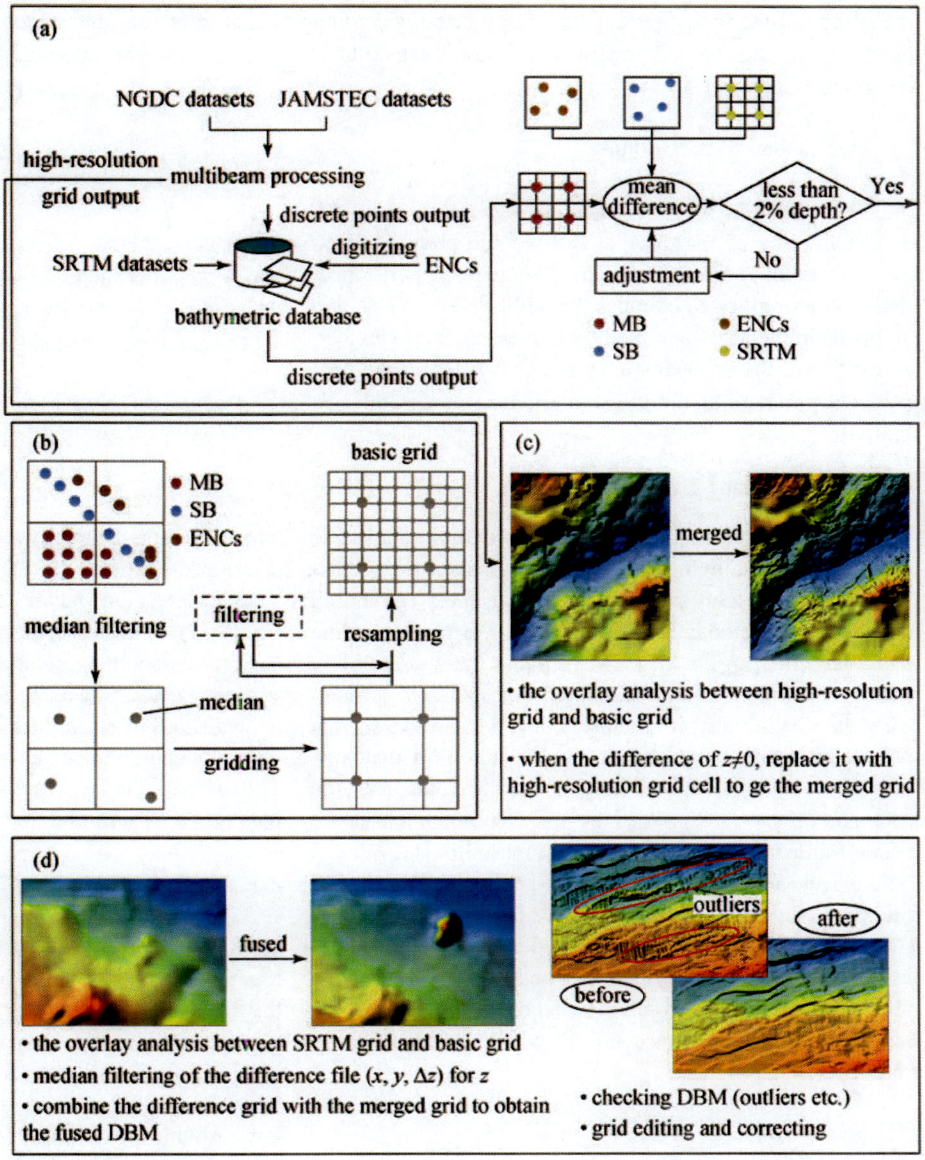

Fig. 7.2 Principle and workflow of the MN method. MB: multi-beam data. **a** Data pre-processing and accuracy assessment; **b** gidding; **c** multi-resolution grid merging; **d** nomalization

multi-beam swath data (Salzmann et al. 1994; Bjorke et al. 2005). The error is represented as follows:

$$F(x,y) = a_0 + a_1x + a_2y + a_3x^2 + a_4xy + a_5y^2 \tag{7.13}$$

where (x, y) is the coordinate of the measuring point and a_0, a_1, a_2, a_3, a_4 and a_5 are coefficients to be determined.

Let the correction of the bathymetric observation value Z be V. Based on the central beam of the multi-beam swath-bathymetric data, the following equation for the error is established:

$$-(V_{S_i} - V_{M_i}) = (a_0 + a_1x + a_2y + a_3x^2 + a_4xy + a_5y^2)_{S_i} - (a_0 + a_1x + a_2y + a_3x^2 + a_4xy + a_5y^2)_{M_i} - (Z_{S_i} - Z_{M_i}) \tag{7.14}$$

where V_{S_i} and V_{M_i} are the corrections of the multi-source data and multi-beam data at point (x, y), respectively, and Z_{S_i} and Z_{M_i} are their respective bathymetric values.

The error correction coefficients are obtained by solving Eq. (7.14) as follows:

$$\sum (V_{S_i} - V_{M_i})^2 = \min \tag{7.15}$$

Finally, all the bathymetric data are corrected according to Eq. (7.13), and the bathymetric values are obtained after the above adjustment.

(2) Multi-source Data Gridding

To avoid the influence of potential outliers in the grid, median filtering of the discrete bathymetric points derived from the multi-source bathymetric database is performed. Median filtering first determines the odd cells of grid G, and the pixels in the grid are sorted by numerical value. Finally, the median of the original values $s(x, y)$ is taken as the final bathymetry $d(x, y)$ in the grid node (Mann et al. 2001) as follows:

$$d(x, y) = \text{median}\{\ s(x-u,\ y-v),\ (u,\ v) \in G\} \quad (7.16)$$

To ensure the accuracy of data interpolation in the sparse region, the filtered multi-source bathymetric data are gridded to form a low-resolution grid. One of the most commonly used gridding methods in the geosciences is bicubic spline interpolation (Briggs et al. 1974; Hell and Jakobsson 2011). This method can obtain smoothed interpolated surfaces using the given points accurately, but it is easy to generate outliers in terrain mutations or sparse data areas that may obscure real topographic details. In this chapter, the multi-source data are gridded by the continuous curvature tension spline interpolation algorithm (Smith and Wessel 1990). This algorithm eliminates large fluctuations in the minimum curvature surface fitting due to the introduction of a tension factor. Regardless of the deformation point, the results after gridding are smoother and more accurate than other methods. The algorithm is based on the principle of the minimum curvature surface fitting, which is calculated as follows:

$$\begin{cases} \nabla^2(\nabla^2 z) = \sum_i^N f_i \delta(x - x_i,\ y - y_i) \\ z = z(x_i,\ y_i) \\ \nabla^2 = \frac{\partial^2 z}{\partial x^2} + \frac{\partial^2 z}{\partial y^2} \end{cases} \quad (7.17)$$

where (x_i, y_i, z_i) are the coordinates of a given point, N is the number of given points, f_i is the coefficient in a linear combination solution of Green's function at point (x_i, y_i), $\delta(x - x_i,\ y - y_i)$ is the given response function, and ∇^2 is the Laplacian.

The boundary conditions are as follows:

$$\begin{cases} \frac{\partial^2 z}{\partial n^2} = 0 \\ \frac{\partial(\nabla^2 z)}{\partial n} = 0 \\ \frac{\partial^2 z}{\partial x \partial y} = 0 \end{cases} \quad (7.18)$$

where n is the unit vector of the normal direction of the surface and $\frac{\partial}{\partial n}$ is a derivative normal to an edge.

On this basis, tension factor T is introduced to obtain

$$(1 - T_I)\nabla^2(\nabla^2 z) - T_I \nabla^2 z = \sum_i^N f_i \delta(x - x_i,\ y - y_i) \quad (7.19)$$

where T_I is the internal tension factor of the surface that is taken in the interval [0, 1).

The boundary conditions are as follows:

$$\begin{cases} (1 - T_B)\frac{\partial^2 z}{\partial n^2} + T_B \frac{\partial z}{\partial n} = 0 \\ \frac{\partial(\nabla^2 z)}{\partial n} = 0 \\ \frac{\partial^2 z}{\partial x \partial y} = 0 \end{cases} \quad (7.20)$$

where T_B is the boundary tension factor of the surface that is taken in the interval [0, 1).

When tension factor $T = 0$, Eq. (7.19) simplifies to Eq. (7.17). In this case, there is no tension, and the minimum curvature smoothing solution is obtained. When $T = 1$, the tension reaches its maximum and the harmonic spline solution is obtained. For ocean bathymetric data, the tension factor is generally chosen to be between 0.32 and 0.4 (Jakobsson et al. 2012). Then, the low-resolution grid is filtered and is resampled to the target resolution to obtain a "basic" grid.

(3) Multi-resolution Grid Data Merging

To merge the multi-resolution grid data and preserve the topographic details of the high-precision bathymetric areas, the low-resolution grid data are replaced with the high-resolution grid data. First, the high-resolution grid obtained in the pre-processing step is superimposed on the basic grid to obtain the difference at each position. The low-resolution data point is replaced by the corresponding values in the multi-beam data grid points unless the difference is zero. Then, we obtain a "merged" grid.

To eliminate the edge effect in the grid merging process (Zhao et al. 2014), a buffer is defined on the side of the high-resolution grid data. In addition, the width of the buffer is determined according to the joint areas between the high-resolution and the low-solution grid, and must be at least three grid points. Its principle is based on the hyperbolic weighting function (Arndt et al. 2013):

$$W = \frac{1}{L^2} \quad (7.21)$$

where L is the distance to next constraint unit of the high-resolution data or constraint unit outside of the buffer and W is the weight.

It can be seen from Eq. (7.21) that the closer to the high-resolution data grid unit, the smaller the influence of the depth from the low-resolution data, which weakens the probability that an edge effect will occur.

(4) Normalization

In the data gaps, the SRTM grid and the merged grid are "differenced." That is, each data point in the SRTM grid is subtracted from the value of the interpolated data point at the same position to obtain a difference dataset, which includes the difference in longitude, latitude, and depth. The difference data are median filtered and gridded to obtain a "differenced" grid. At last, the differenced grid is integrated into the merged grid to form the fused DBM.

The DBM obtained through the above steps inevitably has various outliers such as "glitches", so a 3D terrain browsing method is used to find these problem areas. The artifacts are highlighted by adjusting the illumination and edited by a surface fitting algorithm to obtain the best DBM (Ware et al. 1991; Calder and Mayer 2001). The principle of the surface fitting algorithm is as follows: the seabed surface is fitted according to the beam points, the depth differences between the measured bathymetric data and the surface are calculated, and the outliers are eliminated by incorporating error processing theory. The general form of the surface fitting function is as follows:

$$z = \{\Psi(x, y) | \sum_{l=0}^{k}\sum_{m=0}^{l} h_{lm} x^{l-m} y^{m},\ (x, y, z) \in Q(x_c, y_c, z_c)\} \tag{7.22}$$

where $\Psi(x, y)$ is the surface fitting function, (x, y, z) denotes the space coordinates of the beam points, h_{lm} is a polynomial coefficient, k is the total order of the polynomial, (x_c, y_c, z_c) is the point c to be detected, and $Q(x_c, y_c, z_c)$ denotes the local surface centered on c that is fit by the surface fitting function.

7.2.3 Implementation of the MN Data Fusion Method

According to the above steps, the multi-source bathymetric data were fused to construct the DBM of the Mariana Trench. First, the multi-source bathymetric data were transformed into a discrete data format. The projection and vertical data were unified with respect to the Universal Transverse Mercator projection and mean sea level, respectively. Then, systematic error processing was performed, and outliers in the multi-source bathymetric data were cleaned using the combined uncertainty and bathymetric estimator (CUBE) algorithm (Calder and Wells 2007). Finally, these data were re-exported into a discrete format to form a multi-source bathymetric database of the Mariana Trench.

To evaluate the accuracy of the multi-source bathymetric data, the depth differences of the multi-source bathymetric data of region A in the northwestern part of Guam (Fig. 7.1) were analyzed. This area is fully covered by multi-beam data with high precision, so it is suitable for an accuracy evaluation. The discrete bathymetric points of electronic navigational charts and single-beam data were compared with the central beam sounding results of multi-beam swath-bathymetric data with a resolution of 100 m (Fig. 7.3). Every data point is combined with a multi-beam data point to form a pair such that the distance between the two points is less than 50 m (Jakobsson et al. 2012). A total of 115 pairs of electronic navigational chart data and 207 pairs of single-beam bathymetric data were selected, and the results are compared in Table 7.1 and Fig. 7.3 according to Eq. (7.12).

According to the *International Hydrographic Organization Standard for Hydrographic Surveys*, for electronic navigational chart data, the mean bathymetric difference is better than 2% of the bathymetry when the mean depth is 1 039 m. For the single-beam data, the mean bathymetric difference is 23.2 m when the mean depth is 1 278 m, which is also better than 2% of the bathymetry. The accuracies of the single-beam data and electronic navigational chart data are within the tolerance and hence can be used as datasets in the data fusion experiment.

Based on the results of the central beam sounding of the multi-beam swath-bathymetric data in Region A, the accuracy of the SRTM_15 data was evaluated. A spatial distribution of the bathymetric differences between SRTM_15 and multi-beam data was obtained at a resolution of 100 m (Fig. 7.4). A comparison of 29 605 pairs of bathymetric data points is presented in Table 7.1 as well as a pie chart (Fig. 7.4, inset). When the mean depth is 1 185 m, the mean depth difference is less than 2% of the bathymetry. The data accuracy of SRTM_15 is also within the tolerance, so it can be used as a dataset in the fusion experiment for data gaps.

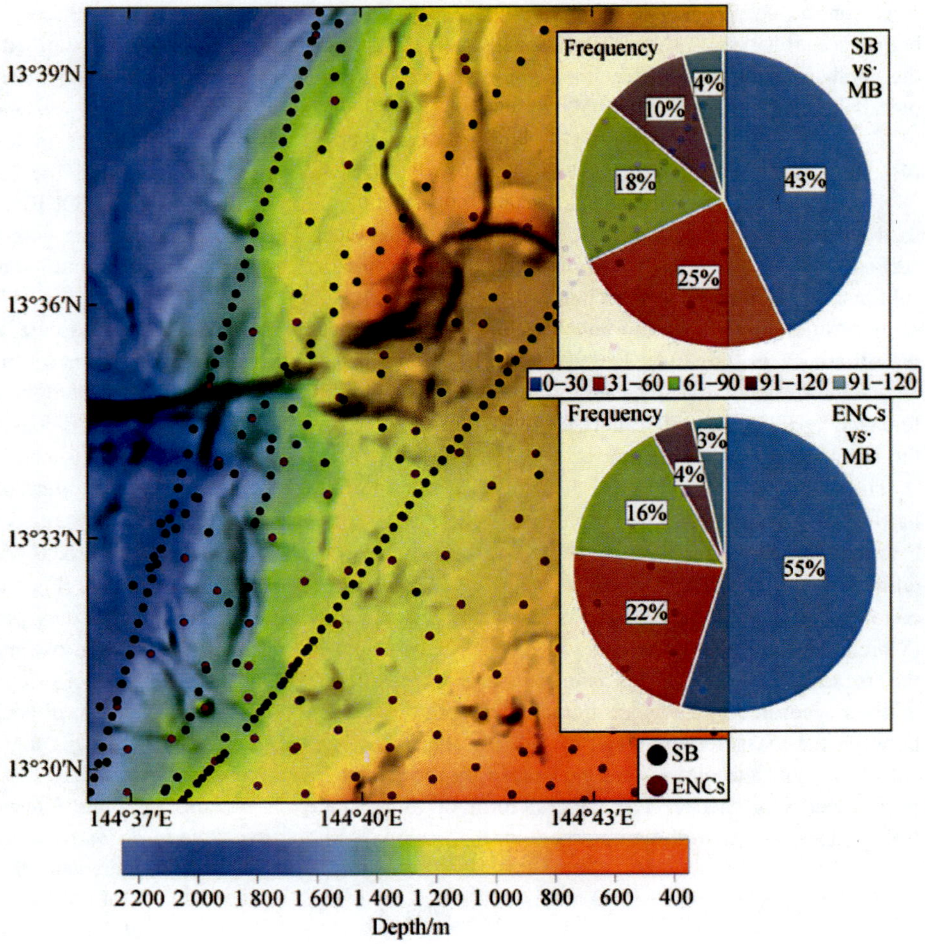

Fig. 7.3 Comparison of the bathymetry of multi-source data

Table 7.1 Comparison of multi-source data accuracy assessment

Object	Mean difference/m	Mean depth/m	±2%*depth/m
ENCs versus MB	18.9	1 039	±20.7
SB versus MB	23.2	1 278	±25.5
SRTM_15 versus MB	10.7	1 185	±23.7

Region B (Fig. 7.1) is used as an example for the multi-source bathymetric data fusion experiment. Median filtering was performed on the multi-source bathymetric data with a grid cell width of 1 000 m (Fig. 7.5a). It was then gridded to form a low-resolution grid at a resolution of 1 000 m. The low-resolution grid was smoothed and filtered simultaneously. Finally, the low-resolution grid was resampled to 100 m to form the basic grid (Fig. 7.5b).

A high-resolution grid with a resolution of 100 m was obtained by processing multi-beam data with steps such as sound velocity profile correction and outlier rejection (Calder and Mayer 2003; Zhao et al. 2015; Yang et al. 2017) (Fig. 7.5c). Then, the high-resolution grid and the basic grid were superimposed to obtain the difference between them at the same position. If the difference of z at a point was non-zero, we replaced it with the corresponding bathymetric value in the

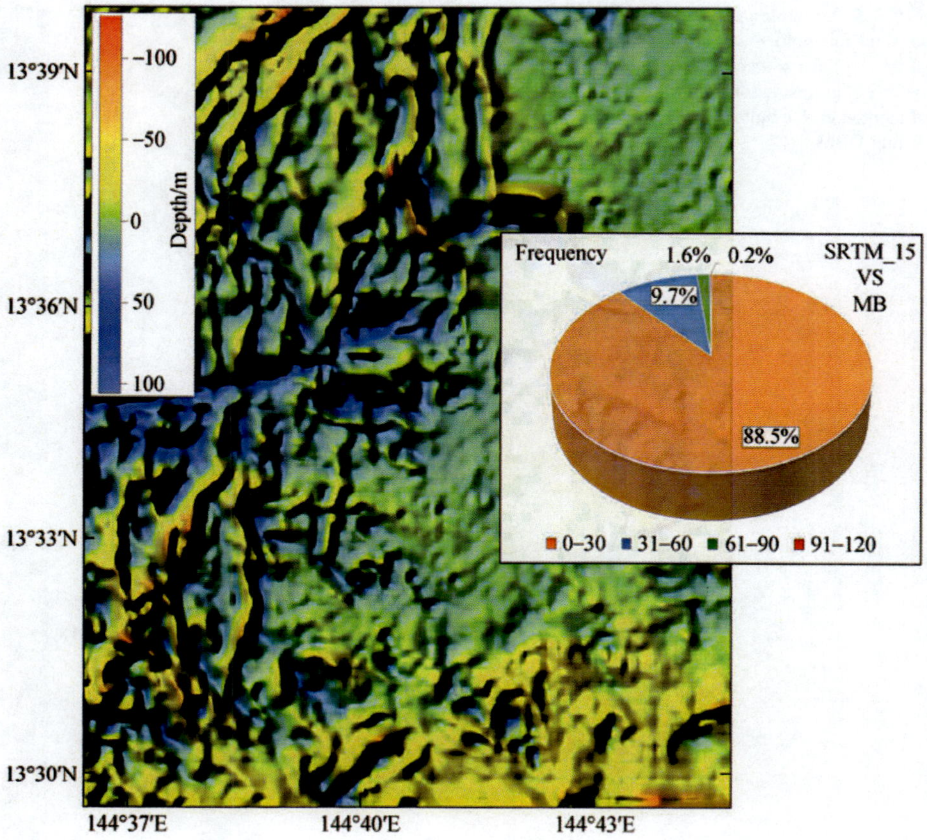

Fig. 7.4 Comparison of the bathymetry obtained using SRTM_15 and multi-beam data

high-resolution grid to obtain the merged grid (Fig. 7.5d). To eliminate edge effects between multi-resolution grids, we defined a 5 km buffer near the high-resolution grid data boundary (Fig. 7.5e), which greatly reduces the probability of edge effects (Fig. 7.6).

SRTM_15 data were used to fill the data gaps. First, the SRTM_15 grid was fused with the merged grid to obtain the bathymetric difference dataset at the same position within the data gaps. The difference dataset includes the longitude, latitude, and depth difference, and then it was subjected to median filtering with a grid size of 1 000 m. The filtered difference file was then gridded to create a difference grid with a resolution of 100 m. Finally, the difference grid was integrated with the merged grid to obtain the fused DBM. The last step is to visualize the combined grid to identify and highlight outliers. After that, problem areas of the grid were edited and errors were processed to obtain a satisfactory DBM (Fig. 7.5f).

Figure 7.7 shows the high-resolution DBM of the Mariana Trench obtained using the MN bathymetric data fusion method with a grid resolution of 100 m. The coverage rates of the multi-beam bathymetric data, single-beam bathymetric data, and electronic navigational chart data in this area are 43.3%, 0.6% and 0.2%, respectively, and the remaining 56% of the data gaps are filled with SRTM data. The final DBM not only preserves the detailed seabed features of the high-resolution data area but also provides a smooth topography for sparse data areas and data gaps.

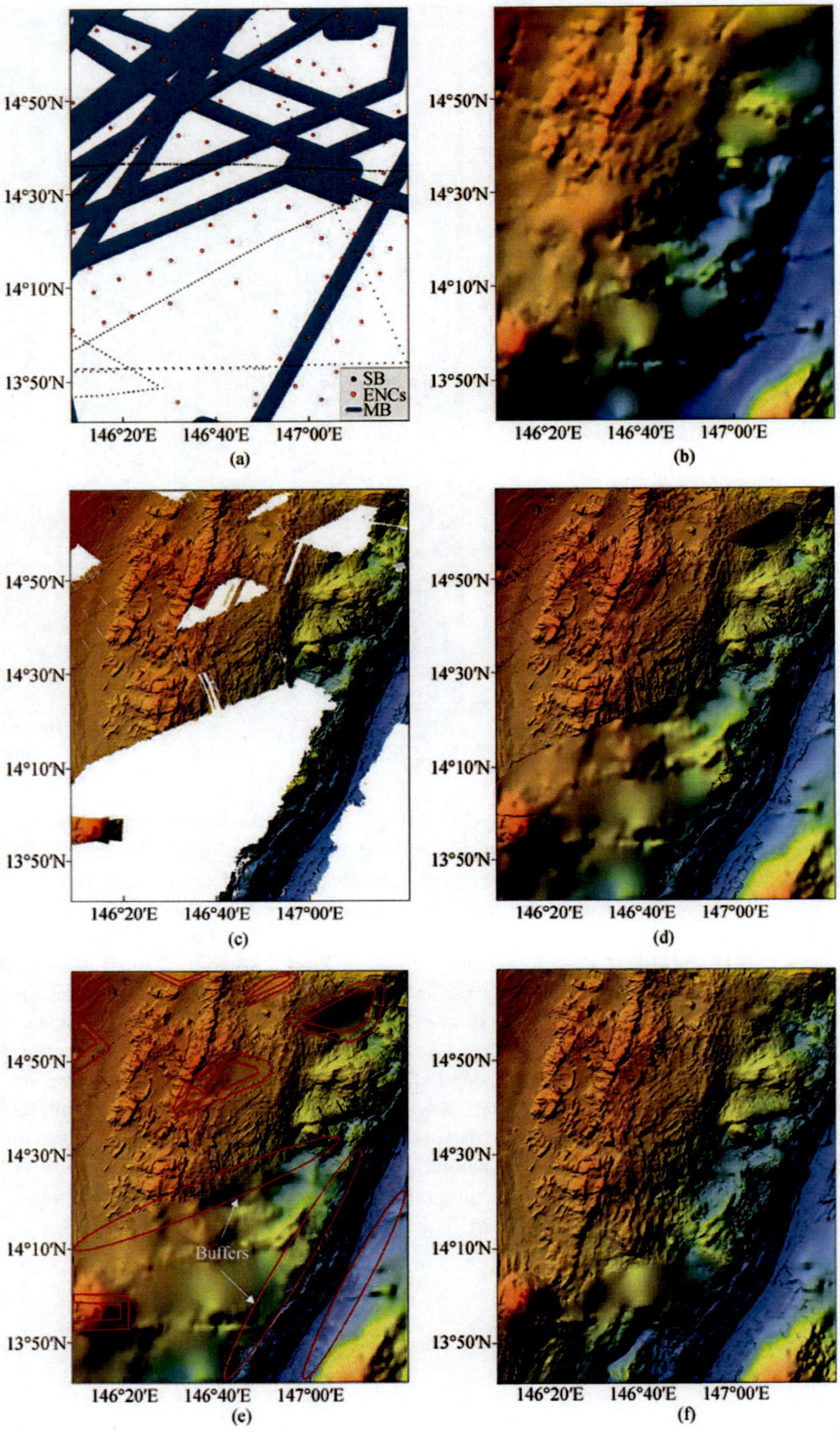

Fig. 7.5 Example of the MN method for multi-source data fusion. **a** Data sources; **b** basic grid; **c** high-resolution grid; **d** merged grid; **e** buffered grid and **f** final DBM

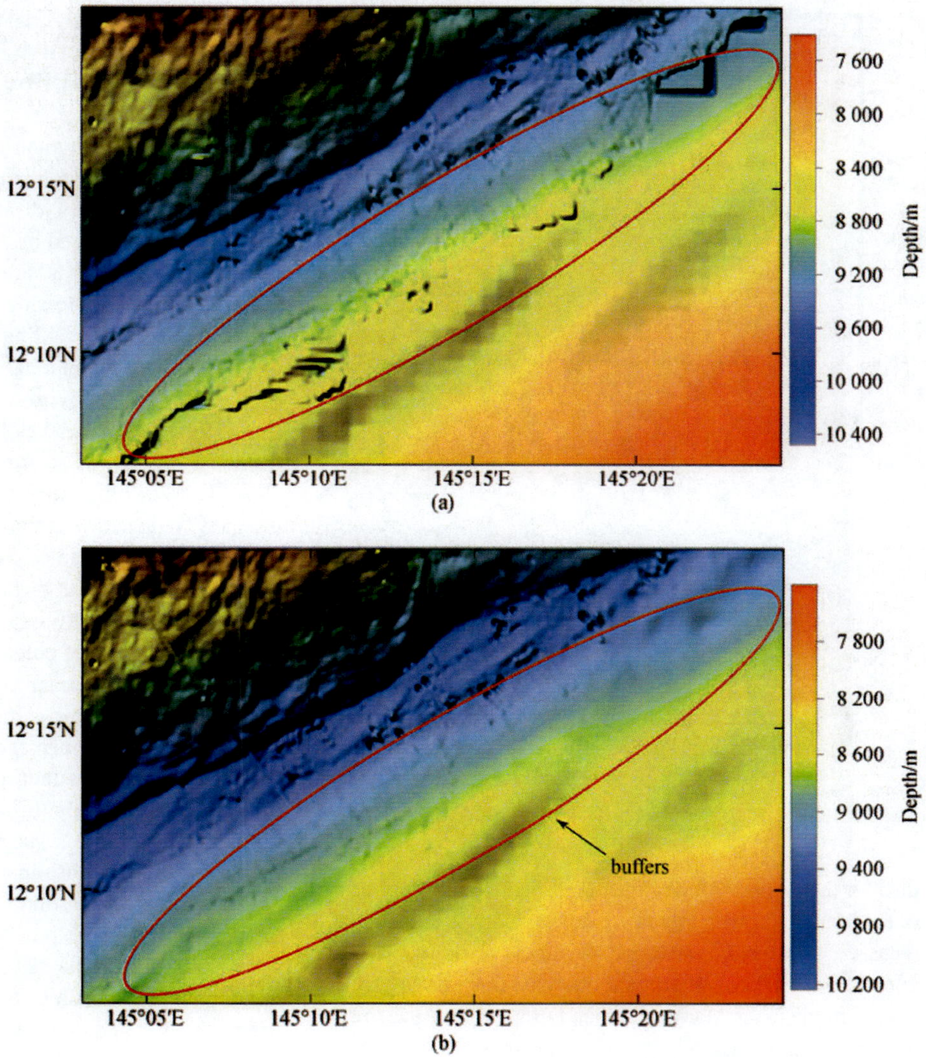

Fig. 7.6 Comparison of the DBM before defining the buffer **(a)** and after defining the buffer **(b)**

7.2.4 Comparison of Data Stitching Methods and Data Details

Except for the DBM built in this article, the area of the Mariana Trench is only covered by global datasets. This article compares the Mariana Trench DBM with the 2014 version of the General Bathymetric Chart of the Oceans (GEBCO) DBM, which is a bathymetric dataset widely used in the geosciences. The results show that the MN method achieves ideal results at the data stitching area. The overall tolerance of the joint is in accordance with the *International Hydrographic Organization Standard for Hydrographic Surveys*, which states that when the depth is greater than 100 m, the error of the depth measurements should be less than 2% of the bathymetry. In the experimental area, Region C (Fig. 7.1) with a large difference in the junction of the bathymetric data is selected, and Profiles a-a' (Fig. 7.8a) and b-b' (Fig. 7.8b) are obtained for the conventional and MN data fusion methods, respectively. It can be seen that the difference between the two methods is mainly concentrated in the region within 2-4 km of the beginning of the section line (the blue sections in Fig. 7.8). Sections A and B are located at a distance between 2.6 and 2.8 km of Profiles a-a' and b-b', respectively, with the water depth of about 3 600 m and a limit of the water depth error of 50 m. For the traditional method, the mean slope is 72.32°, whereas for the MN method, a mean slope of 37.63°, so the topography at the data stitching area of the latter method is more smoothly.

In addition, it is also found that a higher proportion of high-resolution bathymetric data leads to a higher grid resolution. Using the MN method, the Mariana Trench DBM with a resolution of 100 m now contains the most accurate

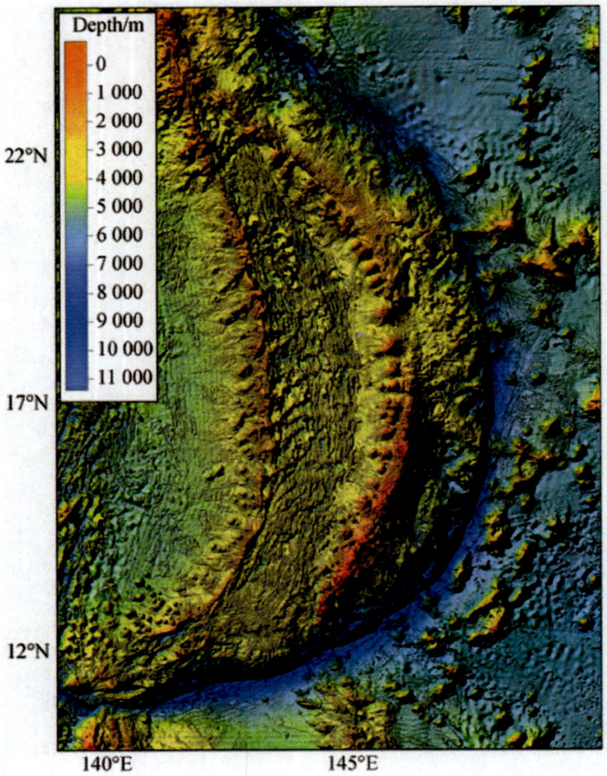

Fig. 7.7 Final DBM of the Mariana Trench

data for this region. At the same time, the topographical details of the Mariana Trench region are substantially improved in this dataset. For example, seamounts with more obvious shapes and contours as well as submarine canyons and ridges can be clearly found in the Mariana Trench DBM (Figs. 7.9a2, b2 and c2). In contrast, such topographical features are vague in the GEBCO_2014 dataset (Fig. 7.9 a1, b1 and c1). In the sparse data areas, the low-resolution grid obtained using processed single-beam or electronic navigational chart data avoids the occurrence of outliers (Fig. 7.9d1 and d2). However, there are still areas with poor data density and quality. These areas contain potential outliers and noise points, which need further investigation to identify and process.

7.3 Topological Construction Method of a Submarine Topographic Map Based on the DTM Boundary

The multi-beam bathymetric data obtained from the seabed have high density, high precision, and full coverage. It is difficult to intuitively display the high density and high precision data surveyed by a multi-beam bathymetric system with a traditional isoline map. Hence, the common multi-beam post-processing systems generally use a graphic model or 2D color filling maps to draw the submarine topography. Although a 3D figure can intuitively display the submarine topography, it cannot accurately locate the water depth point, which makes the accurate use of the map difficult. At present, 2D color filling maps are widely used.

A 2D color filling map renders the map using a color that is exactly matched with the water depth based on the original isoline map. A 2D color filling map not only preserves all the line characteristics of the isoline map, but the change in submarine topography can also be displayed intuitively by the gradual change of color. However, because of the stages and particularity of ocean surveys, there are no data in some areas of the map frame at the time of mapping, which results in many irregular inner and outer boundaries of the DTM (Fig. 7.10). These boundary lines make the implementation of a vector filling algorithm more difficult. For the above reasons, we propose a vector filling algorithm based on the full topology of the isoline based on DTM boundaries.

The essence of isoline vector filling is to establish a topological relationship between isolines, that is, to establish a tree-structured relationship. It is difficult to establish such a topological relationship between isolines in the case of irregular inner and outer boundaries and arbitrary equivalent spacings. In this section, based on the accurate tracking of DTM boundary lines under the condition of a regular grid, the topological relationship between unclosed contours is determined by the DTM boundaries, and closed polygons are constructed using a sequential tracking method to track unclosed isolines along a DTM boundary line. Then, according to the order of the isoline, closed polygons are constructed using the minimum border and rotation angle (or

Fig. 7.8 Comparison of data stitching areas. Profiles obtained by the traditional method (a) and the MN data fusion method (b)

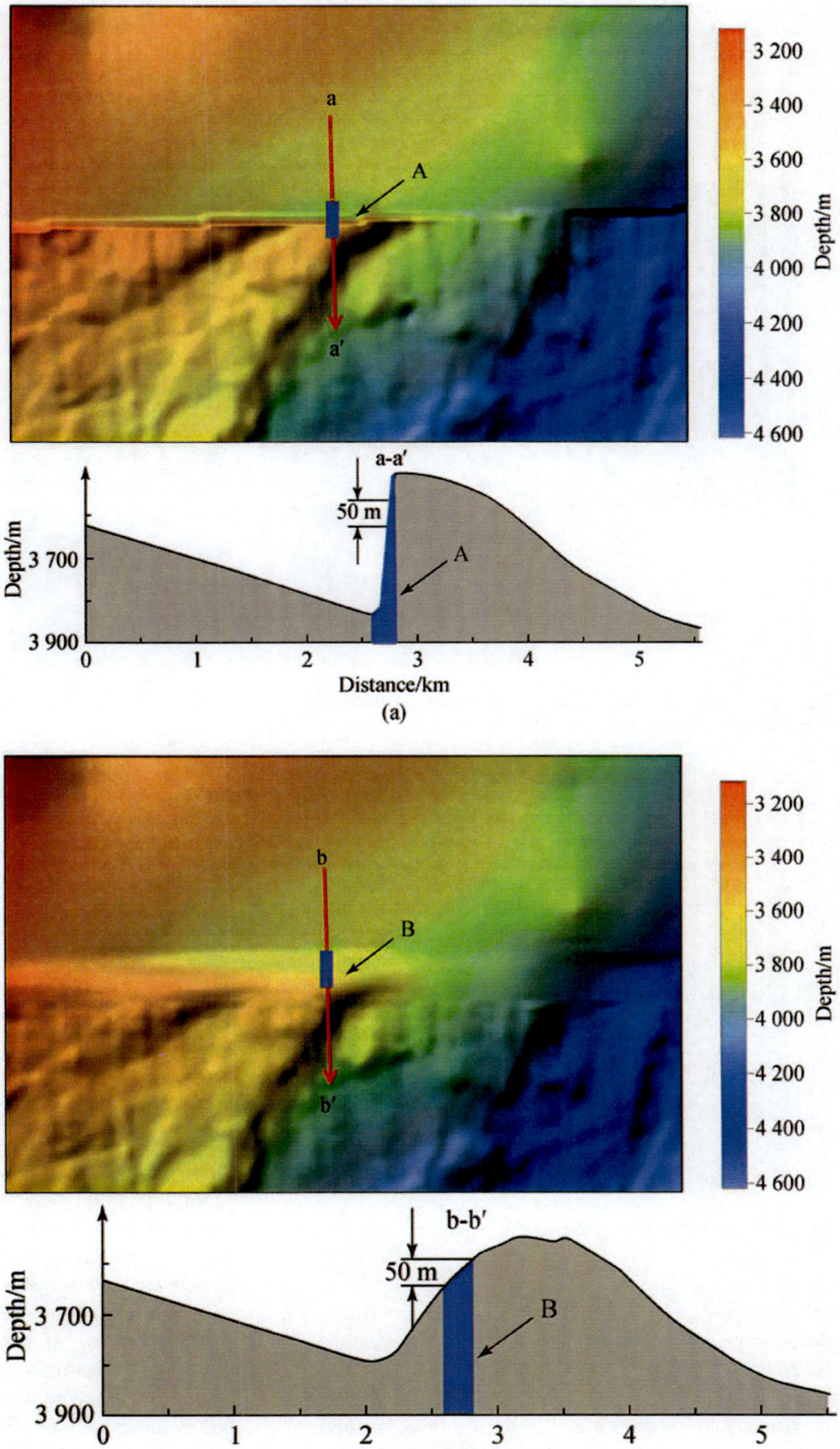

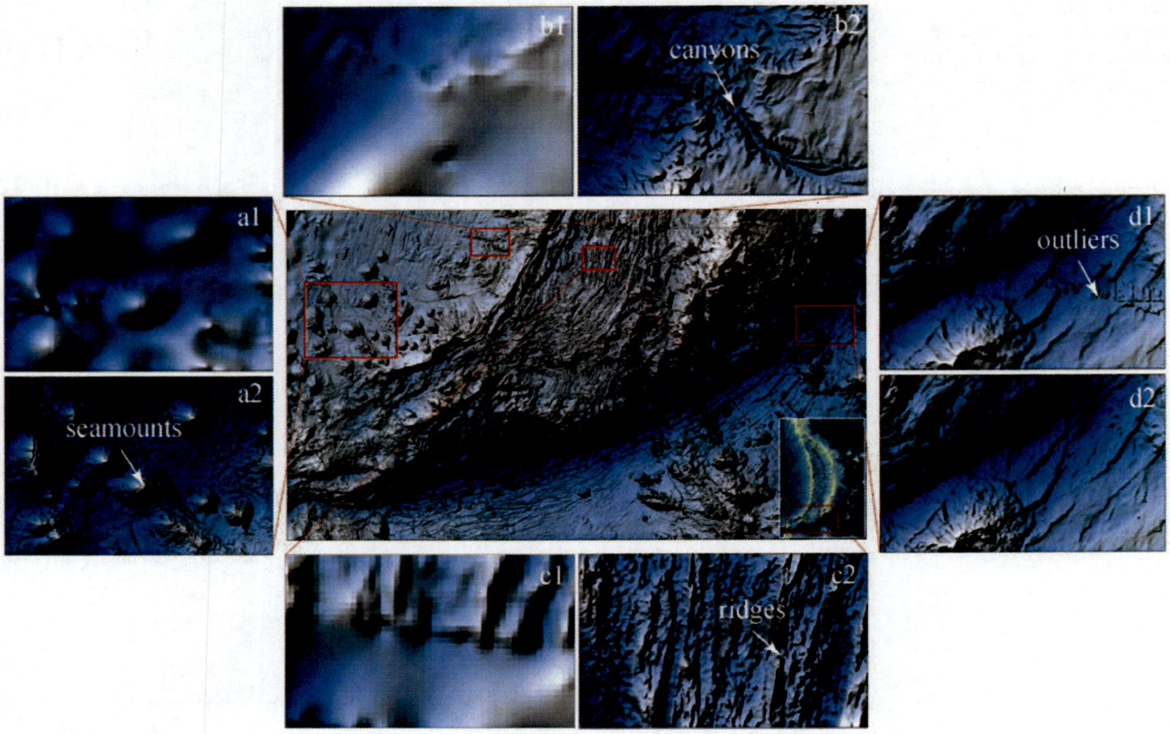

Fig. 7.9 Comparison of data details of the Mariana Trench DBM and GEBCO_2014 DBM. Index 1 indicates GEBCO DBM and Index 2 indicates the DBM obtained by the MN method

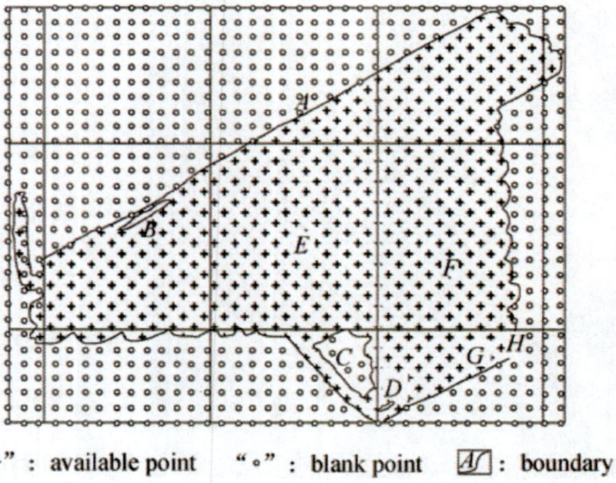

Fig. 7.10 Tracked DTM boundary lines

ray) methods, and the nesting relationship between the closed polygon, closed isoline, and inner boundary line is determined. Finally, the corresponding color or pattern is matched according to the isoline level of the closed polygons, and isoline vectors are filled for any boundary and any equivalent spacing. The algorithm is divided into the following steps: (1) tracking the DTM boundary line; (2) sequential inserting the unclosed isoline into the DTM boundary line; (3) establishing the topological relationship of the unclosed isoline based on the DTM boundary line; (4) tracking closed polygons and establishing nesting relations based on the DTM boundary; and (5) filling the isoline map in the nesting order.

7.3.1 Tracking of the DTM Boundary Line

In a multi-beam survey, because of the frequent loss of a beam due to sea conditions or man-made factors, the collected data do not fully cover the sea floor locally. In addition, because the topography of the survey area can be too complex, the survey line distance may not be adjusted in time for the survey. As a result, the data over-cover the areas in deep water area and do not completely cover shallow water areas. In the construction of the seabed DTM, these under-covered areas will leave some blank grid points. An island is also an area that cannot be measured by a multi-beam survey. If there is no elevation data for the island, there will also be a blank area in the construction of the seabed DTM. The boundaries of these blank areas are called the inner boundaries of the DTM. In a standard map, a large amount of data is needed, and sometimes the multi-beam data obtained is not sufficient. The maximum outer limit of these data is called the outer boundary. In

Fig. 7.10, *A*-*H* are boundary lines, where *A* is the outer limit of the survey area. Here, boundary *B* exists because the distance between the measuring lines is too narrow to cover the blank area left behind; *C* occurs because of the large blank area left behind by the turning of the vessel; and *D*-*H* mark blank areas caused by the loss of the beam. Boundaries *B*-*H* are called inner boundaries.

To track a boundary line, it is necessary to distinguish the boundary line points (the so-called boundary line points, that is, points where there is at least one empty point or non-empty boundary point around a certain lattice point) and then track the closed boundary line. The tracking of the boundary line is not the same as the tracking of the isoline. Isoline tracking inserts isoline points on the basis of the lattice points, and then constructs the isoline from the sequence, tracking the isoline from the border. When the value of a lattice point is exactly the same as the equivalent level of tracking, a small offset of the grid point value is generally adopted, and the boundary line point is just a lattice point. Hence, the boundary line cannot be traced completely according to the tracking sequence of the isoline.

It is necessary to deal with several special types of lattice points before tracking the boundary line (Figs. 7.11a-d). Figure 7.11a illustrates a case of isolated blank points, that is, the eight adjacent points of the blank points are effective points. The proposed algorithm uses a Gaussian distance weighted method to fill them. Figure 7.11b shows a case of isolated effective points, that is, the valid lattice point is surrounded by blank points, and the isolated effective points cannot construct an isoline, so the algorithm eliminates it to make it blank. Figure 7.11c shows the T-shaped boundary case, which generally does not play a role in the construction of an isoline, and the algorithm also eliminates it. Figure 7.11d shows the case of a common point boundary. According to the actual situation, the algorithm will eliminate or fill the blank points near it to avoid the occurrence of a common point in different boundary lines. Employing the above special point processing, the algorithm uses sequential tracking to track the boundary line and stores it into a boundary-line linked list, which is a triple linked list: BOUNDARIES→BOUNDARY→BOUNDARYPOINT. Here, BOUNDARIES is a linked list to store all the boundary lines, the BOUNDARY linked list is used to store a certain boundary line, and the BOUNDARYPOINT linked list is used to store the end points of an unclosed isoline.

7.3.2 Sequentially Inserting the Unclosed Isoline into the DTM Boundary Line

According to the tracking rules for an isoline, the end point of an unclosed isoline must be in contact with a boundary line and be arranged in increasing or decreasing order. There is no direct relationship between the end point of an isoline and the point of a boundary line, but they are closely related to the grid. The coordinates of the end point of an equivalent line can be easily converted to the position of the grid point, that is, located at the end point of the isoline. The boundary line is composed of non-empty grid points. If a direct connection between the boundary line linked list and the grid is established, the boundary line can be directly located from the end point of the unclosed isoline to the target boundary line. The method we employ is to create a temporary sparse matrix with the same number of grid cells as the grid after tracking the boundary line. In this matrix, we can easily store pointers to the boundary line points. The two-way relationship between the end point of an unclosed isoline and the boundary line is established quickly, laying the foundation for the creation of the topological relationship (Figs. 7.12 and 7.13). The end point of the unclosed isoline is inserted in turn according to the grade of the isoline and the order of the boundary line points, that is, after the end point of the unclosed isoline is inserted into the boundary line linked list, if it is sorted, it should be arranged in turn according to the rotation direction of the boundary line. The sequential arrangement of the end points facilitates the correct tracking of the closed polygons. Figure 7.12 shows the process of establishing a relationship between the end point of a closed isoline and the boundary through a grid. Figure 7.13 is a multi-beam depth data surveyed in the East China Sea, which was processed by a mapping system to form a grid. Then, the relationship between a boundary line and the end point of an unclosed isoline is established accordingly.

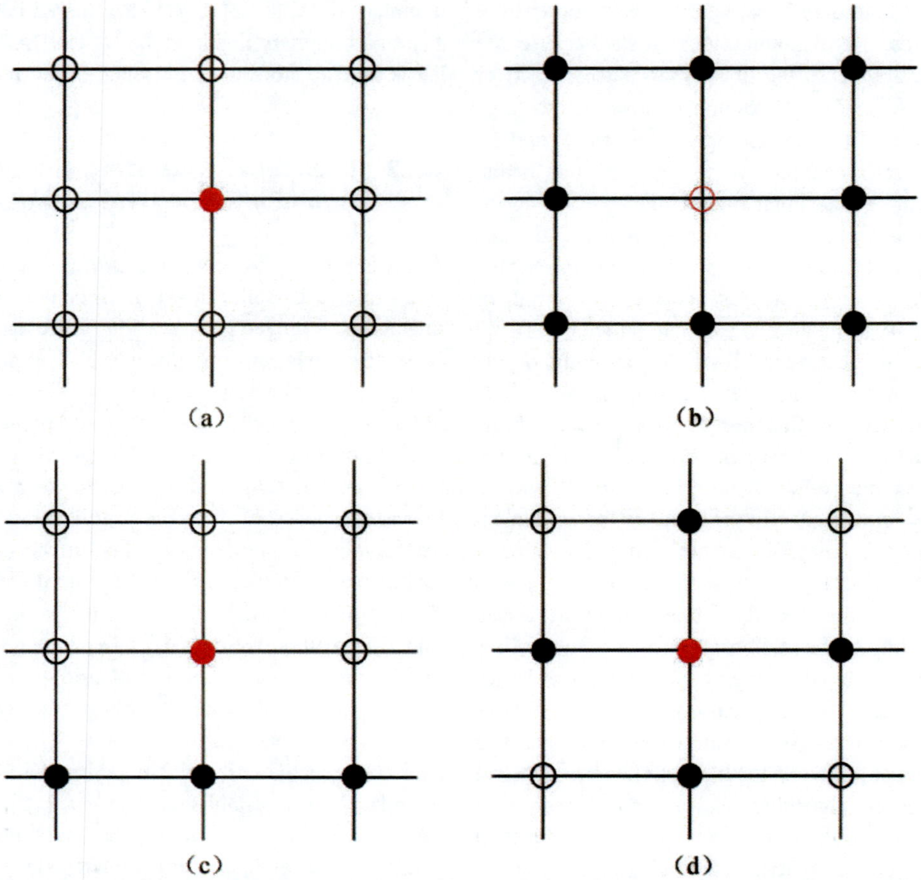

Fig. 7.11 Processing of special lattice points: four cases of isolated blank points (a), isolated effective points (b), T-shaped boundary (c), and a common point boundary (d)

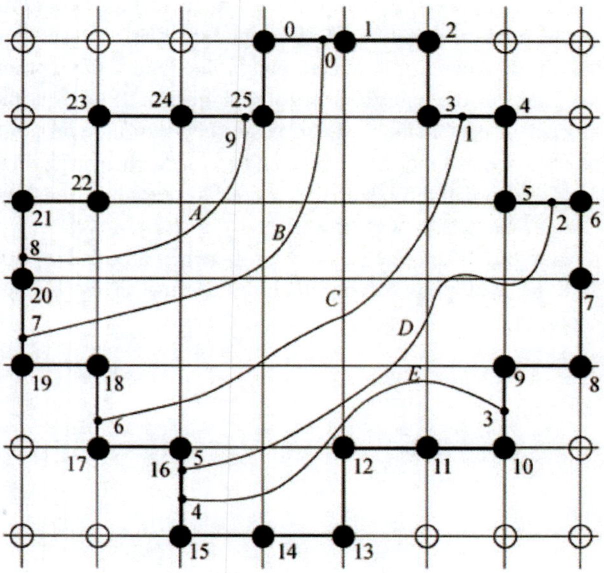

Fig. 7.12 Inserting the end points of unclosed isolines

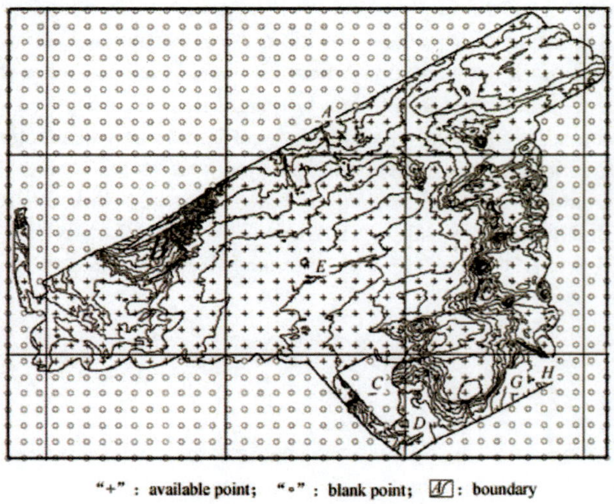

Fig. 7.13 Example of multi-beam data in the East China Sea

7.3.3 Establishing the Topological Relationships of Unclosed Isolines

The topological relationships of unclosed isolines are established to find adjacent unclosed isolines, that is, to query left and right unclosed isolines. For an unclosed isoline, there are up to four unclosed isolines with a topological relationship, with two end points corresponding to two contours, which may also point to this isoline (as shown in Fig. 7.12, for Isolines *A* and *E*). Because the unclosed isoline end points have been inserted sequentially into the boundary lines in the DTM boundary sequence and the relationships between the unclosed isolines and the boundary lines has been established, the isoline end points are queried in order of the boundary lines. The end point pointer of the next unclosed isoline is queried forward or backward along the boundary line to establish the topological relationship between the unclosed isoline and the boundary line. Even if the two end points of the unclosed isoline are not on the same boundary line, the topological relationship between the unclosed isoline can be established according to this idea.

7.3.4 Tracking Closed Polygons and Establishing Nesting Relations Based on the DTM Boundary

7.3.4.1 Tracking Closed Polygons in Boundary Line Sequence

Section 7.3.2 described how to insert an unclosed isoline into the boundary line point linked list BOUNDARYPOINT in an orderly manner, and the bidirectional relationship between the two elements was established. That is, the corresponding unclosed isoline can be directly queried from the boundary line point, and the corresponding boundary line point can be directly queried from the unclosed isoline. Section 7.3.3 described how the topological relationship between unclosed isolines is established using a boundary line, that is, the left and right unclosed isolines of each unclosed isoline are located. Therefore, it is feasible to track closed polygons in sequence according to the sequence of unclosed isolines in the boundary line. When constructing a closed polygon, each unclosed isoline can be related to up to four unclosed isolines, and each unclosed isoline can construct up to two closed polygons, that is, left and right polygons.

The basic idea of the algorithm to track polygons is as follows. Consider the boundary line points in the following order. First track clockwise, from the top of the isoline end point, and determine the end point of the unclosed isoline which has a topological relationship. Then, the next isoline is queried from the end point until the algorithm returns to the starting point. Then, from the end point, track counter-clockwise until the starting point is returned to so that the tracking of the left and right polygons of an isoline is completed. Then, in the order of the end point of the isoline in the boundary line linked list, the left and right polygons formed by the next unclosed isoline are traced from the end point of the next unclosed isoline. The task of the algorithm in this step is not completed until all the unclosed isoline end points have been traced. Tracking in the order of the end points of the isoline in the boundary line linked list can avoid repeated tracking (Fig. 7.11). Each isoline has a flag attribute to show whether it is a closed isoline and, if it is an unclosed isoline, whether it constructs a closed polygon when an unclosed isoline has completed the tracking of the left and right polygons. If the program automatically sets this flag value to indicate that left and right polygons have been constructed, the isoline will not participate in the construction of later polygons. Generally speaking, if the boundary line tracking is correct, the end point of the unclosed isoline is correctly inserted into the boundary line linked list, the correct topological relationship is established, and the construction of the closed polygons will be very smooth (Fig. 7.12).

7.3.4.2 Establishing Nesting Relationships Among Closed Polygons, Inner Boundary Lines, and Closed Isolines

To accurately fill an isoline, it is essential to determine the nesting relationship among closed polygons after the construction of the closed polygons. This is because the filling is drawn using polygons as the basic graphic elements. If the polygons with small areas are drawn first, large polygons will be stacked on top of them, which will not accurately color the map. In this algorithm, polygons, inner boundary lines and closed isolines constructed by unclosed isolines and boundary lines are treated as polygons. After tracking the closed polygons, the nesting of all three kinds of polygons is determined together to establish a complete nesting relationship. In this algorithm, the nesting relationship among polygons is built in two steps. First, the initial nesting arrangement is determined using the minimum border judgment method, and then the point-polygon inclusion relations are determined for accurate sorting. When tracking each isoline and boundary line, the algorithm determines the minimum bounding rectangle for each isoline according to its corner coordinates as well as for the closed polygons when tracking the closed polygons. The algorithm first sorts all polygons according to their size, and the inclusion relationship of the minimum bounding rectangle and creates a separate terrain unit. It then releases the nesting polygons in the order of the extended box size of the minimum bounding rectangle, so as to achieve a preliminary sorting. Then, the interior points are found in the smallest polygon of the extended box, and the inclusion relationship between the interior points and the polygons with nesting relationships are determined by the

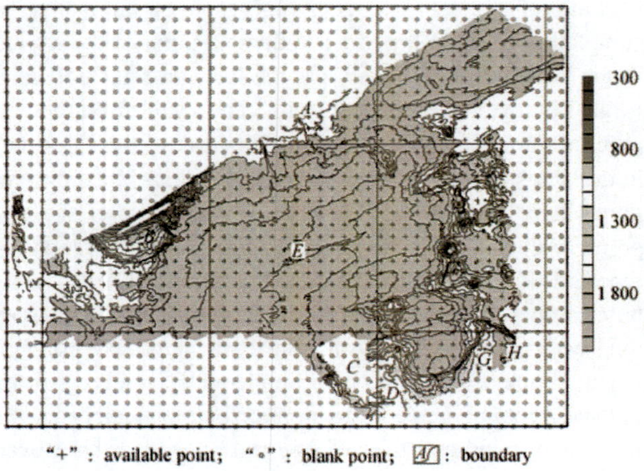

Fig. 7.14 Filling the isoline map in nesting order

"rotation angle" or "ray" method to exclude false nesting relationships caused by concave polygons (Fig. 7.14).

7.3.5 Filling the Isoline Map in Nesting Order

After constructing the closed polygons and establishing the nesting relationship among them, the inner boundary lines, and the closed isolines, it is necessary to draw the polygons in sequence according to the nesting order to avoid overlapping them. When drawing polygons, it is also necessary to match the correct color or pattern. When constructing the isoline tree, the program has a predetermined color or pattern for each level of isoline, and different matching schemes should be adopted for different types of polygons. If it is a closed isoline, the process can directly query the isoline tree to search for the corresponding color or pattern. In the case of closed polygons constructed from unclosed isolines and boundary lines, the filling color and pattern can be determined by the grid water depth (elevation) value of the boundary line points on the polygons. Polygons constructed by inner boundary lines are filled with white.

Figure 7.12 shows the basic process of tracking the DTM boundary line, constructing the topological relationship between the unclosed isolines based on the DTM boundary lines, constructing the equivalent tree based on the DTM boundary sequence, and finally realizing the vector color filling algorithm. To store the closed areas during tracking, the following triple linked-list structure is used: BLOCKNUM→BLOCKS→BLOCK. BLOCK linked lists are used to store closed areas, BLOCKS linked lists are used to store a series of closed areas of a terrain unit (e.g., seamounts or concave features), and BLOCKNUM linked lists are used to store blocks.

7.4 Establishing the Topology of a Submarine Geomorphologic Map

Submarine geomorphology, which is a discipline that studies the morphological characteristics, genesis, distribution and evolution of the seabed, is a branch of geography and a part of geology. The research content of submarine geomorphology mainly includes the characteristics of the submarine morphology and the dynamics of its formation, the origin and development of submarine morphology, and the research of sediments making up an accumulation landform. The submarine geomorphologic map is an essential form of submarine morphology interpretation and expression. It is also an important means of studying the characteristics of submarine geomorphology. It consists of three types of graphic features: the geomorphic unit and range (represented by polygonal symbols), the topographical unit boundary (represented by linear symbols), and the internal form of the geomorphic unit (represented by dotted symbols). A submarine geomorphologic map conveys information with surface symbols filled with color or pattern, the essence of which is to construct a topological relationship among graphical objects.

A five-step strategy is proposed to build a topological tree that incorporates the design of submarine geomorphologic maps. (1) Topological error detecting. Here, the self-intersection and mutual intersection of curves are eliminated through topological error detection. Moreover, dangling nodes and arcs that do not participate in the topological construction are deleted. (2) Establishing of an arc node network in the order of angles. (3) Tracking closed polygons in node order and establishing topology. (4) Nesting sorting to determine the nested relationship between topological and island surfaces to avoid overlaps in the topological surface. (5) Dynamic selecting the topological surface and interactive filling.

7.4.1 Automatically Detecting Topological Error

The purpose of preprocessing before topological tree construction is to eliminate the error information in the topological layer, check for arc segments that do not meet the topological construction requirements, and perform interactive man-machine modification to meet the topological construction requirements and delete primitives that do not participate in the topology. In the topological error detecting process, the following main tasks are completed (Fig. 7.15): finding arc self-intersections and mutual-intersections, finding common nodes, and deleting dangling arcs. Figure 7.15a–d show common problems encountered in the actual mapping process. In Fig. 7.15a,

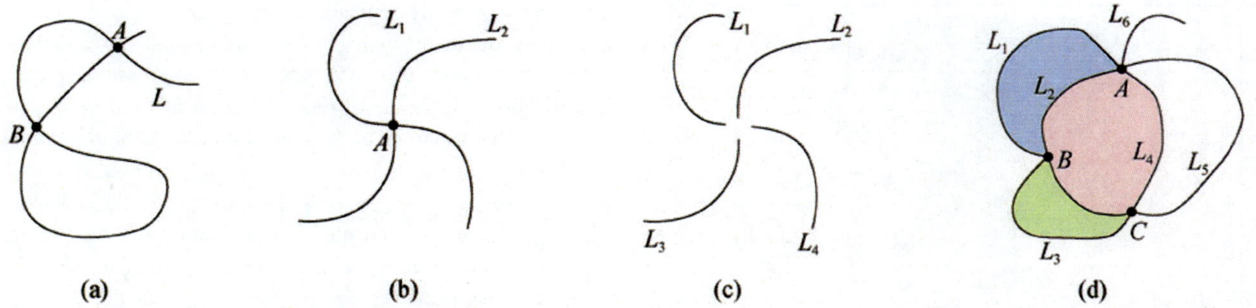

Fig. 7.15 Detecting Topological error before topological tree construction. **a** Finding self-intersection nodes; **b** finding mutual-intersection nodes; **c** finding common nodes; **d** deleting dangling arcs

Arc segment L has self-intersecting nodes A and B. The system will automatically find the intersecting nodes and automatically cut the arc segments. In Fig. 7.15b, Arc segments L_1 and L_2 have intersection A, and the system will automatically divide the relevant arc segments at the intersection points during topological error detecting. In Fig. 7.15c, the Arc segments $L_1 - L_4$ should have a common node according to the curve shapes. In Fig. 7.15d, Arc segments $L_1 - L_5$ can participate in the construction of the topological tree, whereas L_6 does not have a relationship with the other arcs. Hence, is it a dangling arc and must be deleted before the topological tree is built.

7.4.2 Establishing Node Network

Before constructing a topological tree, the relationships among the processed arcs need to be established (Fig. 7.15). There are two types of arcs that remain after detecting topological error: One is self-closing arc segments, as illustrated by Arc segments *i-n* in Fig. 7.16. These arcs connect at the end points of the arc to form a closed circle and do not need to participate in topological tracking. The other is open arc segments, which do not constitute a closed arc segment by themselves but are connected to other open Arc segments, and the closed relationship can be built through topological tracking, as shown in Arc segments *a-h* in Fig. 7.16. By analyzing the spatial distribution of these open arcs, it is not difficult to find that the arcs are connected end to end. (The difference between arcs and general curves is that a segment is a directed line segment.) In other words, there are some virtual nodes connecting these open segments (i.e., Nodes 1-4 in Fig. 7.16). Through these nodes, the network relationship is established. By traversing the open segment list, all the open segments can be found and saved with a linked list. In the process of establishing linked lists, the end points with the same location are merged to form a common linked list of nodes. Each node structure should keep the pointer of the arc linked list with which the relationship occurs. At the same time, the system should add the pointer of the linked list of the first and last nodes in the arc attribute structure so as to establish a two-way connection between the arc segments and the nodes. The key problem is the ordering of the arcs in the nodes. Although the arcs are ordered in space, there is no sequential relationship in the linked list. Our strategy is to establish the ordering relationship of arcs in nodes according to angle (Fig. 7.16). In short, a virtual ray is drawn from the node to the nearest corner of the arc segment, and the angle between the ray and the horizontal line is calculated. The arc segments inserted into the node linked list are sorted according to the angle to establish an ordering relationship between the arc segment and node. There should be at least three arc segments associated with each node; otherwise, it is a redundant node, and the arc segment at the node should be merged and the node deleted.

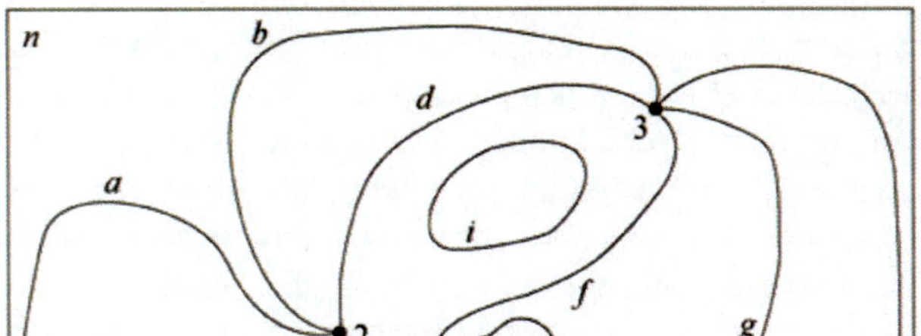

Fig. 7.16 Diagram of establishing an arc node network

7.4.3 Automatically Tracking Closed Polygons

After establishing the node network, although the two-way connection between the arcs and the nodes has been

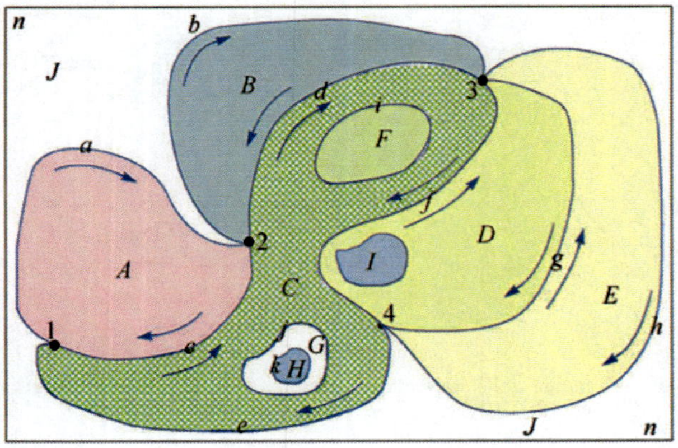

Fig. 7.17 Tracking a topological surface

established, the relationship between the arcs is not clear, and the topological surface has not yet been formed (Fig. 7.16). It is necessary to track the relevant arcs along the nodes to build the topological relationships among the arcs (Fig. 7.17). In the mapping system, the node linked list and the arc linked list were used to double-track the topological surface. First, we track the topological surface in the order of nodes, and then we track the topological surface in the order of arcs arranged in the order of nodes (clockwise or counterclockwise).

We use the tracking of Topological surface *C* in Fig. 7.17 as an example to illustrate the process of establishing a topological surface: starting from Node 1 and going clockwise, Arc segment *c* is traced through the node linked list, and Arc segment *c* is traversed. The arc linked list can be traced to Node 2, and Node 2 can be traversed to Arc segment *d*, which is the closest to Arc segment *c* (the directional angle between these arc segments is the smallest). We then traverse to Node 3 and Arc segment *f* in turn. Then, Node 4 and Arc segment *e* are traversed and we finally return Node 1, so that the tracking of the Topological surface *C* is completed, and the tracked arc segments are sequentially stored in the topological surface linked list to facilitate drawing.

There are three arcs that are directly related to Node 1: *a*, *c* and *e*. After tracing the Topological surface *C* from Arc segment *c*, then tracing the Topological surface *A* from Arc segment *a*, because e is the most peripheral arc segment, tracing along Arc segment *e* will not constitute a topological surface. Tracking in the order of nodes, each arc segment is used twice to construct the left and right topological surfaces respectively (for example, the left and right topological surfaces of Arc segment *c* are *A* and *C*, respectively), but it is important to note that only one related topological surface exists in the outer arc segment, such as for Arc segments *a*, *b*, *h* and *e* in Fig. 7.17. In the tracking process, it is necessary to identify the arcs of the constructed topological surface to avoid unnecessarily repeated tracking. It is also critical to store the tracked topological surface using a reasonable structure because this will affect the expression, query, modification, storage, and reconstruction of the topological surface. The topological surface structure should contain information such as the size of the expansion box, the node pointer, the arc pointer, and the pointer to the adjacent topological surface. The file storage structure of the system is also important because a reasonable file structure will affect the openness, scalability, and storage of the topological relationships as well as fast and accurate reconstruction of the system. Self-closing arc segments do not participate in topological tracking and directly form independent topological surfaces, such as *F-J* in Fig. 7.17. These surfaces are often called island-shaped surfaces.

7.4.4 Automatic Nesting Sorting

After the topological surfaces are tracked, it is necessary to establish a nesting sorting relationship with the island surface to avoid superimposing topological surfaces with smaller areas so that they are not visible (the relationship between *G*, *H* and *C* in Fig. 7.17). We adopt a two-step strategy for nesting sorting: first, the preliminary nesting relationship is established in order of the minimum bounding box size of the topological surface; then, the ray or corner method is used to accurately sort and exclude false nesting relationships. As shown in Fig. 7.17, it is difficult to discriminate the nesting relationship between closed surface*I*and Topological surfaces *C* and *D* using the minimum bounding boxes. In contrast, when using the ray or corner methods for further judgment, it is easy to eliminate the false nesting relationship between *I* and *C*. Through nesting sorting, a complete topological tree structure is built (Fig. 7.18). For Fig. 7.17, a four-tier tree structure is established. Here, Topological surface *J* is located at the top (root) of the tree, under which there are five first-level child topological surfaces: *A*, *B*, *C*, *D* and *E*. Child *C* has two children: *F* and *G*, and Children *I*, *F*, *G* and *I* belong to the same level of child topological surface. Child *H* belongs to the fourth-level child topological surface. When drawing graphics, it is necessary to follow this topological tree structure order; otherwise, some topological surfaces will be superimposed and not be visible. In this topological tree structure, the arc combination relationship of each topological surface, the included nodes, and the left and right topological surfaces of each arc segment are also saved. Using the topological tree, most of the information about the graphics that the user needs can be retrieved. Through simple calculation, the area and length of a specific area can be obtained, and even the regional connectivity and the minimum path of the arc can be discerned.

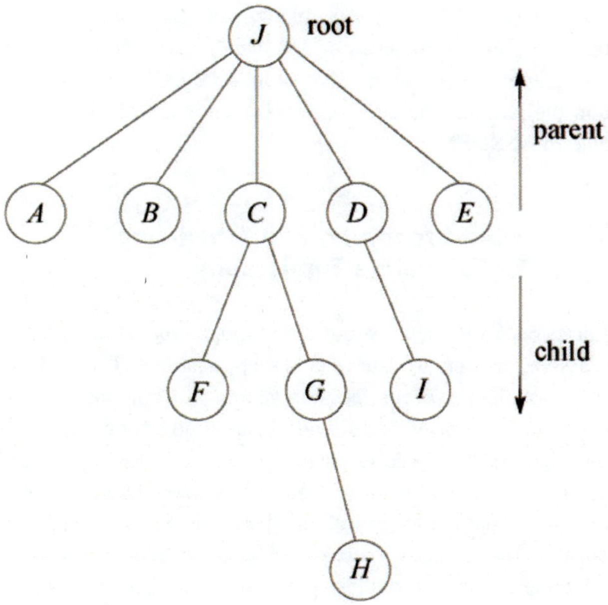

Fig. 7.18 Nesting sorting to construct a topological tree

7.4.5 Human-Computer Interaction Filling

After the above four steps are complete, the topological surfaces have been tracked on the basis of the arc segments, and the nesting has been sorted to construct a complete topological tree structure. The last step is to fill the established topological surfaces. In a geomorphological map, the color, pattern, and line type of each geomorphological type differ according to the geomorphological map specifications. These are then filled into the topological surface by human-computer interaction (Fig. 7.19). The key issue in this step is interactive operation, dynamic selection of the target topology, and editing. A complete geomorphological map consists of several topological surfaces that are connected to each other. It is difficult to select the target topological surface using a general retrieval method. The solution we adopted is to first use the minimum bounding box method to perform an initial investigation, determine the topological surface that has the inclusion relationship with the queried point, and then use the point-to-arc relationship to determine the topological surface closest to the queried point and the smallest area. However, there may be multiple topological surfaces that simultaneously meet the requirements. In this case, the graphics user interface flashes to prompt the user to perform the selection operation, and the landscape map is edited until the requirements are met. Figure 7.19 shows the interactive operation of the mapping system to edit the submarine geomorphologic map.

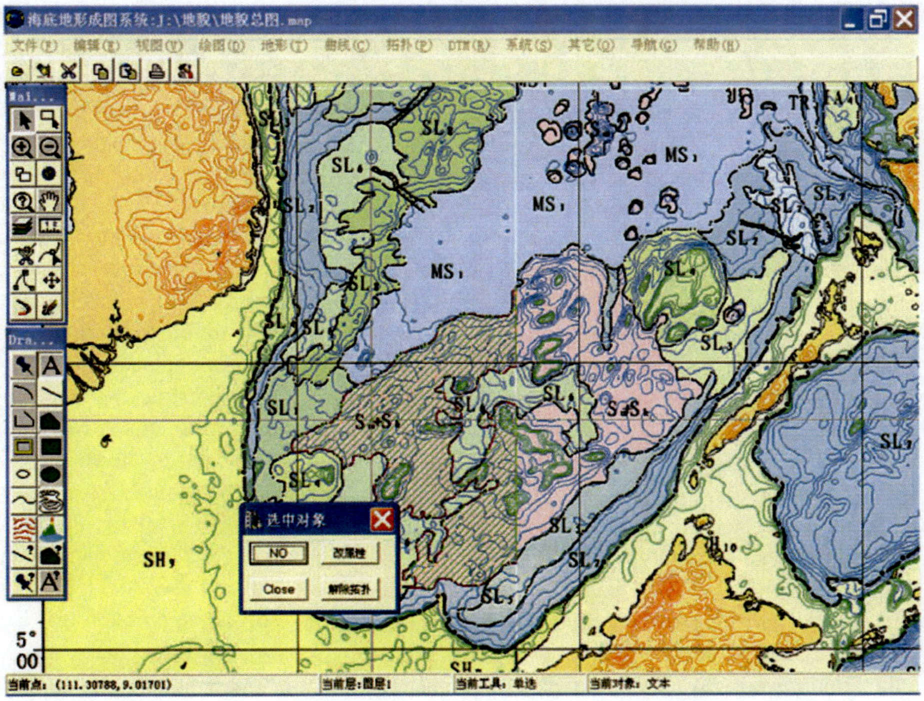

Fig. 7.19 Human-computer interactive geomorphologic mapping

7.5 Constructing 3D Submarine Topography Maps

Traditionally, submarine topography is mainly expressed by 2D methods such as marking water depth points on charts and drawing isobath lines. Although these forms can meet the needs of nautical or marine engineering construction to a certain extent, they are not intuitive and not conducive to terrain analysis. With the development of computer and graphics technology, 3D graphics processing technology of computer has made great progress. The 3D visualization of submarine topography has become the main form and key research topic of submarine topography expression.

3D visualization of submarine topography is a 3D solid construction technology employing the storage, display, simplification, and simulation of a seabed digital elevation model (SDEM). Its aim is to use graphics processing technology of computer to visualize the submarine topography in 3D for browsing, query, analysis, and a series of interactive operations while ensuring the smoothness of the operation system and visual realism. The fundamentals are to transform the 3D submarine topography into a 2D plane according to a perspective projection and calculate the color of each pixel. Alternatively, an appropriate image can be chosen as a texture to attach to the submarine topography surface according to the position and color of the light, submarine topography surface shape, and reflective properties. The aim is to enhance the authenticity of the submarine topography (Tang et al. 2010). When the scale of a submarine topography is small, topographic data can be read into memory all at once, and the organization and visualization of the data are relatively simple. The key steps of this process include DEM data organization, perspective projection transformation, blanking, and cropping, illumination modelling, texture mapping, graphics plotting and rendering. As the size of the submarine topographic data increases, the amount of data to be processed poses a challenge to the performance of the computer system. Many research topics on large-scale submarine topography visualization have been carried out, such as out-of-core, level-of-detail (LOD), geometry mipmap and GPU acceleration technologies.

MBES technology not only produces high accuracy, high spatial and temporal resolution, sounding data but also causes a large increase in the amount of data taking on characteristics of massive data. Although current computer performance has improved, memory capacity, calculation and plotting performance are still limited, and cannot meet the requirements brought about by the rapid growth of submarine topography. The massive data characteristics of multi-beam data have raised a series of problems in the construction of an SDEM and its subsequent 3D visualization. This section mainly discusses the relevant technologies and methods of 3D visualization for submarine topography given the massive submarine DEM data sources and the large scope and refinement tendency of 3D visualization of SDEM.

7.5.1 Data Organizing and Scheduling of Submarine Topography

Large-scale fine-detail submarine topography often employs massive amounts of data in actual applications. The scale of the data often exceeds the storage limit of computer memory, so it is impossible to load all the topographic data into the memory for rendering at one time. Therefore, it is necessary to segment the large-scale submarine topography data and divide them into several smaller topographic sub-blocks. Data are dynamically loaded as needed for rendering without affecting the terrain rendering performance. This method of segmenting and rearranging large-scale topographic data not only reduces memory consumption and speeds up input/output (I/O) operation as well as CPU processing time, but also avoids unnecessary rendering costs. Therefore, how to organize and schedule data scientifically is an important step to improving the rendering efficiency of the whole scene.

7.5.1.1 Hierarchical Blocking of Submarine Topographic Data

In the process of submarine topographic 3D visualization, there are two means of reducing topographic data volume in the memory from the aspect of data organization: (1) submarine topographic data blocking, that is, dividing the same level of data segmentation into suitable topographic sub-blocks and then determining which topographic sub-blocks need to be added into memory for display on the basis of the view display range; and (2) submarine topographic data hierarchy, that is, creating multiple data layers with different resolutions for submarine topographic data in the same spatial range. The aim is to build a multi-resolution pyramid structure, and then invoke the hierarchy model with the corresponding resolutions for rendering according to the viewpoint or the adaptive performance of hardware.

The simplest method of dividing topographic data is uniform blocking, that is, the whole topographic is divided into $M \times N$ sub-blocks and the size of each sub-block is same (Zhao 2011). This method is easy and conveniently solves the problem that large-scale topographic data cannot be transferred into memory at once, but the number of storage files and data redundancy is greatly increased when the original topographic data volume is very large and divided into too many topographic blocks. This approach also increases the time the CPU needs to determine which topographic sub-blocks should be loaded into memory. Therefore, the most commonly used method of large-scale topographic data blocking is a combination of hierarchical and blocking methods, that is, the original topographic data are divided into smaller topographic sub-blocks in a hierarchical form. In this method, the original topographic data are taken as the 0th layer, the divided topographic sub-blocks constitute the first layer, each topographic sub-block of the first layer is further divided into the second layer, and so on.

The hierarchical blocking of topographic data is usually achieved by a binary tree or quadtree structure. Non-leaf nodes of the tree structure only preserve the index relationship of the topographic sub-blocks, whereas the leaf nodes store specific topographic data. As shown in Fig. 7.20, the structure of the hierarchical blocking looks like a pyramid, so this scheme is also called the pyramid model. This model is a multi-resolution hierarchy model most commonly used in large-scale topographic data organization. It generally forms multi-resolution levels using the multiplier method.

The resolution becomes increasingly higher from the top to the bottom of the pyramid model, but the representation range remains unchanged, so it can meet the real-time requirements of 3D visualization of a submarine topography. To achieve the level of detail in a real-time display of submarine topographic data, topographic sub-blocks with different resolutions are required for different positions. Then, the topographic data pyramid can directly provide topographic data without "real-time" resampling. If there is no pyramid model, the data must be simplified in real time on the basis of original topographic and texture data to achieve the required level of detail. Although the pyramid model increases the storage space needed for the data, it can reduce the total time required to complete each frame of rendering. The tiled pyramid model can further reduce the amount of data access and improve the I/O efficiency of the system, thus improving its overall performance.

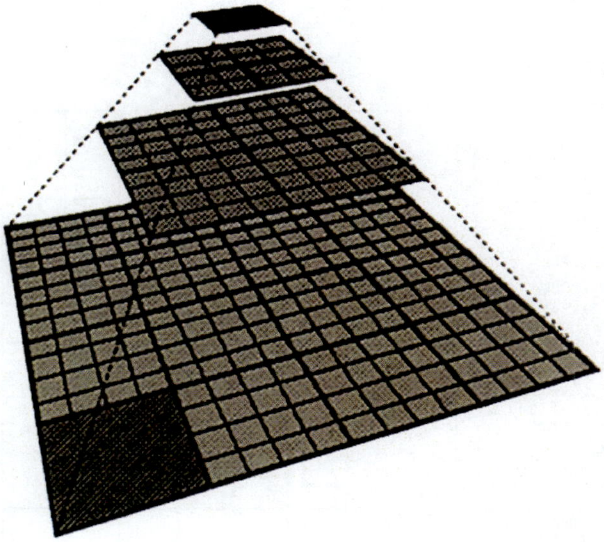

Fig. 7.20 Topography data pyramid model

Fig. 7.21 Multi-resolution topographic representation based on a quadtree organization

A key factor of the pyramid model is spatial blocking, that is, how the pyramid is divided vertically and horizontally. A spatial blocking mode directly determines the storage mode and index mode of large-scale topographic data. It also affects the scheduling efficiency of the topographic database. Common division approaches are equal-interval spatial blocking and equal-area spatial blockings. A typical example of the equal-interval spatial blocking is the quadtree algorithm. The basic idea is to divide topographic data into equal-interval patches. The intervals of the patches on one layer are equal, and the interval ratio to the adjacent layer is two. The well-known open source graphics system Open Scene Graph (OSG) supports massive topographic data by building a pyramid model using a quadtree. Figure 7.21 shows the display results after building a pyramid model for a submarine topography (Meng 2013). The bottom right quarter of the topographic block is shown as a more refined topographic block because it is the closest to the viewpoint, whereas the other three quarters of the topographic data are shown as slightly rougher topographic blocks and the grid lines are relatively sparse.

7.5.1.2 Dynamic Scheduling Strategy for Submarine Topographic Blocks

After the submarine topographic data have been organized, the next problem is how to efficiently load the necessary topographic blocks and release unnecessary data blocks according to the demand of the visual field of view to balance the memory data during the operation of the whole visualization system. This problem is called dynamic data scheduling, and is restricted by the data storage model. Its efficiency will directly affect the I/O efficiency of the system, thus ultimately affecting the efficiency of scene rendering. The data dynamic scheduling process generally includes the following stages: viewing area-intersection test, data preloading, and multi-thread acceleration.

1. **Viewing area-intersection test**

When roaming in a microtopographic scene, only topographic blocks within the view frustum are rendered. Therefore, to determine which data blocks need to be added or released, it is necessary to test in real time which data blocks fall within the view frustum. In computer graphics, the viewable field an observer can see is usually defined as a view frustum that changes with the observer, as shown in Fig. 7.22. When the position of the observer moves or changes the direction of the sight-lines, it is necessary to calculate the topographic sub-block intersecting with the observer's view frustum. At the same time, topographic sub-blocks that should be loaded or loaded in this frame are determined according to the topographic sub-blocks loaded in the memory of the previous frame.

It is a complicated and time-consuming process to precisely intersect the topographic sub-blocks with the view frustum. To simplify the calculation and improve its efficiency, a common strategy is to project the view frustum and topographic sub-block to the 2D plane and test the 2D graphics intersection. The sight-lines can be arbitrarily changed when the observer roams in the topographic scene, so the spatial posture of the view frustum is arbitrary in that the projected shape of the view frustum onto the horizontal plane would be very different.

2. **Data preloading**

In the roaming process in the topographic scene, every frame that needs to be rendered updates the topographic sub-block in the view field through the loading or unloading process according to the change in the observer's current view field. Because the CPU reads data in memory much faster than in external memory, if the topographic block is loaded directly from external memory while the view field is updating, this will inevitably lead to a long waiting time for the CPU or GPU, affecting the refresh speed of the topographic scene. Therefore, it is necessary to use a buffer mechanism for data preloading in the data scheduling policy. The preloading principle defines a larger area outside the observer's view field, in addition to reading in the data block needed for the current frame from the external memory, surrounding topographic blocks are read into the memory cache in advance. Thus, when the system needs to update the data for changes in viewpoint, it does not need to read the new topographic blocks directly from external memory. Instead, data can be directly read from the memory, which speeds up the rendering efficiency.

As shown in Fig. 7.23, the rectangular unit represents the segmented topographic blocks, the triangle represents the projection of the view frustum on the 2D plane, the solid side triangle represents the current view field, and the virtual side triangle represents the view field of the previous frame. If the data prefetch strategy is not used (Fig. 7.23a), it is necessary to read and load a topographic sub-block data from the hard disk in real time and unload two topographic sub-blocks simultaneously when the triangle's position is moved. The process of loading the topographic sub-block in real time is the reason the CPU delays each frame. When the rectangular prefetch strategy is adopted (Fig. 7.23b), all

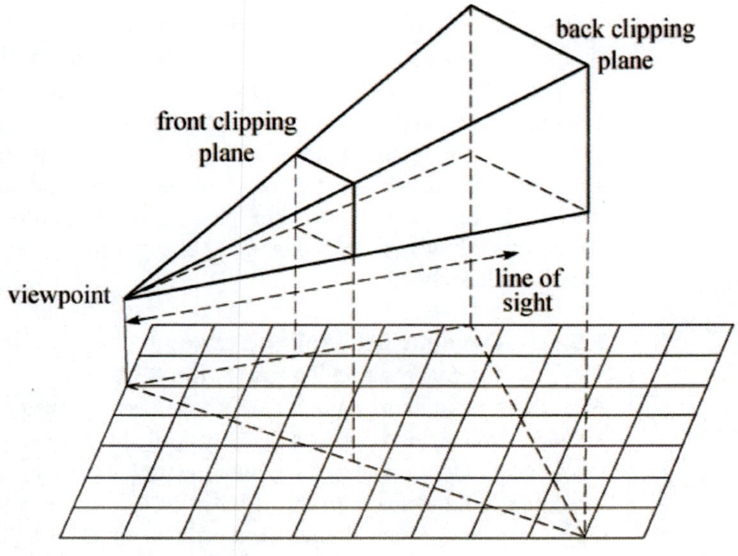

Fig. 7.22 View frustum and projection

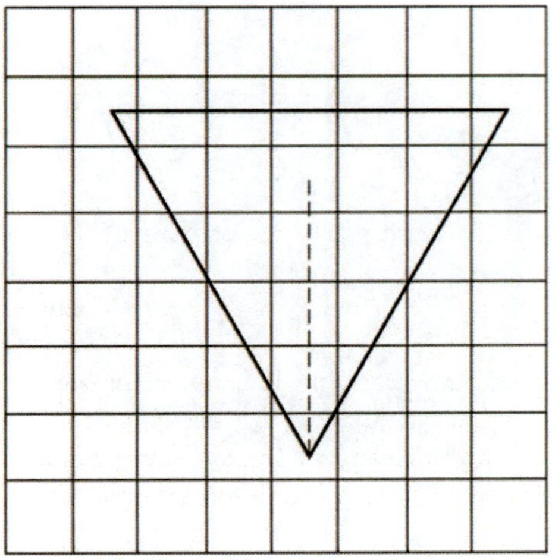

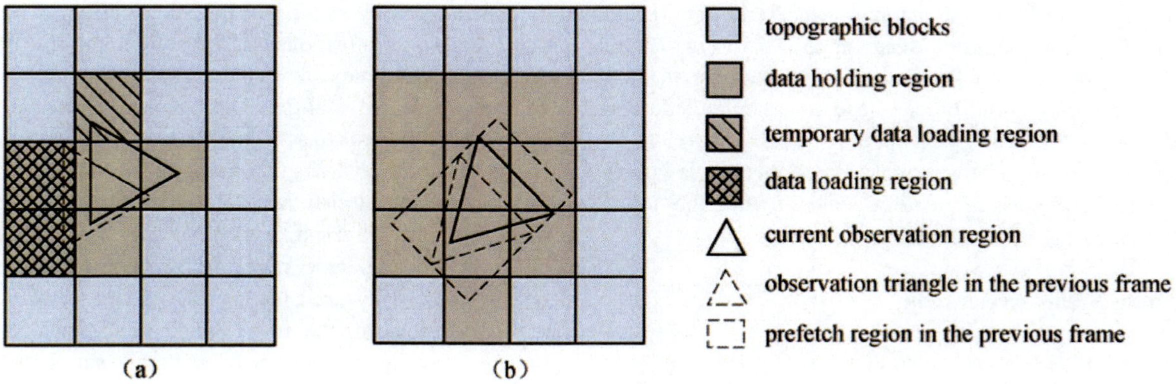

Fig. 7.23 Principle of data preloading (Zhao 2011). **a** No Prefetch strategy; **b** Rectangle Prefetch strategy

topographic sub-blocks within the view field of the current frame have been prefetched in the previous frame. Removing the need for real-time data loading substantially improves the rendering efficiency of the current frame and improve the refresh rate of the system. In addition to rectangular prefetch regions, circles, triangles, and other self-defined shapes are commonly used in prefetch strategies. Larger prefetching ranges mean that more topographic sub-blocks are preloaded, and the memory footprint and load time increase accordingly. Therefore, the shape and range of a prefetch region directly affect the data scheduling. The efficiency of real-time loading and data prefetch should be comprehensively considered, and only by finding a balance between them can the performance of the whole scheduling strategy be improved. An example of a scratchable latex buffer area preprocessing method to illustrate the principle of dynamic data preloading is shown in Fig. 7.24.

The scratchable latex algorithm segments the topographic block into sub-block using the following principles to determine which data blocks to preload in memory. (1) When the view triangle falls over a topographic unit in full, the topographic unit and the eight adjacent topographic units need to be loaded (Fig. 7.24a). (2) When the view triangle intersects multiple topographic units, it is necessary to load data for each topographic unit according to principle (1), but prevent the repeated loading of topographic units in overlapping areas. Figure 7.24b shows the topographic units that need to be preloaded when the view triangle intersects with four topographic units. According to the above principles, the topographic scheduling strategy reads the topographic blocks that may be processed in the next step into the cache in advance, which ensures that no matter which direction the viewpoint moves, the topographic blocks to be rendered in the next frame are all in the cache, thus greatly reducing the dropped frame phenomenon. In the actual roaming process, when the viewpoint moves and the topographic blocks covered by the view triangle change, the system needs to load or unload

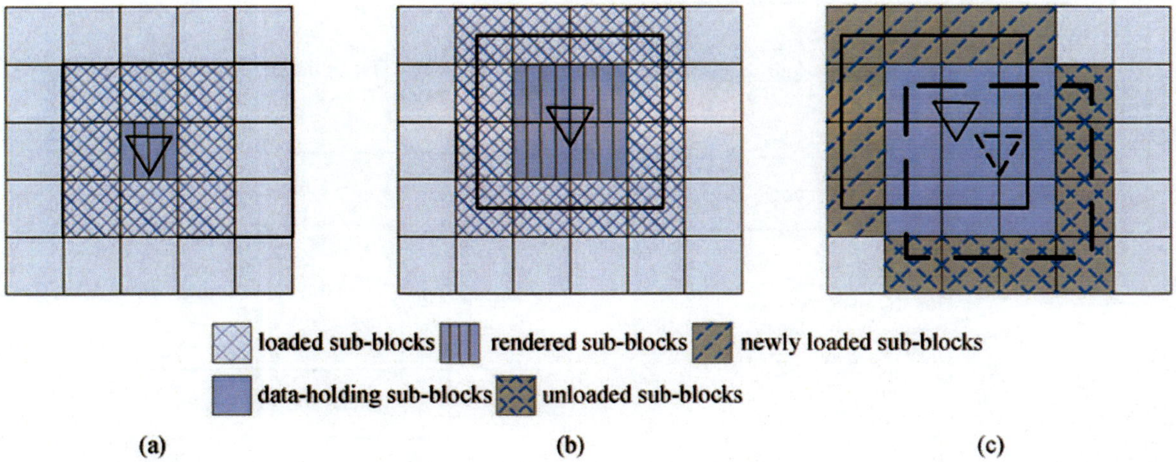

Fig. 7.24 Buffer preprocessing strategy based on the scratchable latex algorithm (Zhao 2011). **a** Scratchable latex data prefetching; **b** prefetching of a twelve-block sub-block and; **c** data updating when viewpoint moves

the topographic blocks dynamically in the buffer. In the case of the view triangle movement shown in Fig. 7.24c, seven topographic blocks need to be loaded in real time and seven topographic blocks need to be unloaded. The scratchable latex strategy completely covers the area to be displayed, and no matter which direction the viewpoint moves, this processing method ensures that the topographic block to be displayed next is cached.

3. **Multi-thread acceleration**

Multi-thread is the ability of an operating system to support the execution of multiple threads in a process. It is often used in multi-task software development, that is, the whole software process contains multiple threads to complete different functions, such as the data collecting thread, preprocessing thread, real-time data displaying thread, graph curve generating thread, and user interface thread. In this way, multiple threads execute simultaneously and accomplish more tasks in parallel in a period of time, which speeds up the response of the system and improves its execution efficiency.

In the process of real-time topographic rendering, the program not only needs to schedule the topographic data that may be needed for rendering according to the viewpoint and other information but also the data needed for specific rendering operations. To improve the continuity and efficiency of the whole topographic rendering process, multi-thread technology is usually used to run two threads at the same time in the program: the main thread is responsible for real-time topographic rendering, and the sub-thread is responsible for data scheduling. The idea is as follows: first, a thread pool is opened up. The main thread is mainly responsible for calculating the visual space, determining the topographic block that needs to be scheduled, updating the rendering queue, and rendering the topographic scene. The sub-thread is mainly responsible for invoking the required topographic block data from external memory (Fig. 7.25).

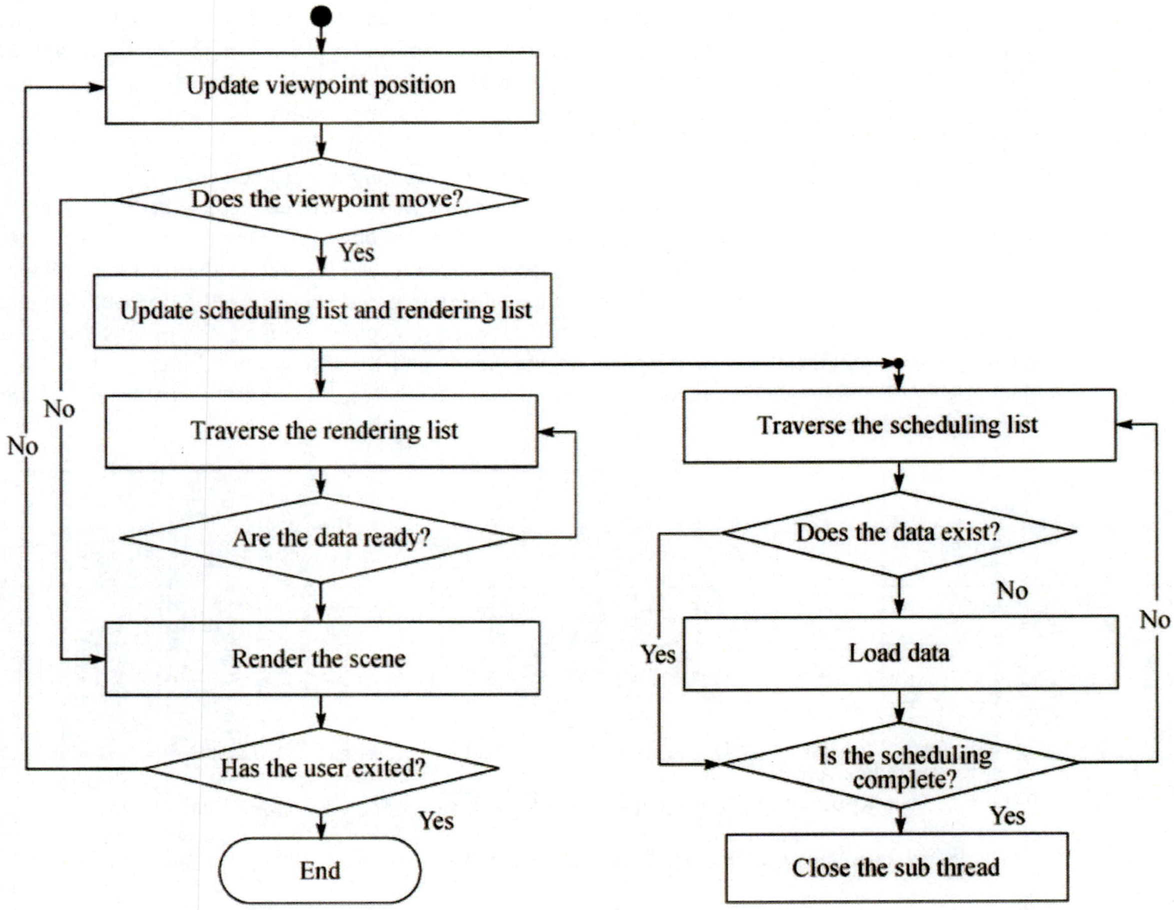

Fig. 7.25 Flow diagram of multi-threaded data scheduling policy

7.5.2 Real-Time Rendering Acceleration Technology for Submarine Topography

7.5.2.1 Acceleration of Graphics Rendering Based on the LOD

Before GPU graphics acceleration technology, the key method for improving the rendering efficiency of complex scenes is to reduce the triangle mesh data needed to construct scenes as much as possible. In addition to the simplification of complex models, Level of Detail (LOD) technology is the most commonly used method to reduce the triangle mesh data in a scene. This technique creates multiple levels of models with different details for different parts of the same object according to the visual characteristics of human eyes, which make near objects clearly and distant objects less clearly. It then selects a rough- or fine-detailed model for display according to the distance of the observer. When constructing a topographic scene with LOD technology, data redundancy, as well as the number of triangles needed for scene rendering, can be reduced to ensure that the topography display results change little under this premise.

1. **Classification of topography LOD models**

According to their construction principles, topography LOD models are usually divided into two categories: discrete LOD and continuous LOD (CLOD) models. The discrete LOD model constructs several detailed models with different resolutions for the same object in advance. The internal details of each detailed model are of the same degree. We select the appropriate LOD model for rendering according to the distance of the observer. Because the geometric topologies of different LOD hierarchical models are different, the switching of different LOD hierarchical models causes an obvious visual jump or mutation.

When building scenes, the CLOD model does not need to establish the topographic detail hierarchy in advance but dynamically calculates the resolution required by each region of the scene in real time through the corresponding rendering algorithm during the rendering process. A fine-detail model is used for the parts that are close to the viewpoint and have strong topographic fluctuations, and a rough-detail model is used for the parts that are far from the viewpoint and have a smoother topography. In this case, the LOD model related to the observer's position is automatically generated to complete the scene rendering (Fig. 7.26). The CLOD method can reduce the mutation phenomenon effectively during scene roaming. However, because splicing the whole topographic scene needs topographic block models at different resolutions, obvious cracks will appear between these topographic blocks. It is easier to eliminate fractures than to solve the mutation problem of discrete LODs.

2. **Generation of a CLOD model**

The real-time generation algorithm of a CLOD model is closely related to the data structure of the topography. Data structures for DEMs include regular square grids (RSGs) and triangulated irregular networks (TINs). Large-scale topography often uses an RSG. For RSG DEM, we often use topography refinement and subdivision based on a quadtree or binary tree to construct the CLOD model in real time. For a TIN, because of its irregular geometric structure, its refined subdivision is relatively complex, and the commonly used method is similar to the triangular mesh simplification algorithm of general 3D geometric objects. It simplifies the mesh through edge folding, vertex deletion, edge deletion, triangle deletion, and a series of geometric simplifications. The classic algorithms include the progressive mesh algorithm presented by Hoppe and the improved view-dependent progressive mesh algorithm.

A triangular binary tree is similar to a quadtree, but the rectangle's quadtree division is replaced by a binary division of the triangle (Fig. 7.27). The advantage of constructing a CLOD model using a binary tree structure is that each binary tree node

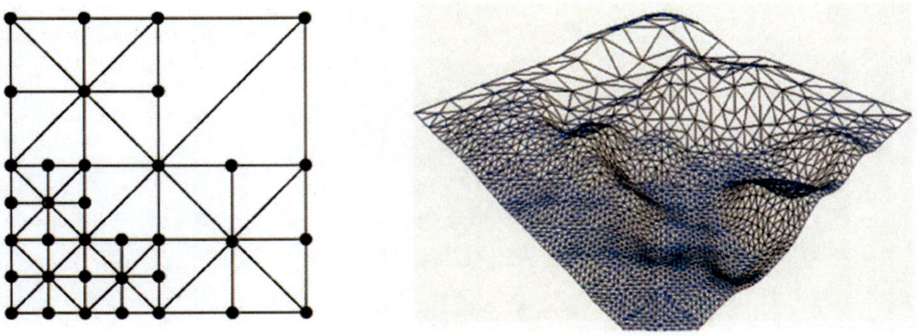

(a) (b)

Fig. 7.26 CLOD structures and examples of topography. **a** CLOD structure of topography; and **b** CLOD instance of topography

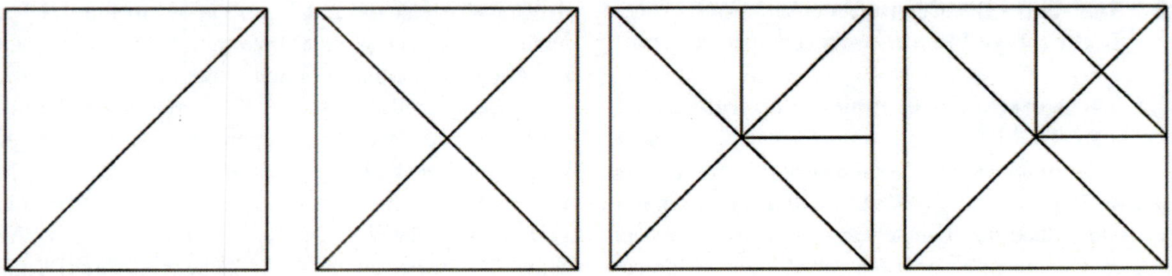

Fig. 7.27 Binary tree-based CLOD division

happens to be a grid rendering unit, and a topographic scene can hence be constructed more quickly in the rendering stage. However, the division level of the binary tree structure is deep, the construction of a binary tree scene is slow, and the subdivision process will also increase the number of the same vertices, so redundant triangles will be generated during rendering. The number of divisions used by a quadtree to construct a CLOD model is less than that of the binary tree structure, and the model construction time is relatively short.

3. **Crack treatment**

In CLOD topography rendering, cracks are usually generated between topographic blocks with different resolutions because different topographic blocks are rendered at different levels of detail. As shown in Fig. 7.28, at the joining edge of two topographic blocks with different resolutions, the midpoint of the high-resolution topographic block uses the true height of the vertex, but there is no height data for the vertex on the low-resolution topographic block, so it needs to be obtained by the interpolation of two border vertices, which will lead to the occurrence of two high values at the same location, resulting in a topographic crack.

Crack elimination is one of the key problems to be solved by LOD technology. The commonly used solutions include vertex addition/deletion methods, the forced subdivision method, and the nepotism method (Meng 2013).

7.5.2.2 GPU-Based Graphics Rendering Acceleration

In the late 1990s, with the rapid development of GPUs and the graphics hardware, a new generation of graphics cards provided vertex shaders and fragment shaders capable of being programmed. Moreover, a geometry shader was added to the Shader Model 3.0, so programmable graphics hardware now contains efficient graphics rendering algorithms based on programmable shaders.

A GPU has a better calculation performance than a CPU. GPU-supported programmable pipelining allows complete control of the specific stages of the pipeline by creating shader programs, and each specific stage is replaced by compiled code fragments to provide standard vendor code for execution (Jin et al. 2010). A vertex shader is a program that performs vertex and normal transitions, texture coordinate generation, and vertex lighting calculation. A fragment or pixel shader is a program that performs calculations during the pixel processing phase of a graphics pipeline. It

Fig. 7.28 Formation of a topographic fracture. **a** Crack locations; and **b** diagram of crack occurrence

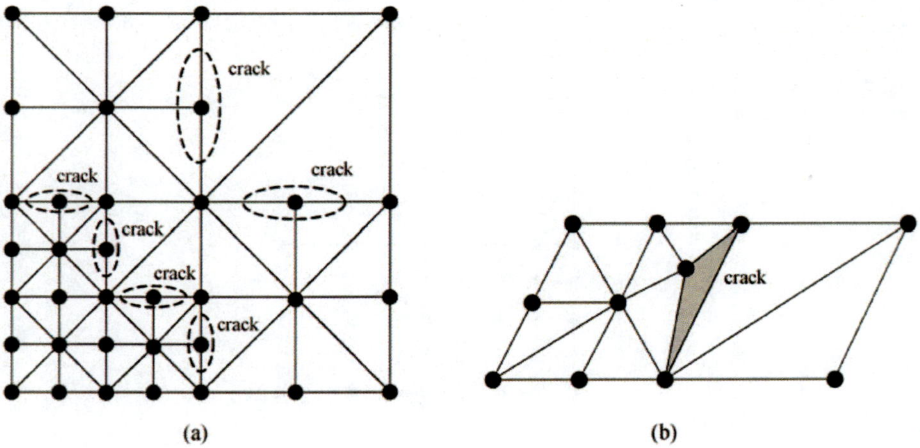

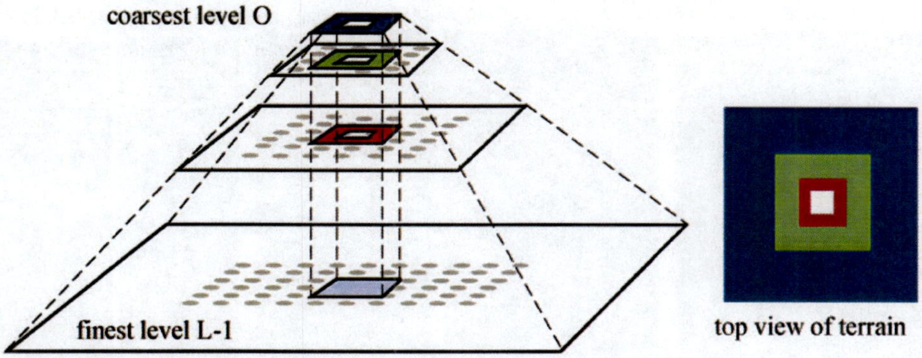

Fig. 7.29 Operating principle of geometry clipmaps algorithm

determines exactly how each pixel is colored, how the texture is applied, and whether a pixel is rendered or not. When rendering, the GPU first receives the geometric data sent by the CPU in the form of triangle vertices and then processes them using the programmable vertex shader unit to complete the geometry and vertex attribute calculations as well as other functions. Then, these 3D triangles are converted into pixels on a 2D screen by a fixed-function raster generator. The final color value of each pixel is calculated by a small program running on the pixel shader.

In the field of large-scale topography rendering, typical algorithms for graphics rendering acceleration using GPU are the geometry mipmap algorithm (DeBoer 2000), chunked LOD algorithm (Ulrich 2002), and geometry clipmaps algorithm (Losasso and Hoppe 2004). This chapter only introduces the geometry clipmaps algorithm.

The geometry clipmaps algorithm is a LOD algorithm based on a GPU optimization proposed by Losasso and Hoppe in 2004. The idea of this algorithm is completely different from the traditional LOD algorithm. It focuses on maximizing the batch rendering capability of GPU instead of focusing on reducing the number of triangles per rendering. The algorithm treats the topography as a 2D height map and prefilters it into a mipmap pyramid containing an L layer. For large-scale topography, a complete mipmap pyramid cannot be put into memory all at once. Therefore, the algorithm takes $n \times n$ samples for each layer and caches them, forming a nested high clipmap structure in memory (Fig. 7.29).

In the process of rendering, the structure forms a set of nested regular grids centered on the viewpoint. The hollow part of each rough layer is occupied by a higher layer, and only the finest-detailed layer is rendered as a complete square grid, while the other layers are rendered as a hollow ring grid (Fig. 7.30). When displayed on the screen, the triangles of all layers will have approximately the same size as projected onto the screen because the center of the viewpoint adopts the finest grid, and the rougher grid is gradually adopted as it moves away from the viewpoint.

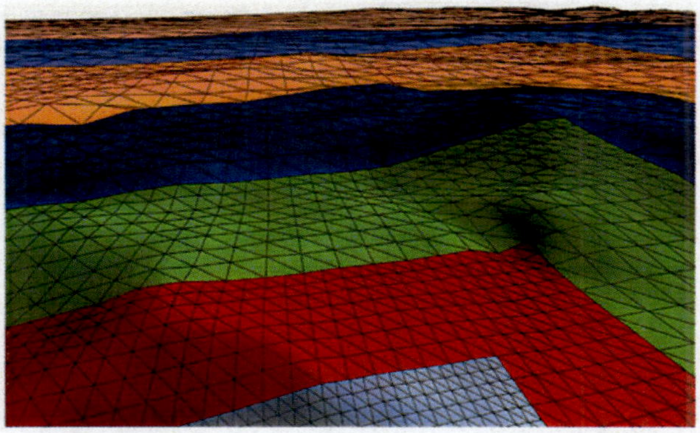

Fig. 7.30 Topography rendering using rough geometry clipmaps

As the viewpoint moves, the render window needs to be changed and the data must be updated. To improve the updating efficiency, we use a ring array to store the data blocks displayed in each layer. Each layer only needs to update one "L" type area with the help of a modular operation. For the rougher layers, the change in position is exponentially reduced compared to the movement of the viewpoint, so updates are rarely required. Because the final rendering topography grid is composed of different LOD levels from inside to outside that range from fine to rough, the uneven grid space will lead to the appearance of cracks in the topography. The algorithm eliminates cracks by introducing a transition zone near the outer edge of each layer (Fig. 7.31). The size of the transition zone can be specified in advance. Using the vertex shader and pixel shader, geometric data and textures can be interpolated smoothly to the next lower level of detail.

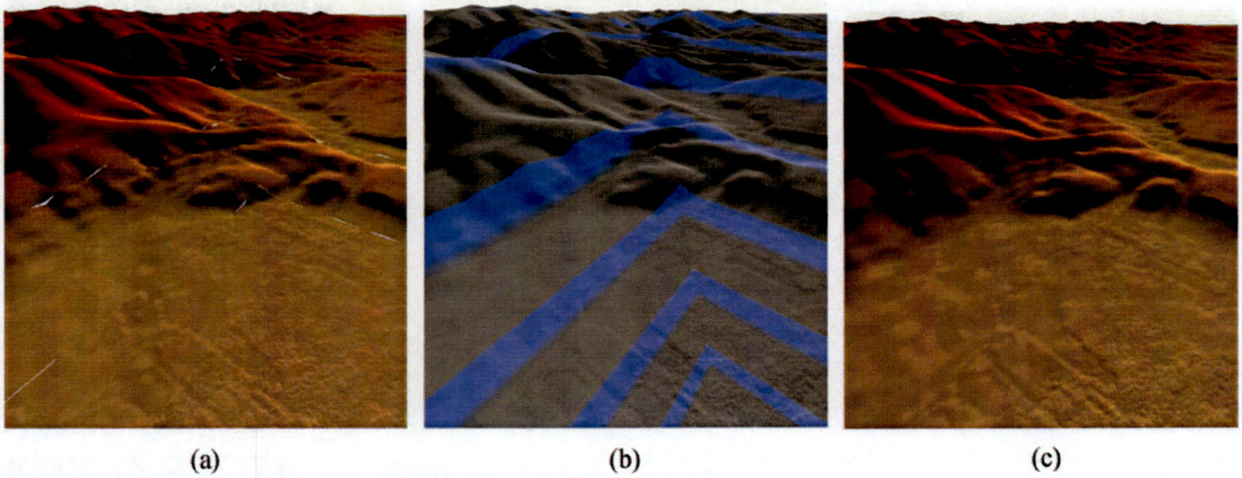

Fig. 7.31 Cracks and crack elimination at different LODs (Losasso and Hoppe 2004). **a** Results with no transition zones; **b** transitional fusion zones; and **c** fusion result

References

Arndt JE, Schenke HW, Jakobsson M et al (2013) The international bathymetric chart of the Southern Ocean (IBCSO) Version 1.0—a new bathymetric compilation covering Circum-Antarctic waters. Geophys Res Lett 40(12):3111–3117. https://doi.org/10.1002/grl.50413

Beaman RJ, O'Brien PE, Post AL et al (2011) A new high-resolution bathymetry model for the Terre Adélie and George V continental margin. East Antarctica. Antarct Sci 23(1):95–103. https://doi.org/10.1017/S095410201000074X

Bjorke JT (2005) Computation of calibration parameters for multibeam echo sounders using the least squares method. IEEE J Oceanic Eng 30(4):818–831. https://doi.org/10.1109/JOE.2005.862138

Bolmer ST, Beardsley RC, Pudsey C et al (2004) A high-resolution bathymetry map for the Marguerite Bay and adjacent west Antarctic Peninsula shelf for the Southern Ocean GLOBEC Program. Center for Financial Institutions Working Papers 28(11):2603–2639

Briggs IC (1974) Machine contouring using minimum curvature. Geophysics 39(1):39–48. https://doi.org/10.1190/1.1440410

Calder B, Mayer L (2001) Robust automatic multi-beam bathymetric processing. Paper presented at the US Hydrographic Conference, Norfolk, Virginia, 21–24 May 2019

Calder B, Mayer L (2003) Automatic processing of high-rate, high-density multibeam echosounder data. Geochem Geophy Geosy 4(6):1048. https://doi.org/10.1029/2002gc000486

Calder B, Wells D (2007) CUBE user manual. Available via DIALOG. http://ccom.unh.edu/sites/default/files/publications/Calder_07_CUBE_User_Manual.pdf. Accessed 25 June 2019

DeBoer WH (2000) Fast terrain rendering using geometrical mipmapping. Available via DIALOG. http://www.flipcode.com/archives/article_geomipmaps.pdf. Accessed 2 July 2019

Gardner JV, Armstrong AA, Calder BR et al (2014) So, how deep is the Mariana Trench? Mar Geod 37(1):1–13. https://doi.org/10.1080/01490419.2013.837849

Hell B, Jakobsson M (2011) Gridding heterogeneous bathymetric data sets with stacked continuous curvature splines in tension. Mar Geophys Res 32(4):493–501. https://doi.org/10.1007/s11001-011-9141-1

Jakobsson M, Mayer L, Coakley B et al (2012) The international bathymetric chart of the Arctic Ocean (IBCAO) Version 3.0. Geophys Res Lett 39(12): L12609. https://doi.org/10.1029/2012GL052219

Jin HL, Lu XP, Liu HJ (2010) Large-scale terrain realistic rendering based on programmable GPU hardware (in Chinese with English abstracts). Geomat Inf Sci Wuhan Univ 35(2):143–146

Losasso F, Hoppe H (2004) Geometry clipmaps: terrain rendering using nested regular grids. Paper presented at the international conference on computer graphics and interactive techniques, ACM SIGGRAPH, Los Angeles, 08–12 August 2004

Mayer L, Jakobsson M, Allen G et al (2018) The Nippon foundation—GEBCO seabed 2030 project: the quest to see the world's oceans completely mapped by 2030. Geosciences 8(2):63. https://doi.org/10.3390/geosciences8020063

Mann M, Agathoklis P, Antoniou A (2001) Automatic outlier detection in multibeam data using median filtering. In: IEEE Pacific Rim Conference on Communications, Computers and Signal Processing (IEEE Cat. No. 01CH37233), vol 2, pp 690–693

Meng JX (2013) Rapid display and rendering of large-scale submarine topography DEM (in Chinese with English abstracts). Dissertation, Shandong University of Science and Technology

Nakanishi M, Hashimoto J (2011) A precise bathymetric map of the world's deepest seafloor, challenger deep in the Mariana Trench. Mar Geophys Res 32(4):455–463. https://doi.org/10.1007/s11001-011-9134-0

Salzmann MA (1994) A real-time quality control procedure for use in integrated navigation systems. Hydrograph J 72:25–30

Smith WH, Sandwell DT (1997) Global sea floor topography from satellite altimetry and ship depth soundings. Science 277 (5334):1956–1962. https://doi.org/10.1126/science.277.5334.1956

Smith WH, Wessel P (1990) Gridding with continuous curvature splines in tension. Geophysics 55(3):293–305. https://doi.org/10.1190/1.1442837

Tang GA, Li FY, Liu XJ (2010) The tutorial of DEM, 3rd edition (in Chinese). Science Press, Beijing

Tobler WR (1970) A computer movie simulating urban growth in the Detroit region. Econ Geogr 46(sup1):234–240. https://doi.org/10.2307/143141

Ulrich T (2002) Rendering massive terrains using chunked level of detail control. Paper presented at the International Conference on

Computer Graphics and Interactive Techniques, Acm Siggraph, New York, 2002

Wang MW, Wu ZY, Yang FL et al (2018) Multifeature extraction and seafloor classification combining LiDAR and MBES data around Yuanzhi Island in the South China Sea. Sensors 18(11):3828. https://doi.org/10.3390/s18113828

Ware C, Knight W, Wells D (1991) Memory intensive statistical algorithms for multibeam bathymetric data. Comput Geosci 17 (7):985–993. https://doi.org/10.1016/0098-3004(91)90093-S

Wu ZY, Yang FL, Luo XW et al (2017) High-resolution submarine topography—theory and technology for surveying and post-processing. Science Press, Beijing

Wu ZY, Milliman JD, Zhao DN et al (2018) Geomorphologic changes in the lower Pearl River Delta, 1850-2015, largely due to human activity. Geomorphology 314:42–54. https://doi.org/10.1016/j.geomorph.2018.05.001

Wu ZY, Saito Y, Zhao DN et al (2016) Impact of human activities on subaqueous topographic change in Lingding Bay of the Pearl River estuary, China, during 1955-2013. Sci Rep 6(1):37742. https://doi.org/10.1038/srep37742

Yang FL, Bu XH, Ma Y et al (2017) Geometric calibration of multibeam bathymetric data using an improved sound velocity model and laser tie points for BoMMS. Ocean Eng 145:230–236. https://doi.org/10.1016/j.oceaneng.2017.09.010

Zhou JQ, Wu ZY, Jin XL et al (2018) Observations and analysis of giant sand wave fields on the Taiwan Banks, northern South China Sea. Mar Geol 406:132–141. https://doi.org/10.1016/j.margeo.2018.09.015

Zhao JH, Yan J, Zhang HM et al (2014) A new method for weakening the combined effect of residual errors on multibeam bathymetric data. Mar Geophys Res 35(4):379–394. https://doi.org/10.1007/s11001-014-9228-6

Zhao DN, Wu ZY, Zhou JQ et al (2015) A new method of automatic SVP optimization based on MOV algorithm. Mar Geod 38(3):225–240. https://doi.org/10.1080/01490419.2015.1006798

Zhao Q (2011) Research and implementation of large-scale terrain data scheduling and rendering technology (in Chinese with English abstracts). Dissertation, University of Electronic Science and Technology of China

Zhou ZY, Lin J, Behn MD et al (2015) Mechanism for normal faulting in the subducting plate at the Mariana Trench. Geophys Res Lett 42 (11):4309–4317. https://doi.org/10.1002/2015GL063917

Acoustic Seafloor Characterization

8

In the past three decades, rapid promotion and application of new, high-end technology in the field of marine exploration have led to a deeper understanding of the world's oceans (Zhai and Huang 2009). The recent global population explosion and shortage of terrestrial resources make the strategic significance of the oceans ever more prominent. Oceanographic, biological, nuclear and aerospace engineering are among the cutting-edge sciences and technologies, and interest in ocean studies continues to increase (Liu 2003). Seafloor sediment is a major component of the marine environment and it plays an important role in activities such as marine engineering construction, seafloor resource development, marine fisheries, port channel construction and surveying for seafloor pipeline routes. Consequently, surveying and mapping investigations of this component of the marine environment constitute the basis of all marine activities. US compiled a 1:1 000 000-scale map of the geological structure of the continental shelf as early as 1969, and the work of compiling large-scale geological maps of coastal and continental shelves was mostly completed by 2000. Japan began seafloor geological surveying in 1974, compiling and publishing a 1:500 000-scale map of the seafloor geological structure in 1984. To date, Japan has completed four campaigns of regional marine geological surveys, and various seafloor sediment maps (covering the seafloor geological structure, surface sediment and sediment type) have been published (Cui et al. 2003).

As ocean exploitation continues, marine geologists and engineering experts are increasingly demanding a comprehensive and detailed understanding of the properties of seafloor sediment, and the acoustic method is the generally considered the most effective and quickest means of detection (Breslau 1965). In addition, with the development of modern sonar technology, hydro acousticians need to understand the impact of seafloor acoustic characteristics on the propagation of sound waves in the ocean (Hampton 1974; Guo 2004). Using an acoustic method to determine the relationship between the acoustic parameters (e.g., reflection coefficient, sound velocity, attenuation and scattering) and the physical properties (e.g., sediment type and particle size distribution) of seafloor sediment, to achieve automatic classification and identification of the sediment, is an important aspect of marine acoustic remote sensing development (Zhu 2000).

Research on acoustic seafloor characterization began in the 1960s (Collins and Voulgaris 1993; Dyer et al. 1997), and sonar technologies such as the single beam echo sounder (SBES), multi-beam echo sounder (MBES), side-scan sonar (SSS) and subbottom profiler have experienced rapid development in the recent several decades. The use of sonar data in the characterization of seafloor type has been shown to have wide potential in the fields of underwater acoustics, marine geology, marine surveying and oceanic engineering applications.

8.1 Development Status of Seafloor Classification Approaches

Direct sampling and indirect detection are two methods often used to obtain information on seafloor sediment properties. Direct sampling involves the use of a sampler to collect specimens of seafloor sediment that are analyzed subsequently using technical means to obtain sediment information (e.g., sediment type, particle size, physical and chemical properties, biological component content and age). This method can determine seafloor sediment characteristics intuitively and accurately. However, direct sampling is often a time-consuming labor-intensive process, especially in a deep-water setting. In addition, direct sampling can often cause substantial disturbance or damage to the seafloor sample, resulting in a considerable decrease in its organization and representativeness, thereby affecting the accuracy and credibility of the sample data. Indirect detection methods employ non-contact means to obtain seafloor sediment characteristics. Common indirect approaches include optical, acoustic and biochemical methods. Detailed features (e.g., surface texture) of sediment can be obtained using optical

Z. Wu et al., *High-resolution Seafloor Survey and Applications*,
https://doi.org/10.1007/978-981-15-9750-3_8

methods such as underwater photography. However, owing to the significant absorption of electromagnetic waves in seawater, it is often difficult to use optical methods widely in seafloor sediment characterization. In contrast, the acoustic method provides a practical means with which to infer category information and physical characteristics of seafloor sediment. Being an efficient and cost-effective approach, the acoustic method has gradually become the principal technique used for detecting sediment.

The principle of the acoustic method is to use an acoustic transducer to direct sound waves toward the seabed, and then to determine the type and characteristics of the seabed sediment by recording and analyzing the characteristics of the backscatter strength (BS) of the returned signals. Since the 1950s, many studies have investigated the relationship between the acoustic characteristics and the physical and mechanical properties of seafloor sediment, laying the foundation for further study of acoustic detection of seafloor sediment type (Lu 1997; Du et al. 2006). The rapid development of both acoustic technology and data acquisition and processing technology has allowed the consequent development of seafloor sediment detection technologies based on SBES, SSS and MBES, which offer the great potential of acoustic information in detecting seafloor sediment (Anderson et al. 2002; Tegowski et al. 2003).

Among the various acoustic methods, the MBES, with its wide coverage and ability to record bathymetry and BS simultaneously, has become an attractive choice for seafloor mapping. In particular, BS data acquired by MBES can provide effective information for the characterization of seabed sediment type. Thus, seafloor characterization based on MBES data is the major focus of this section. Although the recording methods and content of BS data might differ slightly among different systems, the classification process generally includes three aspects: data processing, feature extraction, and classifier design (Zhou 2005).

8.1.1 Research on BS Data Processing

The purpose of processing MBES BS data is to obtain data that reflect only the nature of the seafloor sediment. Owing to the complexity of factors affecting BS, existing data processing methods are all developed for specific instruments. For example, Beaudoin et al. (2002) analyzed the Reason Seabat 8101 MBES system, focusing on the geometric and radiometric correction of BS values. Lurton et al. (1994) analyzed the influence of seafloor topography on the BS measurements of the Simrad EM12D and they constructed a model for local seafloor slope correction. Hellequin et al. (2003) analyzed various factors affecting the BS data of the Simrad EM1000 MBES, with particular focus on the processing method of the central beam BS outliers under the equidistant beam mode. Augustin et al. (1997) both analyzed the BS anomalies of multiple sectors of the Simrad EM300 MBES and studied the method of filtering noise. Gonidec et al. (2003) discussed the influence of the absorption coefficient, seabed topography and acoustic line trace bending on BS, as well as the corresponding correction methods (based on Simrad EM300 MBES data).

8.1.2 Feature Extraction and Selection

The purpose of feature extraction is to elicit information on seafloor sediment characteristics from BS data or sonar images. Similar to the process of feature extraction using satellite remote sensing imagery, existing algorithms mainly include basic statistics, quantiles, histograms, power spectrum ratios, gray-level co-occurrence matrices and fractal dimensions. Although many previous studies have investigated the correlation between characteristic parameters and substrate types based on specific environments, the obtained results are not applicable in a general context because of the complexity of the BS generation mechanism. Normally, to improve classification performance, feature information is extracted to the maximum extent. Consequently, not only is the feature dimension increased, but considerable correlation and redundancy might exist among the extracted features. Therefore, following feature extraction, optimization of feature selection is needed to reduce the dimension of the feature space and to remove potential collinearity. For example, the Quester Tangent's QTC MULTIVIEW system can extract 132 feature parameters and reduce the extracted feature vector to three main components using principal component analysis (PCA) (Quester Tangent 2005). Similarly, Zhou (2005) used a fuzzy neural network combined with a genetic algorithm to achieve feature selection and optimization.

8.1.3 Acoustic Classification Methods

At present, statistical identification, neural networks and cluster analysis are methods commonly used in acoustic seafloor classification. Simrad Inc. developed the Triton seafloor classification software module based on the Bayes maximum likelihood estimation criterion, which is a supervised learning method. The disadvantage of this method is that the probability distribution function (PDF) of each bottom type must be known in advance, but there are no such statistical models available at present. Michalopoulou et al. (1995) used back propagation (BP) neural networks to classify seafloor sediments based on BS data. The QTC MULTIVIEW system uses a cluster analysis method to achieve automatic segmentation of seafloor sediment type (Quester Tangent 2005). To obtain improved characterization results, Zhou (2005) and Tang et al. (2007) combined a neural network algorithm with a genetic algorithm to classify multi-beam BS data.

8.2 Properties of Seafloor Sediment and Principles of Acoustic Classification

The term "seafloor sediment" usually refers to seafloor surface material with a density greater than 1.20 g/cm^3. Information on the composition, physical/structural characteristics and evolutionary history of seafloor material can be studied using methods such as acoustic, physical sampling and other test analyses. Generally, the obtained parameters can be categorized into two groups: physical and acoustic parameters. Physical parameters include sediment density, porosity, permeability and other parameters (e.g., related to sediment particles, pore fluids and pore space). Acoustic parameters include acoustic impedance, velocity and attenuation of both longitudinal and shear waves and seafloor roughness (a key input for most acoustic scattering models and acoustic transmission models).

8.2.1 Physical Properties of Seafloor Sediment

Some parameters of the physical properties of sediment (e.g., volume density) can be used directly as input for acoustic theory and models, and other physical properties (e.g., sediment type and average particle size) can be used to predict sediment acoustic properties indirectly.

8.2.1.1 Composition of Particle Size of Sediment

The composition of the particle size of seafloor sediment is an important physical property that not only reflects the formation mechanism and environmental characteristics of seafloor sediment, but also establishes its relationship with the acoustic characteristics. Thus, studying particle composition is the principal method for the classification of seafloor sediment in marine surveying. Particle size is usually described in logarithmic form:

$$\phi = -\log_2 d \tag{8.1}$$

where d is the particle size (unit: mm).

The reason for taking the logarithm is that the scale of variation of particle size of seafloor sediment is reasonably large. Thus, following the computation of the logarithm, the distribution of ϕ approximately follows the normal distribution, which is convenient for further analysis of particle size composition. To characterize composition, a weighted frequency distribution diagram of each particle size is plotted, following which statistical data (e.g., the average, mode, sorting coefficient, standard deviation (SD), skewness and kurtosis) of the sediment composition can be depicted graphically.

At present, the most widely used method of classification of sediment type is the Udden-Wentworth classification standard, which is based on the equivalent particle size (Table 8.1).

Based on particle size, sediment can be divided into five major groups: rock, gravel, sand, silt and clay, and most of these groups can be subdivided further (Table 8.1). In fact, seafloor

Table 8.1 Udden-Wentworth classification ϕ standard

Particle type	Particle name		Particle size range		$\phi = (-\log_2 d)$		Code
	Simple taxonomy	Detailed taxonomy	Millimeter	Micrometer			
Rock (stone)	Rock (stone)	Rock (stone)	>256		256	−8	R
Gravel (G)	Coarse	Coarse	256–128		128	−7	GG
			128–64		64	−6	
	Middle	Middle	64–32		32	−5	MG
			32–16		16	−4	
			16–8		8	−3	
	Fine	Fine	8–4		4	−2	FG
			4–2		2	−1	
Sand (S)	Coarse	Very coarse	2–1	2 000–1 000	1	0	VCS
		Coarse	1–0.5	1 000–500	1/2	1	CS
	Middle	Middle	0.5–0.25	500–250	1/4	2	MS
	Fine sand	Fine	0.25–0.125	250–125	1/8	3	FS
		Very fine	0.125–0.063	125–63	1/16	4	VFS
Silt (T)	Coarse	Coarse	0.063–0.032	63–32	1/32	5	CT
		Middle	0.032–0.016	32–16	1/64	6	MT
	Fine	Fine	0.016–0.008	16–8	1/128	7	FT
		Very fine	0.008–0.004	8–4	1/256	8	VFT
Clay (Y)	Clay	Coarse	0.004–0.002	4–2	1/512	9	CY
			0.002–0.001	2–1	1/1 024	10	
		Fine	<0.001	<1	1/2 048	>11	FY

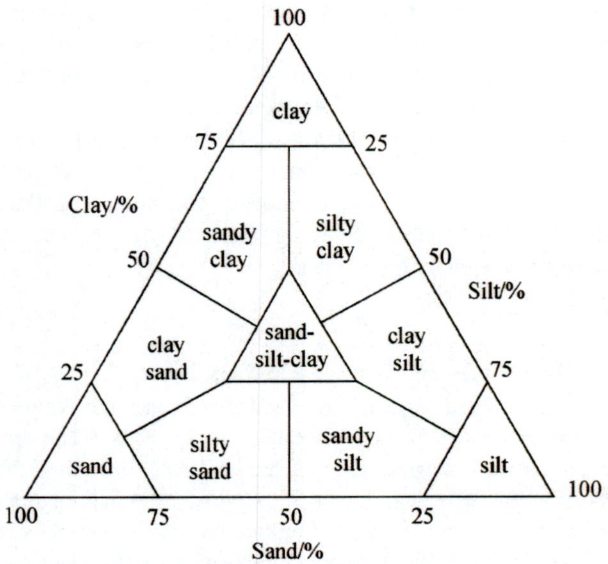

Fig. 8.1 Schematic description of the Shepard classification

sediment is often composed of several different particle sizes. For example, silt might be present in sand, and vice versa. The Shepard classification method has long been employed to classify seafloor sediment according to the relative mass ratio of components of different grades (Zhao 2009).

In the Shepard method, a triangle is used to describe the sediment composition of sand, silt, and clay (Fig. 8.1). When the gravel composition is greater than 5%, the sediment is named according to the dominant particle size of sand and gravel (e.g., gravel, sand gravel, gravel sand); the sand grade is divided further according to the dominant grain size (e.g., medium coarse sand and medium fine sand). Common types of sediment in the shallow seas off China (codes in parentheses) include the following: gravel (G), sand gravel (SG), gravel sand (GS), coarse sand (CS), medium coarse sand (MCS), medium sand (MS), medium fine sand (MFS), fine sand (FS), sand (not divided) (S), silty sand (TS), sandy silt (ST), silt (T), clay silt (YS), silty clay (TY), clay sand (YS), sandy clay (SY), sand-silt-clay (STY) and clay (Y). Here, MCS and MFS refer to CS and FS, respectively, in the third column of Table 8.1. The difference between sand-silt-clay and clay-silt-sand is that the percentage of each component is different. The Shepard classification method states that the first part of the name is the one with the smallest ratio and the final part of the name is the one with the greatest ratio. Only when the relative ratio is greater than 25% can the corresponding substance participate in the naming of the seafloor type.

8.2.1.2 Volume Density of Sediment

The volume density of sediment is defined as the sediment mass per unit volume, which is expressed as

$$\rho = \frac{m_0}{V} \tag{8.2}$$

where m_0 is the sediment mass (unit: g) and V is the sediment volume (unit: cm^3).

Because of the variation in the mineral composition, pore volume and water content of sediment, the volume density of terrestrial sediment changes considerably under natural conditions, generally varying in the range 1.60–2.20 g/cm^3. However, for seafloor sediment, the volume density is relatively smaller because of the high water content.

8.2.1.3 Water Content of Sediment

The water content of sediment is defined as the ratio of the water mass to the mass of solid particles in the sediment. It is determined as the ratio of the water weight loss and the dry sediment weight when a sediment sample is dried to a constant amount at a certain temperature (105–110 °C). It is expressed in terms of percentages:

$$w = \frac{m_w}{m_d} \times 100 = \left(\frac{m_0}{m_d} - 1\right) \times 100 \tag{8.3}$$

where m_0 is the weight of the sediment, m_w is the weight of the water in the sediment and m_d is the weight of the solid particles in the sediment.

Water content is an important indicator of sediment moisture. Generally, its value varies widely depending on the type of sediment, burial conditions and geographical environment. The water content of dry coarse sand sediment is close to zero and that of saturated sandy sediment is up to 40%. Coarse sand generally has water content lower than 30%, and the water content of saturated soft viscous sediment (e.g., silt) is greater than 80%. The water content of fine clay sediment in the deep ocean can reach more than 200%.

8.2.1.4 Porosity

Fractional porosity is equal to the ratio of the pore water volume V_w to the total volume V:

$$\beta = \frac{V_w}{V} \tag{8.4}$$

The void ratio is equal to the ratio of the pore water volume to the solid particle volume V_s:

$$e = \frac{V_w}{V_s} = \frac{\beta}{1-\beta} \tag{8.5}$$

Porosity is also usually expressed in terms of percentages.

8.2.2 Geoacoustic Properties of Seafloor Sediment

Geoacoustic parameters such as the velocity of sound and attenuation of sound waves propagating in sediment are necessary input for the acoustic wave propagation theory of seafloor sediment. Because of the difficulty in obtaining a calculation model for the geoacoustic parameters of sediment and the inaccuracy of the results of such a model, direct measurement is the best method for obtaining geoacoustic parameters.

Laboratory-based and in situ measurements have established the information necessary for determining the velocity of sound and the range of the acoustic attenuation coefficient of shallow superficial sediments, and for establishing predictive models of the sound velocity and physical properties of sediment (Hamilton 1956, 1963; Shumway 1956, 1958, 1960; McCann and McCann 1969; McCann 1972). It has been determined that the velocity of sound in sediment is related to porosity and average particle size. Silt sediment (with the porosity of 70–90%) has the lowest sound velocity (approximately 3% lower than seawater), while coarse sand sediment (with porosity of 35–40%) has the highest sound velocity. Sediment with a median diameter of 30 μm and porosity of 40% has the lowest acoustic attenuation coefficient; sediment with a median diameter of 125 μm and porosity of 60% has the highest acoustic attenuation coefficient. Hamilton (1972, 1980), Hamilton and Bachman (1982) and Bachman (1985, 1989) established multiple regression equations to predict the sediment sound velocity and the acoustic attenuation coefficient based on the physical properties of sediment.

Different types of acoustic detector can be employed for in situ measurements. The velocity of sound is usually calculated directly based on the measured sound wave propagation time and the distance between detectors. The acoustic attenuation coefficient is usually obtained by comparing the loss of acoustic energy between detectors at different distances to a standard energy level. Based on the spherical wave assumption, geoacoustic properties are often measured by laying a hydrophone and detector on the seafloor and placing a series of sound sources in the water and sediment.

8.2.3 Seafloor Roughness

Seafloor roughness is the main influencing factor in high-frequency acoustic scattering. It can cause acoustic reverberation through backscattering, and change the propagation characteristics of sound waves through forwarding scattering. In terms of seafloor roughness, marine geologists and geophysicists are primarily concerned with large-scale seafloor features, while researchers on high-frequency acoustics are primarily concerned with small-scale features comparable with the acoustic wavelength. Roughness characteristics comparable with the wavelength of a sound wave are considered the main reason for acoustic scattering, but characteristics on a larger scale can also cause a change in the grazing angle between a sound wave and the seafloor. Therefore, research on roughness is limited primarily to scales with a resolution of 0.1 mm to several meters.

The statistical characteristics of seafloor roughness are often expressed in the form of root mean square roughness or power spectral density (roughness spectrum), which are also the main input parameters for most acoustic scattering models and acoustic transmission models.

Seafloor fluctuation can be expressed as

$$z = f(\boldsymbol{R}), \boldsymbol{R} = (x, y) \tag{8.6}$$

where $f(\boldsymbol{R})$ is the interface fluctuation function, the mean of which is expected to be zero:

$$\langle f(\boldsymbol{R}) \rangle = 0 \tag{8.7}$$

The mean square error is given as

$$h^2 = \langle f^2(\boldsymbol{R}) \rangle \tag{8.8}$$

where h is the root mean square roughness. For acoustic scattering models such as the Kirchhoff approximation and small-slope approximation, the covariance of interface fluctuations is required as an input parameter. The relevant expression is

$$B(\boldsymbol{R}) = \langle f(\boldsymbol{R}_0 + \boldsymbol{R}) f(\boldsymbol{R}_0) \rangle \tag{8.9}$$

where $\langle \cdot \rangle$ represents the mean value. The result of the covariance divided by the mean square roughness is the autocorrelation value. When the delay parameter $\boldsymbol{R}$ is equal to 0, the autocorrelation value is 1.

The assumption of a spatial stationary process is inherent in Eqs. (8.7)–(8.9). In other words, the variance, covariance and higher moment are independent of where their values are estimated and they depend only on the difference between two points. This means that the seafloor should have similarities within a certain scale.

The acoustic scattering model based on the small roughness perturbation theory requires a roughness spectrum, which is the Fourier transform of the seafloor relief covariance:

$$W(\boldsymbol{K}) = 1/(2\pi)^2 \int B(\boldsymbol{R}) e^{-i\boldsymbol{K}\cdot\boldsymbol{R}} d^2\boldsymbol{R} \tag{8.10}$$

where the variable $\boldsymbol{K} = (K_x, K_y)$ is a two-dimensional wave vector, the magnitude of which ($K = \sqrt{K_x^2 + K_y^2}$) is called the wave number. The relationship between the wave number and wavelength is

$$K = 2\pi/\lambda \tag{8.11}$$

It can be seen that a small wave number corresponds to a large wavelength. The wave number integral over the entire range of K_x and K_y is equal to the mean square roughness h^2 after normalization of the roughness spectrum:

$$h^2 = \int W(\boldsymbol{K})\mathrm{d}^2\boldsymbol{K} \tag{8.12}$$

The dimension of the two-dimensional roughness spectrum is the fourth power of the length. The measuring results of roughness are usually expressed by the spectrum format. Given a roughness spectrum, the covariance of seafloor fluctuation can be obtained using the inverse Fourier transform.

In addition to the assumption of stability, it is often assumed that seafloor roughness follows a Gaussian statistical distribution. Thus, the first- and second-order statistical parameters (or covariance spectrum) could provide a complete representation that could determine any desired statistical property.

If seafloor roughness exhibits a nondirectional characteristic, its roughness spectrum is known as isotropic, and then the roughness depends only on the wave number ($\boldsymbol{K}$).

Scattering models typically employ a few simple functional forms to represent the roughness spectrum. The simplest and most commonly used form for the isotropic roughness spectrum is the "power law" representation, i.e.,

$$W(\boldsymbol{K}) = \frac{w_2}{K^{\gamma 2}} \tag{8.13}$$

where w_2 is the spectral intensity and γ^2 is the spectral index. Because the power law spectrum form extending to zero and of infinite scale can lead to ill-conditioned behavior, it can be effective only on a limited scale. For example, if the power law were effective on a sufficiently wide scale, the terrain could be seen as a "fractal", which is scalable and exhibits "self-similarity". According to this characteristic, regardless of the factors used to reduce or amplify the roughness, they all exhibit statistical similarity, i.e., the same scaling ratio in both horizontal and vertical directions.

8.2.4 Acoustic Classification Principle of Seafloor Sediment

Acoustic scattering by the seafloor is a random process due to fluctuation in the seafloor interface and spatial variation in the physical properties. The study of seafloor acoustic scattering can be divided into two principal areas. The first explores the variation law of the average BS level from the perspective of energy, while the second uses the mathematical method of the stochastic process to model the statistical properties of the scattering intensity, e.g., the distribution function and the energy spectrum. This section focuses on two key issues: the use of seafloor BS data for sediment classification, and the characteristics that should be extracted from the BS data to describe the change in seafloor sediment type. To provide the hydroacoustic theoretical basis for seafloor sediment classification focusing on the above two issues, this section discusses the following topics: (i) the various factors affecting seafloor BS level; (ii) the correlation of BS level with acoustic wave frequency, seafloor sediment type and seafloor incident angle; and (iii) the statistical characteristics of seafloor BS values.

8.2.4.1 Seafloor Reflection and Scattering

Although reflection and scattering are difficult to separate, they are normally considered as two separate concepts. Reflection is generally used to describe the interaction of sound waves with a uniform flat seafloor, in which case the seafloor is considered an acoustic mirror that reflects the sound wave at an angle equal to the incident angle. Scattering means the seafloor exhibits obvious roughness and non-uniform characteristics that result in the redistribution of energy in all directions. When this happens, the scattering angle is no longer equal to the incident angle.

When discussing seafloor reflection and scattering, an appropriate coordinate system should first be defined. The coordinate system discussed here is an angular coordinate system that is often used in hydroacoustic scenarios (Fig. 8.2). Taking the horizontal direction as the reference, the angle between the direction of the incident acoustic wave and the horizontal is known as the incident grazing angle θ_i. The direction of the scattered sound wave is determined by the scattering grazing angle θ_s and the scattering azimuth angle ϕ_s measured by non-concentric arrays. For a non-random flat seafloor, seafloor reflection is in the direction of specular reflection, i.e., the scattering grazing angle is equal to the incident grazing angle and the scattering azimuth is zero. To measure the scattering intensity of the non-concentric arrays, the scattering sound field is measured at the same position as the sound source ($\theta_s = \theta_i$, $\phi_s = \pi$).

1. **Reflection**

Although the seafloor is not exactly flat, it is useful to regard it as an idealized flat surface. It is a reasonable approximation of the actual situation, and it can help define the reflection coefficient and scattering coefficient that are often employed in reflection and scattering models. The incident sound field is assumed to be a plane wave with amplitude of

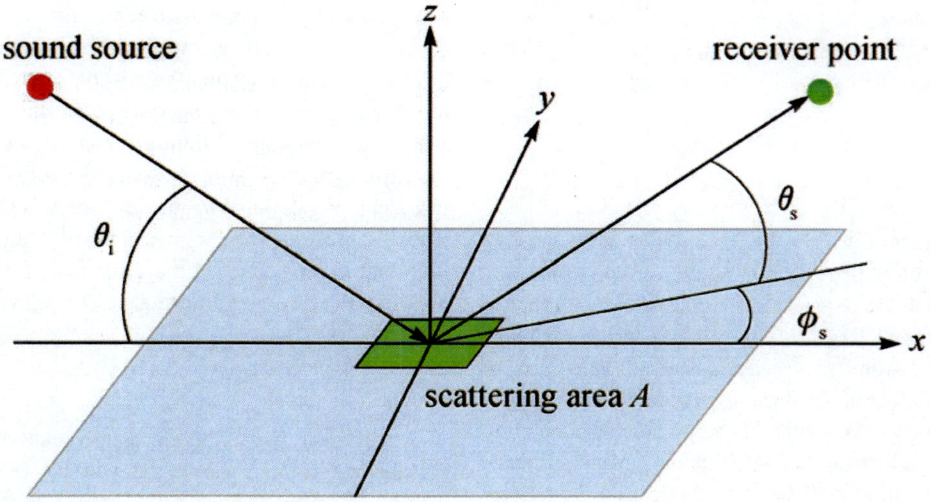

Fig. 8.2 Schematic of coordinate system for seafloor reflection and scattering angle definition (if the seafloor has a trending direction such as sand waves, an azimuth angle ϕ_i for the incident sound waves should be specified)

1 and angular frequency of ω. At the interface, the plane wave can be expressed approximately as

$$P_i = P_{i0}e^{iK_{ir}} \tag{8.14}$$

where P_{i0} is the complex sound pressure amplitude at the initial position and K_{ir} is the wave vector, the direction of which is the propagation direction of the plane wave. The z-axis is perpendicular to the seafloor (Fig. 8.2). For convenience, a suitable coordinate system is selected such that the direction of the plane wave propagation lies in the x-z plane. Thus, the wave vector, which has components in the directions of x, y and z, can be expressed as

$$\boldsymbol{K}_i = \frac{\omega}{c_w}(\cos\theta_i,\ 0,\ -\sin\theta_i) \tag{8.15}$$

where c_w is the seawater sound speed. As reflection involves only one angle, for simplicity, subscript i in Eq. (8.15) is omitted. Although the reflected wave has the same grazing angle as the incident wave, its amplitude generally becomes smaller and it undergoes a phase shift. Thus, the reflected sound pressure field has the following expression:

$$P_r = V_{ww}(\theta)P_{i0}e^{i\boldsymbol{K}_r\cdot r} \tag{8.16}$$

In the above formula, the expression of $\boldsymbol{K}_r$ is

$$\boldsymbol{K}_r = \frac{\omega}{c_w}(\cos\theta,\ 0,\ \sin\theta) \tag{8.17}$$

where $V_{ww}(\theta)$ is the reflection coefficient (dimensionless) that is a function of the grazing angle θ.

The reflection and refraction of plane waves at the interface of an infinite uniform medium may be used as a basis and guide for discussing complex scattering.

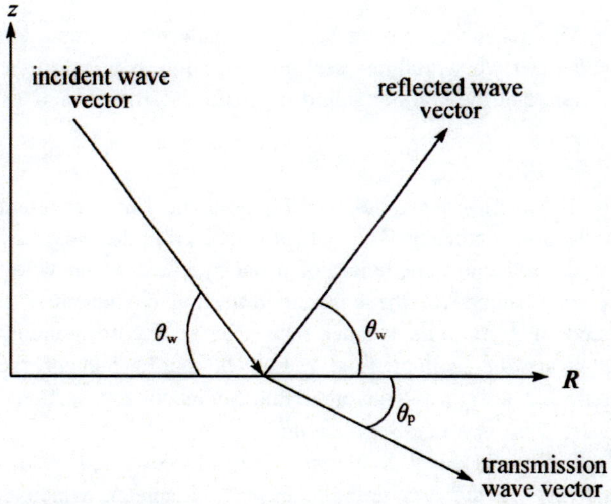

Fig. 8.3 Schematic of incident, reflected and transmitted plane waves at an interface

A schematic of the reflection and transmission of a plane wave incident at a grazing angle θ_w to a completely flat sea bottom is shown in Fig. 8.3. A more detailed explanation of the acoustic reflection problem can be found in Brekhovskikh and Godin (1990).

The reflection and transmission coefficients of a plane wave are

$$\begin{aligned} V_{ww} &= \frac{\rho_p c_p \sin\theta_w - \rho_w c_w \sin\theta_p}{\rho_p c_p \sin\theta_w - \rho_w c_w \sin\theta_p} \\ V_{wp} &= \frac{2\rho_p c_p \sin\theta_w}{\rho_p c_p \sin\theta_w + \rho_w c_w \sin\theta_p} \end{aligned} \tag{8.18}$$

where $\rho_w c_w$ and $\rho_p c_p$ are the acoustic impedances in water and sediment, respectively, V_{ww} is the reflection coefficient and V_{wp} is the transmission coefficient. According to the energy conservation law, $V_{ww} + V_{wp} = 1$. However, as no energy is returned in the direction of the sound source in this simple scenario, the impact of seafloor roughness must be considered.

2. **Scattering**

Discontinuity in the actual physical properties of the seafloor occurs because of the roughness of the seafloor interface, spatial variation in the physical properties of the seafloor and irregularities of shell fragments or bubbles in the sediment. Therefore, when sound waves are projected onto such an inhomogeneous medium, scattering will occur (Fig. 8.4).

Studying the average value of the complex sound pressure field $\langle P \rangle$ and its fluctuation P_s is the classical method with which to examine seafloor scattering caused by random media. The sound field P can be decomposed into the following:

$$P = \langle P \rangle + P_s \tag{8.19}$$

The mean square fluctuation, which is an important parameter when studying seafloor scattering, is equal to the variance in the seafloor sound field (Eq. (8.20)):

$$\langle |P_s|^2 \rangle = \langle |P|^2 \rangle - |\langle P \rangle|^2 \tag{8.20}$$

The average wave field can be represented by a coherent reflection coefficient V_{wwc}, which is defined in the same way as the reflection coefficient of a flat interface sound wave. The magnitude of the coherent reflection coefficient of a random seafloor is smaller than that of a corresponding non-random seafloor. The reason is that a part of the reflected energy is converted into an incoherent scattered wave field in a random medium.

Seafloor acoustic scattering is usually described quantitatively by the "scattering intensity" parameter, which is defined in Fig. 8.2. The area of seafloor scattering in the far field of the sound source is A. If the scattering area is rough and non-uniform compared with the wavelength of the acoustic wave, acoustic energy will be scattered in multiple directions. The scattered sound pressure field shown in Fig. 8.2 is measured at a specific point; thus, it is scattered in a specific direction. If multiple measurements were taken, the mean square field (average of the set of measurements) would be proportional to the seafloor scattering region area and the square of the incident field $|P_i|^2$, and inversely proportional to the square of the distance r_s from the scattering region to the receiving transducer. The relevant expression is

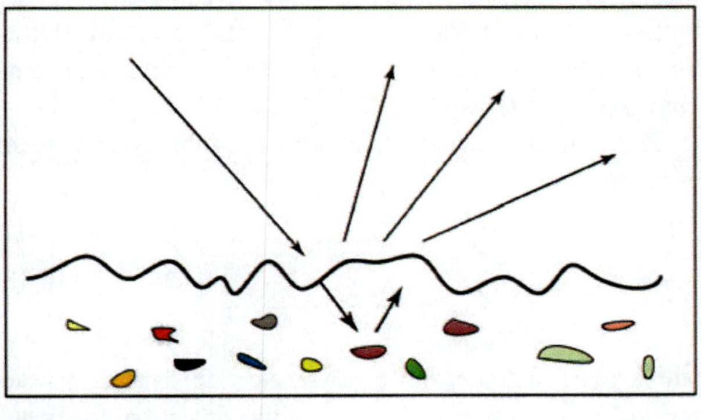

Fig. 8.4 Seafloor scattering

$$\langle |P_s|^2 \rangle = |P_i|^2 A\sigma \frac{1}{r_s^2} \tag{8.21}$$

Attenuation and refraction of the seafloor are ignored in the above formula. The factor σ in Eq. (8.21) is a dimensionless quantity, whose solid angle integral on the upper hemisphere is the total average scattering energy. Therefore, σ is called the scattering cross section per unit area and per unit angle and the relevant expression is

$$U_s = \frac{A|P_i|^2}{2\rho_w c_w} \int_{2\pi} \sigma(\theta_i, \theta_s, \phi_s) d\Omega_s \tag{8.22}$$

where ρ_w is the density of seawater, the product of $\rho_w c_w$ is acoustic impedance, Ω_s is the infinitesimal element, and $\sigma(\theta_i, \theta_s, \phi_s)$ is the scattering cross section. Seafloor scattering energy is usually expressed as "seafloor scattering intensity" (Urick 1983) and the expression of seafloor scattering intensity (in decibels) is

$$S_b = 10\log_{10}\sigma \tag{8.23}$$

It should be noted that the scattering cross section is defined by a statistical average.

3. **Seafloor backscattering intensity**

Seafloor scattering occurs in various directions in space, and the scattering energy returning to the source sonar system is known as seafloor backscattering. Multi-beam sonar systems operate in an active manner (i.e., transmitting and receiving in approximately the same position in space) to receive the backscattered signals.

Seafloor backscattering is defined as the ratio of the intensity of the sound scattered by unit area or unit volume to the intensity of the incident plane wave at a reference distance of 1 m. The scattered sound intensity is returned to the unit distance after the far-field measurement (expressed in decibels) (Fig. 8.5). The sound intensity emitted by the sound source is I_i, the propagation distance is R, the seafloor incident angle is θ, the backscattering coefficient is $s(\theta)$, the instantaneous sound area is A and the received sound intensity I_s (expressed in decibels) is

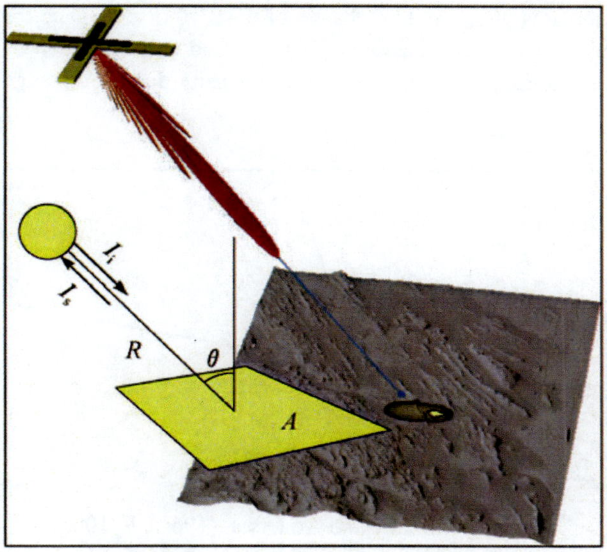

Fig. 8.5 Sketch of seafloor backscattering intensity measurement

$$I_s = I_i \frac{As(\theta)}{R^2} \quad (8.24)$$

The seafloor backscatter coefficient calculated from the far-field measurement is

$$s(\theta) = \frac{I_s R^2}{I_i A} \quad (8.25)$$

Expressed in decibels, the seafloor backscattering intensity (BS) is given as

$$\mathrm{BS}(\theta) = 10\log_{10}(s(\theta)) \quad (8.26)$$

The seafloor backscattering intensity is a function of the incident angle. For a given frequency, it reflects the seafloor sediment properties, constituting the hydroacoustic theoretical basis for seafloor sediment characterization based on BS values.

8.2.4.2 Correlation Between Seafloor BS and Sediment Type

The seafloor BS of measured data indicates that its value is affected by many factors, e.g., acoustic frequency, grazing angle (residual angle of the seafloor incident angle) and seafloor sediment.

1. **Frequency effect**

Variation in seafloor BS caused by different sound wave frequencies is known as the frequency effect. Measured BS data as a function of frequency at the same grazing angle for a muddy, sandy, gravel and rocky seafloor are shown in Figs. 8.6 and 8.9, respectively. The muddy seafloor data (Fig. 8.6) show that (i) seafloor BS increases with increasing frequency over a wide frequency range; and (ii) BS varies with frequency by approximately 3 dB/(10 octave) (i.e.,

Fig. 8.6 Frequency effect of mud seafloor

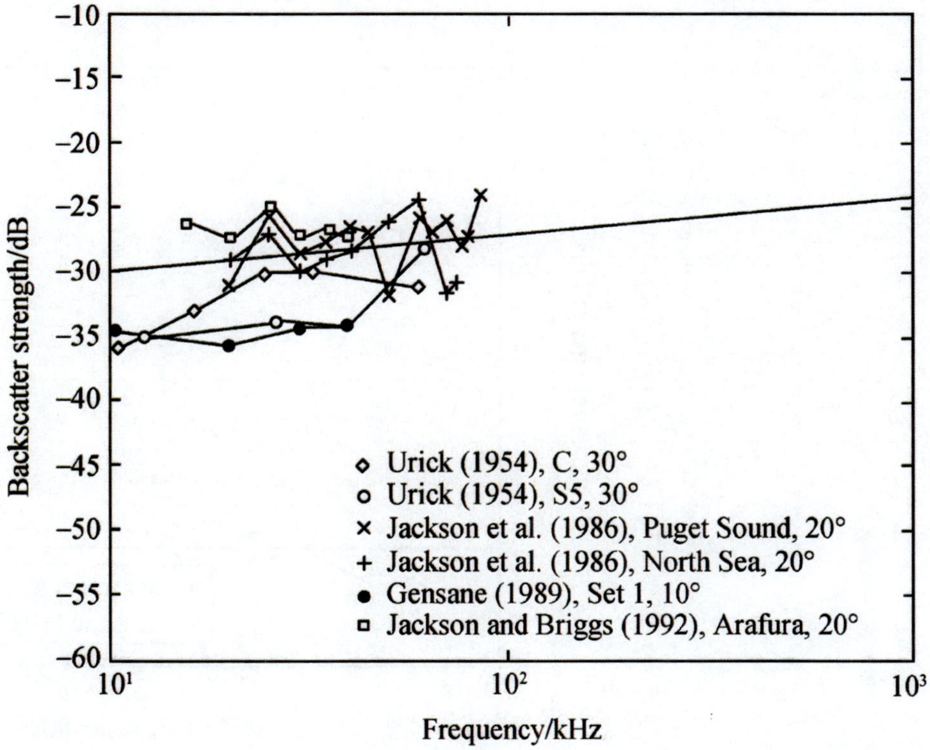

seafloor BS increases by 3 dB when the acoustic frequency changes from 10 to 100 kHz). Data for most sandy seafloor sites (Fig. 8.7) show that as frequency increases, BS increases with a rate of change of 6 dB/10 octave. Data for gravel and rocky seafloors (Figs. 8.8 and 8.9) are independent of frequency (i.e., scattering intensity does not change

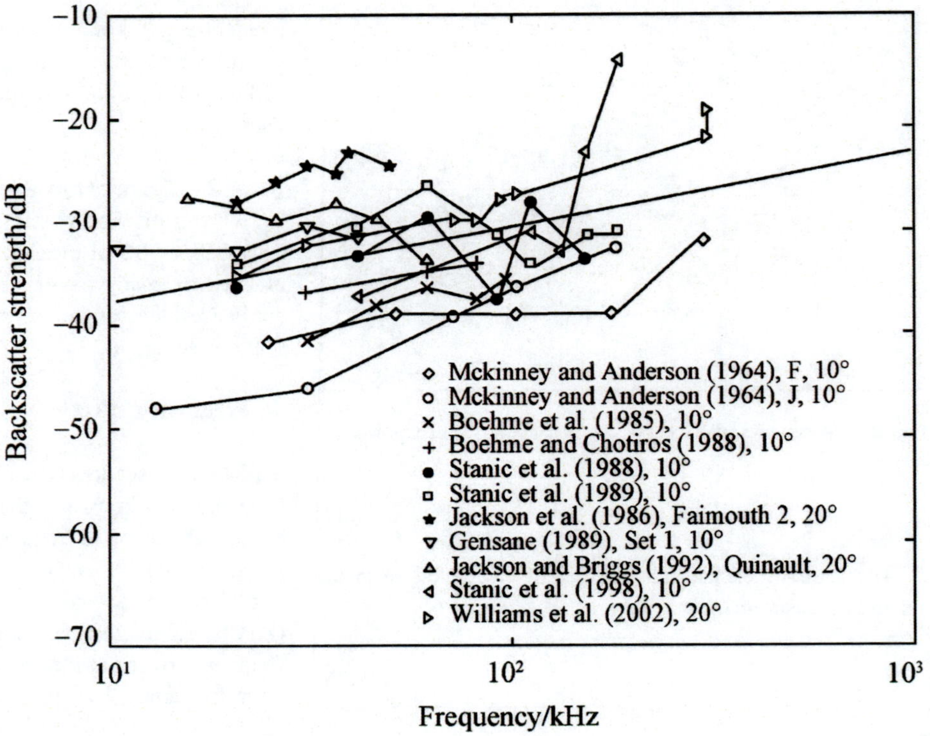

Fig. 8.7 Frequency effect of sandy seafloor

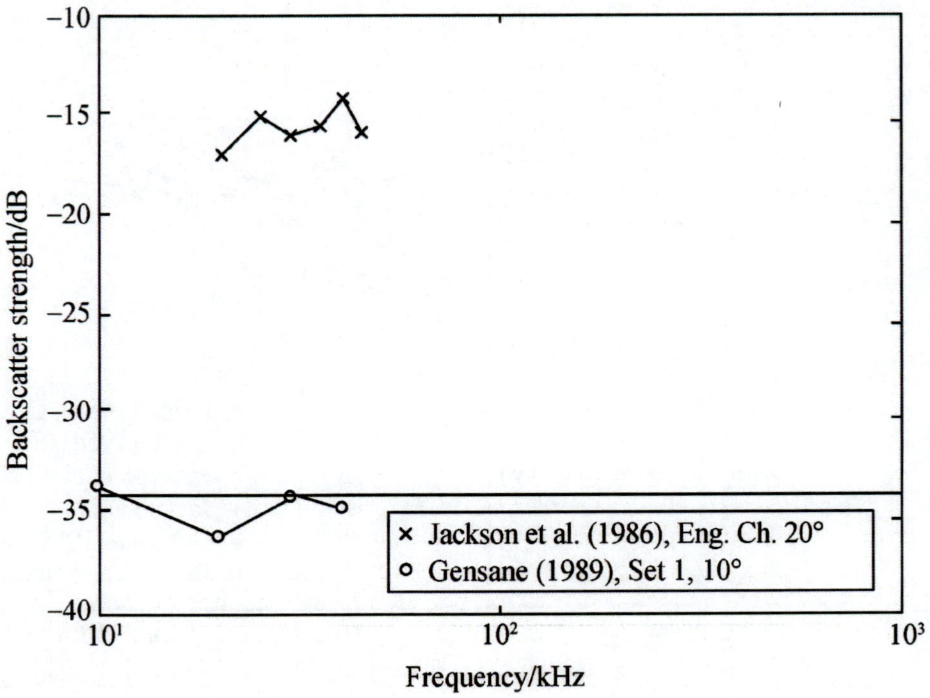

Fig. 8.8 Frequency effect of gravel seafloor

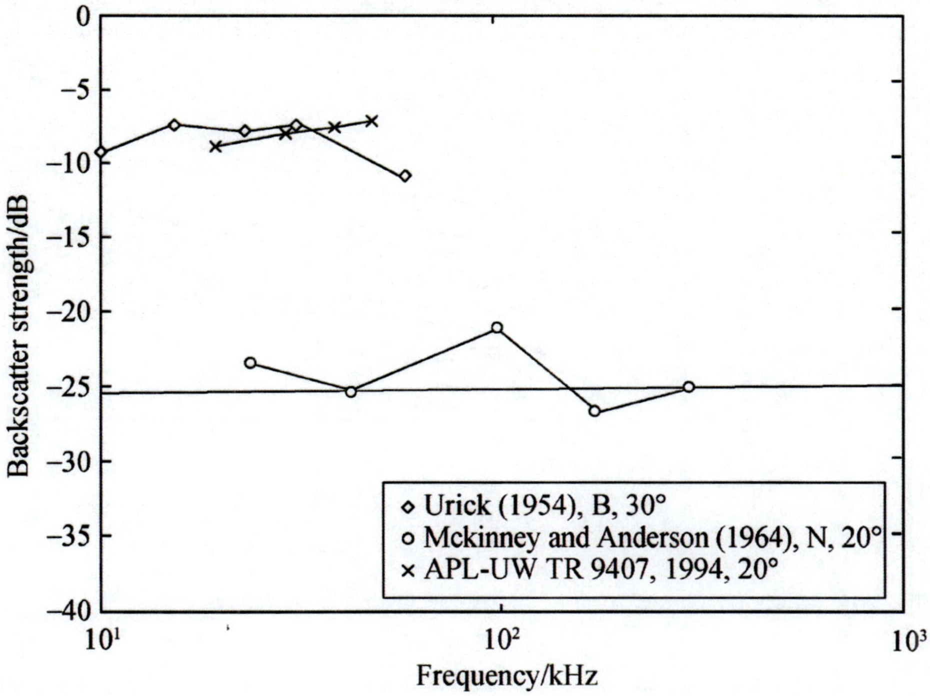

Fig. 8.9 Frequency effect of rocky seafloor

with frequency). With respect to the test results, the following physical explanation can be made: (i) seafloor roughness affects the seafloor acoustic scattering process; (ii) seafloor backscattering is independent of frequency when roughness is greater than wavelength; and (iii) when a considerable part of the seafloor roughness is smaller than the wavelength, BS increases with frequency. Thus, it is difficult to eliminate the influence of frequency on BS.

When mapping the seafloor with an MBES, the general emission frequency is in the range 10–500 kHz, and the variation in the collected BS values can be as high as 10 dB. Because the frequency effect is difficult to remove, current research on acoustic seafloor classification is usually based on instruments of a specific frequency. If fusion processing of seafloor BS data acquired using different measurement systems is undertaken, attention should be paid to the variation in seafloor BS caused by the different instruments.

2. **Grazing (incident) angle effect**

The variation in seafloor BS caused by the difference in the grazing angle (residual angle of the seafloor incident angle) of the sound wave reaching the seafloor is called the grazing (incident) angle effect. Figure 8.10 shows the measured seafloor BS with a grazing angle at high frequency (24–100 kHz).

As the grazing angle increases, seafloor BS generally increases (Fig. 8.10). According to the relationship between BS and incident angle, the grazing angle can be divided roughly into three regions: (i) the low grazing angle zone (grazing angle $\theta < 10°$), where BS increases rapidly as the grazing angle increases; (ii) the intermediate zone ($10° \leq \theta \leq 65°$), where BS changes gradually as the grazing angle increases; and (iii) the high grazing angle zone ($\theta > 65°$), where BS again increases rapidly as the grazing angle increases.

The variation in seafloor BS with grazing angle is essentially the result of the spatial redistribution of the acoustic energy projected onto the seafloor. Lambert's law quantitatively describes this redistribution process (Fig. 8.11).

The incident wave with sound intensity I_i is projected onto a rough surface with unit area dA at grazing angle θ (Fig. 8.11). According to Lambert's law, the incident power will be scattered in all directions, and the scattering intensity in each direction is

$$I_s = \mu I_i \sin\theta \sin\varphi dA \quad (8.27)$$

where μ is the proportionality constant, I_i $\sin\theta$ dA is the incident sound intensity and φ is the grazing angle. According to the definition of backscattering intensity, we can obtain

$$\mathrm{BS_B}(\theta) = 10\log_{10}\mu + 10\log_{10}(\sin^2\theta) \quad (8.28)$$

This is the relationship between the rough surface scattering intensity and grazing angle according to Lambert's law, which is known as the Lambert model. Let BS_o= 10 $\log_{10}\mu$, then Eq. (8.28) can be converted to

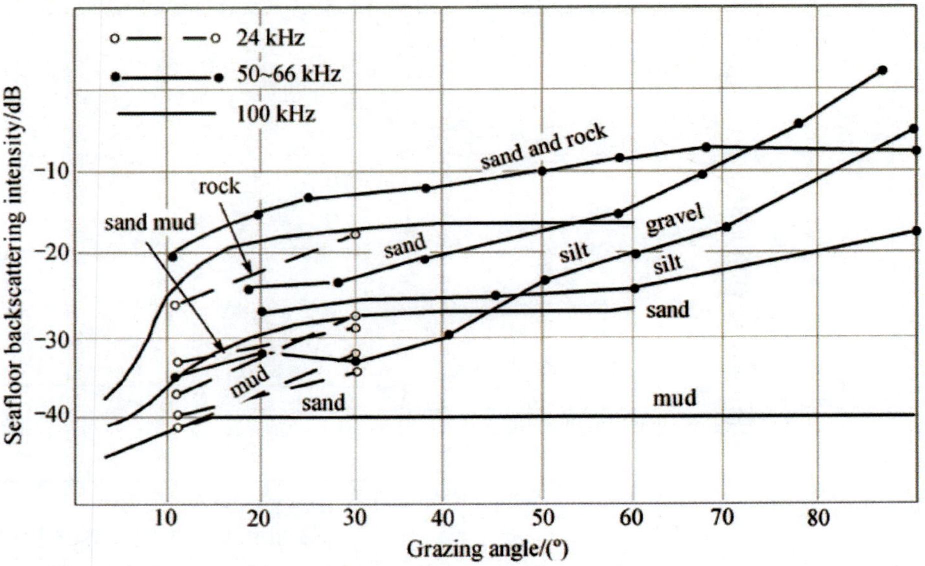

Fig. 8.10 Seafloor BS measured at various stations with different incident angles

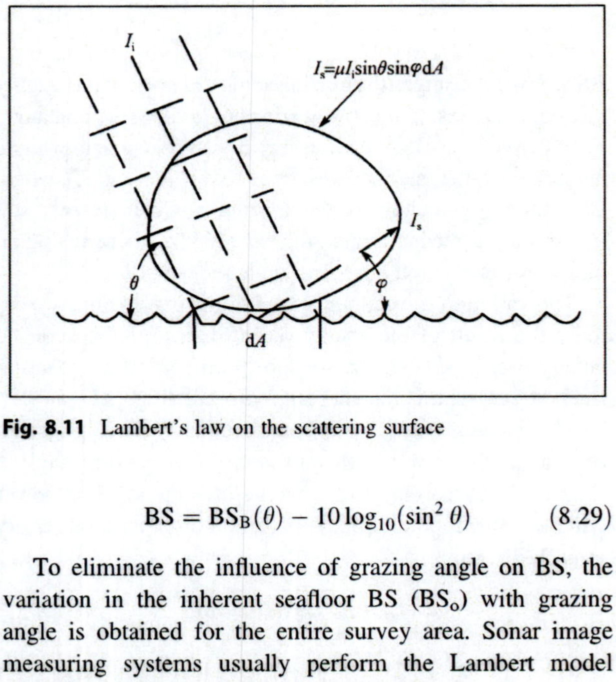

Fig. 8.11 Lambert's law on the scattering surface

$$\mathrm{BS} = \mathrm{BS_B}(\theta) - 10\log_{10}(\sin^2\theta) \quad (8.29)$$

To eliminate the influence of grazing angle on BS, the variation in the inherent seafloor BS ($\mathrm{BS_o}$) with grazing angle is obtained for the entire survey area. Sonar image measuring systems usually perform the Lambert model correction during backscatter gain processing.

Although the Lambert model can describe the backscattering process of a rough seafloor well, it is not applicable in all environments. Figures 8.12, 8.13, 8.14 and 8.15 show measured BS data for a muddy, sandy, gravel and rocky seafloor, respectively (with different grazing angles), where the solid line represents the fitted Lambert model curve. The inherent BS of the muddy, sandy, gravel and rocky seafloor is –19.7, –20.2, –11.7 and –5.6 dB, respectively.

The following observations can be determined from Figs. 8.12, 8.13, 8.14 and 8.15.

(1) When the grazing angle is <65°, the Lambert model can describe well the change rule of BS with grazing angle for muddy and sandy seafloors (Figs. 8.12 and 8.13). When the grazing angle is > 65°, the measured BS data are larger than the value calculated by the Lambert model. In the near-vertical incident region ($\theta = 90°$), the BS value increases rapidly. It indicates that the Lambert model is valid in the middle region of the grazing angle.
(2) The dispersion of the measured seafloor BS data is large. For muddy and sandy seafloors, the difference is approximately ±(10–15) dB at the same grazing angle, which means it is impossible to predict the BS value based simply on sediment type.
(3) The inherent BS values of muddy and sandy seafloors are equivalent. The figures indicate that the type of seafloor sediment cannot be predicted based solely on the inherent BS value.
(4) Despite the small amount of data on the BS of gravel and rocky seafloors (Figs. 8.14 and 8.15), it is obvious that these seafloor types have strong scattering ability. Thus, in general terms, the seafloor can be divided easily into "hard" and "soft" types based on BS data.

8.2.4.3 Statistical Characteristics of Seafloor BS

Seafloor scattering is a random process attributable to fluctuation of the seafloor interface and irregularity of the physical properties of the seafloor. When studying seafloor

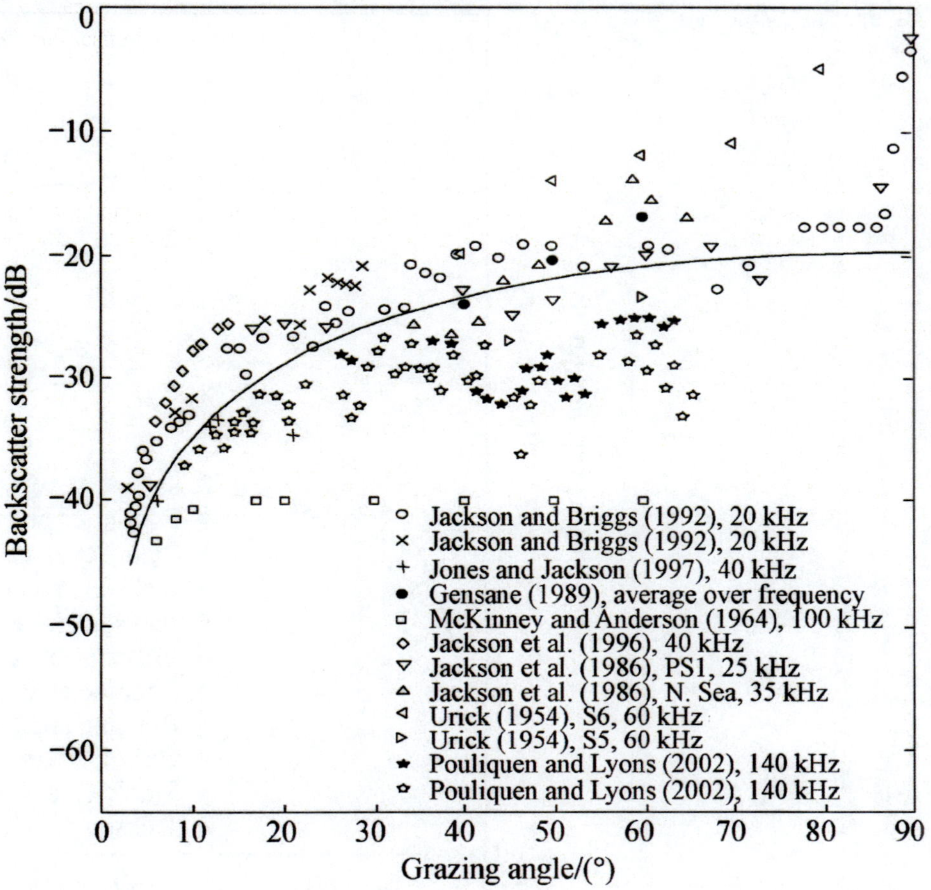

Fig. 8.12 BS data for muddy seafloor

acoustic scattering, in addition to exploring the variation in average seafloor BS, the statistical properties of the scattering intensity (e.g., the distribution function and the energy spectrum) can also be examined using mathematical methods.

The seafloor backscatter signal is formed by superimposing scattered sound waves generated by a large number of independent seafloor scatterers at the receiving position:

$$V(t) = \sum_{i=1}^{N} a(t_i)v(t - t_i) \tag{8.30}$$

where $a(t_i)$ is the random amplitude corresponding to the ith scatterer, function $v(t - t_i)$ represents the signal of a single scatterer, and N is the number of scatterers. To describe the distribution of the amplitude of the scattered signal, the backscatter signal is expressed as

$$V(t) = Z(t)\cos\left[\omega t + \varphi(t)\right] \tag{8.31}$$

where $Z(t)$ is the backscatter amplitude, $\varphi(t)$ is the phase of the backscatter signal and ω is the central angular frequency of the spectrum. Here, $Z(t)$ is a function with a slower time than ωt.

For convenience of analysis, we can rewrite Eq. (8.31) as

$$V(t) = V_c(t)\cos\omega t - V_s(t)\sin\omega t \tag{8.32}$$

where the stochastic processes $V_c(t)$ and $V_s(t)$ are determined by

$$\begin{aligned} V_c(t) &= Z(t)\cos\varphi(t) \\ V_s(t) &= Z(t)\sin\varphi(t) \end{aligned} \tag{8.33}$$

Thus, the amplitude and phase of the backscatter signal can be expressed as

$$\begin{aligned} Z(t) &= \left[V_c^2(t) + V_s^2(t)\right]^{1/2} \\ \varphi(t) &= \arctan\left[V_s(t)/V_c(t)\right] \end{aligned} \tag{8.34}$$

The phases of the scattered waves are generally considered independent of each other and distributed uniformly within the range 0–2π. If the number of effective scatterers within a sounding area is sufficiently large ($N \to \infty$), according to the central limit theorem, the instantaneous

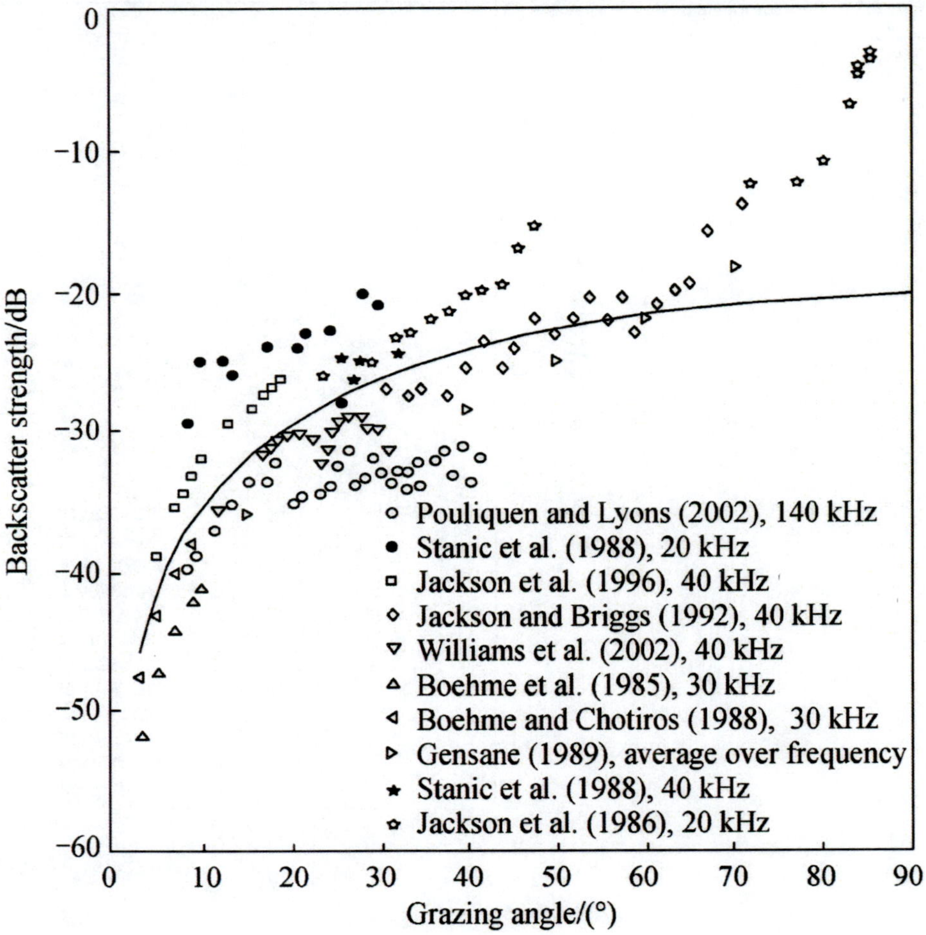

Fig. 8.13 BS data for sandy seafloor

value V of the backscatter signal satisfies the normal distribution law, amplitude Z obeys the Rayleigh distribution and the BS obeys the exponential distribution.

Many modern high-resolution sonar systems use beam-forming techniques to increase the azimuth resolution and to reduce the transmitted pulse width to increase range resolution, both of which result in a reduction in the instantaneous sounding area of the seafloor (equivalent to reducing the effective scatterers). In this case, the central limit theorem is not satisfied. The main phenomenon is that the probability density function has a heavier tail than the Rayleigh distribution. The models that describe a non-Rayleigh distribution mainly include the lognormal distribution, Weibull distribution, Rice distribution and K-distribution (Skolnic 2001; Lyons and Abraham 1999; Jackson and Richardson 2007). Among these, the K-distribution model not only describes the measured sonar data well, but also provides a physical explanation of the relevant backscattering phenomenon. Thus, the K-distribution model is widely used in acoustic research.

1. **Rayleigh distribution model**

If the number of scatterers in a scattering unit is reasonably large and non-random, according to the central limit theorem, the instantaneous value of the backscatter signal is a Gaussian random process with a zero mean value, which can result in its amplitude being a Rayleigh distribution. The PDF of the Rayleigh distribution is

$$f_Z(z) = \frac{2z}{\lambda_0} e^{-(z^2/\lambda_0)} \tag{8.35}$$

The cumulative distribution function is

$$F_Z(z) = 1 - e^{-(z^2/\lambda_0)} \tag{8.36}$$

The mathematical expectation and variance of random variable Z can be expressed as Eqs. (8.37) and (8.38), respectively:

$$E(Z) = \sqrt{\pi\lambda_0}/2 \tag{8.37}$$

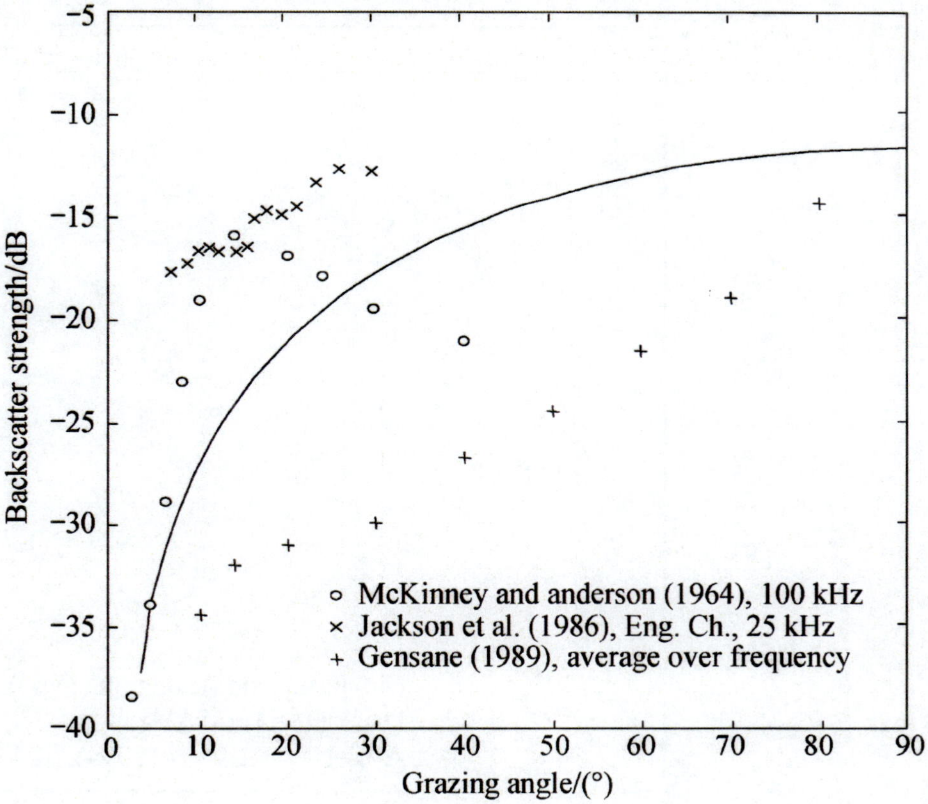

Fig. 8.14 BS data for gravel seafloor

$$Var(Z) = \left(1 - \frac{\pi}{4}\right)\lambda_0 \tag{8.38}$$

Thus, the BS can be obtained, which obeys the exponential distribution with $I = Z^2$. The PDF is

$$f_I(I) = \lambda_0 e^{-\left(\frac{1}{\lambda_0}I\right)} \tag{8.39}$$

The valuation of parameter λ_0 is

$$\hat{\lambda}_0 = \mu_I \tag{8.40}$$

where μ_I is the mathematical expectation of the backscatter intensity.

Figures 8.16 and 8.17 show the PDFs of the Rayleigh distribution and the exponential distribution ($\lambda_0 = 1$), respectively.

2. **K-distribution model**

The K-distribution model was originally proposed to describe the statistical properties of sea clutter backscatters (Jakeman and Pusey 1978). In this model, the amplitude of sea clutter backscatter can be seen as the product of two factors: (i) the spot component, which is produced by the multipath scattering properties of the clutter; and (ii) the basic amplitude modulation component of sea clutter, reflecting the spatial variation of the average energy of the scattered beam.

Lyons and Abraham (1999) analyzed the relationship among the number of scatterers, shape parameters of the K-distribution and characteristics of the sonar system. This led to an expression of the physical interpretation of the distribution of BS. Oliver (1984) derived the statistical moments of BS by studying the correlation between the number of scatters and the size of the finite scattering elements. Practical application of radar and sonar data has shown that the model can reflect the statistical properties of experimental data well.

According to Oliver's product model, the K-distribution reverberation envelope can be expressed as a combination of two different physical processes:

$$Z = X \cdot Y \tag{8.41}$$

In Eq. (8.41), X is a slow-varying component that is related to the physical properties of each scatterer at the seafloor interface, which obeys the generalized χ distribution; Y is a fast-varying component (also known as speckle) that obeys the Rayleigh distribution.

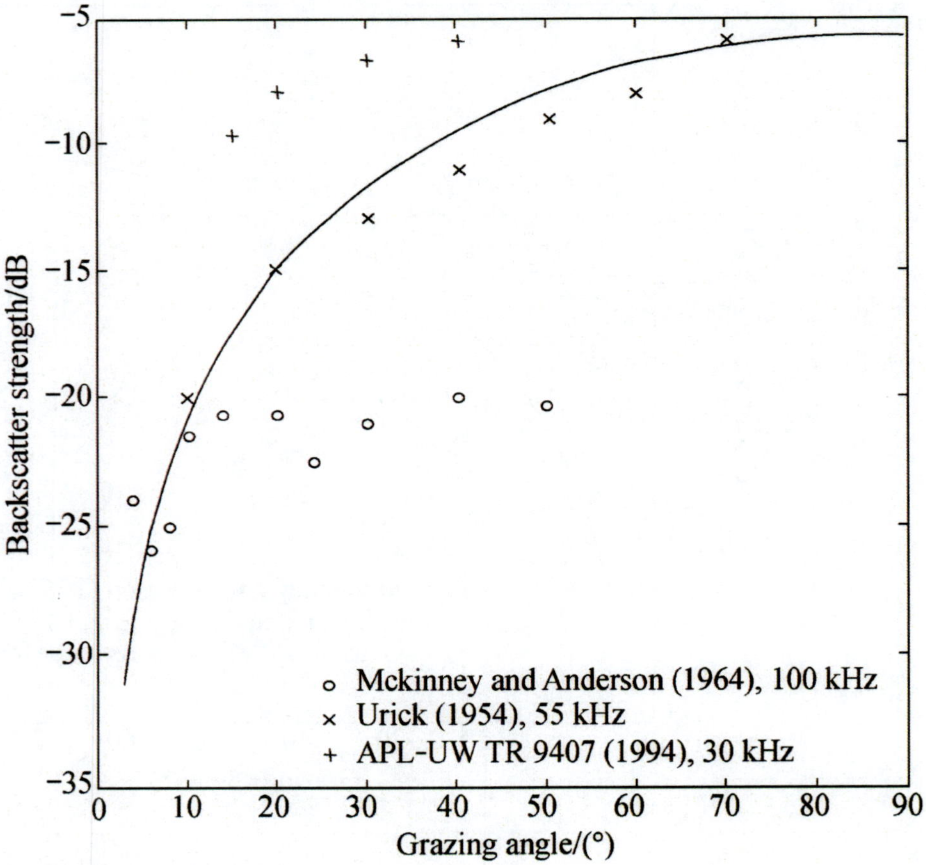

Fig. 8.15 BS data for rocky seafloor

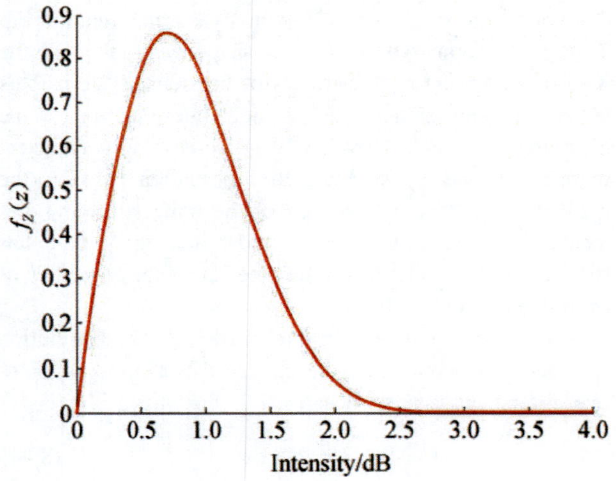

Fig. 8.16 Rayleigh distribution PDF

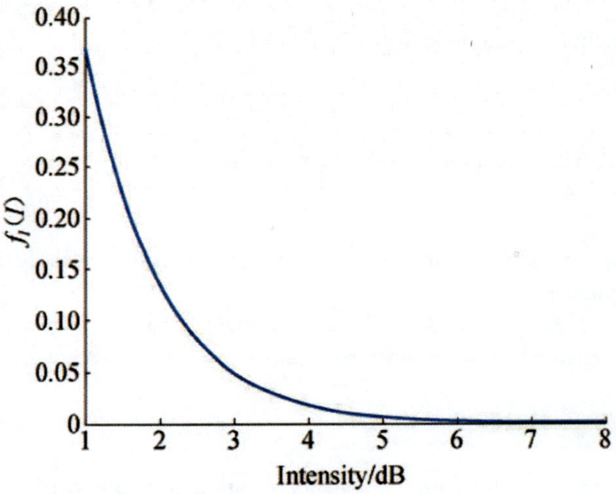

Fig. 8.17 Exponential distribution PDF

According to the literature (Lyons and Abraham 1999), the PDF of the K-distribution can be written as

$$f_Z(z) = \frac{4}{\sqrt{\lambda}\Gamma(v)}\left(\frac{z}{\sqrt{\lambda}}\right)^v K_{v-1}\left(\frac{2z}{\sqrt{\lambda}}\right) \tag{8.42}$$

The corresponding cumulative distribution function is

$$F_Z(z) = 1 - \frac{2}{\Gamma(v)}\left(\frac{z}{\sqrt{\lambda}}\right)^v K_v\left(\frac{2z}{\sqrt{\lambda}}\right) \tag{8.43}$$

where $K_{v-1}(\cdot)$ is the second-type v–1 order modified Bessel function, v is the shape parameter and λ is the scale parameter. The shape parameter reflects the extent to which the K-distribution deviates from the Rayleigh distribution. The steepness of the kurtosis of the K-distribution increases in tandem with both the decrease in size of the v value and the increase in the deviation from the Rayleigh distribution. The shape parameter has a value range of $0.1 < v < \infty$ and when $v \to \infty$, the distribution converges to a Rayleigh distribution. The scale parameter reflects the strength of the backscatter and the larger the value is, the stronger the backscatter power is.

By defining $I = Z^2$, the distribution of BS can be obtained:

$$f_I(I) = \frac{2}{\Gamma(v) \cdot I}\left(\frac{I}{\lambda}\right)^{\frac{v+1}{2}} K_{v-1}\left(2\sqrt{\frac{I}{\lambda}}\right) \tag{8.44}$$

The moment estimation method is often adopted to estimate the K-distribution parameters:

$$\hat{\lambda} = \frac{\langle I^2 \rangle - 2\langle I \rangle^2}{2\langle I \rangle} = \frac{\langle I \rangle}{\hat{v}} \tag{8.45}$$

$$\hat{v} = \frac{2\langle I \rangle^2}{\langle I^2 \rangle - 2\langle I \rangle^2} \tag{8.46}$$

Figures 8.18 and 8.19 present the K-distribution PDFs for the backscatter envelope and intensity, respectively (λ = 4, v = 0.5, 1, 2.5, 5).

3. **Analysis of test data**

MBES data measured by the Simrad EM3000 system (in Jiaozhou Bay, Qingdao, China) are used to express visually the statistical characteristics of BS. The seafloor sediment types in the survey area include bedrock, sand and clay-silt. After processing, the raw measurement data BS_0 are

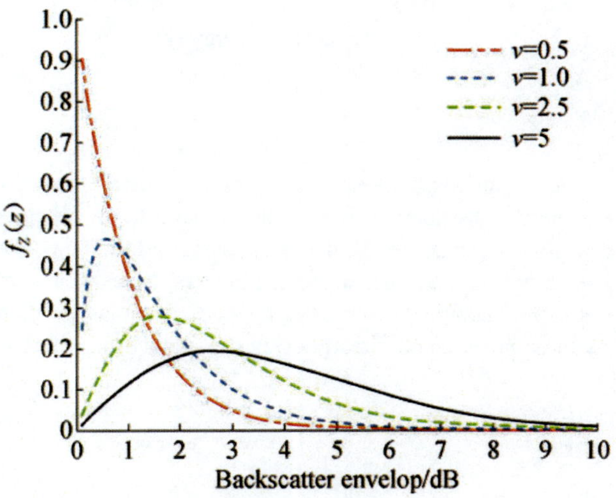

Fig. 8.18 K-distribution PDF for the backscatter envelope

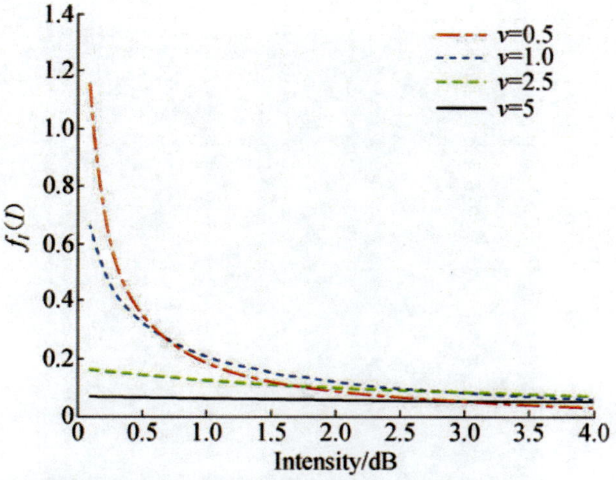

Fig. 8.19 K-distribution PDF for the backscatter intensity

obtained with values in the range –64 to 0 dB, which can be converted into energy form I:

$$I = 10^{\frac{BS_0}{10}} \tag{8.47}$$

Backscattering images of bedrock, sand and silty-sand are shown in Figs. 8.20, 8.21 and 8.22, respectively, and their corresponding histograms are also presented. The parameters of the exponential distribution and the K-distribution are estimated using Eqs. 8.39, 8.45 and 8.46. The values of $\hat{\lambda}_0$ are 0.034, 0.013 and 0.002, respectively. Among the three types of sediment, the BS of the bedrock seafloor is strongest, followed by sand; the BS of the muddy seafloor is relatively weak. The K-distribution shape parameter is estimated as 111.34, 12.75 and 65.18, respectively. Although the value varies, it is not reflected in the fitting curve (i.e., the K-distribution and the exponential distribution fitting curves are consistent), which is mainly attributable to the large valuation in the shape parameters. According to the description of the K-distribution shape parameter, when $v \to \infty$, the distribution converges to the Rayleigh distribution. In fact, when $v > 5$, the K-distribution is essentially the same as the Rayleigh distribution. Once v is estimated, the scale parameter λ can also be computed based on Eqs. (8.40) and (8.45).

Experimental results show the backscattering data obey the exponential distribution (corresponding to the K-distribution with $v \to \infty$). Obviously, the shape of the exponential distribution is determined by $\hat{\lambda}_0$. Once $\hat{\lambda}_0$ is known, the variance, kurtosis, skewness and other statistics of the distribution can be calculated directly. Note that these statistics are usually extracted from sonar images for seafloor classification. This study shows that if the seafloor backscattering energy obeys the exponential distribution, the statistical characteristic parameters (e.g., mean, variance, kurtosis, skewness and quantile) are often inter-correlated.

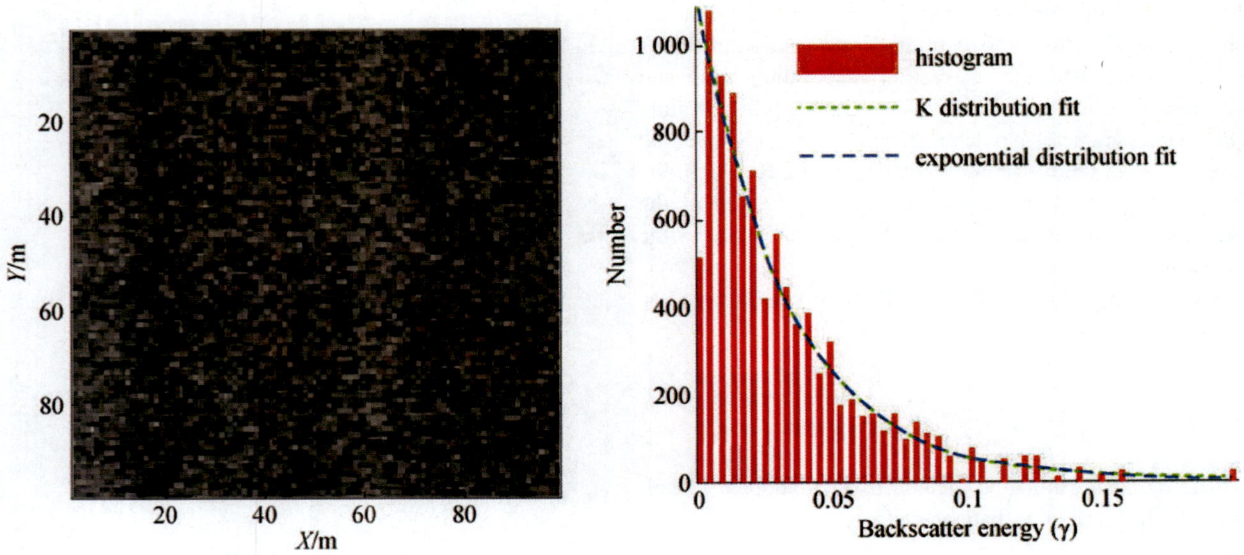

Fig. 8.20 Statistical characteristics of backscattering energy in bedrock seafloor

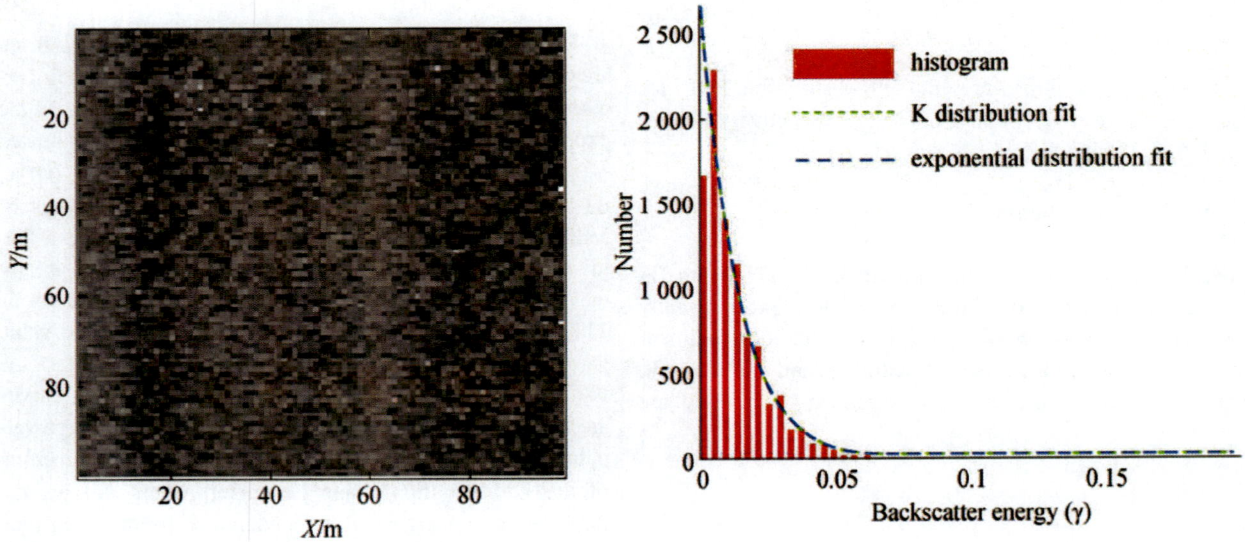

Fig. 8.21 Statistical characteristics of backscattering energy in sandy seafloor

Therefore, changes in the type of seafloor sediment could be described effectively by solely extracting the mean feature. In contrast, if features such as kurtosis and skewness are also extracted for classification, the classification accuracy is often degraded because of the high impact of outliers on high-order moment statistics.

The K-distribution has one parameter more than the exponential distribution, which means that it has a structure that is more flexible. If the backscatter values obey the K-distribution and the shape parameters have values of $v < 5$, the feature vector used for sediment classification could be constructed directly using the λ and v parameters.

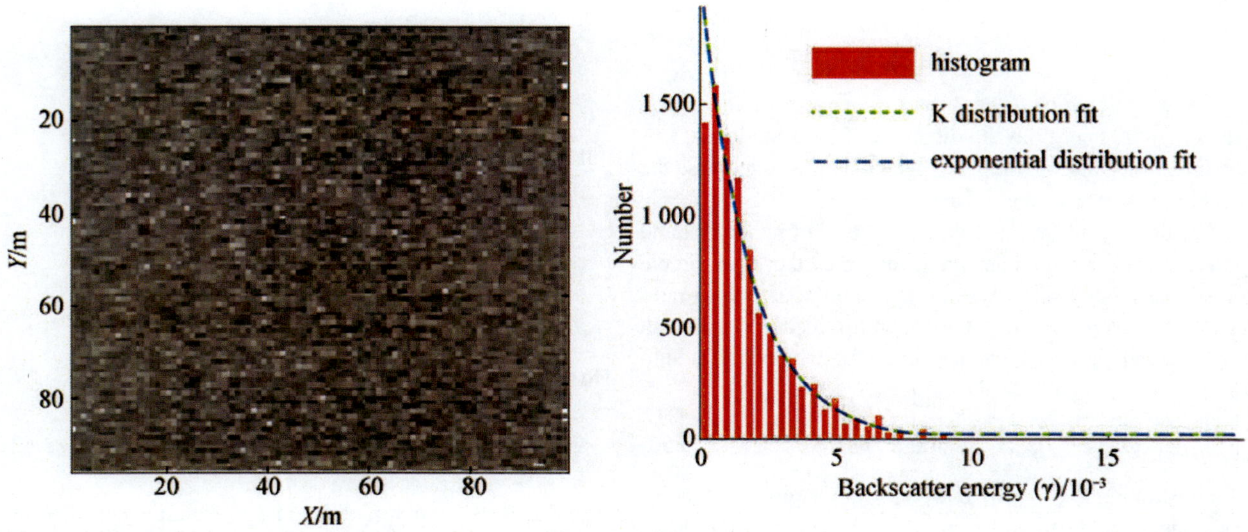

Fig. 8.22 Statistical characteristics of backscattering energy in muddy seafloor

8.3 Backscatter Data Processing

Using acoustic data to classify a seafloor into different regions is a practical method for seafloor characterization. Acoustic instruments currently used for seafloor classification include the SBES, SSS and MBES (Dyer et al. 1997; Collins and Rhynas 1998; Hamilton et al. 1999; Davis et al. 2002). The purpose of processing BS data is to remove the components that are independent of seafloor type and to obtain normalized data reflecting only the properties of the seafloor sediment.

8.3.1 Interpretation of Seafloor BS Data

8.3.1.1 Sonar Equation

The information relevant in performing a sonar measurement is similar regardless of system type. It consists of three principal elements: the seawater medium, detected target and sonar equipment. The state, characteristics and performance of these elements will directly affect the measurement quality. Further analysis shows that each element contains several factors (known as sonar parameters) that affect the operation of the sonar equipment (Urick 1983). These sonar parameters are described in the following sections.

1. **Sonar parameter**

(1) Source level

The source level (SL) is used to describe the strength of the acoustic signal emitted by an active sonar instrument (Urick 1962). It can be defined as

$$\mathrm{SL} = 10\log_{10}\frac{I}{I_0}\bigg|_{r=1} \tag{8.48}$$

where I is the sound intensity at 1 m from the source center of the transmitting transducer (in the direction of the sound axis) and I_0 is the reference sound intensity. In underwater acoustics, the plane wave sound intensity with a root mean square sound pressure of 1 μPa is usually taken as the reference sound intensity I_0, which is approximately equal to 0.67×10^{-22} W/cm^2.

To increase the operating distance of the active sonar, the transmitter is always fabricated with certain emission directivity. Thus, the emitted sound energy is concentrated mainly in one direction. An example of emission directivity is shown in Fig. 8.23. Emission directivity is generally described by the transmission directivity index (I_{TD}), which is defined as the number of decibels by which the sound level on the sound axis is higher than that in a non-directivity direction over the same distance (as given in Eq. 8.49):

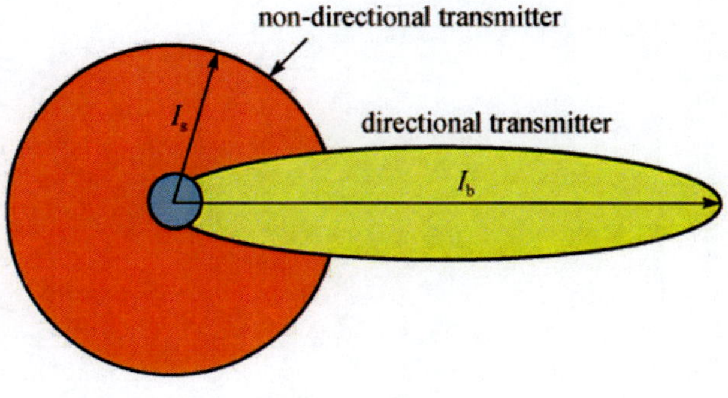

Fig. 8.23 Emission directivity pattern

$$I_{TD} = 10\log_{10}\frac{I_b}{I_s} \quad (8.49)$$

The magnitude of I_{TD} indicates the degree of concentration of sound energy in the direction of the sound axis, i.e., the larger the value is, the more beneficial it is to increase the working distance of the device.

The SL of a transmitter reflects the level of radiated sound power emitted by the transmitter, for which there is a simple functional relationship. Assuming a point source with radiated sound power P_a in a non-absorbing medium, and with basic knowledge of acoustics, the sound intensity at unit distance of the center of the source is

$$I|_{r=1} = P_a/4\pi(W/m^2) \quad (8.50)$$

Substituting Eq. (8.50) into Eq. (8.48), given the reference sound intensity is 0.67×10^{-22} W/cm^2 and the directivity index of the transmitter is I_{TD}, then

$$\mathrm{SL} = 10\log_{10}P_a + 170.8 + I_{TD} \quad (8.51)$$

The magnitude of SL is usually given by the instrument manufacturer (e.g., the SL of the RESON SeaBat 8101 is 210 dB).

(2) Transmission loss

Seawater is a non-homogeneous and non-ideal medium. Absorption occurs when sound waves propagate in the medium because of both expansion of the wave front during propagation and non-uniform scattering in the seawater. Thus, acoustic intensity in the direction of propagation will gradually decrease during propagation. The transmission loss (TL), which quantitatively describes the attenuation of sound intensity after a certain distance of propagation, is defined as

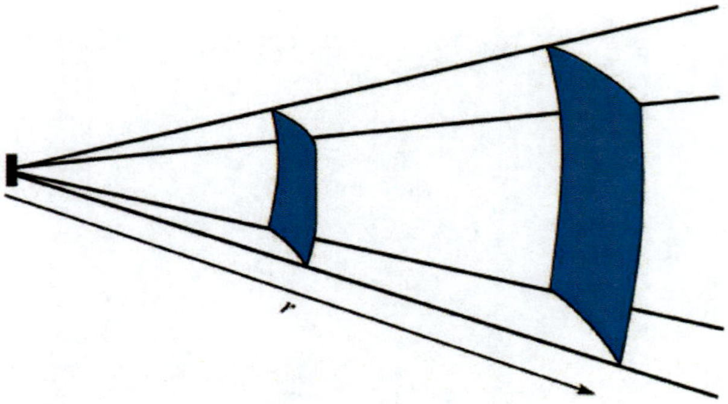

Fig. 8.24 Spherical wave expansion loss

$$\mathrm{TL} = 10\log_{10}\frac{I_1}{I_r} \quad (8.52)$$

where I_1 is the sound intensity at 1 m from the center of the sound source and I_r is the sound intensity at a distance(r, unit: m) from the source. Propagation loss consists of two parts: attenuation of the sound intensity caused by expansion of the wave front with distance (expansion loss) and attenuation caused by water viscosity and the chemical relaxation process related to divalent salts ($MgSO_4$) in water (absorption loss). For expansion loss, the rate of attenuation is proportional to the surface area of the wave front. For a spherical wave emitted by a multi-beam sonar system, the rate of attenuation is proportional to the propagation distance r^2 (Fig. 8.24); thus, the entire attenuation of energy is (Hughes Clarke 2005):

$$\frac{1}{r^2} \times \frac{1}{r^2} = \frac{1}{r^4} \quad (\text{logarithmic form}: \ 40\log_{10} r) \quad (8.53)$$

Absorption loss is usually described by the absorption coefficient a. According to the model established by Francois and Garrison (1982a, b), a is a function of sonar frequency f and the propagation medium properties (salinity, temperature, depth and pH):

$$a = \frac{A_1 f_1 f}{f^2 + f_1^2} + \frac{A_2 p_2 f_2^2 f^2}{f^2 + f_2^2} + A_3 p_3 f^2$$
$$A_1 = \frac{8.86 \times 10^{(0.78\mathrm{pH}-5)}}{C} \quad A_2 = \frac{21.44S(1 + 0.025T)}{C}$$
$$A_3 = \begin{cases} T \le 20^\circ\mathrm{C}: 4.937 \times 10^{-4} - T(2.59 \times 10^{-5} - T(9.11 \times 10^{-7} - 1.5 \times 10^{-8}T)) \\ T > 20^\circ\mathrm{C}: 3.964 \times 10^{-4} - T(1.146 \times 10^{-5} - T(1.45 \times 10^{-7} - 6.5 \times 10^{-10}T)) \end{cases} \quad (8.54)$$
$$P_2 = 1 - H(0.137 - 0.006\,2H) \quad P_3 = 1 - H(0.038\,3 - 4.9 \times 10^{-4}H)$$
$$f_1 = 2.8\sqrt{S/35} \times 10^{(4 - \frac{1245}{273+T})} \quad f_2 = \frac{8.17 \times 10^{(8 - \frac{1990}{273+T})}}{1 + 0.001\ 8(S - 35)}$$

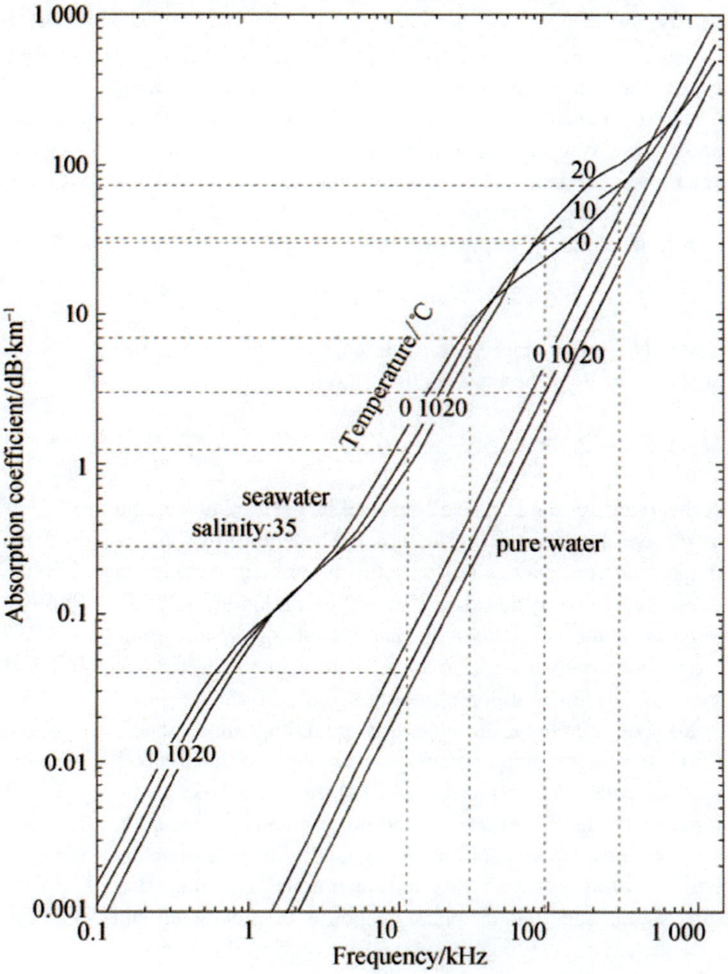

Fig. 8.25 Absorption coefficient (a) plotted against frequency

The units of C, T, Z, S and f in the above expressions are m/s, °C, km, 10^{-12} and kHz, respectively.

Figure 8.25 presents curves of the absorption coefficient with frequency. Obviously, as the frequency increases, so does the absorption coefficient. In freshwater, the absorption coefficient varies linearly with frequency; as the water temperature increases, the absorption coefficient decreases. In seawater, the absorption coefficient is higher than in freshwater owing to the influence of chemical relaxation processes.

Once a is determined, the TL of an acoustic wave can be determined as

$$2\text{TL} = 40\log_{10} r + 2ar \quad (8.55)$$

Table 8.2 shows the values of absorption coefficients of typical MBESs in seawater and freshwater. It can be seen that the absorption coefficient value depends mainly on the frequency variation. The difference in the absorption coefficient value between the EM3000 and EM1002 multi-beam sonar systems is 41 dB/km and thus their influence on BS cannot be ignored. In addition, for the same systems, the absorption coefficients in seawater and freshwater might differ considerably. For example, the difference in the absorption coefficient of the EM1002 instrument between seawater and freshwater is 29 dB/km, which makes it necessary to pay particular attention to the influence of the absorption coefficient in an estuarine environment.

Table 8.2 Absorption coefficients of Simrad system

Multi-beam sonar system	Frequency (kHz)	Seawater (dB · km^{-1})	Freshwater (dB · km^{-1})
EM3000	300	73	30
EM1002	95	32	3
EM300	30	7	0.28
EM120	12	1.2	0.04

(3) Noise level

In seawater, there are various sources of noise and their respective sound waves constitute marine environmental noise. This type of environmental noise is interference in sonar measurements. The environmental noise level (NL) is used to measure the intensity of environmental noise, and its energy level can be expressed as:

$$\mathrm{NL} = N_c + 10\log_{10} W \quad (8.56)$$

where N_c is the noise spectral intensity at the center frequency and W is the bandwidth of the receiver.

(4) Reception directivity index (I_{RD})

A receiving system is typically placed in a certain noise field to receive the target acoustic signal. The output of a directional receiver depends not only on the directivity function of the receiving system but also on the directional characteristics of the noise field. In other words, given the same directional receiver and the same target signal, a receiving system will have different outputs under different noise fields. For simplicity, the following conditions are assumed: (i) in the sound field, there are many uncorrelated noise sources with random phases; and (ii) the noise sources are far from the receiving system and noise intensity is equal in all directions (i.e., an isotropic noise field). Suppose there are two receiving systems: one non-directional and the other directional, and that the axial response of the directional system (axial receiving sensitivity) is equal to the response of the non-directional system. When the two systems are placed in the same isotropic noise field, the ratio of their output noise power (in decibels) is defined as the directional exponent I_{RD} of the directional receiving system, which is

$$I_{RD} = 10\log_{10}\frac{N_0}{N_d} \quad (8.57)$$

where N_0 is the output noise power of the non-directional receiver and N_d is the output noise power of the directional receiver. From the above analysis, we know that the role of the reception directivity index is to suppress noise.

(5) Backscatter strength

The BS is used to describe quantitatively the magnitude of the reflection power of a target. It depends not only on the characteristics of the seafloor but also on the instantaneous insonified area A of the sound wave on the seafloor, which can be expressed (Lurton 2002) as:

$$\mathrm{BS} = \mathrm{BS_B} + 10\log_{10} A \quad (8.58)$$

where $\mathrm{BS_B}$ is the seafloor backscatter intensity, which is a function of seabed type and incident angle θ. It has been shown that when $\theta \approx 0°$, $\mathrm{BS_B}$ is usually approximated to a constant ($\mathrm{BS_n}$); when $\theta \geq 25°$, the variation of $\mathrm{BS_B}$ with incident angle approximately obeys Lambert's law; and when $0° < \theta < 25°$, $\mathrm{BS_B}$ varies linearly with incident angle:

$$\begin{aligned} &\mathrm{BS_B} = \mathrm{BS_n} && \theta \approx 0° \\ &\mathrm{BS_B} = \mathrm{BS_o} + (\mathrm{BS_n} - \mathrm{BS_o}) \cdot (25 - \theta)/25 && 0° < \theta < 25° \\ &\mathrm{BS_B} = \mathrm{BS_o} + 10\log_{10}\cos^2\theta && \theta \geq 25° \end{aligned} \quad (8.59)$$

The insonified area is the area that contributes to the BS measurement at the time of backscatter reception, i.e., all backscattered echoes from the area arrive at the receiving transducer simultaneously. For multi-beam sonar systems, near the central beam ($\theta < \theta_{\lim}$), the insonified area is equal to the beam footprint (Fig. 8.26), which is determined based on the transmitted beam width θ_T (along the track direction), received beam width θ_R (vertical to the track direction), incident angle θ and sound wave propagation distance R:

$$A = \frac{\theta_R \theta_T R^2}{\cos\theta} \qquad \theta \leq \theta_{\lim} \quad (8.60)$$

As the incident angle increases, the pulse length τ controls the lateral distance of the instantaneous sound zone (Fig. 8.27), and its area is

$$A = \frac{c\tau\theta_T R}{2\sin\theta} \qquad \theta_{\lim} < \theta < 90° \quad (8.61)$$

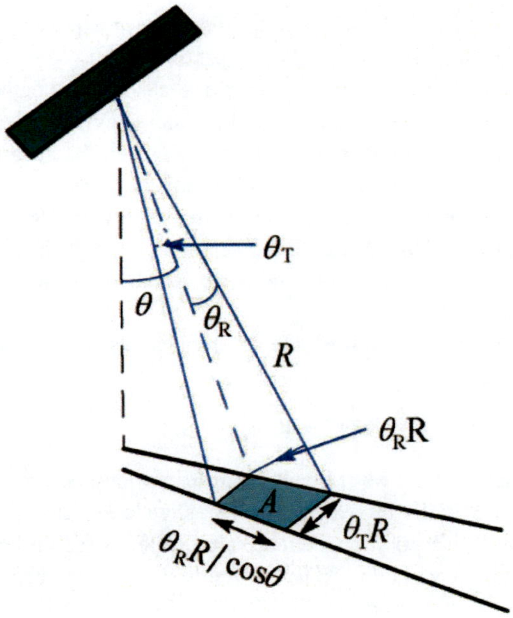

Fig. 8.26 Instantaneous sound zone near the central beam

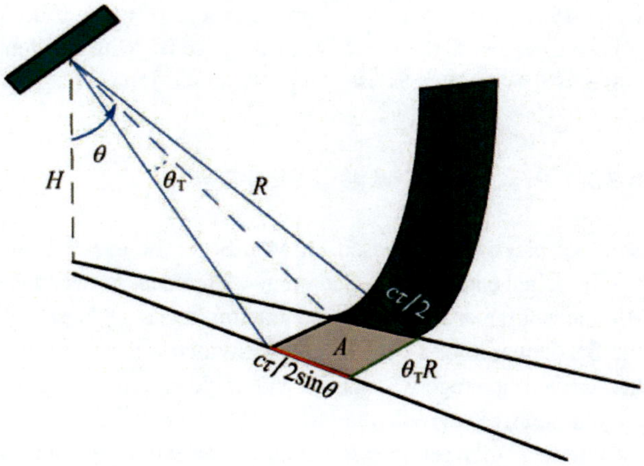

Fig. 8.27 Instantaneous sound zone of oblique incident area

To calculate the insonified area correctly, the first step is to determine the defined angle θ_{lim} of the formula; at the defined angle, the areas calculated by Eqs. (8.60) and (8.61) should be equal:

$$\frac{\theta_R R}{\cos\theta_{\text{lim}}} = \frac{c\tau}{2\sin\theta_{\text{lim}}} \tag{8.62}$$

If the sea floor is flat and the water depth is H, then $R = H/\cos\theta_{\text{lim}}$ and the formula for calculating angle θ_{lim} is

$$\sin\theta_{\text{lim}} = -\frac{\theta_R H}{c\tau} + \sqrt{\left(\frac{\theta_R H}{c\tau}\right)^2 + 1} \tag{8.63}$$

(6) Reception threshold

The receiver of the sonar device operates in a noisy environment, receiving both the sonar signal and the noise. Only when the signal-to-noise ratio of the power is sufficiently high can the device operate effectively. The signal-to-noise ratio (in decibels) at the processor input, which is usually called the reception threshold (RT), is defined as (Lurton 2002):

$$\text{RT} = 10\log_{10}\frac{\text{Signal power}}{\text{Noise power}} \tag{8.64}$$

2. **Sonar equation**

Seabed acoustic detection equipment belongs to the class of active sonar systems. When an acoustic signal reaches a target, its SL is reduced by the TL. If the scattering intensity of the target is BS, the energy level at the receiving array is SL–2TL + BS, which is often referred to as the BS level. However, the NL is also received by the receiving transducer. Because of suppression of by the I_{RD}, the NL is reduced (i.e., NL-I_{RD}). It should be noted that because the acoustic axis of the transducer always points toward the target, the BS level is not depressed by the I_{RD}. The received BS and noise are converted into an electrical signal by the transducer and sent to the processor. The process of sonar signal strength variation is depicted in Fig. 8.28, and the signal-to-noise ratio (SN, in decibels) of the electrical signal is:

$$\text{SN} = \text{SL} - 2\text{TL} + \text{BS} - \text{NL} + I_{\text{RD}} \tag{8.65}$$

If the signal-to-noise ratio is greater than the RT, then the sonar device can operate effectively and the sonar equation can be expressed as

$$\text{SL} - 2\text{TL} + \text{BS} - \text{NL} + I_{\text{RD}} > \text{RT} \tag{8.66}$$

For acoustic seafloor classification research, the variable of interest is the BS level. After amplification by the received processed gain (PG), the output backscatter intensity echo level (EL) is

$$\text{EL} = \text{SL} - 2\text{TL} + \text{BS} + \text{PG} \tag{8.67}$$

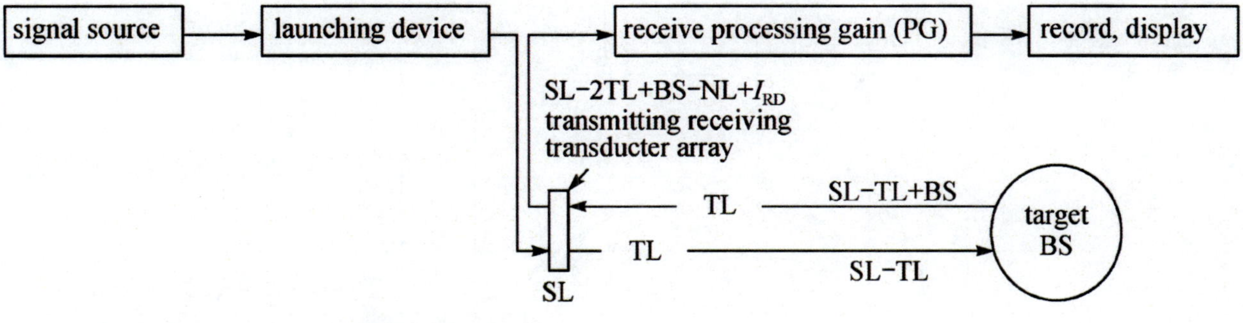

Fig. 8.28 Schematic of sonar signal strength variation process

8.3.1.2 System Gain

Seafloor acoustic measurement system gain generally comprises two parts: fixed gain (FG) and time varied gain (TVG). As the dynamic range of the received BS is limited, the main function of the gain is to avoid backscatter signal overload or the BS being masked by noise.

The TVG is designed such that the average signal level is maintained at the optimum level in the receiver to adapt to the random variations in seafloor reflectivity. Moreover, it is used to flatten the beam sampling amplitude for an MBES. Its configuration is mainly concerned with the contrast in seafloor backscatter intensity.

The system gain process varies slightly in different types of equipment. For example, in the Simrad EM MBES, the specific implementation process (Hammerstad 2000) is as follows.

(1) Set the FG such that the received BS level has the largest dynamic range in the receiver.
(2) Use the models in Eqs. (8.55) and (8.58) to correct the effects of TL and the insonified area ($10\log_{10}A$) on BS.
(3) Based on previous pings, estimate the BS_n of the beam at normal incidence and the BS_o at oblique incidence (crossing angle). The crossing angle means that the BS changes linearly with incident angle to the critical angle, which obeys Lambert's law. The angle value set in Eq. (8.59) is 25°; however, this value can be modified and it may be any value between 5° and 40° depending on sediment type. After estimating the values of BS_n and BS_o, if the crossing angle is 25°, a linear curve is fitted in the inner region ($0° < \theta < 25°$) and a Lambert curve is fitted in the outer region ($25° \leq \theta$ 90°) (Fig. 8.29, heavy dashed line). The correction value produced by the curve (shown by the arrow) is used for correction of the subsequent ping.

After the above steps, based on the flat seafloor assumption, the output BS is expected to reflect the change in the seafloor under an oblique incident angle.

For the Simrad EM series MBES, the information BS_B is stored in the deep data packet (not processed in Steps 3–4), while BS_0 is stored in the sonar image data packet. The former is used to generate the curve of the average BS with incident angle, while the latter is used to generate the sonar image.

8.3.2 Processing of MBES BS Data

Seafloor classification based on MBES can be divided into two principal categories (based on the differences between the BS data source and the feature extraction method). Based on the BS data recorded by a MBES bathymetric data package, the curve of average BS change with incident angle (i.e., the angular relative curve; hereafter, ARC) can be calculated, and the features of the seafloor sediment can be extracted from the curve for subsequent classification. While based on the BS data recorded by an MBES side-scan data package, the normalized BS can be calculated and sonar image mosaic correction adopted to generate a sonar mosaic map reflecting the change in seafloor sediment. The relevant features of the seafloor sediment are extracted from the mosaic map for subsequent classification. This section focuses on the correction of the ARC effect and the near-nadir BS anomaly. A flow chart of MBES data processing is shown in Fig. 8.30.

8.3.2.1 Influence and Correction of ARC Effect

When processing the gain, a basic assumption is that the departure angle of the beam is equal to the incident angle on the seafloor. However, in practical applications, this ideal situation is often not satisfied because of the influence of ship attitude, sound line bending and the impact of seabed topography. Thus, it is necessary to calculate the incident angle of the beam reaching the seafloor and to correct the ARC effect.

1. **Space vector model for calculating seafloor incidence angle**

The beam incident angle θ_f refers to the angle between the beam incidence vector $\vec{V}_i$ and the normal vector $\vec{V}_n$ of the incident point (Fig. 8.31), which is expressed as

$$\theta_f = \arccos(\vec{V}_i \cdot \vec{V}_n / (\|\vec{V}_i\| \|\vec{V}_n\|)) \tag{8.68}$$

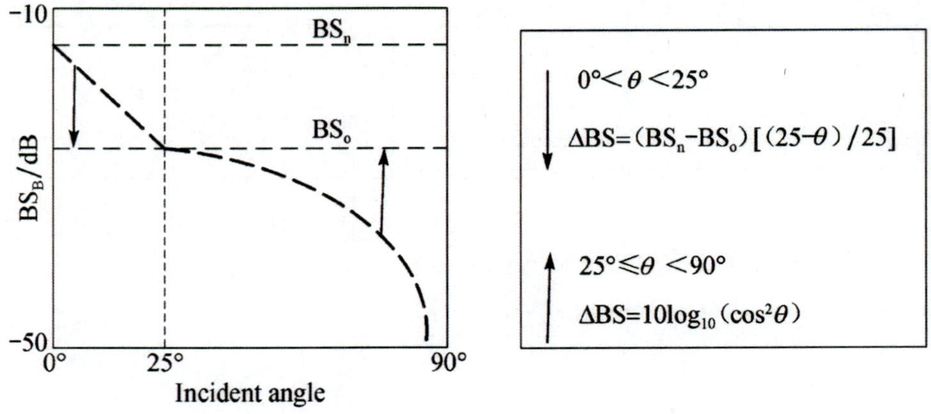

Fig. 8.29 Correction of the BS of different incident angles

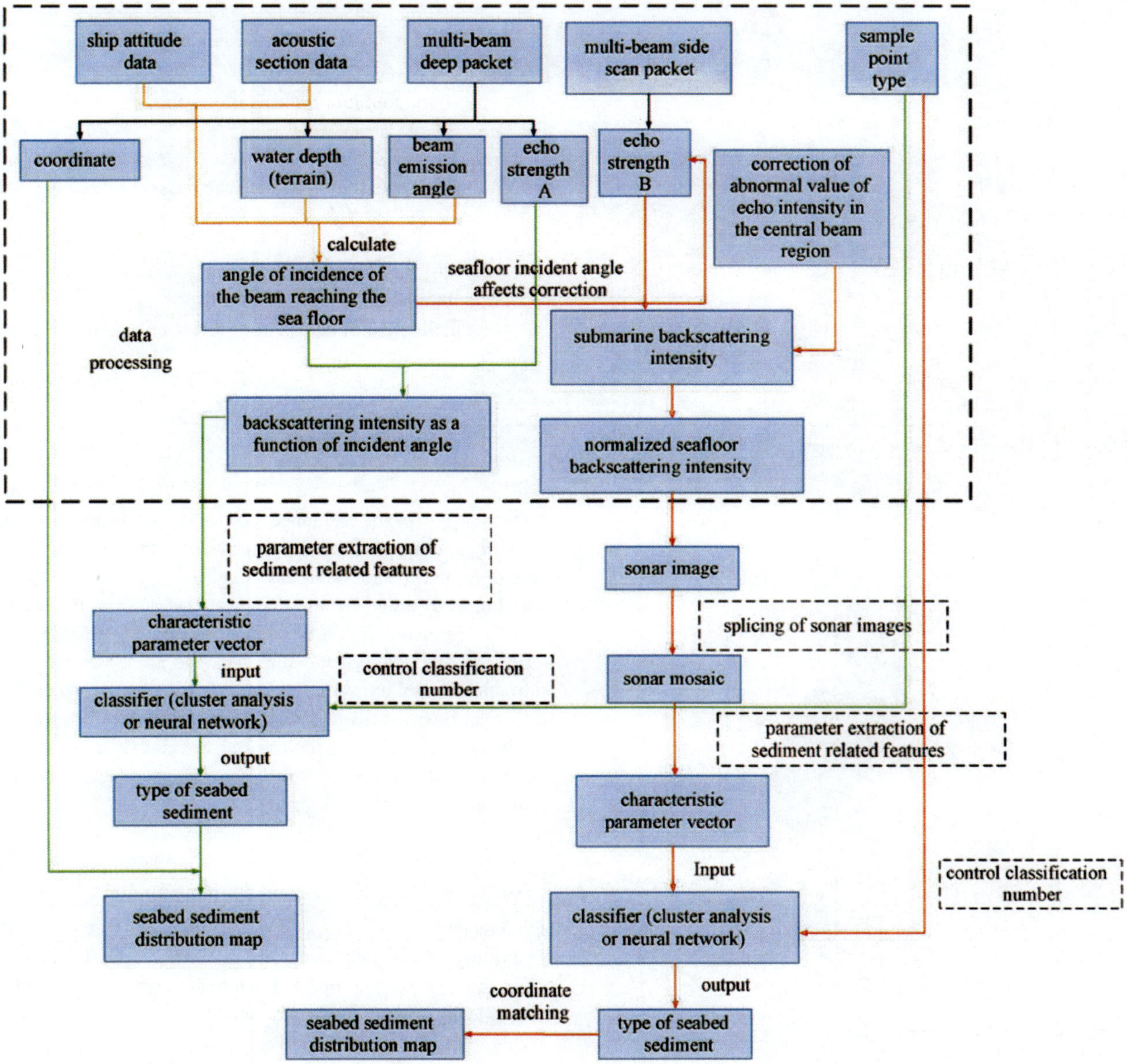

Fig. 8.30 MBES data processing flow

where • represents the inner product, and $\|*\|$ represents the modulus of the computed vector. Although the form of the space vector model for calculation of the incident angle of the seafloor is reasonably simple, determination of $\vec{V}_i$ and $\vec{V}_n$ is a complicated process in actual operation because of the influence of ship attitude, sound line bending and seabed topography.

2. **Calculation model of seafloor incident angle**

For the convenience of the discussion, the vessel coordinate system is introduced first. The origin of the coordinate system of the vessel coincides with the center of the sonar (Fig. 8.32). The coordinate system used throughout is left-handed, i.e., the positive x-axis is oriented toward starboard, positive y-axis points toward the bow and the z-axis is perpendicular to the OXY plane. Both the vessel coordinate system and the instantaneous heading change as the ship moves.

(1) Influence of ship attitude

Because of the influence of wind, currents, waves and other factors, a ship will inevitably roll and pitch during navigation, which will change the beam incidence direction. To

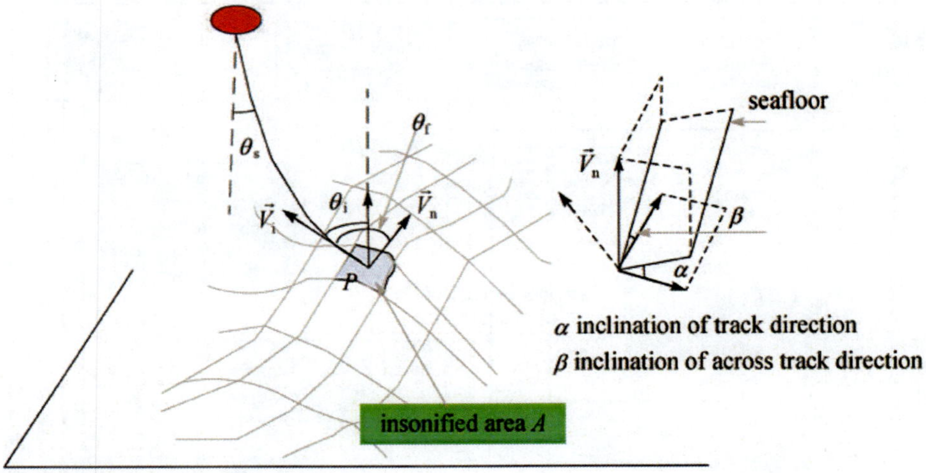

Fig. 8.31 Schematic of seafloor incident angle

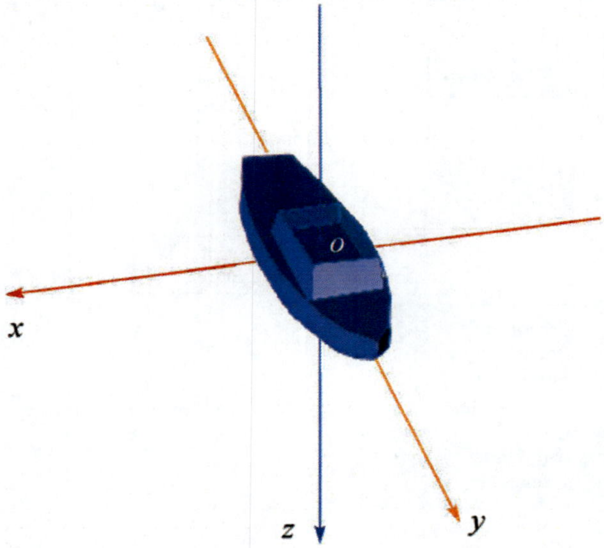

Fig. 8.32 Vessel coordinate system

determine the beam incidence vector $\vec{V}_i$, the beam steering angle θ_{steer} controlled by the transducer array must be converted to the beam departure angle θ_{sector} in the vessel coordinate system. In most cases, the roll is the main influencing factor. It can be seen from Fig. 8.33 that θ_{roll} is set to the roll angle of the ship (i.e., the angle between the ship's horizontal axis and the x-axis). It is positive when the ship's starboard is tilted upward. The parameter $\theta_{installation}$ is the installation angle of the transducer relative to the transverse axis of the survey ship. The departure angle of the beam in the xoz plane is then:

$$\theta_{sector} = \theta_{steer} + \theta_{roll} + \theta_{installation} \tag{8.69}$$

When the influence of pitch on the beam vector is considered (Fig. 8.34), θ_{pitch} denotes the ship pitch angle. According to navigation custom, parameter θ_{pitch} is positive when the bow is tilted upward, the beam vector is rotated by an angle θ_{pitch} and the beam departure angle θ_s (i.e., the

Fig. 8.33 Influence of roll on the beam incidence direction

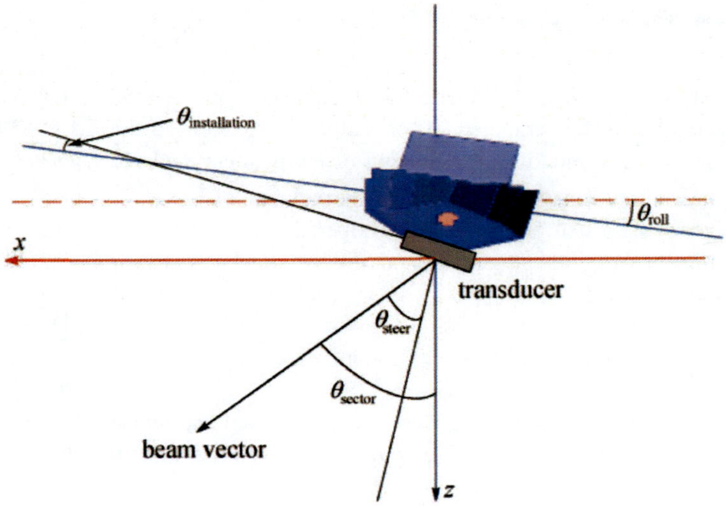

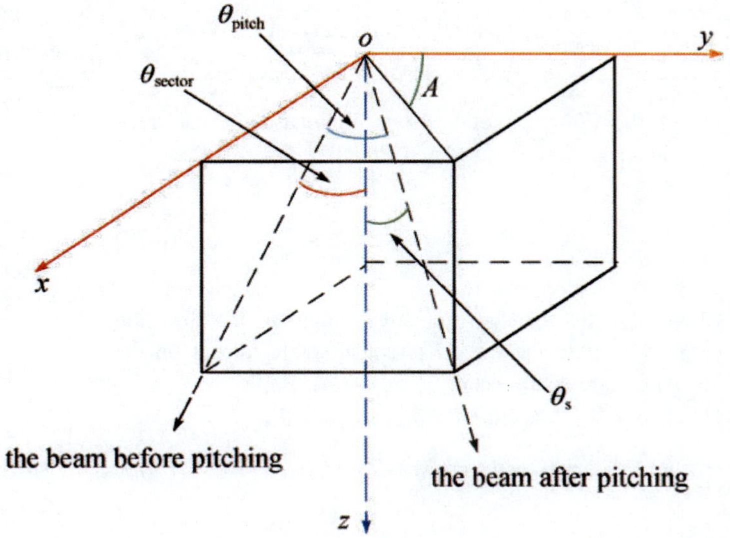

Fig. 8.34 Influence of pitch on beam incidence direction

angle between the beam vector and the z-axis) in the vessel coordinate system is

$$\theta_s = \arccos(\cos\theta_{pitch}\cos\theta_{sector}) \tag{8.70}$$

To describe the orientation of the beam vector, it is also necessary to define the azimuth A of the beam relative to the heading:

$$A = \arccos(\sin\theta_{pitch}/\sin\theta_s) \tag{8.71}$$

At present, most multi-beam systems are equipped with a horizontal and vertical tilt compensation device; thus, the beam departure angle θ_s and the azimuth angle A can be measured and recorded in real time. It means the data can be used directly for subsequent research, which simplifies the calculation process.

(2) Ray-tracing correction

Because of the effect of the refraction of sound waves in water, the incident angle θ_i of the sound wave reaching the seafloor is not the beam departure angle θ_s. If the sound velocity profile (SVP) is measured, the incident angle of the beam can be determined by the Snell rule (Lurton et al. 1994):

$$\theta_i = \arcsin((c_i/c_s)\sin\theta_s) \tag{8.72}$$

where c_s and c_s are the sound speed at the transducer and at the seafloor, respectively. Given azimuth angle A, the incident vector $\vec{V}_i$ of the beam in the vessel coordinate system is

$$\vec{V}_i = \{\sin\theta_i \sin A, \sin\theta_i \cos A, \cos\theta_i\} \tag{8.73}$$

(3) Correction of seafloor topography effect

If the seafloor surface is represented as $z = f(x, y)$ (Fig. 8.31), the normal vector $\vec{V}_n$ of the beam at the seafloor incident point P is

$$\vec{V}_n = \left\{ \frac{\partial f}{\partial x}\bigg|_P, \frac{\partial f}{\partial y}\bigg|_P, -1 \right\} \tag{8.74}$$

In the vessel coordinate system, the inclination angle of the seabed in the track direction (y-axis) and in the track vertical direction (x-axis) is α and β, respectively, and the normal vector of the seabed at the measuring point is

$$\vec{V}_n = \{\tan\beta, \tan\alpha, -1\} \tag{8.75}$$

Substituting Eqs. (8.73) and (8.75) into Eq. (8.70), and considering that the incident angle is < 90°, the incident angle θ_f of the beam at the seabed is

$$\theta_f = \arccos\left(\frac{\cos\theta_i - \sin\theta_i \cdot \sin A \cdot \tan\beta - \sin\theta_i \cdot \cos A \cdot \tan\alpha}{\sqrt{1 + \tan^2\alpha + \tan^2\beta}}\right) \tag{8.76}$$

Taking the influence of the ship's attitude, sound line bending and seabed topography into account, the model for correction of the incident angle relative to the seafloor can be derived rigorously.

(1) In an MBES with a horizontal and vertical automatic compensation device, it is unnecessary to consider the influence of roll and pitch on the incident angle. Thus, Eq. (8.76) is converted into

$$\theta_f = \arccos\left(\frac{\cos\theta_i - \sin\theta_i \tan\beta}{\sqrt{1+\tan^2\alpha + \tan^2\beta}}\right) \quad (8.77)$$

(2) If the influence of seafloor inclination in the track direction is not considered ($\alpha = 0°$), Eq. (8.77) is simplified as

$$\theta_f = \theta_i + \beta - 75° < \theta_i < 75° \quad (8.78)$$

Obviously, the seafloor inclination angle in the direction perpendicular to the track has considerable impact on the incident angle of the beam.

(3) If $\beta = 0°$, then Eq. (8.76) is simplified as:

$$\theta_f = \arccos(\cos\theta_i \cdot \cos\alpha) - 75° < \theta_i < 75° \quad (8.79)$$

The influence of seafloor obliquity in different directions on seafloor incident angle is shown in Fig. 8.35. Obviously, with an increase of the incident angle, the influence of seafloor obliquity generally decreases.

3. **Correction of ARC effect**

(1) Extraction of the relationship between BS and incident angle

In a MBES bathymetry packet, BS is paired with the departure angle θ_s of the beam. By computing the true incident angle θ_f relative to the seafloor, the relationship between BS_B and θ_f can be established.

(2) Correction of BS data of oblique incidence sector

An MBES side-scan data packet records the BS of the oblique incidence sector. From Eqs. (8.58), (8.59) and (8.61), the following relation is obtained:

$$BS(\theta) = BS_o + 20\log_{10}\cos\theta + 10\log_{10}\frac{c\tau\theta_T R}{2\sin\theta} \quad \theta \geq 25° \quad (8.80)$$

To estimate the influence of the error on the incident angle θ, we differentiate Eq. (8.80):

$$dBS = (-20\tan\theta - 10c\tan\theta)\log_{10} e d\theta \quad \theta \geq 25° \quad (8.81)$$

where e = 2.718 28 (note the unit of $d\theta$ is degrees), then:

$$dBS = -0.076(2\tan\theta + c\tan\theta)d\theta \quad \theta \geq 25° \quad (8.82)$$

The relationship between the accuracy of the incident angle against the incident angle itself, when the BS error remains unchanged, is shown in Fig. 8.36. To meet a given

Fig. 8.35 Influence of seafloor inclination on incident angle in different directions

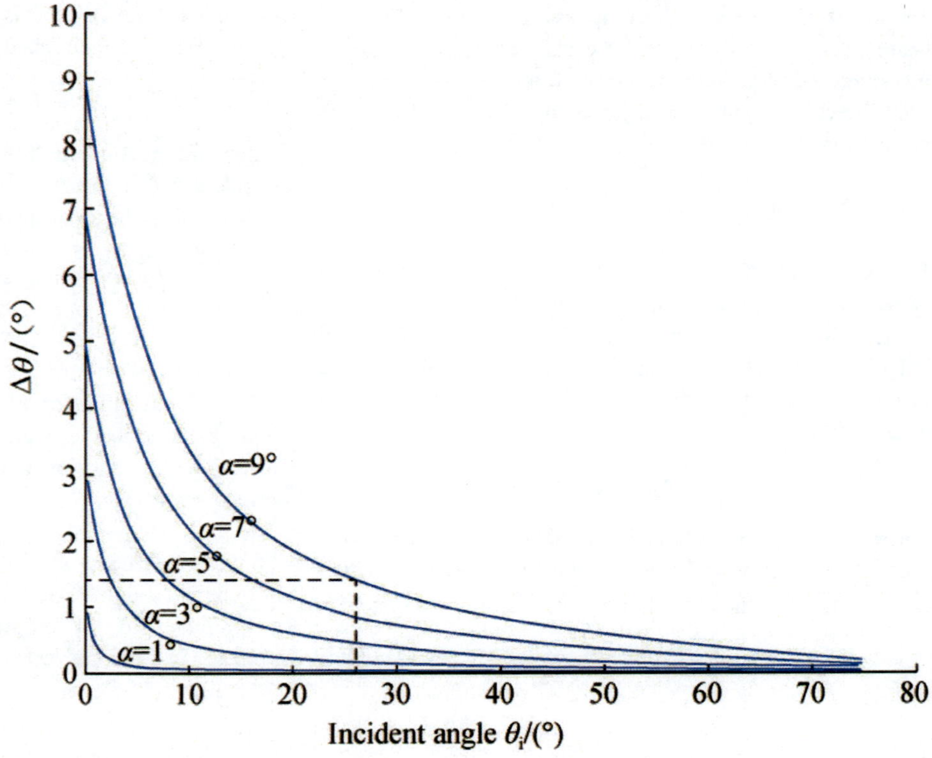

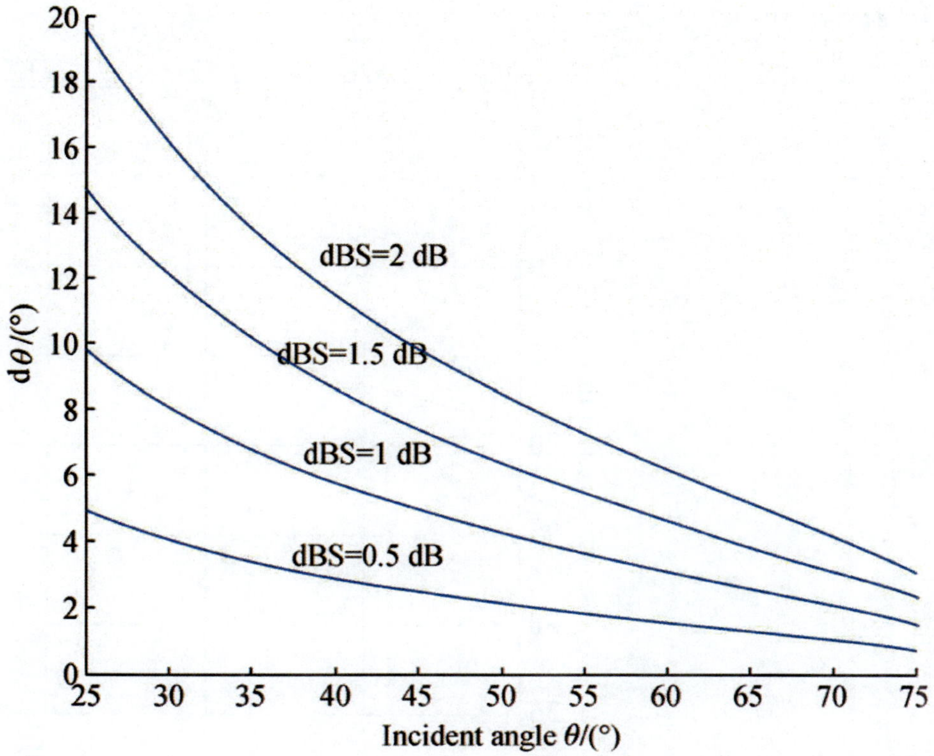

Fig. 8.36 Estimation accuracy of incident angle

requirement of BS error, a higher level of estimation accuracy is required when the incident angle increases.

During data acquisition, to obtain BS_0 values and to eliminate the influence of incident angle on BS, an MBES usually works under the assumption that the sound velocity is constant and that the seafloor is flat. In this case, the difference between the assumed incident angle θ_s and the actual incident angle θ_f is

$$\Delta BS = 20\log_{10}\left(\frac{\cos\theta_s}{\cos\theta_f}\right) + 10\log_{10}\left(\frac{\sin\theta_f}{\sin\theta_s}\right) \qquad \theta \geq 25° \tag{8.83}$$

Thus, the corrected BS data in the oblique sector is

$$BS_o' = BS_o + \Delta BS \tag{8.84}$$

4. **Calculation example**

(1) Performance of multi-beam BS correction

The experiment used BS and bathymetry data measured using a Simrad EM3000 MBES on July 14, 2010, in Jiaozuo Bay, Qingdao, China. Because the attitude compensation device was installed, the influence of ship attitude is not considered here. The SVP data are shown in Fig. 8.37, the maximum variation in sound velocity in the survey area is 2 m/s. Hence, the associated maximum influence on the incident angle is only 0.3° (Eq. (8.72)), which can be ignored.

To facilitate the illustration, bathymetric measurements from 50 pings of an area with complex topographic changes are selected. Figure 8.38 shows the rear view of the bathymetric data. The red lines represent the measurements of port beams and the blue lines represent measurements of starboard beams. Because of the randomness of the BS data, the validity of the BS correction cannot be shown in the measurement of a single ping. Thus, the average ARC is studied. Theoretically, when the seafloor sediment type is consistent and the seabed is flat, the curve should be symmetrical with respect to the nadir direction. However, before the correction, the center of the curve is located near the incident angle $\theta = -18°$ (Fig. 8.39), whereas after the correction, the curve becomes symmetrical with respect to the nadir direction.

(2) Performance of the correction of BS of oblique incidence sector

The BS of the oblique incidence sector is used to generate sonar images. Figure 8.40 shows the rear view of

Fig. 8.37 The SVP data

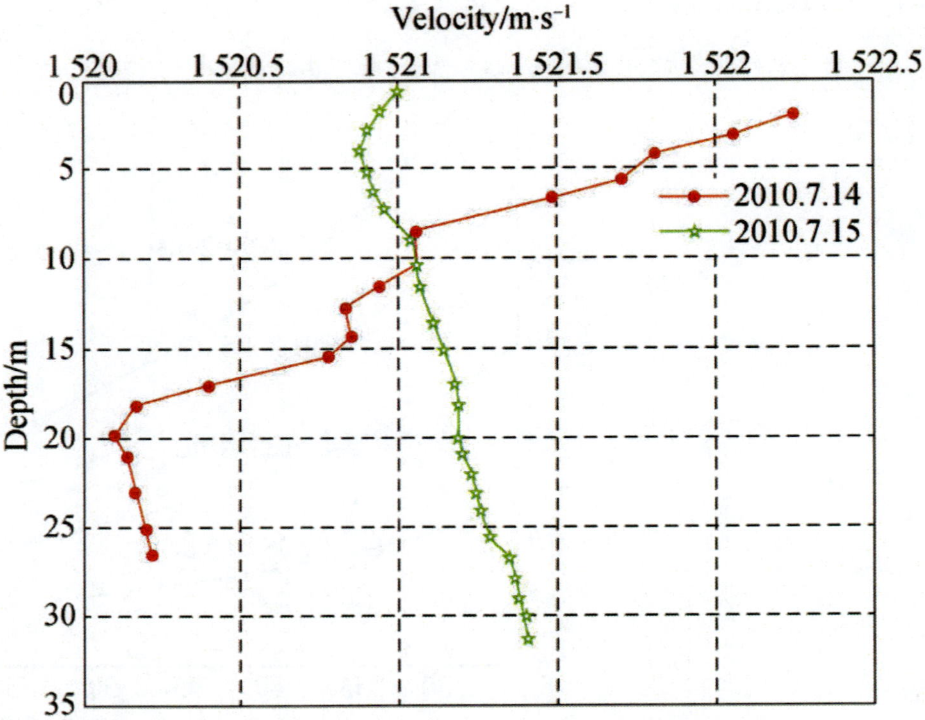

Fig. 8.38 Seafloor rear view

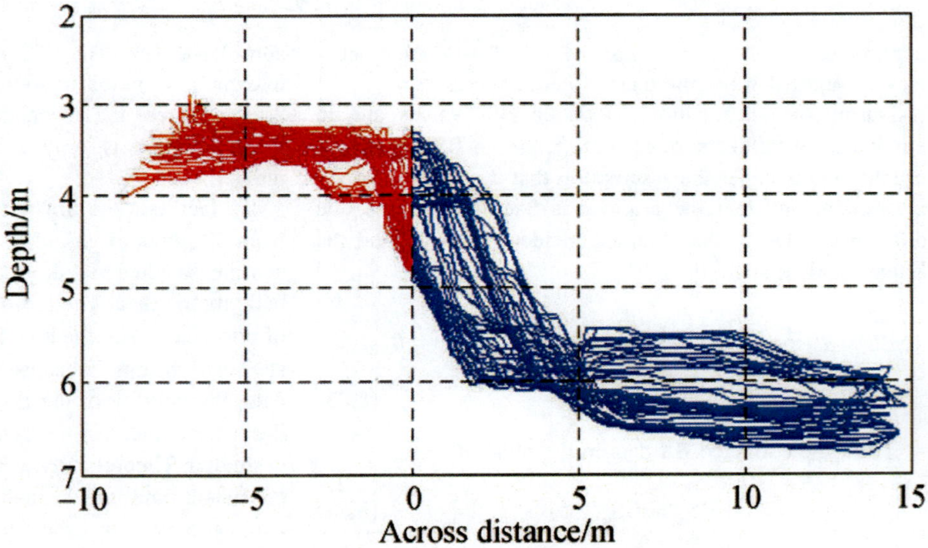

multiplying bathymetry data of seafloor with a large degree of variation. The drastic variation in the seabed topography causes obvious changes in the grayscale of the sonar images (Fig. 8.41a). Figure 8.41b shows the sonar image after the seabed incident angle is corrected. Obviously, the method described above effectively corrects the impact of seabed topography on the sonar image (reducing the impact of seabed incident angle on the BS data).

The model used here can reduce the influence of seabed topography on BS, but it cannot eliminate it completely (Fig. 8.41b). This is because the Lambert model cannot fully describe the pattern of the ARC. In addition, the BS of the data near the nadir area is significantly higher than in other areas, which results in abnormal gray values along the track in the sonar image. If not corrected, such interference could significantly degrade the classification result.

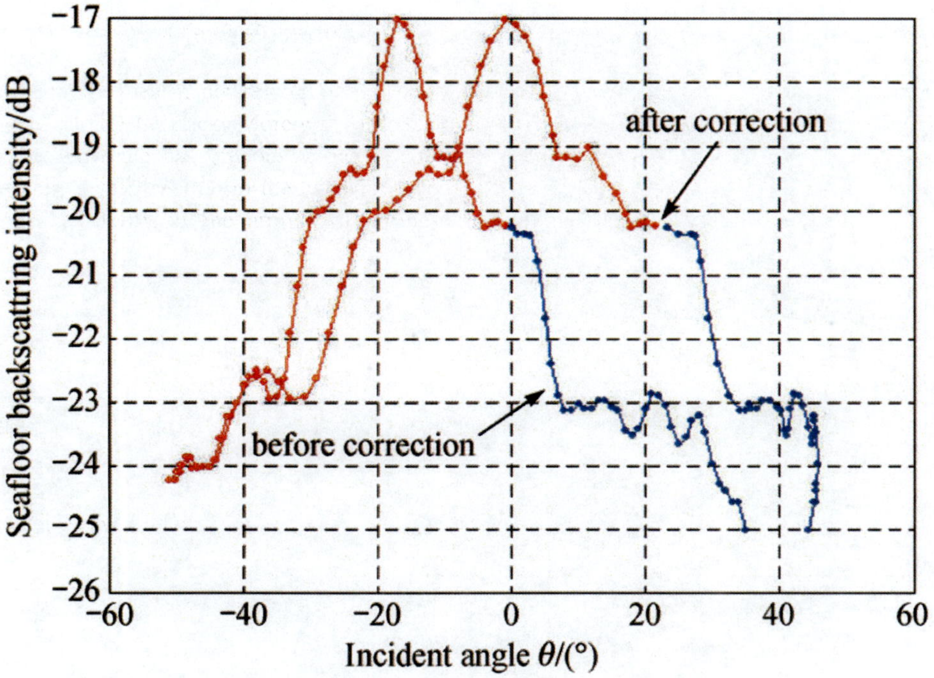

Fig. 8.39 ARC information before and after correction

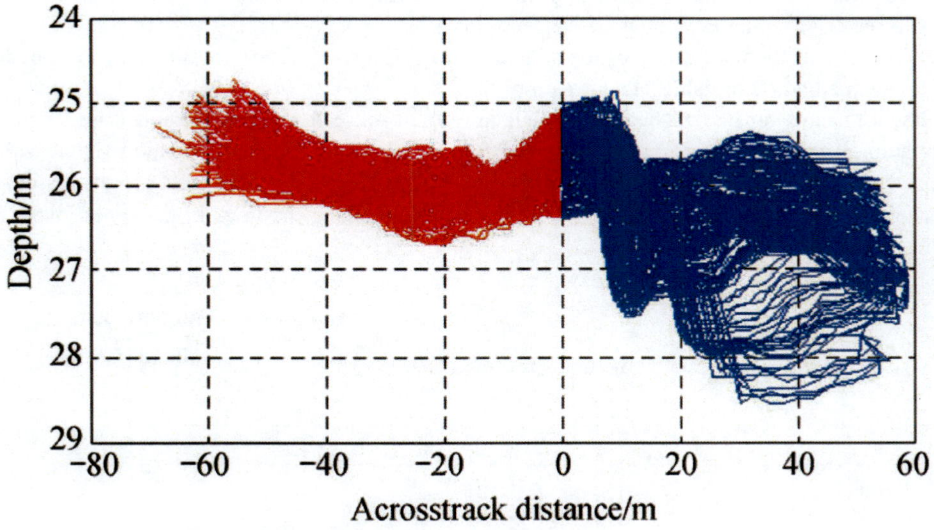

Fig. 8.40 Sea floor rear view of multiping bathymetry data

8.3.2.2 Correction of Outliers in the Near-Nadir Region

Specular reflection in the near-nadir region often causes abnormal BS values along the track. Moreover, at certain incident angles, the grayscale has obvious variation compared with other areas, which is due to the effect of beam directivity and the sensitivity of the array. All of these factors will influence the accuracy of the acoustic seafloor classification.

Tang (2003) introduced two methods to address these biases in BS data. One choice is to eliminate the specular reflection signal, which would result in a loss of effective data. Another method is to use the Gaussian weighted average algorithm that is developed in MB-system software. This

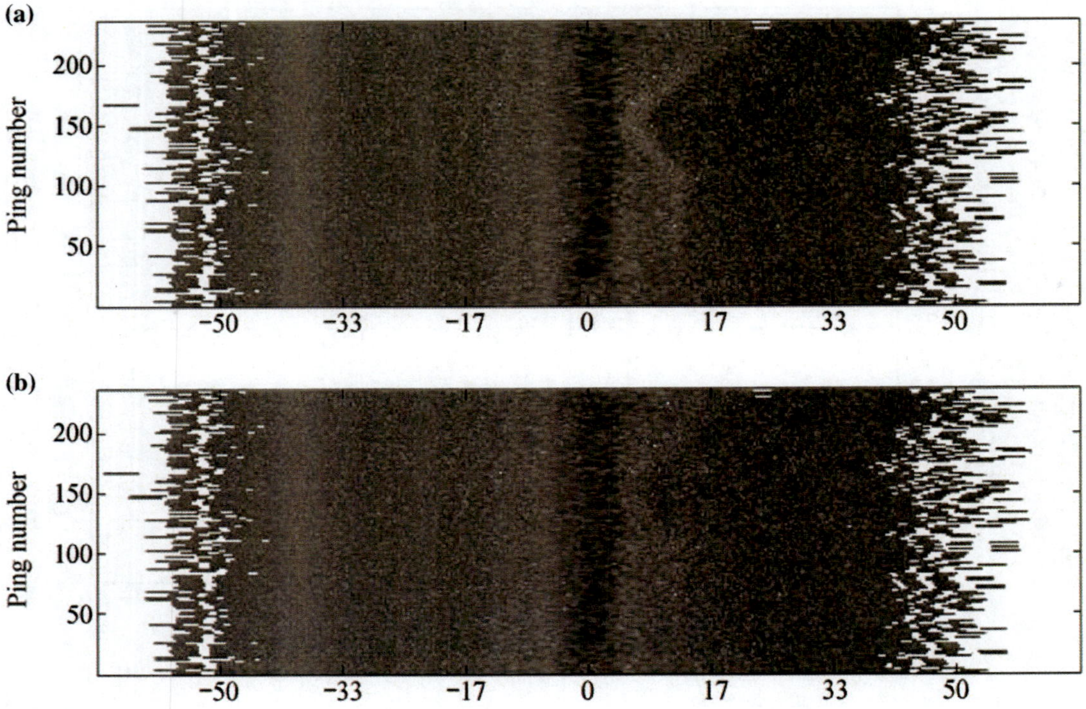

Fig. 8.41 Effects of seabed topography on sonar images (**a**) and the performance of the correction (**b**)

method assigns different weights to the BS data of different incident angles and recalculates the BS values of the central beams. However, because of the uncertainty of the weights, the performance of the method is not stable. Here, we propose a correction method for near-nadir BS data based on weighted least square estimation. It exploits the fact that the average BS of multiple pings varies systematically with the change in incident angle when the seafloor type is consistent.

1. **Analysis of abnormal BS values**

(1) Cause of abnormal BS in the near-nadir area

The backscatter signal in the near-nadir area of a transducer is mainly the specular reflection signal with distinctly large values (Fig. 8.42). The data are expressed as obvious gray outliers along the track in the sonar image (Fig. 8.43).

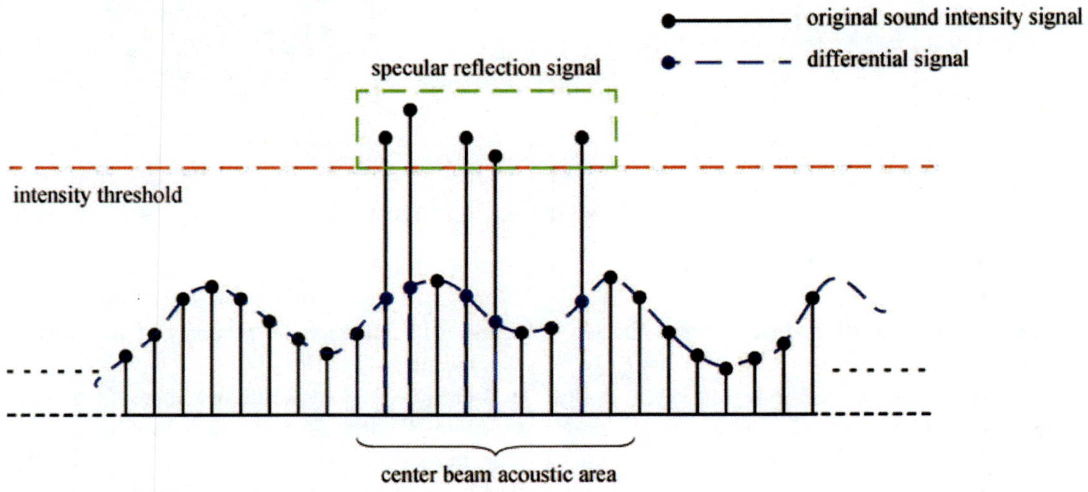

Fig. 8.42 Specular reflection signals in the near-nadir area

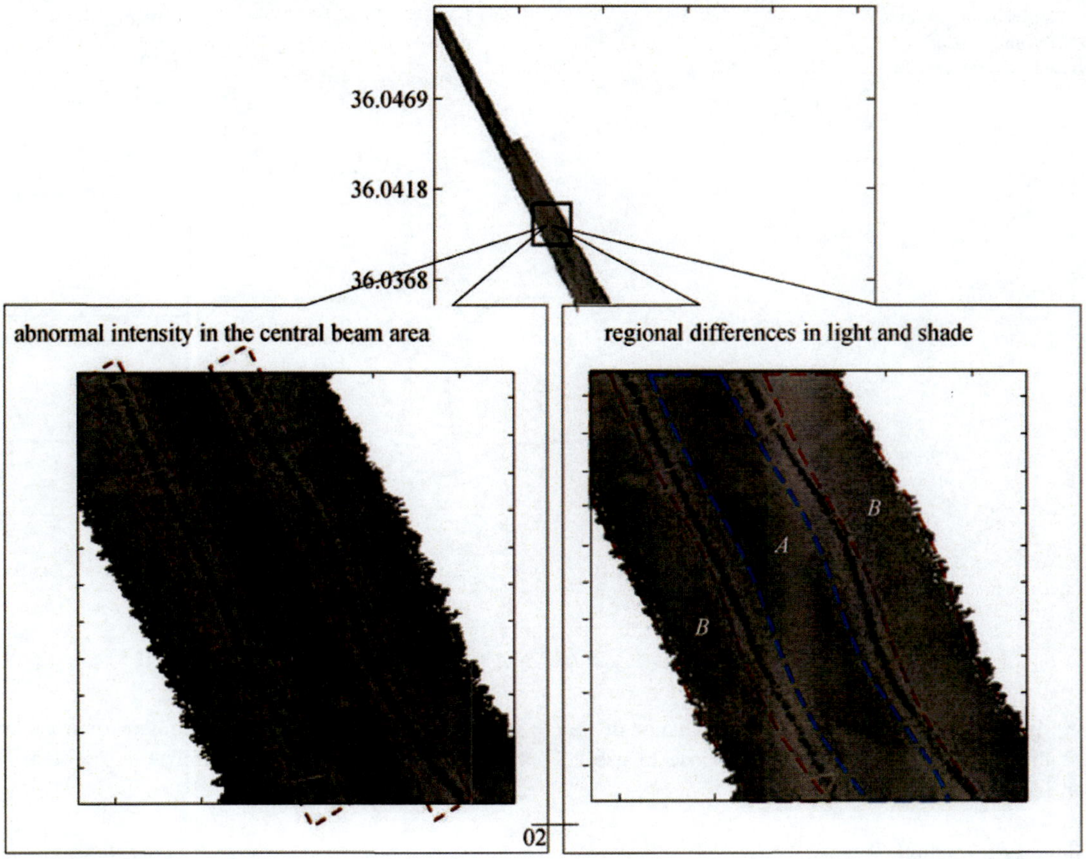

Fig. 8.43 Abnormal BS in sonar images

(2) Analysis of brightness difference between overlapped and non-overlapped data

Figure 8.43 shows a sonar image of two survey lines with an overlap rate of approximately 40%. There is an obvious change in the brightness between the data in the overlapping area *A* (BS data of two survey lines overlapped) and the non-overlapping area *B* (BS data of only one survey line). This artifact is also referred to as "false terrain" and the reason for its appearance is shown in Fig. 8.44. The BS data of the edge beams roughly increase with the incident angle (because of factors such as the non-uniform beam directivity pattern, non-uniform array sensitivity, and measurement noise; however, the BS profile might also present other forms), as shown by curve *a* (Fig. 8.44). When two sets of BS data are merged, the usual approach is to simply average the overlapped data, which results in the evident "bulge" phenomenon (Fig. 8.44c).

2. **Least squares (LS) method correction**

(1) Gaussian weighted average method

The weights of the data are set according to the incident angle of the beam (Table 8.3), and the Gaussian weighted mean algorithm is used to calculate the BS near the central beams (Boulinguez and Quinquis 2002):

$$\mathrm{BS} = \sum_{i=1}^{n} W_i \mathrm{BS}_i / \sum_{i=1}^{n} W_i \tag{8.85}$$

where BS is the backscatter intensity value near the central beam, BS_i is that of other regions and W_i is the weight value.

After the Gaussian weighted mean method has been applied, obvious gray outliers remain near the central beam and between the overlapped and non-overlapped areas

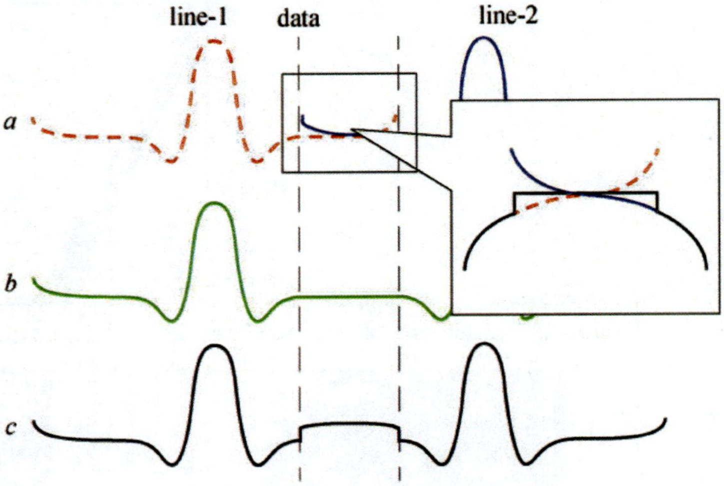

Fig. 8.44 Causes of brightness difference between areas of overlapped and non-overlapped data

Table 8.3 Weight at different incident angles

Incident angle/(°)	–60.0	–45.0	–15.0	–14.9	14.9	15.0	45.0	60.0
Weight	0.2	1.0	0.8	0.1	0.1	0.8	1.0	0.2

(Fig. 8.45). In consideration of the systematic change in gray outliers at the same incident angle, we propose to use LS estimation instead.

(2) Weighted LS method

The basic principle of the correction based on weighted LS estimation is as follows. Note that the BS signal is composed of the trend signal and the high-frequency fluctuation signal. First, BS data of consecutive pings are used to extract the initial BS sequence. Then, the weighted LS estimation is used to extract the trend of the initial BS. Meanwhile, the high-frequency component is also obtained. Next, BS data of a high-quality area (other than the near-nadir area) is used to extract the trend information, which is used as the reference trend. Finally, the reference trend and the extracted high-frequency component are added together to realize the reconstruction of the BS data of the near-nadir area, which effectively reflects the change in seafloor sediment.

In the LS computation, the problem of trend extraction can be transformed into an optimization problem (Farbman et al. 2008):

$$\min \sum_{p} \left((u_p - g_p)^2 + \lambda a_{x,p}(g) \left(\frac{\partial u}{\partial x} \right)_p^2 \right) \tag{8.86}$$

where p is the BS sample point number, g is the BS signal to be processed, u is the trend sequence of the desired BS, $\left(\frac{\partial u}{\partial x}\right)_p$ is the partial derivative of u at point p in the x direction and $a_{x,p}(g) = \left(\left| \frac{\partial l}{\partial x}(p) \right|^{\alpha'} + \varepsilon \right)^{-1}$ is a smoothing weight that determines the tracking effect of the trend of the BS sequence. Here, ε is a minimum quantity (to prevent the denominator becoming zero), $l = \log_{10} g$ and λ is a smoothing factor. The larger the value of λ, the smoother the contour curve will be. Parameter α' determines the sensitivity of the trend when reflecting changes in the BS signal (it is generally between 1.2 and 2.0). For the convenience of

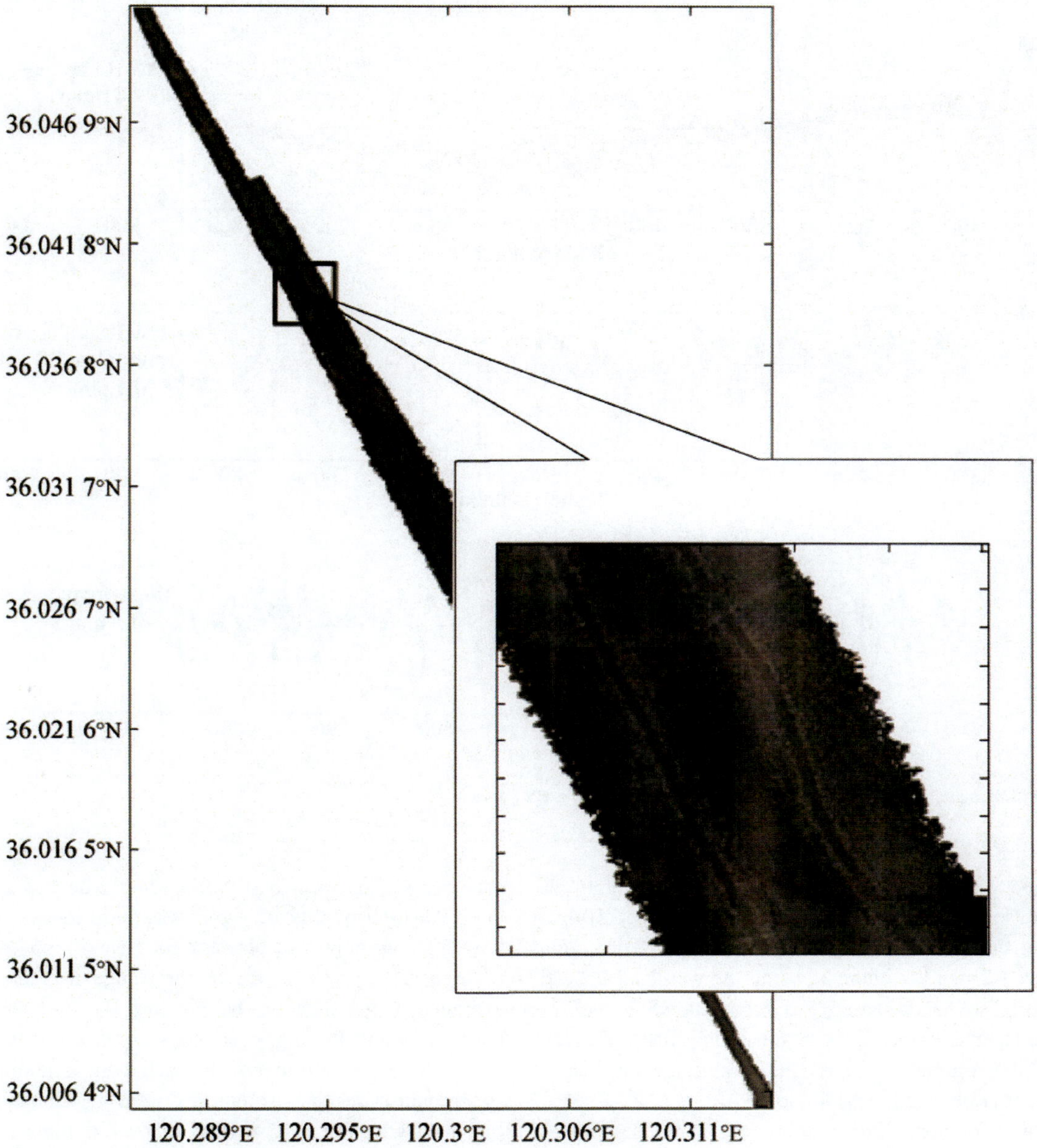

Fig. 8.45 Result of Gaussian weighted average method gray anomaly correction

calculation, Eq. (8.85) can be converted into an optimization problem in the matrix form:

$$\min f = \min\left[(\boldsymbol{u} - g)^{\mathrm{T}}(\boldsymbol{u} - g)\right] + \lambda(\boldsymbol{u}^{\mathrm{T}}\boldsymbol{D}^{\mathrm{T}}\boldsymbol{A}_x\boldsymbol{D}_x\boldsymbol{u}) \quad (8.87)$$

where $\boldsymbol{A}_x$ is a diagonal matrix composed of $a_{x,\,p}(g)$, and the $\boldsymbol{D}_x$ matrix is a discrete partial differentiation operator (composed of 0, 1 and −1). Based on Eq. (8.87), we have

$$\boldsymbol{U} = (\boldsymbol{I} + \lambda\boldsymbol{L}_g)^{-1}g \quad (8.88)$$

where $\boldsymbol{L}_g = \boldsymbol{D}_x^{\mathrm{T}}\boldsymbol{A}_x\boldsymbol{D}_x$, $\boldsymbol{I}$ is the identity matrix and $\boldsymbol{U}$ is the obtained BS trend sequence.

Through superposition of the reference trend value and the corresponding high-frequency signal, the BS sequence of each ping is reconstructed and the corrected sonar image obtained. A segment of BS data of 50 consecutive pings is

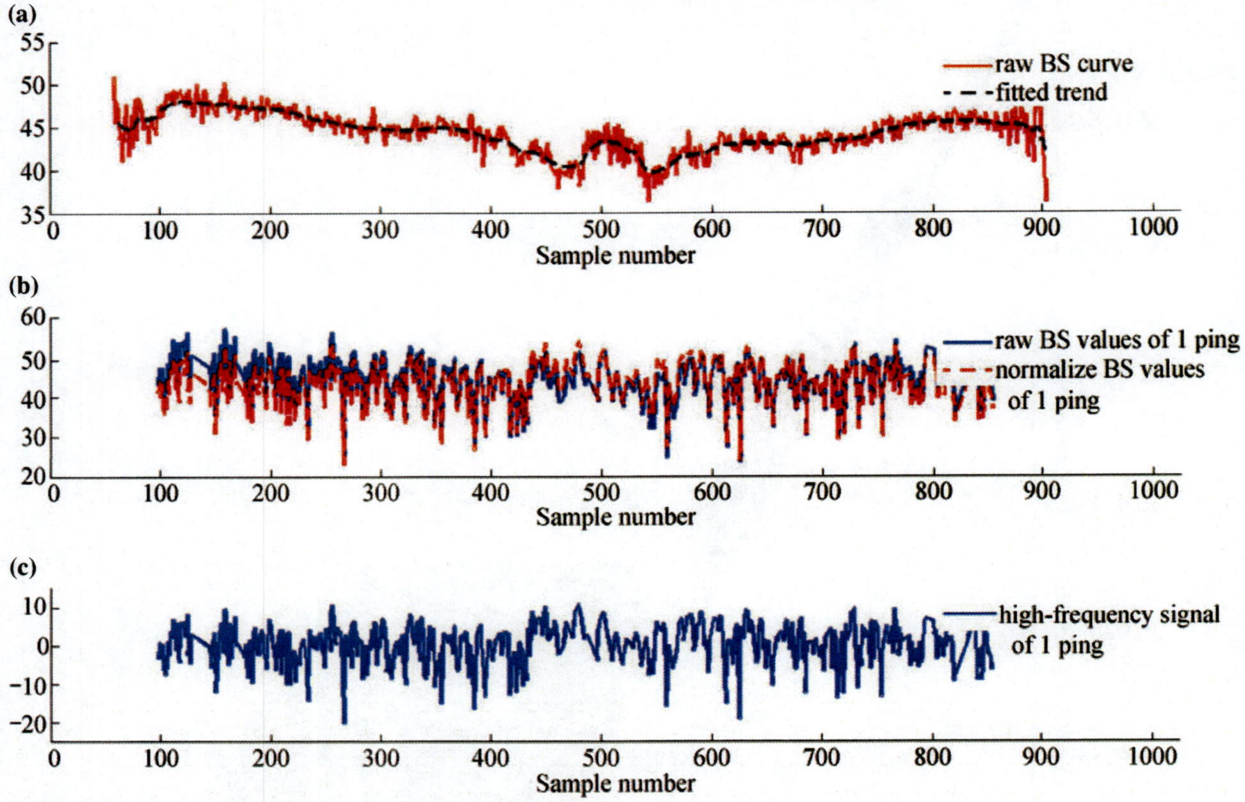

Fig. 8.46 Schematic of the decomposition and reconstruction of a BS curve

considered and the chosen data are averaged along the across-track direction to obtain the BS sequence (50-ping data with the same statistical properties) (Fig. 8.46a, solid line). The BS trend is obtained using the weighted LS estimation (Fig. 8.46a, dotted line). By contrast, BS data of one ping are also extracted (Fig. 8.46b, solid line). By subtracting the estimated trend, the high-frequency signal component is obtained (Fig. 8.46c).

The high-frequency fluctuation component extracted can only reflect the high-frequency variation part of the BS signal. Therefore, to represent the BS of the substrate in each ping, a uniform base value should be added. In BS data, the high quality of signal reception means that data with incident angles of between ±25° and ±65° are often of high quality. Thus, BS in this range in each ping is averaged to generate the base value. By superposing the base value and the high-frequency component, the reconstruction of BS information in each ping is realized. The superposition result is shown in Fig. 8.47. Comparison of Fig. 8.47 with Fig. 8.45 reveals that the outliers in the BS data are eliminated and that the quality of the sonar image is improved.

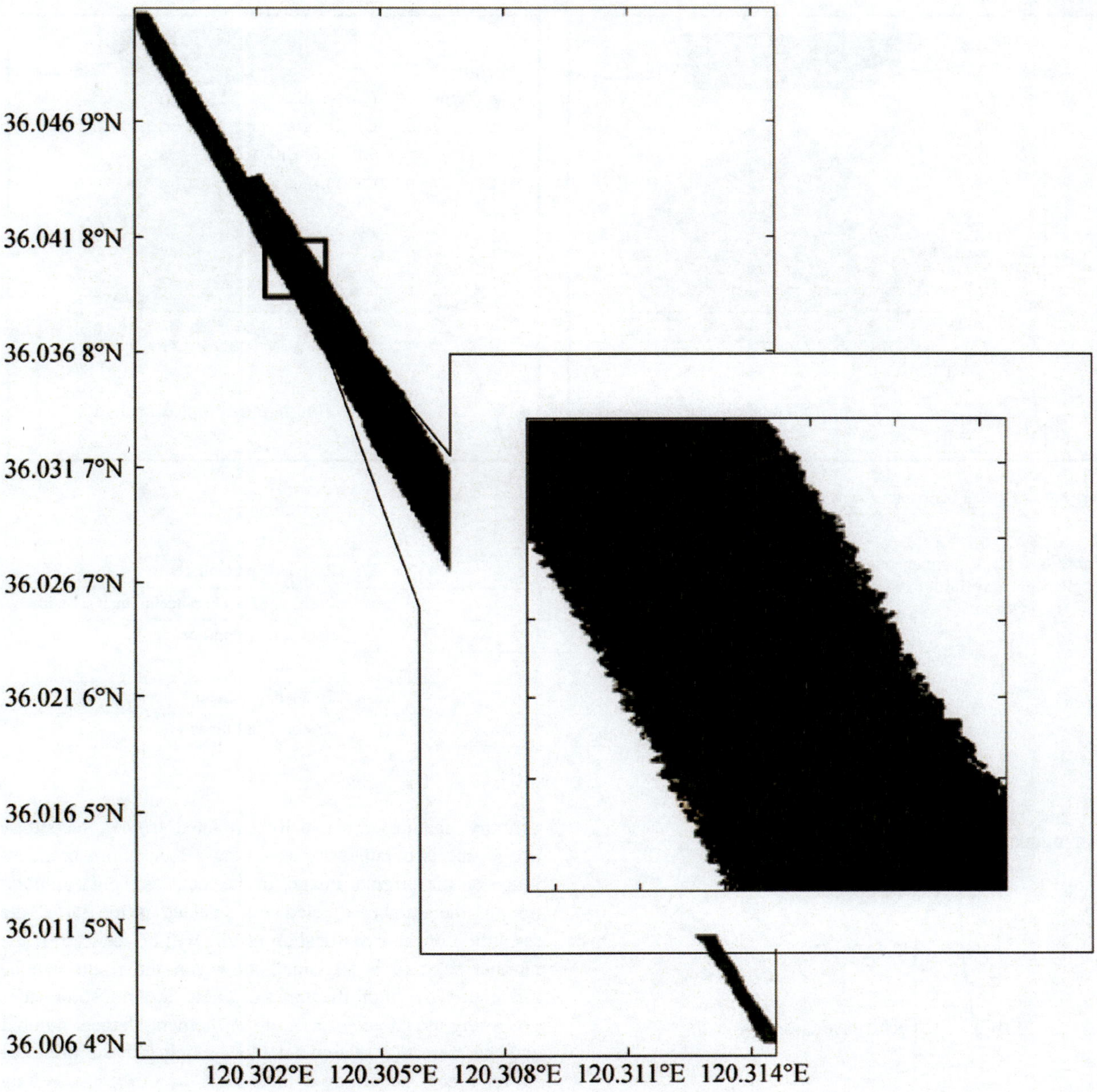

Fig. 8.47 Corrected sonar image

8.4 Acoustic Feature Extraction

8.4.1 Feature Extraction of SSS Time-Series Backscatter Signal

SSS transmits signals to the seabed and receives a continuous time series of BS signal. By analyzing the waveform characteristics of the BS, the type of seafloor sediment can be inferred. For a flat and smooth seabed, the backscatter waveform is reasonably sharp, whereas for a rough and complex seabed, the backscatter waveform is relatively gentle and it persists for a longer time (Fig. 8.48).

In seafloor classification, the characteristic structure of the backscatter waveform from an SSS should be analyzed. For example, QTC software uses five algorithms (Table 8.4) to extract 166 features for sediment classification. A typical backscatter waveform structure is illustrated in Fig. 8.49.

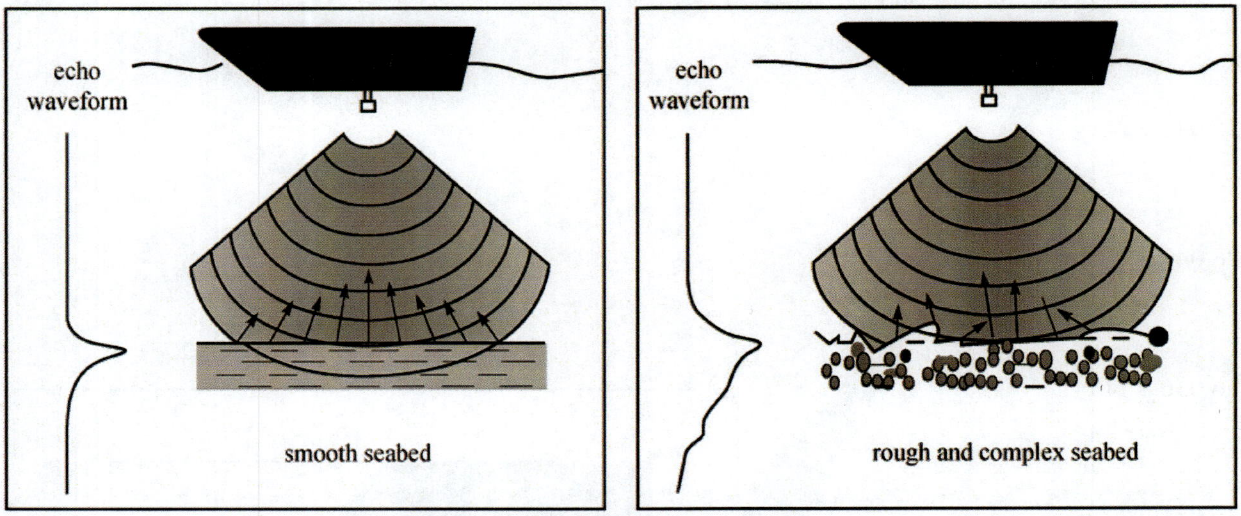

Fig. 8.48 Typical backscatter waveforms from different types of seafloor

Table 8.4 Typical parameters of single beam waveform extracted by QTC software

Feature	Responds particularly to
Cumulative integral of amplitudes	Position of backscatter in analysis window
Amplitude histogram	Backscatter duration
Amplitude quantiles	
Power spectrum	Frequency content
Wavelet decomposition	Rise and fall times

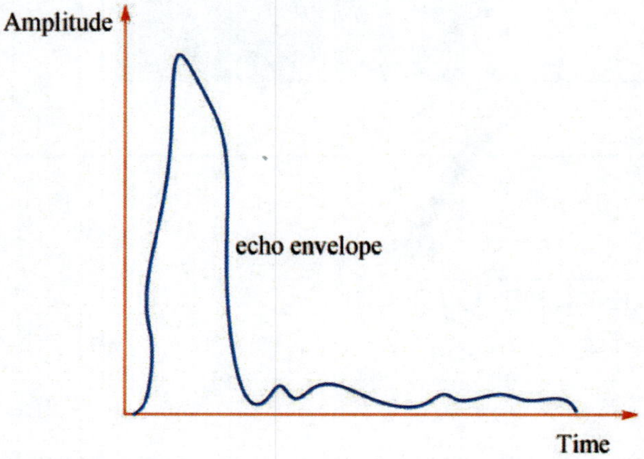

Fig. 8.49 Typical SSS backscatter waveform structure

8.4.2 Feature Extraction Method

High-resolution sonar images can be obtained using SSS, multi-beam sonar or a combination of both (Fig. 8.50). Before image classification, feature parameters should be extracted first, but not directly from the entire image. On the contrary, features are usually extracted from a secondary image that is obtained by executing a local moment calculation on the original image. In the local moment computation, if the number of elements selected is too large, the resolution of the classification results will be reduced. If the number selected is too small, however, the result will be influenced by high fluctuation. There are no strict rules governing the choice of size of the moment element number and therefore this section establishes only a basic principle for reference. According to the central limit theorem, as long as the moment function of a matrix distribution exists, $\bar{x} = \frac{1}{n}(x_1 + \cdots + x_n)$ will converge to the normal distribution $N\left(\xi, \frac{\sigma}{\sqrt{n}}\right)$. Here, ξ is the mean of the distribution and σ^2 is the variance. Thus, for a subsample data set, if the distribution skewness is not too large and the sampled number is not too small ($n > 30$), the distribution of the sampled mean can be approximated by a Gaussian distribution with a standard deviation that converges to $1/\sqrt{n}$. The mean BS (grayscale value) of the local subset is an important characteristic parameter for seafloor sediment classification. The better it approximates to the Gaussian distribution and the

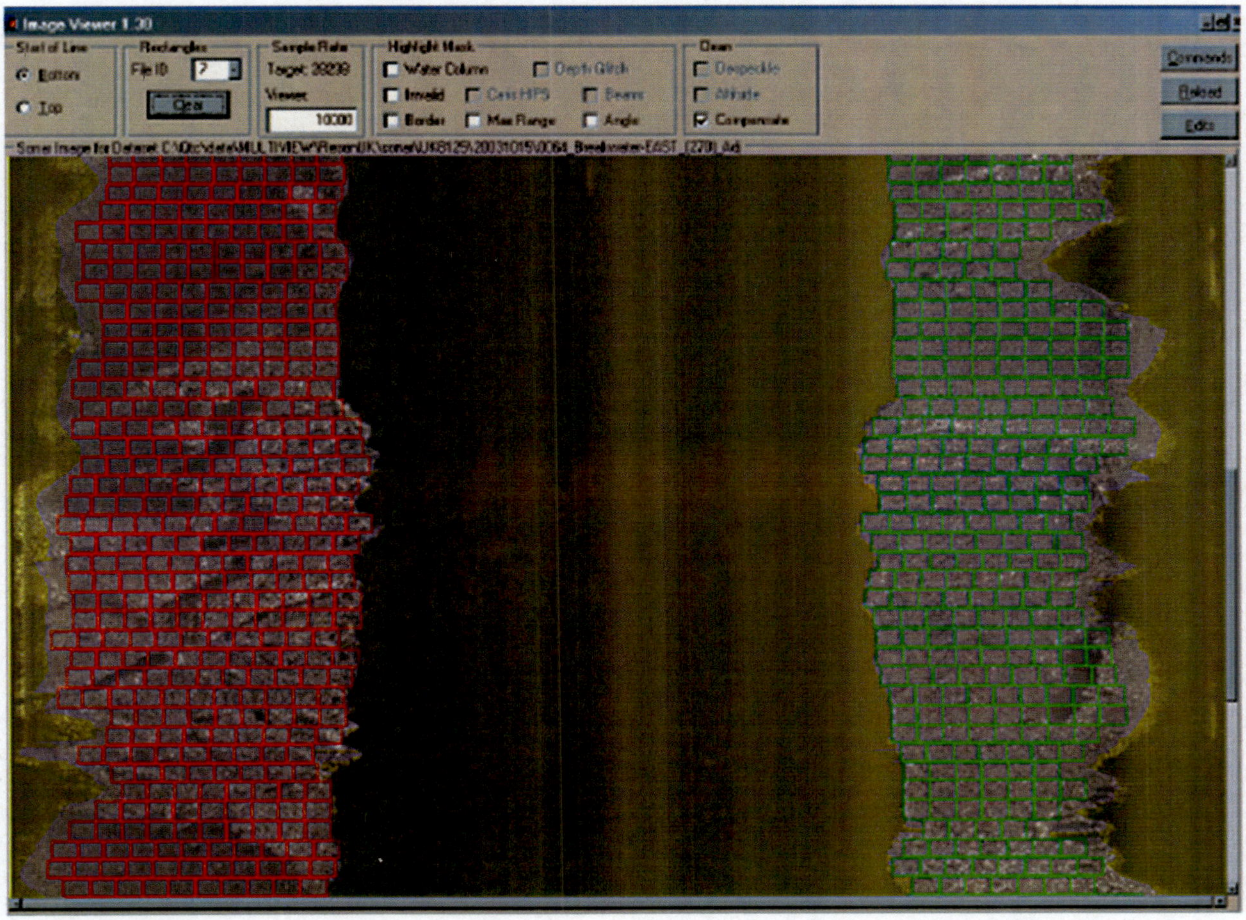

Fig. 8.50 Sonar image and local moment computation

smaller the variance is, the more significant the statistical characteristics are. Therefore, the number of local points for moment computation should be greater than 30.

After the local subset data are selected, feature parameters can be extracted. Feature parameters of interest for seafloor classification mainly include basic statistics, quantiles, histograms, power spectrum ratios, and gray-level co-occurrence matrices.

8.4.2.1 Basic Statistics

1. **Mean**

The mean of BS refers to the mean of the gray value selected, which reflects the average scattering intensity of the seafloor in the image:

$$u = \frac{1}{MN}\sum_{i=0}^{M-1}\sum_{j=0}^{N-1} f(i,j) \tag{8.89}$$

where M and N are the row and column values of the image, respectively.

2. **Standard deviation (SD)**

The SD σ reflects the total dispersion of the gray values selected, and the calculation formula is

$$\sigma = \sqrt{\frac{\sum_{i=0}^{M-1}\sum_{j=0}^{N-1}[f(i,j)-u]^2}{MN}} \tag{8.90}$$

8.4.2.2 Quantiles and Histograms

A quantile is used to describe the probability distribution of random variables. For a random variable x, its quantile Q is the number that satisfies the condition $p(x > Q) = \beta$. Thus, the median corresponds to the quantile of $\beta = 0.5$. Compared with the average value, the quantile statistic is robust against outliers. For a given value of β, Q_β denotes the β

quantile of vector X. For a given image, we often extract the quantile of $\beta = 0.1, 0.2, \ldots, 0.9$.

The histogram is a discrete form of the first-order probability distribution. Its shape indicates general information about the grayscale distribution in the selected area of the image. For example, the occurrence of a histogram with a narrow peak indicates small image contrast, whereas the presence of a double peak indicates the image is composed largely of two areas of different brightness. Apart from the mean and the variance, information commonly obtained from histogram statistics also includes the skewness and the kurtosis. Skewness (S), which represents the deviation of a histogram distribution from a symmetrical distribution, is defined as

$$S = \frac{\eta^3}{\sigma^3} = \frac{1}{MN\sigma^3}\sum_{i=1}^{M}\sum_{j=1}^{N}[f(i,j) - u]^3 \quad (8.91)$$

Kurtosis (K), which describes the degree of dispersion of a histogram, is defined as

$$K = \frac{\eta^4}{\sigma^4} = \frac{1}{MN\sigma^4}\sum_{i=1}^{M}\sum_{j=1}^{N}[f(i,j) - u]^4 \quad (8.92)$$

8.4.2.3 Power Spectral Rate

Given the BS values in a local window $g_i(t)(i = (1, 2, \ldots, n))$, the energy spectrum is defined as

$$P_i(f) = |F[g_i(t)]|^2 \quad (8.93)$$

where F represents a Fourier operator. To improve accuracy, the averaged energy spectrum of n windows is often computed as

$$\bar{P}(f) = \frac{1}{n}\sum_{i=1}^{n}P_i(f) \quad (8.94)$$

The log power spectrum is defined as

$$P_{\mathrm{L}}(f) = 10\log_{10}\left(\frac{A\bar{P}(f)}{P_{\mathrm{m}}} + 1\right) / \log_{10}(A+1) \quad (8.95)$$

where P_{m} is the maximum value of $\bar{P}(f)$ and A is a constant (usually set at 10 000). Then, the normalized logarithmic energy spectrum is

$$P_{\mathrm{NL}}(f) = P_{\mathrm{L}}(f)/f\int_0^{f_{\mathrm{NY}}} P_{\mathrm{L}}(f)\mathrm{d}f \quad (8.96)$$

where f_{NY} is the Nyquist frequency, which is determined according to the pixel number in the window. According to the logarithmic energy spectrum, two spectral features can be obtained:

$$D_{f_1} = \int_0^{f_{\mathrm{BA}}} P_{\mathrm{NL}}(f)\mathrm{d}f \Big/ \int_{f_{\mathrm{BA}}}^{f_{\mathrm{NY}}} P_{\mathrm{NL}}(f)\mathrm{d}f \quad (8.97)$$

$$D_{f_2} = \int_0^{\frac{1}{2}f_{\mathrm{BA}}} P_{\mathrm{NL}}(f)\mathrm{d}f \Big/ \int_{f_{\mathrm{BA}}}^{f_{\mathrm{NY}}} P_{\mathrm{NL}}(f)\mathrm{d}f \quad (8.98)$$

where $f_{\mathrm{BA}} = f_{\mathrm{NY}}/2$, D_{f1} represents the ratio of low frequency to high frequency and D_{f2} is the ratio of very low frequency to high frequency. These two features are effective in distinguishing, for example, a seafloor of sand from that of mud. This is because a sandy seafloor has a wide spectrum, while the spectrum of a muddy seafloor is relatively smooth and without a high-frequency term.

8.4.2.4 Co-occurrence Matrix

Texture information directly reflects the roughness of the seafloor surface structure and it can be used for seafloor classification. Statistical analysis based on a co-occurrence matrix is one of the most widely used methods of texture analysis. In general, a rectangular window is superposed on a region with the same cluster on a sonar image, and the co-occurrence matrix is computed for further statistical analysis.

The co-occurrence matrix studies the relationship between pixels at a given distance along a certain direction (0°, 45°, 90°, 135°). The value of a co-occurrence matrix element (i, j) is equal to the probability p (or frequency) of a pair of pixels with grayscale i and j when their distance is d along the given direction (Subramaniam et al. 1993). There are six basic related statistics: contrast, entropy, uniformity, correlation, energy and variance. Assuming that the matrix size is m × m, then the specific formulas are

$$\text{contrast}: \ f_1 = \sum_{n=0}^{m-1} n^2 \sum_{i=1}^{m}\sum_{j=1}^{m} p(i,j), \quad |i-j| = n \quad (8.99)$$

$$\text{entropy}: \ f_2 = \sum_{i=1}^{m}\sum_{j=1}^{m} p(i,j)\log_2 p(i,j) \quad (8.100)$$

$$\text{homogeneity}: \ f_3 = \sum_{i=1}^{m}\sum_{j=1}^{m} \frac{p(i,j)}{1+(i-j)^2} \quad (8.101)$$

$$\text{relevancy}: \ f_4 = \frac{\sum_{i=1}^{m}\sum_{j=1}^{m} p(i,j) - u_x u_y}{\sigma_x \sigma_y} \quad (8.102)$$

$$\text{energy}: \ f_5 = \sum_{i=1}^{m}\sum_{j=1}^{m} p(i,j)^2 \quad (8.103)$$

$$\text{variance}: \ f_6 = \sum_{i=1}^{m}\sum_{j=1}^{m}(i-u)^2 p(i,j) \tag{8.104}$$

where u_x and u_y are the mean values of each row and column, respectively, σ_x andσ_y are the SD values of each row and column, respectively, and u represents the mean value. Finally, by averaging the values of the four different directions, we can obtain the eigenvector invariant of rotation.

8.4.3 Extraction of Average ARC Feature

8.4.3.1 Feature Extraction by Differential Methods

Various parameters need to be extracted to describe the shape of the ARC. Hughes Clarke et al. (1997) extracted feature parameters by computing the derivation of the curve. Specifically, he divided the curve into three regions (Fig. 8.51): the specular reflection area (Area 1), oblique incidence area (Area 2) and small grazing angle area below the critical angle (Area 3). Furthermore, the region boundary values and the mean value of the BS of each region were taken as the feature parameters. The implementation steps are as follows.

1. Calculate the first and second derivatives of the ARC.
2. Calculate the maximum values B and F of the first and second derivation results, and use the corresponding incident angles as regional boundaries, i.e., C and G.
3. Calculate the mean value of BS in each region (Parameters A, E and I).
4. Calculate the average slope of Regions 2 and 3 on the first derivative curve (denoted as H and J).

Table 8.5 provides a detailed description of the above parameters.

8.4.3.2 Fitting Parameter Using the LS Method

1. **Parametric model**

Hellequin et al. (2003) built a parametric model of BS for the entire range of incident angles according to the characteristics of acoustic scattering at different incident angles:

$$\mathrm{BS}(\theta) = 10\log_{10}(A\exp(-\gamma\theta^2) + B\cos^{\beta}\theta) \tag{8.105}$$

where θ is the incident angle relative to the seabed, and A, B, γ and β are model parameters. The model tends to the

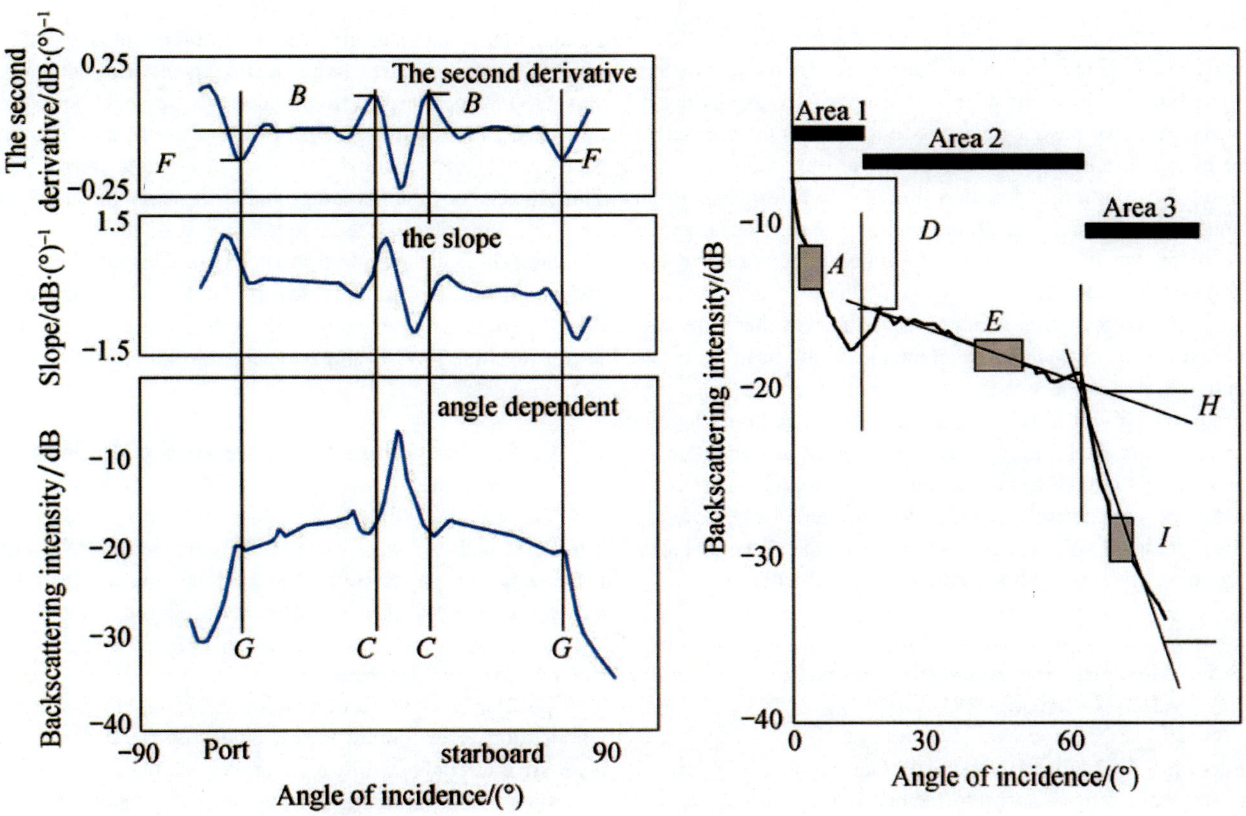

Fig. 8.51 Illustration of ARC feature extraction

Table 8.5 Description of feature parameters

Parameter	Description	Unit
A	average BS near the central beam (0°–5°)	dB
B	maximum value of the second derivative (near Areas 1 and 2)	dB/(°)^2
C	the boundary position of Areas 1 and 2	(°)
D	Area 1 variation range of BS	(°)
E	average BS at center of Area 2	(°)
F	maximum value of the second derivative (near Areas 2 and 3)	dB/(°)^2
G	the boundary position of Areas 2 and 3	(°)
H	the slope of Area 2	dB/(°)
I	average BS at the center of Area 3	dB
J	the slope of Area 3	dB/(°)

Kirchhoff model near the vertical incident angle region and it approximates Lambert's law in other angular regions.

2. **Nonlinear LS fitting model**

The variation model of the average BS against incident angle is given above. For such a nonlinear model, the parameters can be fitted based on the LS principle, for which the mathematical model is

$$\min_{x}\frac{1}{2}\|F(x,x\text{data})-y\text{data}\|_2^2=\frac{1}{2}\sum_{i=1}^{n}(F(x,x\text{data}_i)-y\text{data}_i)^2 \tag{8.106}$$

where *x*data is the data vector constructed by the parameters to be fitted, *y*data is the observation vector corresponding to *x*data, *F*(*x*, *x*data) is the objective function, *x* is the coefficient to be fitted, and *n* is the length of the observation vector. Thus, the solution of function fitting is transformed into an unconstrained nonlinear programming problem. To solve the problem, we use the lsqcurvefit function provided in the MATLAB software.

In the computation result, it is found that there are systematic errors or fluctuations in the measured mean BS curve and the estimation errors of the parameters are relatively large, which will affect the accuracy of the seafloor classification. Thus, the LS fitting parameter method is used based on Hellequin's model. Experimental results show that the characteristic parameters extracted by this method can describe the significant variation characteristics of the curve and they are insensitive to small systematic errors (fluctuations).

8.5 Methods for Acoustic Seafloor Classification

Section 8.4 discussed how to extract acoustic features of the substrate from a single beam time series backscatter signal and sonar image. In this section, we introduce how to segment these features into clusters with different seafloor characteristics. Various seafloor classification methods have been proposed previously, which include among others the Bayes maximum likelihood classification method used in Simrad's software Triton, a neural classification method based on GA-FAMNN and LVQ established by Zhou (2005) and Tang et al. (2007). From the viewpoint of learning methods, the existing methods can be divided into two categories: supervised and unsupervised learning methods.

A supervised learning method learns the target pattern based on a training set with known label tags and it usually has high classification accuracy. However, a large amount of ground truth data is needed for the training step, which is often difficult to realize (e.g., in a deep-sea setting). Conversely, unsupervised learning methods segment the data into clusters without the help of ground truth data. Within each cluster, the statistical properties of the data are similar. Thus, when the number of ground truth sampling sites is small, the unsupervised learning method is often found more suitable. Three classification methods are considered in this section: the self-organizing feature map (SOM) network, neural networks based on genetic wavelet and the support vector machine (SVM) and *k*-means algorithm.

8.5.1 Seafloor Classification Based on a SOM Network

The SOM network was proposed by Kohonen (1982). The network is an unsupervised self-organizing and self-learning network composed of a fully connected neuron array.

8.5.1.1 Network Model

The SOM network has only two layers: an input layer and a competition layer (output layer). Nodes on the competition layer are interconnected by a set of weights that are adjusted according to the minimum (or maximum) distance criterion (Fig. 8.52).

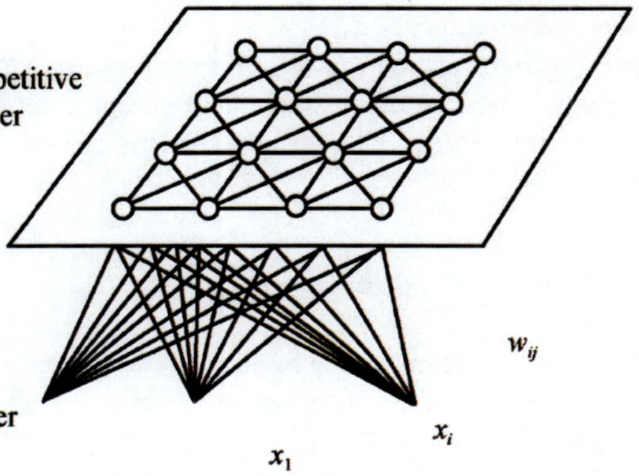

Fig. 8.52 Example of a SOM network

A major advantage of a SOM network is that it can form the characteristic topological distribution of input signals on a one-dimensional or two-dimensional processing array. Therefore, the SOM network has the ability to extract the pattern of input signals.

8.5.1.2 Learning Algorithm

As an unsupervised learning method, the SOM network can transform the input mode of any dimension into a one-dimensional or two-dimensional discrete space in a topological and orderly manner. This transformation from the input space (H) to the output space (A) is known as feature mapping. H is an assembly of input vectors the dimension of which is equal to the input vectors, while A is a two-dimensional self-organizing map.

The SOM network algorithm has three steps: competition, cooperation and adaptation.

(1) In the competition process, the neuron with the largest output is chosen as the winning neuron. Owing to the linear excitation function of the neurons, the maximum output of a neuron depends on the weighted sum of the input $u_i = \sum_{j=1}^{N} w_{ij}x_j$ (the inner product of the input vector $\boldsymbol{X} = [x_1, x_2, \ldots, x_N]^{\mathrm{T}}$ and the weight vector $\boldsymbol{W}_i = [w_{i1}, w_{i2}, \ldots, w_{iN}]^{\mathrm{T}}$ ($i = 1, 2, \ldots, M$)). Given the input vector $\boldsymbol{X}$ and the cth neuron as the winner, the following conditions are met:

$$\|\boldsymbol{X} - \boldsymbol{W}_c\| = \min_i \|\boldsymbol{X} - \boldsymbol{W}_i\|, \quad i = 1, 2, \cdots, M \qquad (8.107)$$

where $\|\cdot\|$ represents the Euclidean distance of input vector $\boldsymbol{X}$ and weight vector $\boldsymbol{W}_i$.

(2) In the cooperation process, the strengthening center of the winning neuron is determined. The center of the topological neighborhood is the winning neuron obtained in the competitive process.

Here, a simple square neighborhood shape is adopted. When the neighborhood radius is 0, the neighborhood contains only the winning neuron. When the radius is 1, the neighborhood contains the eight neighboring neurons near the winning neuron. When the radius increases, the neighborhood area enlarges according to this rule. The topological neighborhood is denoted as $N_c(n)$, which means that the radius of the topological neighborhood is at the nth iteration. Obviously, its value changes with the number of iterations and its change rule can be described by Eq. (8.108):

$$N_c(n) = INT(N_c(0)(1 - n/N)), \quad n = 0, 1, \cdots, N \qquad (8.108)$$

where $N_c(0)$ represents the initial topological neighborhood radius, N is the total number of iterations and $INT(\cdot)$ represents the integer function. Thus, the topological neighborhood shrinks with an increase in the number of iterations.

(3) During the adaptation process, Hebb learning rules are adopted to update the weight vector of the neurons in the topological neighborhood of the winning neuron:

$$W_i(n+1) = W_i(n) + \eta(n)(X - W_i(n)), \quad n = 0, 1, \ldots, N \qquad (8.109)$$

where $\eta(n)$ refers to the learning rate ($0 < \eta(n) < 1$), which decreases with an increase in the number of iterations, as given by Eq. (8.110):

$$\eta(n) = \eta(0)\left(1 - \frac{n}{N}\right), \quad n = 0, 1, \ldots, N \qquad (8.110)$$

where N is the total number of iterations and $\eta(0)$ is the initial learning rate.

8.5.1.3 Classification Process

The classification process using a SOM is shown in Fig. 8.53.

1. **Input the characteristic parameter vector** $\boldsymbol{X}$

Extraction of the characteristic parameters A, a, B and β is based on the averaged ARC, and the input mode of the network is determined as

$$\boldsymbol{X}_i = (10\log_{10} A, \ \gamma, \ 10\log_{10} B, \ \beta)_i \quad i = 1, 2, \ldots, k \qquad (8.111)$$

where k is the total number of samples.

2. **Network structure settings**

The structure of the competitive layer is set as $S = S_1 \times S_2$. The number of neurons (S) represents the number of types

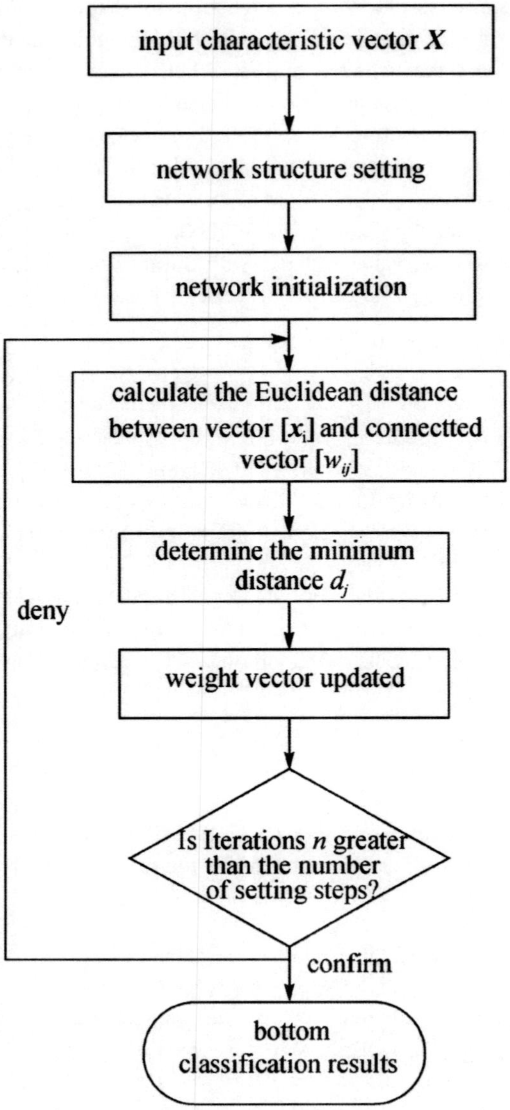

Fig. 8.53 Flow chart of seafloor classification based on a SOM network

into which the input vector could be classified. However, if the number selected is too large, many nodes could be idle after competition, and a large amount of design work and time would be wasted. Therefore, the appropriate number should be determined according to the dimension of the input vector and its estimation. Note that the network structure can be integrated. As the number of different seafloor types is often small in an acoustic seafloor classification application, a one-dimensional network structure is often an appropriate choice, i.e., $S_2 = 1$. In addition, the size of the training steps also affects the clustering performance of the network. Here, the number of training steps is set separately as 10, 100, 500 and 1 000 to assess the classification performance.

3. **Network initialization**

The weight vector with a small random value is initialized using a set of randomly chosen small values. After the initial learning rate $\eta(0)$ is set, the initial weight vector $\boldsymbol{W}_i(0)$ and the input vectors $\boldsymbol{X}$ should be normalized:

$$\boldsymbol{X}' = \frac{\boldsymbol{X}}{\|\boldsymbol{X}\|} \tag{8.112}$$

$$\boldsymbol{W}'_i(0) = \frac{\boldsymbol{W}_i(0)}{\|\boldsymbol{W}_i(0)\|} \tag{8.113}$$

where $\|\boldsymbol{W}_i(0)\| = \sum_{j=1}^{M}\left[w_{ij}(0)\right]^2$ and $\|\boldsymbol{X}\| = \sum_{i=1}^{N}(\boldsymbol{X}_i)^2$. The normalization of weights and vectors is undertaken to ensure that the winning neuron selected based on the minimum Euclidean distance criterion has the maximum output.

4. **Network training competition**

(1) The SOM network calculates the Euclidean distance d_j between w_{ij} and $\boldsymbol{X}$ in the training process:

$$d_j = \sum_{i=1}^{N}(x'_i - w_{ij})^2 \tag{8.114}$$

(2) The minimum distance d_g is found to determine the winning neuron g:

$$d_g = \min[d_j] \tag{8.115}$$

(3) After the competition, the weights of the nodes connected to the winning node are adjusted. The purpose of this step is to decrease the difference between the weights and the input vector such that the weights can represent the characteristics of the corresponding input vector. Thus, similar input vectors are clustered into the same class. The formula for the correction of the SOM weights is

$$\Delta w_{ij} = lr(x'_i - w_{ij}) \tag{8.116}$$

where lr is the learning rate ($0 < lr < 1$, often chosen to be 0.01–0.3) and x'_i is the input vector after normalization.

(4) Iterate Steps (1)-(3). After repeated training and modification, the resultant w_{ij} is the averaged value of several input modes of different types. When there is a different type of pattern input, other nodes in the network competition layer will win and a new weight matrix is generated. By repeating this until the maximum number of training steps is reached, all the different types of patterns can be aggregated into different weight matrices with each weight matrix representing one type.

5. **Sediment classification results**

For each input vector, a SOM network will generate a corresponding classification output. Thus, output with the same categories will be assigned the same color and a seafloor classification map can be obtained.

8.5.2 Seafloor Classification Based on *K*-Means Algorithm

8.5.2.1 Classification Principle

MacQueen (1967) proposed the *k*-means algorithm to distribute samples into clusters with different centers (means). As a popular unsupervised learning algorithm, *k*-means is a repartitioning strategy based on iteration. When the iteration is complete, the data set will be divided into *k* (a predefined number) clusters. In the iteration process, the Euclidean distance between each data point and the clustering center is continuously optimized. For the data point set $X = \{x_1, \ldots, x_N\}$, the *k*-means algorithm gives a partition of data set $\{X_l\}^k\ l = 1{:}k$. Thus, if $\{C_1, \ldots, C_k\}$ is used to represent *k* partition centers, the goal is to optimize the following objective function:

$$E = \sum_{l=1}^{k} \sum_{x_i \in X_l} \|x_i - C_l\|^2 \quad (8.117)$$

The simplest *k*-means algorithm consists of the following three steps.

1. Divide all samples into *k* initial clusters.
2. Allocate a sample in the sample set into the cluster whose center is the closest. Recalculate the centers of clusters.
3. Repeat Step 2 until no samples can be redistributed.

In the above process, Step 1 can also start from the *k* predefined initial centers (without the need to divide *k* initial clusters), following which Step 2 can be carried out. A flow chart of the algorithm is shown in Fig. 8.54.

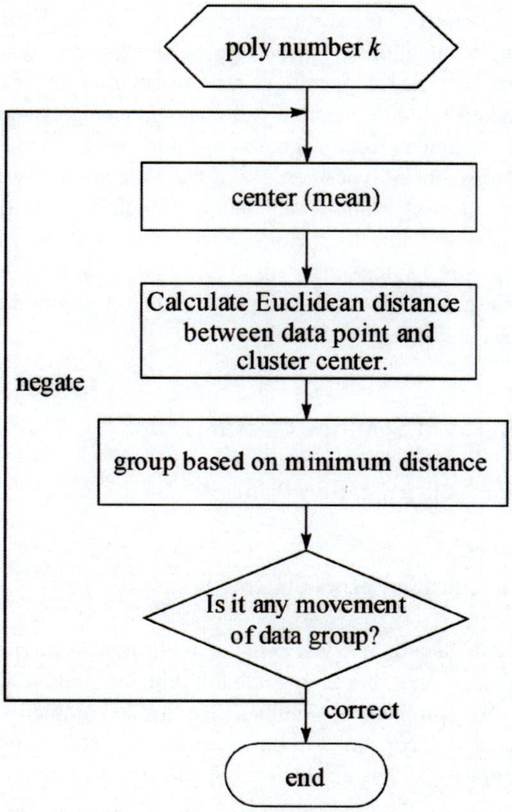

Fig. 8.54 Flow chart of *k*-means algorithm

8.5.2.2 Determination Principle of Clustering Number *K*

In seafloor classification based on the k-means algorithm, the number *k* is a predefined variable that represents the types of seafloor to be segmented. To determine the classification number *k* value accurately, we introduce two principles.

1. **Using the Φ standard grain classification table as reference**

To determine the value of *k*, ground truth data of the survey area are compared with the Φ standard classification table. The *k* value can be chosen according to the number of overlapped seafloor types that occur in both the ground truth data and the first-class groups in the Φ standard classification table.

2. **Assessing regional variation of different *k* settings**

The problem of choosing an appropriate *k* value can be addressed from a phenomenon perspective. For a chosen *k* value, we can evaluate its suitability by observing the regional effect of the seafloor classification results when k is increased intentionally. From the perspective of hydrodynamics and marine geology, the distribution of seafloor sediment should be regional and it should change smoothly. There is often a certain degree of transition between different seafloor sediment types because of the correlation of various factors affecting seafloor sediment distribution. Therefore, if k is increased and the regional effect of the seafloor classification results remains satisfactory, then using a larger k value can be justified; otherwise, it is not appropriate to increase the k value.

8.5.3 SVM Seafloor Classification

8.5.3.1 Classification Principle

1. **Classification processing mode**

The basic idea of an SVM is that classification scenarios can be divided into linear separable problems and nonlinear separable problems. Nonlinear separable problems can eventually be converted to linear separable problems using a hyperplane in high-dimensional space. Suppose we have a linearly separable data set: $\left\{(\vec{X}_1, y_1), (\vec{X}_2, y_2), \ldots, (\vec{X}_N, y_N)\right\}$ and the flag ensemble is $y \in \{-1, +1\}$ (there are only two types of samples). To separate the data set, an optimal hyperplane equation $\vec{W}^{\mathrm{T}} \cdot \vec{X} + b = 0$ should be determined, and we define the sample points closest to the hyperplane to be $\vec{X}_k$ (these points are called support vectors).

Samples in the positive side of the hyperplane are defined as positive examples (denoted as +1, i.e., the sample point of the class label is y_i) and the samples in the negative side of the hyperplane are defined as negative samples (denoted as –1). Therefore, we have

$$\text{ensemble}: \quad (\vec{W}^{\mathrm{T}}\vec{X}_i + b) \geq 1 \tag{8.118}$$

$$\text{counter ensemble}: \quad (\vec{W}^{\mathrm{T}}\vec{X}_i + b) < 1 \tag{8.119}$$

Thus, linear separable problems can be easily classified.

To solve a nonlinear separable problem, the relaxation variable $\varepsilon_i > 0$ is introduced and the constraint condition becomes

$$y_i(\vec{W} \cdot \vec{W}_i + b) \geq 1 - \varepsilon_i, \quad y_i = +1 \ \ (x \text{ is positive}) \tag{8.120}$$

$$y_i(\vec{W} \cdot \vec{W}_i + b) \leq 1 - \varepsilon_i, \quad y_i = -1 \ \ (x \text{ is opposite}) \tag{8.121}$$

When $\varepsilon_i = 0$, the regression is a linearly separable problem. When $0 < \varepsilon_i < 1$, the constraint conditions allow samples to fall within the classification interval, and the samples can be correctly classified by hyperplane. For $\varepsilon_i > 1$, the constraint allows for misclassification. For samples of multiple types, an SVM still adopts a dichotomy, except that the value of the class tag is no longer +1 and –1, but instead multiple class tags are used.

For classification of multiple types, an SVM has three processing methods: one against all, one against one with voting and one against one with elimination. The classification effect of the three methods is similar. Thus, this volume only discusses an SVM with the one against one with voting processing method.

2. **One against one with voting method**

For the above data set, when the samples are paired to form different training sets, there will be $\frac{n(n-1)}{2}$ SVM classifiers. During the test, the test samples are input successively into the $\frac{n(n-1)}{2}$ classifiers for voting. The specific process is as follows.

(1) Initialization: $vote(\vec{X}_1) = vote(\vec{X}_2) = \cdots = vote(\vec{X}_n) = 0$.
(2) If the classifier obtained by training set $(\vec{X}_1, \vec{X}_2)$ classifies $\vec{X}$ into class $\vec{X}_1$, then $vote(\vec{X}_1) = vote(\vec{X}_2) + 1$; otherwise, $vote(\vec{X}_2) = vote(\vec{X}_2) + 1$. If the classifier obtained by set $(\vec{X}_{n-1}, \vec{X}_n)$ classifies $\vec{X}$ into class $\vec{X}_{n-1}$, then $vote(\vec{X}_{n-1}) = vote(\vec{X}_{n-1}) + 1$; otherwise, $vote(\vec{X}_n) = vote(\vec{X}_n) + 1$.
(3) Final decision: classifier with the largest value will be chosen as the result.

8.5.3.2 Classification Process

It is first necessary to correct the BS data to extract features that reflect solely the properties of the seafloor, as discussed in Section 8.3. Note that to achieve fast convergence in the

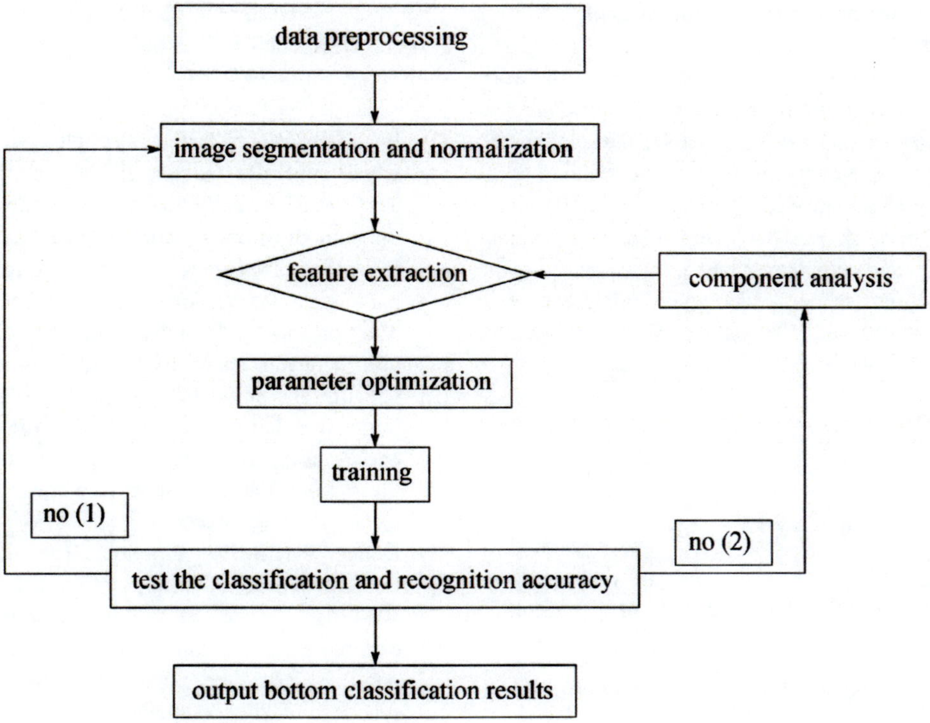

Fig. 8.55 Flow chart of a SVM classification process

training process, sonar images must first be normalized. In this respect, Tang et al. (2006) systematically summarized the influencing factors of sonar data correction. In essence, executing seafloor classification using sonar data directly is equivalent to converting the sonar data into images and then executing the classification subsequently.

In the process of an SVM-based classification, the selection of appropriate parameters is extremely important, and this mainly includes the selection of the error cost coefficient c (penalty parameter) and kernel function parameter g. The selection of sample feature vectors also has considerable impact on training performance. In the training process, the selection of feature vectors should be repeatedly tested. First, the dimension of the training data should be reduced as much as possible. The training speed will decrease markedly as the dimension of the feature vector increases, and the training complexity will be reversed. Second, through PCA and repeated testing, it is possible to determine the contribution of each eigenvalue during the classification. Thus, only those PCA-generated eigenvectors with large eigenvalues should be selected in the following steps. Third, the size of the image segmentation resolution will also affect classification accuracy. It is difficult for a low-resolution map to reflect detailed seafloor features, whereas a map with a resolution that is too high is unable to generate sufficient training samples. The overall classification process is shown in Fig. 8.55. In the process of training, a one-to-one voting strategy is adopted and the algorithm cycle is divided into two steps. The first step tests whether the image segmentation is reasonable. If it is considered acceptable, the computation turns to the second step; otherwise, the segmentation process restarts. The second step aims to test whether the extraction of feature vectors meets the high-quality classification requirements. Step 2 is halted when different seafloor types can be distinguished effectively from the chosen features; otherwise, the extraction of eigen features restarts.

8.5.4 Seafloor Classification Based on Genetic Wavelet Neural Network

A genetic wavelet neural network (G-WNN) effectively combines the advantages of genetic algorithms and wavelet neural networks. G-WNNs have been used successfully in the fields of facial image recognition (Jiang and Li 2010), voice recognition (Han et al. 2010), fault diagnosis (Liu 2009), GPS applications (Lin and Ju 2011), multisensor information fusion technology and submarine pipeline detection (Gao et al. 2007).

8.5.4.1 Construction of a Wavelet Neural Network

The classification problem of a wavelet neural network (WNN) is similar to that of a BP network, i.e., the parameters (weights and wavelet parameters) must be adjusted in the learning process until the network reaches the required accuracy. A WNN has a three-layer structure network, similar to the structure of a BP network (Fig. 8.56), which includes the input layer, hidden layer and output layer. There are N neurons in the input layer, L neurons in the hidden layer and M enterprise neurons in the output layer. Wavelet analysis is performed in the hidden layer, and the output of the hidden layer is

$$H(j) = f_j\left(\frac{\sum_{i=1}^{N} w_{ij}x_i - b_j}{a_j}\right), \quad j = 1,\ 2,\ \cdots,\ L \tag{8.122}$$

The result of the output layer is

$$Y(k) = \sum_{j=1}^{L} w_{jk}H(j), \quad k = 1,\ 2,\ \cdots,\ M \tag{8.123}$$

Weights and wavelet parameters are modified to

$$w(k+1) = w(k) + \Delta w \tag{8.124}$$

$$a(k+1) = a(k) + \Delta a \tag{8.125}$$

$$b(k+1) = b(k) + \Delta b \tag{8.126}$$

where the wavelet basis function used is $f(x) = e^{-x^2/2}\cos(1.75x)$.

8.5.4.2 Construction of a G-WNN

1. **Genetic algorithm**

A genetic algorithm is a random optimization search process (Shi et al. 2011) that is particularly suited to optimization problems. In view of the shortcomings of WNNs, the genetic algorithm is used here to obtain global optimization during seafloor classification. The basic steps of the genetic algorithm are as follows.

(1) Encoding. Binary coding is adopted with a length of 10, which generally satisfies all data length requirements. The size of the individuality is determined by analyzing the number of optimization parameters. The parameters to be optimized include network weights and wavelet parameters (expansion factor and translation factor). The final size of the individuality is 140 in the example used here (the length of the chromosome is 1 400).
(2) Creating the population. After chromosome coding, an initial population is created. According to test experience, the size of the population is generally 2/3 to 4/5 of the training data; the specific number chosen depends on the training effect.
(3) Define the fitness function. The fitness allocation function is used for sorting in this case. When training a WNN, it is expected that the error between the predicted value and the expected value is as small as possible. The fitness value is chosen as the inverse function of the residuals. The smaller the function value, the larger the fitness value will be and the better the corresponding individual. Thus, the calculation formula of the fitness value is $F = 1/f(x)$, where $f(x)$ is the calculated residual value.
(4) Option. In the option step, individuals in the old population are selected for the new population based on a certain probability. The probability of an individual being selected is determined by its fitness value. The greater its fitness value, the greater the probability of that individual being selected. In this study, a generation gap of 0.05 is adopted.
(5) Crossover or recombination. The new individual populations with acceptable adaptability selected above are paired and mapped for hybridization, with a crossover probability of 0.7, to obtain the new population.
(6) Variation. The variation step generates a number of mutation genes based on a certain mutation probability.

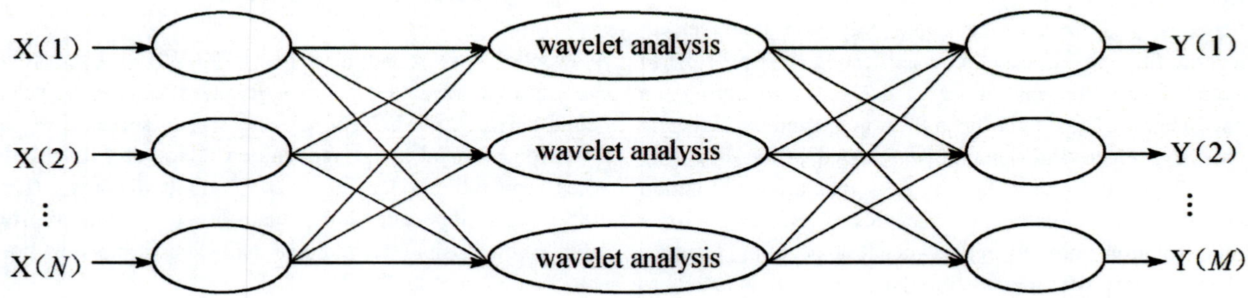

Fig. 8.56 Structural diagram of a WNN

In this study, the mutation gene is selected using a random method with a mutation probability of 0.01. Based on the principle of survival of the fittest, eventually "excellent" offspring are obtained. The replacement of a father by a child is repeated until the inheritance is completed.

2. **Wavelet analysis**

Wavelet analysis can represent the local characteristics of a signal in the domains of both time and frequency. To combine neural network and wavelet analyses, one choice is to use wavelet analysis for data preprocessing and feature extraction, and then to send the extracted feature vector to the neural network for processing. Another choice is the so-called wavelet network, which organically combines the wavelet transform and neural network to make full use of the advantages of both (Chen and Feng 1999). Wavelet analysis can be executed in a multiscale way, by adjustment of the scaling and translation factors, to achieve the purpose of local feature analysis. Thus, different seafloor characteristics can be distinguished effectively at a small scale. The scale factor and shift factor are sometimes learned separately to accommodate the different characteristics of different seafloor types. In terms of choosing a suitable learning step length, two study factors (with the same step length) are tested in this chapter. In the learning process, the gradient correction method is used to correct the wavelet parameters.

In terms of the training function, the wavelet basis function is introduced to ensure strong learning ability and generalization capability. The selection of a suitable wavelet base is currently a focus of research. With respect to the problem of acoustic seafloor classification, experiment results for this chapter indicate that the Morlet mother wavelet represents an acceptable choice.

8.5.4.3 Classification Process

For this study, three layers of network structure are chosen for the construction of a WNN. The output layers all have 6 neurons and each hidden layer has 10 neurons (Figs. 8.56 and 8.57). Figure 8.57 shows the genetic process of the WNN classification. In the process, two circulations are used. The first circulation is the optimization of parameters. The evolution characteristic of the genetic algorithm is used

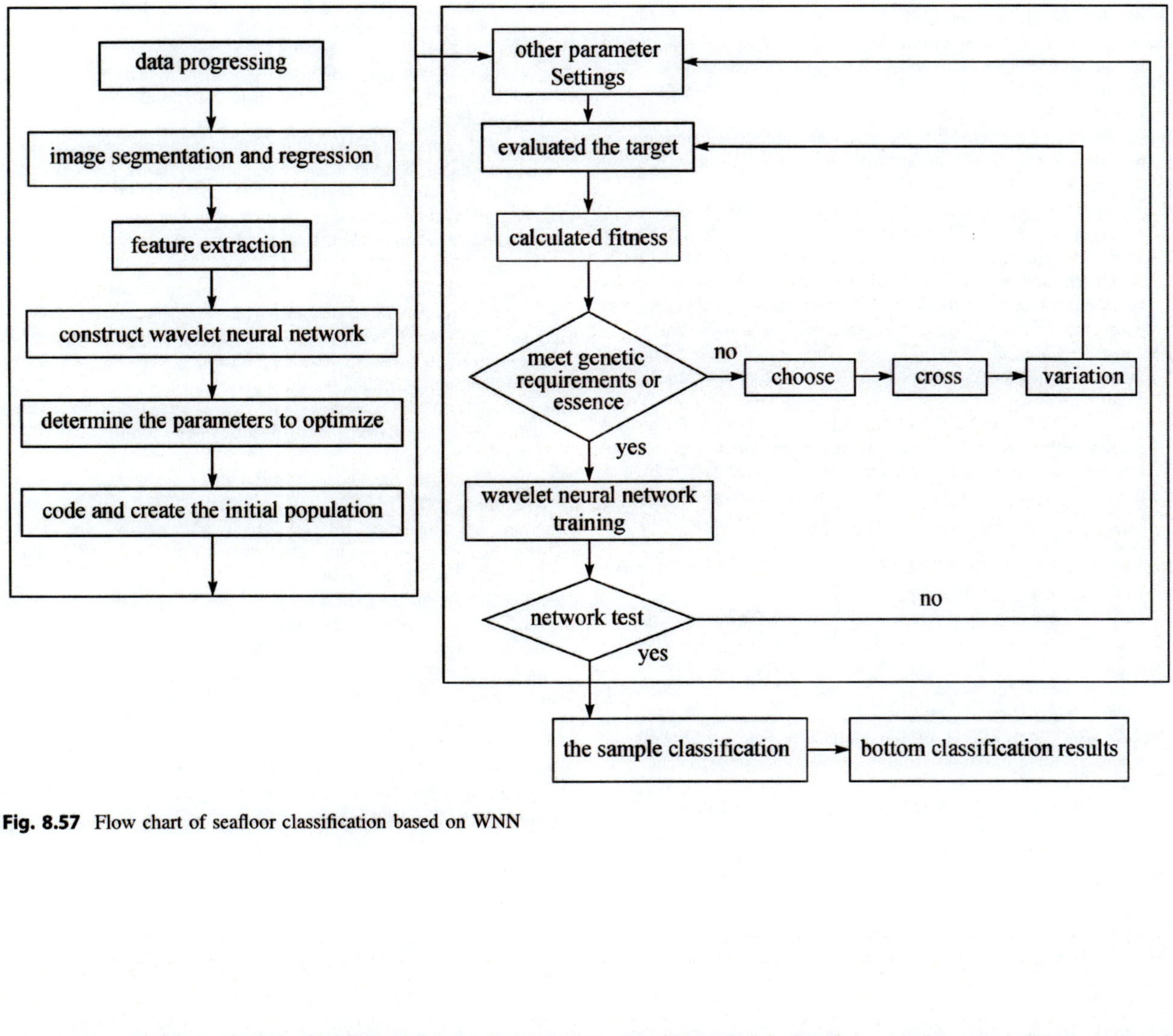

Fig. 8.57 Flow chart of seafloor classification based on WNN

to choose the best network weights and wavelet parameters. The second circulation deals mainly with parameter setting when the test result is inadequate. For example, unsuitable learning rates will result in non-convergence in the network training step. Thus, an appropriate value must be chosen such that ideal training results can be obtained during the training step. Similarly, different substrates require different learning step sizes, different network weights and different wavelet parameters. For example, mud is sensitive to wavelet parameters and therefore the step sizes must be adjusted accordingly. To this end, automatic adjustment is adopted in the training step in this study. At the end of the genetic iteration process, the coding parameters must be decoded and input into the training network. Afterward, the chosen parameters are saved for further seafloor classification applications.

References

Anderson JT, Gregory RS, Collins WT (2002) Acoustic classification of marine habitats in coastal Newfoundland. ICES J Mar Sci 59 (1):156–167. https://doi.org/10.1006/jmsc.2001.1126

APL-UW TR 9407 (1994) High-frequency ocean environmental acoustic models handbook. Applied Physics Laboratory, University of Washington, October 1994

Augustin JM, Dugelay S, Lurton X et al (1997) Applications of an image segmentation technique to multibeam echo-sounder data. Paper presented at the IEEE oceans conference, Halifax, NS, Canada, 6–9 October 1997

Bachman RT (1985) Acoustic and physical property relationships in marine sediment. J Acoust Soc Am 78(2):616–621. https://doi.org/10.1221/1.392429

Bachman RT (1989) Estimating velocity ratio in marine sediments. J Acoust Soc Am 86:2029–2032. https://doi.org/10.1221/1.398585

Beaudoin JD, Hughes Clarke JE, Van Den Ameele EJ et al (2002) Geometric and radiometric correction of multibeam backscatter derived from Reson 8101 systems. Paper presented at the Canadian hydrographic conference, Ottawa, Ontario, 28–31 May 2002

Boehme H, Chotiros NP (1988) High-frequency acoustics bottom interaction. J Acoust Soc Am 84(S1):S120–S120. https://doi.org/10.1121/1.2025720

Boehme H, Chotiros NP, Rolleigh LD et al (1985) Acoustical backscattering at low grazing angles from the ocean bottom. J Acoust Soc Am 77:962–974. https://doi.org/10.1121/1.392064

Boulinguez D, Quinquis A (2002) 3-D underwater object recognition. IEEE J Ocean Eng 27(4):814–829. https://doi.org/10.1109/joe.2002.805097

Brekhovskikh LM, Godin OA (1990) Acoustics of layered media, plane and quasi-plane waves. Springer, Berlin

Breslau LR (1965) Classification of seafloor sediments with a shipborne acoustical system. Le petrole et La Mer 132:1–9

Chen Z, Feng TJ (1999) Research developments and prospects of wavelet neural networks (in Chinese with English abstracts). J Ocean Univ Qingdao 29(4):663–668

Collins MB, Voulgaris G (1993) Empirical field and laboratory evaluation of a real-time acoustic seabed surveying system. Proc Inst Acoust 15:343–351

Collins WT, Rhynas KP (1998) Acoustic seabed classification using echo sounders: operational considerations and strategies. Paper presented at the Canadian hydrographic conference, Victoria BC, Canada, 17–20 April 1998

Cui GS, Liu YC, Wang R (2003) Status and prospect of marine sediment survey (in Chinese with English abstracts). J Dalian Naval Acad 26(5):36–40

Davis A, Haynes R, Bennell J et al (2002) Surficial seabed sediment properties derived from seismic profiler responses. Mar Geol 182:209–223. https://doi.org/10.1016/S0025-3227(01)00235-3

Du DW, Wang N, Zhou XH (2006) Interdisciplinary study of acoustics and marine sedimentology (in Chinese with English abstracts). Prog Mar Sci 24(3):392–396. https://doi.org/10.3969/j.issn.1671-6647.2006.03.019

Dyer C, Murphy K, Heald G et al (1997) An experimental study of sediment discrimination using 1st and 2nd echoes. Paper presented at the SACLANTCEN conference proceedings series CP-45, high frequency acoustics, La Spezia

Farbman Z, Fattal R, Lischinski D et al (2008) Edge-preserving decompositions for multi-scale tone and detail manipulation. Paper presented at the ACM transactions on graphics, Los Angeles, California, 11–15 August 2008

Francois RE, Garrison GR (1982a) Sound absorption based on ocean measurements: part I: pure water and magnesium sulfate contributions. JASA 72:896–907. https://doi.org/10.1221/1.388170

Francois RE, Garrison GR (1982b) Sound absorption based on ocean measurements: Part II: Boric acid contribution and equation for total absorption. JASA 72:1879–1890. https://doi.org/10.1121/1.388673

Gao MJ, Tian JW, Li K (2007) Research on detecting method of submarine oil pipelines corrosion degree based on chaos genetic algorithm neural network. Paper presented at the eighth ACIS international conference on software engineering, artificial intelligence, networking, and parallel/distributed computing, Qingdao, China, 30 July-1 August 2007

Gensane M (1989) A statistical study of acoustic signals backscattered from the sea bottom. IEEE J Ocean Eng 14:84–93. https://doi.org/10.1109/48.16818

Gonidec YL, Lamarche G, Wright CI (2003) Inhomogeneous substrate analysis using EM300 backscatter imagery. Mar Geophys Res 24:311–327. https://doi.org/10.1007/s11001-004-1945-9

Guo YG (2004) Research and application of seabed acoustic parameter inversion (in Chinese with English abstracts). Dissertation, Ocean university of China

Hamilton EL (1956) Low sound velocities in high-porosity sediments. J Acoust Soc Am 28:16–19. https://doi.org/10.1121/1.1908208

Hamilton EL (1963) Sediment sound velocity measurements made *in-situ* from bathyscaph Trieste. J Geophys Res 68:5991–5998. https://doi.org/10.1029/jiz068i021p05991

Hamilton EL (1972) Compressional wave attenuation in marine sediments. Geophysics 37:620–646. https://doi.org/10.1190/1.1440287

Hamilton EL (1980) Geoacoustic modeling of the seafloor. J Acoust Soc Am 68:1313–1340. https://doi.org/10.1121/1.385100

Hamilton EL, Bachman RT (1982) Sound velocity and related properties of marine sediments. J Acoust Soc Am 72:1891–1904. https://doi.org/10.1121/1.388539

Hamilton LJ, Mulhearn PJ, Poeckert R (1999) Comparison of roxann and QTC-View acoustic bottom classification system performance for the Cairns area, Great Barrier Reef, Australia. Cont Shelf Res 16:1577–1597. https://doi.org/10.1016/s0278-4343(99)00020-5

Hammerstad E (2000) EM technical note, backscattering and seabed image reflectivity. Available via DIALOG. https://www.manualslib.com/manual/381545/Simrad-Em-Technical-Note-2000.html#manual. Accessed 2 July 2019

Hampton L (ed) (1974) Physical of sound in marine sediments. Plenum Press, New York and London

Han ZY, Wang J, Lun SX (2010) Design of speech recognition classifier based on genetic wavelet neural network (in Chinese with English abstracts). Comput Sci 37(11):243–246. https://doi.org/10.3969/j.issn.1002-137X.2010.11.058

Hellequin L, Boucher JM, Lurton X (2003) Processing of high-frequency multibeam echo sounder data for seafloor characterization. IEEE J Ocean Eng 28(1):78–88. https://doi.org/10.1109/joe.2002.808205

Hughes Clarke JE (2005) Multibeam training course notes. Paper presented at the swath sonar training course given as part of the Canadian Hydrographic Conference, Ottawa, Ontario, May 2005

Hughes Clarke JE, Danforth BW, Valentine P (1997) Areal seabed classification using backscatter angular response at 95 kHz. Available via DIALOG. https://openlibrary.cmre.nato.int/bitstream/handle/123456789/428/CP-45_HughesClarke_ArealSeabedClassification.pdf?sequence=1&isAllowed=y. Accessed 2 July 2019

Jackson DR, Baird AM, Crisp JJ et al (1986) High-frequency bottom backscatter measurements in shallow water. J Acoust Soc Am 80 (4):1188–1199. https://doi.org/10.1121/1.393809

Jackson DR, Briggs KB (1992) High-frequency bottom backscattering: roughness vs. sediment volume scattering. J Acoust Soc Am 92:962–977. https://doi.org/10.1121/1.403966

Jackson DR, Briggs KB (1996) Tests of models for high-frequency seafloor backscatter. IEEE J Ocean Eng 21(4):458–470. https://doi.org/10.1109/48.544057

Jackson DR, Richardson MD (2007) High-frequency seafloor acoustics. Springer, New York

Jakeman E, Pusey PN (1978) Significance of K distributions in scattering elements. Phys Rev Lett 40(9):546–550. https://doi.org/10.1103/physrevlett.40.546

Jiang YJ, Li X (2010) A Method for Face Recognition Based on Wavelet Neutral Network. In: Paper presented at the Second WRI Global Congress on Intelligent Systems, Wuhan, China, 16–17 December 2010

Jones CD, Jackson DR (1997) Temporal fluctuations of scattered field due to bioturbation in marine sediments. J Acoust Soc Am 100 (4):2713. https://doi.org/10.1121/1.416122

Kohonen T (1982) Self-organized formation of topologically correct feature maps. Biol Cybern 43:59–69. https://doi.org/10.1007/BF00337288

Lin XY, Ju JB (2011) GPS/SINS integrated navigation algorithm based on neural network prediction (in Chinese with English abstracts). Geomat Inf Sci Wuhan Univ 36(5):601–604

Liu MR (2009) The research on method of fault diagnosis for analog circuits based on genetic algorithm, wavelet analysis and neural networks (in Chinese with English abstracts). Dissertation, Central South University

Liu YC (2003) Spatial structure of ocean sounding and data processing (in Chinese). Surveying and Mapping Press, Beijing

Lu B (1997) Preliminary study on physical properties of shallow sediments in Nanshan qundao (in Chinese with English abstracts). Sci China (Ser D) 27(1):77–81

Lurton X (2002) An introduction to underwater acoustics. Praxis Publishing, UK

Lurton X, Dugelay S, Augustin JM (1994) Analysis of multibeam echo-sounder signals from the deep seafloor. Paper presented at the IEEE oceans conference, Brest, France, 13–16 September 1994

Lyons AP, Abraham DA (1999) Statistical characterization of high-frequency shallow-water seafloor backscatter. J Acoust Soc Am 106:1307–1315. https://doi.org/10.1121/1.428034

McCann C, McCann DM (1969) The attenuation of compressional waves in marine sediments. Geophysics 34:882–892. https://doi.org/10.1190/1.1440059

McCann DM (1972) Measurement of acoustic properties of marine sediments. Acoustica 26(2):55–66. https://doi.org/10.1109/tau.1973.1162481

Mckinney CM, Anderson CD (1964) Measurements of backscattering of sound from the ocean bottom. J Acoust Soc Am 36(1):158–163. https://doi.org/10.1121/1.1918927

Michalopoulou ZH, Alexandrou D, De MC (1995) Application of neural and statistical classifiers to the problem of seafloor characterization. IEEE J Ocean Eng 20(3):190–197. https://doi.org/10.1109/48.393074

Oliver C (1984) A model for non-rayleigh scattering statistics. Opt Acta 31(6):701–722. https://doi.org/10.1080/713821561

Pouliquen E, Lyons AP (2002) Backscattering from bioturbated sediments at very high frequency. IEEE J Ocean Eng 27(3):388–402. https://doi.org/10.1109/JOE.2002.1040926

Quester Tangent (2005) QTC multiview user's manual and reference. Quester Tangent, Canada

Shi F, Wang H, Yu L et al (2011) MATLAB intelligent algorithm 30 case analyses. Beijing University of Aeronautics and Astronautics Press, Beijing

Shumway G (1956) A resonant chamber method for sound speed velocity and attenuation measurements in sediments. Geophysics 21:305–319. https://doi.org/10.1190/1.1438231

Shumway G (1958) Sound velocity versus temperature in water saturated sediments. Geophysics 23:494–505. https://doi.org/10.1121/1.1930017

Shumway G (1960) Sound speed and absorption studies in marine sediments by a resonance method. Geophysics 25(2):451–476. https://doi.org/10.1190/1.1438717

Skolnic MI (2001) Introduction to radar systems, 3rd edn. McGraw-Hill, New York

Stanic S, Briggs KB, Fleischer RI et al (1988) Shallow-water high-frequency bottom scattering off Panama City, Florida. J Acoust Soc Am 83(6):2134–2144. https://doi.org/10.1121/1.396341

Stanic S, Briggs KB, Fleischer P et al (1989) High-frequency acoustic backscattering from a coarse shell ocean bottom. J Acoust Soc Am 85:125–136. https://doi.org/10.1121/1.397720

Stanic S, Goodman RR (1998) Shallow-water bottom reverberation measurements. IEEE J Ocean Eng 23(3):203–210. https://doi.org/10.1109/48.701192

Subramaniam S, Barad H, Martinez AB (1993) Seafloor characterization using texture. Paper presented at the IEEE southeastcon, Charlotte, NC, USA, 4–7 April 1993

Tang QH (2003) Classification of multibeam seabed sediments (in Chinese with English abstracts). Dissertation, First Institute of Oceanography, State Oceanic Administration, Qingdao

Tang QH, Liu BH, Chen YQ et al (2007) Application of LVQ neural network combined with genetic algorithm in acoustic bottom classification (in Chinese with English abstracts). Acta Geophys Sin 50(1):313–319. https://doi.org/10.3321/j.issn:0001-5733.2007.01.039

Tang QH, Zhou XH, Ding JS et al (2006) Data processing of multibeam backscattering intensity (in Chinese with English abstracts). Acta Oceanol Sin 28(2):51–55. https://doi.org/10.3321/j.issn:0253-4193,2006.02.007

Tegowski J, Gorska N, Klusek Z (2003) Statistical analysis of acoustic echoes from underwater meadows in the eutrophic Puck Bay (southern Baltic Sea). Aquat Living Resour 16:215–221. https://doi.org/10.1016/s0990-7440(03)00015-9

Urick RJ (1954) The backscattering of sound from a harbor bottom. J Acoust Soc Am 26:231–235. https://doi.org/10.1121/1.1907314

Urick RJ (1962) Generalized form of the sonar equation. J Acoust Soc Am 34:547. https://doi.org/10.1221/1.1918166

Urick RJ (1983) Principles of underwater sound, 3rd edn. McGraw-Hill, New York

Williams KL, Jackson DR, Thorsos EI et al (2002) Acoustic backscattering experiments in a well characterized sand sediments: data/model comparisons using sediment fluid and Biot models. IEEE J Ocean Eng 27:376–387. https://doi.org/10.1109/JOE.2002.1040925

Zhai GJ, Huang MT (2009) Development of marine surveying and mapping in China (in Chinese with English abstracts). Mar Surv Mapp 29(4):74–81

Zhao DB (2009) Discussion on general methods of the grain-size classification and nomenclature of sediments (in Chinese with English abstracts). Mar Geol Lett 25(8):41–44

Zhou XH (2005) An approach to seafloor classification using fuzzy neural networks combined with a genetic algorithm (in Chinese with English abstracts). Dissertation, Hong Kong Polytechnic University

Zhu GW (2000) Review and prospect of 50 years' development of marine exploration technology in China (iii) (in Chinese with English abstracts). Mar Technol 19(1):23–31

9 Intelligent Detection and Recognition of Seabed Targets in Side-Scan Sonar Images

Artificial intelligence (AI) is a new technological science that studies and develops theories, methods, technologies and application systems for simulating, extending and expanding human intelligence. AI research includes robots, language recognition, image recognition, natural language processing, and expert systems. AI can simulate the information process of human consciousness and thinking. One of the most interesting examples of an AI application is Alpha Go and another is the developing driverless car. AI has been widely used in daily life, for example, the automatic recognition of vehicles in parking lots and toll stations, and automatic recognition of faces in important places.

Oceans, which account for about 70% of the earth's surface area, contain abundant resources and a large number of unsolved scientific problems. AI technology is an important approach that can be used to detect items in the ocean quickly. Side-scan sonar is an important seabed geomorphology detection technology, which can provide high-resolution underwater target images in a low-visibility underwater environment. Therefore, it is widely used in marine surveying, underwater search and rescue, mine detection, pipeline survey and other fields. For underwater target detection, side-scan sonar can be used to search for wrecked aircraft, shipwrecks and personnel; locate submarine pipelines; and detect prominent reefs, submarine ores, and military targets, such as submerged mines. At present, artificial methods are used to recognize and study seabed targets. They are inefficient and unreliable, and rely heavily on the level of knowledge of researchers. Based on actual need, the research status of seabed target detection and recognition is systematically discussed in this chapter, a feasible seabed target detection and recognition method is proposed, and an experimental application is developed.

9.1 Research Progress on Target Detection in Side-Scan Sonar Images

Side-scan sonar provides high-resolution images of marine environments, even in zero-visibility water, and thus has been widely used in many marine applications, such as ocean mapping, underwater search and rescue, mine detection, and offshore oil prospecting (Celik and Tjahjadi 2011). To detect objects lying on the seabed, three types of regions can typically be identified in side-scan sonar images (Mignotte et al. 1999, 2000; Ye et al. 2010; Celik and Tjahjadi 2011; Huo et al. 2017): highlight, shadow and sea-bottom reverberation. The highlight and shadow regions correspond to the reflection of the acoustic wave on an object and a lack of acoustic reverberation behind the object, respectively. So-called sea-bottom reverberation is caused by acoustic scattering from a rough sea bottom. Because the shape and size of the shadow or highlight region can be used for object classification and analysis, the extraction of the highlight and shadow regions is typically an important step before object classification and analysis (Mignotte et al. 1999, 2000). However, side-scan sonar image segmentation is a very challenging task (Celik and Tjahjadi 2011; Ye et al. 2010), mainly because of the strong speckle noise of sea-bottom reverberation.

Clustering techniques, such as fuzzy *k*-means, were first used in 1996 in sonar image segmentation by Guillaudeux et al. (1996) and Daniel et al. (1997). To reduce the effect of noise, clustering methods typically process the intensity mean and variance in windows of a specific size. Morphological filtering and a Markov random field (MRF)-based method (Mignotte et al. 1999, 2000), which can incorporate contextual information, have been proposed to further

Z. Wu et al., *High-resolution Seafloor Survey and Applications*,
https://doi.org/10.1007/978-981-15-9750-3_9

improve the segmentation results. The MRF-based segmentation method proposed by Mignotte et al. (1999) uses the Weibull probability density function (PDF) and an MRF to model the noise distribution and segmentation label field, respectively; adopts a maximum likelihood method for the noise model parameters and a least-squares method to estimate the MRF prior model; and implements a two-step process, which uses a scale-causal and spatial model, and an MRF monoscale model to extract the shadow and highlight regions separately. To precisely describe the global and local characteristics of noisy sonar images at different scales, Mignotte et al. (2000) proposed a hierarchical MRF model using a pyramid label field. By combining coarse-to-fine causal interactions with a spatial neighborhood structure, this method cleanly segments sonar images into two types of regions: shadow and sea-bottom reverberation. For specific mine target segmentation, Dura et al. (2008) first used an MRF model to segment the target shadow and then used the hyperelliptic fitting of the shadow shape to identify a mine target. Recently, Acosta and Villar (2015) proposed an accumulated cell average-constant false alarm rate in two dimensions (ACA-CFAR 2-D), which was applied to pipeline detection with great success.

Active contour models can obtain sub-pixel accuracy of object boundaries, easily incorporate various prior information (e.g., shape and intensity distribution) into an energy minimization model for robust segmentation, and provide closed and smooth boundaries that are necessary for shape analysis (Li et al. 2008). Because of these advantages, *X* active contour models have been extensively applied to contour detection and image segmentation (Ye et al. 2010; Li et al. 2008, 2010; Kass et al. 1988; Malladi et al. 1995; Caselles et al. 1997; Chan and Vese 2001; Reed et al. 2003; Lianantonakis and Petillot 2007; Yang et al. 2014, 2015; Gao et al. 2011; Balla-Arabé et al. 2013; Wang et al. 2014; Zhang et al. 2016). According to the representation and implementation of active contours, active contour models can be classified into two categories: parametric active contours (PAC) (or snakes) (Kass et al. 1988; Xu and Prince 1998; Reed et al. 2003) and geometric active contours (GACs) (Li et al. 2008, 2010; Malladi et al. 1995; Caselles et al. 1997; Chan and Vese 2001; Lianantonakis and Petillot 2007; Yang et al. 2014, 2015; Gao et al. 2011; Balla-Arabé et al. 2013). PACs are represented explicitly as parameterized curves, whereas GACs are represented implicitly as level sets (Xu et al. 2000). Using the level set method introduced by Osher and Sethian (1988), GACs easily process topological changes, which are difficult for PACs to cope with. Depending on the type of information used to drive the active contour toward the boundaries of distinct regions in the image, active contours can also be divided into two major classes: edge-based models (Kass et al. 1988; Xu et al. 1998; Caselles et al. 1997; Li et al. 2010; Gao et al. 2011; Yang et al. 2014) and region-based models (Chan and Vese 2001; Lianantonakis and Petillot 2007; Li et al. 2008; Balla-Arabé et al. 2013; Yang et al. 2015). Some popular region-based models are generally less sensitive to noise, but maybe more sensitive to intensity inhomogeneity, for example, the famous Chan-Vese model (Chan and Vese 2001) is sensitive to intensity inhomogeneity because it is based on the assumption that image intensities are statistically homogeneous in each region. To deal with intensity inhomogeneity, Li et al. (2008) proposed a region-scalable fitting (RSF) model. This model first defines local data fitting energy in terms of a contour and two fitting functions that locally approximate the image intensities on both sides of the contour, and then incorporates this energy into a variational level set formulation. A Gaussian kernel function is used for the data fitting energy, and by moving the window of the Gaussian kernel function, all the local regions in the image can be selected and described. The localization property of the RSF model helps it to achieve an accurate segmentation result in the presence of intensity inhomogeneity.

Active contour models have also been used for sonar image segmentation during the last decade. One efficient approach is to combine the MRF model or other techniques that can involve contextual or local information with active contour models. For mine detection and classification using side-scan sonar images, Reed et al. (2003) proposed a statistical snake model. This method first uses an unsupervised MRF model to directly segment the image into three regions and then incorporates a priori information into a statistical snake model to precisely extract the highlight and shadow regions of the mine-like object. For seabed segmentation, Lianantonakis and Petillot (2007) proposed a new method, which integrates a subset of the 14 Haralick features into the Chan-Vese model to efficiently segment sea-bottom reverberation into two distinct regions, such as sand ripples and flat seabed. To improve the accuracy and robustness of side-scan sonar segmentation, Ye et al. (2010) proposed a method that uses the Gauss-MRF (GMRF) model and an improved Chan-Vese model, which uses the GMRF model to remove speckle noise and accentuate the shadow and highlight regions, and introduces a level set regularization term into the multiphase Chan-Vese model to avoid the costly re-initialization of level sets. This method can achieve satisfactory results for extracting the shadow and highlight regions in noisy sonar images.

Because of the inhomogeneity of the sea bottom, and variations of the incident angle between the sonar beam and insonified surface caused by towfish rolling, intensity inhomogeneity might occur in side-scan sonar images and thereby create another difficulty. Geometric correction can only correct intensity inhomogeneity caused by towfish rolling, and it requires the attitude parameters of the towfish, which may be not available (Cobra et al. 1992).

Post-processing that takes advantage of the spatial dependency between echo and shadow areas can also help to eliminate some false segmentation (Mignotte et al. 1999). Another approach is to use more sophisticated segmentation models, such as the RSF model. To better cope with intensity inhomogeneity, the RSF model generally uses a small window, which makes it sensitive to level set initialization and noise. By considering both noise and intensity inhomogeneity, Li et al. (2011) proposed a robust level set model, which is faster and more robust than both the Chan-Vese model and RSF model. Similar to the RSF model, this model also adopts a Gaussian kernel function and is still an RSF model. However, unlike the RSF model, this model simultaneously estimates the bias field, updates the intensity bias-corrected image and uses the corrected image for segmentation at each iteration. The model has been successfully applied to magnetic resonance imaging (MRI) images; however, when it is used in the segmentation of sonar images that are noisier and coarser than MRI images, larger length controlling parameters have to be used, which may also make it sensitive to the initial position of active contours.

By penalizing the length of an active contour, active contour models can smooth the segmentation result. However, this may be insufficient in the segmentation of sonar images with strong speckle noise, which is why the method proposed by Ye et al. (2010) uses the GMRF model to remove speckle noise before segmentation. To achieve accurate segmentation, removing noise effectively without destroying edge details is essential. Wiener filtering, wavelet transform shrinkage (Mallat 2008; Donoho 1995), and a nonlocal filtering approach (Buades et al. 2005) are three typical image denoising methods. The classical Wiener filter is the optimum linear minimum mean-square-error estimator for noisy images that are wide-sense stationary. The wavelet transform can provide a sparse representation of images in the transform domain, where large detail coefficients correspond to object boundaries, whereas small detail coefficients mostly contain noise (Mallat 2008). Therefore, noise can be rejected with good preservation of edges using some form of coefficient thresholding, such as wavelet shrinkage (Donoho 1995). The nonlocal filtering approach, which depends on the observation that most images have clear self-similarities, was recently proposed for denoising from a different perspective (Buades et al. 2005). Image patches with similar features are identified and grouped together, and then filtering can be performed on the patch groups to effectively remove noise and better preserve edge details. This strategy leads to competitive results when compared with state-of-the-art methods, and has inspired several extensions, such as the block-matching and 3D filtering (BM3D) method (Dabov et al. 2007), where nonlocal filtering is combined with wavelet shrinkage and Wiener filtering. At present, BM3D is regarded as the best method to remove additive white Gaussian noise (Chatterjee and Milanfar 2010; Burger et al. 2012; Parrilli et al. 2012; Cozzolino et al. 2014). To better deal with speckle noise in synthetic aperture radar or ultrasound images, several nonlocal despeckling methods have also been developed (Coupé et al. 2008, 2009; Parrilli et al. 2012; Cozzolino et al. 2014). Based on the Bayesian formulation and general speckle model, Coupé et al. (2009) proposed a nonlocal means-based speckle filtering (NLMSF) method for ultrasound images, which can preserve the accurate edges and structural details of the image.

In this paper, to improve robustness in term of dealing with speckle noise and intensity inhomogeneity, and also accelerate the overall segmentation process, a new method based on nonlocal despeckling and an improved RSF model is proposed for segmenting side-scan sonar images. First, speckle noise is a major disturbance that leads to false segmentation and slow convergence; therefore, NLMSF is used to eliminate speckle noise, and thereby improve the accuracy and accelerate the convergence speed. The use of block matching and the consideration of the speckle model makes NLMSF suitable for sonar image denoising because sufficiently similar blocks can be grouped together to achieve better restoration. Most active contour models generally converge slowly when the initial positions of the active contours are far from boundaries of objects, and are therefore computationally costly (Yang et al. 2014). Second, an initial coarse segmentation is obtained using *k*-means clustering on the denoised image. Because noise has been removed by the nonlocal despeckling method, the coarse segmentation used as the initial segmentation is typically not far from the final segmentation, and therefore can reduce the number of iterations and reduce the processing time. Third, the RSF model is adopted to cope with intensity inhomogeneity, and an edge-driven constraint is added to the RSF model to accelerate the convergence speed and drive the active contour to adhere to the desired boundary. Using BM3D instead of NLMSF, the proposed method can also be applied to segmenting common images that are mainly affected by Gaussian noise.

9.2 Active Contour Model and Variational Method

9.2.1 Active Contour Model and Its Development

The active contour model is essentially a model based on a variational method and partial differential equation (PDE), which regards the boundary to be segmented as a movable contour under the combined action of multiple constraints. Under the guidance of energy functional minimization, the contour is continuously deformed in the direction of the target's boundary until it reaches the target's boundary. The general method of solving this type of problem is as follows: firstly, an appropriate energy function is used to define the boundary to be segmented, and the global minimum of the energy functional is used to represent the optimal segmented boundary; secondly, a PDE called the evolution equation, which minimizes the energy functional, is obtained using the variational method and gradient descent method (or steepest descent method); and finally, the evolution equation is solved iteratively using a specific numerical algorithm. Kass proposed the first active contour model, that is, the snake model (Kass et al. 1988). The problem of image segmentation is described as minimizing the energy function of closed curve V(s). The energy function of the snake model consists of three parts:

$$E^{*}_{\text{snake}} = \int_0^1 E_{\text{snake}}\text{d}s = \int_0^1 (E_{\text{int}}(V(s)) + E_{\text{image}}(V(s)) + E_{\text{con}}(V(s)))\text{d}s \tag{9.1}$$

where s is a free parameter and

$$E_{\text{int}}(V(s)) = \frac{1}{2}(\alpha(s)|V_{\text{s}}(s)|^2 + \beta(s)|V_{\text{ss}}(s)|^2) \tag{9.2}$$

where α and β are the elastic and rigid force parameters of the curve, respectively, and $V_{\text{s}}(s)$ and $V_{\text{ss}}(s)$ are the first and second derivatives, respectively. Minimizing these derivatives means making the curve shrink inward and as smooth as possible, respectively. $E_{\text{image}}(V_{\text{s}}(s))$ is an image energy item, which can be composed of line segments, edge constraints, and other constraints, and the corresponding external constraints of $E_{\text{con}}(V_{\text{s}}(s))$ can be added artificially, such as adding motion information, to strengthen the constraints on curves. The snake model has many limitations, such as a dependence on free parameters, small capture range of edges, falling into local extremum points, an unstable numerical solution, and inability to detect depression boundaries and deal with the topological changes of contours. To overcome the disadvantage of the snake model's inability to detect the edge of a depression, Xu and Prince (1998) proposed the gradient vector flow model. To reduce the influence of the local extremum, Cham and Cipolla (1997). proposed a double active contour model.

All the models mentioned above regard the contour as a spline curve with parameter variables, so they are collectively called PAC models. A PAC model expresses a curve explicitly, but this characteristic is not conducive to describing the change of the topological structure of contour lines, particularly the splitting and merging of contour lines, which is very important for actual segmentation. The level set method (Osher and Sterian 1998), originally proposed by Osher and Sterian, is an implicit representation of curves that embeds low-dimensional curves into high-dimensional curves, and can deal with the changes of curve topology well. To overcome the disadvantage that the PAC model is difficult to adapt to a change of topology, Caselles et al. (1993) proposed a GAC model based on curve evolution theory and a level set method. The evolution equation of the GAC model does not involve other parameters independent of the geometric structure of the curve, except the unit normal vector and curvature. However, the evolution equation of this model is derived directly from the theory of curve evolution rather than minimizing it (used by the snake model); hence, there is no explicit criterion to support it. To solve this problem, Caselles et al. (1997) and Yezzi et al. (1997) proposed a geodesic active contour (GAC) model based on the snake model and curve evolution theory, which does not contain free parameters. With the help of a variational method and level set method, the evolution equation of the geometric active contour model can be obtained from the energy functional of the GAC model. Therefore, it can be said that the geodesic active contour model is a generalized geometric active contour model (note: at present, the term geometric active contour model has two meanings, geometric active contour model refers specifically to the geodesic active contour model when referring to a specific model; by contrast, to distinguish it from the PAC model, which uses an explicit method to describe curves, the active contour model, which introduces the level set method, is generally referred to as the GAC model and also known as the level set segmentation model). The geodesic active contour model is a new breakthrough in the application of PDEs in image segmentation. Based on the arc length parameters of the curve itself, the active contour is determined by minimizing the following energy functional:

$$E(C) = L_{\text{R}}(C) = \oint_C g(C)\text{d}s = \int_0^{L(C)} g(|\nabla I(C(s))|)\text{d}s \tag{9.3}$$

where s is the arc length parameter, $L(C)$ is the arc length of closed curve C, $L_{\text{R}}(C)$ is the weighted arc length, $|\nabla I(C(s))|$ represents the gradient modulus of each point on the curve, and g is the edge indicator function, which is inversely proportional to the gradient modulus of the image. Because the gradient modulus at the edge of the image is relatively large, minimizing Eq. (9.3) means that closed curve C eventually moves to the edge of the image. Calculating the gradient directly using a finite difference method is very sensitive to noise. Therefore, the GAC model uses Gauss smoothing to preprocess the image before calculating the gradient. Based on the variational method and gradient descent method, the curve evolution equation that corresponds to the minimization Eq. (9.3) can be obtained as follows:

$$\frac{\partial c}{\partial t} = g(C)\kappa N - (\nabla g \cdot N)N \tag{9.4}$$

where κ is the curvature, N is the unit normal vector, • represents the point product operation of the vector, and ∇ is the Hamiltonian operator. The snake model, GAC model, and improved model are all active contour models based on edge information, which depend on the local gradient of the contour to detect the boundary of the target area. If the boundary of the target is weak, then the model is at risk of failure. To solve this problem, the active contour model based on regional information emerged as the times require. Inspired by the Mumford-Shah functional (Mumford and Shah 1989), Chan and Vese (2001) proposed the classical active contours without edges model (C-V model). Simultaneously, Tsai et al. (2001) proposed a similar model. The C-V model assumes that the two regions to be segmented have different mean values, and defines the energy function as follows:

$$\begin{aligned} E(c_1, c_2, C) &= \mu \cdot \text{Length}(C) + \nu \cdot \text{Area}(\text{inside}(C)) \\ &+ \lambda_1 \int_{\text{inside}(C)} |u_0(x,y) - c_1|^2 \text{d}x\text{d}y \\ &+ \lambda_2 \int_{\text{outside}(C)} |u_0(x,y) - c_2|^2 \text{d}x\text{d}y \end{aligned} \tag{9.5}$$

where the first term denotes the length of the curve; the second term denotes the area of the region surrounded by the curve; the third and fourth terms are the square errors between the gray values of the internal and external areas and c_1 and c_2, respectively, that is, the deviation between the image and the assumed piecewise constant image; and μ, ν, λ_1 and λ_2 are weighted coefficients; typically ν is set to zero, λ_1 and λ_2 are set to one. The C-V model is a classical and practical model for weak edge image segmentation. However, it is difficult to apply the model to complex images because it only uses the mean information of regions to locate the boundary. Moreover, the model can only be applied to two types of region segmentation, and cannot be used for multi-region image segmentation.

For multi-region segmentation, Zhao et al. (1996) proposed a model using N level set functions to segment N-class regions; Vese and Chan (2002) extended the C-V model to multi-region scenarios, and proposed a multi-phase active contour model, which used N level set functions to segment images with 2^N different regions; Chung et al. (2009) used different levels of a level set function. Based on the idea of regional competition (RC), Mansouri et al. (2006) proposed an N-1 level set function to implement the multi-phase level set segmentation model of N regions.

To improve the fitting degree of data and its adaptability and anti-noise performance, level set segmentation based on a statistical model has attracted the attention of scholars in recent years. Zhu and Yuille (1996) proposed an RC model based on Bayes and minimum description length (MDL) criteria. The RC model is essentially a segmentation model that uses a PDF. Although the RC model uses statistical optimization criteria for modeling, which reduces the sensitivity to noise, it uses the Gauss distribution to fit image data and describes the contour in a parameterized manner. Therefore, the practicability of the RC model is greatly limited. Ayed et al. (2005) proposed a level set segmentation method based on a single parameter Gamma distribution. To promote the adaptability of the statistical model, Ayed et al. (2006) also proposed a level set segmentation method based on the two-parameter Weibull distribution, which has better adaptability to speckle noise. In the process of deriving the model, Ayed et al. first provided the energy functional in the form of a parametric curve, then solved the explicit evolution equation that the curve should satisfy, and finally obtained the corresponding implicit evolution equation using the level set equation, so the convergence speed of the algorithm was slow.

Using the level set method to solve active contour models is convenient; however, we also need to consider how to improve the efficiency of the level set calculation: using an implicit level set function to represent curves increases the dimension of the problem that can be solved. Moreover, we need to re-initialize the level set function to the symbolic distance function regularly. Starting from the level set method itself, Adalsteinsson and Sethian (1995) proposed the narrow band algorithm, which is only computed near the zero level set. Sethian (1996) proposed the fast marching method, in which the contour line only evolves in one direction. Starting from the evolution equation of the contour line, Goldenberg et al. (2001) and Weickert and Kuhne (2003) adopted the additive operator splitting scheme to alleviate the time step limitation of the explicit difference scheme of the evolution equation, whereas Papandreou and Maragos (2004, 2007) used a multigrid technique to accelerate the evolution of the contour line. To avoid the problem of the re-initialization of level set functions, Li et al. (2005) proposed a model in which penalty terms for level set functions were added directly to the energy function, and the re-initialization process of the level set functions was embedded into the evolution equation.

Because of the nonconvexity of the energy function, the active contour model has the risk of falling into a local minimum. To solve this problem, many scholars have proposed various solutions, including two typical solutions: the first scheme is to add a priori knowledge of the target shape to the model (Chen et al. 2002; Gastaud et al. 2004) and the second scheme is to design a convex energy functional with a global minimum solution (Appleton and Talbot 2005; Bresson et al. 2007). The latest development trend has shown that the level set method has the characteristic of easy integration with other methods. Scholars have conducted a variety of studies on the integration of level set methods with other theories and methods, including the level set segmentation model based on watershed, level set segmentation model based on a support vector machine, and level set segmentation model based on a random field (Hodneland et al. 2009; Tsechpenakis and Metaxas 2009).

9.2.2 Variational Method and Gradient Descent Method

It can be seen from the previous subsections that the desired curve of the active contour model is typically determined by minimizing a certain energy functional. Considering the two-dimensional case as an example, the following function is considered to be solved:

$$J(z\ (x,y)) = \iint_{\Omega} F\left(x, y, z, \frac{\partial z}{\partial x}, \frac{\partial z}{\partial y}\right) \mathrm{d}x\mathrm{d}y \tag{9.6}$$

where $z(x, y)$ is the surface that corresponds to the extreme value of the functional. The value of functional $z(x, y)$ on boundary $\partial\Omega$ of region Ω has been given, that is, the space contour is known, and all admissible surfaces must pass through it, which is the variational problem of a fixed boundary.

The extremum surface is denoted by $z(x, y)$. If functional F is differentiable in the third order, then $z(x, y)$ is differentiable in the second order. Let $\overline{z(x,y)}$ be an admissible surface close to $z(x, y)$. $\overline{z(x,y)}$ and $z(x, y)$ are merged into a family of surfaces with small parameter α:

$$z(x,y,\alpha) = z(x,y) + \alpha(\overline{z(x,y)} - z(x,y)) = z(x,y) + \alpha\delta z \tag{9.7}$$

In the family of functionals defined by Eqs. (9.7), (9.6) becomes a function of α, and reaches the extreme value $z(x, y)$ when $\alpha = 0$. Therefore, we have the following:

$$\delta J = \frac{\partial}{\partial \alpha} J(z(x,y,\alpha))\bigg|_{\alpha=0} = 0 \tag{9.8}$$

With the help of variational operations, the following formula can be obtained:

$$\begin{aligned}\delta J &= \frac{\partial}{\partial \alpha}\iint_{\Omega} F\left(x, y, z(x,y,\alpha), \frac{\partial z(x,y,\alpha)}{\partial x}, \frac{\partial z(x,y,\alpha)}{\partial y}\right)\mathrm{d}x\mathrm{d}y\bigg|_{\alpha=0} \\ &= \iint_{\Omega}\left(\frac{\partial F}{\partial z}\delta z + \frac{\partial F}{\partial z_x}\delta z_x + \frac{\partial F}{\partial z_y}\delta z_y\right)\mathrm{d}x\mathrm{d}y = 0\end{aligned} \tag{9.9}$$

Because

$$\begin{aligned}\frac{\partial}{\partial x}\left(\frac{\partial F}{\partial z_x}\delta z\right) &= \frac{\partial}{\partial x}\left(\frac{\partial F}{\partial z_x}\right)\delta z + \frac{\partial F}{\partial z_x}\delta z_x,\ \frac{\partial}{\partial y}\left(\frac{\partial F}{\partial z_y}\delta z\right) \\ &= \frac{\partial}{\partial y}\left(\frac{\partial F}{\partial z_y}\right)\delta z + \frac{\partial F}{\partial z_y}\delta z_y\end{aligned} \tag{9.10}$$

the following can be obtained:

$$\begin{aligned}&\iint_{\Omega}\left(\frac{\partial F}{\partial z_x}\delta z_x + \frac{\partial F}{\partial z_y}\delta z_y\right)\mathrm{d}x\mathrm{d}y = \iint_{\Omega}\left(\frac{\partial}{\partial x}\left(\frac{\partial F}{\partial z_x}\delta z\right) + \frac{\partial}{\partial y}\left(\frac{\partial F}{\partial z_y}\delta z\right)\right)\mathrm{d}x\mathrm{d}y \\ &- \iint_{\Omega}\left(\frac{\partial}{\partial x}\left(\frac{\partial F}{\partial z_x}\right) + \frac{\partial}{\partial y}\left(\frac{\partial F}{\partial z_y}\right)\right)\delta z\mathrm{d}x\mathrm{d}y\end{aligned} \tag{9.11}$$

From Green's theorem, $\int_{\Omega}\left(\frac{\partial N}{\partial x} + \frac{\partial M}{\partial y}\right)\mathrm{d}x\mathrm{d}y = \int_{\partial\Omega} N\mathrm{d}y - M\mathrm{d}x$, by transforming the area integral of the first term of Eq. (9.11) into the contour integral, we obtain

$$\begin{aligned}&\iint_{\Omega}\left(\frac{\partial}{\partial x}\left(\frac{\partial F}{\partial z_x}\delta z\right) + \frac{\partial}{\partial y}\left(\frac{\partial F}{\partial z_y}\delta z\right)\right)\mathrm{d}x\mathrm{d}y \\ &= \int_{\partial\Omega}\left(\frac{\partial F}{\partial z_x}\delta z\right)\mathrm{d}y - \left(\frac{\partial F}{\partial z_y}\delta z\right)\mathrm{d}x = 0\end{aligned} \tag{9.12}$$

Because all allowable surfaces pass through the same spatial contour $\partial\Omega$, variation δz on contour $\partial\Omega$ is equal to zero, that is, $\delta z|_{\partial\Omega} = 0$. Therefore, Eq. (9.12) is zero and Eq. (9.11) becomes

$$\begin{aligned}&\iint_{\Omega}\left(\frac{\partial F}{\partial z_x}\delta z_x + \frac{\partial F}{\partial z_y}\delta z_y\right)\mathrm{d}x\mathrm{d}y \\ &= -\iint_{\Omega}\left(\frac{\partial}{\partial x}\left(\frac{\partial F}{\partial z_x}\right) + \frac{\partial}{\partial y}\left(\frac{\partial F}{\partial z_y}\right)\right)\delta z\mathrm{d}x\mathrm{d}y\end{aligned} \tag{9.13}$$

Substituting Eq. (9.13) into Eq. (9.9) yields

$$\delta J = \iint_{\Omega}\left(\frac{\partial F}{\partial z} - \frac{\partial}{\partial x}\left(\frac{\partial F}{\partial z_x}\right) - \frac{\partial}{\partial y}\left(\frac{\partial F}{\partial z_y}\right)\right)\delta z\mathrm{d}x\mathrm{d}y = 0 \tag{9.14}$$

Because variation δz is arbitrary and $\delta z|_{\partial\Omega} = 0$, Eq. (9.14) satisfies the condition of the variational preparation theorem. On surface $z(x, y)$ where the extremum is reached, the following holds:

$$\frac{\partial F}{\partial z} - \frac{\partial}{\partial x}\left(\frac{\partial F}{\partial z_x}\right) - \frac{\partial}{\partial y}\left(\frac{\partial F}{\partial z_y}\right) = 0 \tag{9.15}$$

Equation (9.15) is a necessary condition for the functional extremum, that is, the Euler–Lagrange equation. The extremum surface $z(x, y)$ obtained is the solution of the Euler–Lagrange equation. The variational method and Euler–Lagrange equation represent the same problem; thus,

$z(x, y)$ can be obtained by solving the Euler–Lagrange equation. However, it is typically difficult to solve the Euler–Lagrange equation directly. By introducing time variable t, the gradient descent method is used to solve the problem, which corresponds to the steepest descent of the functional. The gradient descent method is as follows:

$$\begin{aligned}\frac{\partial z}{\partial t} &= -\left(\frac{\partial F}{\partial z} - \frac{\partial}{\partial x}\left(\frac{\partial F}{\partial z_x}\right) - \frac{\partial}{\partial y}\left(\frac{\partial F}{\partial z_y}\right)\right) \\ &= \frac{\partial}{\partial x}\left(\frac{\partial F}{\partial z_x}\right) + \frac{\partial}{\partial y}\left(\frac{\partial F}{\partial z_y}\right) - \frac{\partial F}{\partial z}\end{aligned} \quad (9.16)$$

9.2.3 Curve Evolution Theory

Considering closed curve $C(p, t)$ evolving over time t, the PDE, typically obtained using the variational method and gradient descent method, is as follows:

$$\frac{\partial C(p,t)}{\partial t} = \boldsymbol{V} = \alpha(p,t)\boldsymbol{T} + \beta(p,t)\mathrm{N} \quad (9.17)$$

where α and β represent the tangential and normal components of the velocity, respectively. Expressing y as a function of x, for example, $y = \gamma(x)$, curve C can be expressed by parameter x as follows:

$$C(p) = C(x(p), y(p)) = C(x,\ \gamma(x)) \quad (9.18)$$

Then the tangent vector of curve C is $C_x = (1, \gamma_x)$, and the unit tangent vector and normal vector are, respectively,

$$\boldsymbol{T} = \frac{(1,\ \gamma_x)}{\sqrt{1+\gamma_x^2}},\quad \mathrm{N} == \frac{(-\gamma_x,\ 1)}{\sqrt{1+\gamma_x^2}} \quad (9.19)$$

The equation of motion of x and y can be obtained by substituting Eq. (9.19) into Eq. (9.17):

$$\begin{aligned}\frac{\mathrm{d}x}{\mathrm{d}t} &= \alpha\frac{1}{\sqrt{1+\gamma_x^2}} + \beta\frac{-\gamma_x}{\sqrt{1+\gamma_x^2}},\quad \frac{\mathrm{d}y}{\mathrm{d}t} \\ &= \alpha\frac{\gamma_x}{\sqrt{1+\gamma_x^2}} + \beta\frac{1}{\sqrt{1+\gamma_x^2}}\end{aligned} \quad (9.20)$$

From $\frac{\mathrm{d}y}{\mathrm{d}t} = \gamma_x\frac{\mathrm{d}x}{\mathrm{d}t} + \gamma_t$, we obtain $\gamma_t = \frac{\mathrm{d}y}{\mathrm{d}t} - \gamma_x\frac{\mathrm{d}x}{\mathrm{d}t}$, and by substituting Eq. (9.20) into it, we obtain

$$\begin{aligned}\gamma_t &= \alpha\frac{\gamma_x}{\sqrt{1+\gamma_x^2}} + \beta\frac{1}{\sqrt{1+\gamma_x^2}} - \alpha\frac{1}{\sqrt{1+\gamma_x^2}} + \beta\frac{\gamma_x}{\sqrt{1+\gamma_x^2}} \\ &= \beta\sqrt{1+\gamma_x^2}\end{aligned} \quad (9.21)$$

Equation (9.21) shows that the change of curve geometry is only related to the normal component β of $\boldsymbol{V}$ and not the tangential component α. Therefore, when discussing curve evolution, only the normal velocity is considered. Based on this, the general equation of curve evolution is obtained:

$$\frac{\partial C}{\partial t} = \beta\mathrm{N} \quad (9.22)$$

Before the level set method was proposed, Eq. (9.22) was mostly solved by the "marker particle method": sufficient points were marked on the interface, and then the coordinates of each point were moved to a new position according to the velocity and direction determined by the equation. By connecting the points of these new positions, a new interface was obtained after an interval of Δt. The traditional snake model first obtains the PDE of the curve evolution that corresponds to Eq. (9.1) using the variational method, and then solves it using the marker particle method.

However, related studies have shown that there are two problems in the solution of Eq. (9.22): (1) the discontinuity of curves leads to singular phenomena in the solution of PDEs, and it is difficult to obtain accurate results; and (2) the parametric model makes it difficult to implement the free topological transformation of curves, thus restricting the scope of use of the model. The solution of Eq. (9.22), a seemingly simple PDE, is actually extremely difficult. It was not until Sethian proposed the level set method that this problem made a breakthrough.

9.2.4 Level Set Method

9.2.4.1 Basic Concept of the Level Set Method

The level set method, originally proposed by Osher et al. in 1988, is an implicit representation of curves that embeds low-dimensional curves into high-dimensional curves, which can deal with the changes of the topological structure of curves well. The key idea of the level set method is to consider the n-dimensional description as the $(n + 1)$-dimensional level set, or the n-dimensional description as the level set of level set function ϕ that contains n-dimensional variables.

In the two-dimensional case, the level set method typically expresses time-varying closed curve $C(p, t)$ on the plane as the zero level set of level set function $\phi(x, y, t)$, that is,

$$C(p,\ t) = \{(x,\ y)|\phi(x,\ y,\ t) = 0\} \quad (9.23)$$

According to the chain rule of derivatives of the compound function, the total derivative of the level set function of Eq. (9.23) is solved as follows:

$$\frac{\mathrm{d}\phi}{\mathrm{d}t} = \frac{\partial\phi}{\partial t} + \nabla\phi\cdot\frac{\partial C(p,\ t)}{\partial t} = 0 \quad (9.24)$$

By substituting Eq. (9.17) into Eq. (9.24), we obtain

$$\frac{\partial\phi}{\partial t} = -\nabla\phi\cdot\boldsymbol{V} = -|\nabla\phi|\frac{\nabla\phi}{|\nabla\phi|}\cdot\boldsymbol{V} = |\nabla\phi|\boldsymbol{N}\cdot\boldsymbol{V} = \beta|\nabla\phi| \tag{9.25}$$

which is the basic expression of the level set method of curve evolution, where $\beta = \boldsymbol{N}\cdot\boldsymbol{V}$ is the normal component of the velocity of motion. Clearly, Eqs. (9.25) and (9.22) correspond to each other. The evolution of closed curve *C*(*p*, *t*) using Eq. (9.22) under given initial value C_0 is equivalent to that of level set function $\phi(x,y,t)$ using Eq. (9.25) under given initial value $\phi_0(x,y)$. Therefore, current curve *C*(*p*, *t*) can be determined by selecting the level set of $\phi(x,y,t) = 0$s at any time. Note that in Eq. (9.25), unit normal vector $\boldsymbol{N} = -\frac{\nabla\phi}{|\nabla\phi|}$, and the level set functions inside and outside the closed curve are considered as positive and negative, respectively, which means that $\boldsymbol{N}$ in this chapter is defined as the external normal vector.

After introducing the level set method, we can solve the evolution equation of the level set function of the GAC model. By comparing Eq. (9.4) with the general Eq. (9.22) of curve evolution, we obtain

$$\beta = g(C)\kappa\mathrm{N} - (\nabla g\cdot\boldsymbol{N}) \tag{9.26}$$

By substituting Eq. (9.26) into Eq. (9.25), the evolution equation of the level sets of the GAC model is obtained:

$$\frac{\partial\phi}{\partial t} = \beta|\nabla\phi| = [g(C)\kappa\mathrm{N} - (\nabla g\cdot\mathrm{N})]|\nabla\phi| \tag{9.27}$$

Substituting $\kappa = -\nabla\cdot\boldsymbol{N}$ and $\boldsymbol{N} = -\frac{\nabla\phi}{|\nabla\phi|}$ into Eq. (9.27), after simplification we obtain

$$\frac{\partial\phi}{\partial t} = \nabla\cdot\left(g\frac{\nabla\phi}{|\nabla\phi|}\right)|\nabla\phi| \tag{9.28}$$

where $\nabla\cdot$ represents the divergence operator. Equation (9.28) can be numerically calculated using half-point discretization and an upwind scheme.

9.2.4.2 Level Set Function

To calculate Eq. (9.25) steadily, it is necessary to select the appropriate form for $\phi(x,y)$ and make the zero level set of the initial value $\phi_0(x,y)$ of the level set function $\phi(x,y)$ correspond to given initial curve C_0. The choice of $\phi(x,y)$ is not unique, but the symbolic distance function, that is, the symbolic distance from point (*x*, *y*) on the plane to curve *C*, is generally adopted, which is defined as follows:

$$\phi(x,y) = \begin{cases} d[(x,y),C] & (x,y)\text{inside } C \\ -d[(x,y),C] & (x,y)\text{outside } C \end{cases} \tag{9.29}$$

where *d*[(*x*, *y*), *C*] denotes the Euclidean distance between point (*x*, *y*) and curve *C*. Because the distance function has the basic property of $|\nabla\phi(x,y)| \equiv 1$, this means that the change rate of $\phi(x,y)$ is uniform everywhere, which is conducive to the stability of the numerical calculation. $\phi(x,y)$ is initialized by calculating the minimum Euclidean distance from each grid point (*x*, *y*) to initial curve C_0 in definition domain Ω, and then assigning a positive sign or negative sign according to whether the point is inside or outside C_0.

To solve Eq. (9.25), it is required that all *β*s of every point (*x*, *y*) in the definition domain Ω are known, not only *β* on the zero level set of $\phi(x,y)$. *β* in Eq. (9.22) is only the zero level set of $\phi(x,y)$ in Eq. (9.25). Specifically, Eq. (9.25) should be rewritten as follows:

$$\frac{\partial\phi}{\partial t} = \beta_{\text{ext}}|\nabla\phi| \tag{9.30}$$

where β_{ext} is a continuation velocity, which can be any function that satisfies the following conditions:

$$\beta_{\text{ext}} = \beta,\ \forall\phi(x,y,t) = 0 \tag{9.31}$$

To calculate Eq. (9.30), the velocity field must be extended by satisfying Eq. (9.31). Sethian (1999) discussed a two-tier system that ran alternately to achieve velocity continuation. This method has the advantage of maintaining $\phi(x,y)$ as a symbolic distance function, but the disadvantage is that the efficiency of the numerical calculation is extremely low. The more commonly used method is natural continuation, which assumes that the given curve evolution equation holds not only for the zero level set of the level set function, but also for all the level sets of the level set function. Natural continuation can be achieved by considering *β* in Eq. (9.25) as the normal rate of all level sets rather than just the normal rate of the zero level set. Natural continuation is simple and easy, but it cannot guarantee that $\phi(x,y)$ is always a symbolic distance function in the evolution process: because of the discreteness of digital images and the influence of noise, some parts may have peaks or deep valleys because $|\nabla\phi|$ is far greater than one, or have flat areas because $|\nabla\phi|$ is far less than one, which leads to an unstable numerical calculation. Therefore, it is necessary to re-initialize $\phi(x,y)$ as a symbol distance function. Sethian (1999) proposed that this can be accomplished using the following formula:

$$\frac{\partial\phi}{\partial t} = \operatorname{sgn}(\phi^n)(1 - |\nabla\phi|) \tag{9.32}$$

where ϕ^n denotes the level set function that has been updated at the current time. The steady- state solution of Eq. (9.32) satisfies $|\nabla\phi|$=1. The steady-state solution of Eq. (9.32) is regarded as a new initial value of the level set function, and the evolution of the level set function continues. Re-initialization does not need to be performed after each update, and can be performed periodically according to the experimental conditions.

9.2.4.3 Variational Level Set Method

The level set method first obtains the motion Eq. (9.22) of curve C (typically obtained by minimizing the energy functional using the variational method or directly derived from the laws of physics), then introduces level set function ss$\phi(x,y)$, and obtains the PDE of the evolution of level set function $\phi(x,y)$ from Eq. (9.25). Considering that when curve evolution is applied to image processing, the equation of motion of the curve often comes from an energy function that minimizes closed curve C, for this type of curve evolution problem derived from the minimization of the energy functional of curves, Zhao et al. (1996) proposed a new level set method called the variational level set.

By introducing level set function ϕ and Heaviside function H, the variational level set method first transforms energy functional $E(C)$ of curve C into $E(\phi)$, and then uses the variational method to minimize energy function $E(\phi)$ so that the PDE of the evolution of level set function ϕ can be obtained directly. Heaviside function H is defined as follows:

$$H(z)=\begin{cases}1 & z\geq 0\\ 0 & z<0\end{cases} \tag{9.33}$$

For the GAC model, with the help of Heaviside function H and level set function ϕ, which is a loop integral along closed curve C, energy functional $E(C)$ of curve C can be formally rewritten as a surface integral, defined as

$$E(C)=\oint_C g(C)\mathrm{d}s=E(\phi)=\iint_\Omega g(x,y)|\nabla H(\phi)|\mathrm{d}x\mathrm{d}y \tag{9.34}$$

Because $\nabla H(\phi)=\delta(\phi)\nabla\phi$, Eq. (9.34) is equivalent to

$$E(\phi)=\iint_\Omega g(x,y)\delta(\phi)|\nabla\phi|\mathrm{d}x\mathrm{d}y \tag{9.35}$$

Based on the variational method and gradient descent method, the evolution equation of Eq. (9.35) can be obtained by Eq. (9.16):

$$\frac{\partial\phi}{\partial t}=\mathrm{div}\left(g\frac{\nabla\phi}{|\nabla\phi|}\right)\delta(\phi) \tag{9.36}$$

Dirac function $\delta(\phi)$ needs to be regularized to make Eq. (9.36) a practical PDE:

$$\frac{\partial\phi}{\partial t}=\mathrm{div}\left(g\frac{\nabla\phi}{|\nabla\phi|}\right)\delta_\varepsilon(\phi) \tag{9.37}$$

where $\delta_\varepsilon(z)=\frac{\mathrm{d}}{\mathrm{d}z}H_\varepsilon(z)$ and $H\varepsilon(z)$ satisfies $\lim_{\varepsilon\to 0}H_\varepsilon(z)=H(z)$, which, for example, can be chosen as follows:

$$H_\varepsilon(z)=\frac{1}{2}\left(1+\frac{2}{\pi}\arctan\frac{z}{\varepsilon}\right) \tag{9.38}$$

Therefore, we have

$$\delta_\varepsilon(z)=\frac{1}{\pi}\cdot\frac{\varepsilon}{\varepsilon^2+z^2} \tag{9.39}$$

Equation (9.37) is the same as Eq. (9.28), except $|\nabla\phi|$ is replaced by $\delta(\phi)$. Moreover, there are essential mathematical differences between them. The former belongs to a parabolic curve and the latter belongs to a hyperbolic curve; thus, the former is more stable. Therefore, in the numerical implementation, the former can adopt a larger time step, and often does not need to re-initialize the level set function, thus improving computational efficiency. By introducing Heaviside function H and level set function ϕ the variational level set method can be used to solve the C-V model. Equation (9.5) can be modified to a function of a level set function, that is,

$$\begin{aligned}E(c_1,c_2,\phi)&=\mu\iint_\Omega\delta(\phi)|\nabla\phi|\mathrm{d}x\mathrm{d}y\\&+v\iint_\Omega H(\phi)\mathrm{d}x\mathrm{d}y+\lambda_1\iint_\Omega H(\phi)|u_0(x,y)\\&-c_1|^2\mathrm{d}x\mathrm{d}y+\lambda_2\iint_\Omega(1-H(\phi))|u_0(x,y)-c_2|^2\mathrm{d}x\mathrm{d}y\end{aligned} \tag{9.40}$$

Fixing parameters c_1 and c_1, the evolution equation of level set function ϕ can be obtained by Eq. (9.16):

$$\frac{\partial\phi}{\partial t}=\left[\mu\mathrm{div}\left(\frac{\nabla\phi}{|\nabla\phi|}\right)-v-\lambda_1(u_0(x,y)-c_1)^2+\lambda_2(u_0(x,y)-c_2)^2\right]\delta(\phi) \tag{9.41}$$

By regularizing $\delta(\phi)$, we obtain

$$\frac{\partial\phi}{\partial t}=\left[\mu\mathrm{div}\left(\frac{\nabla\phi}{|\nabla\phi|}\right)-v-\lambda_1(u_0(x,y)-c_1)^2+\lambda_2(u_0(x,y)-c_2)^2\right]\delta_\varepsilon(\phi) \tag{9.42}$$

Although the variational level set method has the advantage of being easy to solve, it cannot completely replace the level set method. In some cases, the PDE of the curve or surface evolution is not obtained by minimizing the energy functional. For example, the evolution of the interface in the fields of fluid mechanics and materials science is often derived directly from the laws of physics; thus, the variational level set method cannot be used at this time. Therefore, the level set method is more applicable than the variational level set method.

9.3 Region-Scalable Fitting Active Contour Model and Nonlocal Means-Based Speckle Filtering

9.3.1 Region-scalable Fitting Active Contour Model.

The defining characteristic of inhomogeneous images is that in the background region or object region, an intensity difference exists, which may cause considerable difficulties in image segmentation. To overcome this problem, Li et al. (2008) proposed a controlled scalable kernel function to ensure minimum energy in the local region, and built the RSF model. For two-region segmentation, let C be a closed contour that divides image domain Ω into two regions, Ω_1 and Ω_2, which are outside and inside contour C, respectively. For given point $x \in \Omega$, a local intensity fitting energy of contour C at point x is defined as

$$\varepsilon_x^{\text{Fit}}(C,f_1(x),\ f_2(x)) = \sum_{i=1}^{2} \lambda_i \int_{\Omega_i} K(x-y)\ |I(y)-f_i(x)|^2 \mathrm{d}y \tag{9.43}$$

where λ_1 and λ_2 are positive real constants; $f_1(x)$ and $f_2(x)$ are two functions that describe image intensities in Ω_1 and Ω_2, respectively; K is a symmetric, nonnegative and localized kernel function; and $y \in \Omega$ is an arbitrary point in the image domain. Typically, kernel function K can be chosen as a Gaussian kernel:

$$K_\sigma(u) = \frac{1}{(2\pi)^{n/2}\sigma^n} \mathrm{e}^{-|u|^2/2\sigma^2} \tag{9.44}$$

with scale parameter $\sigma > 0$. Because the kernel function is localized, the contribution of intensity $I(y)$ to fitting energy $\varepsilon_x^{\text{Fit}}$ decreases and approaches zero when the point goes away from the center point. Fitting energy $\varepsilon_x^{\text{Fit}}$ with small σ only involves the intensities within a small neighborhood of point x, whereas the fitting energy with large σ involves the image intensities in a large area around x, which means that intensities $I(y)$ for the fitting energy Eq. (9.43) can be in a region of any size: from a small neighborhood to the entire image domain. Therefore, the local intensity fitting energy in Eq. (9.43) is called the RSF energy of a contour at point x.

For given point x, $\varepsilon_x^{\text{Fit}}$ arrives at the minimum value when contour C is on the boundary of the object and $f_1(x)$ and $f_2(x)$ optimally describe the local intensities on the two sides of the contour C. By minimizing the integral of $\varepsilon_x^{\text{Fit}}$ over all center points x in domain Ω the entire object boundary can be obtained. Because it is often necessary to smooth the contour by penalizing its length $|C|$, the total energy functional is defined as

$$\varepsilon(C,f_1(x),f_2(x)) = \oint_{\Omega} \varepsilon_x^{\text{Fit}}(C,f_1(x),f_2(x))\mathrm{d}x + v|C| \tag{9.45}$$

where v is a positive real length controlling parameter. The larger v is, the smaller length $|C|$ is, which therefore provides a shrinkage force to erase some noise points and small regions.

Because topological changes may occur as contour C evolves, it is better to convert the above energy to a level set formulation by implicitly representing contour C using a level set. According to the level set methods, contour $C \subset \Omega$ can be represented as the zero level set of Lipschitz function $\varphi : \Omega \to R$, which is called a level set function. Let level set function ϕ take positive and negative values outside and inside contour C, respectively, and then by introducing Heaviside function H, energy functional $\varepsilon_x^{\text{Fit}}(C,f_1(x),f_2(x))$ is rewritten as

$$\varepsilon_x^{\text{Fit}}(\phi,f_1(x),f_2(x)) = \sum_{i=1}^{2} \lambda_i \int_{\Omega} K_\sigma(x-y)|I(y)-f_i(x)|^2 M_i(\phi(y))\mathrm{d}y \tag{9.46}$$

where $M_1(\phi) = H(\phi)$ and $M_2(\phi) = 1 - H(\phi)$ are masks that correspond to regions Ω_1 and Ω_2s, respectively. Accordingly, total energy ε in Eq. (9.45) can be written as

$$\varepsilon(\phi,f_1,f_2) = \sum_{i=1}^{2} \lambda_i \int_{\Omega} \left(\int_{\Omega} K_\sigma(x-y)|I(y)-f_i(x)|^2 M_i(\phi(y))\mathrm{d}y \right)\mathrm{d}x + v\int_{\Omega} |\nabla H(\phi(x))|\mathrm{d}x \tag{9.47}$$

where the last term $\int_{\Omega} |\nabla H(\phi(x))|\mathrm{d}x$ is used for computing the length of the zero level contour of ϕ.

To ensure that level set function ϕ remains a signed distance function (SDF) during level set evolution, which is necessary for accurate computation and stable evolution, a level set regularization term was proposed in (Li et al. 2005), which is defined as

$$P(\phi) = \int_{\Omega} \frac{1}{2}(|\nabla \phi(x)| - 1)^2 \mathrm{d}x \tag{9.48}$$

which describes the deviation of the function from an SDF. Finally, the energy functional to be minimized is given by

$$F(\phi,f_1,f_2) = \varepsilon(\phi,f_1,f_2) + \mu P(\phi) \tag{9.49}$$

where μ is a positive real constant. To minimize energy functional F, which depends on level set function ϕs, and the two functions $f_1(x)$ and $f_2(x)$, an iterative two-step algorithm is adopted, with the functional decreasing at each step.

A small-scale parameter of the kernel function of the RSF model is typically preferred for accurate edge detection, intensity inhomogeneity elimination, and fast computation. However, this selection makes it sensitive to level set initialization and noise, and determining a proper initialization is a problem. When the initial position of the active contour is far from the boundary of the object, the RSF model may lack sufficient driving force to reach the desired boundary and therefore stop at a local minimum.

9.3.2 Nonlocal Means-based Speckle Filtering.

The nonlocal means approach introduced by Buades et al. (2005) is suited to the additive white Gaussian noise model. To adapt it to spatial speckle patterns, a dedicated speckle model was considered by Coupé et al. (2009). The speckle model is defined as

$$u(x) = v(x) + v^{\gamma}(x)\eta(x) \tag{9.50}$$

where $u(x)$ is the observed noisy image, $v(x)$ is the true image, γ is a constant factor, and $\eta(x) \sim N(0, \sigma^2)$ is zero-mean Gaussian noise with variance σ^2. This model takes the following into consideration: local correlation caused by periodic arrangements of scatters, envelope detection and logarithm amplification of signals, and additive Gaussian noise of sensors. Therefore, this model is more flexible and less restrictive than the usual Rayleigh speckle model.

Blockwise means-based speckle filtering first divides the image into overlapping blocks, whose centers are equally distributed. Then these blocks can be restored. Empirical Bayesian estimator $\hat{\boldsymbol{v}}(B_i)$ of block B_i is defined as (Coupé et al. 2009)

$$\hat{\boldsymbol{v}}(B_i) = \frac{\sum_{j=1}^{|\Delta_i|} \boldsymbol{v}(B_j)p(\boldsymbol{u}(B_i)|\boldsymbol{v}(B_j))}{\sum_{j=1}^{|\Delta_i|} p(\boldsymbol{u}(B_i)|\boldsymbol{v}(B_j))} \tag{9.51}$$

where B_i is a square block centered on pixel $x_i$of size $S = (2\alpha + 1)^{\text{dim}}$ ($\alpha \in N$, dim = 2 or dim = 3 for two-dimensional and three-dimensional images, respectively); Δ_is is the square search volume centered at pixel $x_i$of size $|\Delta_i| = (2M+1)^{\text{dim}}$, $M \in N$;$\boldsymbol{u}(B_i)$ is the vector gathering the intensity values of block B_i; $\boldsymbol{v}(B_j)$ is the unobserved vector of true values of block B_j; and $p(\boldsymbol{u}(B_i)|\boldsymbol{v}(B_j))$ denotes the pdf of $\boldsymbol{u}(B_i)$ given the unknown noiseless patches $\boldsymbol{v}(B_j)$. Because $\boldsymbol{v}(B_j)$ is unknown, Eq. (9.51) can be computed by substituting $\boldsymbol{u}(B_i)$ for $\mathbf{v}(B_j)$, which yields

$$\hat{\boldsymbol{v}}(B_i) = \frac{\sum_{j=1}^{|\Delta_i|} \boldsymbol{u}(B_j)p(\boldsymbol{u}(B_i)|\boldsymbol{u}(B_j))}{\sum_{j=1}^{|\Delta_i|} p(\boldsymbol{u}(B_i)|\boldsymbol{u}(B_j))} \tag{9.52}$$

where $p(\boldsymbol{u}(B_i)|\boldsymbol{v}(B_j))$ denotes the pdf of $\boldsymbol{u}(B_i)$ conditionally to $\boldsymbol{u}(B_j)$.

According to the speckle model Eq. (9.50), the following can be given:

$$p(u(x)|v(x)) \propto \exp\left(-\frac{(u(x)-v(x))^2}{2v(x)^{2\gamma}\sigma^2}\right) \tag{9.53}$$

where r is a constant.

Accordingly, the likelihood can be factorized as

$$p(\boldsymbol{u}(B_i)|\boldsymbol{u}(B_j)) = \prod_{s=1}^{S} p(u^{(s)}(x_i)|u^{(s)}(x_j))$$
$$\propto \exp\left(-\sum_{s=1}^{S}\frac{(u^{(s)}(x_i)-u^{(s)}(x_j))^2}{2(u^{(s)}(x_j))^{2\gamma}\sigma^2}\right) \tag{9.54}$$

With all the blocks known, for pixel $x_i$included in several blocks B_i, the final restored intensity of pixel $x_i$is the mean of all the restored values in the matched blocks. A pixel selection scheme proposed by Coupé et al. (2009), which is based on means of blocks and controlled by threshold μ_1, is adopted to select the most relevant patches for given block B_i. This scheme improves the preservation of the detailed regions while smoothing the flat regions.

9.4 Object Segmentation Based on the Nonlocal Means-Based Speckle Filtering and Edge-Constrained Region-Scalable Fitting Model

By considering both noise and intensity inhomogeneity, which make sonar image segmentation very difficult, a new segmentation method is proposed. Nonlocal despeckling is adopted as a pre-treatment to improve robustness in terms of dealing with noise, and an edge-driven constraint is integrated into the RSF model to avoid it from becoming trapped in the local minima and reduce the number of iterations. Coarse segmentation is also integrated into the proposed scheme to accelerate the speed. As shown in Fig. 9.1, the proposed method comprises the following main stages:

(1) Image preprocessing. To reduce noise interference in the sonar image and improve the processing speed, NLMSF is used to remove the most important speckle noise from the sonar image. The value of each block to be restored is calculated by Eq. (9.52) combined with Eq. (9.54). For pixel x_i contained in multiple blocks, its intensity value is the average of the strength value of all blocks to be restored.

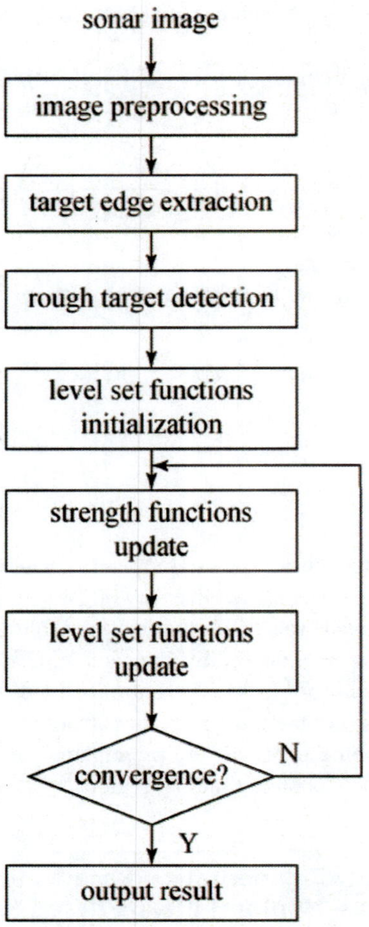

Fig. 9.1 Target detection in a side-scan sonar image based on the edge constraint level set

(2) Target edge extraction. Set $\boldsymbol{E}$ of target edge points is detected by the Canny operator, and the edge-driven constraint is calculated by combining Eqs. (9.61) and (9.62). An edge constraint is added to the RSF model to avoid it falling into the local modulus maxima and reduce the number of iterations.
(3) Rough target detection. As described in the following subsection, the k-means clustering algorithm is used to obtain the initial rough detection results, and the detection results are marked as target area Ω_1, background area Ω_3 and shadow area Ω_2 according to the brightness mean, from large to small.
(4) Initialize level set functions ϕ_1 and ϕ_2. According to the labeling results of rough target detection, Eqs. (9.70) and (9.71) are used to initialize level set functions ϕ_1 and ϕ_2, where number of iterations $n = 0$.
(5) Update the strength functions f_1, f_2 and f_3. Equation (9.64) is used to calculate and update the three regions Ω_1, Ω_2 and Ω_3, and the corresponding strength functions f_1, f_2 and f_3.
(6) Update level set functions ϕ_1 and ϕ_2. The level set functions ϕ_1 and ϕ_2 are calculated and updated using Eq. (9.65) and Eq. (9.66), and the number of iterations is $n + 1$.
(7) Assess the convergence. According to Eq. (9.72), whether the evolution of level set convergences is assessed, and the evolution stops if the convergence condition is satisfied.
(8) Output the result. According to Eq. (9.58), the final three regions Ω_1, Ω_2 and Ω_3 are assigned the values 255, 0 and 125, respectively, to represent the target, shadow and background, respectively, which is easy for human eyes to interpret. The result is output and displayed.

9.4.1 Coarse Segmentation Using *k*-means Clustering

After NLMSF, a coarse segmentation is obtained using k-means clustering. Because the coarse segmentation result is typically close to the final result, it can be used to accelerate the speed, which is shown in Sect. 9.5.

Given an initial set of $k = 3$ intensity means $m_1^{(1)}$, $m_2^{(1)}$ and $m_3^{(1)}$, coarse segmentation using k-means clustering proceeds by alternating between two steps:

Step 1: Assignment step. Assign each $x \in \Omega$ to the region whose intensity mean yields the least within-cluster sum of squares using the following:

$$\Omega_i^{(n)} = \left\{x \in \Omega : \left\| I(x) - m_i^{(n)2} \le \right\| I(x) - m_j^{(n)2} \forall j, 1 \le j \le 3 \right\}, \; i = 1, 2, 3 \tag{9.55}$$

where each $x \in \Omega$ is assigned to region $x \in \Omega$, and n is the number of iterations.

Step 2: Update step. Calculate the new means to be the centroids of the intensities in the new regions using

$$m_i^{(n+1)} = \frac{1}{\left|\Omega_i^{(n)}\right|} \sum_{x \in \Omega_i^{(n)}} I(x) \tag{9.56}$$

When the assignments no longer change, each $x \in \Omega$ has been assigned to region Ω_i, and therefore a coarse segmentation Ω_1, Ω_2 and Ω_3 can be obtained. For three-class segmentation of sonar images, the three regions Ω_1, Ω_2 and Ω_3 are re-labeled according to their intensity means m_1, m_2 and m_3, respectively. The region with the highest intensity mean corresponds to the object region, and is denoted by Ω_1; the region with the lowest intensity mean corresponds to the shadow region, and is denoted by Ω_2; and the remaining region is denoted by Ω_3, whose intensity mean is between that of Ω_1 and $s\Omega_2$.

9.4.2 Fine Segmentation Using the Edge-driven Constraint Region-scalable Fitting Model.

The RSF model described in Section II is a two-phase model. It can be extended to a multiphase model using the method proposed by Vese and Chan (2002), which uses n level set functions to separate 2^n regions. To guarantee an unambiguous segmentation, a multi-region competition scheme was proposed by Mansouri et al. (2006) and is adopted in this paper. For a partition into N regions, a family of closed contours $\{C_i,\ i = 1,\ldots, N\text{-}1\}$ is used. The relationship between the family $\{\Omega_{C_i}\}$ of regions enclosed by the contours $\{C_i\}$ and the segmentation regions $\{\Omega_i\}$ of image domain Ω is defined by

$$\Omega_1 = \Omega_{C_1}$$

$$\Omega_2 = \Omega^c_{C_1} \cap \Omega_{C_2}$$

$$\Omega_N = \Omega^c_{C_1} \cap \Omega^c_{C_2} \cap \cdots \cap \Omega^c_{C_{N-2}} \cap \Omega^c_{C_{N-1}} \tag{9.57}$$

where $\Omega^c_{C_i}$ is the complement of Ω_{C_i}. For a level set implementation, the proposed method represents each contour C_i implicitly using the zero level set of function ϕ_i with the region enclosed by C_i corresponding to $\phi_i < 0$. Considering three-class segmentation, two level set functions ϕ_1 and ϕ_2 are required, and the segmentation regions $\{\Omega_i, i = 1, 2, 3\}$ can be given by

$$\Omega_i = \{x \in \Omega,\ M_i(\phi_1(x), \phi_2(x)) > 0\}, \quad i = 1,\ 2,\ 3 \tag{9.58}$$

where $M_1(\phi_1, \phi_2) = 1 - H(\phi_1)$, $M_2(\phi_1, \phi_2) = H(\phi_1)(1 - H(\phi_2))$ and $M_3(\phi_1, \phi_2) = H(\phi_1)H(\phi_2)$ are three masks that correspond to the three regions Ω_1, Ω_2 and Ω_3, respectively. Extending Eq. (9.47) for three-phase segmentation, total energy ε is given by

$$\varepsilon(\phi_1,\ \phi_2,\ f_1,\ f_2,\ f_3) = \sum_{i=1}^{3} \lambda_i \int_\Omega \int_\Omega (K_\sigma(x-y)|I(y) - f_i(x)|^2 M_i(\phi_1(y), \phi_2(y))dy)dx + \sum_{i=1}^{2} v_i \int_\Omega |\nabla H(\phi_i(x))|dx \tag{9.59}$$

where f_1, f_2 and f_3 are three functions that describe image intensities in Ω_1, Ω_2 and Ω_3, respectively; and λ_1, λ_2, λ_3, v_1 and v_2 are positive real constants. To avoid the re-initialization of ϕ_1 and ϕ_2, the regularization term is defined as

$$P(\phi_1,\ \phi_2) = \sum_{i=1}^{2} \mu_i \int_\Omega \frac{1}{2}(|\nabla\phi_i(x)| - 1)^2 dx \tag{9.60}$$

Considering residual noise and other disturbances, an optimal edge detection operator, that is, the Canny operator (Canny 1986), is adopted to obtain a clean image gradient, which is further integrated into the RSF model to help it jump out of the local minimum. Let E denote the set of coordinates of the edge points detected by the Canny operator. The gradients of those non-edge points are set to zero by

$$|\nabla I(x)| = \begin{cases} |\nabla I(x)|, & if\ x \in E \\ 0, & \text{otherwise} \end{cases} \tag{9.61}$$

To accelerate the convergence speed and drive the active contours to arrive at the desired boundary, an edge-driven constraint based on the gradients of the edge points is introduced as

$$E(\phi_1, \phi_2) = \sum_{i=1}^{2} \alpha_i \int_\Omega gH(-\phi_i(x))dx \tag{9.62}$$

where g is an edge indicator function defined by $g(|\nabla I(x)|) = 1/\ (1 + |\nabla I(x)|^2)$, and α_1 andα_2 are positive real constants. Finally, the energy functional to be minimized is given by

$$F(\phi_1, \phi_2, f_1, f_2, f_3) = \varepsilon(\phi_1, \phi_2, f_1, f_2, f_3) +\ P(\phi_1, \phi_2) + E(\phi_1, \phi_2) \tag{9.63}$$

Energy functional F can be minimized using an iterative two-step algorithm. The first step updates f_1, f_2 and f_3 using the following:

$$f_i(x) = \frac{K_\sigma(x) \times (M_i(\phi_1(x), \phi_2(x))I(x))}{K_\sigma(x) \times M_i(\phi_1(x), \phi_2(x))}, \quad i = 1,\ 2,\ 3 \tag{9.64}$$

The second step updates ϕ_1 and ϕ_2 according to the two following gradient flow equations:

$$\frac{\partial\phi_1}{\partial t} = \delta(\phi_1)(\lambda_1 e_1 - \lambda_2(1 - H(\phi_2))e_2 - \lambda_3 H(\phi_2)e_3) + v_1\delta(\phi_1)\text{div}\left(\frac{\nabla\phi_1}{|\nabla\phi_1|}\right) + \mu_1\left(\nabla^2\phi_1 - \text{div}\left(\frac{\nabla\phi_1}{|\nabla\phi_1|}\right)\right) + \alpha_1\delta(\phi_1)g \tag{9.65}$$

$$\frac{\partial\phi_2}{\partial t} = \delta(\phi_2)H(\phi_1)(\lambda_2 e_2 - \lambda_3 e_3) + v_2\delta(\phi_2)\mathrm{div}\left(\frac{\nabla\phi_2}{|\nabla\phi_2|}\right) + \mu_2\left(\nabla^2\phi_2 - \mathrm{div}\left(\frac{\nabla\phi_2}{|\nabla\phi_2|}\right)\right) + \alpha_2\delta(\phi_2)g \tag{9.66}$$

where δ is the Dirac delta function, and e_1, e_2 and e_3 are given by

$$e_i(x) = \int_\Omega K_\sigma(y-x)|I(x) - f_i(y)|^2 \mathrm{d}y, \quad i = 1,\ 2,\ 3 \tag{9.67}$$

In practice, Heaviside function H and Dirac delta function δ should be slightly regularized. The regularized Heaviside function H_ε is defined as

$$H_\varepsilon(x) = \frac{1}{2}\left(1 + \frac{2}{\pi}\arctan\left(\frac{x}{\varepsilon}\right)\right) \tag{9.68}$$

where ε is a positive real constant. The regularized Dirac delta function δ_ε is the derivative of H_ε and therefore is given by

$$\delta_\varepsilon(x) = H'_\varepsilon(x) = \frac{1}{\pi}\frac{\varepsilon}{\varepsilon^2 + x^2} \tag{9.69}$$

To calculate ϕ_1 and ϕ_2 using the iterative two-step algorithm, ϕ_1 and ϕ_2 need to be initialized. With the coarse segmentation Ω_1, Ω_2 and Ω_3 obtained using the method given in *B*, level set functions ϕ_1 and ϕ_2 can be initialized by

$$\phi_1 = \begin{cases} -2c_0, & if\ x \in \Omega_1 \\ 2c_0, & \text{otherwise} \end{cases} \tag{9.70}$$

$$\phi_2 = \begin{cases} -2c_0, & if\ x \in \Omega_2 \\ 2c_0, & \text{otherwise} \end{cases} \tag{9.71}$$

where c_0 is a positive constant.

9.4.3 Termination Criterion for Contour Evolution

When the active contours finally arrive at the true boundary, they should stop evolving and converge. Because the areas enclosed by the active contours change negligibly when they begin to converge, the number of pixels in the two regions enclosed by active contours C_1 and C_2 tend to be stable. Therefore, the number of pixels in the two regions enclosed by active contours C_1 and C_2 are used to provide the termination criterion for contour evolution. If the absolute differences of the number of pixels in the two regions between successive iterations remain smaller than a given threshold, then the evolution of contours stops, which is defined as

$$|N(C_i(t)) - N(C_i(t-1))| \le n_p \wedge |N(C_i(t-1)) - N(C_i(t-2))| \le n_p,\ i = 1, 2 \tag{9.72}$$

where $N(C_i(t))$, $N(C_i(t-1))$ and $N(C_i(t-2))$ are the number of pixels in the region enclosed by active contour C_i after t, $t-1$ and $t-2$ iterations, respectively; and n_p is the threshold.

9.5 Experimental and Comparative Studies

In this section, the experimental and comparative studies are conducted to demonstrate the effectiveness and efficiency of the proposed method. The experimental results of the proposed method are mainly compared with those of two robust image segmentation methods, namely, a robust fuzzy local information c-means clustering algorithm (FLICM) (Krinidis and Chatzis 2010) and the level set method with intensity bias correction (LSIBC) (Li et al. 2011). All the experiments are conducted on MATLAB R2011b installed in a computer with Intel Core i3 2.27 GHz CPU, 2G RAM, and Windows 7 operating system. Since there is strong speckle noise in sonar images, the NLMSF method is also used with the other two methods. For the NLMSF method, $\gamma = 0.5$ and $\alpha = 3$ are used. The smoothing parameter h^2 is defined by $h^2 = 2\beta\sigma_n^2 S$, where σ_n^2 is the estimated noise variance, $S = (2\alpha + 1)^2$ is the block size, and β is used for adjustment with a default value of 1. The following level set parameters of the proposed method are kept the same: the time step $\Delta t = 0.1$s, the regularization parameter $\varepsilon = 1$, the scale parameter $\sigma = 5$, $c_0 = 1$, $\lambda_1 = \lambda_2 = \lambda_3 = 1$, and $\mu_1 = \mu_2 = 1$. For the Canny operator, the high threshold is set to be 0.5 and the scale parameter of the Gaussian function is set to be 2. The default values of v_1, v_2, α_1 and α_2 are $v_1 = v_2 = 0.03 \times 255^2$, and $\alpha_1 = \alpha_2 = 150$. The length controlling parameters v_1 and v_2 should be a little larger for sonar images with more noise or coarser background, and the edge information controlling parameters α_1 and α_2 should be larger for sonar images with obvious intensity inhomogeneity. For the FLICM method, to achieve better results, the window size is usually set to be 5, the threshold is 0.001 and the maximum iteration number is 500. For the LSIBC method, multiple small circles that are evenly distributed over the image are used as the level set initialization and the diameter of each circle is set to be 10. The initial positions of active contours must be carefully selected (see part *E*). The following parameters are kept the same for the LSIBC method: $\Delta t = 0.1$, $\varepsilon = 1$, $c_0 = 1$, and $\mu = 1$. The default values of σ and v is: $\sigma = 5$, and $v = 0.023 \times 255^2$; and they will be adjusted with different images to achieve better results. The termination criterion given in Eq. (9.72) is used for the LSIBC method and the proposed method, and n_p is set to be 2 for all the images.

For segmenting the sonar images into highlight, shadow, and sea-bottom reverberation, two side-scan sonar image sets are used, which will be described in Sect. 9.5.1. The final segmentation results and a qualitative assessment are presented in Sect. 9.5.2. Quantitative assessment is given in Sect. 9.5.3. Comparison of running time is given in Sect. 9.5.4. Comparison of the evolution of the active contour is shown in Sect. 9.5.5.

In part Sect. 9.5.6, a two-phase segmentation of the proposed scheme is applied to pipeline detection in sonar images and the results are compared with the latest method (Acosta and Villar, 2015).

9.5.1 Sonar Image Sets

The two sonar image sets are given in Figs. 9.2a and 9.3a, respectively. The first sonar image set consists of four sonar images with different levels of noise, while the second consists of four sonar images with both noise and intensity inhomogeneity. In Fig. 9.2a, the first was imaged by a Klein 5000 sonar and is of a Cessna twin-engine airplane; the second was obtained by an EdgeTech 4200-FS system and is of boxes; the third was imaged by a Klein 3900 sonar and is of lobster traps; the last was imaged by a Sea Scan 900 sonar and is of sand ripples. In Fig. 9.3a, the first is a synthetic

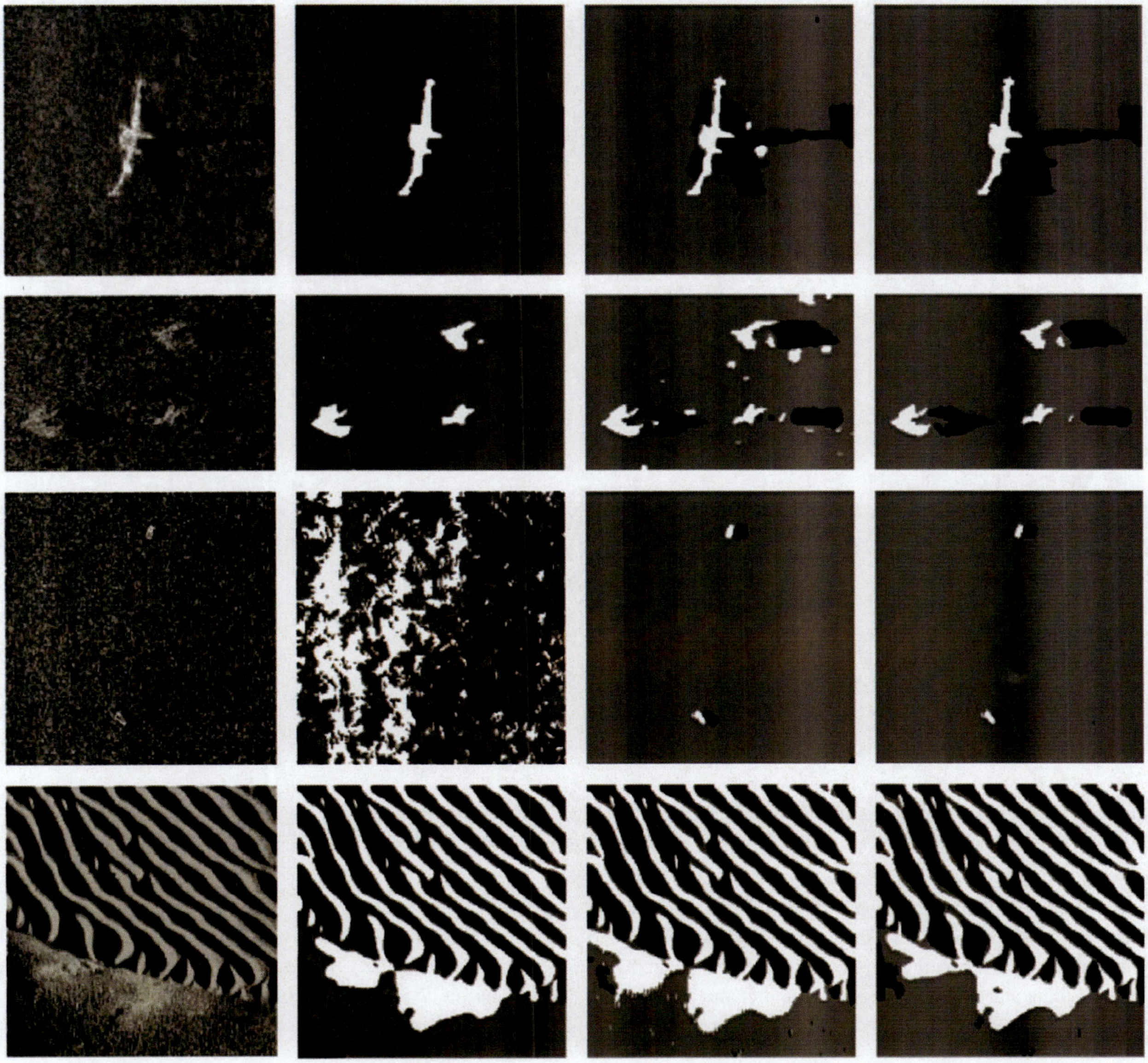

(a) (b) (c) (d)

Fig. 9.2 Comparison of the three methods in segmenting four noisy side-scan sonar images. **a** Original side-scan sonar images; **b** results using the FLICM method; **c** results using the LSIBC method; and **d** results using the proposed method

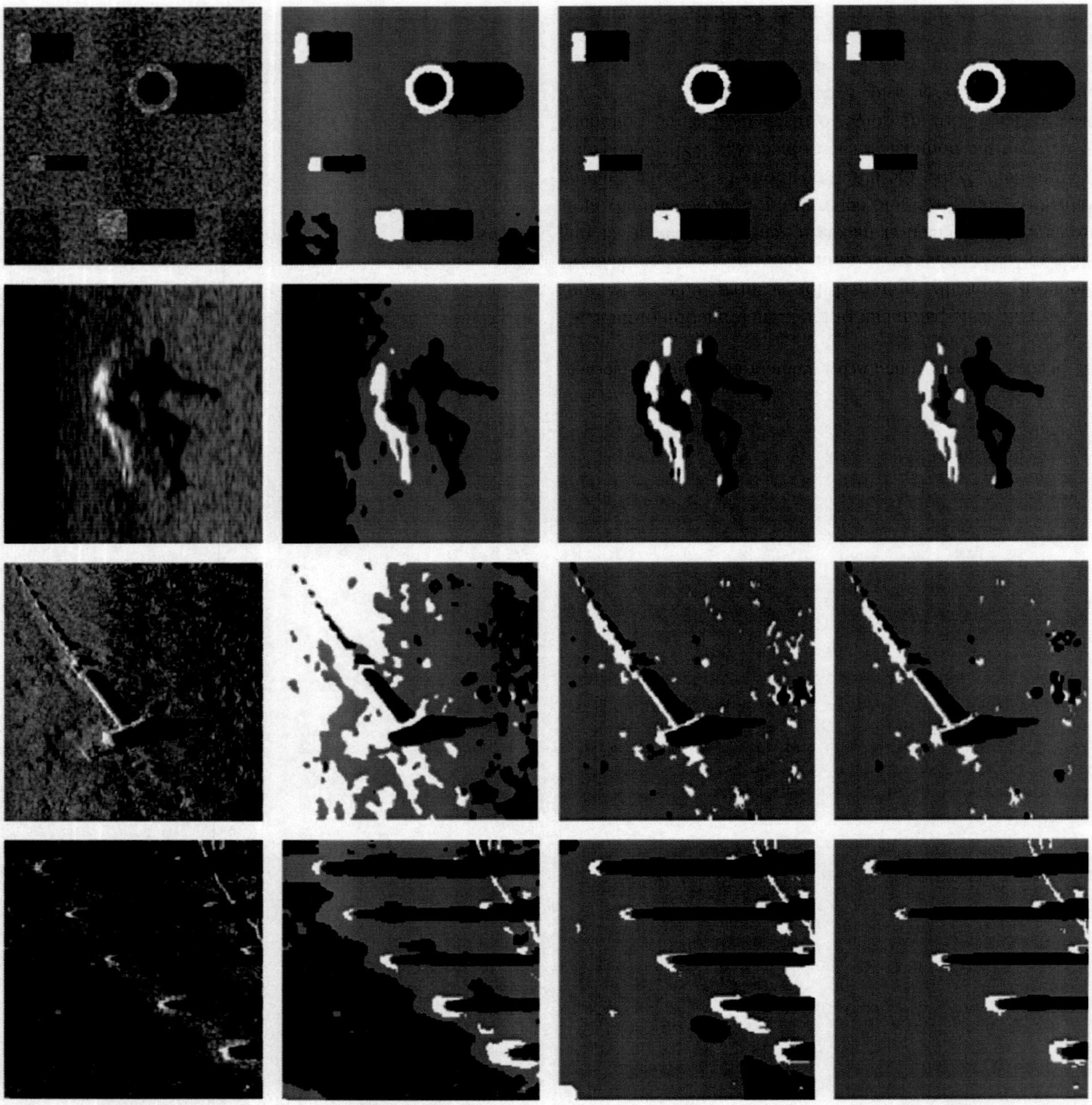

(a) (b) (c) (d)

Fig. 9.3 Comparison of the three methods in segmenting four noisy and intensity inhomogeneous side-scan sonar images. **a** Original side-scan sonar images; **b** results using the FLICM method; **c** results using the LSIBC method; and **d** results using the proposed method

sonar image containing a ring and three rectangles of different sizes; the second was made with an EdgeTech 4125 sonar and is of a drowning victim; the third was imaged by a Klein 3900 sonar and is of an anchor and chain; the last was provided by a CM2 sonar and is of dockside pilings. The sizes of the eight images are 262 × 262, 178 × 274, 250 × 250, 300 × 300, 190 × 190, 260 × 260, 320 × 320 and 135 × 135 pixels, respectively. By dividing an image into multiple 7 × 7 blocks and averaging all the variances, the noise variance can be approximately estimated. The estimated noise standard deviations of the eight side-scan sonar images are 22.3, 32.1, 36.8, 42.5, 24.0, 16.8, 34.2 and 26.9, respectively.

9.5.2 Segmentation Results and Qualitative Assessment

The final segmentation results of the FLICM, the LSIBC and the proposed methods for the first image set are given in Figs. 9.2b–d, respectively. The results are presented as pseudo-image representations. Pixels that belong to the same region are given the same intensity value, i.e., 255, 125 and 0 are assigned to highlight, sea-bottom reverberation and shadow, respectively. The window size of the FLICM method is set to be 4 for the third image. For the LSIBC method, $\sigma = 6$ is used on the first image, and v is set to be 0.035×255^2 for the last image. For the proposed method, v_1 and v_2 are slightly increased according to the estimated noise levels: $v_1 = v_2 = 0.035 \times 255^2$ is used on the second image and $v_1 = v_2 = 0.04 \times 255^2$ is used on the last two images.

By using the NLMSF, the FLICM method can often produce clean segmentation results, except for the third image *Lobster Traps*. Due to the strong disturbance of sea bottom reverberation and the classification bias, the FLICM method fails to extract the two small lobster traps and their shadows. Moreover, the FLICM method cannot give accurate results, e.g., the shadow of the airplane extracted by the FLICM method for the first image *Cessna Airplane* is less complete than those results produced with the other two methods; for the last image *Sand Ripples*, the lower-left corner is incorrectly labeled as shadow, and some of the extracted shadows are discontinuous (e.g., the third, fourth and fifth ripple shadows counting from the left). Compared with the FLICM method and the proposed method, obvious false segmentation exists in the results produced with the LSIBC method, e.g., two false shadows and objects are detected for the image *Cessna Airplane*, multiple false objects are detected for the second image *Boxes*, and an obvious shadow region at the lower-left corner is not detected for the image *Sand Ripples*. The proposed method can always produce clean and accurate results without obvious false segmentation, which can be seen from the results in Fig. 9.2d, e.g., the airplane and its shadow are cleanly and completely extracted; two small lobster traps and their shadows are accurately detected.

The final segmentation results of the second image set are given in Figs. 9.3b–d, respectively. The window size of the FLICM method is set to be 4 on segmenting the last image. For the LSIBC method, $\sigma = 6$ and $v = 0.04 \times 255^2$ is used on the second image, and $v = 0.035 \times 255^2$ is used on the third image. For the proposed method, considering obvious intensity inhomogeneity, $\alpha_1 = \alpha_2 = 400$ is used on the second, third and last images; $v_1 = v_2 = 0.04 \times 255^2$ is used on the second and third images.

Although the FLICM method is robust in dealing with noise, it cannot well segment sonar images with intensity inhomogeneity, which can be seen in Fig. 9.3b. Obvious false segmentation exists in the four results produced with the FLICM method, e.g., a large region is incorrectly labeled as the shadow at the left of the result for the second image *Drowning Victim*, and nearly half of the reverberation region is wrongly identified as highlight and connected to the anchor. Besides, the edges of the results produced with the FLICM method are less accurate, which can be seen from the results for the first image *Ring & Rectangles* and the last image *Dockside Pilings*. Compared with the FLICM method, the LSIBC method can well deal with intensity inhomogeneity. However, obvious false segmentation also exists in the results produced with the LSIBC method, e.g., a false object at the lower-right is detected in segmenting the image *Ring and Rectangles*; the man detected is broken and a false shadow is detected in segmenting the image *Drowning Victim*; and one of the pilings is false detected and the lower-right corner is wrongly identified as shadow in segmenting the image *Dockside Pilings*. Among all the three methods, the proposed method can produce cleaner results without obvious false segmentation. Besides, the proposed method is also accurate, which can be clearly seen from the results for the image *Ring & Rectangles* and the image *Dockside Pilings*.

From the results in Figs. 9.2 and 9.3, it can be seen that side-scan sonar image segmentation is quite a challenge. The FLICM method cannot extract small objects due to the disturbance of sea bottom reverberation and the classification bias, and also cannot deal with intensity inhomogeneity. The LSIBC method can well deal with intensity inhomogeneity. However, it still produces obvious false segmentation due to the ill-posed property of intensity bias correction and the selection of initial contour position. A possible way to improve the segmentation is to use post-processing, e.g., to take advantage of the spatial dependency between echo and shadow areas can help to eliminate some false segmentation (Mignotte et al. 2000). However, this strategy cannot work well on the results produced with the FLICM method in segmenting images *Lobster Traps* and *Anchor and Chain*, because in these results the real object or shadow is connected to the false one. The proposed method is robust in dealing with noise and intensity inhomogeneity and can provide accurate segmentation results.

9.5.3 Quantitative Assessment

To allow a quantitative assessment of the segmentation results, the corresponding ground-truth segmentation maps usually need to be given. For the synthetic sonar image *Ring & Rectangles*, a ground-truth segmentation map is given in Fig. 9.4 based on the reference image. The corresponding ground-truth segmentation maps of the other real sonar images are obtained by manual segmentation (Celik and Tjahjadi 2011), and are also given in Fig. 9.4 for additional quantitative assessment. Segmentation success rate ρ is used to give an overall evaluation of the whole segmentation result and is calculated according to the following:

$$\rho = \frac{N_c}{N_t} \times 100\% \tag{9.73}$$

where N_c is the number of correctly segmented pixels and N_t is the total number of pixels. The higher the value of ρ is, the better the segmentation is. Considering object recognition and analysis, the result is also evaluated through quantitative measures that take into consideration relative aspects of position, size and shape of segmented objects and their shadows. Success of position fit F_{xy}, success of size fit F_n, and success of shape fit F_s, which are first defined in (Lucca et al. 1998) and used in (Marques et al. 2012), are used in this paper to evaluate the accuracy of the extracted objects and their shadows. The mean of F_{xy} from all the objects can be given by

$$F_{xy} = \frac{1}{A}\sum_{i=1}^{A}\left(1 - \frac{d_x^{(i)} + d_y^{(i)}}{2}\right) \tag{9.74}$$

where A is the total number of objects; $d_x^{(i)}$ is the difference in the horizontal direction between the center of the ith object in the result and that of its corresponding object in the ground-truth segmentation map, while $d_y^{(i)}$ is the difference in the vertical direction between the two centers. The values of $d_x^{(i)}$ and $d_y^{(i)}$ can be calculated by

$$d_x^{(i)} = \frac{1}{W}|\langle x_{S_i}\rangle - \langle x_{G_i}\rangle|, \quad d_y^{(i)} = \frac{1}{H}|\langle y_{S_i}\rangle - \langle y_{G_i}\rangle| \tag{9.75}$$

where W is the number of the columns, while H is the number of the rows; S_i and G_i denote the ith object in the result and that in the ground-truth segmentation map, respectively; $\langle x_{S_i}\rangle$ and $\langle y_{S_i}\rangle$ represent the means of the abscissas and the ordinates of the ith object in the result, respectively, while $\langle x_{G_i}\rangle$ and $\langle y_{G_i}\rangle$ represent those means in the ground-truth segment- ation map. The mean of F_n is given by

$$F_n = \frac{1}{A}\sum_{i=1}^{A}\left(1 - \frac{|N_{S_i} - N_{G_i}|}{N_{S_i} + N_{G_i}}\right) \tag{9.76}$$

where N_{S_i} and N_{G_i} are the number of pixels of the ith object in the result and in the ground-truth segmentation map, respectively. The mean of F_s is given by

Fig. 9.4 The ground-truth segmentation maps for the eight sonar images

$$F_s = \frac{1}{A}\sum_{i=1}^{A}\frac{N_{S_i \cap G_i}}{N_{S_i \cup G_i}} \quad (9.77)$$

where $N_{S_i \cap G_i}$ is the number of pixels of the intersection of the sets S_i and G_i, while $N_{S_i \cap G_i}$ is the number of pixels of the union of the two sets. When the segmentation result and the ground-truth segmentation map are the same, F_{xy}, F_n and F_s are 1, and they decrease when the dissimilarity between the segmentation result and the ground-truth segmentation map increases. Therefore, the higher F_{xy}, F_n and F_s are, the better the segmentation is. The quantitative measures of the results are given in Table 9.1.

Table 9.1 shows that the performance of the proposed method is usually better than that of the FLICM method and the LSIBC method in segmentation success rate and accurate extraction of objects and their shadows. It is worth noticing that F_s of the results produced with the FLICM significantly decrease because the shapes of the extracted objects using the FLICM method change a lot, which is confirmed by the results in the first row of Fig. 9.3.

As a state-of-the-art denoising method for additive Gaussian noise, the BM3D can be well applied to many common images. However, in segmenting sonar images that are often disturbed by multiplicative speckle noise, using the NLMSF as the preprocessing method is a better choice. To demonstrate this, a comparison of the denoised images for the image *Cessna Airplane* using the BM3D and the NLMSF is given in Fig. 9.5. It can be seen in Fig. 9.5 that the speckle noise in the reverberation region is reduced more effectively by the NLMSF than that by the BM3D. Using the BM3D instead of the NLMSF, some segmentation results of the three methods are given in Fig. 9.6 and the quantitative assessment of these results are given in Table 9.2. Comparing Table 9.2 with Table 9.1, it can be seen that the NLMSF can usually help to give better segmentation results than the BM3D.

9.5.4 Comparison of the Running Time

Table 9.3 gives a comparison of the iteration numbers and the running times of the three methods when combined with the NLMSF in segmenting the eight images in the same circumstances. To ensure the fairness of comparison, the code of the proposed method is also not optimized. From Table 9.3, it can be seen that the proposed method needs less iterations than the other two methods, and therefore is the fastest among the three methods in almost all cases. The reason why the FLICM method is the fastest in segmenting the image *Dockside Pilings* is that it uses a smaller window size. However, the FLICM method fails to provide a satisfactory result in this case. The iteration numbers of the proposed method in segmenting the eight images are usually smaller than 100, which are very small to our best knowledge.

9.5.5 Comparison of the Active Contours Evolution

Compared with the LSIBC method, the proposed method converges more quickly and is less likely to trap into local minima, which is demonstrated by Figs. 9.7 and 9.8. Figure 9.7 shows a perfect combination of the three steps. After despeckling using the NLMSF, coarse segmentation provides a good initialization, and by integrating the

Table 9.1 Comparison of quantitative measures

Input image	FLICM				LSIBC				Proposed method			
	ρ	F_{xy}	F_n	F_s	ρ	F_{xy}	F_n	F_s	ρ	F_{xy}	F_n	F_s
Cessna airplane	99.24	0.998 3	0.967 2	0.897 8	98.14	0.997 3	0.983 7	0.924 0	99.69	0.999 2	0.991 7	0.954 7
Boxes	98.86	0.996 0	0.986 2	0.867 8	97.75	0.997 8	0.973 3	0.925 8	99.61	0.998 7	0.977 7	0.949 3
Lobster traps	49.95				99.88	0.999 3	0.950 1	0.888 5	99.96	0.999 5	0.981 6	0.948 2
Sand ripples	91.46	0.990 3	0.917 4	0.793 3	94.28	0.997 3	0.959 1	0.912 6	98.01	0.999 1	0.985 7	0.967 8
Ring and rectangles	93.76	0.998 8	0.978 9	0.920 3	99.35	0.999 4	0.990 6	0.960 3	99.60	0.999 5	0.990 0	0.968 8
Drowning victim	66.04	0.988 1	0.882 7	0.753 5	96.31	0.991 4	0.926 4	0.812 9	99.25	0.997 5	0.976 2	0.921 4
Anchor and chain	54.41				94.86	0.965 3	0.863 2	0.735 9	96.69	0.968 2	0.883 1	0.768 5
Dockside pilings	53.26	0.989 8	0.854 1	0.686 6	90.91	0.991 4	0.893 3	0.821 5	98.63	0.996 4	0.943 8	0.880 3

Note ρ- success rate; F_{xy}- success of position fit; F_n-success of size fit; F_s- success of shape fit

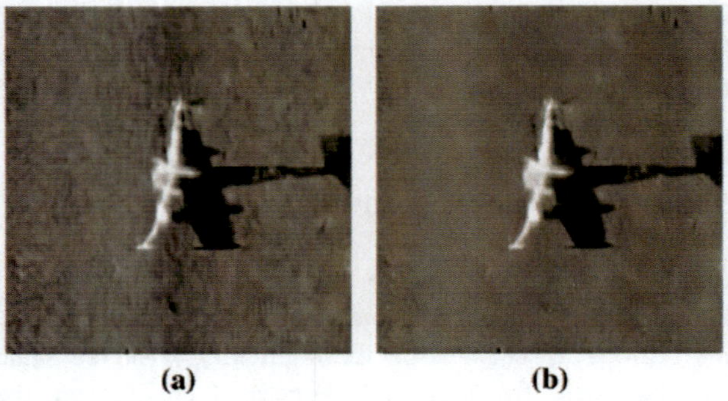

Fig. 9.5 Comparison of the denoised images for the image *Cessna Airplane* using the BM3D and the NLMSF. **a** Denoised image using the BM3D; and **b** denoised image using the NLMSF

edge-driven constraint with the RSF model, the proposed method converges after 24 iterations, which is much faster than the LSIBC method. Figure 9.8 shows that even if the initialization is not very good due to obvious intensity inhomogeneity, by using the edge-driven constraint, the proposed method can still converge quickly. In Figs. 9.7d and 9.8d, it is worth noticing that the LSIBC method is more likely to trap into local minima, which will stop at false positions and result in obvious false segmentation, including false object or shadow detection and coarse edges.

The LSIBC method is also sensitive to noise and the initial position of active contours, because a larger length controlling parameter v has to be used to erase residual noise points, however, a larger length controlling parameter

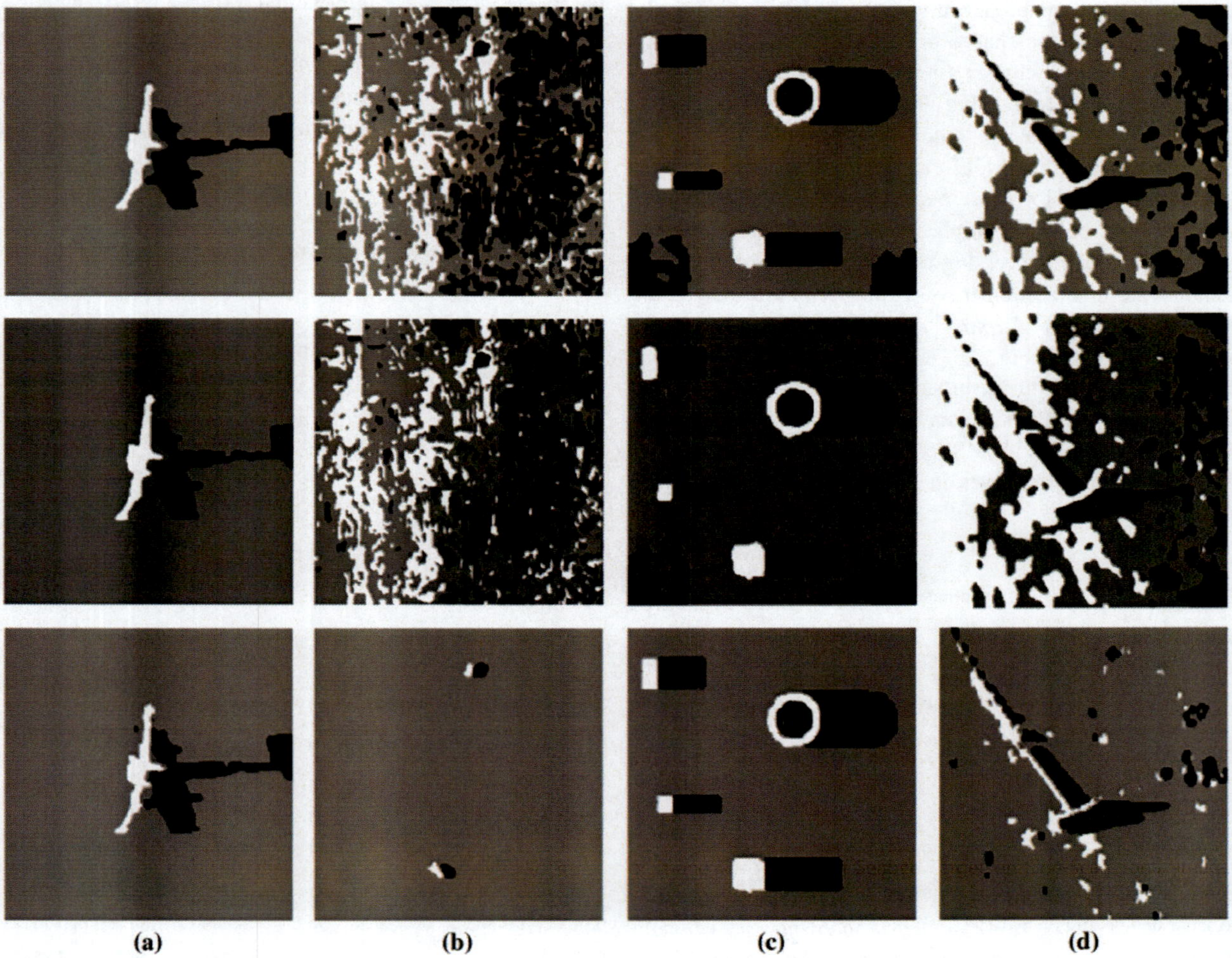

Fig. 9.6 The segmentation results of the three methods using the BM3D instead of the NLMSF. Upper row: results using the FLICM method. Middle row: results using the LSIBC method. Lower row: results using the proposed method. **a** Results for the image *Cessna Airplane*; **b** results for the image *Lobster Traps*; **c** results for the image *Ring and Rectangles*; and **d** results for the image *Anchor and Chain*

Table 9.2 Comparison of quantitative measures

Input image	FLICM (using BM3D)				LSIBC (using BM3D)				Proposed method (using BM3D)			
	ρ	F_{xy}	F_n	F_s	ρ	F_{xy}	F_n	F_s	ρ	F_{xy}	F_n	F_s
Cessna airplane	99.22	0.998 2	0.970 7	0.894 4	97.79	0.996 2	0.977 8	0.911 4	99.67	0.998 4	0.994 4	0.957 3
Lobster traps	50.10				99.87	0.999 3	0.958 8	0.867 8	99.94	0.999 2	0.974 1	0.909 3
Ring and rectangles	93.47	0.998 7	0.978 2	0.920 6	99.25	0.999 0	0.981 7	0.948 0	99.56	0.999 5	0.988 4	0.968 6
Anchor and chain	54.10				93.29	0.958 0	0.849 4	0.724 2	96.04	0.965 5	0.869 2	0.752 7

Note ρ-success rate; F_{xy}-success of position fit; F_n-success of size fit; F_s-success of shape fit

Table 9.3 Iteration numbers and total running time using different methods

Input image	Iteration number			Running time/s		
	FLICM	LSIBC	Proposed method	FLICM	LSIBC	Proposed method
Cessna airplane	52	97	24	29.83	32.02	9.21
Boxes	97	142	61	36.71	27.62	13.48
Lobster traps	49	64	26	10.66	15.61	10.23
Sand ripples	104	285	82	71.80	93.80	27.58
Ring and rectangles	73	79	21	21.73	10.66	5.98
Drowning victim	279	233	121	219.80	109.59	43.05
Anchor and chain	424	205	111	213.71	51.98	27.48
Dockside pilings	101	93	79	5.88	8.34	6.60

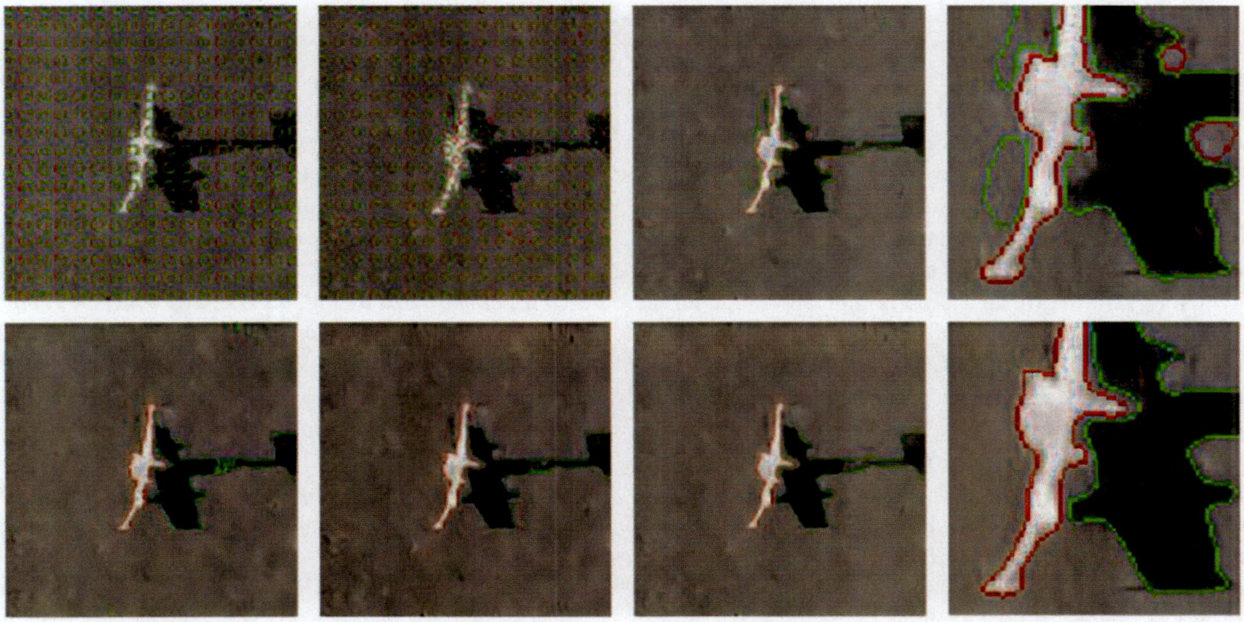

(a) (b) (c) (d)

Fig. 9.7 Comparison of the evolution of the active contour of the LSIBC and the proposed methods in segmenting the image *Cessna Airplane*. The active contours C_1 and C_2 of the two methods are plotted as the solid red and green contours, respectively. Upper row: results using the LSIBC method. Lower row: results using the proposed method. **a** Initial contours; **b** active contours after 15 iterations; **c** final contours, upper contours after 97 iterations, and lower contours after 24 iterations; **d** enlarged local parts of (c)

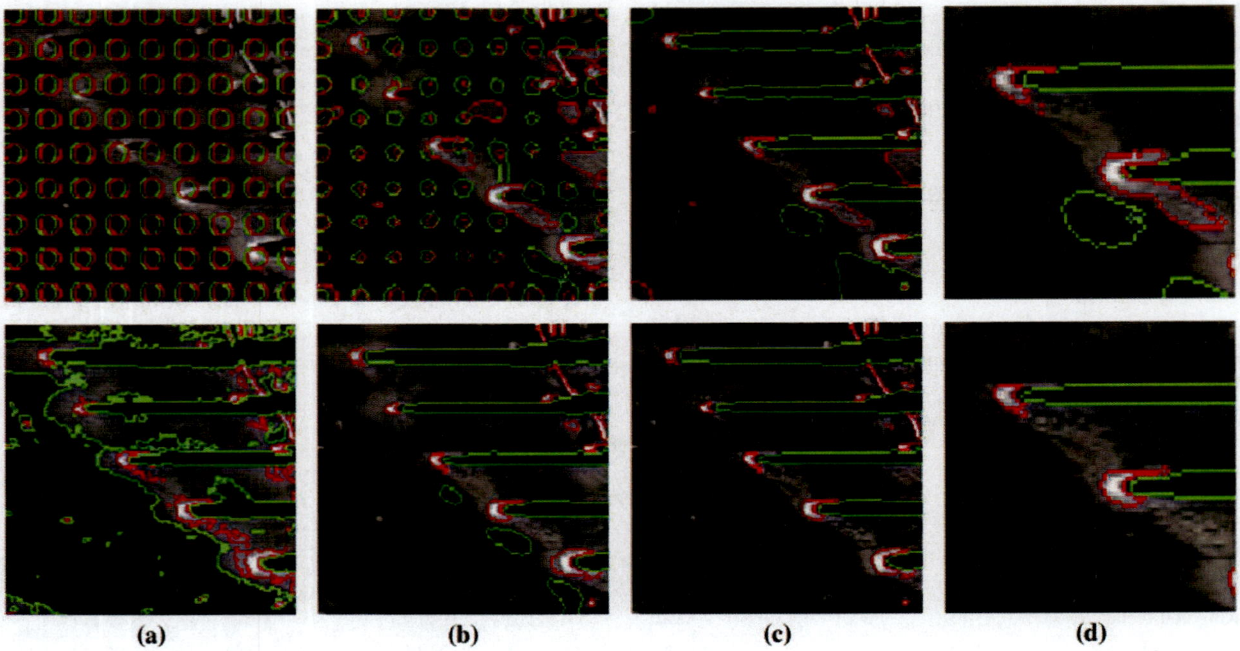

Fig. 9.8 Comparison of the evolution of the active contour of the LSIBC and the proposed methods in segmenting the image *Dockside Pilings*. The active contours C_1 and C_2 of the LSIBC method are plotted on the denoised image as the solid green and red contours, respectively; while those of the proposed method are plotted as the solid red and green contours, respectively. Upper row: results using the LSIBC method. Lower row: results using the proposed method. **a** Initial contours; **b** active contours after 25 iterations; **c** final contours, upper contours after 93 iterations, and lower contours after 79 iterations; **d** enlarged local parts of **(c)**

v provides a strong shrinkage force and therefore may erase some initial circles when they are small and lack enough force from the boundary of object to overcome this shrinkage force. The sensitivity of the LSIBC method to the initial position of active contours can be demonstrated by Figs. 9.9 and 9.10. Even if a slight position shift will result in a quite different result, e.g., one of the two contours finally disappears and only the objects are extracted in Figs. 9.8 and 9.9. Therefore, how to give an appropriate initial position for the LSIBC method is a problem, and it has to be solved by trial-and-error. For the LSIBC method, the intervals in the horizontal and vertical directions between the two adjacent circles that belong to the same active contour are usually set to be 6. The initial contour position for producing the results in Figs. 9.2 and 9.3 are {(2, 2), (2, 0)}, {(3, 3), (3, 0)}, {(3, 3), (2, 0)}, {(2, 2), (2, 0)}, {(1, 1), (2, 0)}, {(3, 3), (2, 0)}, {(1, 1), (2, 0)}, and {(3, 3), (2, 0)}, where the first item in each { } represents center coordinates in the horizontal and vertical directions of the first circle in the upper-left corner, and the second item includes the rightward and downward offsets of C_2 with respect to C_1. Compared with the LSIBC method, the proposed method automatically gives the initialization, and by combining the edge-driven constraint with the RSF model, it can finally produce a good result even if the initial segmentation is not very good.

It is also worth mentioning that by using the coarse segmentation proposed in the paper, the combination of the edge-driven constraint with the RSF model is essential, which is demonstrated by Fig. 9.11. In Fig. 9.11, the results are produced with the extended multiphase RSF model using the NLMSF and the coarse segmentation. The parameters are the same except that $\alpha_1 = \alpha_2 = 0$, which means that the edge-driven constraint will not be used in the active model. It can be seen from Fig. 9.11 that using the coarse segmentation, the original multiphase RSF model alone lacks enough force to overcome the initial big area false segmentation caused by intensity inhomogeneity and stops at an obvious local minimum, which results in obvious false segmentation in the reverberation and a not straight shadow of the third piling.

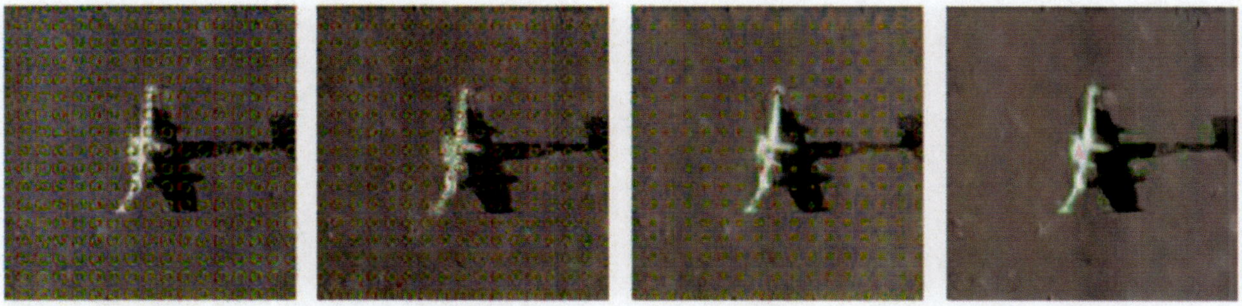

(a) (b) (c) (d)

Fig. 9.9 The active contours evolution of the LSIBC method using a different initial position from that in Fig. 9.7 in segmenting the image *Cessna Airplane*. The active contours C_1 and C_2 are plotted as the solid red and green contours, respectively. **a** Initial contours with {(1, 1), (2, 0)}; **b** active contours after 15 iterations; **c** active contours after 25 iterations; **d** final contours after 90 iterations

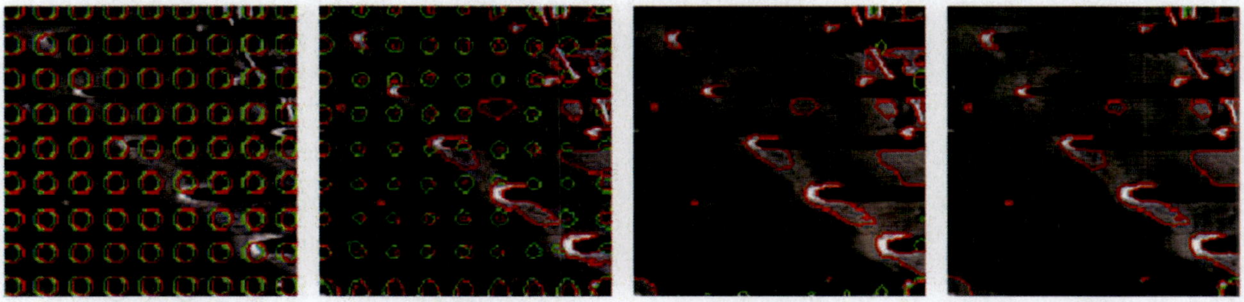

(a) (b) (c) (d)

Fig. 9.10 The active contours evolution of the LSIBC method using a different initial position from that in Fig. 9.8 in segmenting the image *Dockside Pilings*. The active contours C_1 and C_2 are plotted as the solid green and red contours, respectively. **a** Initial contours with {(2, 2), (2, 0)}; **b** active contours after 25 iterations; **c** active contours after 50 iterations; **d** final contours after 81 iterations

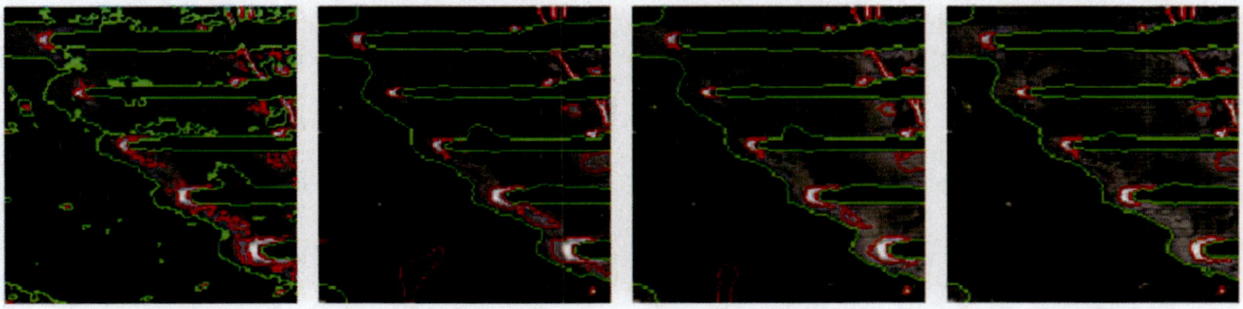

(a) (b) (c) (d)

Fig. 9.11 The active contours evolution of the original multiphase RSF model using the NLMSF and the coarse segmentation in segmenting the image *Dockside Pilings*. The active contours C_1 and C_2 are plotted as the solid red and green contours, respectively. **a** Initial contours; **b** active contours after 25 iterations; **c** active contours after 50 iterations; **d** final contours after 80 iterations

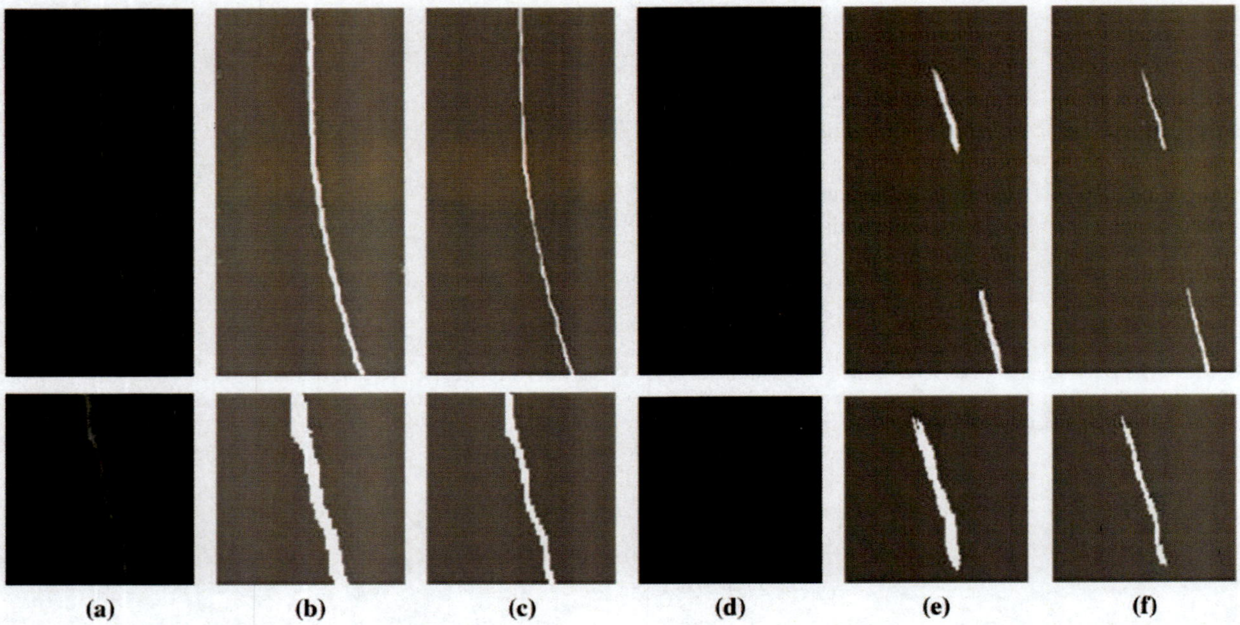

Fig. 9.12 Comparison of the proposed method and the ACA-CFAR 2-D method in pipeline detection. Upper row: **a** Sonar image of a curved pipeline; **b** detected pipeline of (a) using ACA-CFAR 2-D; **c** detected pipeline of (**a**) using the proposed method; **d** sonar image of a pipeline partially buried; **e** detected pipeline of (**d**) using ACA-CFAR 2-D; **f** detected pipeline of (**d**) using the proposed method. Lower row: (**a**)–(**f**) are local parts of those in the upper row, respectively

9.5.6 Pipeline Detection Using the Proposed Method

Using two-class clustering in the coarse segmentation and only one level set function in the fine segmentation, a two-phase model of the proposed method can be applied to pipeline detection of side-scan sonar images. The results produced with the proposed method using the default parameters and by the ACA-CFAR 2-D method are given in Fig. 9.12. Both methods can well extract the whole pipelines. However, the pipelines detected by the ACA-CFAR 2-D are bigger than the real ones, while those detected by the proposed method are almost the same, which can be clearly seen in Fig. 9.12.

9.6 Target Detection Software Interface and Description

The proposed sonar image target detection method has been developed and applied, which has been tested in cooperation with Shanghai AoTuo Deepwater Equipment Technology Development Co., Ltd. The two parties cooperated in underwater sonar imaging of Shallow Caosha Reservoir and detected fish and other targets. The interface of the developed sonar image segmentation software is given in Fig. 9.13.

When the software is used, the user first clicks the "Open Image" button to open the sonar image of the target to be detected. Because of the large number of parameters in the level set, the software provides an adaptive detection parameter configuration, and the default detection parameters can be read by clicking the "Read Parameters" button. The denoising method, the number of regions and the level set segmentation parameters can be directly modified by the user in the interface. After reading the parameters, click the "Target Detection" button. The software uses the encapsulated fast level set segmentation method to detect the target. The intermediate detection results can also be displayed. When the convergence condition is reached or the maximum number of iterations is reached, the software outputs the final target detection results and displays them on the upper right of the interface. Users can click on "Save Result" to save the processing results, and at the same time, they can click on "Target Measurement" to calculate the size and size of the target quantitatively.

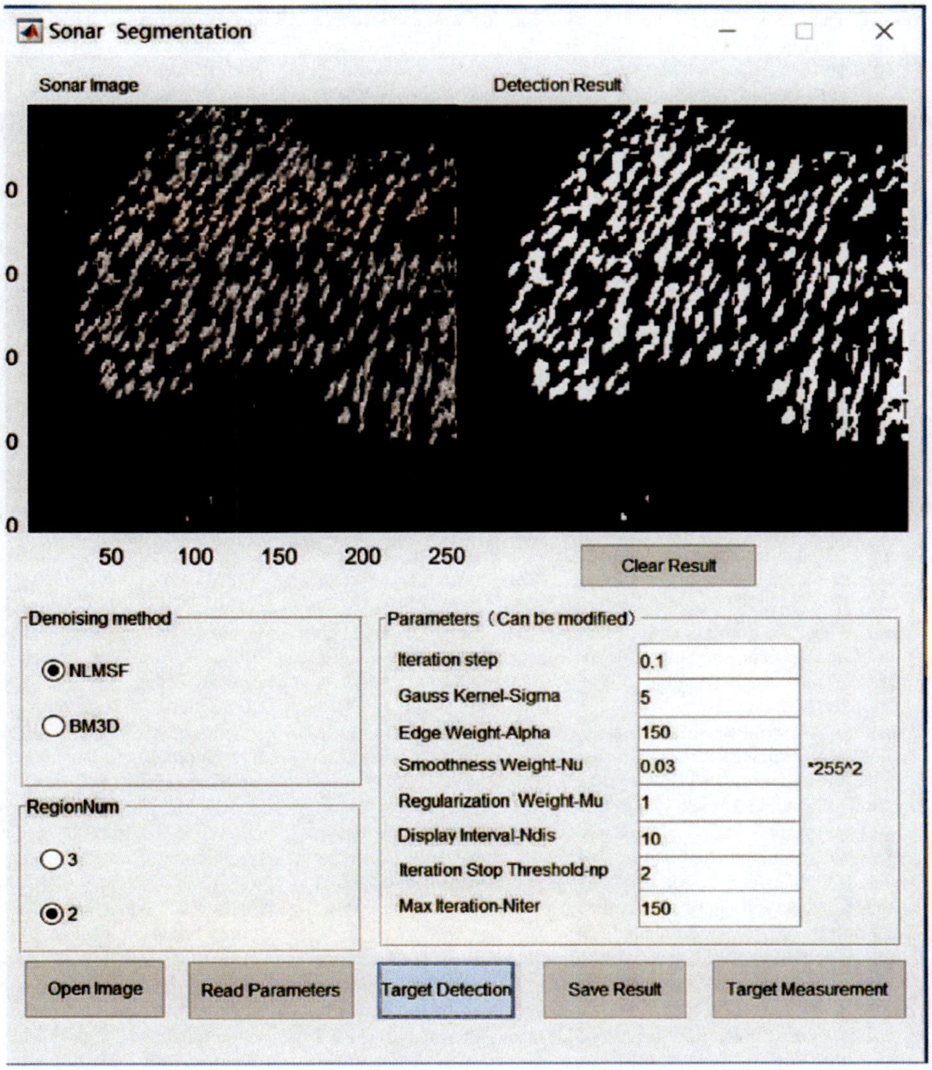

Fig. 9.13 Software interface of sonar image target detection

References

Acosta G, Villar S (2015) Accumulated CA-CFAR process in 2-D for online object detection from side-scan sonar data. IEEE J Oceanic Eng 40(3):558–569. https://doi.org/10.1109/JOE.2014.2356951

Adalsteinsson D, Sethian JA (1995) A fast level set method for propagating interfaces. J Comput Phys 118(2):269–277. https://doi.org/10.1006/jcph.1995.1098

Appleton B, Talbot H (2005) Globally optimal geodesic active contours. J Math Imaging Vis 23(1):67–86. https://doi.org/10.1007/s10851-005-4968-1

Ayed IB, Hennane N, Mitiche A (2006) Unsupervised variational image segmentation/classification using a Weibull observation model. IEEE Trans Image Process 15(11):3431–3439. https://doi.org/10.1109/TIP.2006.881961

Ayed IB, Mitiche A, Belhadj Z (2005) Multiregion level set partitioning of synthetic aperture radar images. IEEE Trans Pattern Anal Mach Intell 27(5):793–800. https://doi.org/10.1109/TPAMI.2005.106

Balla-Arabé S, Gao X, Wang B (2013) A fast and robust level set method for image segmentation using fuzzy clustering and lattice Boltzmann method. IEEE Trans Cybern 43(3):910–920. https://doi.org/10.1109/TSMCB.2012.2218233

Bresson X, Esedoglu S, Vandergheynst P et al (2007) Fast global minimizes of the active contour/snake model. J Math Imaging Vis 28:151–167. https://doi.org/10.1007/s10851-007-0002-0

Buades A, Coll B, Morel JM (2005) A review of image denoising algorithms, with a new one. Siam J Multiscale Model Simul 4 (2):490–530. https://doi.org/10.1137/040616024

Burger HC, Schuler CJ, Harmeling S (2012) Image denoising: can plain neural networks compete with BM3D. Paper presented at the IEEE conference on computer vision and pattern recognition, Providence, RI, USA

Canny J (1986) A computational approach to edge detection. IEEE Trans Pattern Anal Mach Intell 8(6):679–698. https://doi.org/10.1109/TPAMI.1986.4767851

Caselles V, Kimmel R, Sapiro G (1997) Geodesic active contours. Int J Comput Vision 22(1):61–79. https://doi.org/10.1109/ICCV.1995.466871

Celik T, Tjahjadi T (2011) A novel method for side-scan sonar image segmentation. IEEE J Oceanic Eng 36(2):186–193. https://doi.org/10.1109/JOE.2011.2107250

Cham TJ, Cipolla R (1997) Stereo coupled active contours. Paper presented at the IEEE conference on computer vision and pattern recognition, San Juan, Puerto Rico, USA, 17–19 June 1997

Chan TF, Vese LA (2001) Active contours without edges. IEEE Trans Image Process 10(2):266–277. https://doi.org/10.1109/83.902290

Chatterjee P, Milanfar P (2010) Is denoising dead? IEEE Trans Image Process 19(4):895–911. https://doi.org/10.1109/TIP.2009.2037087

Chen Y, Tagare HD, Thiruvenkadam S et al (2002) Using prior shapes in geometric active contours in a variational framework. Int J Comput Vision 50(3):315–328. https://doi.org/10.1023/a:1020878408985

Chung G, Vese LA (2009) Image segmentation using a multilayer level-set approach. Comput Vis Sci 12(6):267–285. https://doi.org/10.1007/s00791-008-0113-1

Cobra DT, Oppenheim AV, Jaffe JS (1992) Geometric distortions in side-scan sonar images: a procedure for their estimation and correction. IEEE J Oceanic Eng 17(3):252–268. https://doi.org/10.1109/48.153442

Coupé P, Hellier P, Kervrann C et al (2009) Nonlocal means-based speckle filtering for ultrasound images. IEEE Trans Image Process 18(10):2221–2229. https://doi.org/10.1109/TIP.2009.2024064

Coupé P, Yger P, Prima S et al (2008) An optimized blockwise non local means denoising filter for 3D magnetic resonance images. IEEE Trans Med Imaging 27:425–441. https://doi.org/10.1109/TMI.2007.906087

Cozzolino D, Parrilli S, Scarpa G et al (2014) Fast adaptive nonlocal SAR despeckling. IEEE Trans Geosci Remote Sens Lett 11(2):524–528. https://doi.org/10.1109/igrs.2013.2271650

Dabov K, Foi A, Katkovnik V et al (2007) Image denoising by sparse 3-D transform-domain collaborative filtering. IEEE Trans Image Process 16(8):2080–2095. https://doi.org/10.1109/TIP.2007.901238

Daniel S, Guillaudeux S, Maillard E (1997) Adaptation of a partial shape recognition approach. Available via DIALOG. https://www.sci-hub.shop/. Accessed 2 July 2019

Donoho DL (1995) Denoising by soft-thresholding. IEEE Trans Inf Theory 41(3):613–627. https://doi.org/10.1109/18.382009

Dura E, Bell J, Lane D (2008) Superellipse fitting for the recovery and classification of mine-like shapes in Side-scan sonar images. IEEE J Oceanic Eng 33(4):434–444. https://doi.org/10.1109/JOE.2008.2002962

Gao XB, Wang B, Tao DC et al (2011) A relay level set method for automatic image segmentation. IEEE Trans Cybern 41(2):518–525. https://doi.org/10.1109/TSMCB.2010.2065800

Gastaud M, Barlaud M, Aubert G (2004) Combining shape prior and statistical features for active contour segmentation. IEEE Trans Circuits Syst Video Technol 14(5):726–734. https://doi.org/10.1109/TCSVT.2004.826758

Goldenberg R, Kimmel R, Rivlin E et al (2001) Fast geodesic active contours. IEEE Trans Image Process 10(10):1467–1475. https://doi.org/10.1109/83.951533

Guillaudeux S, DanieL S, Maillard EP (1996) Optimization of a sonar image processing chain: a fuzzy rules based expert system approach. In: OCEANS 96 MTS/IEEE Conference Proceeding, Fort Lauderdale, FL, USA, 23–26

Hodneland E, Tai XC, Gerdes H (2009) Four-color theorem and level set methods for watershed segmentation. Int J Comput Vision 82(3):264–283. https://doi.org/10.1007/s11263-008-0199-4

Huo GY, Yang SM, Li QW et al (2017) A robust and fast method for side-scan sonar image segmentation using nonlocal despeckling and active contour model. IEEE Trans Cybern 47(4):855–872. https://doi.org/10.1109/TCYB.2016.2530786

Kass M, Witkin A, Terzopoulos D (1988) Snakes: active contour models. Int J Comput Vision 1(4):321–331. https://doi.org/10.1007/BF00133570

Krinidis S, Chatzis V (2010) A robust fuzzy local information C-means clustering algorithm. IEEE Trans Image Process 19(5):1328–1337. https://doi.org/10.1109/TIP.2010.2040763

Li CM, Huang R, Ding ZH et al (2011) A level set method for image segmentation in the presence of intensity inhomogeneities with application to MRI. IEEE Trans Image Process 20(7):2007–2016. https://doi.org/10.1109/TIP.2011.2146190

Li CM, Kao C, Gore JC et al (2008) Minimization of region-scalable fitting energy for image segmentation. IEEE Trans Image Process 17(10):1940–1949. https://doi.org/10.1109/TIP.2008.2002304

Li CM, Xu CY, Gui CF et al (2005) Level set evolution without re-initialization: a new variational formulation. In: Paper presented at the IEEE conference on computer vision and pattern recognition

Li CM, Xu CY, Gui CF et al (2010) Distance regularized level set evolution and its application to image segmentation. IEEE Trans Image Process 19(12):3243–3254. https://doi.org/10.1109/TIP.2010.2069690

Lianantonakis M, Petillot Y (2007) side-scan sonar segmentation using texture descriptors and active contours. IEEE J Oceanic Eng 32(3):744–752. https://doi.org/10.1109/joe.2007.893683

Lucca EV, Freitas CC, Frery AC et al (1998) Comparison of SAR segmentation algorithms. In: Paper presented at the Anais IX Simpósio Brasileiro de Sensoriamento Remoto, Santos, Brasil

Malladi R, Sethian JA, Vemuri BC (1995) Shape modeling with front propagation: A level set approach. IEEE Trans Pattern Anal Mach Intell 17(2):158–175. https://doi.org/10.1109/34.368173

Mallat S (2008) A Wavelet Tour of Signal Processing, 3rd edn. Academic Press, Salt Lake City, The Sparse Way

Mansouri AR, Mitiche A, Vazquez C (2006) Multiregion competition: a level set extension of region competition to multiple region image partitioning. Comput Vis Image Underst 101:137–150. https://doi.org/10.1016/j.cviu.2005.07.008

Marques RCP, Medeiros FN, Nobre JS (2012) SAR image segmentation based on level set approach and model. IEEE Trans Pattern Anal Mach Intell 34(10):2046–2057. https://doi.org/10.1109/tpami.2011.274

Mignotte M, Collet C, Prez P et al (1999) Three-class Markovian segmentation of high-resolution sonar images. Comput Vis Image Underst 76(3):191–204. https://doi.org/10.1006/cuiu.1999.0804

Mignotte M, Collet C, Prez P et al (2000) Sonar image segmentation using an unsupervised hierarchical MRF model. IEEE Trans Image Process 9(7):1216–1231. https://doi.org/10.1109/83.847834

Mumford D, Shah J (1989) Optimal approximation by piecewise smooth functions and associated variational problems. Commun Pure Appl Math 42:577–685. https://doi.org/10.1002/cpa.3160420503

Osher S, Sethian JA (1988) Fronts propagating with curvature-dependent speed: Algorithms based on Hamilton-Jacobi formulations. J Comput Phys 79(1):12–49. https://doi.org/10.1016/0021-9991(88)90002-2

Papandreou G, Maragos P (2004) A fast multigrid implicit algorithm for the evolution of geodesic active contours. In: Proceedings of the 2004 IEEE Computer Society Conference on Computer Vision and Pattern Recognition, Washington, DC, USA, 27 June–2

Papandreou G, Maragos P (2007) Multigrid geometric active contour models. IEEE Trans Image Process 16(1):229–240. https://doi.org/10.1109/TIP.2006.2006.884952

Parrilli S, Poderico M, Angelino CV et al (2012) A nonlocal SAR image denoising algorithm based on LLMMSE wavelet shrinkage. IEEE Trans Geosci Remote Sens 50(2):606–616. https://doi.org/10.1109/tgrs.2011.2161586

Reed S, Petillot Y, Bell J (2003) An automatic approach to the detection and extraction of mine features in side-scan sonar. IEEE J Oceanic Eng 28(1):90–104. https://doi.org/10.1109/joe.2002.808199

Sethian JA (1996) A fast marching level set method for monotonically advancing fronts. Proc Natl Acad Sci 93(4):1591–1595. https://doi.org/10.1073/pnas.93.4.1591

Sethian JA (1999) Level set methods and fast marching methods. Cambridge University Press, London

Tsai A, Yezzi A, Willsky AS (2001) Curve evolution implementation of the Mumford-shah functional for image segmentation, denoising, interpolation, and magnification. IEEE Trans Image Process 10 (8):1169–1186. https://doi.org/10.1109/83.935033

Tsechpenakis G, Metaxas DN (2009) CoCRF deformable model: a geometric model driven by collaborative conditional random fields. IEEE Trans Image Process 18(10):2316–2329. https://doi.org/10.1109/tip.2009.2026624

Vese LA, Chan TF (2002) A multiphase level set framework for image segmentation using the mumford and shah model. Int J Comput Vis 50(3):271–293. https://doi.org/10.1023/A:1020874308076

Wang B, Gao X, Tao D et al (2014) A nonlinear adaptive level set for image segmentation. IEEE Trans Cybern 44(3):418–428. https://doi.org/10.1109/TCYB.2013.2256891

Weickert J, Kuhne G (2003) Fast methods for implicit active contour models. Geometric Level Set Methods in Imaging, Vision, and Graphics. Springer, New York, pp 43–57

Xu CY, Prince JL (1998) Snakes, shapes, and gradient vector flow. IEEE Trans Image Process 7(3):359–369. https://doi.org/10.1109/83.661186

Xu C, Yezzi A, Prince J (2000) On the relationship between parametric and geometric active contours. Paper presented at the 34th asilomar conference on signals, systems and computers, Pacific Grove, CA, USA. https://doi.org/10.1109/ACSSC.2000.911003

Yang X, Gao XB, Tao DC et al (2014) Improving level set method for fast auroral oval segmentation. IEEE Trans Image Process 23 (7):2854–2865. https://doi.org/10.1109/TIP.2014.2321506

Yang X, Gao XB, Tao DC et al (2015) An efficient MRF embedded level set method for image segmentation. IEEE Trans Image Process 24(1):9–21. https://doi.org/10.1109/TIP.2014.2372615

Ye XF, Zhang ZH, Liu PX et al (2010) Sonar image segmentation based on GMRF and level-set models. Ocean Eng 37(10):891–901. https://doi.org/10.1016/j.oceaneng.2010.03.003

Yezzi A, Kichenassamy S, Kumar A (1997) A geometric snake model for segmentation of medical imagery. IEEE Trans Med Imaging 16 (2):199–209. https://doi.org/10.1109/42.653665

Zhang KH, Zhang L, Lam KM et al (2016) A level set approach to image segmentation with intensity inhomogeneity. IEEE Trans Cybern 46(2):546–547. https://doi.org/10.1109/TCYB.2015.2409119

Zhao HK, Chan T, Merriman B et al (1996) A variational level set approach to multiphase motion. J Comput Phys 127:179–195. https://doi.org/10.1006/jcph.1996.0167

Zhu SC, Yuille A (1996) Region competition: Unifying snakes, region growing, and Bayes/MDL for multiband image segmentation. IEEE Trans Pattern Anal Mach Intell 18(9):884–900. https://doi.org/10.1109/34.537343

10 Applications of Submarine Geomorphology

Submarine geomorphology is a large branch of the submarine geosciences that carries out studies of the seabed geomorphology worldwide using high-resolution multi-beam echo sounding and other data sources. The study of seabed geomorphology has regional characteristics; it uses common research methods and addresses specific regional scientific problems. In this chapter, we present examples that illustrate how to carry out geomorphology studies using detection data such as multi-beam echo soundings, sub-bottom profiles and side-scan sonar images. Three of the most common types of geomorphic units in the seafloor are selected: sand ridges in the continental shelf, structural geomorphology in the marginal basin and structural geomorphology in a mid-oceanic ridge. Additionally, some technical methods of underwater feature naming and examples of name selection for underwater features discovered in the China seas are introduced in this chapter.

10.1 Sand Ridges on Continental Shelf of the East China Sea

Submarine sand ridges are typically developed on tide control continental shelves around the world. Since Off (1963) study of linear sand bodies, tidal sand ridges on tide-dominated shelves have been studied extensively. Currently, the North Sea (Anthony and Leth 2002), the offshore Atlantic (Twichell et al. 2003) and the shelves of the China's seas (Liu and Xia 2004; Wu et al. 2017) are still the hot spots for the Linear Sand Ridges (LSR) research. It has gradually become a tendency to study the fine structures of sand ridges utilizing high-precision and high-density multi-beam echo sounding (MBES) data and high-resolution, single-channel seismic profiles (Li and King 2007; Zhou et al. 2018).

Studies of the sand ridges in China were based on a discussion of the origin of radial submarine terrain in Qionggang, Jiangsu (Li and Li 1981). Yang (1989) studied the internal structures of sand ridges in the horn-like topography near the Changjiang River Estuary; Chen et al. (2003) studied the mud ridges in the outer Changjiang River Estuary. Many achievements have been made in the study of the structure, classification, and development of models of the buried sand ridges in the East China Sea (ECS) shelf (Berné and Vagner 2002; Liu et al. 2007b).

There are many theories explaining the origin of sand ridges, including the secondary-flow theory, the long-period wave theory, the seabed stability analysis theory, and the theory of ebb and flood channels in estuary mouths (Dyer and Huntley 1999). Most researchers have accepted the theory that submarine sand ridges were caused by tidal currents or storms. The North Sea is a typical region for the study of tidal sand ridges (Anthony and Leth 2002), and the offshore shelves in the North Atlantic Ocean are the focal areas for the study of storm sand ridges (Twichell et al. 2003). The depositional pattern and evolution of the sand ridges in the continental shelf are the result of the interaction between the multi-phase ancient tidal current fields and the bed sediment (Uehara et al. 2002), which is an important basis for studying the evolution of sand ridges (Uehara and Saito 2003).

Since the 1990s, a large number of multi-beam bathymetric surveys have been carried out in the ECS. Based on these high-resolution bathymetric data, this section will analyze the fine structure of the sand ridge, together with the high-resolution single-channel seismic profiles, the distribution, formation and evolution of LSR on the ECS shelf are finally analyzed and discussed.

During the Last Glacial Maximum (LGM), the sea level of the ECS continually decreased to 140 m below the present level of the continental shelf. After that, it rose quickly and by 7 000 a BP it approached or even rose above the current sea level. Here, we carry out the simulation of ancient tide field. The hydrodynamic model used here is the Princeton Ocean Model (POM), a three-dimensional, non-linear primitive equation model with Boussinesq and

Z. Wu et al., *High-resolution Seafloor Survey and Applications*,
https://doi.org/10.1007/978-981-15-9750-3_10

hydrostatic approximations (Mellor et al. 2002). The POM has been applied to many regional and shelf ocean tidal applications (Niwa and Hibiya 2004; Koropitan and Ikeda 2008). Our model uses a Cartesian coordinate system in the horizontal and a sigma coordinate system in the vertical. The model domain covers from 8°S to 56°N and from 99°E to 145°E with a resolution of (1/12)°. In the vertical, the domain is divided into 10 equal sigma levels ($\sigma = 0$ at the sea surface and $\sigma = 1$ at the sea bottom). The bathymetry adopted by the model is taken from ETOPO2 (Smith and Sandwell 1997). This dataset is inaccurate in the coastal regions and the shelf of the ECS; hence, we correct it by using some published bathymetry charts. Tidal elevations are specified by the linear interpolation of the output from the global tidal model TPXO7 (Egbert and Erofeeva 2002), which is assimilated Topex/Poseidon altimeter data with Root Mean Square(RMS) error less than 1.43 cm for M2 tidal at open boundary with bathymetry above 3000 m. Harmonic constants derived from the model are compared with the tide gauges listed in literature (The Hydrographer of Navy 1990) and crossover points of Topex/Poseidon trace (AVISO 1996). Until now, Paleo-tides in the ECS shelf have been estimated by applying paleo-bathymetries for every 10 m from −150 m, and the +5 m case has also been investigated. The formulae containing critical bed shear stress are proposed by Soulsby and Whitehouse (1997), Uehara and Saito (2003). The sediment grain size data are obtained from geological survey results.

10.1.1 Statistics of Sand Ridge Strikes

Based on MBES data and some historic topography maps, submarine topographic maps and 3D topographic maps are drawn, the crests of sand ridges on the ECS shelf are traced, and the crest lines are also marked (Fig. 10.1). Thirty-two segments of crest lines with a total length of 2 178 km are indentified using historic topographic maps (the dotted red lines in Fig. 10.1), whereas 218 segments with a total length of 10 101 km are identified using surveyed data (the continuous red lines in Fig. 10.1). The entire length of the crest lines of the sand ridges totals about 12 279 km.

The statistical results of the strikes and lengths of the LSR are obtained through the weighed method of five intervals

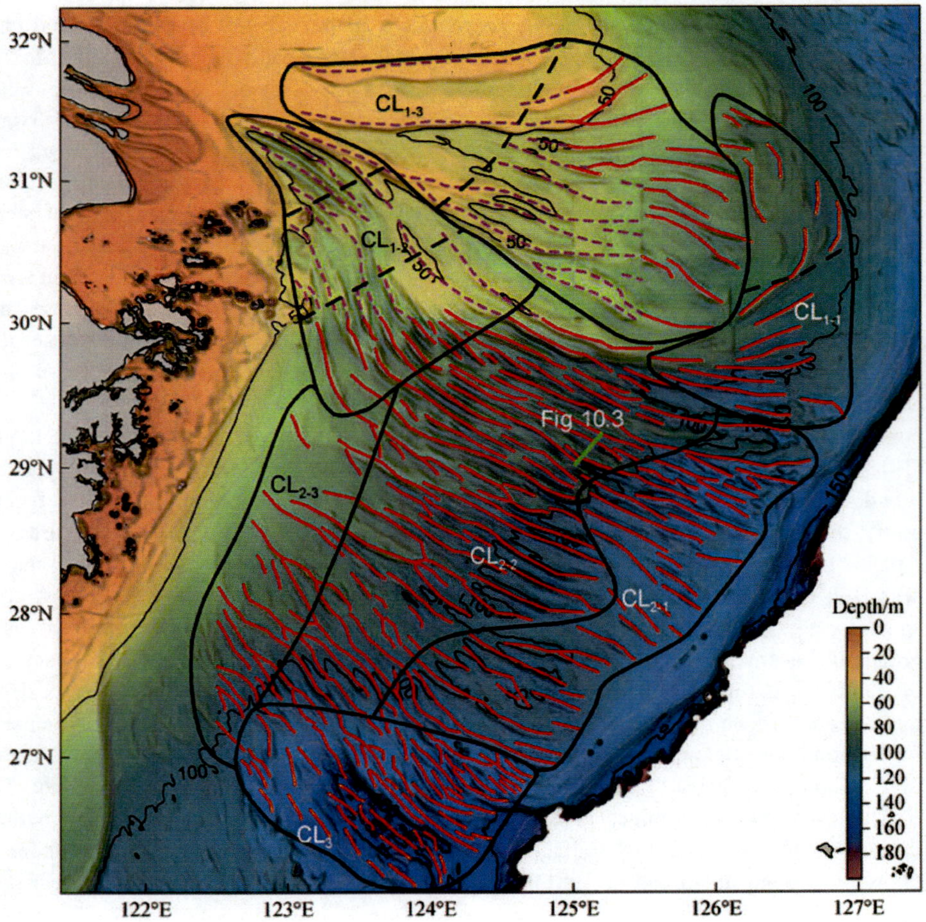

Fig. 10.1 Spectrum and subzones of LSR on the ECS shelf. Red lines represent the crest lines of sand ridges. The solid lines are based on surveyed data, the dotted lines are based on historic data, black lines are the boundaries among subzones

Table 10.1 Statistics of the crest strikes (weighed statistics at 5° intervals)

Azimuth /(°)	Length /km	Percentage /%	Azimuth /(°)	Length /km	Percentage /%	Azimuth /(°)	Length /km	Percentage /%
60–65	0.0	0.0	120–125	395.9	3.2	180–185	153.0	1.2
65–70	0.0	0.0	125–130	630.1	5.1	185–190	97.2	0.8
70–75	0.0	0.0	130–135	739.2	6.0	190–195	149.7	1.2
75–80	13.0	0.1	135–140	824.6	6.7	195–200	83.8	0.7
80–85	21.5	0.2	140–145	879.0	7.2	200–205	72.6	0.6
85–90	39.7	0.3	145–150	1232.2	10.0	205–210	72.9	0.6
90–95	98.0	0.8	150–155	1507.1	12.3	210–215	54.9	0.4
95–100	151.3	1.2	155–160	1392.9	11.3	215–220	48.3	0.4
100–105	137.3	1.1	160–165	829.8	6.8	220–225	14.3	0.1
105–110	225.6	1.8	165–170	673.2	5.5	225–230	15.9	0.1
110–115	288.8	2.4	170–175	498.6	4.1	230–235	0.0	0.0
115–120	381.9	3.1	175–180	531.7	4.3	235–240	25.7	0.2

from 1° to 5° azimuth angle. The results of 5° azimuth are listed in Table 10.1. The distribution profile of the crest strikes is finally formed (Fig. 10.2) wherein the horizontal axis represents the crest strikes of the sand ridges, while the longitudinal axis represents the percentage of the length of sand ridges in a particular crest strike range compared to the total length of the sand ridges on the continental shelf. We changed the statistic range of crest strikes from 0–180° to 60–240°, as the value around 180° is one of the concentrated regions for crest strikes.

The distribution on the profile is a normal curve and the normal characteristics become more apparent as the statistic intervals of the crest strikes increase. The percentage of central peak value domain reaches beyond 12% when the interval reaches 5° (Table 10.1 and Fig. 10.2). The main distribution domain of the crest strikes is in region of 110°–180° and the length of sand ridges in this region is 88% of the total length of the ECS shelf. The azimuth 155° is the center point of the normal distribution but there are several other centralized regions of crest strikes: 125°, 130°, 140° and 180° azimuth. Despite the differences in the strikes of the sand ridges in the different sea areas, as influenced by the direction of tidal currents during a sea level period, the total number of strikes of the sand ridges on the ECS shelf is characterized by normal and centralized distribution. This proves that the sand ridge bodies in different regions of the ECS shelf have been controlled by the direction of the dominant tidal current and the main flow of tidal currents migrating from the outer shelf to the inner, along with the rising sea level.

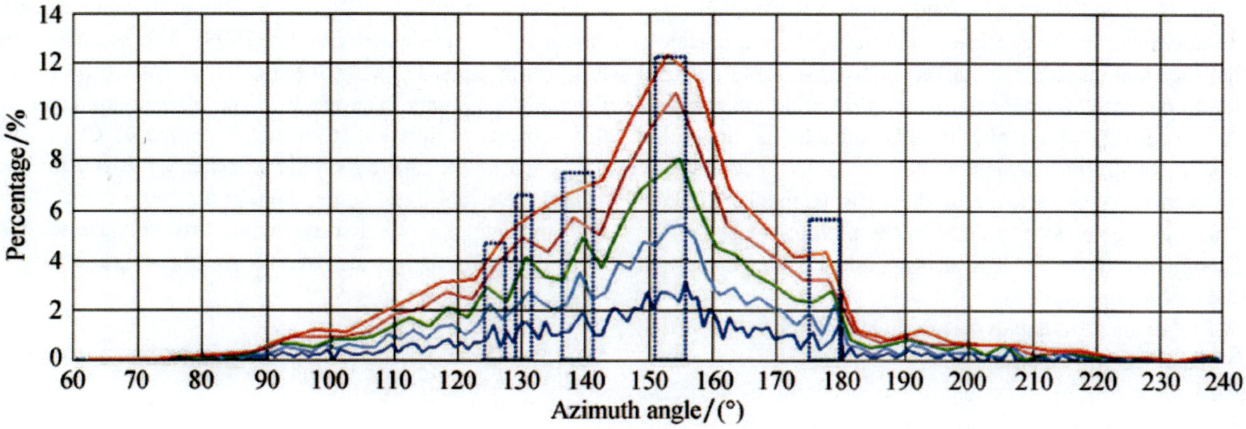

Fig. 10.2 Statistical chart of crest strikes

10.1.2 Subzones of Sand Ridges Based on Their Strikes

The LSR on the ECS shelf are characteristically intensive in the central part, sparse at the south and north ends, disperse and bifurcated to the east, and convergent to the west (Fig. 10.1). Seven subzones, $CL_{1\text{-}1}$-CL_{1-3}, $CL_{2\text{-}1}$-$CL_{2\text{-}3}$ and CL_3 as listed, are delineated according to their strikes features and the distribution of LSR on the ECS shelf. The general distribution features of these four subzones, $CL_{1\text{-}1}$-$CL_{1\text{-}3}$ and CL_3, are similar to the estuary mouth ridges (Dyer and Huntley 1999), which show convergent characteristics from southeast to northwest. Subzones from $CL_{2\text{-}1}$ to $CL_{2\text{-}3}$ are normal, open shelf ridges (Dyer and Huntley 1999).

Subzone $CL_{1\text{-}1}$ is further divided into two parts: south and north. The general strikes of sand ridges in the southern part are in accordance with the main strikes of the shelf, which disperse from the top of the western bay to the northeast. The geomorphology of this subzone appears as a horn-like topography, which was probably always an estuary of Paleo-Changjiang River (PCR). In the northern part of this subzone are six short sand ridges trending from north to south, which are perpendicular to the strikes of the sand ridges on Subzone CL_{1-3}. This suggests the existence of different hydrodynamic conditions, as the strikes of these sand ridges are similar to those found in the central-eastern part of the Yellow Sea shelf (Jin and Chough 1998, 2002) and deviate from all the strikes of the LSR on the ECS shelf.

The strikes of the sand ridges in Subzone $CL_{1\text{-}2}$ are S-shaped and convergent from SE to NW. The strikes of sand ridges on the central part of the ECS shelf apparently begin to change at the south end of the subzone from NW–SE to NNW-SSE, and gradually revert to NW–SE. The troughs of the LSR in the northern part are apparent, while its crests are indistinct. The sand ridges in the southern part are not consistent with those in the northern part, revealing that the tidal current field of the ECS most likely changed during the early development of sand ridges in the subzone. The horn-like topography in this subzone is noticeable though its origin remains in debate. It has been widely believed to be an estuary of the PCR (Uehara and Saito 2003), as the subzone is close to the present estuary of the Changjiang River, but it is also regarded as a scour trough caused by tidal currents.

Ridges are abundant to the east, while scarce to the west of Subzone $CL_{1\text{-}3}$, decreasing sharply from 9 to 2 respectively and with spaces between their crests widening as well. The ridges in the western part are similar to the PTSRs (pseudo-tidal sand ridges) of Ganghwa Bay in the Yellow Sea (Jung et al. 1998), the difference being that the linear terrain of this region is characterized by narrow ridges and wide troughs, while the PTSR on Ganghwa Bay are characterized by wide platforms and narrow troughs. The region is adjacent to the Changjiang River Estuary and mainly consists of complicated deltas of PCR shown on seismic profiles (Yang 1996). So these troughs of linear terrain may be part of the PCR. However, more research is needed to determine whether or not these linear terrains are actually sand ridges.

Most of the LSR are distributed from 27 °N to 30 °N with a water depth of 60–120 m on the ECS shelf. These LSR are characterized by their extensive distribution, straight relief, lesser divergence and basic strikes from NW to SE, which indicates that the shelf of the ECS has experienced an approximation of tidal current environment. The distinguishing spatial attributes of these LSR can be described as dense and straight to the north; sparse and winding to the south. The three subzones, $CL_{2\text{-}1}$, $CL_{2\text{-}2}$ and $CL_{2\text{-}3}$, have been divided according to appearance and the spatial continuity of the sand ridge distribution. The LSR on Subzone $CL_{2\text{-}1}$ are characterized by their short length and strikes that disperse from NW to SE. The area between the LSR of Subzones $CL_{2\text{-}1}$ and $CL_{2\text{-}2}$ is apparently discontinuous as these subzones are divided by the 100-m isobath. The LSR of Subzone $CL_{2\text{-}2}$ are flat, straight and continuous; the spaces between the sand ridges are enlarged with several LSR crossing one another in the central part. The LSR on Subzone $CL_{2\text{-}3}$ have enlarged spaces and widened troughs and they are typically referred to as comb-like sand ridges as the 60-m isobath convergence somewhat resembles a comb.

There is a boat-like deep depression with a maximum water depth of approximately 180 m in the southwest region of this area (near 27°N) where partially developed LSR can be observed from MBES. This area is designated as Subzone CL_3 because the LSR of this area differs from the normal, open shelf ridges. The distribution of LSR on this subzone is similar to that of Subzones $CL_{1\text{-}1}$ to $CL_{1\text{-}3}$, and these ridges, convergent from SE to NW, are similar to wide estuary mouth ridges (Dyer and Huntley 1999). The sediments in this region are starved, for the area is far from the Changjiang River Estuary. The origin of negative topography of this subzone is different from that of Subzones $CL_{1\text{-}1}$ and $CL_{1\text{-}2}$. These LSR have multiple branches (at least 2 to 3 or 4) and their root-like features (Wu et al. 2005) from SE to NW indicate that the hydrodynamic environment in this subzone during the process of LSR formation is mutative.

10.1.3 Overlapped Sand Ridges After Multi-Stages

In previous studies, the LSR on the ECS shelf are defined as sand ridges of U2 (Wu et al. 2005), which are characterized by multi-stage growth. Based on seismic stratigraphy (Fig. 10.3), U2 sand ridges can be divided into four

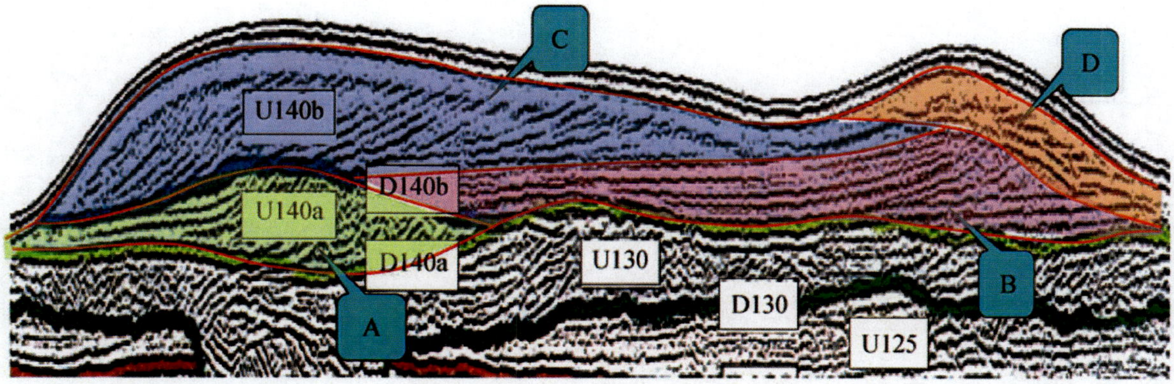

Fig. 10.3 Multi-stage buried sand ridges shown on a typical seismic profile, modified from Berné and Vagner (2005). Refer to Fig. 10.1 for the position of the profile

sub-stage buried sand ridges from A to D. The seismic units of Sand ridges A and C correspond to the units U140a and U140b defined by Berné and Vagner (2002). Sand ridges in U140a, which are called cored sand ridges by Berné and Vagner (2002), maybe the earlier core of the sand ridges in U140b based on the relationship of their seismic units. The seismic beddings of Sand ridges A and C are characterized by inclined, high angles, while Sand ridge B is characterized by inclined, small angles. Sand ridge D is immature and its structure, obscure, while the boundary between Sand ridges D and B is distinct. Sand ridge B can be seen as the core of Sand ridge D. In other words, Sand ridge D is developed on the base of Sand ridge B and overlaps it.

10.1.4 Discussion on the Origin of Sand Ridges

10.1.4.1 Factors Influencing the Formation of Sand Ridges

The growth of sand ridges is related to water depth, sediment supply and hydropower. The growth and termination of the sand ridges are determined by variations in water depth, while the linear shape of sand ridges is determined by hydrodynamics. Sufficient sediments are the foundation for the growth and burying of sand ridges (Wu et al. 2005). The most favorable conditions for the development of sand ridges are a water depth near 30 m, 1–3 kt tidal velocity, and an ellipticity of less than 0.4 for the M2 tidal component (Liu and Xia 2004).

Storm tide is a short-term regional climatic change formed by hurricanes or typhoons and it can easily form active submarine sand waves or sand ridges in shallow seas with frequent storms. It is the typical study area for storm ridges in offshore Atlantic (Twichell et al. 2003) and around Hainan Island, China (Dong et al. 2004). Although the sand wave is a structure, which is vital to studying the activity of sand ridges, it is seldom reported on the ECS shelf, with the exception of the Changjiang River Estuary shoal (Ye et al. 2004). Long-term global climatic changes such as the glacial-interglacial cycle, can lead to high-amplitude fluctuations of sea level, which can, in turn, cause a large-range transition of sea and land on the shelf (e.g., coastline changes, flow path changes of rivers and environmental changes). During a period of falling sea level, coastlines migrate seaward, continental shelves gradually outcrop into land, rivers go through the wide, flat shelf and the loose sediments formed earlier show a typical V-shape on seismic profiles (Liu and Xia 2004; Berné and Vagner 2002) because they have been eroded by these changed rivers (Berné and Vagner 2002). The bottom boundary of the V-shape channel is regarded as transgressive marine surface erosion (TMSE) (Berné and Vagner 2002) with a braided feature resulting from the swing of the channel (Li et al. 2005). The shelf topography is altered by fluvial erosion and the river simultaneously transports terrestrial sediments to the outer shelf, which can provide material for the growth of sand ridges. During a period of rising sea level, the water depth of a shelf increases and channels and coastlines recede landward. Previously formed V-shape channels are gradually buried by later deposits, and loose sediments on the shelf gradually developed into sand ridges through the action of tidal waves. The top boundary of the sand ridges is considered to represent the maximum flooding surface (MFS) (Berné and Vagner 2002). The tidal component is influenced by three important factors: coastline appearance, water depth and submarine topography. The tidal component changes along with the sea level changes for the coastline. Water depth and submarine topography are also influenced by the quick change of sea level.

During the process of sand ridge formation, a complicated response system is established by multiple factors, such as climate, sea level, water depth, topography, coastline, rivers and tidal currents. These factors are associated and influence each other (Fig. 10.4). Global climatic change is the basic cause of the formation of sand ridges, acting as a domino-effect. Suitable depositional dynamic is

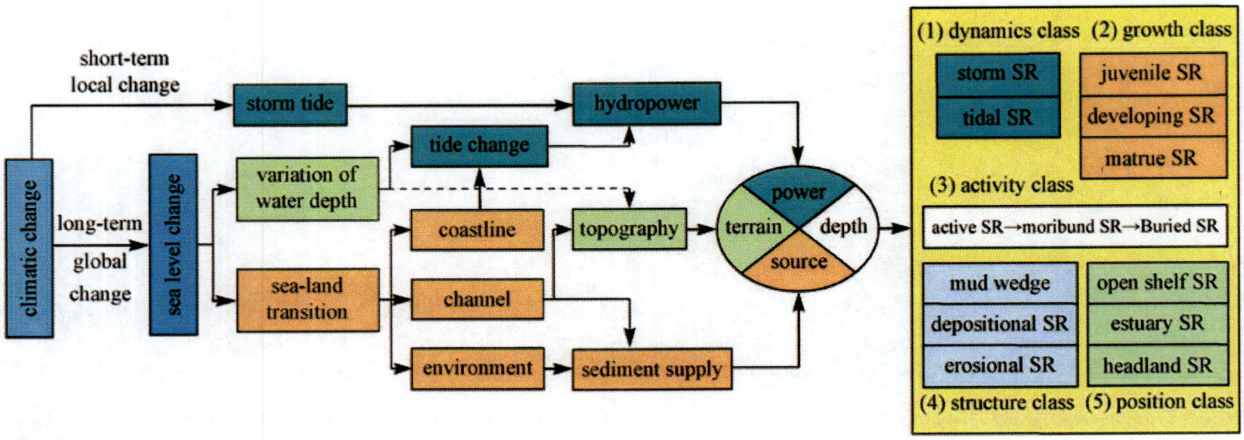

Fig. 10.4 Factors influencing the formation of SR (sand ridges)

the most direct reason for the growth of sand ridges. The developing process and type of sand ridges are influenced by sediment supply, variation of water depth and submarine topography (Fig. 10.4). Storm sand ridges or tidal sand ridges reflect the difference of depositional dynamics. Abundant sediment supply is advantageous for the formation of depositional sand ridges (e.g., the ECS shelf (Liu and Xia 2004)) while lack of sediment or strong hydrodynamics can lead to erosional sand ridges (e.g., offshore Korea Peninsula (Jin and Clough 1998; Jung et al. 1998). Three stages are usually experienced during the evolution of depositional sand ridges: the juvenile stage, the developing stage, and the mature stage. Open shelf ridges, estuary mouth ridges or headland associated banks can only be formed in special submarine topographies (e.g., the North Sea shelf (Anthony and Leth 2002; Dyer and Huntley 1999). Periods of stable or slowly rising sea level are advantageous to the development of sand ridges. During the period of rapidly rising sea level, the activity of sand ridges will weaken because water depth is quickly increasing and the tidal current weakens synchronously. The active sand ridges will gradually weaken to near moribund as the sea level rises further. Conditions advantageous for the formation of buried sand ridges include sufficient depositional time, continual subside, and abundant sediment supply (e.g., sand ridges buried about 100 m under the seafloor of the ECS shelf (Yang 1989; Berné and Vagner 2002).

10.1.4.2 Influence of Sea Level Change on the Growth of Sand Ridges on the ECS Shelf

The Milankovitch Cycles, with a major cycle of 100 ka and secondary cycles of between 20 and 40 ka, led to the global glacial-interglacial change with a cycle of approximately 100 ka. The high amplitude change of the local sea level is the response to glacial oscillation. Sedimentary strata on the shelf are dominated by sea level change and six strata have been formed under the seafloor of the ECS shelf in the last 100 ka due to periodical sea level changes. FRST (forced regressive systems tract), LST (lowstand systems tract), terrestrial deposit, TST (transgressive systems tract) and HST (highstand systems tract) are formed during periods of falling sea level, low sea level, droughts, transgressive sea level and high sea level. Sand ridges belong to TST and are formed during transgressive periods.

During the LGM, the sea level of the ECS shelf descended to 140 m below its present level (Zhu and Li 1979). Subsequently, the general sea-level of the ECS shelf changed in accordance with global changes. The sea level rose from its lowest point at 15 ka BP to its highest point at 7 ka BP. The fall of the sea level was not simple-gradient, but underwent multiple fluctuations in the process (Fairbanks 1989; Liu et al. 2004, 2006, 2007a). Two important meltwater pulses (MWP) (Fairbanks 1989) in the Atlantic Ocean led to the quick changes of sea level. MWP-1A occurred during the period of 14.5 to 13.7 ka BP, resulting in a rapid rise of sea level ranging between 95 and 78 m. MWP-1B occurred during the period of 11.5 to 11.2 ka BP, resulting in a rapid rise of sea level ranging from 60 to 40 m (Liu et al. 2004). The changed hydrodynamics of the shelf resulting from the rapid changes of sea level was not conducive to the development of sand ridges. The present water depth located by seismic profiles in Fig. 10.3 is about 90 m. However, it was only about 30 m during the stable sea level period between WMP-1A and WMP-1B, which was suitable for the growth of sand ridges. The sand ridges in Substage C (Fig. 10.3) were likely developed during this time. The sand ridges in Substages A and B are very small. These sand ridges could be found only in some areas and were possibly formed before WMP-1B and buried by later deposits. After WMP-1B, conditions were not suitable for the development of sand ridges in most areas of the ECS shelf as the water depth further increased. Only some immature sand ridges in Substage D were formed on the base of the sand ridges of Substage C (Fig. 10.3).

10.1.4.3 Influence of Negative Topography of the Seafloor and Variations in Sediment Supply on Growth of Sand Ridges on the ECS Shelf

Dyer and Huntley (1999) divides the tidal deposits in the North Sea into three types (according to geographical position): (1) open shelf sand ridges; (2) estuary mouth ridges, which can be subdivided into wide mouth ridges and narrow mouth tidal deltas (wherein the narrow tidal delta can be further divided into ebb tidal delta without recession and shoreface-connected ridges with recession); and (3) headland associated banks, which can be subdivided into banner banks with non-recessional headland and alternating ridges with recessional headland. This classification method shows that special submarine topography can influence and even restrict the development of tidal deposit.

The formation of some sand ridges on the ECS shelf is affected by submarine negative topography as well. Seven subzones of the ECS shelf have been divided according to the strikes and distribution of the sand ridges (Fig. 10.1). Sand ridges in Subzones CL_{1-1}, CL_{1-2}, CL_{1-3} and CL_3 resemble estuary mouth ridges, while the sand ridges in CL_{2-1}, CL_{2-2} and CL_{2-3} are normal, open shelf sand ridges. Although they are formed under same conditions on the continental shelf and through the same process of changing sea level and similar hydrodynamics, the sand ridges in the seven subzones differ widely in terms of strikes and features because of the influence of submarine topography on the formation of sand ridges. During the period of rising sea level, coastline appearance changes along with submarine topography. The power of the tidal current is influenced by water depth and the tidal current direction is affected by coastline appearance. During the process of rising sea level, the sea-land transition of negative topography materializes earlier than that of adjacent areas (Fig. 10.5). Tidal current can come into negative topography early and interact with existing sediments, and sand ridges will gradually form under suitable conditions. Although Subzones CL_{1-1}, CL_{1-2} and CL_3 with negative topography are adjacent to other subzones, the strikes of sand ridges in these Subzones are different from their adjacent subzones. For example, the strikes of some sand ridges in Subzones CL_{1-1} and CL_3 are even orthotropic. Therefore, the seafloor negative topography must play an important role in the formation of these sand ridge formations.

The PCR also played an important role in the development of sand ridges on the ECS shelf. Since the Eocene, the Indian Plate collided with the Eurasian Plate, leading to the uplift of the Tibetan Plateau and the rifting of the marginal basin. Hence, the Asian river system, which has a radial shape from the center of the Asian continent, was formed. The Changjiang River and the Huanghe River, which originated from the Tibetan Plateau, continuously poured into the ECS, carrying lots of sediments. During the postglacial period, the amount of sediment transported by the Changjiang River Estuary ranged from 2.36×10^8 to 4.86×10^8 t/a, and approached 3.54×10^{12} to 7.08×10^{12} t in total (Li et al. 2003). About half of the sediments were transported to the outer sea or adjacent coast by the longshore current and the tidal current. During the process of sea-land transformation caused by the glacial- interglacial cycle, flow paths of the PCR changed synchronously, and abundant terrigenous sediment was carried to the ECS shelf. However, the flow-paths of the PCR are still under debate (Li et al. 2005). Huge submerged deltas were gradually formed during these changes of the PCR (Hori et al. 2001, 2002). The thickness of the delta depocenter had been over 60 m just since the Holocene (Huang et al. 1996). The estuary deltas migrated with the change of sea level, retreated landward during the period of sea level rise, and HST was formed (Liu et al. 2007a). However, it prograded seaward during the period of falling sea level and the FRST was formed. Two delta sedimentary layers with an apparent depositional break had been formed on the ECS shelf since the latest sea level change of the 100 ka cycle (Berné and Vagner 2002). The topography of the central part of the northern ECS shelf was a typical uplift as it was formed by huge congeries of submerged deltas (Fig. 10.1). Some negative topography should correlate with changes in the PCR (such as the Subzones CL_{1-1} and CL_{1-2}) as the shelf topography had been changed along with the migration and undercut of channels. During the LGM, the ECS shelf was once exposed. As a result, loose terrestrial deposits and weathered deposits were formed (Zhao et al. 1997) and abundant amounts of sediment were carried by the rivers. These sediments formed an important basis for the development of the LSR on the ECS shelf.

10.1.5 Evolution of Sand Ridges on the ECS Shelf

The hydrodynamics of the ECS shelf is suitable for the growth of sand ridges as multi-phase sand ridges (Fig. 10.3) are developed in the same region. The sea-level change is the basic cause of the development of multi-phase sand ridges. The formation of sand ridges is closely related to multiple factors such as water depth, topography and tidal current (Fig. 10.4), and it takes a long time to form large sand ridges. Periods of stable sea level rather than the periods of rapidly rising sea level are conducive to the growth of sand ridges. The sand ridges distributed widely on the ECS shelf

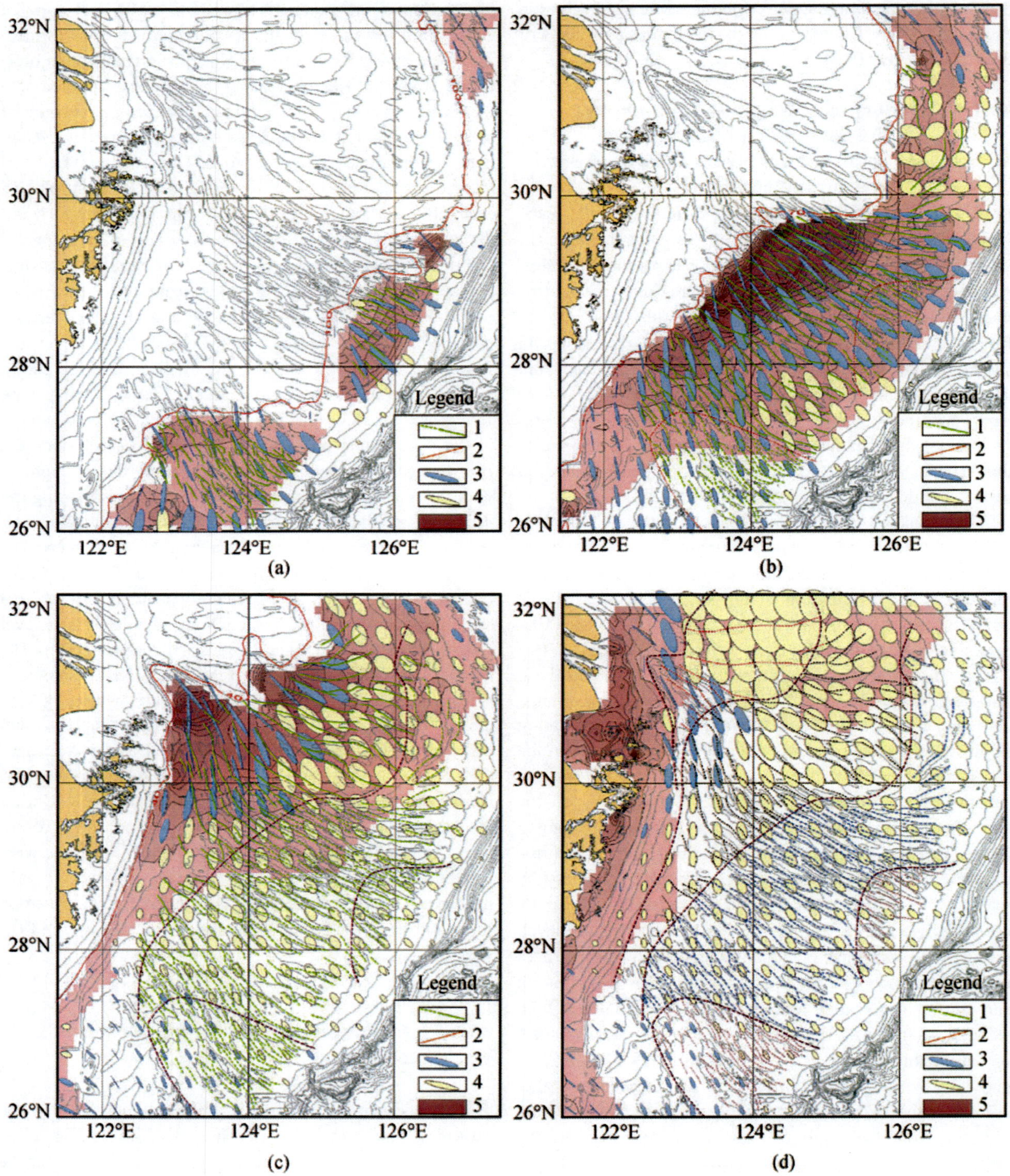

Fig. 10.5 Evolution of LSR on the ECS shelf. **a** LSR in Substage I developed before 14.5 ka BP; **b** LSR in Substage II developed before 12 ka BP; **c** LSR in Substage III developed before 9.5 ka BP; **d** LSR in Substage IV developed after 9 ka BP. 1, Crest lines of sand ridges; 2, boundaries; 3, ellipticity of M2 less than 0.4; 4, ellipticity of M2 greater than 0.4; 5, active bedform areas

were not formed over short periods of time because the ECS shelf is very broad (Fig. 10.1), the topography is complicated, and the water depth and tidal currents differ greatly in many regions (Fig. 10.5). The development of the LSR on the ECS shelf is divided into four main stages according to the integrated results of water depth variation, sea level change, ancient tide evolution and stratigraphy (Fig. 10.5).

Sand ridges in Stage I (Fig. 10.5a) were formed before 14.5 ka BP when the water depth was 100 m lower than its present level (Zhu and Li 1979; Liu et al. 2004). It was fit for the development of sand ridges on the southern boat-like depression and some regions on the outer shelf of the ECS (Fig. 10.5a). At this stage, some small-scale sand ridges were developed by weak tidal currents. The typical sand ridges in the southern depression (Subzone CL_3, see Fig. 10.1) were root-like and multi-branched and on the outer shelf (parts of Subzone $CL_{2\text{-}1}$), were small-scale sand ridges. Several small-scale sand ridges might have been formed in local parts in the center of the shelf (e.g., sand ridges in Substage A in Fig. 10.3), but these were later destroyed or buried by fluctuations of sea level.

Sand ridges in Stage II (Fig. 10.5b) were formed during the period from 12 to 14 ka BP. The MWP-1A, which occurred between 14.5 and 13.7 ka BP (Fairbanks 1989) after the LGM, led to the steep rise in sea level that, after 13.7 ka BP, rose to 70 m below the present water depth (Liu et al. 2004). The water depth of the depression on the southern ECS shelf gradually deepened and some sand ridge strikes changed concurrently with the change in tidal direction. The increased water depth was not suitable for the growth of sand ridges, so the sand ridges of this depression gradually changed into recessionary sand ridges (Fig. 10.5 b). This period served as the main developing stage of sand ridges on the ECS shelf. It was suitable for sand ridge growth in most parts of the central outer shelf (e.g., sand ridges in Substage C on Fig. 10.3 may have been formed during this time). Straight line sand ridges with NW–SE strike began to form (Subzone $CL_{2\text{-}1}$ and parts of Subzone $CL_{2\text{-}2}$). While there were suitable conditions for the growth of sand ridges in some parts of the northern ECS shelf, the tidal direction of these areas (near NS trend, see Fig. 10.5a, b) was vastly different than that of the ECS shelf (near NW–SE trend), and some sand ridges with NS strike were formed in the region with a similar ancient estuary on the north-outer shelf (Subzone $CL_{1\text{-}1}$).

Sand ridges in Stage III (Fig. 10.5c) were formed during the period from 9.5 to 11.5 ka BP. The MWP-1B (Fairbanks 1989) occurred during the rapid rise of sea level from 11.5 to 11.2 ka BP, which led to the steep rise of sea level from 60 to 40 m (Liu et al. 2004). After the MWP-1B, the water depth at the inner-central shelf was suitable for the growth of sand ridges, and sand ridges on the outer shelf gradually regressed into moribund sand ridges. Some former sand ridges were buried (e.g., sand ridges in Subzones A and B in Fig. 10.3) and became the core of subsequent sand ridges (Berné and Vagner 2002). The southwest area of the ECS (Subzone $CL_{2\text{-}3}$) was suitable for the development of sand ridges, which gradually decreased and wide-range linear topography was subsequently formed in the horn-like region in the middle ECS (Subzone $CL_{1\text{-}3}$) and the outer edge of the north Changjiang River Estuary delta (Subzone $CL_{1\text{-}2}$). The strike of the sand ridges in the horn-like region of the shelf (Subzone $CL_{1\text{-}3}$) apparently changed from NW–SE to NNW-SSE.

The sand ridges in Stage IV (Fig. 10.5d) were formed after 9 ka BP. A fast melt water pulse occurred again in the period from 9.5 to 9 ka BP (Liu et al. 2004) leading to the steep rise of sea level from about 40 to 15 m. The water depth of most of the ECS regions with their rapidly rising sea levels was not suitable for the development of sand ridges and sand ridges only developed continuously in the shallow water of the inner shelf (Fig. 10.5d). The most apparent linear topography was found at the top of the bell-mouthed region (Subzone $CL_{1\text{-}3}$), where the deep trough feature is more extensive than that of the ridges, and the strike of the ridges changes from NNW-SSE at the outside of the bell mouth to NW–SE, similar to that of the shelf. These are typically mud-core ridges (Chen et al. 2003). Some linear terrain has also been found in the northern part of the ECS (Subzone $CL_{1\text{-}2}$), but it is characterized by wide troughs and narrow crests, unlike other regions. Ascertaining whether or not it is a sand ridge should be the subject of further research. Simultaneously, the Changjiang River Estuary delta evolved in multiple stages from the beginning of the Holocene (Hori et al. 2001, 2002), with the Changjiang River carrying a gigantic amount of sediments, forming a sigmoidal clinoform, which formed enormously thick mud wedges along the coast of Zhejiang and Fujian under the action of the coastal current (Liu et al. 2006, 2007a).

10.2 Sand Wave Migration in Monterey Submarine Canyon, California

10.2.1 Study Area

Monterey Canyon is the largest submarine canyon in the west coast of the US. The region is supplied with a large amount of sediment every year from the river mouth at the northwest side of the canyon. This steady supply of sand and gravel moves gradually along the narrow pathway which is restricted to the canyon axis. As a result, extinct migration of sandwave organization can be observed in the area. The study area is located in the head of the canyon (the left part of Fig. 10.6). The size of the area is around 900 m × 500 m, and the water depth of the area ranges from 30 to 80 m. Benefit from the

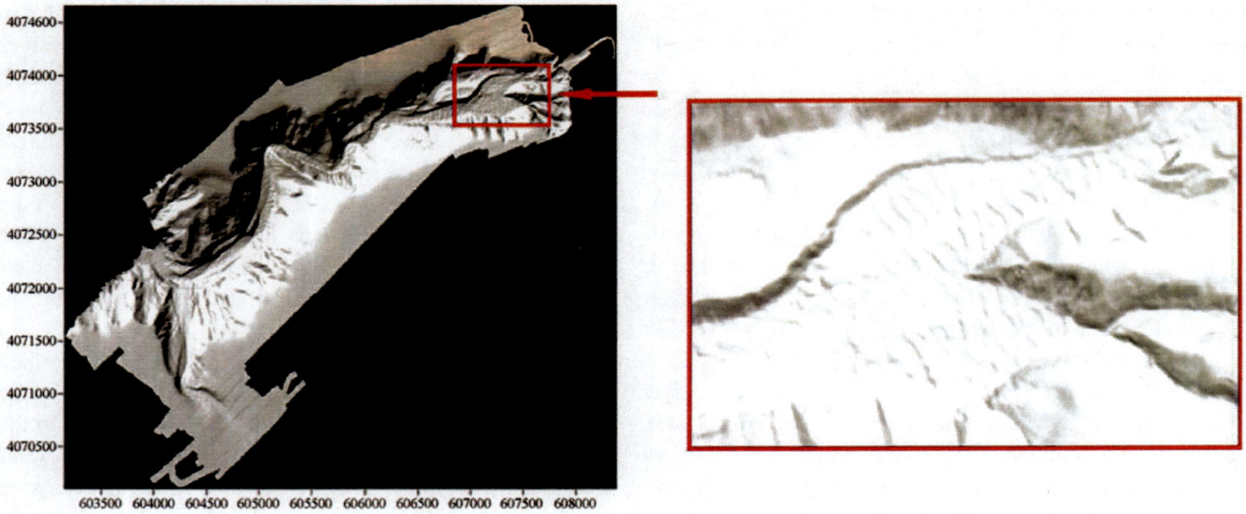

Fig. 10.6 Left panel: The shaded relief image of the bathymetric data at the head of the Monterey Submarine Canyon, California. Right panel: the bathymetry of the study area

large amount of annual sediment input, distinct sandwave organization has been formed in the region (the right part of Fig. 10.6).

The seafloor mapping Laboratory of the California State University Monterey Bay conducted a series of bathymetric surveys at the head of the Monterey Submarine Canyon at the frequency of about a year (Xu et al. 2008; Paull 2010). In the survey, the bathymetric data were collected with a Reson 8101 multibeam sonar system. The horizontal position of the boat was measured with dual-frequency differential GPS, which provided submeter horizontal positioning precision. Meanwhile, the tidal variation was also corrected with the same GPS systems so that the vertical bias was less than 0.2 m (Smith et al. 2005, 2007). After the manual editing, the bathymetric data were gridded into 3-m interval XYZ format. The multi-temporal DTMs used in this paper were collected in September and November of 2004, respectively. These two DTMs have the smallest temporal gap among the whole DTM time series (around two months), which is ideal for sandwave migration estimation.

10.2.2 Methodology

10.2.2.1 Correlation Matching for Migration Estimation

Multi-temporal underwater DTM provides valuable detailed information about the change of bedforms. Based on these data, the spatial correlation technique can be applied to infer the migration of a particular patch of bedform during different periods. To be more specific, for a particular patch of bedform P in the first DTM (termed foreground DTM), the most similar patch Q with the same size is searched in the second DTM (termed background DTM). Thus, by computing the horizontal displacement between P and Q, the migration information can be obtained.

To find the most similar pair between the foreground and the background DTMs, the matching degree is used to quantify the similarity of 2 seafloor patches of different DTMs. To simplify the discussion, the DTM is assumed to be in the grid format in this paper. In the foreground DTM, let P denote the terrain grids in a window with dimension (W_x, W_y). Likewise, Q is a patch with equal window size in the background DTM. Thus, to quantify the similarity degree between P and Q, the matching degree S between them is given as

$$S = \sum_{l=1}^{n} \frac{(d_l - z_l)^2}{\sigma_{d,l}^2 + \sigma_{z,l}^2} \tag{10.1}$$

where $\{d_l, l = 1{-}n\}$ and $\{z_l, l = 1{-}n\}$ are the grid vectors of P and Q respectively, $\sigma_{d,i}^2$ and $\sigma_{z,i}^2$ are the corresponding variances. The squared difference form given in Eq. (10.1) is a bit different from the criterion in Duffy and Hughes-Clarke (2005), but is used extensively in geostatistics and terrain navigation because of its analytical tractability (Nygren 2005). To simplify the expression, a weight term is introduced in the equation:

$$S = \sum_{l=1}^{n} w_l (h_l - z_l)^2 \tag{10.2}$$

where the weight $w_l = \frac{1}{\sigma_{d,l}^2 + \sigma_{z,l}^2}$.

Hence, for a particular patch P in the foreground DTM, the objective is to find a patch Q with equal window size from the background DTM based on the criterion Eq. (10.2). For this purpose, the position of the window is sliding over the background DTM, so as to find a horizontal position with the smallest S. When plotting the correlation sum Eq. (10.2) for every candidate position, the so-called correlation surface appears. Then, the minimum point on the correlation surface corresponds to the displacement $\hat{D}$ that gives the best match:

$$\hat{D} = \arg\min_{D} \sum_{l-1}^{n} w_1 \left(h_l - z_{l,x}\right)^2 \tag{10.3}$$

where the term D denotes the position of Q in the background DTM. After the matching, the temporal displacement of a certain terrain patch can be deduced by subtracting the horizontal positions of the matching pair.

Two implicit assumptions exist in the above procedure. Firstly, success of estimation with correlation matching is based on the presence of morphological features in the different survey. Generally, the seafloor seldom preserve exact morphological expressions during different survey periods. However, general shapes of features are often preserved, which make the estimation possible. Secondly, the validity of measuring the similarity with Eq. (10.2) is based on the premise that there is little bias in depth dimension between multi-temporal DTMs. Nevertheless, this is seldom the case for real data. On the one side, the inaccurate tidal correction could lead to depth bias. On the other side, terrain changes happen during different survey periods. To deal with the depth bias, a modified algorithm, named the profile matching, is a natural choice. In the profile matching, the foreground grid vector $\{d_l, l = 1{-}n\}$ is replaced by

$$dc_l = d_l - \overline{d} \tag{10.4}$$

where $\overline{d}$ denotes the mean of the foreground grid vector, given by $\overline{d}_l = \frac{1}{n}\sum_1^n d_l$. Similarly, the background grid vector is replaced by

$$zc_l = z_l - \overline{z} \tag{10.5}$$

Therefore, by subtracting the mean depth from the grid vector, the absolute depth information is removed, and only the contour information is used for correlation matching. With the replacement, the influence of the bias in depth can be removed.

It is noteworthy that the seafloor terrain is generally nonlinear. In consequence, the distribution of the matching result is seldom Gaussian (often multimodal), which prevents a straightforward interpretation of the matching result. However, as the number of grid points in the window increases, the distribution of the matching result will converge to Gaussian, owing to the central limit theorem (CLT) (Nygren 2005). Moreover, variance R of the matching result can be approximated by the Cramer-Rao lower bound (CRLB):

$$R = E\{[\hat{x} - x]^2\} = \sigma_e^2 \begin{bmatrix} \sum_{l=1}^{N} \left[\frac{\partial h^i(\hat{x})}{\partial \hat{x}_i}\right]^2 & \sum_{l=1}^{N} \frac{\partial h^i(\hat{x})}{\partial \hat{x}_i}\frac{\partial h^i(\hat{x})}{\partial \hat{x}_j} \\ \sum_{i=1}^{N} \frac{\partial h^i(\hat{x})}{\partial \hat{x}_j}\frac{\partial h^i(\hat{x})}{\partial \hat{x}_i} & \sum_{i=1}^{N} \left[\frac{\partial h^i(\hat{x})}{\partial \hat{x}_j}\right]^2 \end{bmatrix}^{-1} \tag{10.6}$$

where x denotes the horizontal position of the window $[x_{\text{north}}, x_{\text{east}}]$, and $h(\cdot)$ the background DTM for searching. Therefore, when the number of the grid nodes in the window is large, the distribution of the matching error can be estimated with Eq. (10.6).

10.2.2.2 Robust Correlation Matching Algorithm

Obviously, the above procedure work efficiently under Gaussian noise assumption, because the matching converges CRLB as the number of matching points increases to infinite. However, outlier arises naturally in the terrain data, which violates the assumption. On the one side, outliers are introduced inevitably during the survey process, due to poor data collection and poor survey environment. On the other side, seafloor morphology features may be distorted between consecutive survey, owing to irregular erosion/deposit, sediment gravity flow, or even the sunken boats. All these factors will perform a disturbance on our investigation. Therefore, to estimate the bedform migration, the impact of outliers and terrain distortion has to be excluded.

To this end, a robust procedure, termed robust correlation matching (RCM) algorithm is developed in this paper. The objective of the procedure is to estimate the displacement of the seabed in the presence of outliers, based on the robust estimation theory. Many robust estimator have been proposed during the last decades, and the most famous one is probably the M-estimator proposed by Huber (2011). When applying robust theory to correlation matching in the context discussed, two problems should be taken into consideration. Firstly, robust theory access the robustness of an estimator with the breakdown point (BP) criterion, which is defined as the largest fraction of outliers that the algorithm can tolerate. Since extensive variation of seafloor may occur between different survey, the developed matching algorithm should possess a high BP. Secondly, the estimator should have high efficiency at the same time.

Based on the above considerations, the robust matching algorithm with both high BP and simple framework is introduced. To achieve these qualities, the method consists of 3 steps.

Firstly, the least median square (LMS) matching (Rousseeuw 1984), an analog to the correlation matching in Eq. (10.2), is executed to provide a robust initial matching result. Different from the least square form in Eq. (10.2), LMS method estimates the displacement based on least median square rule:

$$\hat{D}_0 = \arg\min \ \mathrm{median}(dc_l - zc_l)^2 \tag{10.7}$$

By applying Eq. (10.7), LMS method possesses very high BP (up to 0.5) but poor efficiency, which makes it a suitable tool for the outlier diagnose (Rousseeuw 1984).

During the computation, it is noteworthy that the mean values $\overline{d}$ and $\overline{z}$ involved in Eqs. (10.3) and (10.4) are also very sensitive to outliers. Hence, instead of computing $\overline{d}$ directly, we estimate its value by the trimmed mean, which is given as

$$\overline{d}_t = \frac{1}{U-L}\sum_{i=L+1}^{U} d_{(i)} \tag{10.8}$$

where $L = \lfloor n\delta \rfloor$ is the lower trimming bound, $U = n\text{-}L$ the upper trimming bound, and δ the trimming ratio (20% for example). By trimming $\delta/2$ percent data on both sides, the trimmed mean can give a robust and efficient estimate of the mean value (Olive 2008). Thus, by replacing $\overline{d}$ and $\overline{z}$ with $\overline{d}_t$ and $\overline{z}_t$, the impact of outliers can be accounted for.

Secondly, the contribution of each measurement pair is adjusted according to the initial estimate, so as to remove the impact of the potential outlier. Based on $\hat{D}_0$, the residual values $v_i = h_i - z_i(\hat{D}_0)$ are the main information for outlier identification. Under the no-outlier assumption, the distribution of residuals will converge to Gaussian distribution N (0, σ_v). Thus, when the absolute value of a certain residual is extremely large comparing to σ_v, it is reasonable to identify the corresponding grid pair as outlier. To handle the influence of outliers, the weight w^i of each measurement pair is adjusted according to a threshold value M. Following the idea of (Gervini and Yohai 2002), the empirical probability of each large residual is compared with its probability under the distribution N(0, σ_v).

To be more specific, for the value of v_i which is larger than a certain value η ($3\sigma_v$ for example), a measure of the proportion of outlier is defined as

$$d_n = \sup_{v_i \geq \eta}\{P(v_i) - F_n(v_i)\}^+ \tag{10.9}$$

where $P(v_i)$ is the probability of v_i under N(0, σ_v), $F_n(v_i)$ is the empirical probability of v_i, and $\{\cdot\}^+$ denotes the positive part. Note that if $x_1 \leq \cdots \leq x_n$ are the order statistics of the residuals and $i_0 = \max\{i : x_i < \eta\}$, then

$$d_n = \max_{i > i_0}\left\{P(v_i) - \frac{(i-1)}{n}\right\}^+ \tag{10.10}$$

At this stage, the $\lfloor nd_n \rfloor$ observations with the largest residuals will be recognized as outlier and eliminated. As a result, the cutoff value is given as

$$t^+ = \min\{t : F_n(v_i) \geq 1 - d_n\} \tag{10.11}$$

Instead of decreasing the weight to a decimal as in M-estimator, or setting it to zero directly when the residual exceeds a given threshold, the above procedure identifies the measurement pair as an outlier not only when its residual exceeds the threshold, but also when the residual is sufficiently larger than the corresponding order statistic. By adaptively calculating the cutoff threshold from the data, the method can achieve full efficiency at the Gaussian distribution, but retain a high BP at the same time (Gervini and Yohai 2002).

Obviously, the premise of the above procedure is the knowledge of σ_v. Nevertheless, this information is seldom available because the uncertainty information of the bathymetric data is rarely available. To deal with this problem, we estimate σ_v directly from the residual vectors. To consider the effect of outliers, instead of using sample standard deviation, the mean absolute deviation is used to estimate σ_v, which is given as

$$\hat{\sigma}_v = 1.4826 \cdot \mathrm{median}(\mathrm{abs}(v_i - \mathrm{median}(v_i)) \tag{10.12}$$

Note that the median of the residual vector is 0. Thus, with $\hat{\sigma}_v$, the outliers can be identified based on the adaptive trimming rule.

Thirdly, ordinary correlation matching in Eq. (10.2) is executed again with the updated weights, and the improved matching result can be computed. By iterating weight updating based on previous estimated position and position updating based on updated weights until convergence, the final matching result is obtained. It is noteworthy that the matching result of the LMS matching in Step 1 may be multimodal, because of the inherent nature of LMS estimation as well as correlation matching. To deal with this problem, it is necessary to track multiple local minimum values in LMS matching result simultaneously, so as to grantee the global minimum of the final result. For example, suppose that there are m minimum values in the LMS result. We execute the outlier detection based on every local minimum independently. Thus, after the correlation matching with the updated weights, we will have m matching results. To find the global minimum from these results, the residual vector $\{v_i,\ i = 1\text{-}n\}$ is used as the indicator. Define the weighted sum of the residual square as

$$R = \frac{n}{n - n_t} \sum_{i=1}^{n} w_i (v_i)^2 \qquad (10.13)$$

where $\{w_i, i = 1\text{–}n\}$ are the weights derived by the outlier detection, and n_t is the number of weights which are adjusted to 0 during the outlier detection. Thus, when multimodality occurs, the matching result with the smallest R value will be recognized as the final result.

Finally, the procedure of the robust correlation matching is summarized as.

1. utilize LMS matching to calculate the initial value D_0 based on Eq. (10.7);
2. calculate the cutoff value based on the estimate D_0, and adjust the weights accordingly;
3. execute ordinary correlation matching with the updated weights to obtain the updated estimate;
4. iterate Step 2 and Step 3 until convergence;
5. execute Step 2–4 on every local minimum in the LMS matching result simultaneously, and choose the matching result with the smallest R value in Eq. (10.13) as the final matching result.

10.2.3 Results and Analysis

In the computation, a first attempt is to estimate the migration of sandwaves with the ordinary weighted correlation matching algorithm based on Eqs. (10.1)–(10.3). During the matching, to make a compromise between matching accuracy and resolution, the window size is 20 grids by 20 grids (60 m × 60 m). Figure 10.7 depicts the migration estimates of the correlation matching. As can be seen, for the most part of the area, the migration amplitude of the bedform is relatively small. While for some part of the area, the migration amplitude exceeds 10 m. In general, the main tendency of the bedform movement is in the down-canyon direction. While for a small portion of area, the migration shows an up-canyon tendency, which is an attracting feature when comparing to the neighboring area. To evaluate the quality of the matching result, the mean absolute value (MAV) of the residual vector is used. Clearly, when the matching result in the background is dissimilar to the terrain in the foreground window, MAV will increase. Figure 10.8 reveals that MAV changes dramatically when the foreground window slides over the region. When the matching results are correlated with MAV, it turns out that high MAV (marked with red circles in Fig. 10.8) tempts to be related to the large migration amplitude (marked with red color in Fig. 10.7). Obviously, for a portion of large migration estimates, MAV indicates that their credibility is relatively low.

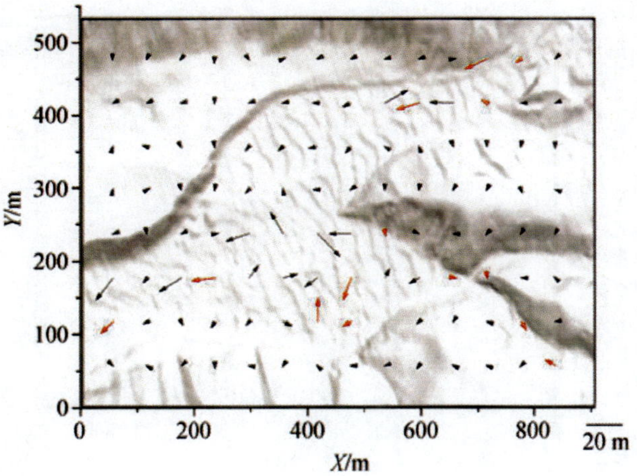

Fig. 10.7 Migration estimates derived by OCM. The red ones denote the migration estimates with large MAV

Secondly, to evaluate the performance of RCM presented in Sect. 10.2.2.2, robust correlation matching is executed on the same foreground window as above. Figure 10.9 shows the differences between the results of RCM and OCM. Generally, the main migration tendency is similar to the result of OCM. While for some local regions, the substantial

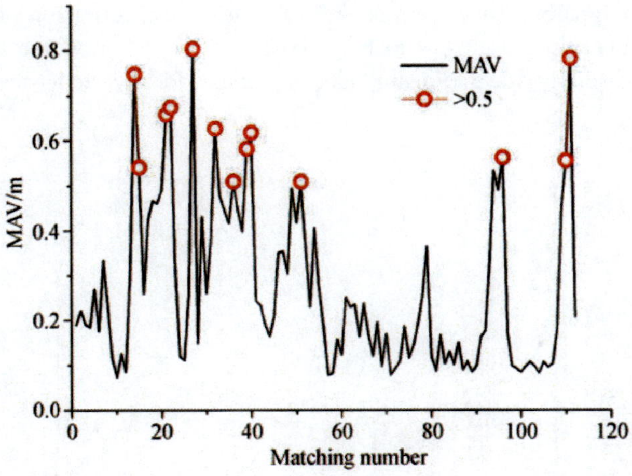

Fig. 10.8 MAV of the matching results derived by OCM. The red circles indicate MAV which are larger than 0.5

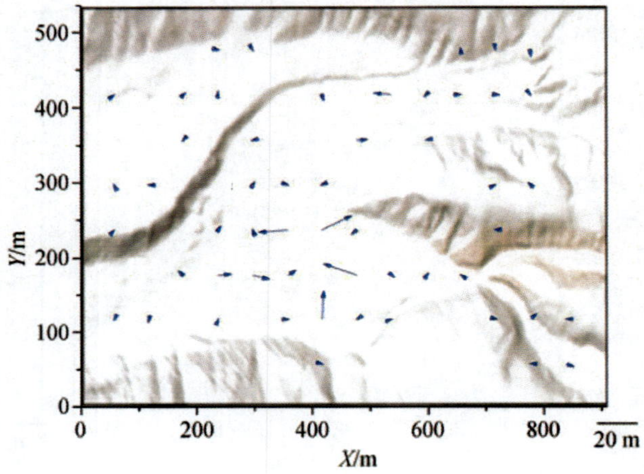

Fig. 10.9 Difference in the migration estimates of RCM and OCM

differences (exceeds 10 m) show in the RCM results. This indicates that in these local regions, the impact of outliers is not negligible. To compare the matching quality of RCM relative to OCM, MAV derived by two matching methods are compared (Fig. 10.10). It is obvious that MAVs of RCM are lower than OCM in general, which indicates the validity of RCM under the outlier environment.

A detailed analysis of the relationship between MAV and RCM estimates shows a similar situation to OCM scenario. The bulk of large MAV correspond to large horizontal migration, which makes these large migration suspicious. Obviously, these suspicious migration estimates severally hinder the accurate characterization of the sandwave migration. To solve this problem, the error information of matching result has to be estimated, so as to provide the credibility degree of the matching result. Ideally, under the Gaussian noise environment, the theoretical error distribution of the matching can be computed based on Eq. (10.6). Thus, the theoretical distribution of the residual vector can be inferred by integrating this information with DTM. By comparing the actual residual vector with its theoretical distribution, the credibility can be assigned to the matching result. Nevertheless, the direct computation with Eq. (10.6) involves the uncertainty information of the bathymetry data, which are seldom available. Therefore, instead of estimating the theoretical distribution of the residuals, the empirical distribution of the actual residual vectors is analyzed. For this purpose, the mean value of the residual vector (MRV) is used to access the estimates empirically.

Figure 10.12 shows the empirical probability distribution of MRV serial derived by RCM. As can be seen, the major part of MRV distribution shows a Gaussian shape. While for the tail section, the heavy tail phenomena is obvious, which indicates the existence of the dubious matching results. Based on the empirical distribution information, the adaptive trimming rule described in Sect. 10.2.2.2 is used to identify the suspicious matching results. To be more specific, the mean absolute deviation given in Eq. (10.12) is used to estimate the stand deviation of MRV. Next, MRV with large absolute values is tested according to equations Eqs. (10.9)–(10.11). For that MRV which fail the test, the corresponding matching result will be regarded as error. The spatial distribution of such migration estimates is depicted in Fig. 10.11 (marked in red). Although some of these marked results seem reasonable when compared with the nearby estimates, the corresponding residual vectors are too large to be trusted. Meanwhile, it is noteworthy that some irregular migration estimates pass the test (marked with the blue circle), which indicates the unique behavior of the local sandwave.

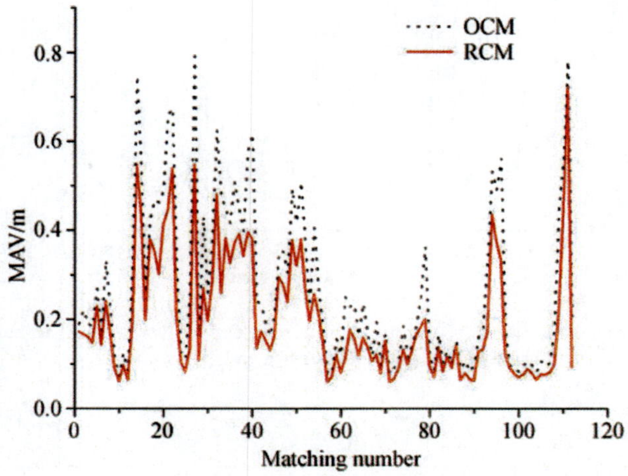

Fig. 10.10 Comparison of MAV derived by RCM and OCM

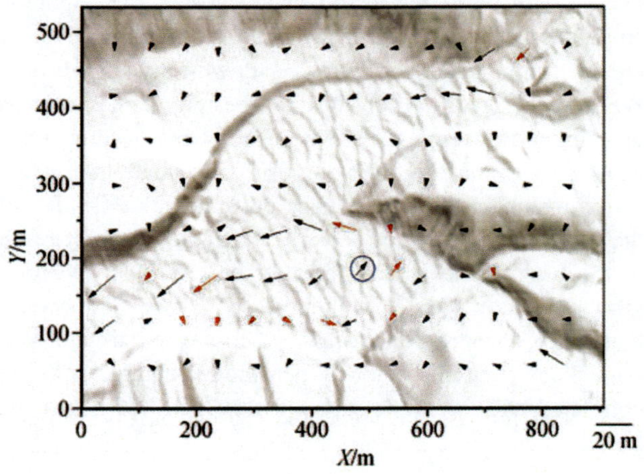

Fig. 10.11 Migration estimates derived by RCM and tested by the adaptive trimming rule. The estimates which don't pass the test are marked in red

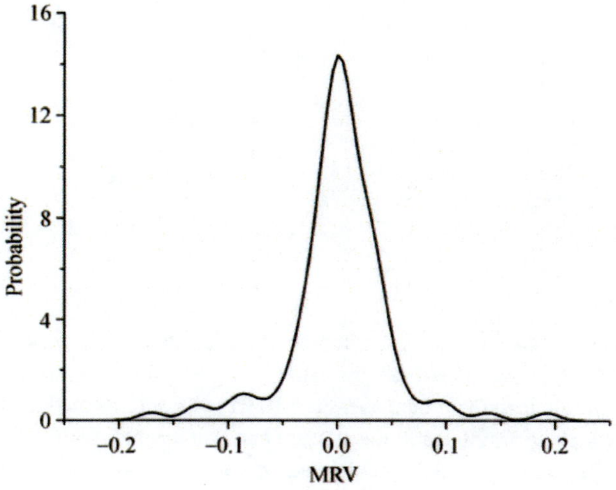

Fig. 10.12 Migration estimates derived by RCM and tested by the adaptive trimming rule. The estimates which don't pass the test are marked in red

10.3 Study on Gas Hydrate Geomorphology Identification Marks on the Northern Slope of the South China Sea

10.3.1 Study Area Overview

The SCS is one of the largest marginal seas in the western Pacific Ocean. It is part of the West Pacific Trench-arc-basin system, where the Indian-Australian Plate, Eurasian Plate and Pacific Plate meet. Complex tectonic movement resulted in the special geophysical field and geological boundaries of the SCS. Many varied geological structures formed in the complex tectonic environment on the northern slope of the SCS, providing ideal conditions for gas hydrate accumulation. The northern slope of the SCS has become a primary area for gas hydrate studies and exploration, aimed at discovering considerable amounts of stored gas hydrate.

The study area is located in the Zhu II depression of the Zhujiang River Mouth Basin; the water depth in the area is between 100 and 3000 m (Fig. 10.13). The depth increases from NW to SE. The upper slope is flat, with water depths of less than 500 m, and the average gradient is 5%. The middle slope is steep, with canyons and nearly vertical walls. The water depth ranges from 700 to 1500 m, and the gradient is 30%. Several studies suggest that the distribution of underwater cliffs, clay diapirs and protuberances is correlated with the decomposing of gas hydrate (Chen et al. 2006). There are many crater-like features in the middle slope, with the water depth ranging from 1300 to 2000 m. Recent research has shown that these features were formed by large amounts of escaping gas causing the sinkage of sediments, which led to the formation of the craters. In the lower slope the topography becomes smoother, the water depth is more than 1500 m and the gradient is 13%.

The sub-bottom and single-channel profiles used in this study (Fig. 10.13) were collected during the Dayang115-20 South China Sea trial cruise in 2008. To achieve high resolution and obtain deep sub-bottom profiles in the Shenhu

Fig. 10.13 Bathymetry, survey lines and stations in the study area; isobaths in meters. Blue lines indicate synchronous sub-bottom and single-channel profiles; red dots (HS) indicate locations of gas hydrate wells; red dots (C–F) indicate locations of carbonate samples from the 2006 cruise; number 1148 indicates the ODP site mentioned in the text, the insert (left corner) shows the tectonics and position of the study area; the black areas represent depressions; the gray areas represent uplift

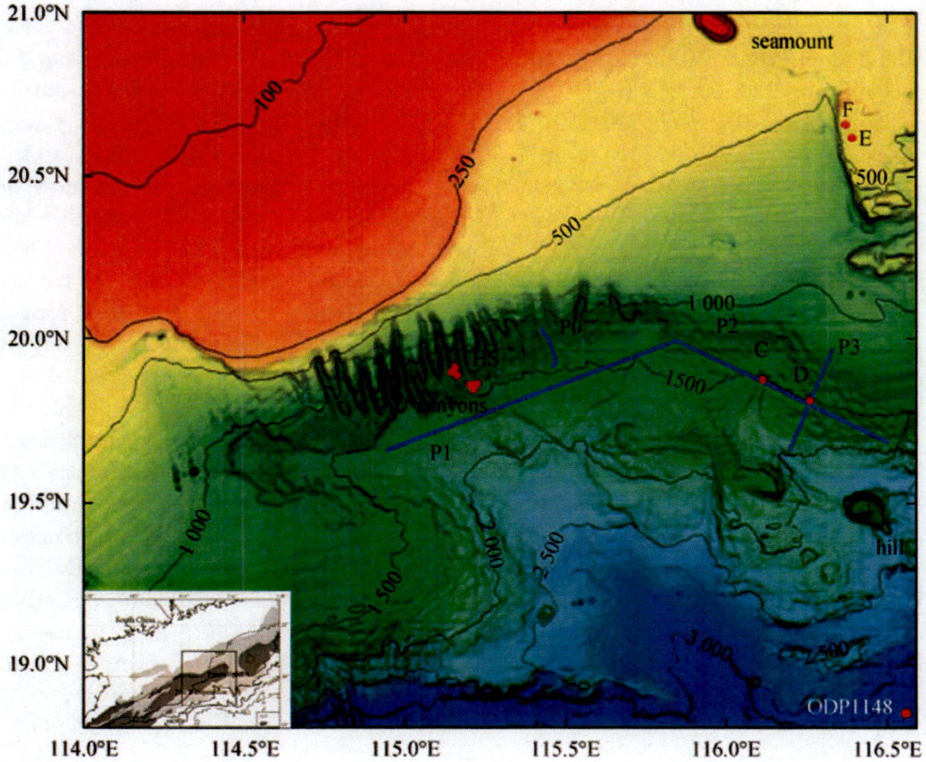

Table. 10.2 Details of sub-bottom and single-channel seismic profiles

Name of profile	Start point			End point			Length of profile/km
	Latitude (N)	Longitude (E)	Water depth/m	Latitude (N)	Longitude (E)	Water depth/m	
P0	20°01′30″	115°25′36″	1087	19°54′48″	115°27′24″	1432	12.8
P1	19°41′06″	114°58′12″	1627	19°59′18″	115°54′00″	1249	109.0
P2	19°58′24″	115°55′12″	1249	19°40′18″	116°30′36″	1860	73.4
P3	19°40′00″	116°15′00″	1930	19°51′00″	116°21′00″	1300	23.2

area on the north slope of the SCS, a sub-bottom profiler and single-channel seismic system were utilized simultaneously. The high-frequency sub-bottom profiler (Topas PS018 3.5 kHz, Kongsberg) can identify thin sedimentary strata in sub-surface layers, with a horizontal resolution of several decimeters. The SIG low-frequency single-channel seismic system (SIG, 30–360 Hz) can detect strata in the lithosphere. The total length of the synchronous survey lines of the sub-bottom profile and single-channel seismic profile is 218.4 km. The survey line details are shown in Table 10.2.

Based on time-depth conversion and depth correction, we digitized the profiles by a combination of human and machine interpretation. We plotted the strata and set up the chronology of the gas hydrate sediment in the Shenhu area in the northern slope of the SCS according to thickness, sedimentary rate and geological age of Ocean Drilling Project (ODP) Site 1148.

10.3.2 Acoustic Characteristic of Sub-Surface Layers and Genetic Mechanism

Sub-bottom Profile P0, lies NW–SE, perpendicular to the isobaths; it is 12.8 km long (Fig. 10.13). The water depth is 1 087 m at the NW point (20°01′30″N, 115°25′36″E) and 1 432 m at the SE point (19°54′48″N, 115°27′24″E). The profile was divided into two sections according to the gradient and terrain features (Fig. 10.14a). The terrain of the upper section (Fig. 10.14b) is very rugged and rich in protuberances while the terrain of the lower section (Fig. 10.14c) is flat. The acoustic reflective properties of the two sections are conspicuously different. The acoustic properties of the geological formations and structure, which are correlated to the gas hydrate, are described below.

10.3.2.1 Protuberance

Three prominent hummocks (elevation 40 m from the seabed) appear in the sub-bottom profile (Fig. 10.14b). The protuberances are not symmetrical in shape, with the lower slope is longer than the upper slope. Blurred reflections become blank reflections from the protuberance's edge towards the center. The blurred reflections are caused by the anisotropy of the stratum in the sediment. Geothermal free gas, such as methane, which accumulates in the stratum can rise through the faults. Gas hydrate crystals, composed of methane and water, can form in the stable zone in the shallow sediment layer. Tectonic movement can cause sudden decomposition of gas hydrate, leading to the release of large volumes of gas, resulting in protuberances in the sediment area.

10.3.2.2 Shallow Fault

Figure 10.14c shows a small fault 60 m below the seabed. Recent research showed several deep faults cutting through the sediment layer and reaching the seafloor in the Shenhu area of the SCS. The small fault in Profile P0 may be the terminal segment of the deep fault in the lithosphere. The fault and rupture seen in the profile provide favorable channels for liquid and gas flow in the shallow layer.

10.3.2.3 Shallow Gas-Rich Layer

The lower section of Profile P0 is flat with clearly recognizable features. The average acoustic penetration is 96 m. Two continuous enhanced reflectors (R1 and R2) can be observed in the parallel stratum, including a series of acoustically blank patches. Partial enhanced reflection occurs at the interface of the sediment and the acoustically blank patches, where the acoustic impedance is high.

There is an almost horizontal acoustically blank zone below Reflector R2. The characteristics of this zone are similar to those of the acoustically blank patches. A few remnant geological forms in the acoustically blank zone indicate the asymmetry of the stratum. The 40 m-thick gas-rich layer comprises continuously enhanced reflectors, acoustically blank patches, partial enhanced reflection and an acoustically blank zone. Well logging data in the sediment layer in this area revealed that the samples from China's first gas hydrate drilling expedition are 153–225 m below the seabed; the gas hydrate stable zone was found to be 18–34 m thick. The shallow gas-rich layer in Profile P0 is 34–82 m below the seabed. Judging from its geographical

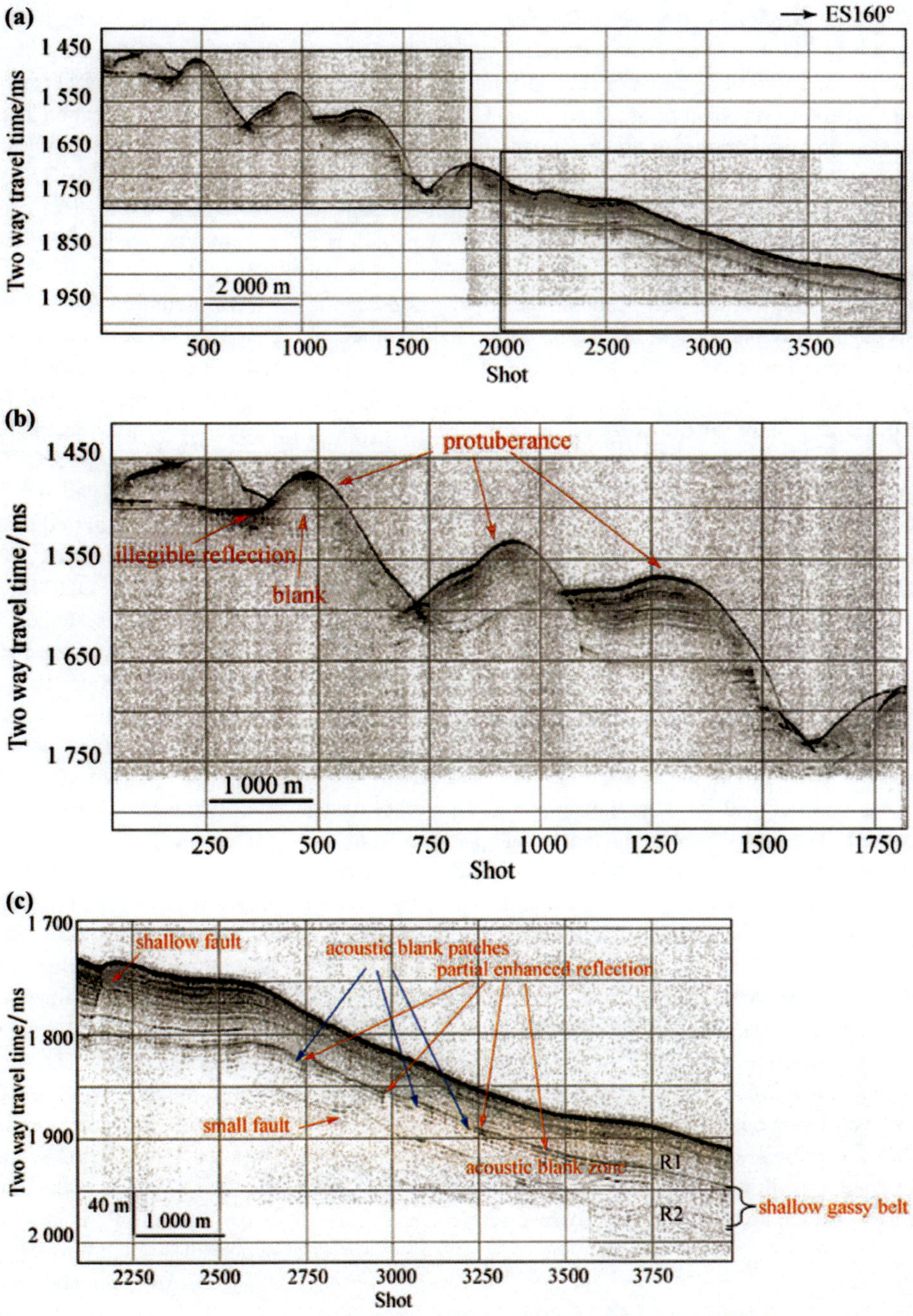

Fig. 10.14 Sub-bottom profile of survey line P0. **a** Some gas hydrate existing evidences on Profile P0; **b** englarged view of P0 upper part; and **c** englarged view of P0 lower part

position, we think the gas may come from the decomposing of deep gas hydrate in the deep sediment.

10.3.3 Characteristics of the Stratum in the Single-Channel Seismic Profile

Single-channel Profile P2 lies NW–SE, almost parallel to the isobaths (Fig. 10.13); it is 73.4 km long. The water depth is 1 249 m at the NW point (19°58′24″N, 115°55′12″E) and 1 860 m at the SE point (19°40′18″N, 116°30′36″E). The seismic data are of good quality and contain rich information of the deep structure. Seismic layers such as T0, T2, T4 and T6 can be identified, which is important for studies of deep crust and gas hydrate evolution in this region. The acoustic properties of the geological forms, which are correlated to gas hydrate, are described below.

10.3.3.1 Indentations and Seepage

A series of U-shaped or V-shaped craters can be identified at the top of Profile P2; they are several hundred meters long and several to several tens of meters deep. Craters are closely related to gas hydrate in the continental margin regions. Large volumes of escaping gas, from the decomposition of gas hydrate, causes sinking of sediments and forms craters in the seabed. The existence of many craters in Fig. 10.15 is a strong indication of escaping gas in the Shenhu area in the northern South China Sea. Another means of gas escape is seepage. Pole-shaped light reflections in the sediment are an indication of seepage, as shown in Fig. 10.15.

10.3.3.2 Folds and Faults

The sediment in Profile P2 at shot numbers 150–350 forms an anticline fold about 40 m in height. Some researchers believe that such folds form when gas is released and accumulates in sediment in the continental slope. The fold also creates favorable conditions for capturing and storing gas for the formation of gas hydrate. Moreover, seepages and faults cut through the folds in Profile P2, increasing the complexity of the stratum.

The oceanic crust beneath the northern SCS contains an extensive system of faults and ruptures after two periods of seafloor spreading in the Late Cretaceous and Early Tertiary. Some faults extend from the lithosphere to the surface of the seabed. The faults become channels for gas flow from the deep strata to the shallow stable zone while folds provide favorable conditions for gas accumulation. These two factors promote the formation of gas hydrate.

10.3.3.3 Bottom Simulating Reflector (BSR)

An intermittent negative strong reflector can be observed 225–240 m below the seabed in Profile P2. The strong reflector is parallel to the seabed, and the longest section is 6 km long. We interpret it as a bottom simulating reflector (BSR). The BSR is just below the level of the samples found in China's first gas hydrate expedition in this region, suggesting that the BSR represents the bottom of the gas hydrate stable zone.

10.3.4 Chronology of the Gas Hydrate Sediment

Deep-sea sediments from depths down to 859 m below the seafloor were recovered from Site 1148 in ODP Leg 184. The age of these cores spans the last 32.8 Ma in the northern SCS, which is the longest record known so far covering the detailed evolutionary history of the SCS since the Oligocene. The effective detection depth of single-channel Profile P2 is 800 m, which roughly corresponds to the drilling depth of ODP Site 1148. The strata derived from Profile P2 are plotted in Fig. 10.15 along with the chronology of the gas hydrate sediment according to the data from ODP Site 1148.

We identified four reflectors under the seabed in Profile P2. The deepest reflector, T6, is a continuous strongly reflective surface, buried 440–500 m below the seabed. The uppermost reflector, T0, is almost parallel to the seabed, buried 120–150 m deep. The middle reflectors, T2 and T4, are 270–300 m and 350–390 m deep, respectively. Generally, the SCS began its rifting activity 32 Ma. However, recent research inferred that the process of depression of the northern SCS occurred 23 Ma, corresponding to the 460–480 m stratum in ODP Site 1148 and reflector T6 in profile P2. The previous studies found a discontinuous sediment layer formed by intense denudation 1–3 Ma between the Oligocene and Miocene at depths of 458–472 m. This suggests that reflector T6 represents the beginning of the Miocene. Using the astronomically tuned time scale methodology Tian et al. (2005) studied the chronology of the Middle Miocene (18–12 Ma) at ODP Site 1148, and the stratum at depths of 280–370 m. Judging from the buried depth of T2 and T4 in Profile P2, the two reflectors correspond to the beginning and end of the Middle Miocene, respectively. The BSR and the fold in Profile P2 are at depths of 210–270 m, while China's first gas hydrate expedition found samples 153–225 m under the seabed. The BSR and gas hydrate samples are in the late Miocene and Pliocene strata (12–2.6 Ma). After two periods of seafloor spreading, in 42–35 Ma and 32–23 Ma, respectively, the tectonic movement in the SCS weakened. In the northern slope of the SCS, the sedimentary rate remained high from the Late Miocene to the Pliocene. The depth of the sediment layer in Profile P2 is more than 300 m; this corresponds to a period of high biological productivity in the SCS (10.6–4.8 Ma). An abundance of sediment and marine organisms provided favorable conditions for the formation of gas hydrate; thus, the strata formed during the Miocene and Pliocene in the northern slope of the SCS has been the focus of gas hydrate studies and exploration.

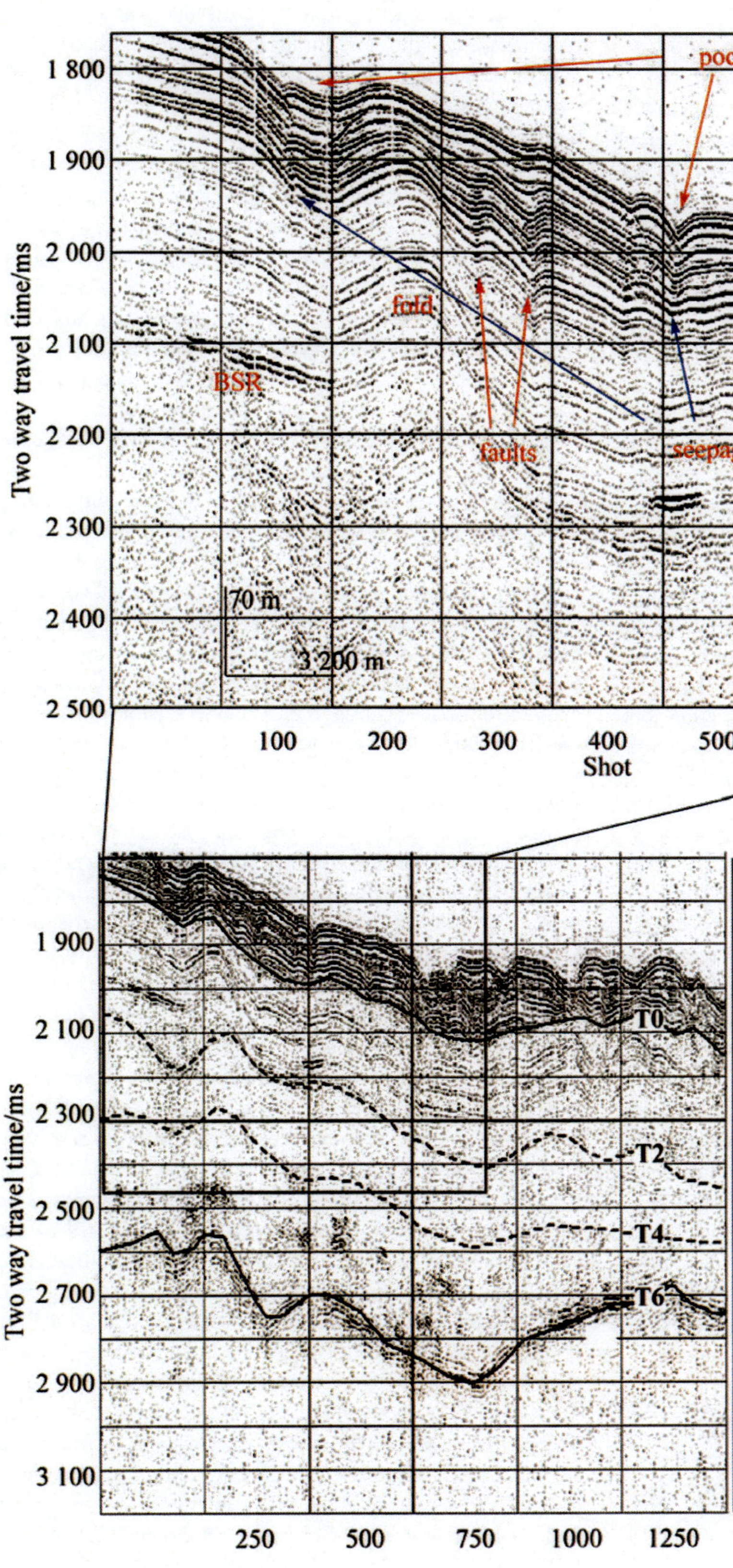

Stratum			Era/Ma	Interfaces	Buried depth/m	Tectonic movement
Q			2.6	T0	120-150	
Neogene	N_2		5	T1		Dongsha
	N_1	N_1^3	12	T2	270-300	
		N_1^2	17	T4	350-390	
		N_1^1	23	T6	440-500	Baiyun
Paleogene	E_3	E_3^2				
		E_3^1	32	T7		Nanhai

Fig. 10.15 Single-channel seismic profile of survey Profile P2

10.4 Study on the Morphotectonics of the Manila Subduction Zone

Tectonic geomorphology is an interdisciplinary field that combines geomorphology, geodynamics and tectonic geology. Tectonic geomorphology explores the dynamic and active landforms (or terrain) directly formed by crustal tectonic movement, called Tectonic Landform or Morphotectonics (Shi and Du 2006). Tectonic geomorphology studies the formation process and development of tectonic landforms (including deposition rate) and deduces the characteristics of crustal movement under certain tectonic stress fields, solving practical application problems (Han 1992). Submarine tectonic geomorphology specializes in the study of specific submarine geomorphology (or topography) that was formed, developed or reformed by tectonic activities. In recent years, with the rapid development of ocean bathymetric technology, especially the development of highly efficient, wide-ranging, high- precision multi-beam technology, a large number of fine seabed landforms have been revealed. The study of tectonic geomorphology has been gradually extended to the field of marine geology, especially in studies of seabed tectonic evolution in areas of submarine hydrothermal activity (Humphris et al. 1996; Kleinrock and Humphris 1996), subduction zones (Soh and Tokuyama 2002; Kopp et al. 2006), marginal basins (Wu et al. 2003; Li et al. 2012) and mid-ocean ridge areas (Carbotte and Macdonald 1992; Grindlay et al. 1992) for example. Submarine morphotectonics is widely used and plays an important role in all these fields.

Compared with traditional geophysical methods such as gravity, magnetism and seismic investigation, multi-beam bathymetric data offer both high resolution and widespread coverage. Fine topographic data provide more reliable information for researchers discussing the causes, formation and evolution of the research target. Thus, a more accurate model of the origin and development mechanism of related morphotectonic characteristics in similar tectonic environments can be derived and combined with other geophysical information such as seismic profiles, ocean bottom seismometer (OBS) data, gravity anomaly data, magnetic anomaly data and even historic drilling data. In this section, we illustrate the application of tectonic geomorphology in submarine scientific research by presenting a tectonic geomorphology study of the accretionary wedge in the Manila subduction zone at the eastern boundary of the South China Sea as an example.

10.4.1 Overview of the Study Area

The Manila subduction zone is a key component of the western Pacific subduction system; it was formed by the interaction between the South China Sea Plate and Philippine Sea Plate. As a young subduction zone deviating from the general subducting direction on the western Pacific continental margin, the northern part of the Manila subduction zone collided with the eastern margin of China's continental shelf, while the central part was compressed into the Huangyan Chain Seamounts, which is the middle ridge of the expansion of the ancient South China Sea. This led to significant changes in many topographic features and structural states in the study area, and a complex area with many unique structural features was formed. A thorough study of such a young subduction tectonic belt, especially the relationship between its initial characteristics and deep tectonic dynamics, will enhance our understanding of the plate tectonic evolution processes and mechanisms in this region and in other subduction belts in the world.

In recent years, the implementation of a large number of National Major Projects in China has enabled scientists to carry out research on the tectonic problems in the areas surrounding the Taiwan Island, systematic studies on tectonic dynamic processes and their variations in the South China Sea Basin and the plate interaction region around the Taiwan Island. Many recent advances have been made in the study of the dynamic mechanism of the Manila subduction zone as the eastern boundary of the South China Sea. Based on comprehensive comparison and analysis of multi-channel seismic profiles and regional topography, Lacombe et al. (2001) and Deffontaines et al. (2016) found that the crust and surface of the fan-shaped subduction front in southwestern Taiwan Island had regularly deformed because they were controlled by the collision and orogeny activities of Taiwan Island. Li et al. (2012) and Wang and Bilek (2014) studied and simulated the phenomenon of seamount "crowding" and the deformation characteristics of the accretionary wedge at northwest Luzon Island based on high-resolution topographic data. They found that the aggregation of seamounts in the upper region of the subduction plate causes a series of changes such as radial uplift of the accretionary wedge to the side of the island and sea-sliding to the side of the trench (Li et al. 2004). Through comprehensive analysis of gravity, magnetic and seismic data Li and King (2007) and Kao et al. (2000) found that the change of inner stress in a deep plate can bring about remarkable changes in the surface fractures and other geomorphological states as the subduction state transforms into a collision state from south to north along southern Taiwan Island. Many scholars constructed geophysical models of this region and analyzed the morphotectonic characteristics of the Luzon double volcanic chains (Yang et al. 1996; Liu et al. 2007a; Fan and Wu 2014; Tang et al. 2017). Ku and Hsu (2009) and Shang et al. (2010) defined the zones of the accretionary wedge at the North Manila Trench according to different multi-channel seismic profiles, and summarized the characteristics of the geological units and their evolution. Chen et al. (2009) and Hirtzel et al. (2009) compared the morphological details and origins of the northern and central regions of the Manila Trench. Huang et al.

(2012), Huang (2017), Chen et al. (2012) and Kuo et al. (2016) identified the deep crustal velocity structure and the location of the Moho under Taiwan Island by using the latest OBS data obtained in the TAIGER program and the drilling data from ODP 148 in the South China Sea.

10.4.2 Data Description

Regional morphotectonic research focused on the acquisition of high-precision topographic data for submarine targets. The study of tectonic geomorphology in the Manila subduction zone is also based on high-resolution regional bathymetry data. The high- resolution multi-beam bathymetry data for the northern part of the Manila Trench subduction zone used in this study are from ship surveys conducted in the past by our supporting institutions. The SeaBeam 2112 deep-water multi-beam system with an operating frequency of 12 kHz was used in the investigation. 12-channel SeaStar 3000 L wide-area differential global positioning system was used for positioning. The data were real-time sound-velocity profiles recorded simultaneously by 18 stations. The accuracy of the fixed-point repetition and cross-line surveying was evaluated after various corrections. The overall bathymetric error is less than 0.3% of the bathymetric value. The study of regional tectonic geomorphology should be based on topographic and geomorphological data, but not limited to these. Other geophysical data, such as seismic profiles, shallow strata profiles, OBS data or gravity and magnetic anomaly data are usually used to help infer the internal dynamic process of special seabed geomorphological features in the study area and to provide other essential information. The study of the dynamic process of deep subduction in the Manila subduction zone also relies on a single-channel seismic profile that passes through our study area that was published in previous articles.

10.4.3 Regional Morphotectonic Features

In our case study, we investigate the north-central part of the Manila subduction zone, located in the South China Sea off the northwest corner of Luzon Island, the main island of the Philippines (Fig. 10.16). The Manila subduction system generally lies NNE in this region. The subduction system can be divided into four secondary tectonic units lying approximately parallel to the axis of the Manila Trench. From west to east they are: the Manila Trench with the deepest water depth near the eastern basin of the South China Sea, the subduction accretion wedge represented by imbricated thrust faults or linear troughs and ridges arranged parallel to each other, the North Luzon Deep-water Trough—a belt of relatively flat terrain located in the middle of the island slope and the Upper Luzon Island slope where the water depth on the eastern side of the profile decreases sharply until it reaches the shoreline.

The edited high-resolution topographic shadow map shows that the seabed to the west of the trench is relatively flat, with scattered small hills and seamounts in some areas. There is no conspicuous V-shaped subsidence at the axis of the trench, which transits gently to the western basin and separates from the eastern accretionary wedge by abrupt thrust slivers. The trench gradually deepens towards the NW, reaching a depth of more than 4000 m.

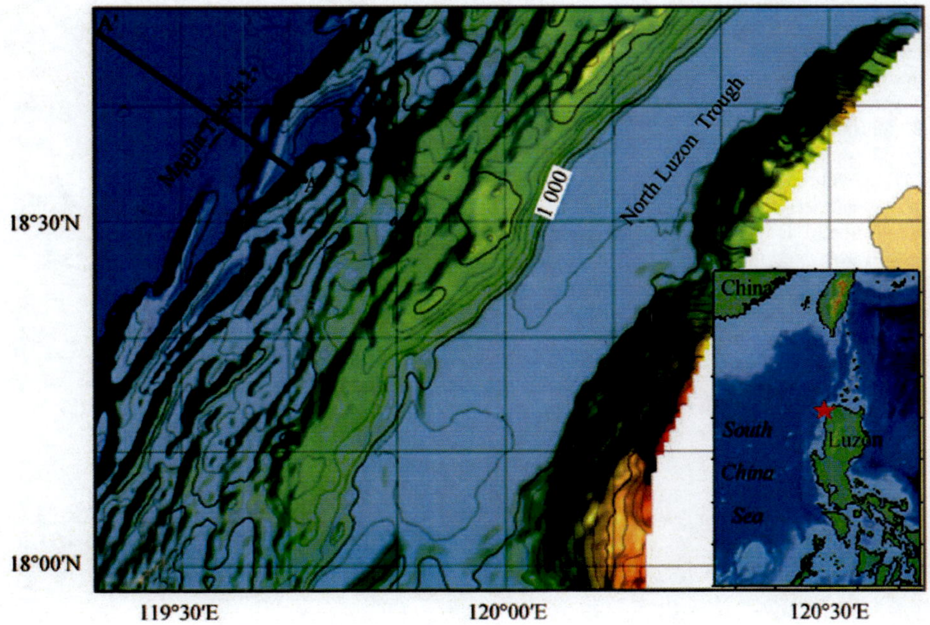

Fig. 10.16 Bathymetric chart of the study area. The solid line represents a single-channel seismic profile acquired in a previous study, the red star in the insert indicates the location of our study area

The transition zone between the trench and the North Luzon Trough is an accretive wedge developing zone. It is characterized by subparallel banded NNE-trending troughs and ridges arranged linearly and alternately. The overall water depth gradually shallows to the east, reaching the shallowest depth (about 1800 m) near the forearc basin, and then drops sharply to 1400 m and connects with the bottom of the North Luzon Trough to the east of the accretionary wedge. The regularly arranged thrust slivers are well developed and orderly, showing typical imbricate thrust faults commonly found in subduction zones around the world. They contain many very small trapped basins formed by arc- or cross-shaped ridges.

The North Luzon Trough is the forearc basin of the subduction zone; it is a "U" type deep-water trough developed on the western slope of Luzon Island. The bottom of the trough is flat and smooth with only individual small hills or depressions in some areas. The average water depth in the trough is about 3200 m. The west side of the trough is bound by wide-angle thrust sliver sets with steep smooth slopes. To the east rises the upper half of the Luzon Island slope, which is nearly vertical. The uneven and gravelly distribution of the slope bedding near the trough bottom indicates different geological composition from that of the accretionary wedge. Looking eastward, the western slope of Luzon Island is formed by the sharp rise of the North Luzon Trough up to the water surface. The island slope is steep, causing many landslides and the faults are concentrated.

The morphotectonic interpretation map of the study area illustrates the special structural and geomorphological features in the region (Fig. 10.17). According to the spatial density and characteristics of the fractures in the region, three main tectonic faults (LMT, MMT and UMT) were identified and the accretive wedge region was divided into four morphotectonic regions correspondingly: the lower tectonic region (LTZ), middle tectonic region (MTZ), upper tectonic region (UTZ) and ridge zone of the accretionary wedge (RZW). The detailed geomorphological features and distribution rules of the four morphotectonic regions will be summarized in following section and compared with adjacent research areas mentioned in Li et al. (2004).

With accurate multi-beam seabed bathymetry, many small trapped basins were identified in the middle and lower tectonic regions of the accretion wedge, especially in the seaward deformation area near the Manila Trench, where the basins are more conspicuous and dense. To further investigate this special tectonic unit, its deep structural state, fracture characteristics and evolution were analyzed by combining the bathymetry with the seismic profile in the region (Fig. 10.18).

The continuous and stable reflection waves west of the Manila Trench on the seismic profile correspond to the flat area between the SCS basin and Manila Trench in the morphotectonic map. The smooth seabed and the undistorted subparallel reflections on the lower part of the profile indicate that the deposits in this region were not affected by deep subduction activity; thus, this region can be defined as a pre-subduction deformation area.

From the axis of the trench, although there is no notable undulation of the seabed on the morphotectonic map, the folded deformation between the lower part of Layer L1 and the detached surface on the seismic profile suggests that the

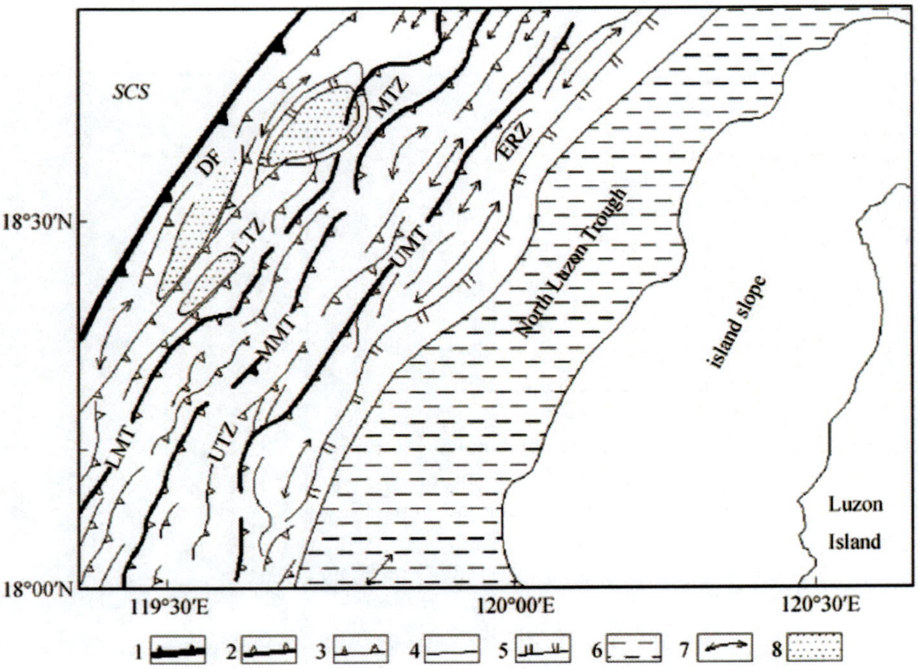

Fig. 10.17 Morphotectonic interpretation map of the study area. 1. DF-deformation front; 2. main tectonic fracture zone; 3. compressive fracture zone; 4. unknown fracture zone; 5. steep cliff; 6. trough region; 7. fold belt; 8. minisize trapped basin; LTZ-lower tectonic zone; MTZ-middle tectonic zone; UTZ-upper tectonic zone; RZW-ridge zone of the wedge; LMT-lower main fracture zone; MMT-middle main fracture zone; UMT-upper main fracture zone

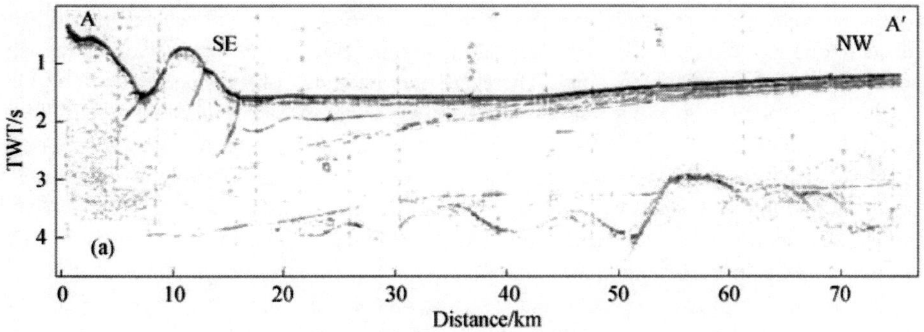

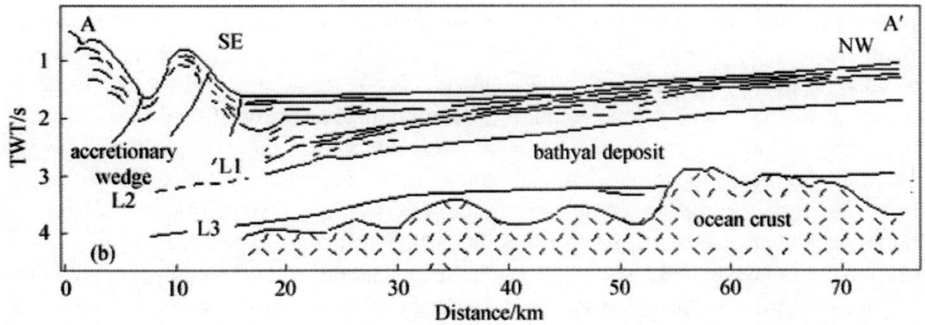

Fig. 10.18 Seismic profile and its interpretation (profile line shown in Fig. 10.16). **a** original seismic Profile AA′near Luzon island on Fig. 10.16; and **b** corresponding interpretation of the Profile AA′

subduction activity has begun to affect the upper sedimentary sequence. The first set of thrust faults on the profile shows conspicuous imbricated NW-trending dip characteristics. The dip angle of the fault increases gradually as the water becomes shallower, and the fault profile is generally concave to the inner side of the trench. The seismic profile reflects the dynamic processes of deformation, accretion, compression and even thrust of the upper deposition layer caused by the deep subduction. The deformation front then becomes the initial position of the subduction activity on the seabed. Although the subduction activity causes deformation and fracturing of the shallow sediments ahead of the deformation belt of the accretionary wedge, the direction of the thrust nappe varies due to the different geological conditions and structural characteristics of the sediments. During this process some small volumes of sedimentary depressions extending along the trench can be trapped by these thrust nappes, forming small trapped basins.

In the middle and upper part of the lower tectonic region, as the subduction continues, more linear folds and thrust faults are formed nearly parallel to the trench. At the same time the thrust nappes, which constitute the small trapped basins are continuously compressed by subsequent sediments and begin to deform. The small trapped basins formed earlier become long and narrow.

With the increasing compression forces perpendicular to the trench axis, the dip angle of the imbricate thrust slivers in the middle tectonic region increases gradually. The direction of the outcrop thrust fragments is pushed into a more regular NNE alignment. The area of the small trapped basins is greatly reduced by sustained compression, and some of the basins completely disappear into trench-like landforms between two adjacent fault sets.

In the upper tectonic region, the dip angle of the thrust slivers is nearly vertical, and the adjacent imbricate strata have merged. All the previously formed small trapped basins have disappeared, and even the trench-like landforms between adjacent slivers have been filled with fresh sediment and disappeared.

In the top area of the accretionary wedge ridge, with the deep subduction activities the roots of the accretionary wedge have merged under the tremendous transverse compressive stress. Longitudinal tension stress causes a sharp uplift of the strata at the top of the accretionary wedge, thus forming a large uplifted ridge area. The thrust slivers with large dip angles even reverse and collapse in this region, forming local landslides and sediment accumulation.

10.4.4 The Law of Geomorphotectonic Development

The small trapped basins (also be called "minisize trapped basins" or MTBs) reflecting the growing process of the accretionary wedge were first identified on the fine geomorphotectonic map in the middle region of the Manila Trench. Similar basins are scattered within the study area, especially near the deformation front on the offshore side of the accretionary wedge. The basins are generally small, long and narrow areas lying parallel to the extension direction of the trench. They are surrounded by thrust or reversed fragments in different directions in the accumulating deformation zone west of the subduction zone. There are no obvious tensional fractures around the basins and they are always found in a stable, loose sedimentary environments. Thus, the basins were formed by passive traps, unlike the down-faulted basins formed by tensional structures such as grabens or semi-grabens, or depression basins which are commonly seen in stable continental margins.

The seismic profiles show that the deposits above the subduction detachment surface have been scraped and peeled off while the subducting plate carried some of the sediments into great depths. Under the effect of friction, the scraped deposits continue to move slowly along the direction of plate subduction, gradually accumulating on the edge of the upper plate. The continuous input of material eventually leads to the formation of a large imbricate thrust-nappe structural accretional wedge. The horizontal displacement of the upper sediment nearest to the subduction detachment surface is much larger than that of the submarine surface, and the deep sediment is more directly impacted and blocked by the frontal collision of the overlying plates, thus leading to more severe extrusion and bending deformation. The displacement of the shallow and middle sediments is relatively small, and partial pressure can be released by collision activities perpendicular to the direction of the subduction, resulting in relatively weak deformation. However, in the uppermost seabed layer, because the top of the sediment layer is higher than the overlying plate, more growth space is available and the tension stress perpendicular to the seabed can be completely released. Under the lateral compression of the accumulated surface sediments, the typical imbricated nappe structure can be rapidly compressed and formed. In this way, because of the difference in the transverse compressive stresses and release spaces, the overlapping thrust-nappe structure with a shovel-shaped depression is controlled and formed in the accretion wedge.

In the horizontal direction perpendicular to the trench axis, the accretionary process is most active near the deformation front in the lower tectonic region under the effects of deep subduction motion. The folds at the bottom of the Manila Trench indicate that the deep compressive stress has begun to transfer upward. Outcrops of several thrust slices in the deformation front area indicate that the compressive forces in the upper deposits are gradually increasing, while the folds have developed into faults and emerged from the seafloor. Because the upper extrusion environment is relatively loose at this time, the thrust fragments in different directions are interlaced, hence the small trapped basins can be closed and form locally. In the middle tectonic region, the dip angle of the thrust faults that were previously formed increases and the gaps between the fault sets are compressed because of the extrusion of the thrusting slivers, which are accumulating and amalgamating in front. The area of the small trapped basins also begins to decrease and they are compressed into long and narrow troughs extending along the trench axis. In the upper tectonic region, the density of the compressed thrust faults increases sharply. The adjacent tectonic slivers are closely joined together, and the dip angle of the thrust slices is nearly vertical. Due to the shortage of horizontal expandable space in this region, the previous small trapped basins have completely disappeared, and even the residual groove-like linear landforms have been filled by new sediments at the top of the accretionary wedges and disappeared. At the top accretionary wedge, the pressured thrusting faults are no longer evident. Many arch-shaped folds appear at the top of the rapidly uplifted ridge. Some areas are starting to collapse under the influences of continuous sedimentation and bottom flows, and even tensional landslides and fractures in random directions may occur.

Generally, the growth process of the small trapped basins with the development of the accretionary wedge in the subduction zone can be divided into four main stages: accretion and fracturing stage, entrapment and basin-formation stage, extrusion and extinction stage and uplifting and nappe stage (Fig. 10.19). Each growing stage has different structural characteristics and is complementary to and inseparable from the whole development of the accretionary wedge. It is a complete process and represents a relatively new formation and development pattern for accretionary wedges. In the lower tectonic zone, a large number of small trapped basins formed. In the middle tectonic zone, the small trapped basins were compressed and the basins have become long and narrow. In the upper tectonic zone, the small trapped basins have transformed into terrains with alternating troughs and ridges and even disappeared completely. In the ridge zone of the accretionary wedge there are many bulges, folds and collapsed formations. The generation and collapse of small trapped basins as a special morphotectonic unit is a structural expression of the dynamic deep subduction process and the tectonic stress environment on the surface of the accretionary wedge, which reflects the increase of the transverse compressive stress along the subduction direction at the root of the accretionary wedge. Although there are no absolute boundaries between the four development stages of the small trapped basins, the growth processes of the accretionary wedge correspond to the structural arrangement of the tectonic regions. The structural features and geomorphological characteristics are distinct and orderly. This is essentially a manifestation of the compressive stress generated by the deep plate subducting activities on the seabed, reflecting the inner dynamic mechanism of the development of accretionary wedges to some extent.

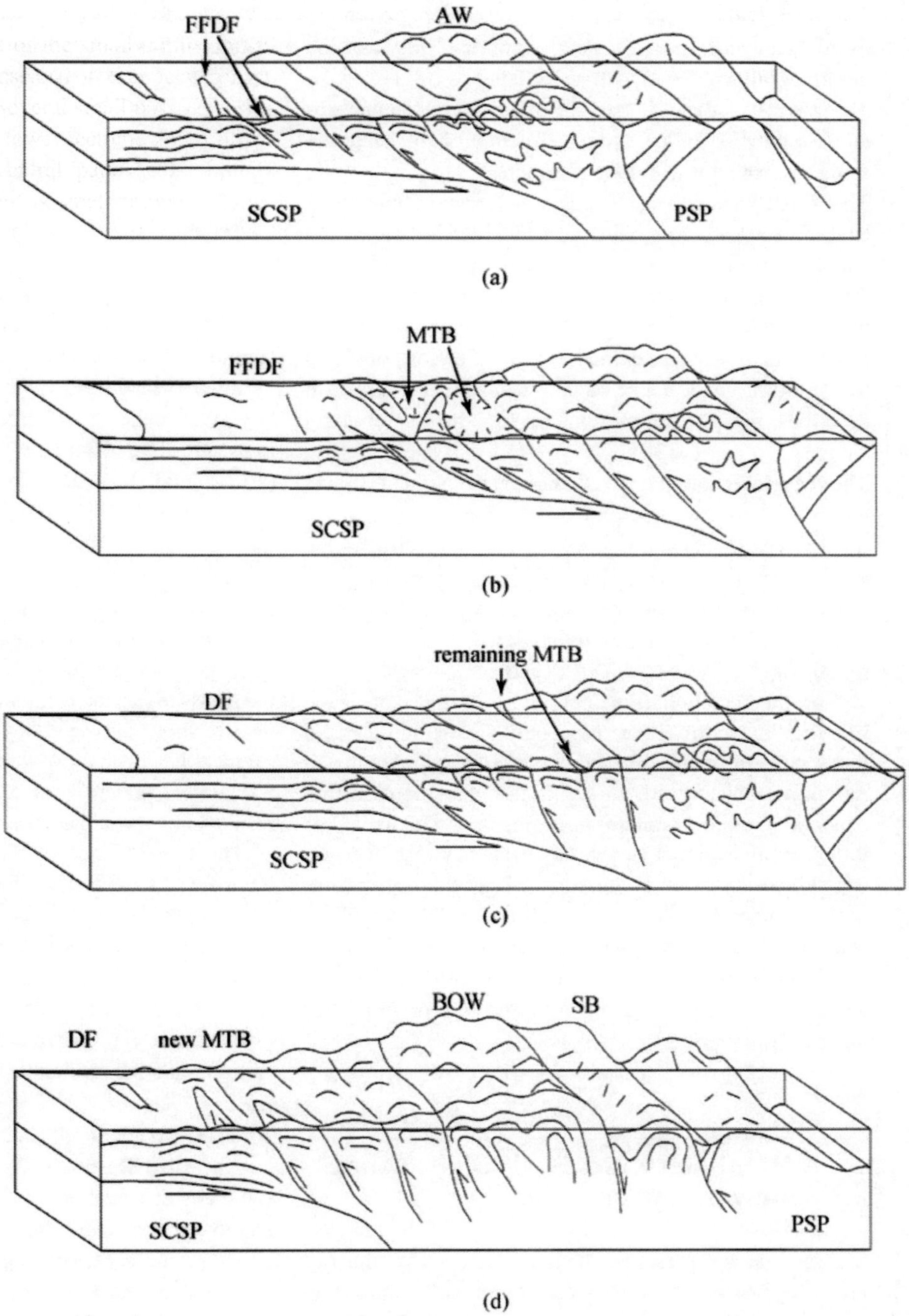

Fig. 10.19 The four main stages of the development of a small trapped basin. **a** Accretion and fracture stage; **b** entrapment and basin-formation stage; **c** extrusion and extinction stage and **d** uplifting and nappe stage. DF-deformation front; FFDF-folds and failures at deformation front; AW-accretionary wedge; MTB-small trapped basins; SCSP-south china sea plate; PSP-philippine sea plate; BOW-bulges on top of wedge; SB-slumping bodies

10.4.5 Conclusions

1. For the first time, the concept of the small trapped basin was introduced to the discussion of the accretionary wedge development pattern. Several small passive basins discovered in the middle and lower tectonic zones of the accretionary wedge in the central part of the Manila subduction zone were defined as small trapped basins, enclosed and trapped by sets of thrusting slivers. They are generally small and elongated parallel to the trench direction.
2. The surface morphotectonic features of the accretionary wedge in the Manila subduction zone vary greatly from the deformation front to the top of the Hengchun Ridge. In the deformation front area, the seabed morphology exhibits loose folds or heterogeneous faults. Toward the

peak of the ridge, the dip angle, density and height of the thrust faults increase. At the top of the ridge, nappe deformation and even partial collapse were observed. The process indicates that the effect of the deep subduction process on the surface structure of the accretion wedge is different in different regions.

3. The small trapped basins begin to form in the lower tectonic zone, decrease in the middle tectonic zone, disappear in the upper tectonic zone and uplift at the top of the ridge; this corresponds to the development process of the accretion and abruption stage, entrapment and basin-formation stage, extrusion and extinction stage and the uplifting and nappe stage, respectively. Thus, it describes a new growth pattern of the accretionary wedge.
4. The development stages of the small trapped basins correspond to the structural arrangement of the accretionary wedge tectonic region, and are controlled by the subduction of the deep plate. The deep basins are the seabed expression of the compressive stress release caused by deep subduction dynamics.
5. Small trapped basins with distinct features were identified on the accretionary wedge in the central part of the Manila subduction zone, especially in the middle and lower parts of the Manila accretionary wedge. Are they a morphotectonic feature unique to the Manila subduction zone or a more common phenomenon, which can be found in other global subduction regions? This question is yet to be answered because there is no fine bathymetry data for other subduction belts around the world. However, the development pattern has distinct characteristics and is a complete process. A thorough understanding of this process can provide a good reference for studies of other subduction belts in the world.

10.5 A Morphotectonics Study of the Central Southwest Indian Ridge

In this section, we present a detailed analysis of the full coverage high-resolution multi-beam topographic data from the Southwest Indian Ridge (SWIR) between 49°E and 51°E using morphotectonics analysis. The tectonics and magmatism information is extracted from the bathymetry and seafloor geomorphology data, and the morphological features of 49−51°E are analyzed in detail to obtain their geological significance. Then, combined with geophysical data such as OBS data from the area, we discuss the tectonics and magmatism of this section from the seabed to the deep crust, and from the ridge axis to the flanks.

10.5.1 Regional Geological Background

SWIR (Fig. 10.20) extends from the Rodriguez triple point in the northeast to the Bouvet triple point in the southwest, with a total length of about 8 000 km. It marks the border between the African and Antarctic plates. Generally, it spreads obliquely, about 60° from its running direction, at a steady rate of 14–16 mm/a (Macdonald 1998). The overall trend of the ridge is NE-SW, and the 16°–25°E segment is close to SEE. According to its spreading history and geometric features, Georgen et al. (2001) divided SWIR into seven segments from west to east. The study area, SWIR 49–51°E, is located between the Indomed and Gallieni fault zones. The Indomed-Gallieni segment is located in the middle SWIR; it has the highest terrain with an average axial depth of 3180 m. To the west, the average depth of the Discovery-Indomed segment increases to 3530 m, while the deepest part of SWIR lies to the east, from the Melville transform fault to the Rodrigues triple point, with an average axial depth of 4730 m. The Indomed-Gallieni segment lies in shallow water and has thick crust and low Na 8.0 content (Sauter et al. 2009), indicating that it may have resulted from an enhanced magmatic event.

The 49–51°E part of SWIR includes, from east to west, Segments 27, 28 and 29 (Fig. 10.21). The axial rift is absent at Segment 27, and a high-temperature hydrothermal activity zone at the ultra-slow spreading ridge was first discovered at Segment 28 (Tao et al. 2011). A hydrothermal vent was also found at the south flank of Segment 29 (Han et al. 2010). From January to March 2010, Chinese cruise *DY*115-19 carried out a 3D artificial source OBS experiment in the area (Li and Chen 2010). The OBS inversion results (Li et al. 2015; Niu et al. 2015; Zhao et al. 2013) show that the crust thickness of Segments 28 and 29 decreases from the volcanic center (thickness of 7–8 km) to both sides (3–4 km), and that this change originates mainly from the lower crust. The volcanic center of Segment 27 can reach a crust thickness of about 9.5 km, which can be tracked 4–5 Ma off-axis, and a low-speed anomaly exists in the lower crust. Sauter et al. (2004) also found a significant low magnetic anomaly zone near 50.5°E, which is thought to be similar to that of the slow spreading MAR. They believe that the mantle temperature and magmatism in this region are high, and the volcanic rocks are not well crystallized, resulting in the low

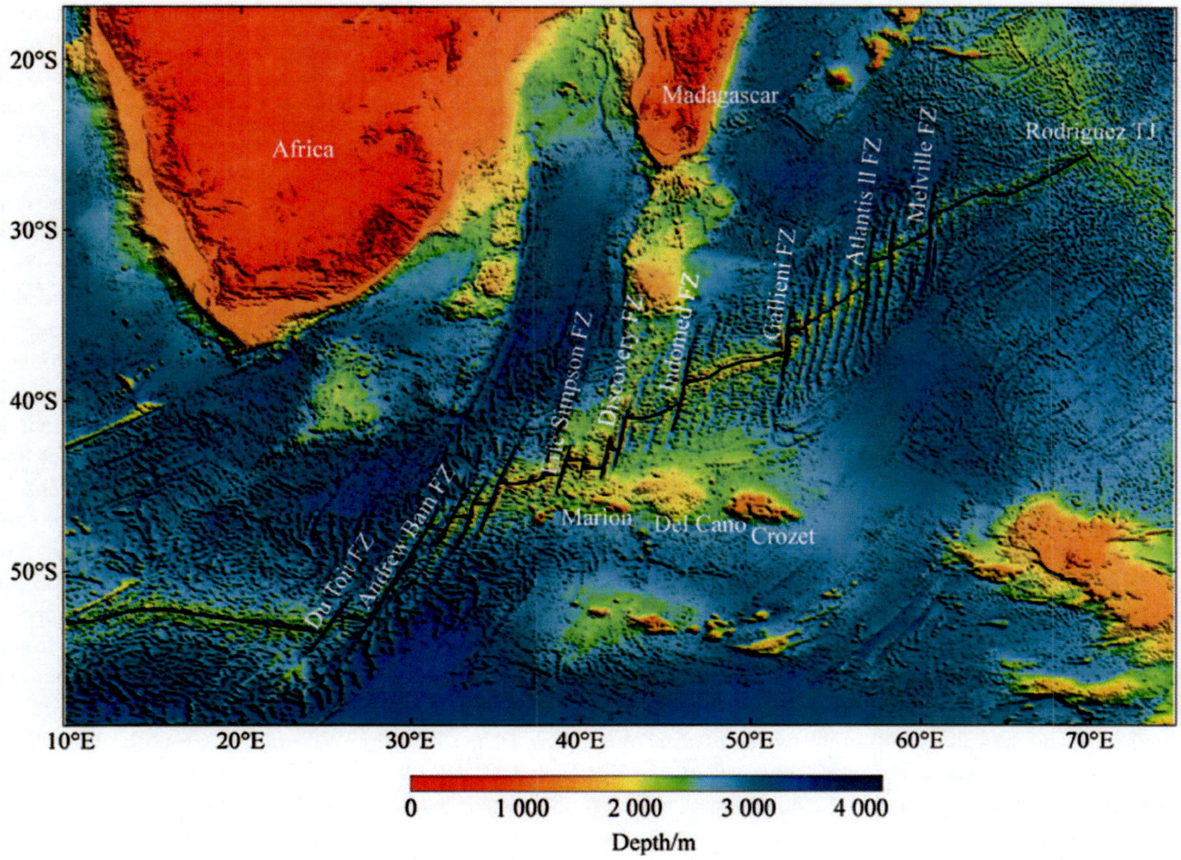

Fig. 10.20 Submarine topographic map of Southwest Indian Ridge

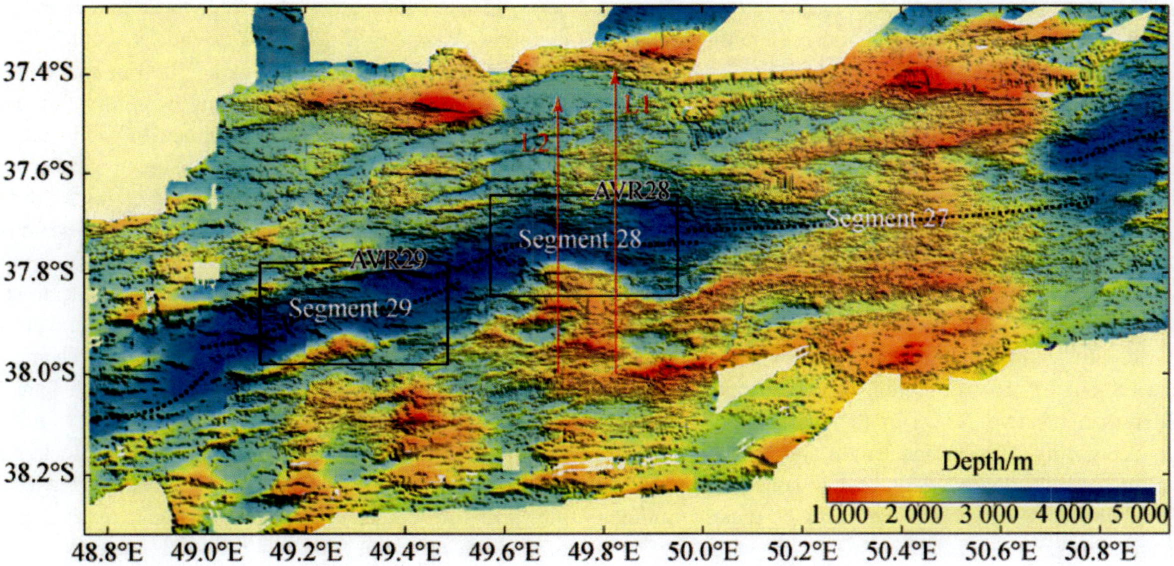

Fig. 10.21 Bathymetry of SWIR along 49°–51°E

magnetic anomaly zone. Various phenomena indicate that Segment 27 has extremely intense magmatic activity compared with the adjacent sections.

10.5.2 Bathymetry Analysis

10.5.2.1 Segments 28 and 29

There is a rift at the axis of Segments 28 and 29, with axial volcanic ridges (AVRs) at the bottom (denoted AVR28 and AVR29, respectively). The AVRs and non-transform discontinuities (NTDs) are arranged alternately in an en-echelon pattern.

As shown in Fig. 10.22a, AVR29 is nearly E-W trending, and its ends partially overlap with the NTDs, where the depth suddenly increases. AVR29 is about 20 km long and 4–5 km wide. It rises more than 500 m from the bottom of the rift valley, with the highest point in the middle, where the water depth is only about 2700 m. Linear structures extending almost E-W and conical volcanoes are scattered on AVR29. The north and south sides of AVR29 are asymmetrical within the rift valley (Fig. 10.22a–c). The bottom of the north side is significantly higher than that of the south side, and the valley walls of the two sides also have very different morphology. The south valley wall is a continuous surface with a slope of about 20°, similar to a fault plane, while the northern wall developed linear structures in an almost E-W direction and conical volcanoes, similar to that of the AVR. As shown in Fig. 10.22d–f, generally, AVR28 has a similar morphology to AVR29, except for a small branch trending ESE. AVR28 is about 20 km long and 2–3 km wide, and rises more than 500 m from the bottom of the rift valley.

There are three NTDs in the study area. The two western NTDs have an "S" plane shape and trough terrain, and extend ENE for nearly 30 km, with no trace of volcanic activity and conspicuous linear structures leading to the smooth valley bottom, where the water depth is more than 3700 m. The width of the rift, with a roughly V-shaped profile, is about 15 km, and a stepped normal fault can be observed on both walls. The north–south terrain profile is also roughly V-shaped; however, the NTD between Segments 28 and 27 shows different characteristics. First, it is smaller than the two NTDs in the west, with its deepest water depth only 3500 m and length about 15 km; furthermore, it extends nearly E-W, being consistent with the AVRs, with no S-shaped plane form or clearly outlined volcanic cones.

The geomorphology of the south and north flanks shows distinct features at Segments 28 and 29. Overall, the south flank has higher elevation, more developed fractures and no volcanic activity, while the north flank has more obvious traces of volcanic activity.

As shown in Fig. 10.23, the north flanks of Segments 28 and 29 have 3–4 groups of long faults about 20–40 km long within the study area, each formed by short E-W and ENE-trending faults lying end to end alternately. Such a multi-segment line shape is very similar to the present rift boundary, suggesting it represents an early rift boundary. A single short fault can be about 10 km long, with a vertical fault displacement of between 300 and 500 m and a slope between 30° and 40°.

The morphology of the northern flank can be divided into two terrain categories. The first has shallow water and well-developed clusters of near E-W linear structures (Fig. 10.23), which may be secondary fractures. These structures create a rugged seafloor and divide the terrain into several phases in the N-S direction, cut by the long faults mentioned above. In addition, volcanic cones can be observed (Fig. 10.23). This type of seabed morphology may be similar to the "volcanic seafloor" of the flank of the eastern SWIR (about 61°–66°E) described by Cannat et al. (2006). The second terrain is much flatter, as it has no volcanic cones and basically has no fractures or linear structures, apart from the above long fault that cuts through it. While the seafloor in this region is flat, it is different from the "smooth seafloor" of the flank in the eastern SWIR (about 61–66°E) described by Cannat et al. (2006). Generally, it is a very gentle slope at the off-axis side, while the axis-facing side is the fault plane of the long fractures above it. Such seafloor topography may have been formed by seafloor spreading with very weak magma, and differs from the "smooth seafloor" produced from non-magmatic spreading. These two types of terrains are arranged alternately along the north flank in a roughly NE direction. A large E-W uplift belt extends at about 37°26′S in the north flank of Segment 29, rising 1000 m above the surrounding area. The axis-facing side of the uplift belt is a large fracture surface, with a slope of about 20° and secondary normal faults; the off-axis side is a much gentler slope. A conjugated uplift belt may have formed on the south flank; however, this lies beyond the scope of our multi-beam data.

The southern flank is characterized by the development of a large number of large fault blocks, which indicate strong tectonic tension (Fig. 10.24a). Their fault planes are all facing the axis and have slopes between 10° and 25°, which are much gentler than those of the north flank. The south flank also developed some smaller breaks, usually secondary faults that developed on the large fault blocks. These fault blocks do not show a regular distribution pattern or a clear correlation with the current rift boundary. This may be due to the strong effect of the high structural tension at the south flank.

An N-S trending terrain with typical oceanic core complex (OCC) morphology was discovered on the southern side of Segment 28 (Fig. 10.25). The northern edge of its corrugated surface is 10 km from the axial ridge; the OCC extends southwards for 10 km, its E-W width is 3–7 km,

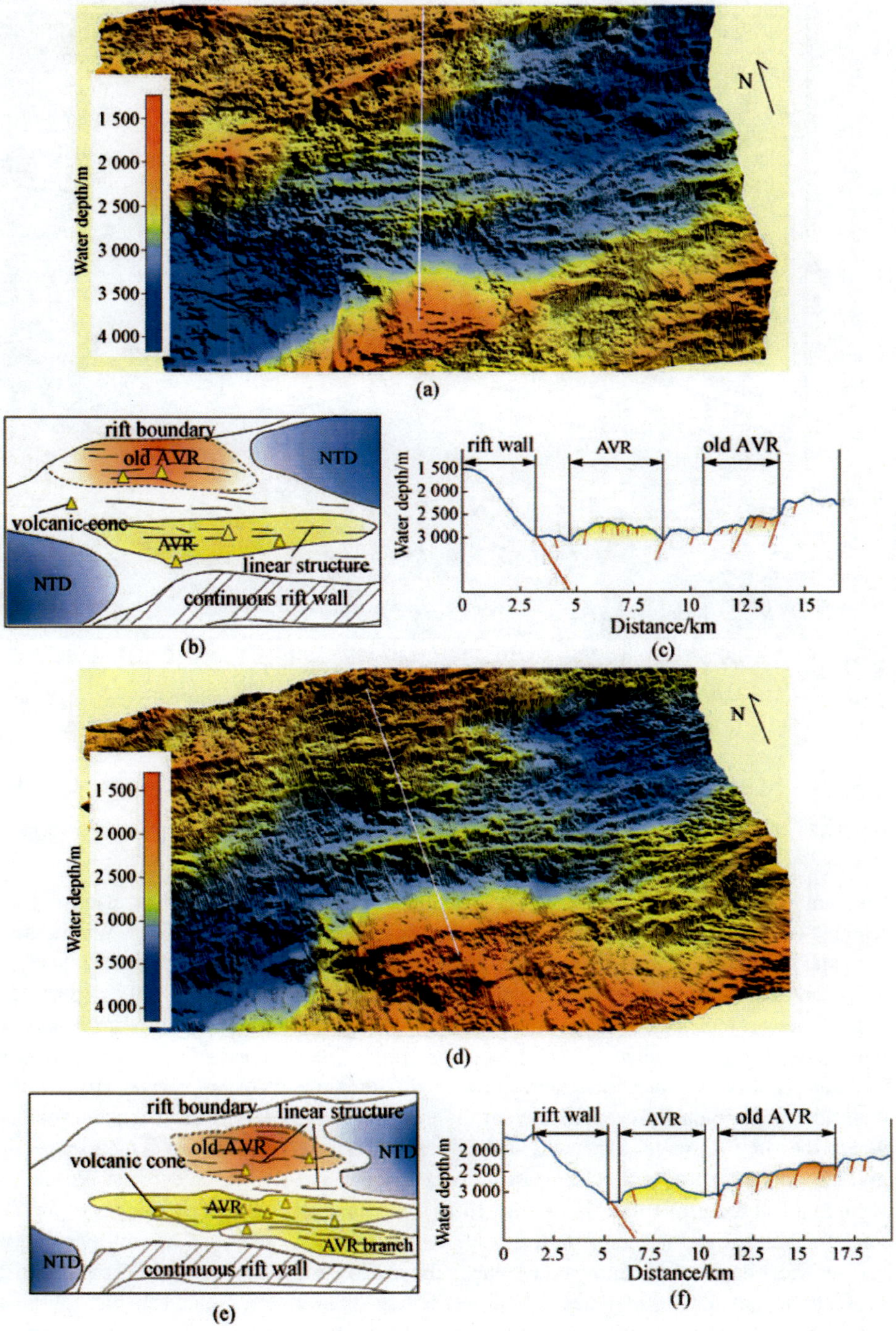

Fig. 10.22 Contrast of the basic topography, landform and their interpretations between two areas of AVR 28 and AVR29. **a** 3D bathymetry of AVR29, the position is shown in the black frame in Fig. 10.21; **b** AVR29 area geomorphology interpretation map; **c** AVR29 profile, along the white line in (**a**); **d** 3D bathymetry of AVR28, shown in the black frame in Fig. 10.21; **e** AVR28 area geomorphology interpretation map; and **f** AVR28 profile along the white line in (**d**)

decreasing towards the north, and it covers an area of about 40 km^2. The top of the formation is about 200 m higher than its base, and the surface is smooth compared with the surrounding terrain, dipping to the north at a low angle. Several steep fractures dipping to the north may have developed within the OCC. The linear ridge on its southern side is partially collapsed, making the initial ruptured surface less continuous, and the blocks on its western side may have been subjected to the same detachment fault (Fig. 10.24). The seismic velocity model shows that the crust here is thin, and the seismic velocity of the core is 4.5 km/s, which is significantly higher than that of the surrounding rock mass.

10.5.2.2 Segment 27

Segment 27 is very different from the other segments of the Indomed-Gallieni segment of SWIR; denoted the

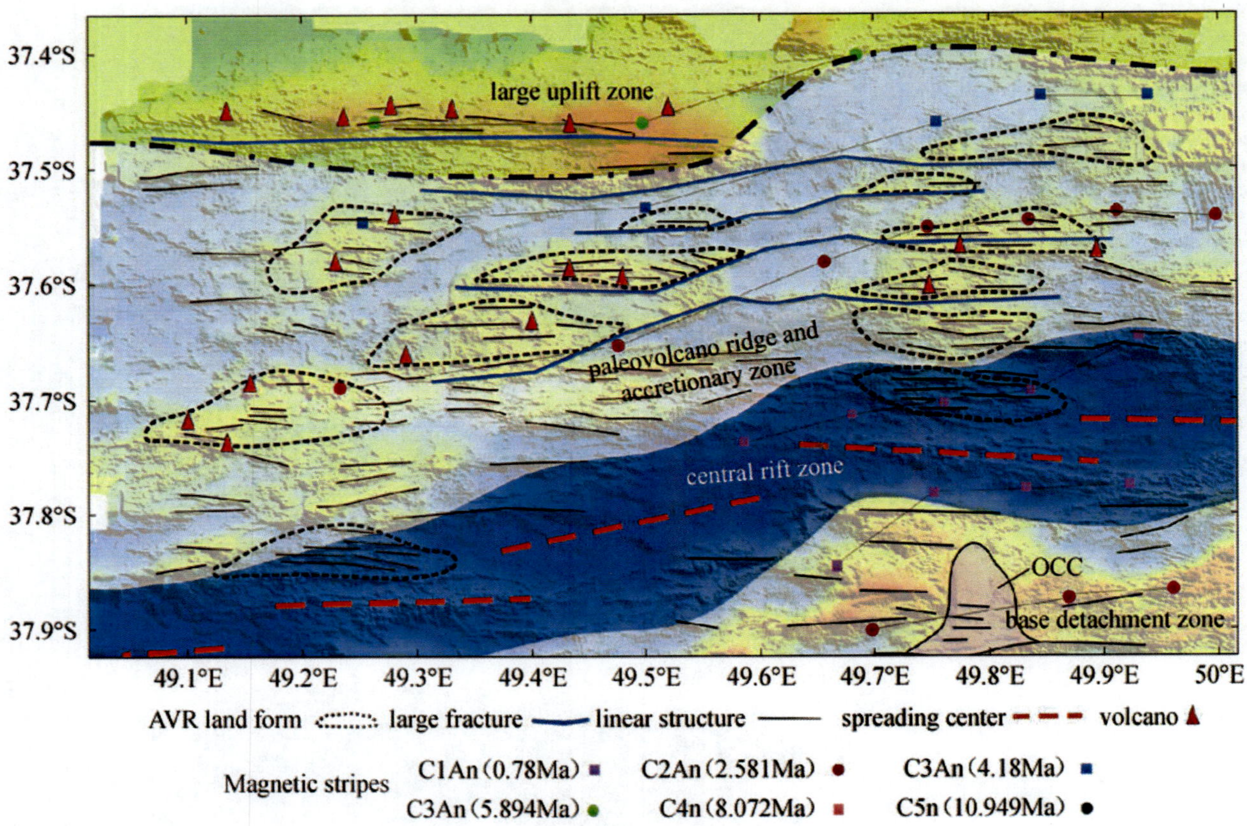

Fig. 10.23 Bathymetry of the north flanks of Segments 28 and 29 and its geomorphology interpretation

"anomalous zone" here, it shows the following distinct features (Fig. 10.26). Firstly, the overall terrain is high, and the axial rift valley is absent. Secondly, the two flanks have two axisymmetric conjugate lifts. Thirdly, traces of volcanic activity are very conspicuous and the spatial density of the volcanoes is high. Segment 27 is the only section of the Indomed-Gallieni segment that lacks an axis rift for more than 40 km. It gradually deepens toward the east and west, where the rift re-emerges. The area of Segment 27 can be divided into three zones according to the water depth and topographical features (Figs. 10.26 and 10.27): the anomalous, transitional and normal zones.

The water depth of the anomalous zone is the shallowest, generally less than 2000 m in depth. Two sets of large uplifts lie parallel to the axis on each side of it, about 13–15 km from the axis (Fig. 10.26). Each uplift extends from about 50°00′E to about 50°42′E for more than 60 km in an almost E-W direction with a N-S width of 3–7 km. The axis-facing slopes of both uplifts are very steep, suggesting they were a fracture surface that split at the axis. Secondary gravity-induced normal faults developed along the steep slopes. The off-axis side of the uplift slopes are much gentler. Another set of conjugated uplifts, much larger than the first set, lies 25–28 km from the rift axis (Fig. 10.26). Similar to the smaller uplifts, the larger lifts also have steep axis-facing slopes and gentle off-axis slopes. Within the anomalous zone, a cluster of volcanoes lies approximately along the 50°30′E line, where the shallowest water depth is less than 1400 m. The highest points of these uplifts are all located near 50°30′E (as shown by white line in Fig. 10.26), which suggests that this area has been a volcanic center since 4 Ma (the age of the outermost uplift based on magnetic anomaly analysis (Sauter et al. 2009)). According to their shape, these volcanoes can be divided into three categories: conical, crater and flat-topped volcanoes. They are mostly between 1 and 2 km in diameter and between 100 and 200 m in height, with the height of the conical volcanoes reaching more than 200 m. The density of the volcanic rock between the uplifts is slightly larger than that of the uplifts. This may be attributed to the magma activity, which changed from a scattered eruption to an axial overflow as the magma supply increased (Behn et al. 2004). The east and west edges of the anomaly zone are asymmetrical. Starting from 50°30′E, the eastern end of these lifts is generally at 50°45′E, while westward they extend farther, with the western end forming a V-shaped trough pointing east (Fig. 10.26).

The transitional zone is an area where the water depth gradually deepens away from the magma center (water depth between 2000 and 3000 m) where volcanic activity is scarcely

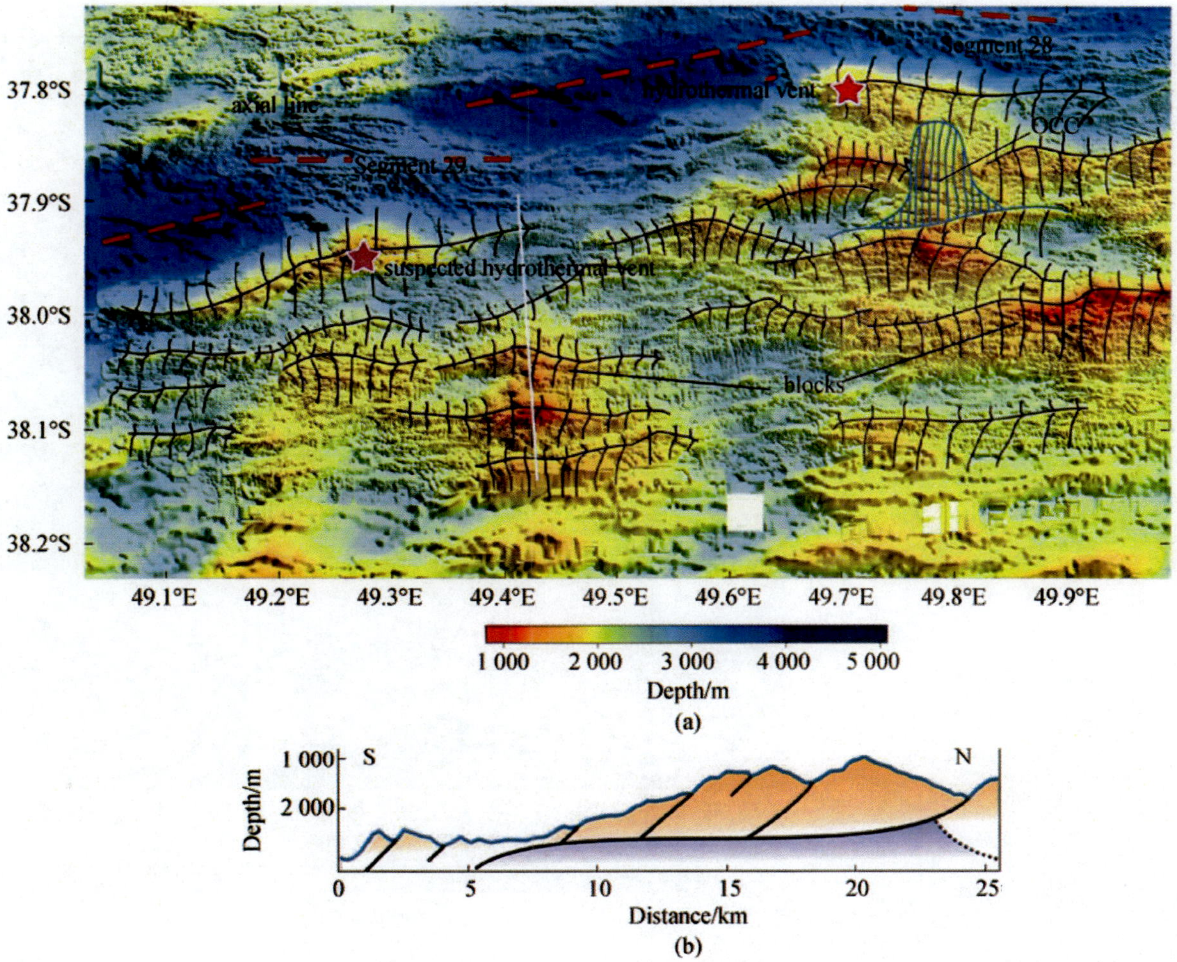

Fig. 10.24 Bathymetry of the south flank of Segments 28 and 29 (**a**) and its geomorphology interpretation (**b**)

seen. The west transitional zone intrudes into the anomalous zone, as the rift terrain gradually drops and fades to the north in a V-shaped form. The water depth becomes gradually shallower and the valley narrows, with its apex at 37.68°S, 50.30°E where the rift valley terrain fades completely. The V-shaped rift terrain intruding into the anomalous zone may indicate that the magmatism of Segment 27 is gradually weakening, and the axial rift terrain is gradually re-appearing. There is no such V-shaped formation at the east transitional zone, which is likely related to the oblique spreading of SWIR and the northward shift of the axis.

The water depth of the normal zone (Fig. 10.27) generally reaches more than 3000 m, with little evidence of volcanic activity. It is considered to be basically unaffected by the 50.5°E magma center, having returned to the general state of the ultra-slow spreading ridge.

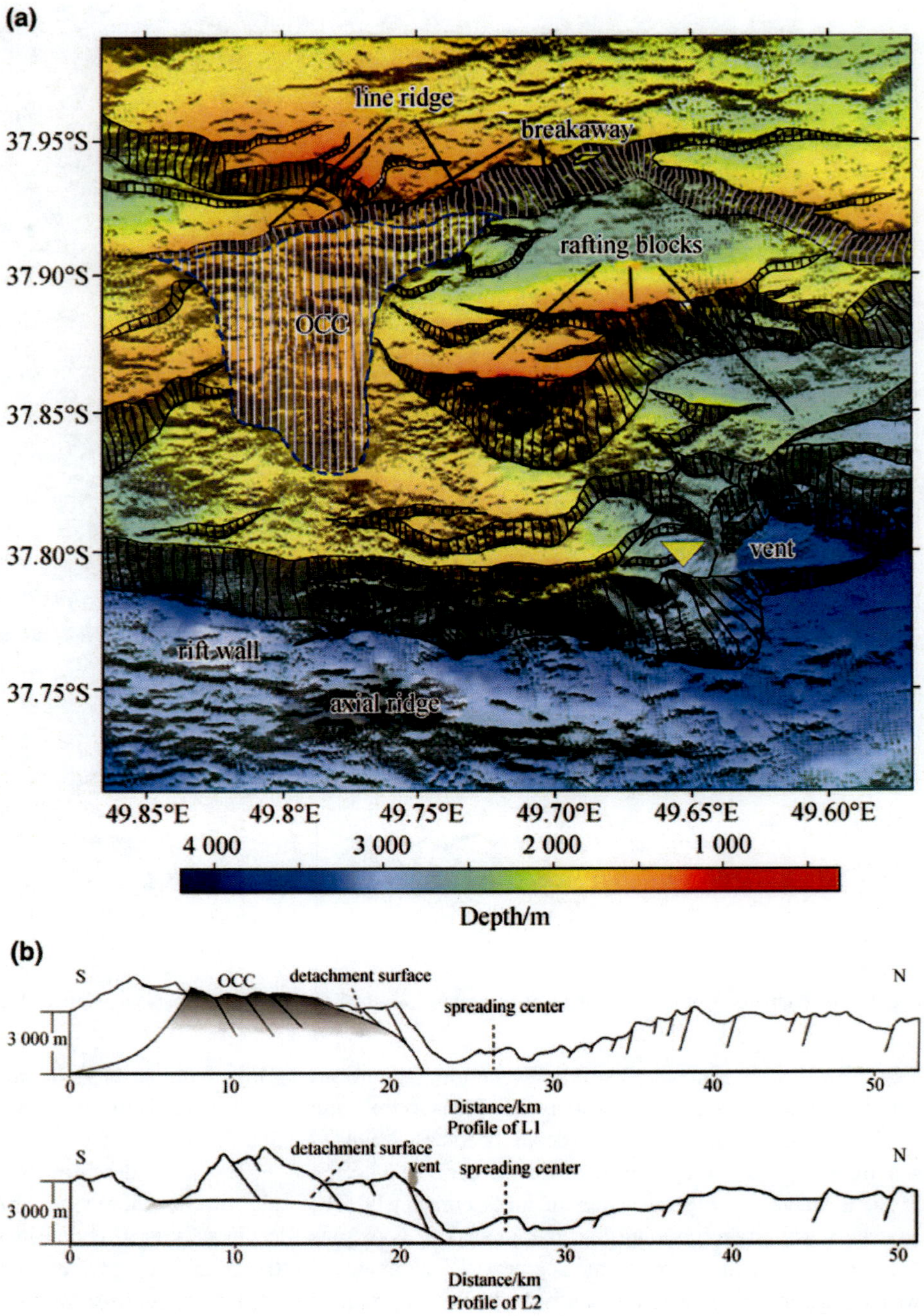

Fig. 10.25 Tectonic interpretation of the detachment zone, the position of L1 and L2 shown in Fig. 10.21

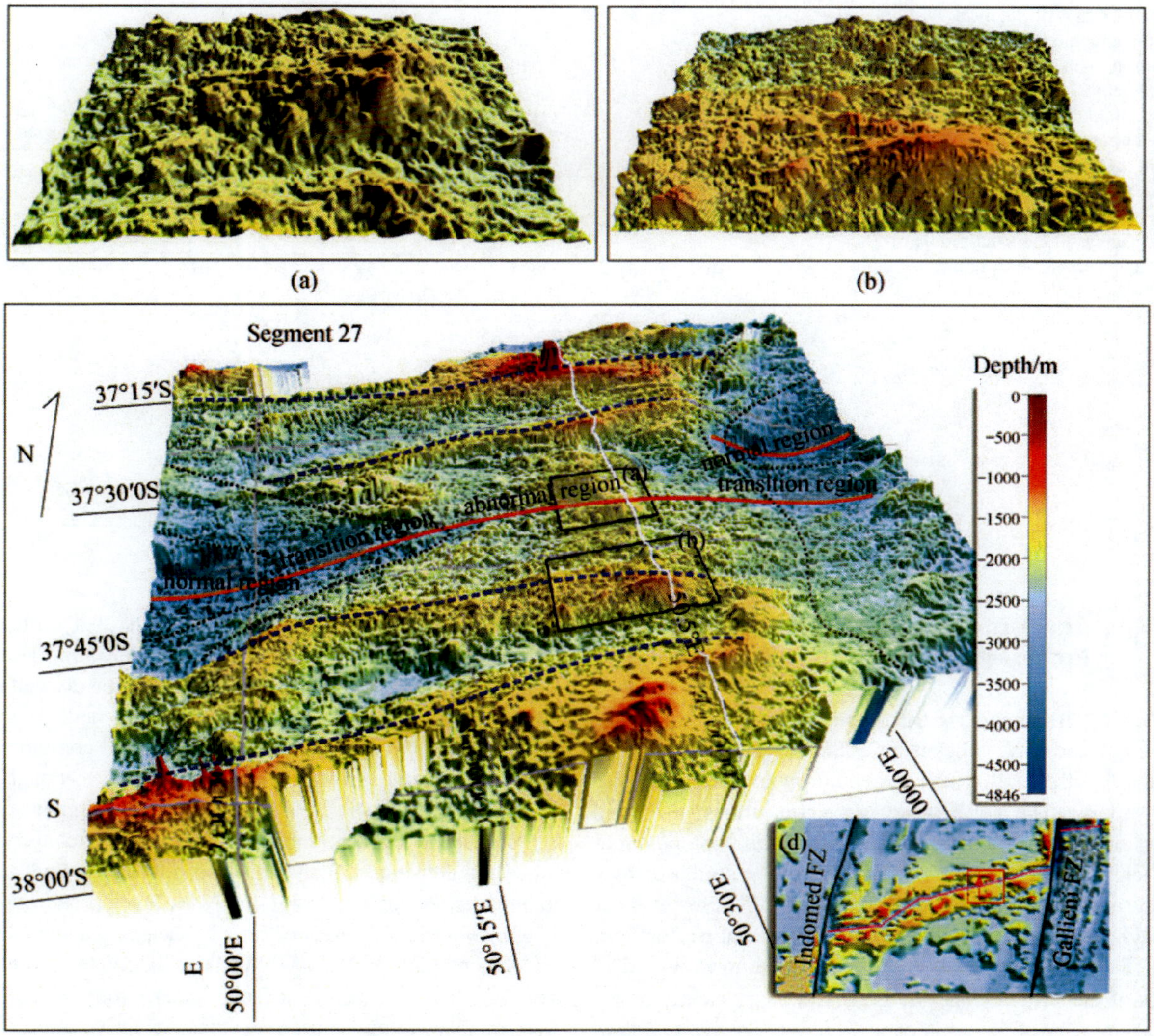

(c)

Fig. 10.26 3D bathymetry of Segment 27. The position of panels (**a**) and (**b**) is outlined in black in panel (**c**), and **d** shows the location of this study area

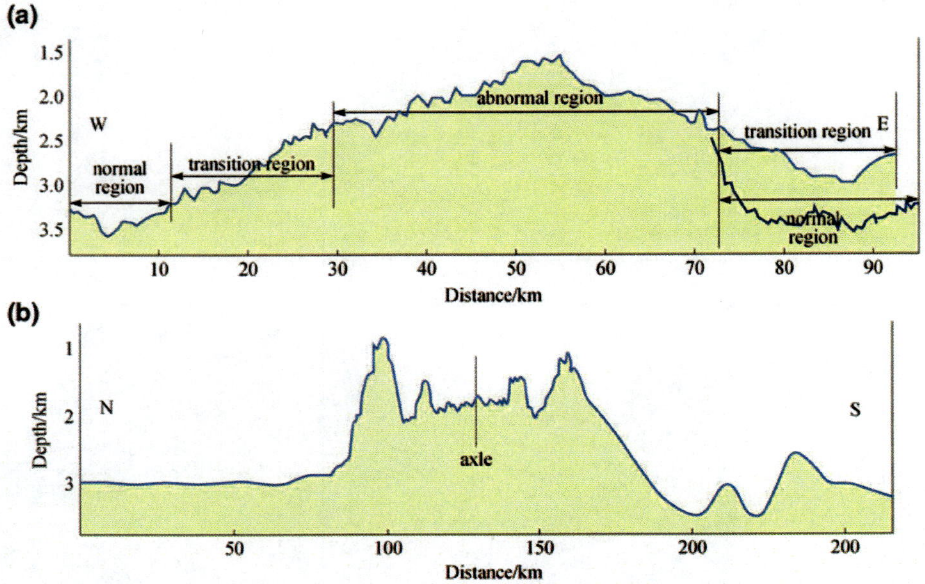

Fig. 10.27 SWIR Segment 27 axis topographic profile and cross section. **a** Axial elevation profile of Segment 27, along the red line in Fig. 10.26c; the dark blue curve corresponds to the short red line in the eastern end of the terrain block in Fig. 10.26c; **b** Elevation profile of Segment 27 perpendicular to the axial profile, along the straight black line in Fig. 10.26d

10.5.3 Analysis of Tectonic and Magma Processes of the Oceanic Ridge

10.5.3.1 Asymmetric Spreading

The south and north valley walls of Segments 28 and 29 show distinct structural and geomorphological features (Fig. 10.20). The north valley walls run almost E-W and traces of volcanic activity can be observed in the topographical features, similar to those of the AVRs. The north wall may be the northward migration of the old AVR. However, as mentioned above, the south valley wall is a flat fracture surface. The asymmetry between the south and north rift valley walls of the structure and the geomorphology clearly reflect the asymmetry of the seafloor spreading at Segments 28 and 29. At present, the fresh oceanic crust migrates mainly to the north flank, while the seafloor spreading of the south flank occurs mainly by structural extension.

The geomorphology features of the two flanks are very different at Segments 28 and 29, which also reflects the asymmetric spreading of this zone. The northern flank has rugged terrain and a near E-W linear structure, and volcanic cones can be observed (Fig. 10.23). As mentioned above, the geomorphological features are very similar to those of AVRs and their north valley walls; this suggests that the north flank was built by the northward migration of the AVRs as part of the seafloor spreading process. Therefore, we can refer to this oceanic crust as "volcanic seafloor". At the south flank, a large number of large fault blocks have developed, with no obvious traces of volcanic activity. Correspondingly, only a regular flat fracture surface developed at the south rift valley wall, illustrating the important position of the tectonic extension during the seafloor spreading in the south. The development of the OCC in the southern part of Segment 28 also shows that structural extension is strong here. According to the "Rolling Hinge" model of Buck (1988), detachment faults can be divided into two categories based on their lifespan: if the fault does not rotate to the angle of being "locked", it remains continuously active; then, if the amount of extension is large enough, an OCC can develop—a so-called long-lived detachment fault. If the hanging wall is "locked" when rotated to a certain angle, the detachment fault will stop developing and under continuous tensioning, new faults will develop and cut the hanging wall into a series of fault blocks, called short-lived detachment faults. The south flank is likely to develop a large number of detachment faults under strong tensioning, most of them of the short-lived type, or OCC cones, thus creating a large number of fault blocks. Based on this analysis, the OCC mentioned above was a result of a long-lived detachment fault (Figs. 10.24, 10.25).

Strong structural tension leads to the thinning of the oceanic crust. Notably, the OBS inversion velocity model of Segments 28 and 29 shows that the oceanic crust of the south flank is significantly thinner than that of the north flank. Cannat et al. (2003) also found that the crust thicknesses of the two flanks were different at the eastern end of the Melville fault zone in SWIR. A similar asymmetric spreading pattern was also found at about 64°E and 28°S in SWIR (Searle and Bralee 2007): the ridge axis and the north flank exhibit volcanic topographic features, while the south flank reflects traces of structural extension, where a large number of fault blocks developed, in addition to the development of the FUJI Dome OCC. Gravity inversion analysis found that the north flank crust is 2 km thicker than the south flank crust. Therefore, such asymmetric spreading may be prevalent in ultra-slow spreading ridges that have a limited supply of magma.

Cannat et al. (2003) proposed a seafloor spreading model of SWIR under a limited magma supply. This model presents an ultra-slow spreading SWIR controlled by deep large asymmetric normal faults, with a rotating and uplifting hanging wall, leading to such asymmetric spreading. Based on the large fracture surface of the south valley wall of Segments 28 and 29 and the location of hydrothermal vents (utilizing the heat and flow channels provided by the deep faults) found by Tao et al. (2011), a large, deep north-facing axial fault very likely developed at Segments 28 and 29 (Fig. 10.22c, f). This asymmetrical deep fault, directly controlling the asymmetric spreading, led to the difference between the south and north flanks.

10.5.3.2 Magmatic Activity

For Segments 28 and 29, based on the topographical features of the ridges (AVRs) and troughs (NTDs) arranged alternately in an en-echelon pattern, the magmatic model is a typical model of an ultra-slow spreading ridge, which has an overall limited magma supply and erupts in concentrated bursts. This mode of magmatism and corresponding axial topography are common in ultra-slow spreading ridges (Dick et al. 2003; Searle and Bralee 2007).

Mendel et al. (2003) discussed the magma-tectonic cycle of the SWIR, considering the evolution of the ridge morphology as a process of alternating action between the magma active stage and the tectonic active stage. During the magma active stage the strong magmatic activity builds the axial volcanic ridge (the AVR), while during the tectonic active stage, the strong tectonic activity splits the AVR and transports the two segments away from the axis on both sides as seafloor spreading. Such a process can be seen as a series of periods where the magma supply changes rather than an intrinsic magma-tectonic cycle. During each period of strong magma supply an AVR develops, and during the weak magma supply period, the AVR is split and transported by tectonic movement. Geomorphological analysis indicates that the morphology of Segments 28 and 29 can be attributed to such a magma supply period. Firstly, the volcanic seafloor of the north flanks of the two segments had developed in a number of stages. According to the long faults representing the early rift boundary, the data suggest 4–5 stages, indicating that the axial magmatic activity also occurred in stages. Furthermore, it is difficult to explain the development of the deep fault shown in the profile of Fig. 10.22c and f during a strong magma supply period because such a deep fault cannot develop in a state of hot and abundant magma supply. Thus, a reasonable explanation is that the segment is currently in the tectonic active stage, which the magma supply is limited. Based on the 5 AVRs within the north flank of Segments 28 and 29, combined with the paleomagnetic age of the rocks, the magma supply period in this area lasted about 1.5 Ma.

Based on the above analysis, the magma-tectonic dynamics model of Segments 28 and 29 is shown in Fig. 10.28. Segments 28 and 29 are under a magma supply period, during which the magma supply gradually decreases. Within this period, when the magma supply is high, AVRs are built from the magmatic activity. When the magma supply is limited, the tectonic activity increases and the asymmetric deep fault develops on one side of the axis, resulting in asymmetric spreading of the oceanic ridge, which leads to the topographical difference between the two flanks.

Segment 27 has high elevation, a dense distribution of volcanoes, and no rift valley, indicating an abundant magma supply. In contrast, Segments 28 and 29, who have a low magma supply, have a very different geomorphology. However, the magma supply of Segment 27 is limited in time, and the constructed topography is also under the control of the corresponding magma-tectonic dynamics mode, but in a different form.

The dense volcanic activity on or around the lifts of Segment 27 indicates these uplifts' volcanic origin. Their gentle off-axis slope and steep axis-facing fracture surface and the symmetry relative to the spreading axis (Fig. 10.29) all indicate that these lifts are conjugated, formed at the axis and then split and shifted with seafloor spreading. During the

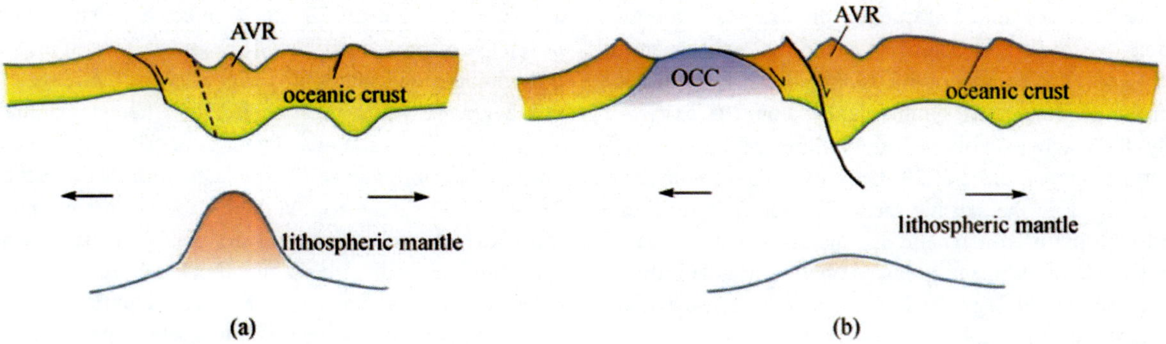

Fig. 10.28 Model of magma supply cycle and spreading mode of Segments 28 and 29. **a** Magma active stage; and **b** tectonic active stage

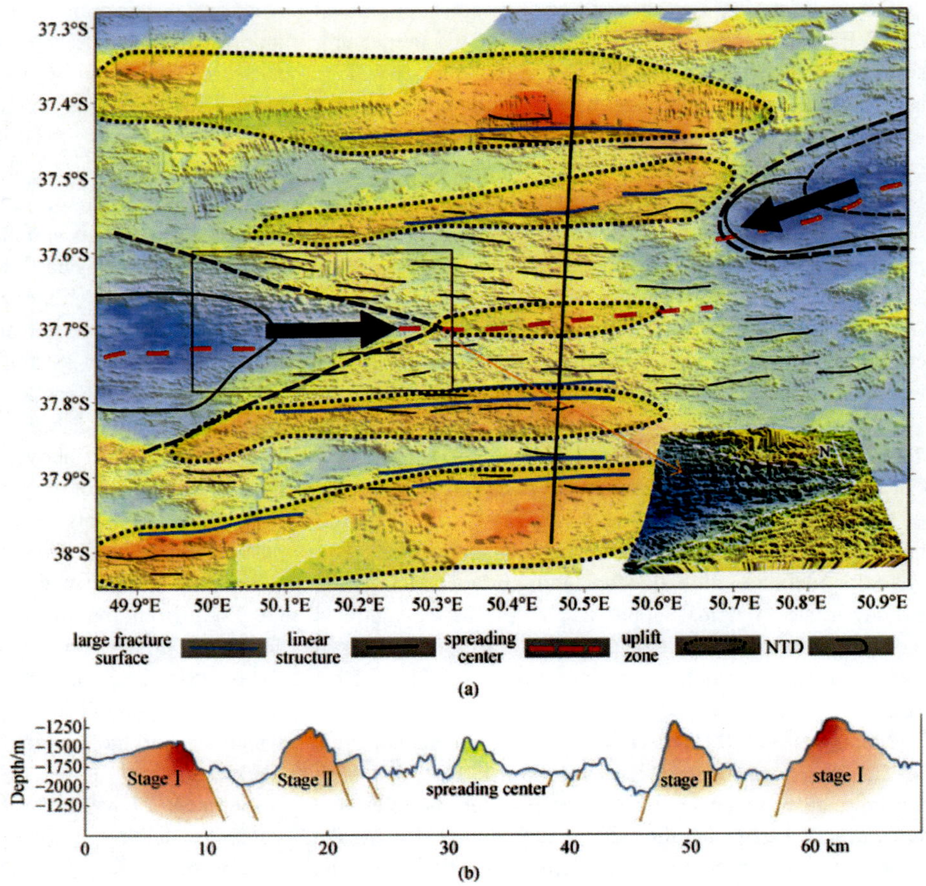

Fig. 10.29 Geomorphology interpretation of Segment 27. **a** Bathymetric map of Segment 27 of SWIR and its morphotectonic interpretation; and **b** topographic profile of Segment 27 of SWIR and its interpretation

high magma supply period, these ridge-shaped lifts were built along the axis, and during the weak magma supply period, the lifts split and migrated to the flanks, with the seafloor continuing to spread between the lifts (Fig. 10.29). At present, a ridge-like uplift exists at the axis of Segment 27 with a volcanic center at about 50.5°E; this indicates that Segment 27 is currently experiencing a period of ample magma supply during, which lifts are constructed. However, within the general magma supply cycle and corresponding magma-dynamics mode, Segment 27 shows its own distinct characteristics. Firstly, during the weak magma supply period, the axial volcanic construction process along the three sections was different: Segments 28 and 29 split asymmetrically under the control of the asymmetric deep fault while Segment 27 developed symmetrically along its axis. Secondly, the magma supply period was longer for Segment 27 than for Segments 28 and 29. Only two sets of conjugated uplifts appear in the seismic section of Sect. 27; this, combined with the paleomagnetic age indicates that the magma supply cycle of Segment 27 lasted about 2 Ma. Thirdly, the tectonic activity of Segment 27 is much less intense than that of Segments 28 and 29.

Despite the similar magma supply cycle and corresponding magma-dynamics mode, Segment 27 shows its own characteristics, distinct from those of Segments 28 and 29. This indicates that as well as being controlled by this model, the topography and construction of Segment 27 are also affected by another factor. The ridges (AVRs) and troughs (NTDs) arranged alternately in an en-echelon pattern, appear in Segments 28 and 29 and beyond, but are interrupted in Segment 27. They then reappear at east of 37.55°S, 50.72°E. This also indicates an additional factor influencing Segment 27. The large faults developed at the axial part of Segments 28 and 29 indicate the current period of limited magma supply, while Segment 27 is in the period of building a volcanic uplift, which means that Segment 27 and Segments 28 and 29 are fed by different magma systems. Thus, the additional factor is most likely the higher magma flux of Segment 27 compared with that of Segments 28 and 29. Different magma fluxes lead to different physical forms of axial volcanism and different patterns of seafloor spreading. Secondly, the magma flux is related to the volume of melt-extracted magma. The large volume of melt can supply magma to Segment 27 for a longer period. Thirdly, the strong magmatic activity of Segment 27 weakens the role of tectonic activities in its seafloor spreading.

10.6 Study of Supporting Techniques for Naming of International Undersea Features

In recent years, the Sub-Committee on Undersea Feature Names (SCUFN) has been responsible for setting standards and approving the names of undersea features that are entirely or mainly (more than 50%) outside the territorial limits of sovereign states, adhering to the principles of not involving sovereignty disputes between countries. The SCUFN proposed that the names of undersea features, once selected, will be added directly to "GEBCO-Gazetteer of Undersea Feature Names", and could be applied to bathymetric charts as well as other oceanic scientific research. With rapid progress in undersea feature naming, more countries are paying close attention to this work. Leading oceanic national authorities (such as the USA, Russia, Germany, Japan, Korea and New Zealand) set up special committees for undersea geographic naming to carry out research on undersea feature naming in sea areas of interest. Consequently, 53 and 81 names for undersea features were proposed and submitted at the 23rd (2010) and 24th (2011) SCUFN conferences, respectively. More than 100 of these proposed names were from Japan and Korea. In addition, the proposed names were studied by some developing countries, including Brazil, Ecuador, Peru, Indonesia and Vietnam.

To participate in this international effort, China initially submitted seven naming proposals for undersea features in 2011: Niaochao Hill, Tonggong Seamounts, Baiju Guyot, Xufu Guyot, Yingzhou Seamount, Penglai Seamount and Fangzhang Guyot. All the names were accepted by the SCUFN committee at the 24th SCUFN conference in September in 2011. Located in the Eastern Pacific Rise, Niaochao Hill, as the first Chinese undersea feature name recognized by the world, fills up the gap in this field for China.

10.6.1 Current Description of International Undersea Feature Naming

SCUFN, established in 1993, is a professional organization guided by the joint IOC-IHO GEBCO guiding committee, where IOC is the Intergovernmental Oceanographic Commission, IHO is the International Hydrographic Organization and GEBCO stands for General Bathymetric Chart of the Ocean. SCUFN aims to carry out research on guidance and principles of global undersea feature naming and also set related standard criteria. As a highly authoritative and influential international organization in this field, SCUFN also provides support for publication of global bathymetric charts, scientific research and other open applications.

SCUFN comprises the Chair, Vice-Chair, Secretary and commissioners. At present, the Sub-Committee consists of 12 regular members from Germany, the USA, Russia, Japan, Korea, India, Pakistan, Brazil, New Zealand, Monaco France and China. In 2011, China was appointed as a regular member. The various roles of SCUFN include (i) defining the nomenclature used for undersea features, e.g., canyon, plateau, fracture zone, as well as naming guidelines; (ii) considering and approving names submitted to the Sub-Committee.

Before SCUFN was established, several special institutes involving undersea feature naming had existed around the world, such as the Advisory Committee on Undersea Features (ACUF) in the US, the Hydrographic and Oceanographic Department of Japan (JHOD) in Japan and the Korean Committee on Marine Geographic Names (KCMGM) in Korea. One important function of SCUFN is to consult with governmental organizations of undersea feature naming in the involved national authorities on name selections. Generally, SCUFN follows the principle of selecting or agreeing with existing names. For instance, SCUFN referred to ACUF recommendation when naming undersea features close to the US coastline, and referred to JHOD for a series of guyots and seamounts in the western Pacific named by Japan. Consequently, these countries have great advantages in this field.

As of August 2019, a total of 4 371 names of undersea features around the world had been added to the GEBCO Gazetteer. With the growing interest and awareness of the role of the oceans in our environment and rapid progress in science and technology, humans are striving to further investigate the oceans and utilize their resources. As exploration horizons expand, more terrain features like plateaus, ridges, rises, trenches and seamounts are being discovered. These are essential geographic elements in oceanic mapping and important features in submarine scientific research. Proposing names for the newly discovered undersea features and participating in setting standards for undersea feature names helps to promote the international influence of a nation. In addition, undersea feature naming has a potential influence on outer continental shelf boundary definition and plays an important role in protecting marine rights and interests.

10.6.2 Standardization and Supporting Techniques for Undersea Feature Names

10.6.2.1 Standardization of Undersea Feature Naming

SCUFN has produced a series of documents involving the principles, rules and standardizations for naming of undersea features (IHO 2019a, b). The major function of SCUFN is to discuss and review the names of undersea features located outside the 12-n mile limit of national authorities. Through continuous revision and improvement, the standardization for undersea feature names and undersea feature terms and definitions currently contains 88 items. The standard version of this series is B-6 (IHO 2019b).

In SCUFN standardization of undersea feature names is a combination of qualitative description and quantitative parameters used to define submarine topography. For instance, a seamount is defined as a discrete (or group of) large isolated elevation(s), greater than 1000 m in relief above the seafloor. A hill is an elevation smaller than a seamount and of a rounded profile, less than 500 m above the seafloor. A knoll is an elevated terrain, characteristically isolated or as a cluster greater than 500–1000 m in relief above the seafloor (IHO 2019b). The standardization requires that the undersea geographic features to be named should have measurable relief and are delineated by relief. Furthermore, the submarine topographic features should be distinguishable and recognizable and have accurate data to support these features.

10.6.2.2 Procedures for Submission and Deliberation of Undersea Feature Naming

SCUFN encourages the naming of undersea features. Individuals or agencies are encouraged to apply names to newly discovered undersea features, to subdivide and rename originally named topographic features within a geographic area, and to apply names to a group of undersea geographic features. However, proposals of undersea naming will not be accepted in the following cases: (1) the name of the submarine geographic feature already exists in the GEBCO Gazetteer; (2) the submarine geographic feature is not independent and has no clear boundaries; (3) the proposed name of the undersea feature is politically sensitive.

Proposals to be considered at SCUFN meetings must be submitted 30 days before the meeting in electronic form, or 60 days prior if in analogue form. In the procedure, the naming proposals should be submitted first by individuals and agencies to their national names authority and be reviewed, and then reported to SCUFN. If no national authority exists, then the naming proposal can be submitted directly to SCUFN. The secretary and commissioners of SCUFN collect and inspect the proposals. Discussions, reviews and voting on the proposed names is conducted collectively at the SCUFN annual meeting and the final decisions are reported. Finally, the selected names of the undersea features are added to GEBCO Gazetteer.

10.6.2.3 Methods and Supporting Techniques for Undersea Feature Naming

The name of an undersea feature consists of a specific term followed by a generic term. The generic term should reflect the physiographic description of the feature. When selecting the specific term two categories are common: commemoration of persons or ships (names of living persons will normally not be accepted) and names of lands or islands nearby. Groups of similar features may be named collectively for specific categories of historical persons, mythical features, constellations, animals, vegetables and so on (Bouma 1990).

SCUFN requires that each undersea feature naming proposal is accompanied by measured and reliable information on the topography, including depth data by single or multi-beam sounding, navigation positioning data, geographic position and bathymetric chart and profile. The naming proposal has a strict format for the content and references. The undersea feature naming proposal form should contain the following information: geographic coordinates, maximum/minimum depth, total relief, steepness, dimension/size, discovery data, discoverer, survey ship, sounding equipment, type of navigation system, positioning accuracy and proposer(s) details.

The restrictions stated above were set to ensure standardization and compliance with the rules of SCUFN undersea naming proposals; therefore, the following guidelines should be followed to support the proposal. (1) Data acquisition of submarine terrain. (2) Processing of submarine terrain data and establishment of digital terrain model. (3) Identification, accurate positioning and boundaries of the undersea geographic features. (4) Parameter measuring, computing and visualization of the undersea topographic feature. (5) Classification of topographic feature according to SCUFN criteria, selection of generic terms and specific terms. (6) Study of the SCUFN operation principles, rules and standardization and preparation and submission of proposals according to SCUFN's rules. (7) Database establishment under SCUFN undersea feature names, names search and comprehensive management of names.

10.6.3 Research on International Undersea Feature Naming in China

To participate in undersea feature naming and embody its international responsibility, obligation and influence of the state, the Chinese State Oceanic Administration submitted seven undersea naming proposals to SCUFN in 2011 for the first time. The Niao Chao proposal utilized data from multi-beam sounding acquired by the Chinese research vessel *Dayangyihao* at the East Pacific Rise. The material preparation, figure plotting and submission of the proposal were conducted according to the standardization and procedure of SCUFN.

On the 19th cruise of the Chinese ocean investigation (in August 2008), a survey of polymetallic sulfides was conducted at the East Pacific Rise in latitudes 1°–3°S. During this expedition a geographic feature was discovered at latitude 1°22′S when using full coverage multi-beam sounding to survey the submarine terrain. This feature, which has a rounded profile with a collapsed crater structure at its top, is 250 m above the sea floor with a water depth contour of 2 875 m. The feature is shown in Fig. 10.30. Figure 10.31 illustrates the terrain profile, which is perpendicular to the mid-ocean ridge and across the collapsed carter. According to SCUFN rules of naming and definitions of a generic term, "hill" is the generic term of the feature. At the time of the discovery, the 28th Olympic Games were held in Beijing, China. As the undersea collapsed carter resembled an the olympic Stadium, the Bird's Nest, or Niaochao in Chinese, the name Niaochao was used as the specific term in the undersea naming proposal (Tao et al. 2011).

At the 24th SCUFN annual meeting in 2011, seven names for undersea features (including Niaochao Hill) proposed by China were accepted. This was the first time China had submitted names to SCUFN. It reflected China's active participation in and contribution to the collaborative international work on undersea feature naming, and also reflected China's comprehensive national power and national attention to marine affairs.

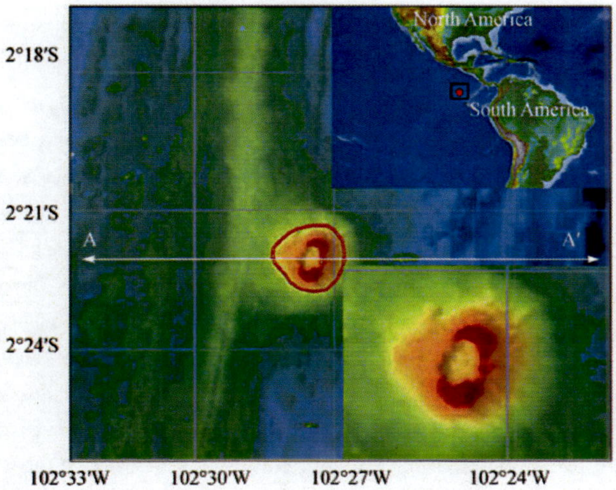

Fig. 10.30 Bathymetry chart of Niaochao Hill on the East Pacific Rise (collapsed caldera). The upper right panel shows the location of Niaochao Hill, lower right panel shows the collapsed caldera of Niaochao. The black arrow AA′marks the location of the terrain profile (Fig. 10.31) and the red line denotes the range of the collapsed caldera of Niaochao

In 2015, the Sub-Committee on Undersea Feature Names of the China Committee on Geographical Names (CCUFN) was formally established to organize and implement undersea feature naming in China. Each unit of the sub-commission prepares naming proposals and related material by using the submarine topography information from the Special National

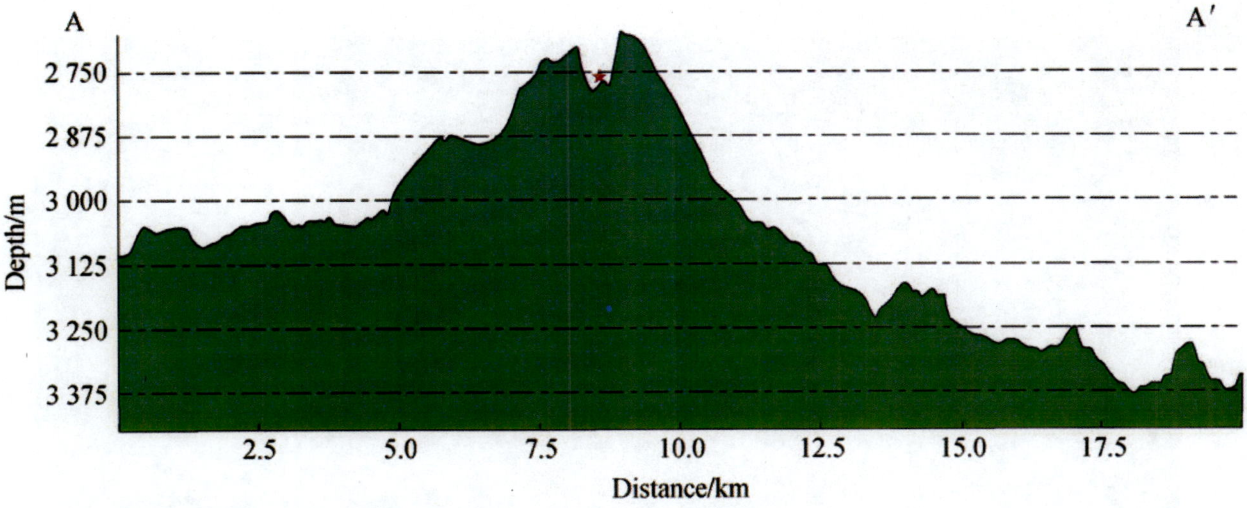

Fig. 10.31 Terrain profile of Niaochao Hill (collapsed caldera) on the East Pacific Rise

Table 10.3 Names and details of some undersea features named by China

Number	Name	Center/Vertex coordinates	Sea area
1	Niaochao Hill	01°22.0′S, 102°27.5′W	East Pacific
2	Baiju Guyot	17°53.9′N, 178°58.7′E	Northwest Pacific
3	Xufu Guyot	19°32.3′N, 157°56.0′E	Northwest Pacific
4	Tonggong Seamounts	14°13.8′N, 165°51.6′E	Northwest Pacific
5	Yingzhou Seamount	09°57.8′N, 157°27.3′E	Northwest Pacific
6	Penglai Seamount	19°12.3′N, 158°14.0′E	Northwest Pacific
7	Fangzhang Guyot	19°46.3′N, 157°22.8′E	Northwest Pacific
8	Risheng Guyot	20°42.6′N, 127°44.1′E	Northwest Pacific
9	Ritan Knoll	21°09.4′N, 127°45.2′E	Northwest Pacific
10	Yuetan Ridge	21°11.7′N, 127°55.7′E	Northwest Pacific
11	Weihan Seamount	00°05.2′S, 101°24.2′W	Southeast Pacific
12	Weiyuan Seamount	09°48.2′N, 154°31.8′W	Southeast Pacific
13	Qianyu Guyot	22°58.4′N, 175°38.5′E	Northwest Pacific
14	Zhinǚ Guyot	19°39.2′N, 160°09.4′E	Northwest Pacific
15	Niulang Guyot	20°44.3′N, 161°11.6′E	Northwest Pacific
16	Qiaoyue Seamount	37°20.0′S, 052°07.0′E	South Indian Ocean
17	Xiaozheng Seamount	16°12.9′S, 013°06.5′W	South Atlantic
18	Kaifeng Seamount	22°56.6′S, 013°25.9′W	South Atlantic
19	Caifan Seamount	14°03.1′S, 014°21.1′W	South Atlantic

Oceanographic Survey. Until 2016, China had submitted 125 naming proposals to SCUFN, covering the Pacific Ocean, Indian Ocean, Atlantic Ocean, Philippine Sea and the South China Sea. The names and details of some undersea features in the Pacific, Indian and Atlantic oceans are shown in Table 10.3. Materials used in the proposal for the naming of the undersea Pingfeng Ridge in the Philippine Sea are shown in Figs. 10.32, 10.33 and 10.34.

Another important work implemented by SCUFN is reviewing undersea feature names and reporting to the State Council for approval so that the undersea feature name and information can be announced. On October 9, 2015, through a press conference convened by the State Oceanic Administration, the names of 124 submarine geographical entities named by China were announced to the public. The generic names of the undersea features include seamounts, ridges, hills and fault zones. Of these, 101 are in the Pacific Ocean, 15 in the Indian Ocean and 8 in the Atlantic. In 2016, CCUFN improved the undersea feature naming and review system in the South China Sea.

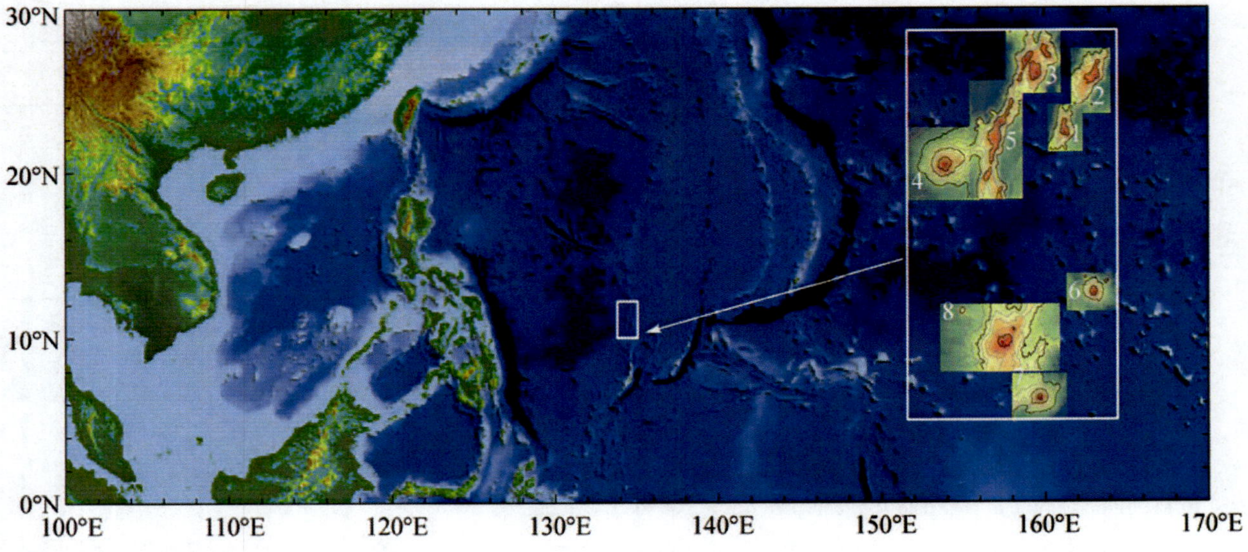

Fig. 10.32 Philippine Sea Pingfeng Ridge location map (2016). 1-Jiali Seamount, 2-Jiayang Seamount, 3-Yize Seamount, 4-Xiangyang Seamount, 5-Pingfeng Ridge, 6-Qilai Seamount, 7-Nanhua Seamount, 8-Taguan Seamount

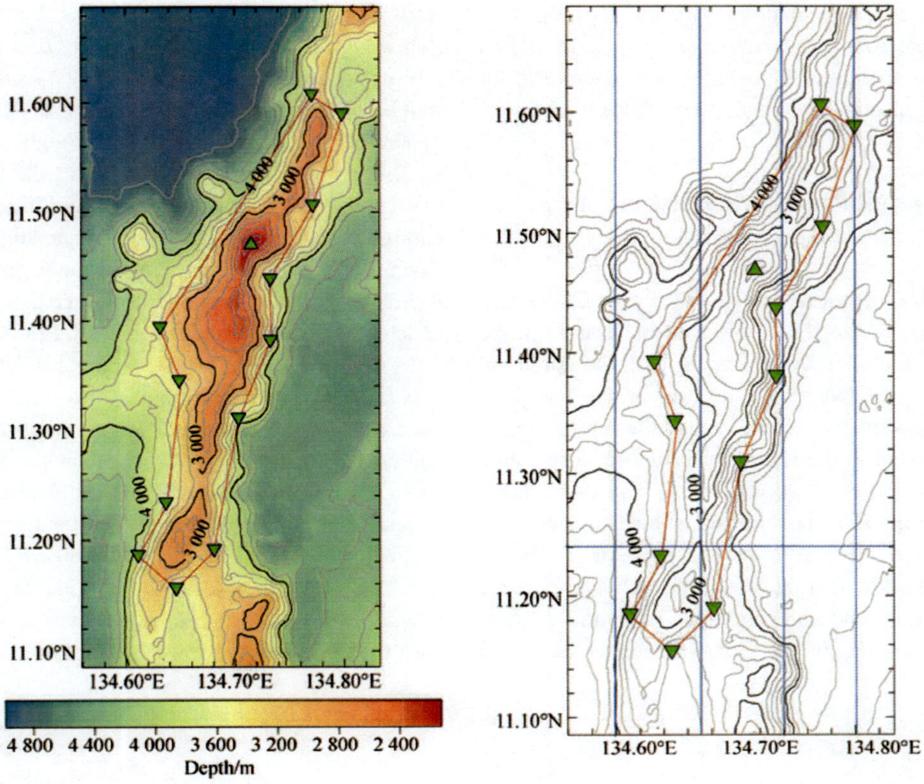

Fig. 10.33 Pingfeng Ridge bathymetry chart and survey line chart

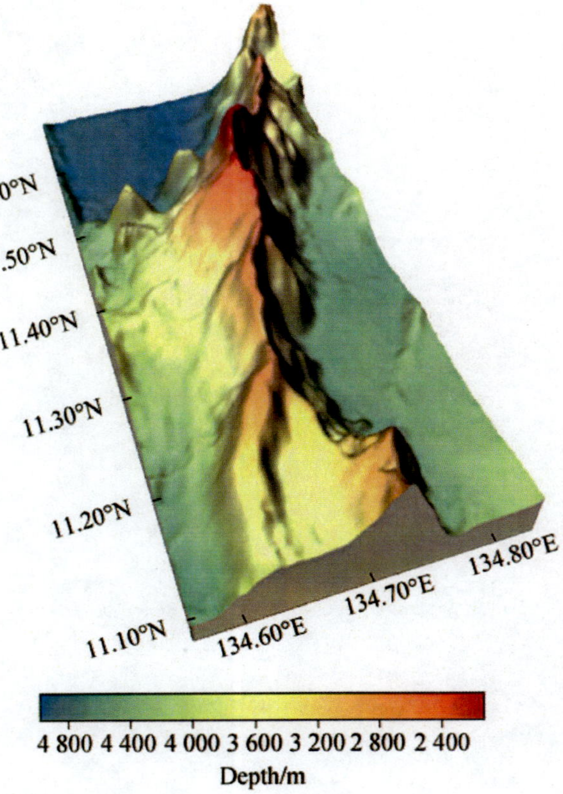

Fig. 10.34 Pingfeng Ridge 3D topographic map

10.7 Study on the Identification of Submarine Topographic Boundaries of the Okinawa Trough in the East China Sea

The region of the East China Sea (ECS) and Okinawa Trough (OT) has become a focal area for worldwide marine science in recent years. However, there are still many unsolved scientific questions in this region. The Okinawa Trough, a back-arc basin that is still in tension, is an important region for studying interactions between the Pacific and Eurasian plates, and also provides a window for understanding the features of the Pacific-type continental margins (Lee et al. 1980; Kimura 1985; Sibuet 1987; Qin et al. 1987; Xu and Le 1988; Jin 1992; Li 2008). Even in the global marginal sea basins, it also has its unique representativeness. It is possible to "analyze a typical case" to reveal the evolution law of the marginal sea and even reconstruct the initial evolutionary history of the marginal seas around the Pacific Rim.

The break point of continental shelf (BOS) line, the foot point of the continental slope (FOS) line and the central axis lines are critical for maritime delimitation and submarine science research, because of their close relationship with the topographical and geomorphological features of the OT. Based on multiple sources of bathymetric data, we aim to finely analyze the submarine topographical characteristics of the OT and then identify and characterize these boundaries.

10.7.1 Topographical Characteristics of the Okinawa Trough

The OT, located at the eastern edge of the East China Sea, extends from the Yilan Plain of Taiwan Island in the southwest to the island of Kyushu in Japan in the northeast. The southern part of the middle section of the trough is slightly convex towards the southeast, and the central axis of the trough is parallel to the Ryukyu Island Arc (also called the Ryukyu Arc) and the Ryukyu Trench. The strike direction of the southern OT is nearly E-W, whereas it is NE-SW in the middle and NNE-SSW to the north (Fig. 10.35). Bathymetric data show that the trough has relatively deeper water in the middle and southern parts, with the 2000 m isobath extending south of the Gonggu-Diaobei Fault and nearly reaching Jiumi Island and the Yushan Fault to its northwest (Jin 1992). The deepest position is located in the E-W extended central depression in the southern part of the trough with a measured maximum depth over 2300 m (generally known as a "trough in trough" or "central graben") that gradually decreases towards the north (Li 2008). The central graben shows relief and spatial distribution features that are similar to slow spreading ocean ridges (Luan and Yue 2007). Inside the trough, isolated seamounts of different heights, linear seamount chains, and en echelon navicular depressions have developed. Profiles perpendicular to the OT have shown that the trough has a typical U-shape, with increasing depth from west to east. This is closely related to the abundant sedimentary supply from the ECS continental shelf, which suggests that the sediment source has a significant influence on sedimentary patterns and submarine terrain in the trough. The continental slope of the ECS is rugged. In the middle and southern part of the slope, several well-developed incised submarine canyons extend upward to the outer margin of continental shelf and downward to the trough bottom. During glacial lowstands in sea

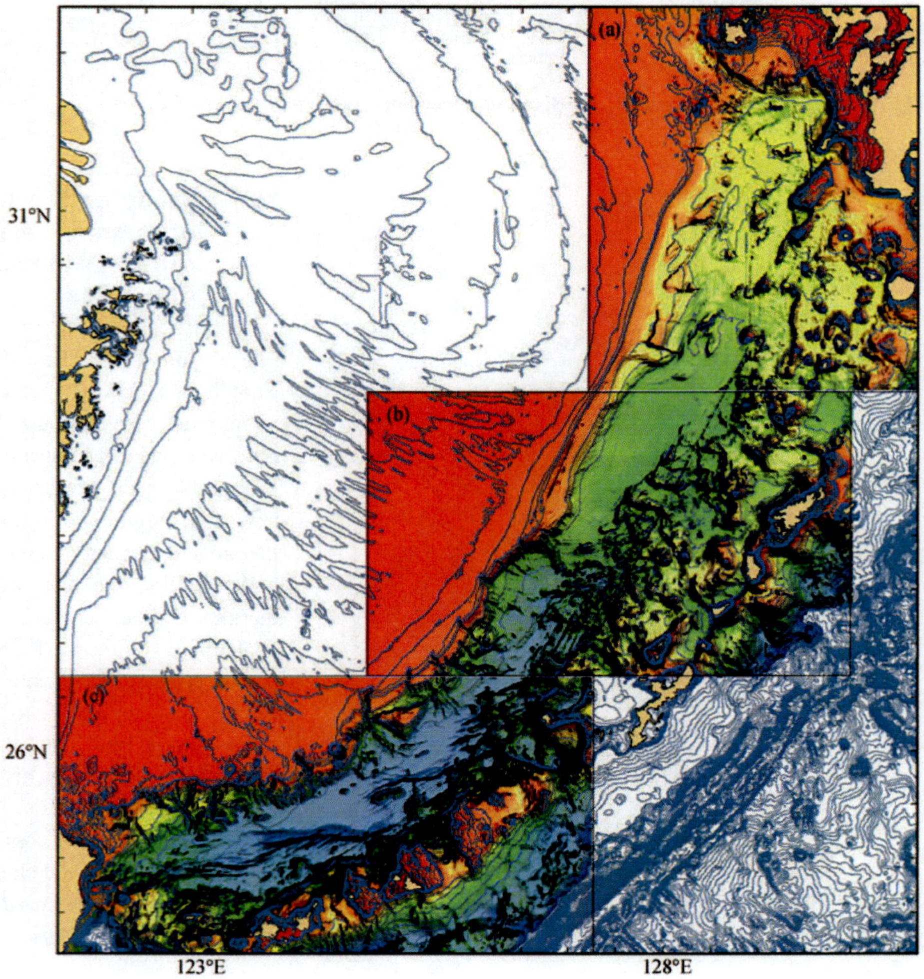

Fig. 10.35 Submarine topography in the East China Sea. **a–c** respectively refers to the approximate location of northern, middle and southern part of the Okinawa Trough

level, these canyons were important channels connecting the shelf and the trough (Zhao et al. 2011) Such topography not only molded the appearance of the ECS continental slope, but also affected the submarine topography in the western part of the OT. A series of fault blocks and ridges formed in the fore-slope area of the northern part of the ECS shelf, probably as a result of Miocene tectonic movements (Fan et al. 2000).

The ECS continental shelf and slope, and the OT, together form a typical continental margin (Fig. 10.35). Water depth on the shelf differs greatly from the trough, and has totally different topographical features. The continental slope connecting the shelf and the trough is steep. Geological and geophysical evidence indicates that the crust in the southern part of the OT thins sharply to < 3 km (Hao et al. 2004) and new oceanic crust might have been generated locally (Huang et al. 2006). Linear magnetic stripes have been identified in the middle and southern parts of the trough (Liang et al. 2001). Topographical, lithological and geophysical evidence demonstrates that the OT is a back-arc basin in an early stage of expansion (Han et al. 2007). As the ECS continental shelf extends naturally to the east and terminates at the OT, the OT becomes the natural separation between the ECS continental shelf and Ryukyu Island Arc to its east.

10.7.2 Determination of Okinawa Trough Terrain Boundaries

10.7.2.1 FOS and BOS Lines

A total of 48 topographical profiles crosscutting the ECS continental slope were constructed (Figs. 10.36 and 10.37) and 48 BOS were determined using the method mentioned in Wu et al. (2016). The average water depth for the BOS points is about 200 m (Fig. 10.38), with 19 of them having a water depth <200 m (mostly distributed at the northern and middle part of ECS continental slope), 23 of them with water depths between 200 and 250 m, and six of them > 250 m. Collectively, most of these points are concentrated at a water depth of 200 m, with a deepening tendency from north to south and more "jumpy" transitions in the south. The BOS line (the green line in Fig. 10.36) interlaces with the 200 m isobath (the blue line in Fig. 10.36); they nearly merge together at the northern part of continental shelf but the BOS locally extends outboard of the 200 m isobath in the middle and southern parts of the margin.

This distribution pattern of the BOS is closely related to the submarine terrain between the continental shelf and slope. In the northern part of the trough, even though some fault blocks and ridges have developed (Fan et al. 2000), the terrain of the continental slope is relatively smooth and the isobaths are generally parallel to the strike of the slope. Thus the BOS line generally is parallel to the isobaths. In the middle and southern parts of the trough, many canyons are found on the slope (Wu et al. 2004; Zhao et al. 2011) that have strongly incised the continental slope and altered the original relief of the continental shelf. This has led to a cracked marginal terrain on the outer shelf and frequent fluctuations in the BOS, but basically in water depths of around 200 m (Fig. 10.38). Therefore, we take the 200 m depth contour to be the BOS line for the ECS, instead of 160 m as previously used (Liu et al. 2005).

From the 48 topographical profiles used above, 57 FOS were obtained using the method of Wu et al. (2016) (Fig. 10.36). For nine profiles, foot points on both the upper and lower continental slope were obtained (shown as blue and red solid circles in Fig. 10.36, respectively). An analysis of these FOS points shows that the water depth of the FOS increases gradually from north to south (Fig. 10.38), with a value at about 800 m in the northern trough (Fig. 10.37a) increasing to 1000 m or more in the middle section of the trough (Fig. 10.37b). In the southern part of the middle trough, the water depth of the FOS is near 2000 m (Fig. 10.37c) and is generally >1500 m in the southern trough (Fig. 10.37d). In the middle part of southern trough, the water depth of the FOS tends to fluctuate with localized values decreasing to 1400 m (Fig. 10.38) and with even more reduction near the Taiwan Island. In the southern part of the trough, submarine canyons are extensively developed and interrupt the terrain of the continental slope. Turbidity currents flow along the canyons and form turbidite fans inside of the trough (Zhao et al. 2011). The canyons and turbidite fans change not only the terrain of the sea floor but also the position of FOS. Both the upper and lower continental slope can be identified in most profiles in the southern trough (Figs. 10.36 and 10.37d), together with their FOS.

If we define the continental slope as the region lying between the BOS and FOS, then the continental slope of the ECS can be characterized by wide features in the north and south, and narrow features in the middle (Fig. 10.36). The northern continental slope is about 40–70 km wide (Fig. 10.37a) with a water depth drop from 600 to 800 m (Fig. 10.38) from BOS to FOS. The middle continental slope has a minimum width about 10 km at the narrowest position (Fig. 10.37b) and a depth drop of 600–1800 m from BOS to FOS (Fig. 10.38). However, the width of the southern continental slope gradually increases to a range of 30–50 km (Fig. 10.37c, d) accompanied by a depth drop of 1000–1800 m (Fig. 10.38). Connecting the points of slope, the position of the continental slope follows a broad S-shape that is relatively flat in the middle part of the trough. In the northern part of the trough, a similar slope shape is also seen, but overall it is slightly broken with some small terrain fluctuations (Fig. 10.37a). In the southern part of the trough, the continental slope is severely incised by many canyons and has the appearance of being totally smashed (Fig. 10.37d).

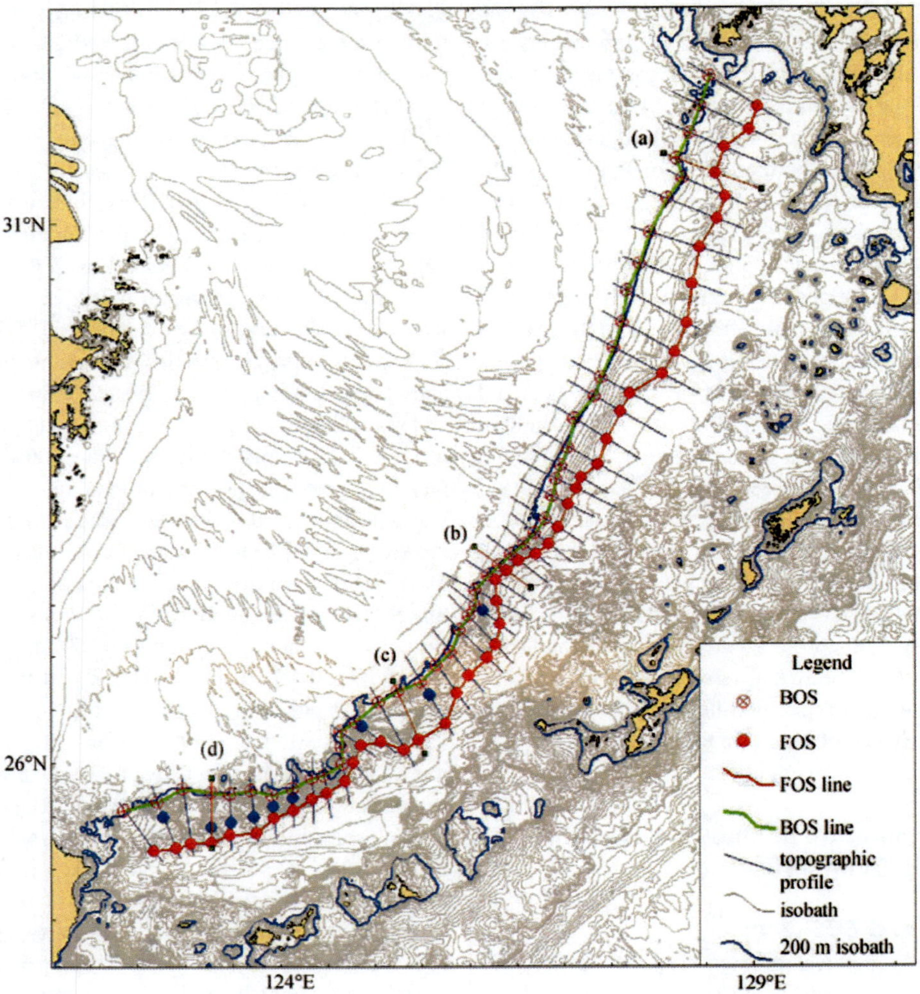

Fig. 10.36 Okinawa Trough and the profiles that determine the BOS and FOS lines. **a–d** are typical topographic profiles. BOS: break point of continental shelf; FOS: foot point of the slope

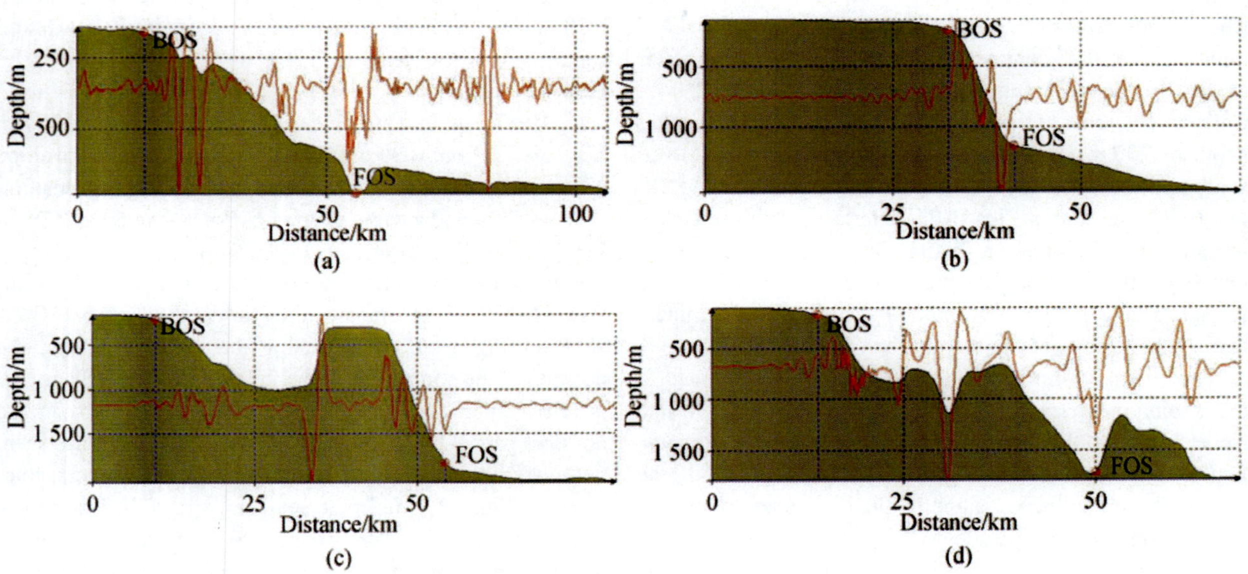

Fig. 10.37 Typical profiles with recognized BOS and FOS. **a** profile from the northern trough; **b** profile from the middle part of the trough crosscutting the narrow continental slope; **c** profile from the middle part of the trough crosscutting a fault block on continental slope; and **d** profile from the southern trough across a submarine canyon. Black curve is topographical profile, and the red curve is second derivative to the terrain; positions of these profiles are shown in Fig. 10.36

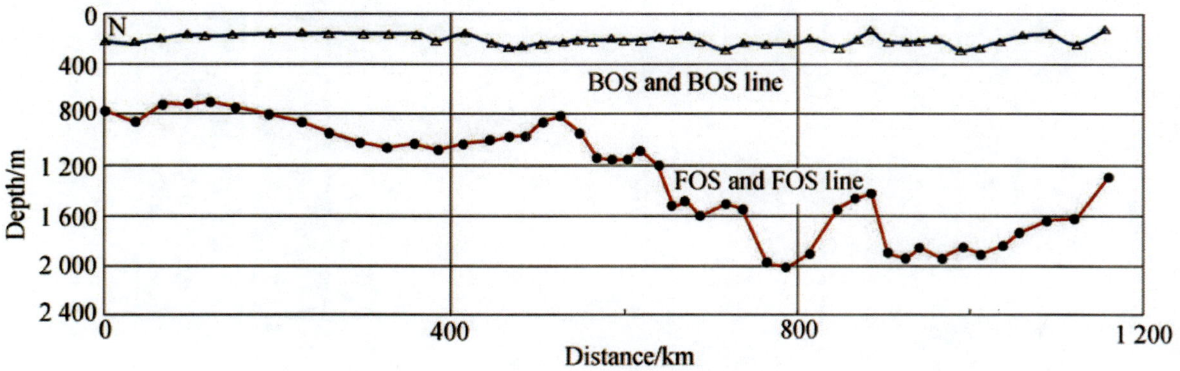

Fig. 10.38 Water depth profile of connecting line between BOS and FOS. Spatial distribution is shown in Fig. 10.36, and two points with the same distance are related to the same profile

10.7.2.2 Central Axis of the Okinawa Trough

Determination Method

The OT has an overall U-shaped groove appearance (Fig. 10.35) with the continental slope of the ECS to its west and the Ryukyu Arc slope to its east. The continental slope of the ECS has a regular FOS, which is generally distributed at the lower part of the continental slope (Fig. 10.37). Although locally cut and disconnected by channels, the Ryukyu Arc is continuous on the whole; the FOS along its west side can be readily traced. Therefore, we can use the method discussed in the previous section to determine the FOS at both the eastern and western sides of the OT; the boundary of the OT axis area can be formed by connecting these FOS in order (Fig. 10.39).

The determination of a trough axis is based on 3D submarine topographical maps. Some typical tectonic units, such as the central axis of an echelon depressions at the bottom of the trough, or a ridgeline associated with linear ocean mountains, can be considered as basic evidence for the determination of a trough axis (Fig. 10.35). By analyzing the topographical profile in the axial area, and combining this with the 3D topographical features, we can confirm the positions of central axial points quantitatively. If linearly oriented seamounts are mapped in 3D, they can be used as the central axis for the profile. If there are obvious linear depressions in such a 3D map, the points with maximum depths should be used to select central axial points in the trough. In addition, geological and geophysical evidence is also critical for the determination of central axes, e.g., the shallow or deep tectonic features of the trough revealed by single- or multi-channel seismic profiles can sometimes directly determine the position of central axial (Sibuet 1987; Luan and Yue 2007).

Central Axial Point and Central Axis

The axial region of the OT has been characterized by the construction of 39 typical topographical profiles perpendicular to the axis (Fig. 10.39) using the method discussed above. After analyzing the topographical characteristics of each profile, a series of topographical feature points on each profile has been determined quantitatively, such as the minimum, maximum, and secondary depth points. Comprehensive analysis of these topographic profiles and maps has led to the positioning of a central axial point on each profile; then, after connecting all these points in order, the central axis in OT axial region is formed (Figs. 10.39 and 10.41).

From the analysis of the 3D map and typical topographical profiles (Figs. 10.35 and 10.40), we find the general appearance of the trough is similar at the northern, middle, and southern parts of the trough with a U-shaped groove terrain. However, there are significant differences in detail that control the position of the central axial points.

In the northern OT (Fig. 10.35a), the continental slope of the ECS is not flat and has developed numerous fault blocks (Fan et al. 2000). In contrast, on the eastern side of the trough, numerous islands exist, which makes the west flank of the Ryukyu Island Arc relatively broken. There are many low sea-knolls and depressions developed in the trough. The terrain in the axial region of the trough is characterized by the features of that high relief in the middle and deep water along the two sides, and most profiles are W-shaped features of symmetry (Fig. 10.40a). The maximum and secondary depth points are always distributed on the two sides of the W-shaped profile. As a result, the maximum depth point in a few profiles jumps between the eastern and western sides of the axis, with most of the maximum depth points distributed east of the axis but a few to the west (Fig. 10.39). The maximum water depth ranges from 800 to 1200 m (red line in Fig. 10.41). This distribution pattern of maximum depth points in the northern OT is consistent with that of boundary points defined in the submission to CLCS (the Commission

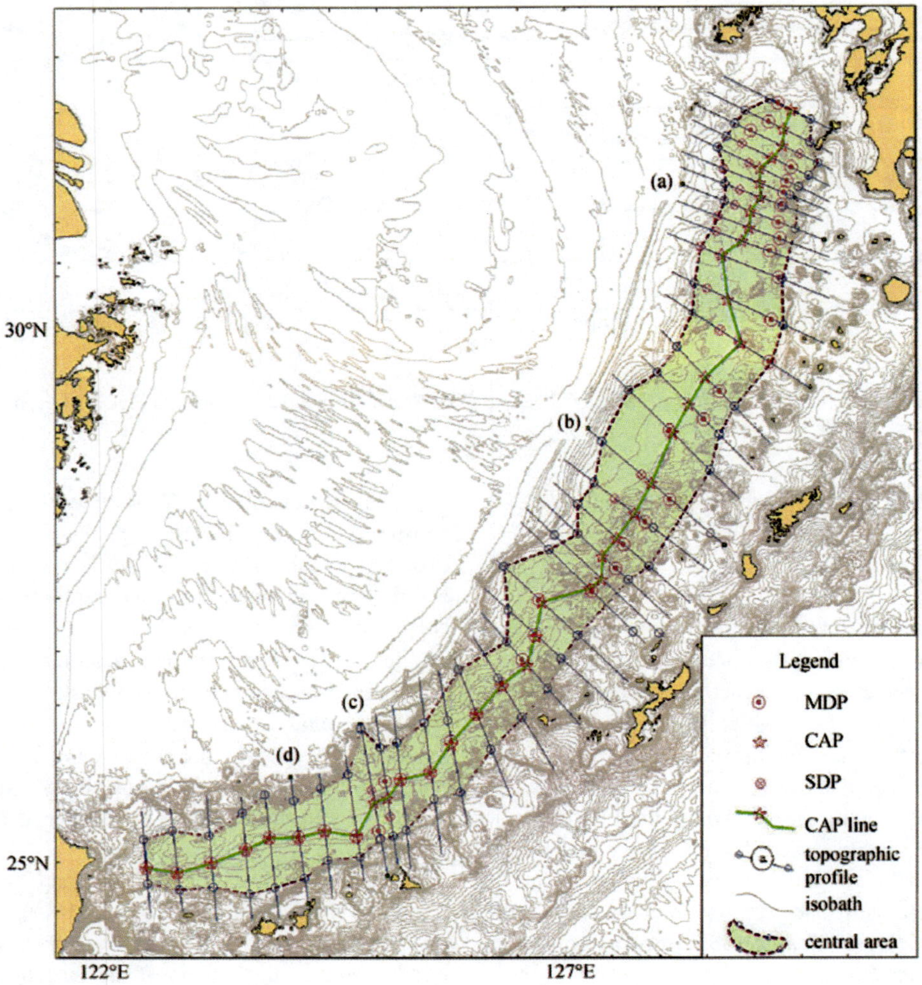

Fig. 10.39 Topographical profiles to determine the central axial points. **a–d** are typical topographic profiles. MDP: maxium depth points; CAP: central axis points; SDP: secondary depth points

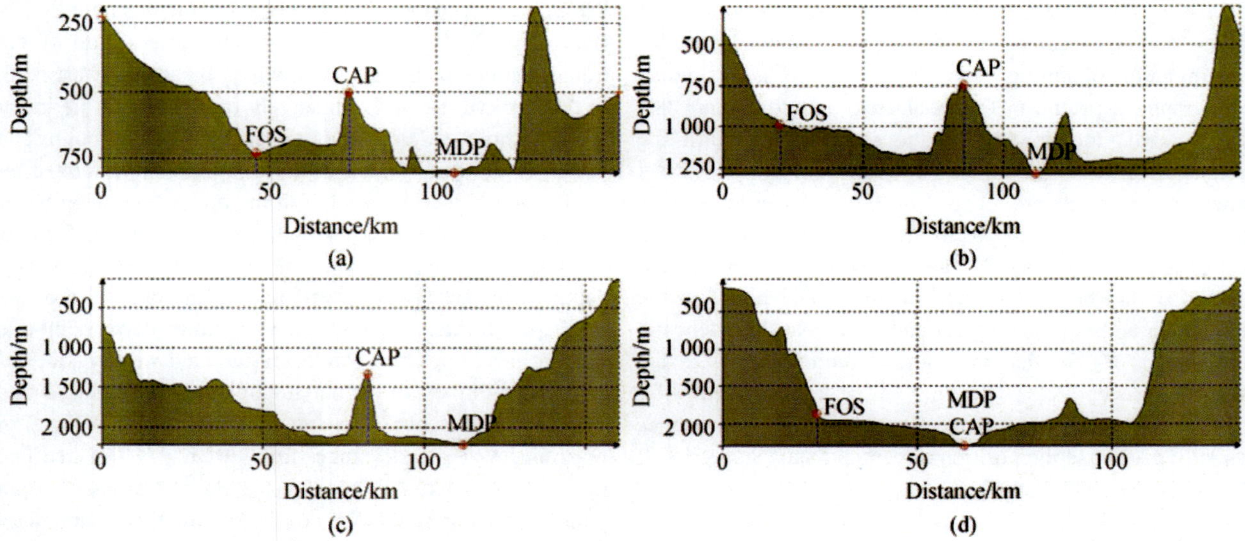

Fig. 10.40 Typical profiles and their central axial points. **a** A profile in the northern trough; **b** a profile in middle trough; **c** a profile in southern trough; and **d** a profile in southern trough. FOS: foot point of the slope, MDP: maximum depth points; CAP: central axial points; positions of the four profiles shown in Fig. 10.39

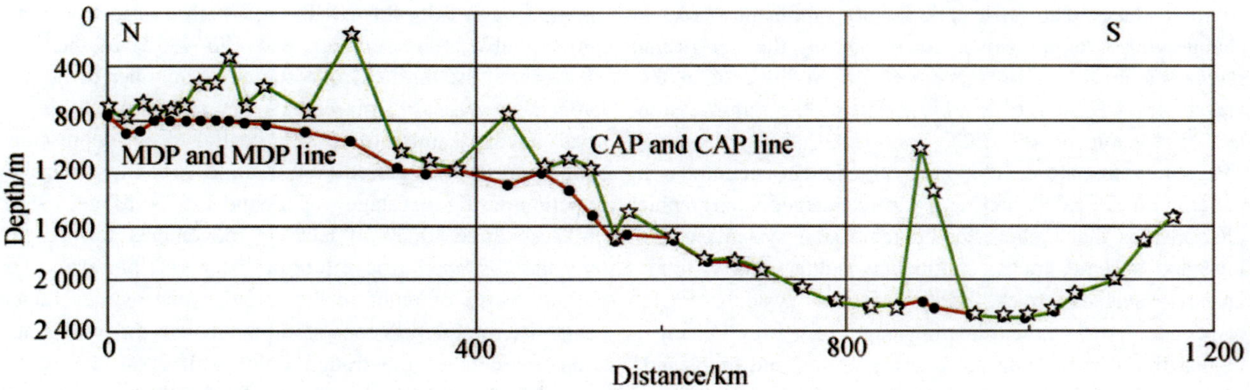

Fig. 10.41 Connecting line of central axial points and that of the maximum depth points in the Okinawa Trough. Spatial distribution is shown in Fig. 10.37, and two points with same distance are corresponding to a same profile

on the Limits of the Continental Shelf). For this type of topography, the shallow point should be chosen as the central axial point, i.e., the middle shallow point on a W-shaped profile. In summary, the central axial points in northern OT are essentially located in the middle part of the axial region with fluctuating water depth. The shallowest water depth of the point is only 200 m but this point has a large depth drop of 400 m on the same profile, and even to 1000 m locally (Fig. 10.41).

In the middle part of the OT (Fig. 10.35b), the continental slope of the ECS is narrow and flat. By contrast, the western slope of the Ryukyu Island Arc is quite uneven, and there are linear seamount chains, conical seamounts and hills distributed from the eastern part of the trough to the Ryukyu Arc (Li 2008). Influenced by linear seamounts, the trough as a whole shows the characteristics of composite W-shaped topography (Fig. 10.40b). The maximum depth points are distributed in the strap-like depression area among linear seamounts, and additionally, this occurs mostly in the central area of the trough axial region. Given the combined features of the topographical profile, a ridge of the linear seamounts is distributed regularly in the middle of the axial region, which can be used as the central axial line of the trough axial region. In contrast, in the region where these amounts are not linear, the maximum depth points in the central depression can be taken as the central axial points. In the middle part of the trough, the maximum depth points and the central axial points gradually gather together in spatial distribution, and even overlap each other in southern part of the middle trough (Fig. 10.39), which is distinct from dispersion in northern part of the trough. The maximum depth in the middle section gradually increases from 1 200 to 2000 m or more (Fig. 10.41). The water depths of axial points have a small-amplitude fluctuation in northern part of the middle section and coincide with the maximum depth points in the south.

In the southern Okinawa Trough (Fig. 10.35c), the continental slope of the ECS is fragmented, whereas the west slope of the Ryukyu Arc is relatively flat. Topography inside the trough is relatively simple with most en echelon depressions (half grabens) located along the trough (Luan and Yue 2007; Li 2008) along with some local linear seamounts. Most of the topographical profiles crosscutting the trough have a standard U-shape (Fig. 10.40d), with some occasionally exhibiting a W-shape (Fig. 10.40c). The maximum depth points of U-shaped profiles are coincided with the central axial points, whereas those of the W-shaped profiles are in the linear depressions and the central axial points are always located at central shallow position of the profiles (Fig. 10.40d). Spatially, all of the maximum depth points were located inside the linear depressions (Fig. 10.39), where they coincide with the majority of central axial points. However, where there is a linear seamount, the central axial point exists on the ridge of the seamount, and the water depth drop can exceed 1000 m between a central axial point and the corresponding maximum depth point. Generally, in the southern OT, the depth of maximum depth points is between 1 600 and 2400 m, whereas the minimum water depth of the central axial point is about 1000 m (Fig. 10.41).

10.7.3 Factors Influencing Boundary Distributions

10.7.3.1 Influence of Sea-Level Fluctuation on Boundary Characteristics

Significant sea-level fluctuation is the most important factor affecting continental shelf deposition and seafloor topography. Milankovich cycles, with a main cycle interval of 100 ka, lead to the global cycle of glacial and interglacial stages (Lin and Sun 1987). The growth and decay of continental ice sheets leads to regionally significant sea-level change with a range of over 100 m. During the last glacial stage, sea level fell to a position 140 m lower than the present level (Zhu and Li 1979), causing the transgression and regression of the ECS continental shelf. Sea-level changes control the development of sediment accumulation patterns on the ECS shelf, and form sequence stratigraphic relationships that include forced regression systems tracts, lowstand systems tracts, continental sedimentation, transgressive systems tracts and highstand systems tracts, respectively, during the following stages of sea-level change: regression, lowstand, drought, transgression and highstand. Multiple stages of comparable strata are imaged by high-resolution single-channel seismic data (Wu et al. 2002). Additionally, changes in hydrodynamic environmental conditions, brought about by sea-level change, also are a major factor in the development of relief on the continental shelf. Overland rivers that crosscut the shelf during lowstands (Li 2008) flow into the ocean directly at the front of the continental slope, or alternatively into canyons on the continental slope (Zhao et al. 2011), providing the source and dynamics for changes in the relief of shelf outer margins and the continental slope of the ECS.

Sea-level fluctuations, because of their role in promoting changes in sedimentary sequences and the dynamic environment during very low sea levels, play a critical role in shaping the form of the BOS. This in turn affects the shape of the slope and the position of the FOS from place to place on the margin.

10.7.3.2 Influence of Sea-Level Fluctuation on Boundary Characteristics

The Ryukyu Trench, Ryukyu Arc and OT make up a typical "trench-arc-basin" system. This specific tectonic pattern is the result of the interaction between the Eurasian and Pacific plates (Lee et al. 1980; Kimura 1985; Sibuet 1987; Qin et al. 1987; Xu and Le 1988; Jin 1992; Li 2008). Three stages of tectonic movements have been identified as being important for the evolution of the OT.

During the late stage of the Early Miocene, the Philippine Sea plate subducted eastward along the eastern side of the Diaoyu Islands folding zone, and initiated the Ryukyu Trench, Ryukyu Arc and OT basin system (Xu and Le 1988). A series of beaded rift basins formed along the normal fault because of the extensively developed rifts in the OT and Ryukyu Island region (Jin 1992). The expansion direction during this stage was 135°–155°N (Sibuet 1987). The second episode of OT tectonic movement started in the Pliocene/Pleistocene. This stage was characterized by the tilting of fault blocks and formation of normal faults with a rough strike of 45°N and approximate displacements of 50 m (Wu et al. 2004). The expansion direction during this stage was 155°–170°N (Sibuet 1987). The OT evolved into a unique trough from this rift basin, which tended to expand preferentially at its southern end (Jin 1992) In the Late Pleistocene, the third episode started. A number of normal faults with vertical displacements >10 m and newly developed linear seamounts in the central deep depressions formed during this period (Wu et al. 2004). The expansion direction during this stage was around 175°N (Sibuet 1987).

These three stages of tectonic movements formed not only the U-shaped groove terrain of the OT, but also other various forms of relief in the trough, such as, several en echelon central depressions along the axis in the middle and southern parts of the trough (Wu et al. 2004; Li 2008), normal faults with displacements of a few meters to tens of meters along the two sides of linear depressions, and new seamounts with apparent linear features in the southern part of the central depression (Wu et al. 2004). Fresh olivine tholeiite samples obtained from the central depressions show geochemical characteristics similar to those of mid-ocean ridge basalts from mantle plume sources. Additionally, there are several linear seamounts scattered through the trough, such as the Ono Temple seamount in the southern part of the trough and other multiple linear seamounts chains aligned in parallel in the middle and northern parts of the trough. All of these linear seamounts are the product of trough expansion (Wu et al. 2004; Li 2008). The seismic profile across the trough has confirmed the existence of a spreading center with a spreading axis extending along the trough axis (Sibuet 1987).

Morphotectonic (Fan et al. 2000; Wu et al. 2004; Liu et al. 2005; Luan and Yue 2007; Li 2008), petrologic (Li and Wang 1997; Huang et al. 2006) and geophysical (Liang et al. 2001; Hao et al. 2004; Han et al. 2007) evidence has proven that the southern OT came into the spreading stage as a result of rifting and subsidence. The linear seamounts and the central depressions in the center of the OT are the results of seafloor rifting and expansion, whereas the seamount ridge and depression axis are the central axis of OT. The three stages of tectonic movements have not only shaped the appearance of the trough but also controlled the distribution of its central axis.

10.7.3.3 Influence of Submarine Canyons on Boundary Characteristics

On the ECS slope, especially to the south, there are numerous variably sized submarine canyons that incise the continental slope and divide it into segments (Fig. 10.35). On the north-central continental slope of the ECS, the submarine canyons are relatively small and streamlined, whereas on the middle and southern continental slope they are usually large. Canyons in the south are generally 10–50 km long, 1–15 km wide and have a gradient of 5−15° with an incised depth of 100–500 m. Many canyons have three or four branches resembling, in plan view, a goose-claw or tree (Li 2008). These canyons extend upward to the shelf, downward to the bottom of the trough and show visible turbidite accumulation along their extent. These deposits indicate that the submarine canyons act as channels for turbidity currents flowing down slope; i.e., the ECS shelf provides sediment sources for deposition on the western side of the OT (Wu et al. 2002; Li 2008; Zhao et al. 2011).

During the Last Glacial Maximum, the sea level of the ECS had dropped to 140 m beneath the present level (Zhu and Le 1979) with the continental shelf completely exposed with widespread desertification. The ancient rivers of the ECS flowed directly across the continental shelf (Li et al. 2004a) into the submarine canyons, which means that the submarine canyons probably acted as the direct channel for sediment flowing from the ECS continental shelf into the OT. The heads of the submarine canyons extending into the continental shelf not only changed the position of the BOS, but also transformed the original shape of the BOS line. In short, the submarine canyons that incised the slope had changed the slope shape. The sediments they carried also changed the relief between the lower continental slope and the trough bottom. They also changed the position of the FOS and FOS line for the ECS shelf. On the southern continental slope, multiple FOS are even available for a single profile.

References

Anthony D, Leth JO (2002) Large-scale bedforms, sediment distribution and sand mobility in the eastern North Sea off the Danish west coast. Mar Geol 182:247–263. https://doi.org/10.1016/S0025-3227(02)00267-0

AVISO (1996) AVISO user handbook for merged Topex/Poseidon products edition 3. Available via DIALOG. https://www.aviso.altimetry.fr/fileadmin/documents/data/tools/hdbk_tp_gdrm.pdf. Accessed 25 June 2019

Behn M D, Sinton JM, Detrick RS (2004) Effect of the Galápagos hotspot on seafloor volcanism along the Galápagos Spreading Center (90.9°-97.6°W). Earth Planet Sci Lett 217(3–4):331–347. https://doi.org/10.1016/S0012-821X(03)00611-3

Berné S, Vagner P (2002) Pleistocene forced regressions and tidal sand ridges in the East China Sea. Mar Geol 188:293–315. https://doi.org/10.1016/S0025-3227(02)00446-2

Bouma AH (1990) Naming of undersea features. Geo-Mar Lett 10:119. https://doi.org/10.1007/BF02085926

Buck WR (1988) Flexural rotation of normal faults. Tectonics 7 (5):959–973. https://doi.org/10.1029/tc007i005p00959

Cannat M, Rommevauxjestin C, Fujimoto H et al (2003) Melt supply variations to a magma-poor ultra-slow spreading ridge (Southwest Indian Ridge 61° to 69°E). Geochem Geophys Geosyst 4(8):27–35. https://doi.org/10.1029/2002GC000480

Cannat M, Sauter D, Mendel V et al (2006) Modes of seafloor generation at a melt-poor ultraslow-spreading ridge. Geology 34 (7):605–608. https://doi.org/10.1130/G22486.1

Carbotte S, Macdonald K (1992) East Pacific Rise 8°-10°30′N: evolution of ridge segments and discontinuities from SeaMARC II and three-dimensional magnetic studies. J Geophys Res 97(B5):6959–6982. https://doi.org/10.1029/91JB03065

Chen ZY, Saito Y, Hori K et al (2003) Early Holocene mud-ridge formation in the Changjiang River Estuary offshore, China: a tidal-controlled estuarine pattern and sea-level implications. Mar Geol 198(3):245–257. https://doi.org/10.1016/S0025-3227(03)00119-1

Chen Z, Yan W, Chen MH et al (2006) Discovery of seep authigenic carbonate nodules on northern continental slope of South China Sea: new evidence of gas hydrate. Chin Sci Bull 51(9):1065–1072. https://doi.org/10.1109/TPSD.2006.5507455

Chen ZH, Li JB, Wu ZY et al (2009) Tectonic evolution implication of geometry shape characteristics for Manila trench (in Chinese with English abstracts). Mar Geol Quat Geol 2:63–69

Chen PF, Craig R, Hao KC et al (2012) Slab-induced waveform effects as revealed by the TAIGER seismic array: evidence of slab beneath central Taiwan. Phys Earth Planet Inter 196–197:62–74. https://doi.org/10.1016/j.pepi.2012.02.004

Deffontaines B, Liu CS, Hsu HH (2016) Structure and deformation of the Southern Taiwan accretionary prism: The active submarine Fangliao Fault Zone offshore west Hengchun Peninsula. Tectonophysics 692:227–240. https://doi.org/10.1016/j.tecto.2016.11.007

Dick HJ, Lin J, Schouten H et al (2003) An ultraslow-spreading class of ocean ridge. Nature 426(6965):405–412. https://doi.org/10.1038/nature02128

Dong ZH, Cao H, Xue RJ (2004) The influence on topography and relief in south Gulf Beibu and suspending of pipe caused by typhoon (in Chinese with English abstract). Ocean Tech 23:24–28

Duffy P, Hughes-Clarke JE (2005) Application of spatial cross correlation to detection of migration of submarine sand dunes. J Geophys Res 110 (F4). https://doi.org/10.1029/2004JF000192

Dyer KR, Huntley DA (1999) The origin, classification and modeling of sand banks and ridges. Cont Shelf Res 19:1285–1330. https://doi.org/10.1016/S0278-4343(99)00028-X

Egbert GD, Erofeeva SY (2002) Efficient inverse modeling of Barotropic ocean tides. J Atmos Oceanic Technol 19(2):183–204. https://doi.org/10.1175/1520-0426(2002)019<0183:EIMOBO>2.0.CO;2

Fairbanks RG (1989) A 17,000-year glacio-eustatic sea level record: influence of glacial melting rates on the Younger Dryas event and deep-ocean circulation. Nature 342(6250):637–642. https://doi.org/10.1038/342637a0

Fan JK, Wu SG (2014) P-wave seismic tomography of the Manila subduction zone(in Chinese with English abstracts). Chinese J Geophys 57(7):2127–2137. https://doi.org/10.6038/cjg20140709

Fan FX, Lin MH, Jiang RH et al (2000) The rift block ridge in north continental shelf front slope of the East China Sea (in Chinese with English abstracts). J Ocean Univ China 30:173–176

Georgen JE, Lin J, Dick HJ et al (2001) Evidence from gravity anomalies for interactions of the Marion and Bouvet hotspots with the Southwest Indian Ridge: effects of transform offsets. Earth Planet Sci Lett 187(3):283–300. https://doi.org/10.1016/s0012-821x(01)00293-x

Gervini D, Yohai VJ (2002) A class of robust and fully efficient regression estimators. Ann Stat 583-616. https://doi.org/10.1214/aos/1021379866

Grindlay NR, Fox RJ, Vogt PR (1992) Morphology and tectonics of the Mid-Atlantic Ridge (25°–27°30′S) from Sea Beam and magnetic data. J Geophys Res-Sol Ea 97(B5):6983–7010. https://doi.org/10.1029/91JB2981

Han MK (1992) Tectonicgeomorphology (in Chinese version). Adv Earth Sci 7(5):61–62

Han B, Zhang XH, Pei JX et al (2007) Characteristics of crust-mantle in East China sea and adjacent regions (in Chinese with English abstracts). Progress Geophys 2:376–382

Han X, Wu G, Cui R et al (2010) Discovery of a Hydrothermal Sulfide Deposit on the Southwest Indian Ridge at 49.2°E. Paper presented at the AGU Fall Meeting, San Francisco, 13–17 December 2010

Hao TY, Liu JH, Guo F et al (2004) Research on crustal structure and lithosphere property in the Okinawa Trough area (in Chinese with English abstracts. Chinese J Geophys 47:462–468

Hirtzel J, Chi WC, Reed D et al (2009) Destruction of Luzon forearc basin from subduction to Taiwan arc-continent collision. Tectonophysics 479:43–51. https://doi.org/10.1016/j.tecto.2009.01.032

Hori K, Saito Y, Zhao QH et al (2001) Sedimentary facies of the tide-dominated paleo-Changjiang (Yangtze) estuary during the last transgression. Mar Geol 177(3–4):331–351. https://doi.org/10.1016/S0025-3227(01)00165-7

Hori K, Saito Y, Zhao Q et al (2002) Architecture and evolution of the tide-dominated Changjiang (Yangtze) River delta. China. Sediment Geol 146(3–4):249–264. https://doi.org/10.1016/s0037-0738(01)00122-1

Huang CY (2017) Geological ages of Taiwan stratigraphy and tectonic events (in Chinese with English abstracts). Scientia Sinica Terrae 47:394–405

Huang HZ, Tang BG, Yang WD (1996) Depositional geology of Changjiang River Estuary Delta (in Chinese with English abstract). Geological Publishing House, Beijing

Huang P, Li AC, Jiang HY (2006) Geochemical features and their geological implications of volcanic rocks from the northern and middle Okinawa Trough (in Chinese with English abstracts). Acta Petrologica Sinica 22:1703–1712

Huang CY, Yan Y, Zhao QH et al (2012) Cenozoic stratigraphy of Taiwan: Window into rifting, stratigraphy and paleoceanography of South China Sea. Chin Sci Bull 57:3130–3149. https://doi.org/10.1007/s11434-012-5349-y

Huber PJ (2011) Robust Statistics. Springer, Berlin Heidelberg

Humphris S, Herzig P, Miller DJ et al (1996) Proceedings, initial reports, Ocean Drilling Program, Leg 158, TAG: drilling an active hydrothermal system on a sediment-free slow-spreading ridge. Available via DIALOG. https://www-odp.tamu.edu/publications/158_IR/158ir.htm. Accessed 8 Feb 2007

IHO (2019a) GEBCO Gazetteer of undersea feature names. Available via DIALOG. https://www.ngdc.noaa.gov/gazetteer/. Accessed 29 June 2019

IHO (2019b) Standardization of undersea feature names. Bathymetric Publication No.6. 4th ed. Available via DIALOG. https://www.gebco.net/data_and_products/undersea_feature_names/documents/b_6_e4_1_0_ce.pdf. Accessed 29 June 2019

Jin XL (1992) Marine Geology of the East China Sea (in Chinese). Science Press, Beijing, pp 303–331

Jin JH, Chough SK (1998) Partitioning of transgressive deposits in the southeastern Yellow Sea: a sequence stratigraphic interpretation. Mar Geol 149(1–4):79–92. https://doi.org/10.1016/S0025-3227(98)00023-1

Jin JH, Chough SK (2002) Erosional shelf ridges in the mid-eastern Yellow Sea. Geo-Mar Lett 21(4):219–225. https://doi.org/10.1007/s00367-001-0082-6

Jung WY, Suk BC, Min GH et al (1998) Sedimentary structure and origin of a mud-cored pseudo-tidal sand ridge, eastern Yellow Sea. Korea. Mar Geol 151(1–4):73–88. https://doi.org/10.1016/s0025-3227(98)00058-9

Kao H, Huang GC, Liu CS (2000) Transition from oblique subduction to collision in the northern Luzon arc-Taiwan region: Constraints from bathymetry and seismic observations. J Geophys Res 105:3059–3079. https://doi.org/10.1029/1999JB900357

Kimura M (1985) Back-arc rifting in the Okinawa Trough. Mar Pet Geol 2(3):222–240. https://doi.org/10.1016/0264-8172(85)90012-1

Kleinrock MC, Humphris SE (1996) Structural control on sea-floor hydrothermal activity at the TAG active mound. Nature 382 (6587):149–153. https://doi.org/10.1038/382149a0

Kopp H, Flueh ER, Petersen CJ et al (2006) The Java margin revisited: Evidence for subduction erosion off Java. Earth Planet Sci Lett 242:130–142. https://doi.org/10.1016/j.epsl.2005.11.036

Koropitan A, Ikeda M (2008) Three-dimensional modeling of tidal circulation and mixing over the Java Sea. J Oceanogr 64(1):61–80. https://doi.org/10.1007/s10872-008-0005-5

Ku CY, Hsu SK (2009) Crustal structure and deformation at the northern Manila Trench between Taiwan and Luzon islands. Tectonophysics 466:229–240. https://doi.org/10.1016/j.tecto.2007.11.012

Kuo YW, Wang CY, Hao KC et al (2016) Crustal structures from the Wuyi-Yunkai orogen to the Taiwan orogen: The onshore-offshore wide-angle seismic experiments of the TAIGER and ATSEE projects. Tectonophysics 692:164–180. https://doi.org/10.1016/j.tecto.2015.09.014

Lacombe O, Mouthereau F, Angelier J et al (2001) Structural, geodetic and seismological evidence for tectonic escape in SW Taiwan. Tectonophysics 333:323–345. https://doi.org/10.1016/S0040-1951(00)00281-X

Lee CS, Shor GG, Bibee LD et al (1980) Okinawa trough: origin of a back-arc basin. Mar Geol 35(1–3):219–241. https://doi.org/10.1016/0025-3227(80)90032-8

Liang RC, Wu JL, Liu BH et al (2001) Linear magnetic anomalies and tectonic development for the middle Okinawa Trough (in Chinese with English abstracts). Acta Oceanol Sin 23:69–79

Li JB (2008) Regional geology in East China Sea (in Chinese). China Ocean Press, Beijing, pp 81–152

Li CZ, Li BC (1981) Studies on the formation of Subei sand cays (in Chinese with English abstracts). Oceanologia Et Limnologia Sinica 12(4):321–331

Li WR, Wang XL (1997) Study on olivine tholeiite in southern Okinawa Trough (in Chinese with English abstracts). Oceanologia Et Limnologia Sinica 28(6):665–672

Li MZ, King EL (2007) Multibeam bathymetric investigations of the morphology of sand ridges and associated bedforms and their relation to storm processes, Sable Island Bank, Scotian Shelf. Mar Geol 243:200–228. https://doi.org/10.1016/j.margeo.2007.05.004

Li J, Chen Y (2010) First Chinese OBS experiment at Southwest Indian Ridge. InterRidge Newsletter 19:16–28

Li BH, Li CX, Shen HT (2003) A preliminary study on sediment flux in the Changjiang Delta during the postglacial period. Sci China Serious D: Earth Sci 46:743–752. https://doi.org/10.1360/03yd9065

Li GX, Liu Y, Yang ZG et al (2004a) Changjiang River Estuary palaeo-river on continental shelf plain of East China Sea during last glaciation (in Chinese with English abstracts). Sci China (Series D): Earth Sci 35:284–289

Li JB, Jin XL, Ruan AG et al (2004b) Indentation tectonics in the accretionary wedge of middle Manila Trench. Chin Sci Bull 49:1279–1288. https://doi.org/10.1360/03wd0412

Li GX, Liu Y, Yang ZG et al (2005) Ancient Changjiang channel system in the East China Sea continental shelf during the last glaciation. Sci China Ser D-Earth Sci 48:1972–1978. https://doi.org/10.1360/04yd0053

Li JB, Ding WW, Wu ZY et al (2012) Progressive expansion of the southwestern basin of the South China Sea (in Chinese with English abstracts). Chin Sci Bull 57(20):1896–1905

Li J, Jian H, Chen YJ et al (2015) Seismic observation of an extremely magmatic accretion at the ultraslow spreading Southwest Indian Ridge. Geophys Res Lett 42(8):2656–2663. https://doi.org/10.1002/2014GL062521

Lin GD, Sun XL (1987) The Sea Level (in Chinese). Geological Publishing Press, Beijing, pp 1–20

Liu ZX, Xia DX (2004) Tidal depositional sand bodies offshore China (in Chinese). China Ocean Press, Beijing, pp 195–210

Liu JP, Milliman JD, Gao S et al (2004) Holocene development of the Yellow River's subaqueous delta. North Yellow Sea. Marine Geol 209(1–4):45–67. https://doi.org/10.1016/j.margeo.2004.06.009

Liu ZC, Liu BH, Huang ZZ et al (2005) Topography of China seas and adjacent regions (in Chinese). China Ocean Press, Beijing, pp 51–67

Liu JP, Li AC, Xu KH et al (2006) Sedimentary features of the Changjiang River Estuary River-derived along-shelf clinoform deposit in the East China Sea. Cont Shelf Res 26(17–18):2141–2156. https://doi.org/10.1016/j.csr.2006.07.013

Liu JP, Xu KH, Li AC et al (2007a) Flux and fate of the Yangtze River sediments delivered to the East China Sea. Geomorphology 85:208–224. https://doi.org/10.1016/j.geomorph.2006.03.023

Liu ZC, Berné S, Saito Y et al (2007b) Internal architecture and mobility of tidal sand ridges in the East China Sea. Cont Shelf Res 27(13):1820–1834. https://doi.org/10.1016/j.csr.2007.03.002

Liu ZF, Zhan WH, Zhang ZQ (2007c) Tectonic implications of the double island arc between Taiwan and Luzon (in Chinese with English abstracts). Geotectonica Et Metallogenia 31(2):45–150

Luan XW, Yue BJ (2007) Topography and distribution of central graben volcanoes in Miyoko section of Okinawa Trough (in Chinese with English abstracts). Oceanologica Et Limnologia Sinica 38:266–271

Macdonald KC, Ridges MO (1998) Exploring the global mid-ocean ridge. Oceanus-Woods Hole Mass 41:2–6

Mellor GL (2002) User guide for a three-dimensional primitive equation numerical ocean model. Available via DIALOG. https://www.ccpo.odu.edu/POMWEB/UG.10-2002.pdf. Accessed 25 June 2019

Mendel V, Sauter D, Rommevauxjestin C et al (2003) Magmato-tectonic cyclicity at the ultra-slow spreading Southwest Indian Ridge: Evidence from variations of axial volcanic ridge morphology and abyssal hills pattern. Geochem Geophys Geosyst 4(5): art.09402. https://doi.org/10.1029/2002GC000417

Niu X, Ruan A, Li J et al (2015) Along-axis variation in crustal thickness at the ultraslow spreading Southwest Indian Ridge (50E) from a wide-angle seismic experiment. Geochem Geophys Geosyst 16(2):468–485. https://doi.org/10.1002/2014GC005645

Niwa Y, Hibiya T (2004) Three-dimensional numerical simulation of M2 internal tide in the East China Sea. J Geophys Res 109 (C04027). https://doi.org/10.1029/2003JC001923

Nygren I (2005) Terrain navigation for underwater vehicles. Dissertation, Royal Institute of Technology

Off T (1963) Rhythmic linear sand bodies caused by tidal currents. AAPG Bull 47(2):324–341. https://doi.org/10.1306/BC743989-16BE-11D7-8645000102C1865D

Olive DJ (2008) Applied robust statistics. Available via DIALOG. https://www.researchgate.net/publication/228392075_Applied_Robust_Statistics_Applied_Robust_Statistics. Accessed 15 June 2019

Paull CK, Ussler IIIW, Caress DW et al (2010) Origins of large crescent-shaped bedforms within the axial channel of Monterey Canyon, offshore California. Geology 6:755–774. https://doi.org/10.1130/GES00527.1

Qin YS, Zhao YY, Chen LR et al (1987) Geology of the East China Sea (in Chinese). Science Press, Beijing, pp 210–287

Rousseeuw PJ (1984) Least median of squares regression. J Am Stat Assoc 79(388):871–880

Sauter D, Carton H, Mendel V et al (2004) Ridge segmentation and the magnetic structure of the Southwest Indian Ridge (at 50°30′E, 55°30′E and 66°20′E): Implications for magmatic processes at ultra-slow- spreading centers. Geochem Geophys Geosyst 5(5):374–378. https://doi.org/10.1029/2003GC000581

Sauter D, Cannat M, Meyzen C et al (2009) Propagation of a melting anomaly along the ultraslow Southwest Indian Ridge between 46°E and 52°20′E: interaction with the Crozet hotspot. Geophys J Int 179 (2):687–699. https://doi.org/10.1111/j.1365-246X.2009.04308.x

Searle RC, Bralee AV (2007) Asymmetric generation of oceanic crust at the ultra-slow spreading Southwest Indian Ridge, 64°E. Geochem Geophys Geosyst 8(5):6. https://doi.org/10.1029/2006gc001529

Shang JH, Li JB, Wu ZY (2010) Discussion on the fine tectonic features of accretionary wedge and developing patterns of minisize trapped basins at middle part of Manila Subduction Zone(in Chinese with English abstracts). Chinese J Geophys 53(1):94–101. https://doi.org/10.3969/j.issn.0001-5733.2010.01.010

Shi XM, Du ZC (2006) Review and Prospect of Tectonic Geomorphology in China (in Chinese with English abstracts). Northwestern Seismol J 28(3):280–284

Sibuet JC (1987) Back arc extension in the Okinawa Trough. J Geophys Res 92:14041–14063. https://doi.org/10.1029/jb092ib13p14041

Smith WHF, Sandwell DT (1997) Global sea floor topography from satellite altimetry and ship depth soundings. Science 277:1956–1962. https://doi.org/10.1126/science.277.5334.1956

Smith DP, Ruiz G, Kvitek R et al (2005) Semiannual patterns of erosion and deposition in upper Monterey Canyon from serial multibeam bathymetry. GSA Bull 117(9–10):1123–1133. https://doi.org/10.1130/B25510.1

Smith DP, Kvitek R, Iampietro PJ et al (2007) Twenty-nine months of geomorphic change in upper monterey canyon (2002–2005). Mar Geol 236(236):79–94. https://doi.org/10.1016/j.margeo.2006.09.024

Soh W, Tokuyama H (2002) Rejuvenation of submarine canyon associated with ridge subduction, Tenryu Canyon, off Tokai, central Japan. Mar Geol 187:203–220. https://doi.org/10.1016/S0025-3227(02)00267-0

Soulsby RL, Whitehouse RJSW (1997) Threshold of sediment motion in coastal environments. In: Proceedings of the Pacific Coasts and Ports. Whitehouse RJSW, Christchurch, pp 149–154

Tang QQ, Zhan WH, Li J et al (2017) Volcanic evidence for slab window induced by fossil ridge subduction at east edge of South China Sea (in Chinese with English abstracts). Mar Geol Quat Geol 37(2):119–126

Tao C, Lin J, Guo S et al (2011) First active hydrothermal vents on an ultraslow-spreading center: Southwest Indian Ridge. Geology 40 (1):47–50. https://doi.org/10.1130/G32389.1

Tian J, Wang PX, Cheng XR et al (2005) Astronomically tuned time scale 12 ma to 18.3 ma odp site 1148 North South China Sea (in Chinese with English abstracts). Earth Sci-J China Geosci 30 (5):513–518

The Hydrographer of Navy (1990) Admiralty Tide Tables and Tidal Stream Tablets. UK Hydrographic Office, London

Twichell D, Brooks G, Gelfenbaum G et al (2003) Sand ridges off Sarasota, Florida: a complex facies boundary on a low-energy inner shelf enviroment. Mar Geol 200:243–262. https://doi.org/10.1016/S0025-3227(03)00185-3

Uehara K, Saito Y (2003) Late Quaternary evolution of the Yellow/East China Sea tidal regime and its impacts on sediments dispersal and seafloor morphology. Sediment Geol 162:25–38. https://doi.org/10.1016/s0037-0738(03)00234-3

Uehara K, Saito Y, Hori K (2002) Paleotidal regime in the Changjiang (Changjiang River Estuary) Estuary, the East China Sea, and the Yellow Sea at 6 ka and 10 ka estimated from a numerical model. Mar Geol 183:179–192. https://doi.org/10.1016/S0025-3227(01)00255-9

Wang K, Bilek SL (2014) Invited review paper: Fault creep caused by subduction of rough seafloor relief. Tectonophysics 610:1–24. https://doi.org/10.1016/j.tecto.2013.11.024

Wu ZY, Jin XL, Li JB et al (2003) Typical tectonic geomorphology of the Okinawa Trough and its geological significance. Paper presented at the Submarine Science Development Strategy Seminar, Hangzhou Zhejiang, China

Wu ZY, Jin XL, Li JB (2002) Seismic stratigraphic interpretation of high-resolution seismic profiles between Changjiang River Estuary Estuary and Okinawa Trough (in Chinese with English Abstract). Marine Geol Quat Geol 22:9–20

Wu ZY, Wang XB, Jin XL et al (2004) The evidences of the backarc spreading and discussion on the key issues in the Okinawa Trough (in Chinese with English abstracts). Marine Geol Quat Geol 24:67–76

Wu ZY, Jin XL, Li JB et al (2005) Linear sand ridges on the outer shelf of the East China Sea. Chin Sci Bull 50(21):2517–2528. https://doi.org/10.1007/BF03183643

Wu ZY, Li JB, Li SJ et al (2016) A new method to identify the foot of continental slope based on an integrated profile analysis. Marine Geophys Res 38(1–2):1–9. https://doi.org/10.1007/s11001-016-9273-4

Wu ZY, Jin XL, Zhou JQ et al (2017) Comparison of buried sand ridges and regressive sand ridges on the outer shelf of the East China Sea. Marine Geophys Res 38(1–2):187–198. https://doi.org/10.1007/s11001-016-9278-z

Xu WL, Le JY (1988) Tectonic movement and evolution of the East China Sea (in Chinese with English abstract). Marine Geol Quat Geol 8:9–19

Xu JP, Wong FL, Kvitek R et al (2008) Sandwave migration in Monterey submarine canyon, centra California. Mar Geol 248 (3):193–212. https://doi.org/10.1016/j.margeo.2007.11.005

Yang WD (1996) Holocene underwater lobes of the Changjiang River Estuary River delta and their development features (in Chinese with English abstracts). Marine Geol Quat Geol 16(3):25–36. https://doi.org/10.1007/BF02029074

Yang CS (1989) Active, moribund and buried tidal sand ridges in the East China Sea and the southern Yellow Sea. Mar Geol 88(1–2):97–116. https://doi.org/10.1016/0025-3227(89)90007-8

Yang TF, Lee T, Chen CH et al (1996) A double island arc between Taiwan and Luzon: consequence of ridge subduction. Tectonophysics 258:85–101. https://doi.org/10.1016/0040-1951(95)00180-8

Ye YC, Zhuang ZY, Lai XH et al (2004) A study of sandy bedforms on the Changjiang River Estuary Shoal in the East China Sea (in Chinese with English abstract). J Ocean Univ China 34:1057–1062

Zhao M, Qiu X, Li J (2013) Three-dimensional seismic structure of the Dragon Flag oceanic core complex at the ultraslow spreading Southwest Indian Ridge (49°39′E). Geochem Geophys Geosyst 14 (10):4544–4563. https://doi.org/10.1002/ggge.20264

Zhao SL, Yu HJ, Liu JP (1997) Origin, development and evoluationary model of shelf desertization environment in late stage of Upper Pleistocene. Sci China, Ser D Earth Sci 40:207–214. https://doi.org/10.1007/BF02878380

Zhao YX, Liu BH, Li XS et al (2011) Sedimentary characters and material transportation of submarine canyon-fan systems in slope of the East China Sea (in Chinese with English abstracts). J Palaeogeogr 1:119–126

Zhou JQ, Wu ZY, Jin XL et al (2018) Observations and analysis of giant sand wave fields on the Taiwan Banks, northern South China Sea. Mar Geol 406:132–141. https://doi.org/10.1016/j.margeo.2018.09.015

Zhu YQ, Li CY (1979) Lowest sea level about the continental shelf of the East China Sea during late Pleistocene (in Chinese with English abstracts). Chin Sci Bull 7:317–320

11 Technology and Practice of Continental Shelf Delimitation

(**Disclaimer**: The views expressed herein are solely those of the author(s) and do not necessarily reflect the views of the Commission on the Limits of the Continental Shelf.)

To define the outer limit of the continental shelf beyond 200 n mile, the following elements are needed: the point representing the foot of the continental slope (FOS), the 2,500 m isobath, the FOS+60 n mile line, the 2,500 m isobath+100 n mile line, the territorial sea boundary, the boundaries of contiguous zones, the limit of the exclusive economic zone and the 350 n mile line. To delineate these lines and boundaries the Law of the Sea uses marine science and the latest graphics techniques. In this chapter, we discuss the basic concepts, technical methods and the practical issues associated with delimitation of the continental shelf and their relationship with the topographical and geomorphological features of the seabed.

11.1 United Nations Convention on the Law of the Sea

11.1.1 Main Topics of the Convention

The United Nations Convention on the Law of the Sea (hereafter referred to as the Convention) was adopted at the final session of the Third United Nations Conference on the Law of the Sea held at the Montego Bay in Jamaica in December 1982 and implemented on November 16, 1994. It has been ratified by more than 150 countries. The Convention comprises 17 parts and 9 annexes, with a total of 446 articles. The main topics of the Convention are: territorial seas and contiguous zones, the exclusive economic zone, the continental shelf, straits used for international navigation, the high seas, the archipelagic states, the regime of islands, enclosed or semi-enclosed seas, rights of access of land-locked states to and from the sea and freedom of transit, international submarine and marine scientific research, marine environment protection and safety, development and transfer of marine technology (Fig. 11.1).

(1) Definition of the territorial sea
Article 2 of the Convention.
  a. The sovereignty of a coastal State extends, beyond its land territory and internal waters and, in the case of an archipelagic State, its archipelagic waters, to an adjacent belt of sea, described as the territorial sea.
  b. This sovereignty extends to the air space over the territorial sea as well as to its bed and subsoil.
  c. The sovereignty over the territorial sea is exercised subject to this Convention and to other rules of international law.

(2) Limits of the territorial sea
Article 3 of the Convention.
Every state has the right to establish the breadth of its territorial sea up to a limit not exceeding 12 n mile, measured from baselines determined in accordance with this Convention.
Article 4 of the Convention.
The outer limit of the territorial sea is the line every point of which is at a distance from the nearest point of the baseline equal to the breadth of the territorial sea.
Article 5 of the Convention.
Except where otherwise provided in this Convention, the normal baseline for measuring the breadth of the territorial sea is the low-water line along the coast as marked on large-scale charts officially recognized by the coastal state.
Paragraph 1 of Article 5 of the Convention.
In localities where the coastline is deeply indented and cut into, or if there is a fringe of islands along the coast in its immediate vicinity, the method of straight baselines joining appropriate points may be employed in drawing the baseline from which the breadth of the territorial sea is measured.

Z. Wu et al., *High-resolution Seafloor Survey and Applications*,
https://doi.org/10.1007/978-981-15-9750-3_11

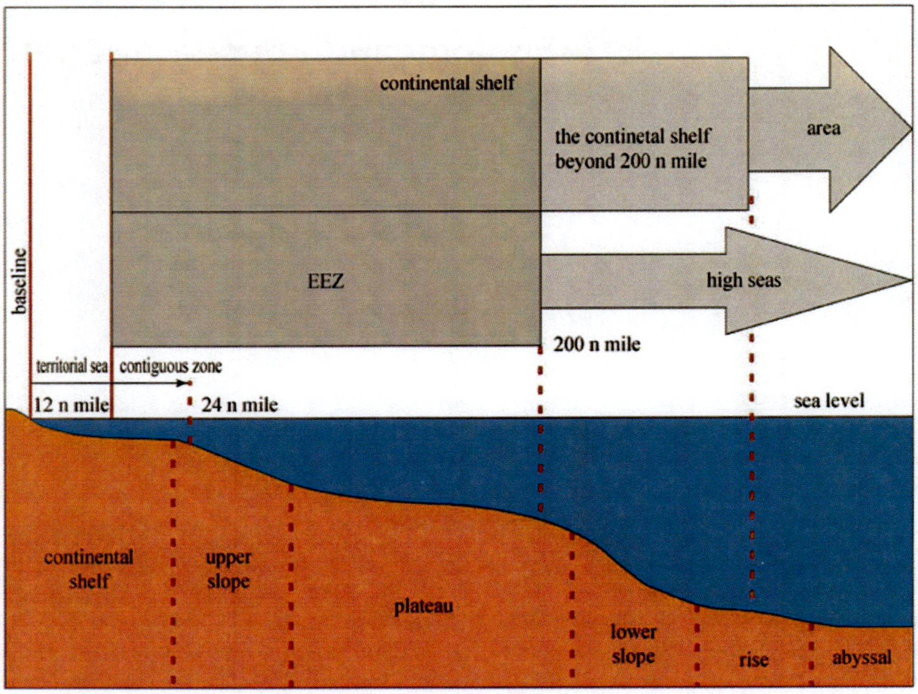

Fig. 11.1 Submarine geomorphology showing continental shelf boundaries and definitions of territorial water limits

(3) Contiguous zone
Paragraph 2 of Article 5 of the Convention.
The contiguous zone may not extend beyond 24 n mile from the baselines from which the breadth of the territorial sea is measured.
(4) Exclusive economic zone
The exclusive economic zone may not extend beyond 200 n mile from the baselines from which the breadth of the territorial sea is measured.
(5) Definition of the continental shelf
Paragraph 1 of Article 76 of the Convention.
The continental shelf of a coastal state comprises the seabed and subsoil of the submarine areas that extend beyond its territorial sea throughout the natural prolongation of its land territory to the outer edge of the continental margin, or to a distance of 200 n mile from the baselines from which the breadth of the territorial sea is measured where the outer edge of the continental margin does not extend up to that distance.
(6) High seas
Article 86 of the Convention.
"High seas" apply to all parts of the sea that are not included in the exclusive economic zone, in the territorial sea or in the internal waters of a state, or in the archipelagic waters of an archipelagic state. This article does not entail any abridgment of the freedoms enjoyed by all states in the exclusive economic zone in accordance with Article 58.
(7) Area
Paragraph 1 of Article 1 of the Convention.
"Area" means the seabed and ocean floor and subsoil thereof, beyond the limits of national jurisdiction.

11.1.2 The Continental Shelf

11.1.2.1 Legal Concept of the Continental Shelf

The legal concept of the continental shelf came into being after the Second World War. On September 28, 1945, the US president Truman published the Policy of the United States with Respect to the Natural Resources of the Subsoil and Seabed of the Continental Shelf (also known as the Truman Proclamation), which stated that "whereas it is the view of the government of the US that the exercise of jurisdiction over the natural resources of the subsoil and sea bed of the continental shelf by the contiguous nation is reasonable and just, since the effectiveness of measures to utilize or conserve these resources would be contingent upon cooperation and protection from shore, since the continental shelf may be regarded as an extension of the land mass of the coastal nation and thus naturally appurtenant to it, since these resources frequently form a seaward extension of a pool or deposit lying within the territory, and since self-protection compels the coastal nation to keep close watch over activities off its shores which are of their nature necessary for utilisation of these resources." On the same day, the US government declared that the extent of the continental shelf will not exceed the water depth of 100 fathoms (183 m). Shortly after the proclamation, many countries, especially those in Latin America, released statements that asserted their rights to the waters, seabed and resources adjacent to their coastline, but the content and scope of their claims varied.

In 1958, the final text of the International Law Commission defined the extent of the continental shelf by combining the water depth with practical exploitability limitations of resources. In the same year, the Continental Shelf Convention, which was adopted at the First UN Conference on the Law of the Sea, accepted the definition. The term "continental shelf" is used as referring: a. to the seabed and subsoil of the submarine areas adjacent to the coast but outside the area of the territorial sea, to a depth of 200 m or, beyond that limit, to where the depth of the superjacent waters admits of the exploitation of the natural resources of the said areas; b. to the seabed and subsoil of similar submarine areas adjacent to the coasts of islands.

In the Third UN Conference on the Law of the Sea, intense and highly technical discussions and negotiations were held to enhance and clarify the continental shelf regime and provide more detailed provisions in the Convention. Paragraph 1 of Article 76 of the Convention: the continental shelf of a coastal state comprises the seabed and subsoil of the submarine areas that extend beyond its territorial sea throughout the natural prolongation of its land territory to the outer edge of the continental margin, or to a distance of 200 n mile from the baselines from which the breadth of the territorial sea is measured where the outer edge of the continental margin does not extend up to that distance.

The paragraph clarifies the legal concept of the continental shelf, and emphasizes the close link between the continental shelf and the natural prolongation of the land territory. It declares that coastal states with a narrow continental shelf (with natural prolongation less than 200 n mile) may claim a continental shelf of the same breadth as its exclusive economic zone. Coastal states with a continental shelf whose natural prolongation is more than 200 n mile shall delineate the outer limits according to Paragraphs 4–7 of Article 76 of the Convention.

The legal continental shelf and the geological continental shelf are closely related but are different. The formation and development of the concept of the continental shelf is based on the geological continental shelf, but is not limited to the extent of the geological continental shelf. The concept of the "continental shelf" set forth in the Convention is not strictly a geological concept; rather, it is a hybrid concept that involves geology, bathymetry, metrology, political diplomacy, law and other fields.

The geographical scope of the "continental shelf" provided in Article 76 of the Convention generally covers the geological "continental margin". Paragraphs 4–7 of Article 76 of the Convention are based on geology and adopt a variety of criteria (including distance constraints and depth constraints) to delineate the outer limits of the continental shelf with natural prolongation of more than 200 n mile. Therefore, the outer limits of the continental shelf of some coastal states may extend beyond the geological continental margin.

11.1.2.2 Geological Concept of the Continental Shelf

Scientific literature defines the continental margin as the transition zone between the continent and the ocean floor; in terms of crustal structure it is the transition zone from continental crust to oceanic crust. In general, most continental margins are composed of the continental shelf, continental slope and continental rise (Fig. 11.2). Depending on the mode of plate movement, there are two geo-tectonic processes that affect the type of continental margin, namely, divergence and convergence, corresponding to passive continental margins and active continental margins, respectively. The third type of continental margin, known as a transform margin, can develop in both divergence and convergence conditions. The passive continental margin is located within the plate that is transiting from continental crust to oceanic crust. It passively follows the plate movement and is characterized by tensile tectonics with a

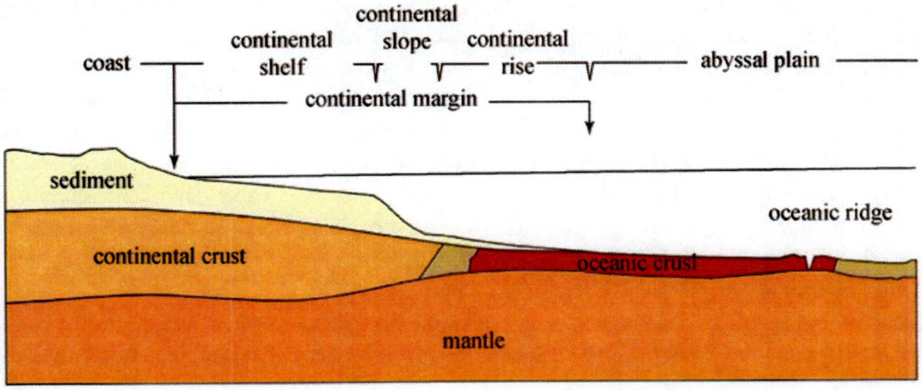

Fig. 11.2 Schematic diagram of the continental margin

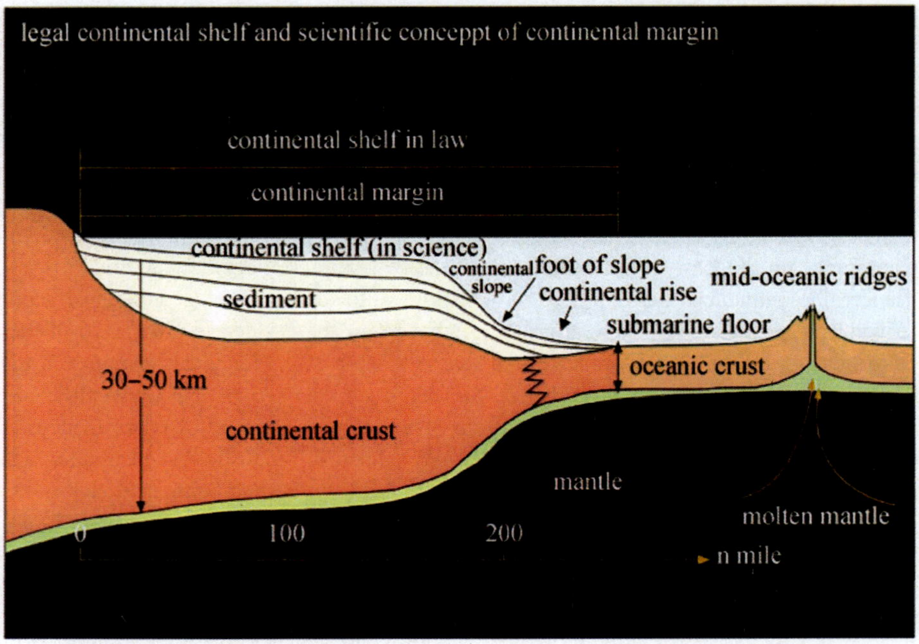

Fig. 11.3 Legal continental shelf and scientific continental margin

well-developed continental shelf, continental slope and continental rise. The active continental margin is found along a convergent lithospheric plate boundary. Its characteristics are related to plate subduction. The typical active continental margin geomorphology is composed of a continental shelf and continental slope, generally without a continental rise. The seaward margins are oceanic trenches or oceanic troughs. Because of the difference between the divergence (low tectonic activity) and convergence (high tectonic activity) conditions most regions with a continental shelf extending more than 200 n mile are related to divergent passive continental margins, except for few regions in the Pacific rim where the expansion of the continental shelf is related to passive continental margins.

The continental shelf is a shallow-water zone around the continent. It extends very gently from the shallow waterline along the shore to the open seas until it reaches the continental slope break where the gradient of the slope suddenly increases. The shallow sea area between the coast and the sharp slope is defined as the continental shelf. The continental shelf has a gentle slope, with an average gradient of about 0.1°. The continental slope is an underwater geomorphological feature on a global scale which separates continents from oceans. Its upper border is the outer edge of the continental shelf (the continental shelf break) and its lower border varies greatly in water depth. The most distinctive feature of the continental slope is its sharp gradient, with an average of 4.3° and a maximum slope of 45° (the offshore slope of Sri Lanka). The surfaces of most continental slopes have topographical formations such as basins, ridges, submarine canyons and abyssal plains. The most common ones are submarine canyons.

The continental rise is a gently sloping submarine protrusion formed by the accumulation of sediment at the foot of the continental slope. It generally extends between water depths of 2,000 to 5,000 m. Its upper half leans against the foot of the continental slope while the lower half covers the oceanic crust. The continental rise appears only at Atlantic-type continental margins, usually appears in the form of abyssal fans. It has a breadth between several hundred to more than 1,000 km, a gentle slope gradient of up to 1° and a thick sediment layer (Feng et al. 1999).

However, the concept of the continental shelf defined in Article 76 of the Convention is a legal concept that extends further than the geological concept. The legal concept of the continental shelf is similar to the geological concept of the continental margin, and includes the continental shelf, continental slope and continental rise as defined in scientific terms. Their relationship is shown in Fig. 11.3.

11.1.3 Relationship Between the Exclusive Economic Zone and Continental Shelf

There are some differences, correlations and similarities between the continental shelf in Part VI and the exclusive economic zone (EEZ) in Part V of the Convention.

A coastal state has inherent rights over the continental shelf of its coast, while the rights of that state over the EEZ of its coast are attained only through proclamation and other legal acts. Paragraph 3 of Article 77 shows that "the rights of a coastal state over the continental shelf do not depend on occupation, effective or notional, or on any express proclamation." However, there is no such provision in Part V. The rights of a coastal state over the continental shelf include sovereign rights and jurisdiction. The coastal state exercises sovereignty rights over the continental shelf for the purpose of exploring the continental shelf and developing its natural resources. The rights of a coastal state over the continental shelf are inherent and do not depend on any occupancy, effective or notional, or on any proclamation, unlike the rights over the EEZ. Without the express consent of the coastal state, no-one can explore the continental shelf along the state's coastline or exploit the natural resources within the continental shelf area. Therefore, rights are exclusive to the coastal state.

According to Paragraph 1 of Article 76, when the natural prolongation of the continental shelf of a coastal state is less than 200 n mile, that is, the outer edge of the continental margin is less than 200 n mile, the coastal state may extend its continental shelf to a distance of 200 n mile. In this case, the outer limit of the continental shelf coincides with the outer limit of the EEZ.

When the natural prolongation of the continental shelf of a coastal state exceeds 200 n mile, the outer limits of the shelf and the EEZ are distinctively different. The coastal state can enjoy an EEZ of 200 n mile at most, but the outer limit of the continental shelf can be beyond the extent of the EEZ of 200 n mile and is determined according to the provisions specified in Paragraphs 4–8 of Article 76. Therefore, in this case, "there may be a continental shelf without an exclusive economic zone, but there will not be an EEZ without a corresponding continental shelf."

All coastal states may claim the rights over an EEZ of 200 n mile, as well as the rights over a maximum continental shelf of more than 200 n mile. In case of an overlap of the claims, the outer limit of the continental shelf and the EEZ will be determined by partitioning in accordance with the provisions specified in Articles 74 and 83. The EEZ and continental shelf are two different regimes; the EEZ regime was established after the Third UN Conference on the Law of the Sea, and many states had already declared the limits of the continental shelf along their coasts. Under certain circumstances, the outer limit of the continental shelf and that of the EEZ are not completely identical, as in the following cases. (1) Only the outer limit of the continental shelf has been set, without EEZ boundaries. (2) Only the outer limit of the continental shelf has been set, but it can also be used as the limit of the EEZ if the two adjacent states laying the claim proclaimed the EEZ according to the provision. Thus, one line is used to define the outer limits of both the continental shelf and the EEZ. (3) The outer limits of the continental shelf and the EEZ are different, following separate guidelines.

Coastal states exercise sovereign rights over the EEZ and continental shelf along their coasts. However, the objects of the two zones are different. The sovereign rights over the EEZ include not only the waters above the seabed, but also the seabed and subsoil in the zone. The sovereign rights over the continental shelf are limited only to the resources in the seabed and subsoil. However, according to the provisions specified in Paragraph 3 of Article 56 of the Convention (Article 56: rights, jurisdiction and duties of the coastal State in the exclusive economic zone), "the rights set out in this article with respect to the seabed and subsoil shall be exercised in accordance with Part VI.", when the outer limit of 200 n mile applies to both the continental shelf and the EEZ, as far as the sovereign rights exercised by the coastal states are concerned, the EEZ actually aims at the living resources, seawater, ocean current and wind energy production. The rights of a coastal state over local living organisms, mineral products and other non-living resources in the seabed and subsoil of the EEZ will be applied according to the regime of continental shelf. However, where the continental shelf extends beyond the EEZ, the part beyond the EEZ boundary will apply only to the continental shelf regime, including the provisions specified in Article 82 concerning the fees and in-kind payment for the development of the continental shelf beyond 200 n mile, while the water areas overlying the continental shelf beyond 200 n mile are considered high seas.

11.2 Continental Shelf Delineation System

11.2.1 Article 76 of the Convention

1. The continental shelf of a coastal State comprises the seabed and subsoil of the submarine areas that extend beyond its territorial sea throughout the natural prolongation of its land territory to the outer edge of the continental margin, or to a distance of 200 n mile from

the baselines from which the breadth of the territorial sea is measured where the outer edge of the continental margin does not extend up to that distance.

2. The continental shelf of a coastal State shall not extend beyond the limits provided for in Paragraphs 4 to 6.
3. The continental margin comprises the submerged prolongation of the landmass of the coastal state, and consists of the seabed and subsoil of the shelf, the slope and the rise. It does not include the deep ocean floor with its oceanic ridges or the subsoil thereof.
4. (a) For the purposes of this Convention, the coastal State shall establish the outer edge of the continental margin wherever the margin extends beyond 200 n mile from the baselines from which the breadth of the territorial sea is measured, by either:
   (i) a line delineated in accordance with Paragraph 7 by reference to the outermost fixed points at each of which the thickness of sediments is at least 1% of the shortest distance from such point to the foot of the continental slope; or
   (ii) a line delineated in accordance with Paragraph 7 by reference to fixed points not more than 60 n mile from the foot of the continental slope.
   (b) In the absence of evidence to the contrary, the foot of the continental slope shall be determined as the point of maximum change in the gradient at its base.
5. The fixed points comprising the line of the outer limits of the continental shelf on the seabed, drawn in accordance with Paragraph 4(a)(i) and (ii), either shall not exceed 350 n mile from the baselines from which the breadth of the territorial sea is measured or shall not exceed 100 n mile from the 2,500-m isobath, which is a line connecting the depth of 2,500 m.
6. Notwithstanding the Provisions of Paragraph 5, on submarine ridges, the outer limit of the continental shelf shall not exceed 350 n mile from the baselines from which the breadth of the territorial sea is measured. This paragraph does not apply to submarine elevations that are natural components of the continental margin, such as its plateaux, rises, caps, banks and spurs.
7. The coastal state shall delineate the outer limits of its continental shelf, where that shelf extends beyond 200 n mile from the baselines from which the breadth of the territorial sea is measured, by straight lines not exceeding 60 n mile in length, connecting fixed points, defined by coordinates of latitude and longitude.
8. Information on the limits of the continental shelf beyond 200 n mile from the baselines from which the breadth of the territorial sea is measured shall be submitted by the coastal state to the Commission on the Limits of the Continental Shelf set up under Annex II on the basis of equitable geographical representation. The commission shall make recommendations to coastal states on matters related to the establishment of the outer limits of their continental shelf. The limits of the shelf established by a coastal state on the basis of these recommendations shall be final and binding.
9. The coastal state shall deposit with the Secretary-General of the UN charts and relevant information, including geodetic data, permanently describing the outer limits of its continental shelf. The Secretary-General shall give due publicity thereto.
10. The provisions of this article are without prejudice to the question of delimitation of the continental shelf between states with opposite or adjacent coasts.

11.2.2 Procedures to Be Followed by a Coastal State When Determining the Outer Limit of the Continental Shelf

The breath of continental shelf is the distance of 200 n mile from the baselines from which the breadth of the territorial sea is measured where the outer edge of the continental margin does not extend to that distance. The coastal state shall establish the outer edge of the continental margin wherever the margin extends beyond 200 n mile from the baselines from which the breadth of the territorial sea is measured, by either: (i) a line by reference to the outermost fixed points at each of which the thickness of sediments is at least 1% of the shortest distance from such point to the foot of the continental slope; or (ii) a line by reference to fixed points not more than 60 n mile from the foot of the continental slope. Meanwhile, the fixed points comprising the line of the outer limits of the continental shelf on the seabed, shall not exceed 350 n mile from the baselines from which the breadth of the territorial sea is measured or shall not exceed 100 n mile from the 2,500 m isobath, which is a line connecting the depth of 2,500 m.

According to the above provisions, the important parameters for determining the outer limits of the continental shelf beyond 200 n mile are the 200 n mile line from the baseline of the territorial sea, the 60 n mile line from the foot of the continental slope, the 100 n mile line from the 2,500 m isobath and the 1% sediment thickness line and 350 n mile line from the baseline of the territorial sea.

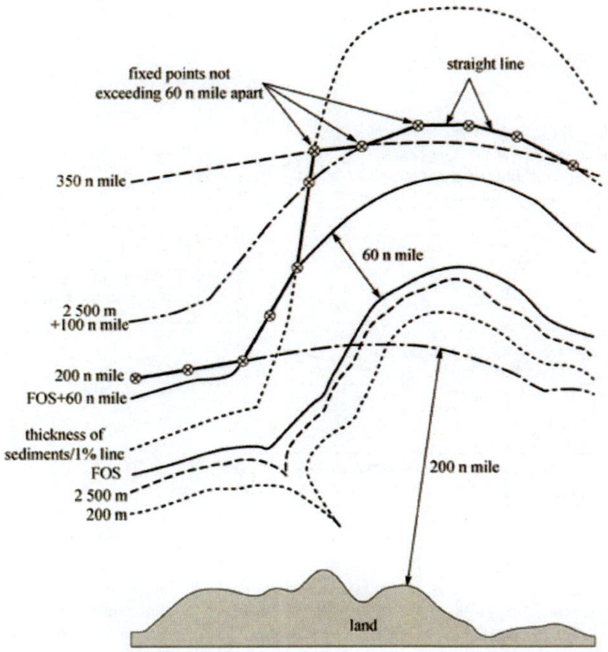

Fig. 11.4 Schematic diagram of the outer limit of the continental shelf established by standards in Article 76

According to the provision of the Convention, the process for determining the outer limit of the continental shelf is as shown in Fig. 11.4.

(1) If a coastal state agrees to set the 200 n mile line from the baseline of the territorial sea as the outer limit of the continental shelf, then the 200 n mile line is determined.
(2) If a coastal state does not agree to set the 200 n mile line from the baseline of the territorial sea as the outer limit of the continental shelf, then the line defining the foot of the continental slope should be determined.
(3) Delineate the 1% sediment thickness line and 60 n mile line from the FOS.
(4) Determine if any of the fixed points along the outer limit line appear on submarine ridges. If any such points are found, only the 350 n mile line from the baseline of the territorial sea shall be delineated; if not, the 350 n mile line from the baseline of the territorial sea, and the 100 n mile line from the 2,500 m isobath should be defined.
(5) The outer limits of the continental shelf will be delineated by straight lines not exceeding 60 n mile in length, connecting the fixed points defined by coordinates of latitude and longitude.

Thus, the key techniques to delineate the outer limits of the continental shelf are: determining the FOS and the 1% sediment thickness line, and setting the continental shelf boundary according to the appropriate distance from the territorial sea baseline.

The entire process for a coastal state to establish the outer limits of its continental shelf under Article 76 of the Convention is presented in the master flowchart in Figs. 11.5 and 11.28 and in the five sub-flowcharts (Figs. 11.29, 11.30, 11.31, 11.32). The sub-flowcharts outline in greater detail the following five procedures are shown in the master flowchart.

I. Establish the 200 n mile limit from the baseline.
II. Determine the location of the foot of the continental slope:
   (a) find the maximum change in the gradient at the foot of the continental slope (Article 76 (4));
   (b) provide evidence that contradicts the general rule (Article 76 (4) (b)).
III. Apply the formulae:
   (a) sediment thickness (Article 76 (4) (a) (i));
   (b) 60 n mile from the foot of the continental slope (Article 76 (4) (a) (ii)).
IV. Determine the outer limit of the continental shelf in the case of submarine elevations and/or ridges (Article 76 (3) and (6)).
V. Application of the constraints:
   (a) the 350 n mile line from the territorial sea baseline (Article 76 (5), first provision);
   (b) the 2,500 m isobath plus 100 n mile boundary (Article 76 (5), second provision).

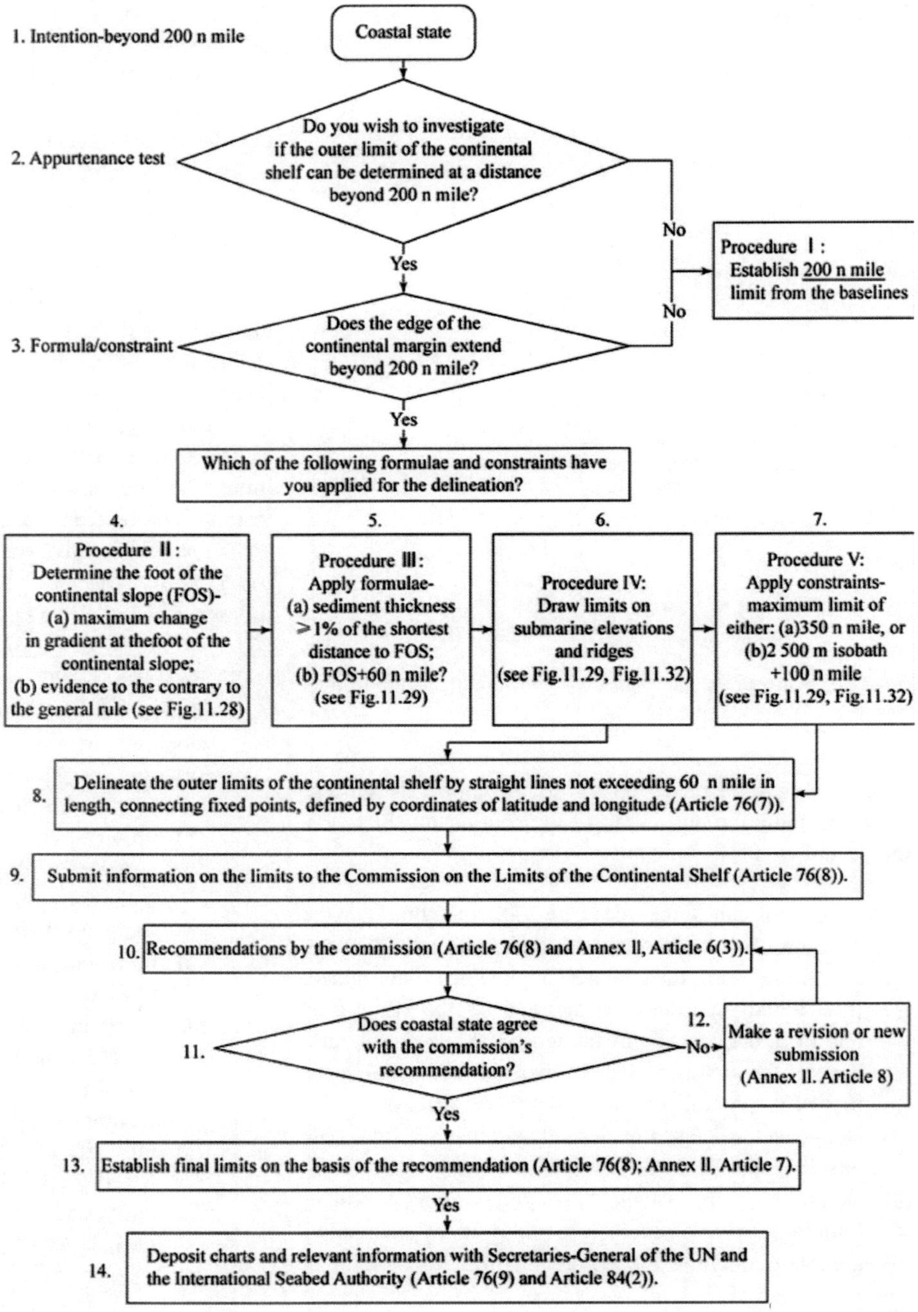

Fig. 11.5 Master flowchart for the establishment of the outer limits of the continental shelf

11.3 Techniques and Methods for Determining the Outer Limit of the Continental Shelf

As can be seen from the submission requirements of the Convention, a complete submission should contain various supporting material and data, including a topographic profile, FOS points, FOS+60 n mile line, 2,500 m isobath, 2,500 m+100 n mile constraint line, 1% sediment thickness line from the FOS line, continent-ocean boundary (COB) and evidence to the contrary, etc. To prepare this evidence and material, survey and research methods of multiple disciplines can be used. These include geodetic surveying, navigation and positioning techniques, marine surveying and mapping, oceanographic charting, seismic reflection and refraction methods, gravity and geomagnetism surveys, geological sampling and marine information processing and management. Among these techniques, the multi-beam detection technology is an extremely important supporting technique. Based on multiple-beam bathymetry data, the following information may be provided: topographic profile, FOS line, FOS+60 n mile line, 2,500 m isobath and 2,500 m+100 n mile constraint line.

11.3.1 Methods and Technical Processes for Determining the Outer Limit of the Continental Shelf

Determining the boundary of the continental shelf beyond 200 n mile is one of the most highly debated issues in marine science at present. Currently, the focus is on the Arctic region and the West Pacific. It is estimated that in the next decade, about 75,000,000 km^2 of sea area will be claimed by coastal states around the world; this amounts to more than half of the land area on the planet; 30,000,000 km^2 of this sea area lies beyond the 200 n mile limit (Lyu et al. 2012).

Article 76 of the Convention on the Law of the Sea states that the continental shelf of a coastal state is composed of the seabed and subsoil of the submarine areas that extend beyond its territorial sea throughout the natural prolongation of its land territory to the outer edge of the continental margin, or to 200 n mile from the baseline from which the breadth of the territorial sea is measured where the outer edge of the continental margin does not extend to that distance. If a coastal state claims a distance of more than 200 n mile from the baseline from which the breadth of the territorial sea is measured, it must delineate the outer limits of the continental shelf beyond 200 n mile according to the requirements of Article 76 and Article 4 of Annex II of the Convention and present a Submission to the Commission on the Limits of the Continental Shelf (CLCS), which shall also be the basis for the coastal state to claim the rights and interests of the continental shelf beyond 200 n mile.

On December 20, 2001, Russia presented the first Submission for the delineation of the outer limits of the continental shelf beyond 200 n mile to CLCS through the Secretary- General of the UN in accordance with the Convention. As of May 2019, the UN website published 84 executive summaries of Submissions, as well as preliminary information of 47 submissions. On May 11, 2009, the permanent delegation of China to the UN submitted Preliminary Information of the People's Republic of China on the Determination of the Outer Limits of the Continental Shelf beyond 200 n mile (Preliminary Information for short) to the UN Secretariat. On December 14, 2012, it presented the Submission of the People's Republic of China on Delimitation of Outer Limits of the Continental Shelf beyond 200 n mile in Part of the East China Sea (Partial Submission for short) to the UN. Republic of Korea and Japan also presented separate submissions for the delimitation of the continental shelf beyond 200 n mile.

The preparation of a complete Submission is a systematic project. It requires topographical, geomorphological, geological and geophysical evidence and also needs to provide relevant scientific evidence and texts, including delimitation texts, supporting data, executive summary and relevant drawings and maps according to Article 76 of the Convention. Additionally, it should comply with the technical requirements of the CLCS for the delimitation of the continental shelf beyond 200 n mile. Of the submitted material, only the executive summaries will be published on the UN website. The key evidence presented to the CLCS for review comprises a series of delineated limits, including the FOS points, formula limit (FOS+60 n mile line and 1% sediment thickness line), constraint line (350 n mile line and 2,500 m+100 n mile line) and outer limits. Thus, maritime delimitation requires interdisciplinary collaboration. It requires the support of marine survey and research data as well as information technology to conduct intensive processing of acquired data and determine the various delineation points.

The preparation of a Submission can be divided into five basic steps (Fig. 11.6). (1) The coastal state shall first demonstrate, in accordance with the Convention, whether the continental shelf of the state extends beyond 200 n mile. States with narrow continental shelves whose breadth is less than 200 n mile cannot apply for a continental shelf limit beyond 200 n mile. (2) The state will determine the foot of the continental slope (FOS) line. (3) The state will establish two

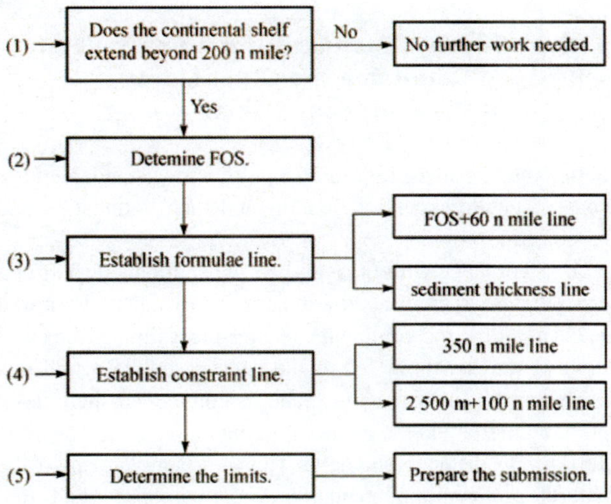

Fig. 11.6 Flow chart showing the basic technical stages for delimitation of the outer continental shelf

formula lines: the FOS+60 n mile line and the 1% sediment thickness line. (4) The state will establish two constraint lines: the 350 n mile line and the 2,500 m+100 n mile line. (5) The state will determine the outer limits of the continental shelf based on the comprehensive study of formulae and constraint lines, and prepare the submission on this basis.

11.3.2 Basis of Technical Methods for Delimitation

According to Article 76, two formulae and two constraint lines may be used to determine the outer limits of the continental shelf based on geodetic surveying, geology, geophysics and mapping.

The formulae are

(1) a line delineated in accordance with Paragraph 7 by reference to the outermost fixed points where the sediment thickness is at least 1% of the shortest distance from such point to the foot of the continental slope;
(2) a line delineated in accordance with Paragraph 7 by reference to fixed points not more than 60 n mile from the foot of the continental slope.

Constraint lines:

(1) the fixed points comprising the line of the outer limits of the continental shelf on the seabed, drawn in accordance with Paragraphs 4 (a) (i) and (ii), should be located either not more than 350 n mile from the baselines from which the breadth of the territorial sea is measured, or;
(2) not more than 100 n mile from the 2,500 m isobath.

11.3.3 The Two Standards for a Coastal State to Determine the Outer Limits of the Continental Shelf

Paragraphs 1 and 2 of Article 76 provide two standards for a coastal state to expand the continental shelf boundary along its coastline.

In accordance with Paragraph 1, the continental shelf of a coastal state is the seabed and subsoil beyond its territorial sea. Therefore, the continental shelf's inner limit is the outer limit of the territorial sea of the coastal state. Beyond this limit, the coastal state can expand its continental shelf based on two criteria: (1) the natural prolongation standard; (2) the 200 n mile distance standard. This is a supplement to the natural prolongation standard. A coastal state shall first determine the extent of its continental shelf according to the natural prolongation standard. If the continental margin exceeds 200 n mile, the outer limit of the continental shelf shall be delineated beyond 200 n mile. Only when part of the natural prolongation is less than 200 n mile measured seaward from the baseline of the territorial sea of the coastal state can the costal state apply the distance standard and extend it to 200 n mile. Therefore, when the 200 n mile distance standard applies, the outer limit of the continental shelf of the coastal state will be a curve parallel to the baseline of the territorial sea with the shortest distance from every point on the curve to the baseline of the territorial sea equal to 200 n mile.

When coastal states are allowed to delineate the outer limits of the continental shelf along their coast based on the natural prolongation of their land territory, some coastal states whose continental shelf is extensive with a gentle slope may expand their continental shelf boundaries beyond 200 n mile. Therefore, Paragraph 2 of Article 76 provides constraints for the excessive extension of the continental shelf of a coastal state. This Paragraph is a supplement to the natural prolongation standard specified in Paragraph 1 and is an additional constraint.

11.3.4 Two Formulae Used to Delineate the Outer Limits of the Continental Shelf of a Coastal State Beyond 200 n Mile

According to Paragraph 4 of Article 76, the outer limits of the continental shelf beyond 200 n mile shall not extend beyond two lines: (1) the sediment thickness line joining the points where the thickness of sediments is at least 1% of the shortest distance from such point to the foot of the continental slope; (2) the line joining the points located not more than 60 n mile from the FOS. The area between these two lines is the zone where the continental shelf of the coastal state beyond 200 n mile may be expanded. The two lines are the formula lines for the coastal state to delineate the outer limits of the continental shelf beyond 200 n mile. However, practically, only after delineation of both formula lines, can we know whether the continental margin of the coastal state extends beyond 200 n mile or not. If both formula lines are within the 200 n mile line, then the 200 n mile line shall be the outer limit of the continental shelf of the coastal state. In this case, it is not necessary to delineate the outer limit of the continental shelf beyond 200 n mile. Only when one or both formula lines are beyond 200 n mile is it necessary to delineate the outer limits of the continental shelf according to Paragraphs 4–6. Therefore, the provisions of Paragraph 4 are the basis and key to the application of Article 76, without which, neither the natural prolongation standard nor the 200 n mile distance standard is applicable.

According to Paragraph 7 of Article 76, the outer limits of the continental shelf of a coastal state is a curve comprising of a series of straight lines connecting the outermost fixed points; therefore, the key to delineate the outer limits of the continental shelf beyond 200 n mile is the selection of the fixed points. In Scientific and Technical Guidelines of the Commission on the Limits of the Continental Shelf, the commission states that the outer limits of the continental shelf of a coastal state can either be expanded to the sediment thickness formula line (that is the appropriate fixed points on the thickness formula line are selected to delineate the outer limit of the continental shelf) or to the distance formula line (that is the appropriate fixed points on the distance formula line are selected to delineate the outer limits of the continental shelf), or be expanded to both formula lines (the fixed points favorable for the coastal state on both lines are selected to delineate the outer limits of the continental shelf).

In addition, the two formula lines provided in Paragraph 4 are used to delineate the outer limits of the continental margin of a coastal state. Therefore, they are established on the basis of the natural prolongation standard provided in Paragraphs 1–3 of Article 76 and are subject to the constraint of the natural prolongation standard. This means that the two formula lines must be within the extent of the geological continental margin of the coastal state, that is, within the scope of the continental margin defined in Article 3, or the area between the two curves must lie within the extent of the geological continental margin of the coastal state. When a coastal state delineates the sediment thickness line, there is no problem; however, when a coastal state draws the distance formula line, it is not always possible to determine the points on the line by reference to the standard of outward prolongation of 60 n mile from the FOS. When the outer edge of the geological continental margin of a coastal state is less than 60 n mile from the FOS, the formula line can only be delineated based on the actual distance. Only when the outer edge of the geological continental margin is at least 60 n mile from the FOS can we delineate the formula line based on the standard of 60 n mile.

11.3.5 Two Constraint Lines for Delineating the Outer Limit of the Continental Shelf Beyond 200 n Mile

Under Paragraph 4 of Article 76, some coastal states with a broad and flat continental margin can still extend their continental shelves a long way offshore, thus characterizing the international seabed area as "the common property of human beings". Therefore, Paragraph 5 of Article 76 provides two constraints lines, one is the distance constraint line, 350 n mile from the baselines from which the breadth of the territorial sea is measured; the outer limit of the continental shelf shall not exceed the distance constraint line. The other line is the depth constraint line, on which all points are on the seaward side of the 2,500 m isobath and no more than 100 n mile from the 2,500 m isobath. The two constraint lines can restrict coastal states from excessively prolonging the continental shelf along their coast. The two constraint lines are totally different from the two formula lines specified in Paragraph 4. The area between the two formula lines is the zone where the coastal state can expand its continental shelf while the area between the two constraint lines is not designated with such authorization. If the two formula lines are within the two constraint lines, then the area within the two formula lines is the extent where the coastal state can extend its continental shelf; if all or part of the two formula lines are outside the two constraint lines, then the coastal state is not allowed to expand the continental shelf beyond the two constraint lines.

In the Scientific and Technical Guidelines of the Commission on the Limits of the Continental Shelf, the commission considers that when determining each fixed point for delineating the outer limits of the continental shelf according to Paragraph 4 of Article 76 under the restrictions in Paragraph 5, the coastal state may combine two constraint lines, that is, either all the fixed points are subject to the 350 n mile standard line or the 2,500 m isobath+100 n mile standard line, or some fixed points are subject to the 350 n mile standard line and the other fixed points are subject to the 2,500 m isobath+100 n mile constraint line. Thus, the outer limits of the continental shelf of a coastal state may be extended beyond the depth constraint, or beyond the distance constraint, as long as it is not extended beyond the two constraint lines at the same time.

11.3.6 Establishing the Outer Limit of the Continental Shelf of a Coastal State Beyond 200 n Mile

According to the provisions of Paragraph 7 of Article 76, the outer limit of the continental shelf of a coastal state beyond 200 n mile comprises of several straight lines not longer than 60 n mile that connect the outermost fixed points. Therefore, to establish the outer limit of the continental shelf beyond 200 n mile, it is necessary first to determine the outermost fixed points. Combining the provisions of Paragraphs 4–7 of Article 76, the fixed points can be determined by the following steps: first, draw the two formula lines according to Paragraph 4 (a) (1) and (2) of Article 76 and select the outermost line if the two lines do not intersect; or select the curve composed of the outermost segments of the both lines if the two lines intersect. Second, draw the two constraint lines according to Paragraph 5 of Article 76 and select the outermost constraint if the two lines do not intersect; or select the curve composed of the outermost segments of both constraint lines if they do not intersect. In case of a submarine ridge, it is necessary to delineate only the 350 n mile line from the baseline of the territorial sea.

Next, compare the curves selected above; the seabed and subsoil between the two lines above are the extents to which the coastal state are allowed to expand the continental shelf boundary. All the points on the outer limits of the seabed and subsoil between the two curves mentioned above that meet the requirements for the outermost fixed points specified in Paragraphs 4–6 can be used by the coastal state as the outermost fixed points for establishing the outer limit of the continental shelf beyond 200 n mile. Finally, select appropriate points among the above-mentioned points as the fixed points so that the distance between two fixed points does not exceed 60 n mile and is the most favorable distance chosen by the coastal state based on available data. Once selected, connect the fixed points by straight lines to establish the outer limit of the continental shelf beyond 200 n mile.

The above-mentioned steps for establishing the outer limits of the continental shelf of the coastal state are only applicable to the areas within the same continental margin of the coastal state. If there are more than two continental margins within the jurisdiction of a coastal state, the above steps shall be applicable to each continental margin to establish its outer limit.

The FOS and the maximum water depth point are determined through the detailed topography profile, the 1% sediment thickness point is determined by the sedimentary rock profile and the FOS+60 n mile line; the 350 n mile line and 2,500 m+100 n mile line derived by extrapolating the fixed points are used to determine the final outer limit through analysis of the various limits. The FOS+60 n mile line, 350 n mile line and 2,500 m+100 n mile line are collectively referred to the extrapolation limits. Please see Fig. 11.10 for the methods for determining the FOS.

11.3.7 Foot of Slope Identification Method

High-precision bathymetric data and the Depth Digital Model (DDM) are the basis for determining the FOS. Extensive research was carried out on bathymetric data processing and DDM construction (Yang et al. 2008, 2009; Huang et al. 2001a, b; Wang et al. 2011; Zhang et al. 2011, 2012), but there are only a few technical studies on maritime demarcation (Peter and Carleton 2000; Liu et al. 2007; Peng and Wang 2002; Ou and Vaníček 1996). Ou and Vaníček (1996) proposed an automatic identification method of the FOS by converting water depth into a maximum curvature surface (MCS). However, the MCS method applies only to simple continental margins, because the same MCS may correspond to different submarine topographies, which can cause inaccurate results. CARIS LOTS is a commercial software used for demarcation. However, because it filters and smoothes the original bathymetric data, it may change the precise position of the FOS. Furthermore, the software does not allow for parameter adjustments; therefore, it is not suitable for conducting in-depth research.

A FOS automatic identification method is proposed here based on a comprehensive analysis of the topography, slope, second derivative and the Douglas-Peucker (D-P) profile. Through experimental and numerical processing methods, multiple direct FOS identification methods are obtained. The detailed technical process is given, and the automatic FOS identification is finally achieved.

11.3.7.1 Basic Definition of FOS

Correct identification of the FOS requires a good understanding of the nature of the continental shelf. The earliest claim of the continental shelf was made in the Truman Proclamation on September 28, 1945. At that time, the continental shelf was defined as the submerged land along the continents with a water depth not exceeding 100 fathoms (about 183 m). The 1958 Convention on the Continental Shelf defined the continental shelf as "seabed and subsoil of the seabed areas at the territorial sea, bordering the coast with depth of not exceeding 200 m, or more than 200 m, extending to such depth as would permit natural development activities." According to the provision of Article 76 of the Convention on the Law of the Sea that came into force in 1982: if a coastal state claims more than 200 n mile from the baseline on which the breadth of the territorial sea is measured, it shall establish the outer limit of the continental shelf beyond 200 n mile according to the relevant requirements. This is also the current legal basis for all coastal states in the world to apply for the continental shelf beyond 200 n mile.

Continental margin comprises continental shelf, continental slope and continental rise (Fig. 11.7a), where the concept of the continental shelf here differs from, but is only a part of the one defined in UNCLOS. The continental margin can be divided into three types (Peter and Carleton 2000): (1) the Atlantic type; (2) the Pacific type; and (3) the transformation type. From the research on the changes from continental crust to oceanic crust, and natural extension boundary markers of coastal states, Hedberg (1976) suggested that "the outer edge of the continental margin should be best defined as the external boundary of the topographical continent, which is usually accurately explained as the bottom of the continental slope".

The determination of the outer limits of the continental shelf beyond 200 n mile, in Article 76 of UNCLOS, mainly refers to the Atlantic-type continental margin, which shows clear submarine morphologic feature. A complete Atlantic-type continental margin comprises continental shelf, continental slope, continental rise, and oceanic basin, where the FOS point is located between the lower part of the continental slope and the continental rise (Fig. 11.7a).

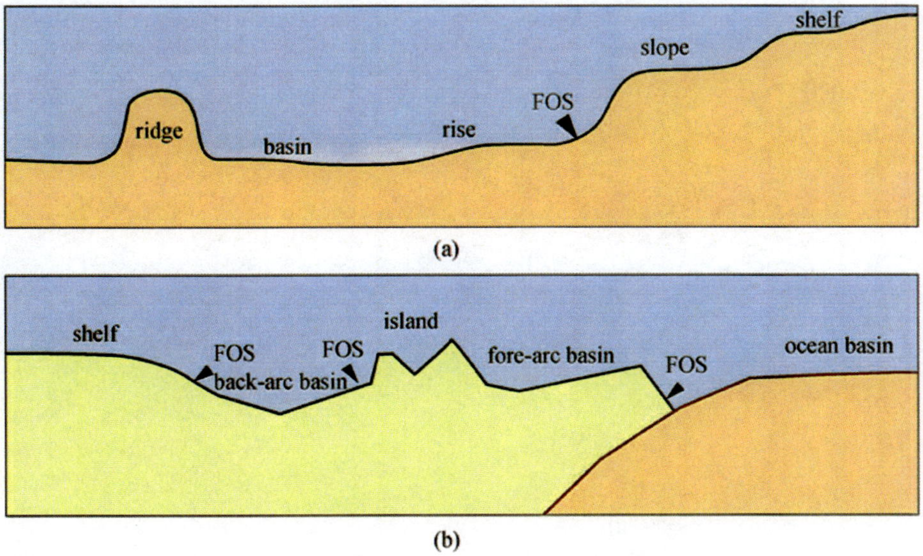

Fig. 11.7 Continental margin model and the theoretical location of the foot of the continental slope. **a** Atlantic type continental margin; **b** Pacific type continental margin; FOS: the foot of the continental slope

Comparatively, a Pacific type continental margin is extremely complicated owing to the effects of plate convergence and subduction compression, and consists of multiple geomorphological units, including continental shelf, back-arc basin, island arc, fore-arc basin, oceanic trench, and abyssal plain from the continent to the ocean. Thus, the FOS point can be found on both sides of the back-arc basin and along the oceanic trench (Fig. 11.7b).

11.3.7.2 Basic Principles of FOS Determination

In the absence of evidence to the contrary, the foot of the continental slope is the point of maximum change in the gradient at its base (Peter and Carleton 2000). It takes two major steps to determine the foot of the continental slope: (i) determine the area defined as the bottom of the continental slope, that is, the base of the slope; and (ii) determine the position of the point of maximum change in the gradient at its base, corresponding to the extremum point of its second derivative.

As submarine topography is complicated on a continental slope, the base region of the continental slope must be determined before determining the FOS points; i.e., the base region of the continental slope, is the area that the FOS point is likely to locate. The FOS point is located along the continent-ocean boundary (COB), where the continental crust transits into the oceanic crust. The determination of the continental slope base region requires multiple evidences, among which submarine topography and landform are the most important. Thus, the region that exhibits the typical topographic feature of "continental shelf-continental slope-oceanic basin" is supposed to be the appropriate region to determine the FOS points.

The submarine topography is often the most intuitive means to determine the base of the continental slope (BOS) (Hedberg 1976). The base region of the slope can be determined by analyzing the water depth, topography, slope and second derivative. The large clinoform region shown in Fig. 11.8a represents a typical continental slope: an oceanic basin topographic transition zone according to the isobaths, where the water depth increases gradually from the northwest to the southeast, from 1,500 to 3,900 m. Based on gradient analysis (Fig. 11.8b), this region is quite rugged, exhibiting a significant gradient change in the submarine topography. Analysis of the second derivative diagram (Fig. 11.8c) shows that the overall characteristics are similar to those of the gradient grid, but with a gentler trend. Moreover, the position of the peak identified by the second derivative grid is different from that of the gradient grid; the latter indicates the maximum change in the topography while the former corresponds to the region with maximum gradient change; this is where the FOS is located. After stacking different layers and analyzing the water depth, topography, gradient, second derivative and other relevant information, the continental slope base region is determined (Fig. 11.8d).

11.3.7.3 Automatic Identification of the FOS

(1) Overall technical ideas

The method used to determine the FOS needs to be accurate, quantitative and verifiable. The present method builds a series of topographic profiles vertical to the strike direction of the continental slope, and then determines the FOS points according to the changes in the submarine topography profile. However, it is difficult to make determination solely on the basis of the topographic profile; hence, other sources of information are needed to perform an integrated analysis of the location of the FOS points.

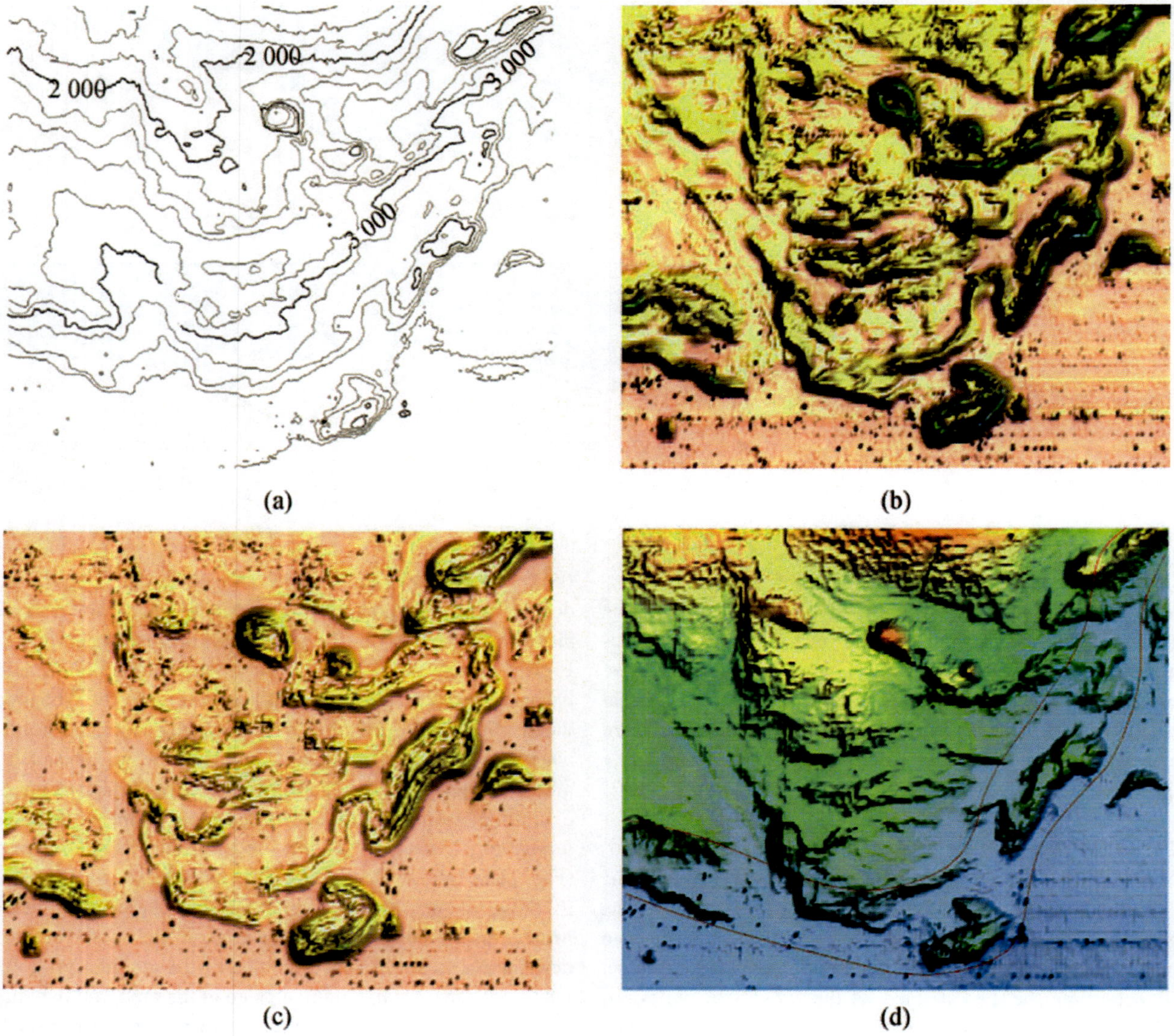

Fig. 11.8 Determination of the base of the continental slope. **a** Topographical map (unit: m); **b** gradient grid; **c** second derivative diagram; and **d** The area between the red curves is the base of the continental slope

The FOS, defined as the topographic point with maximum change in the gradient, corresponds to the extremum point of the second derivative, rather than the zero-value point of the gradient profile; thus, the FOS is often not the extremum point of the topographic profile. Moreover, the comparison result indicates that the extremum point of the second derivative is always near the extremum point of the topographic profile, but normally they do not overlay. The topographic profile where the FOS point located presents convex features (longitudinal axis is displayed in the direction of increasing water depth). Therefore, the point with an extreme and positive second derivative value is a potential FOS point. Owing to the influence of the small-scale topography, a topographical profile might have several second derivative extremum points, and thus, the original topographic profile must be simplified.

A typical topographic profile of the continental marginal comprises three sections: a flat and shallow-water-depth continental shelf, a steep and sharp-change-water-depth continental slope, and a flat and deep-water-depth oceanic basin. The FOS is located at the turning point from the continental slope to the oceanic basin, where the water depth is relatively deep, while the gradient is relatively greater towards the continental slope and smaller toward the oceanic basin, respectively. For an original topographic profile, as influenced by small-scale local topography, there could be several points consistent with the conditions mentioned above. Therefore, the original profile should be simplified to eliminate the interference of local topography.

Filtering can smooth the original topographic profile and make it easier to identify the FOS. However, it has a potential

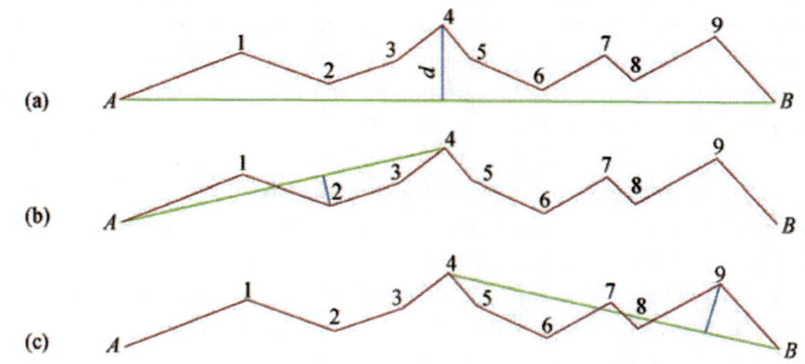

Fig. 11.9 Schematic of the D-P algorithm process. **a**, **b** and **c** show the process of the algorithm to filter redundant points

defect that the basic characteristics of the original topographic profile might be changed. In contrast, by using extremum points and the quadratic fitting of the D-P algorithm, not only the original topographic profile can be simplified, but also the features of the original profile can be retained. The D-P algorithm has the significant advantage of retaining the most basic features of a curve, while it is difficult to determine directly the location of the FOS simply by fitting the original topographic profile. The point selected by the D-P algorithm might not be the extremum point of the second derivative of the curve, i.e., the D-P algorithm might remove the extremum point of the second derivative during the filtering. Therefore, prior to the D-P algorithm fitting, we should fit the original profile based on the extremum point of the second derivative, and then apply the D-P algorithm to perform the quadratic fitting based on the extremum point profile. Thus, it can be ensured that each point obtained is an extremum point of the second derivative, which can avoid a false FOS point.

By calculating the second derivative of the acquired D-P topographic profile, a new gradient profile and a second derivative profile can be obtained. The profile following two processes of simplification retains only the most basic characteristics of the curve, without interference from small-scale topography. Hence, we can first analyze the features of each point in the D-P profile, including the water depth, the gradient, the second derivative, the concave-convex characteristic, and the correlation (continuity) between the point and its neighboring points on the upper slope, lower slope and the water depth, and then judge whether the topographic change correspondent to the typical characteristics (segmentation) of the turning point from the continental slope to the oceanic basin. With these parameters, we can determine accurately the location of the FOS points in the curve.

(2) Topographic profile simplification algorithm

The D-P algorithm proposed by David Douglas and Thomas Peucker (Douglas and Peucker 1973) is an algorithm for curve simplification, which can significantly reduce the number of redundant points while retaining the basic characteristics of the curve. In recent years, the algorithm has drawn increasing attention in China, and has been applied in image compression, redundant point removal and image segmentation (Zhao et al. 2008; Peng et al. 2010; Sun et al. 2012). In short, an initial distance deviation value *D* is given according to the curve dispersion. The beginning and ending points (*A* and *B*) of the curve are connected to form a straight line *AB*, and among all the inflection points of the curve, the point farthest from the straight line *AB* is queried (Point 4 in Fig. 11.9a). If the distance between this point and *AB* (*d*) is less than *D*, all the inflection points on the curve *AB* are deleted and returned. As shown in Fig. 11.9b, Points 1-3 in Segment *A*-4 are directly deleted. If *d* is greater than *D*, the point is retained, and the farthest point between the beginning point and this point is searched again, that is, Point 9 in Segment 4-*B* as shown in Fig. 11.9c. These steps are repeated until all the points of the curve are searched and the feature points are retained.

The D-P algorithm has been found unnecessary in the program implementation process, when a designed data structure can store all the information before and after the query points in the curve. It is worth noting that the value of the initial distance deviation (*D*) will affect the output of the curve simplification. That is, a larger value will remove too many details, whereas a smaller value will add unnecessary computation to the process. The program can automatically adjust the *D* value to produce a quick but good simplification of the curve. The main advantage of the D-P algorithm is an integral algorithm which can preserve the points of maximum change in the curve; thus, the shape of the simplified curve remains unchanged, which meets the requirement of the FOS identification. The FOS points are identified by finding the water depth data points with maximum gradient change, which are also the extremum points of the second derivative in the topographic profile, located at the point where the continental slope meets the oceanic basin.

(3) Detailed technical process

For a given grid digital depth model (DDM), the FOS point can be identified automatically following seven steps (including grid cutting, first derivation, initial topographic simplification, second topographic simplification, second derivation, concave terrain elimination, and comprehensive judgment) and six criteria (Fig. 11.10).

(1) Grid cutting. An original topographic profile g_0 is obtained by using a series of straight lines to cut the digital depth model and then, running the intersection calculation. The topographic profile obtained should match the “continental shelf-continental slope-oceanic basin” feature (Figs. 11.10 and 11.11a).
(2) First derivation. A slope gradient profile and a second derivative profile are obtained by computing the derivative of the topographic profile curve. Then, the distance, the topography, the gradient, and the second derivative of the profile together form the data set G_0 (Figs. 11.10 and 11.11a).
(3) Initial topographic simplification. Only the extremum points of the second derivative profile are retained, the coordinates of which, together with the water depth data points, forming a new simplified topographic profile g_1 and a new data set G_1. Compared with the original topographic profile, it can be found that only a water depth data point that is in accordance with the characteristic of a second derivative extremum point can be retained (Figs. 11.10 and 11.11b).
(4) Second topographic simplification. Applying the D-P algorithm to process the initially simplified topographic profile g_1, we can obtain a new data set G_2, which meets our requirements and forms a new topographic profile g_2. The secondly simplified topographic profile g_2 retains only a small portion of data points that meet the requirements (Figs. 11.10 and 11.11c).
(5) Second derivation. By calculating the derivative of the topographic profile g_2 (the second derivative), a new gradient profile and a second derivative profile are formed (Fig. 11.10).
(6) Concave terrain elimination. Utilizing the second loop to check through all the data points in the topographic profile g_2, the points with concave features can be eliminated to form a new data set G_3. From this, a new topographic profile g_3, a new gradient profile and a second derivative profile are formed (Figs. 11.10 and 11.11d).

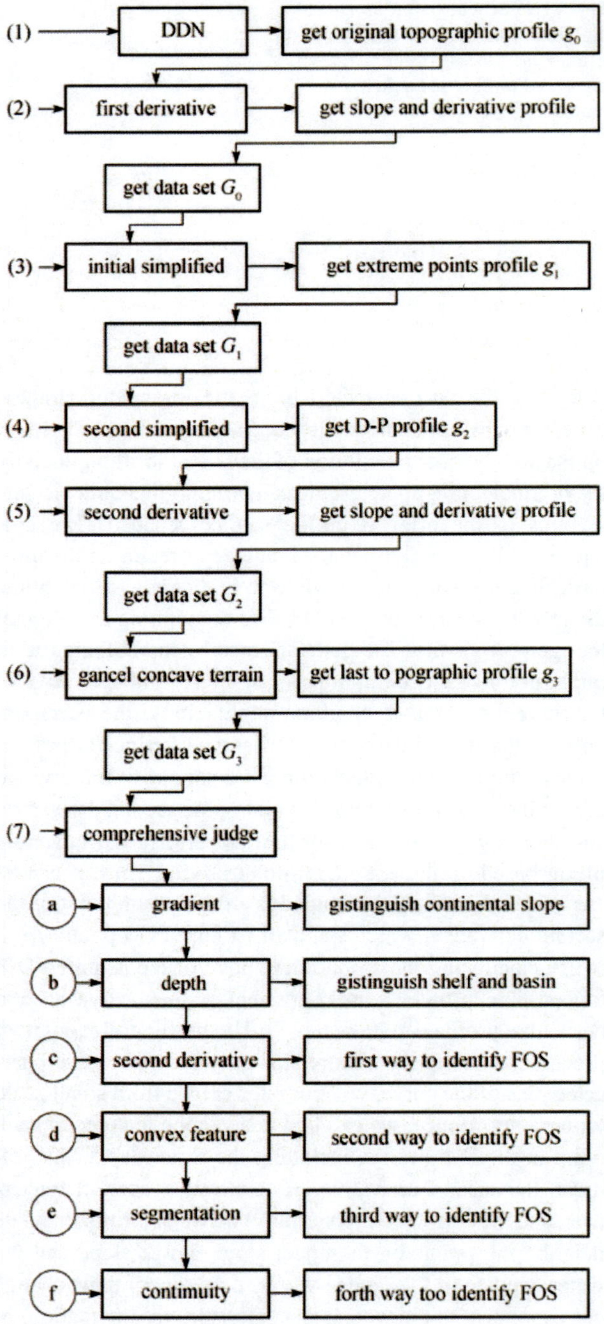

Fig. 11.10 Technical process of FOS identification

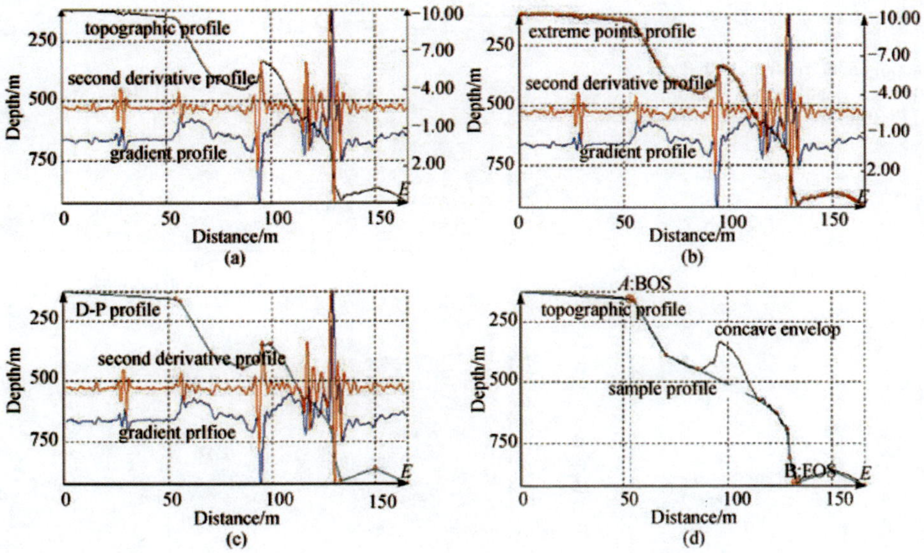

Fig. 11.11 Typical topographic profile and the process of FOS identification. **a** The original topographic profile; **b** the extremum-point profile; **c** the D-P profile; and **d** the concave-eliminated profile and the identified FOS

(7) Comprehensive judgment. Through Steps (1) to (6), a simplified integrated profile is obtained. After two processes of simplification and concave elimination, the topographic profile is simplified considerably, wherein the continental shelf and the oceanic basin present flat topography feature, while the continental slope shows a singular gradient.

Then, the topographic profile g_3 formed in Step 6 can be queried and judged by water depth, slope, second derivative, convex features, continuity and segmentation to identify and judge the FOS (Fig. 11.10).

Criteria a: gradient. By classifying the gradient values of different points in the profile, we can obtain the average gradients of the continental shelf, the oceanic basin and the continental slope, respectively; then, we can identify the region of the continental slope based on the gradient difference.

Criteria b: water depth. By classifying the water depth of different points in the profile, we can obtain the average water depth value of the continental shelf and the oceanic basin; thus, identifying the continental shelf and the oceanic basin.

Criteria c: second derivative. The FOS point is the point with maximum change in gradient from the continental slope to the oceanic basin, which is also the extremum point of the second derivative.

Criteria d: convex feature. Located at the turning point from the continental slope to the oceanic basin, the FOS presents a convex feature, i.e., the FOS point is also a data point with a positive second derivative value.

Criteria e: segmentation. Given that the adjacent points before and after the FOS point are from the continental slope and the oceanic basin, respectively, we can judge preliminarily the location of the FOS according to the segmented gradient differences of the continental slope and the oceanic basin.

Criteria f: continuity. According to the rule that points with a similar gradient are close to each other, each point in a profile will grow towards the starting point and the ending point of the profile, and record its growth distance away from them. Therefore, the point with maximum growth distance is the FOS point.

In the end, the data points that are obtained through the above seven steps, and meet the six criteria, are identified to be the FOS points.

(4) Typical examples to identify FOS

In practice, given that the judgment on the location of the FOS is influenced by various factors, a variety of complex situations should be considered during program design. Figures 11.12a–d show four typical types of topographic profile of continental margin; Figs. 11.12a–c are located at the back-arc basin, while Fig. 11.12d at the spreading continental margin.

Figure 11.12a is a topographic profile of the standard continental margin that comprises "continental shelf-continental

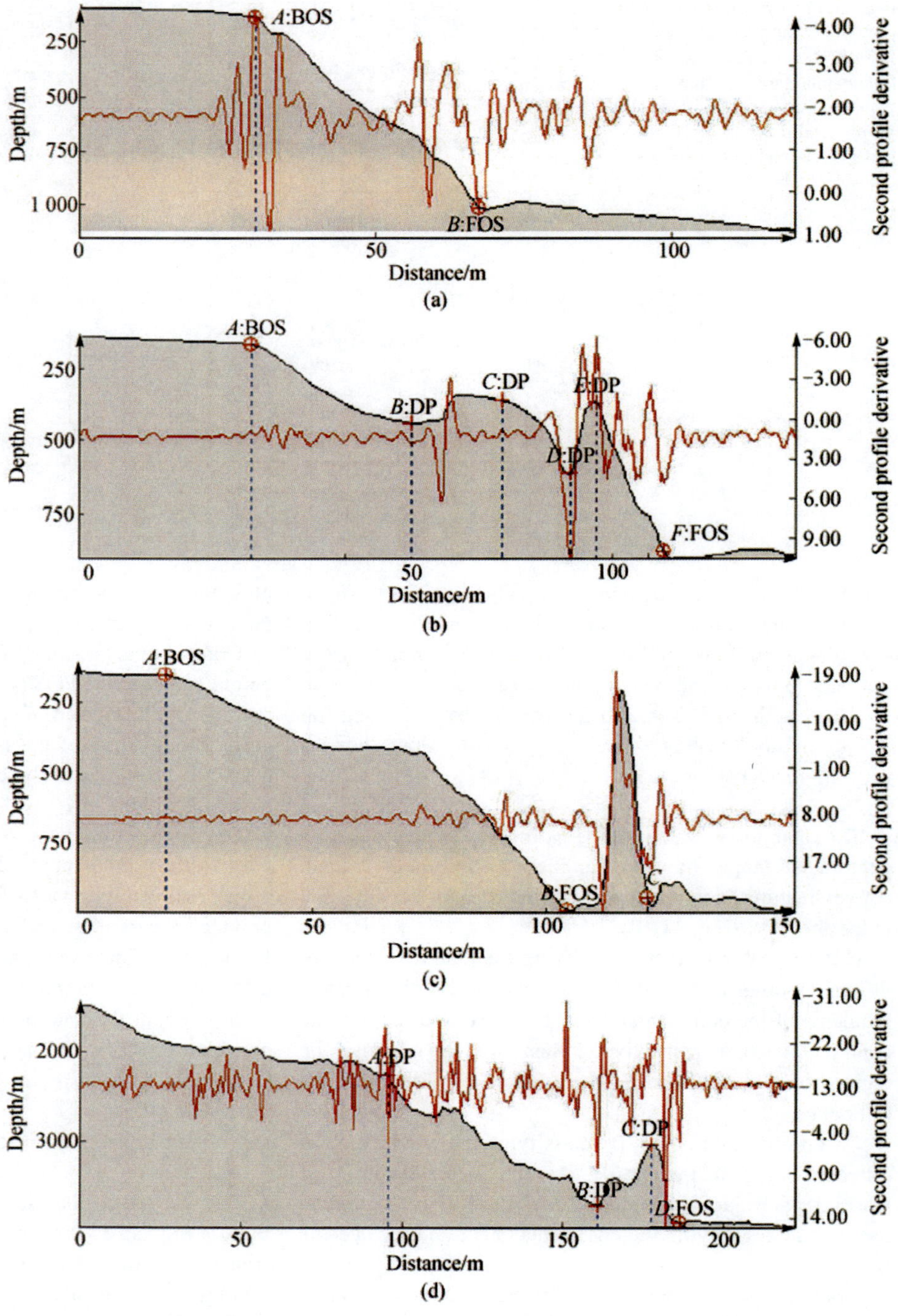

Fig. 11.12 Analysis of typical profiles. The black curve is the topographic profile; the red curve is the second derivative profile; FOS: the foot of the continental slope

slope-oceanic basin", from which it is easy to judge its characteristics. Point *A* is the boundary point between the continental shelf and the continental slope, whereas Point *B* is the dividing point of the continental slope and the oceanic basin, namely the FOS.

Figure 11.12b also has the topographic features of "continental shelf-continental slope- oceanic basin"; however, its continental slope is extremely complex, affected by submarine canyon cutting and tectonic movement, and is fragmented with convex and concave local topography developing. Thus, the identification of the FOS is vulnerable to the influence of local topography; e.g., Points *B* and *D* in this figure could easily be misinterpreted as the FOS by the program. Therefore, we should consider the entire form of the topographic profile to avoid the interference of the local topography by eliminating the concave points.

Figure 11.12c shows the influence of the oceanic-basin-margin seamount on the determination of the FOS. If we only consider the transition characteristics of topography for locating the FOS, then Point *C* is easily misinterpreted as the FOS because both its gradient and second derivative are plotted in the high-value area. Therefore, we should also judge the overall features of the profile and eliminate interference from the characteristics of curve segmentation and continuity.

Figure 11.12d shows a situation when the wide continental slope is overlapped by sea hills. The natural extension of the continental slope towards the oceanic basin is blocked by the relatively low sea hills overlapping onto the outer edge of the continental slope. Thus, Point *B* could easily be misinterpreted as the FOS. However, analyzing of the entire profile, it can be determined that Point *D* is a reasonable location for the FOS, because the submarine topography towards the oceanic basin changes from steep to flat, which agrees with the characteristics of the turning point from the continental slope to the oceanic basin. Here, we can also exclude the interference from local topography on determination of the FOS by eliminating concave points. In summary, the automatic identification of the integral feature of topography profile is the basis of accurate FOS determination, which requires the software program to recognize automatically the features and categories of each data point in the profile.

11.3.8 Establishing the 1% Sediment Thickness Line

The 1% sediment thickness line, also known as the sedimentary rock thickness formula line, was proposed by Cardinar, an Irish engineer, in 1978, and is therefore known as the Cardinar line (Lyu et al. 2012). It is used to determine a certain distance from the FOS. The ratio of the sedimentary rock thickness at a fixed point and the distance from this point to the FOS is 1%. The formula combines geophysical evidence, data on submarine resources and the delimitation of the outer continental shelf. Generally, there is a close relationship between seabed oil and gas resources and the thickness of the sedimentary rocks.

As shown in Fig. 11.13, Point *A* is the foot of the continental slope, the thickness of the sedimentary strata at Point *B* is *d* and the distance from *B* to *A* is *D*. When *d*/*D* is 1%, B is selected as the 1% sediment thickness point. This method

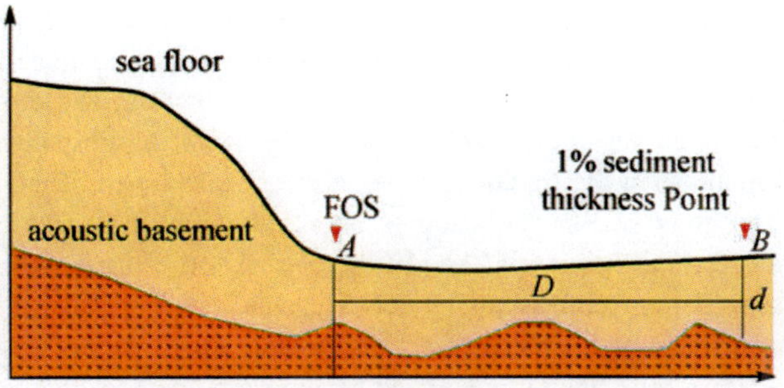

Fig. 11.13 Schematic diagram of the main features used to define the 1% sediment thickness point. Point *A* is the FOS, Point *B* is the 1% sediment thickness point

requires highly accurate stratigraphic acoustic profile data, such as multichannel seismic profiles, to determine the 1% sediment thickness point. However, such seismic data is expensive to obtain. Due to the uncertainty of the acoustic velocity in submarine strata, many Submissions do not provide the 1% sediment thickness line. In an extensive sea area, especially within Atlantictype continental margins, the adoption of the 1% sediment thickness points as the fixed points of the formula line is conducive to obtaining a more favorable outer limit of the continental shelf. If the 1% sediment thickness point is not needed to extend the outer limit of the continental shelf, it is generally unnecessary to provide additional evidence of the 1% sediment thickness.

11.3.9 Establishing the Extrapolation Limits

The FOS+60 n mile line, also known as the distance formula line, is a continuous curve formed after the intersection, cutting and splicing operations of the envelope lines generated by extrapolating the FOS points and adding a radius of 60 n mile as the radius.

As stated in the Convention, no matter which mode of delimitation is adopted, the outermost limits of the continental shelf should not exceed 350 n mile from the baseline of the territorial sea (the 350 n mile line) or the envelop line that extrapolates 100 n mile from the 2,500 m isobath (the 2,500 m+100 n mile line). Therefore, the two limits are also known as the constraint lines for the delimitation of the continental shelf beyond 200 n mile, or the constraint lines. The constraint may consist of one or both of these lines (Lyu et al. 2012; United Nations 1993). The purpose of the constraint lines is clear: to restrict the extent of the outer limits of the continental shelf beyond 200 n mile to avoid excessive prolongation. In continental shelf regions with an extensive scope of sedimentary rock, the points on the boundary line produced by the formulae may extend far out to sea. Excessive prolongation is advantageous to the applicant, but prevents other countries from benefiting from resources in the claimed area. At present, the 2,500 m+100 n mile line is widely used in areas with oceanic ridges. However, abuse of the rules regarding oceanic ridges leads to a substantial reduction in the traditional international seabed area. Therefore, the CLCS is very cautious about the use of the 2,500 m +100 n mile constraint line in the oceanic ridge areas when considering submissions by coastal states.

The FOS+60 n mile line, 350 n mile line and 2,500 m +100 n mile line are generated by the same method, that is, deriving the designated limits through the intersection and combination of extrapolation lines that are generated on the earth's surface with a fixed point as the center and a certain distance as the radius. The calculations are performed in the spherical coordinate system, because the buffer surfaces of the points and lines are severely deformed when using the projected coordinate system (Peng et al. 2005).

The following steps are taken to generate the extrapolation limits (Fig. 11.14). (1) Input the variables: input the initial variables of the starting point of the extrapolation arc *G*, extrapolation radius *r* and extrapolation direction *D* from the seaward side. The starting point of the extrapolation arc is the center of the arc, which is given by the user. The series of fixed points form the initial dataset *G*. The extrapolation radius is determined by the type of extrapolation arc. The extrapolation radius *r* of the FOS+60 n mile line, 350 n mile line and 2,500 m+100 n mile line is 60 n mile, 350 n mile and 100 n mile, respectively. The extrapolation direction *D* is the extension direction of the extrapolation arc. The FOS+60 n mile line, 350 n mile line and 2,500 m+100 n mile line are all extrapolated seaward. (2) Generate the extrapolation arc: generate the original dataset for the spherical extrapolation with each point of the initial dataset *G* as the arc center and the same spherical distance *r* as the radius. (3) Intersection of the

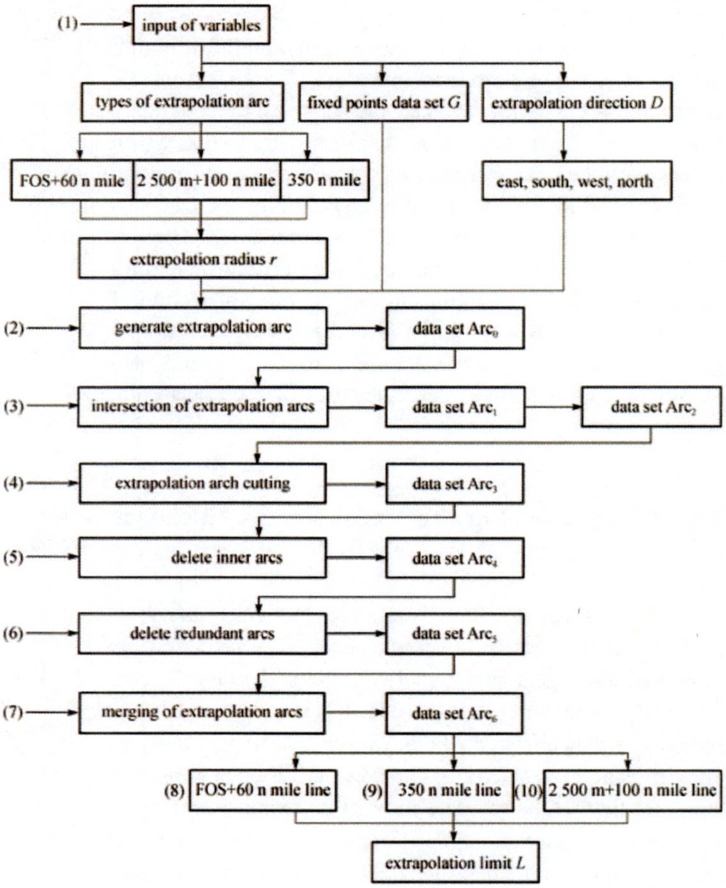

Fig. 11.14 Flowchart illustrating the process of generating the three extrapolation limits

extrapolation arc: double circulation is used to go through the dataset and perform the original extrapolation arc intersection operation. (4) Cut the extrapolation arcs: use double circulation to cut the extrapolation arcs into smaller arcs. (5) Delete the internal arcs: use the double circulation to delete the internal sub-arc segments, that is, the sub-arc segments completely contained within the original arcs. (6) Delete redundant arcs: the FOS+60 n mile line, 350 n mile line and 2,500 m +100 n mile line are all extrapolated seaward and are not closed envelop arcs, unlike the buffer zones applied in computer graphics. Generally, the extrapolation direction *D* of the FOS+60 n mile line, 350 n mile line and 2,500 m+100 n mile line is set as due east, south, west or north. The extrapolation direction *D* is an external input variable. Review the dataset and remove the irrelevant data. (7) Merge the extrapolation arc segments. Double circulation is used to go through the dataset to automatically assess the adjacent extrapolation arcs and form continuous extrapolation lines.

11.3.10 Establishing the Final Outer Limit

The outer limit of the continental shelf beyond 200 n mile is composed of the formula lines and the constraint lines. According to the requirements of Article 76 of the Convention, the points of the outer limit line are defined as longitude and latitude coordinates. The straight line distance between two consecutive points shall not be more than 60 n mile (Lyu et al. 2012; United Nations 1993). The composition of the final outer limit of the continental shelf may vary greatly for different continental shelves in different coastal states. It could be one of the two formula lines or two constraint lines, or a combination of several limits. If multiple complex factors such as politics, neighboring countries and historical rights are taken into account, the process of determining the outer limits under can be extremely complicated.

Figure 11.15 shows the various lines used to define the outer limit of the continental shelf applied by a typical coastal state with a wide continental shelf. The width of the northern shelf is less than 200 n mile, and the width of the southern continental shelf is more than 200 n mile. The 200-m isobath is the boundary between the continental shelf and the continental slope. The sedimentary layer in the northern region is relatively thin. The 1% sediment thickness line is between the FOS limit and FOS+60 n mile line. The sediment in the southern region is much more extensive, and the 1% sediment thickness line is outside the 2,500 m +100 n mile line. After obtaining the two formula lines (i.e., the FOS+60 n mile line and the 1% sediment thickness line) and the constraint lines (i.e., the 350 n mile line and the 2,500 m+100 n mile line), the outer limit of the continental shelf may be obtained through analysis based on the criteria outlined in the Convention. The final boundary line is composed of five curved segments from the different lines, denoted F_1-F_6. Segment F_1-F_2 is part of the FOS+60 n mile line, Segment F_2-F_3 is from the 1% sediment thickness line, Segment F_3-F_4 is from the 350 n mile line, Segment F_4-F_5 is part of the 2,500 m+100 n mile line and Segment F_5-F_6 is on the 350 n mile line. The outer limit of the continental shelf in this example is an ideal scenario. There are no neighboring countries within the declared boundary and the outer limit of the continental shelf may extend far.

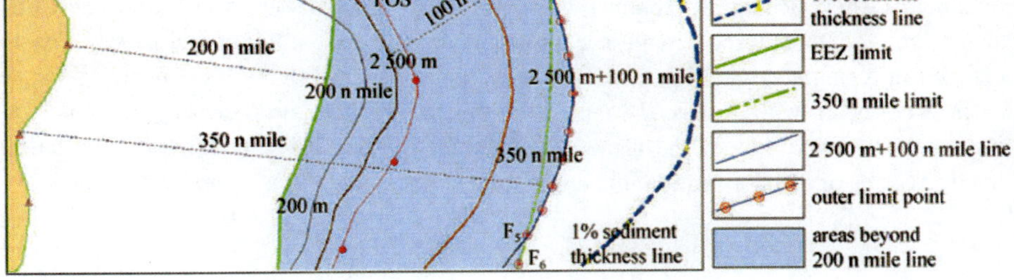

Fig. 11.15 Schematic diagram of the limits lines that may be used to define the continental shelf boundary beyond 200 n mile; EEZ is the exclusive economic zone

11.4 Delimitation of the Continental Shelf-review and Example

11.4.1 Global Overview of the Outer Continental Shelf Submissions

From December 20, 2001, to May 31, 2019, countries around the world presented 84 submissions, 7 amendments (Table 11.1) and 47 preliminary information proposals (Table 11.2). The submissions involved 70 states and the preliminary information involved 42 states. The first state to present a submission was Russia, presenting part of a submission involving the Arctic and Pacific regions. Most recently, on May 23, 2019, Canada presented a partial submission concerning the Arctic. At present, except for the US, all the states that border the Arctic, including Russia, Denmark, Norway, Iceland and Canada, have put forward submissions for outer continental shelf delimitation, with a total of 8 submissions, including one amendment. There are 4 submissions involving the Antarctic regions; namely, submissions on Antarctic territory presented by Australia and by Norway, a submission by the UK for the delimitation of the Falkland Islands, South Georgia Island and South Sandwich Islands and Argentina's submission. There were 23 submissions related to regions surrounding the Indian Ocean; 20 submissions in regions surrounding the Pacific Ocean and 44 submissions in regions surrounding the Atlantic Ocean. Of the 47 preliminary information submissions, 14 related to regions surrounding the Pacific Ocean, 26 related to regions surrounding the Atlantic Ocean and 7 involved regions surrounding the Indian Ocean. Africa comprises 60 countries and regions, including 38 coastal states. At present, 23 countries have presented 19 submissions.

The total area of the outer continental shelf claimed in the 84 submissions is about 3.2×10^7 m^2, some of which was claimed by more than one state. For example, Arctic regions claimed by Russia, Denmark and Canada, claims for the Barents Sea by Norway and Russia and regions in the Bay of Bengal where submission by India, Thailand and several other countries overlap. Australia claims the largest continental shelf area and has also claimed the largest continental shelf area in a single submission. Cuba claimed the smallest continental shelf area, only about 1,400 km^2. The Cuban submission contained the smallest continental shelf area claimed in a single submission.

In accordance with Annex II of the Convention, Resolutions No. 72 and 183 of the General Assembly of Contracting States of the Convention, the coastal states who plan to delineate the outer limits of the continental shelf beyond 200 n mile must present their submissions to the commission within 10 years of the validation of the "UN Convention on the Law of the Sea" by that state. Thus, if a state in which the Convention was validated before 2,000 plans to delineate the outer limit of the continental shelf beyond 200 n mile, it must present the submission or the preliminary information before 2010. At present, 30 coastal states in which the Convention has been validated for more than 10 years have not yet presented submissions or preliminary information. A rough estimate shows that these countries are mainly in the Mediterranean, Baltic and Caribbean Seas.

11.4.2 Submissions Adopted by the Commission

From 2002 when Russia presented its first submission until March 2019, the commission has completed consideration of 32 submissions and published summaries of the suggestions (Table 11.3). According to these summaries, the commission's consideration of a submission is based mainly on the regional geology and geography of the claimed area, the natural prolongation of the landmasses and the rights of countries bordering the continental shelf, determination of the FOS and the outer edge of the continental margin, determination of the outer limit of the continental shelf and other considerations. The key points considered by the commission are whether the base area and the location of the FOS are reasonable, whether the extension of 100 n mile from the 2,500-m isobath is applicable and the points on the 200 n mile limit are reasonable; of these, the most critical is whether the base area and the FOS are reasonable. The FOS is an important parameter to be considered by the commission. There are two methods to determine the FOS, one is to select the point with the maximum change of slope gradient at the BOS based on topographic and geomorphic features; the other is to use evidence to the contrary to support the geological evidence. Most submissions use the first method to determine the. For example, in the partial delimitation submission of Mexico, the position of the continental slope bottom of Campeche Cliff beyond the Yucatan Peninsula is obvious and is very easy to identify. The FOS in that region of the continental slope was easily approved by the commission (Fig. 11.16). However, in other submissions, the determination of the base and foot of the continental slope was more complicated.

Fig. 11.16 Location of the base and FOS of the continental slope of Campeche Cliff beyond Yucatan Peninsula in Mexico

11.4.3 The Submission of Seychelles in the Northern Plateau Region

11.4.3.1 Introduction

On May 7, 2009, the Republic of Seychelles submitted to the commission a request to recognize its claim on the continental shelf along its shores in the northern plateau region. The submission was presented through the UN Secretary-General and contained information on the limits of the continental shelf beyond 200 n mile from the baseline, in accordance with Paragraph 8 of Article 76 of the Convention (the "submission"). The submission was for the northern plateau region in the central Indian Ocean (Fig. 11.17).

The original submission of Seychelles contained three parts: the executive summary, the main body, and scientific and technical data.

Pursuant to the rules of procedure of the commission, the subcommission of Seychelles examined and verified the format and completeness of the original submission and additional data and undertook a preliminary analysis of the Submission, in accordance with Article 76 of the Convention and the guidelines, and ensure that the following items be carried out correctly according to the CLCS requirements.

(1) The test of appurtenance.
(2) The data and methodology to determine the location of FOS.

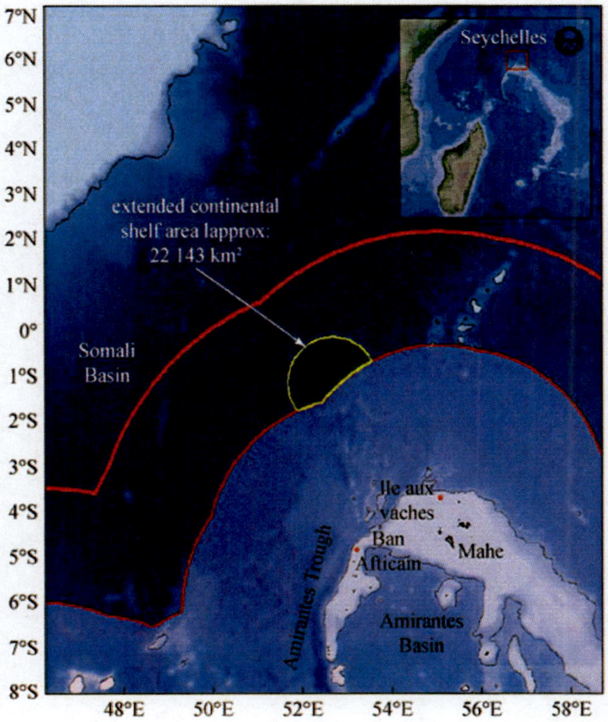

Fig. 11.17 Map of the continental shelf area of the Seychelles Islands beyond 200 n mile in the northern plateau region (executive summary of the submission of May 7, 2009)

(3) The methodology used to determine the FOS+60 n mile.
(4) The data and methodology used to determine the constraint line 350 n mile from the baselines.
(5) Construction of the inner envelope of the formula and constraint lines.
(6) Delineation of the outer limit of the continental shelf by means of straight lines not longer than 60 n mile while ensuring that only the portion of the seabed that satisfies all the provisions of Article 76 of the Convention and the statement of understanding is enclosed.
(7) Estimates of the uncertainties in the methods applied, with a view to identifying the main source(s) of such uncertainties and their effect on the submission.

11.4.3.2 Regional Background

The northern plateau region of Seychelles is located to the north-west of the Seychelles Bank (Fig. 11.18).

The northern plateau region is located at the northern extremity of the Mascarene Plateau (Fig. 11.18). The northern plateau region was described by Seychelles as consisting of three specific morphological features, all connected to the Seychelles Bank (Fig. 11.19): (a) the western pedestal, (b) the central area, and (c) the eastern pedestal. According to the submitting state, the western pedestal is elevated 400–1,000 m above the abyssal plain, while the central area comprises a raised pedestal cut by several longitudinal rib-like ridges spreading out northward. The peaks of the ridges generally rise to more than 2,000 m above the abyssal plain. The eastern pedestal lies at similar elevation as the western pedestal and has NE-SW trending rib-like ridges similar to those in the central area.

In their submission, Seychelles describes the northern plateau region as the northern extension of Mascarene Plateau, underlain by stretched continental crust created during the rifting and eventual separation of the Mascarene Plateau from the east coast of Africa. The Mascarene Plateau was connected to eastern Madagascar and western India prior to 85 Ma. Seafloor spreading during the Late Cretaceous separated the Mascarene Plateau together with India from Madagascar and created the Mascarene and Amirante Basins (Schlich et al. 1990; Dyment 1991). The Mascarene micro-continent subsequently rifted from the western margin of the Indian Plate during the period 83-65 Ma. At approximately 65 Ma, the Mascarene micro-continent further separated from western India with the effusion of the Deccan volcanic, and migrated by generation of basaltic seafloor spreading from the Carlsberg Spreading Ridge.

11.4.3.3 Determining the Foot of the Continental Slope

The FOS should be established in accordance with Paragraph 4 (b) of Article 76 of the Convention. The northern plateau region is dominated by several ridge-like features of

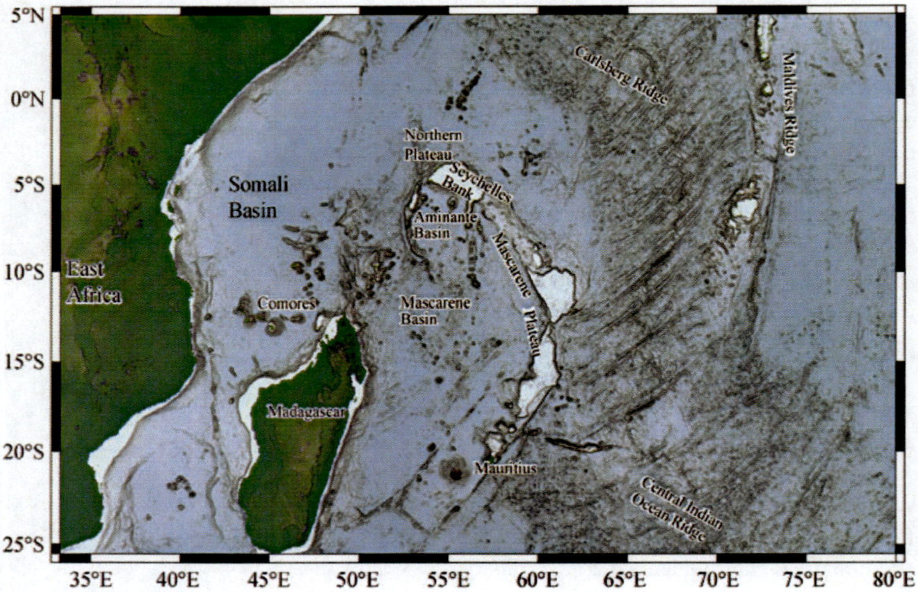

Fig. 11.18 Regional setting of the Mascarene Plateau

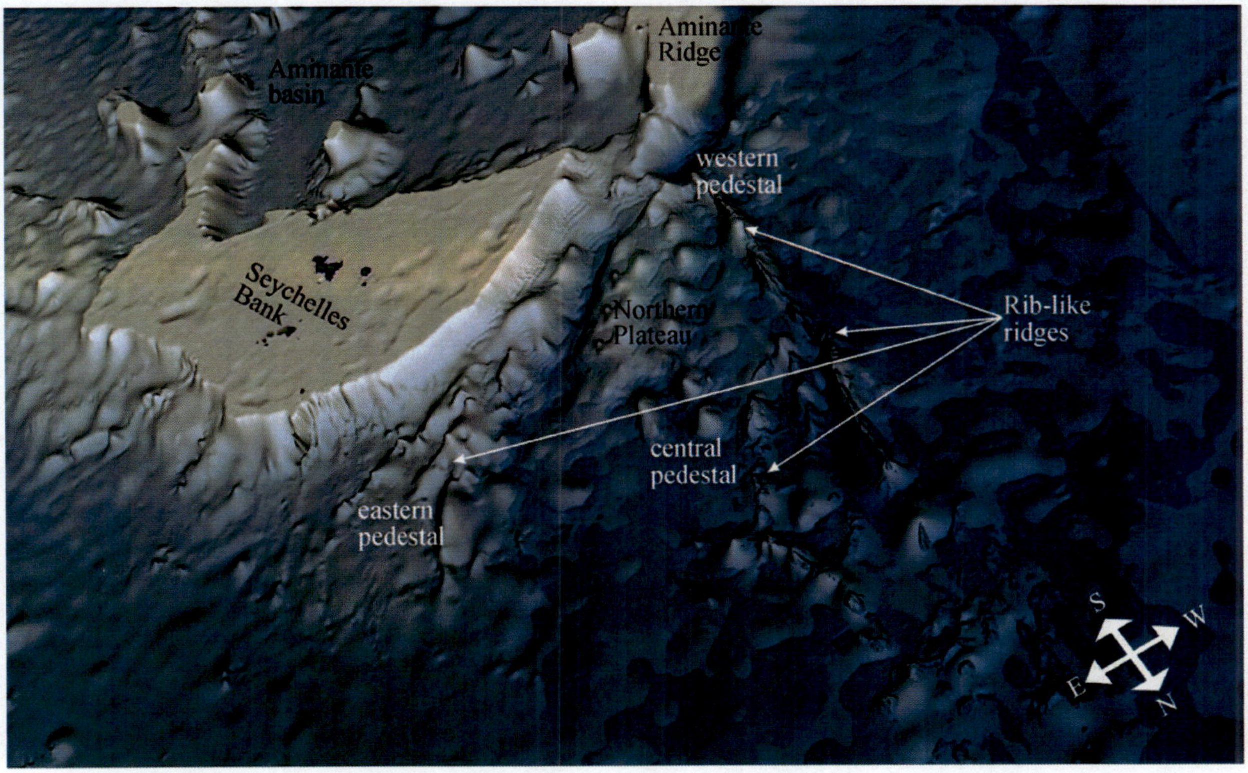

Fig. 11.19 Morphology of the northern plateau region

variable length, orientation and elevation above the surrounding seafloor. The Seychelles claims that the entire area encompassed by these features lies within the FOS envelope.

In July 2016, a survey of topography and geomorphology was carried out in the northern plateau region of Seychelles by the *Xiangyanghong 10*. High-precision bathymetric data was obtained for the regional subcommission of Seychelles (Figs. 11.20 and 11.21).

Based on the analysis of the new bathymetric data, the subcommission decided that Seychelles passed the test of appurtenance using the key FOS-3 point at the northern edge of the westernmost ridge (Figs. 11.22 and 11.23). Based on the new data from the China-Seychelles Joint Cruise, three new FOS points were determined. The subcommission accepted FOS point FOS-3 located at the northern edge of the western ridge (Fig. 11.23). The test of appurtenance was passed on the basis of this FOS point.

The subcommission concluded that the saddles observed along the bathymetric profiles on which FOS-1 and -2 were identified needed further substantiation (Fig. 11.24).

Seychelles subsequently provided more information on the new bathymetric survey, including the preliminary cruise report and reprocessed bathymetry data. Based on the updated grid, the elevation of the saddles above the deep ocean floor was demonstrated by Seychelles to be of the order of 400 m. The subcommission concluded that these saddles were significantly elevated above the very flat deep ocean floor beyond the foot of the continental slope.

In view of the above, the subcommission accepted all three FOS positions (FOS-1, -2 and -3) (Fig. 11.23).

The commission considered the BOS and the three FOS points: FOS-1, -2 and -3, submitted by Seychelles, together with the findings of the subcommission. The commission agreed with the subcommission that the Seychelles submission passed the test of appurtenance based on the location of FOS-3 from which the 60 n mile formula line extends beyond the 200 n mile line (Fig. 11.23).

After examining the technical and scientific data and information submitted by Seychelles, and on the consideration and recommendations made by the subcommission, the commission concluded that the three FOS points in the northern plateau region fulfill the requirements of Article 76 and Chap. 5 of the Guidelines. The commission recommended that these FOS points should form the basis for the establishment of the outer edge of the continental margin in the northern plateau region.

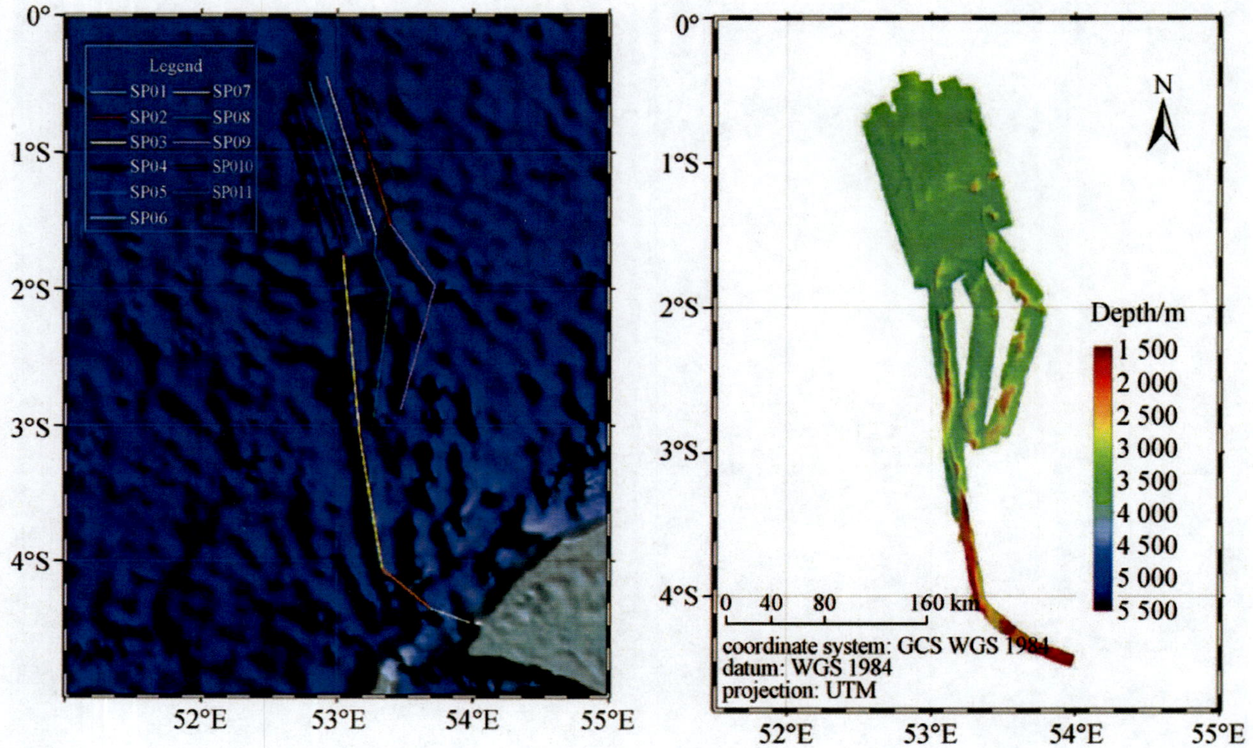

Fig. 11.20 Multi-beam lines of the China-Seychelles Joint Cruise, and bathymetric map derived from the multi-beam bathymetric data collected by the R.V. *Xiangyanghong 10*

Fig. 11.21 Bathymetry and 3D subsurface topographic features in the northern plateau region of Seychelles

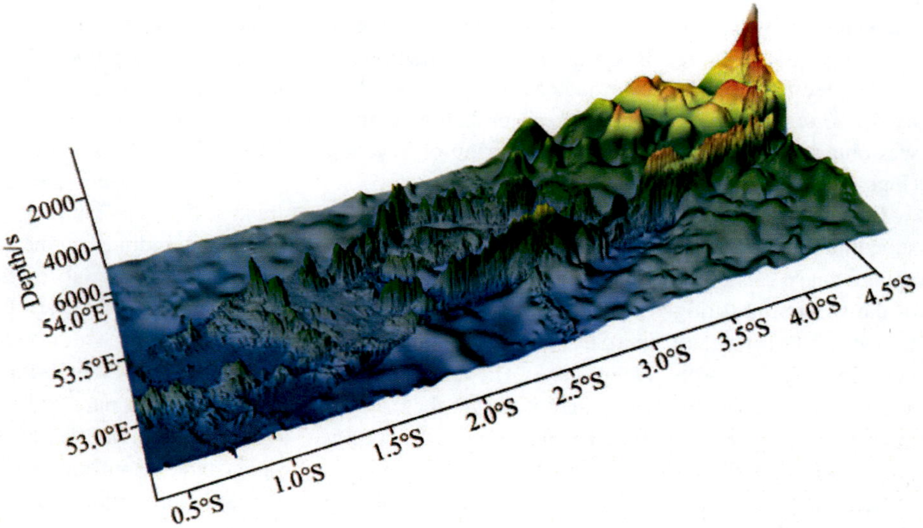

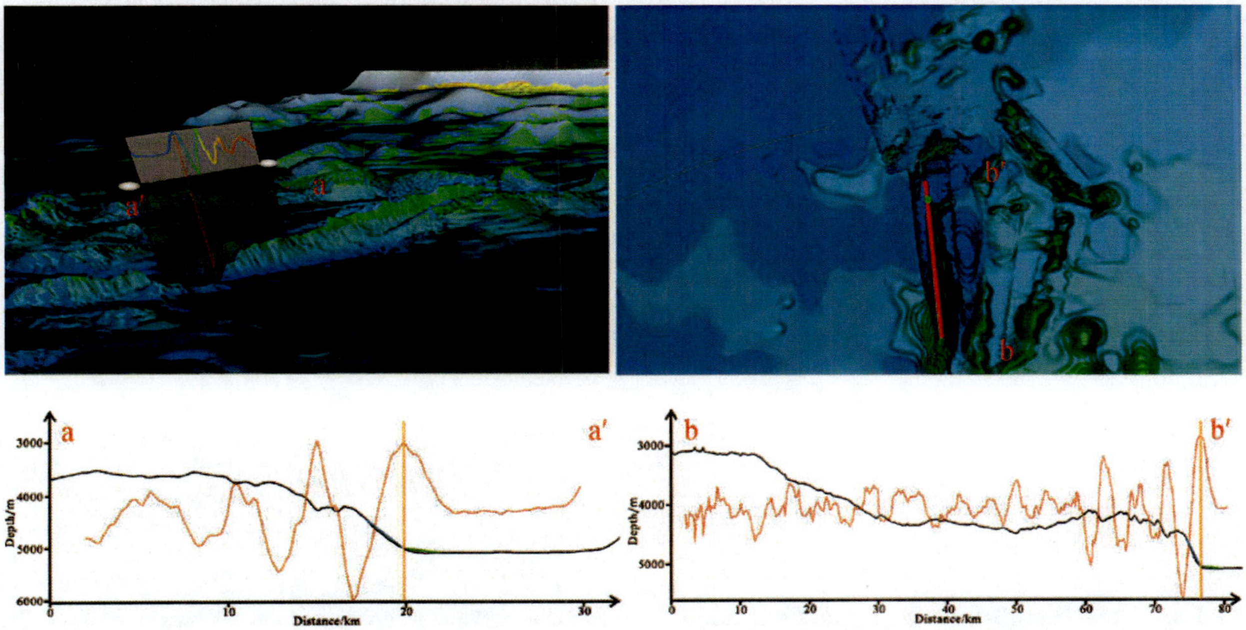

Fig. 11.22 Analysis of the new bathymetric data illustrating the FOS-3 point at the northern edge of the westernmost ridge

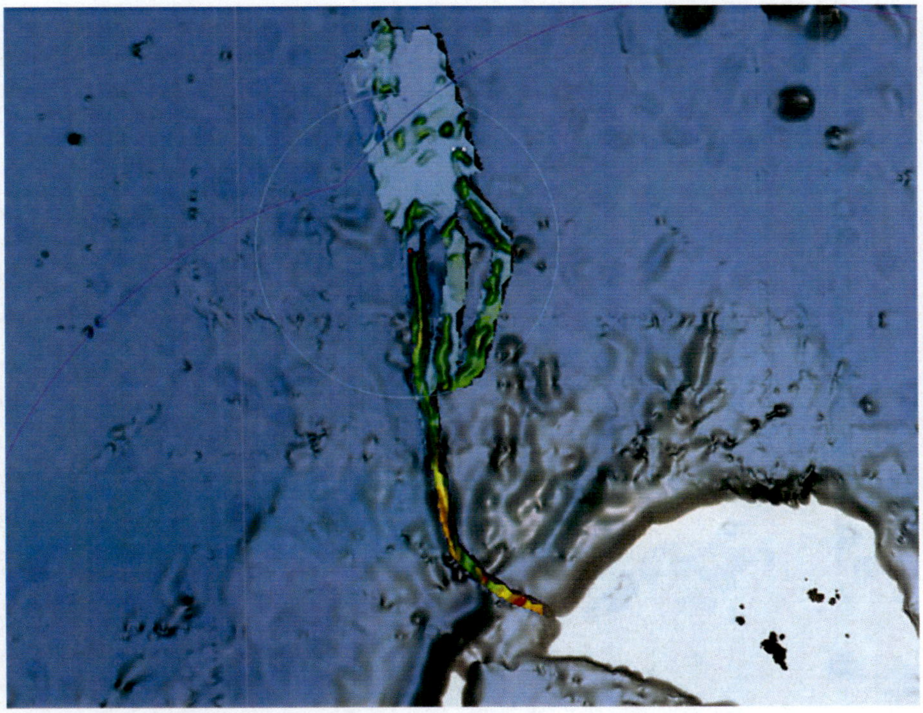

Fig. 11.23 View of the new multi-beam bathymetric data combined with SRTM illustrating the three FOS-3 positions (red dot)

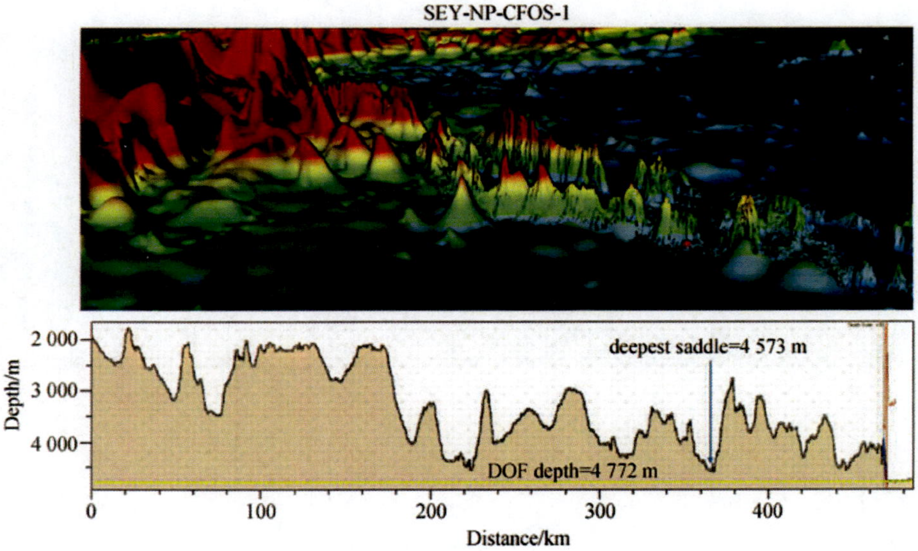

Fig. 11.24 Bathymetric profile derived from new multi-beam bathymetric data submitted by Seychelles, illustrating the position of FOS point FOS-1, indicated by the red dot

11.4.3.4 The Establishment of the Outer Edge of the Continental Margin

The outer edge of the continental margin of Seychelles in the northern plateau region should, for the purposes of the convention, be established in accordance with Paragraph 4 (a) of Article 76 of the Convention.

(1) The Application of the 60 n mile Distance Formula

The outer edge of the continental margin is based on fixed points constructed at a distance of not more than 60 n mile from the FOS points on the continental margin of Seychelles in the northern plateau region, in accordance with the provision contained in Paragraph 4 (a) (ii) of Article 76 of the Convention.

Using the selected FOS points in the northern plateau region (Fig. 11.25), Seychelles established fixed points based on the 60 n mile formula. The outer edge of the continental margin is defined by 360 fixed points (CM_001 to CM_360) connected by straight lines not exceeding 60 n mile in length (Fig. 11.25).

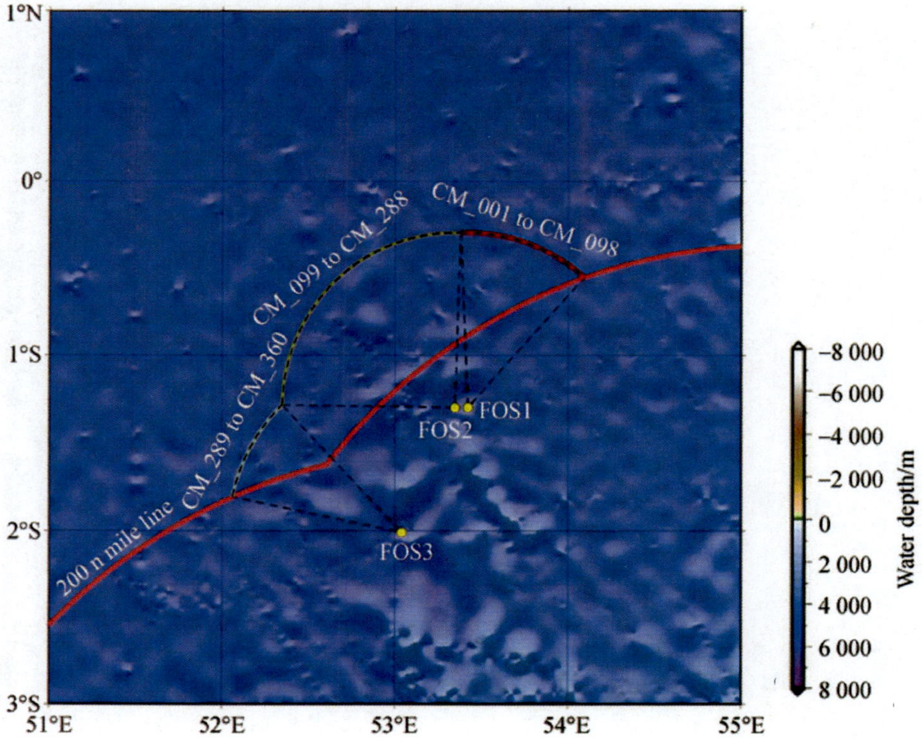

Fig. 11.25 Map illustrating fixed points CM_001 to CM_360, numbered from east to west, composing the outer edge of the continental margin in the northern plateau region

The commission agreed with the procedure and the accuracy by which the points in the northern plateau region were established by Seychelles.

(2) Configuration of the Outer Edge of the Continental Margin

In the northern plateau region, the outer edge of the continental margin extends northwest beyond the 200 n mile line of Seychelles. Fixed points CM_001 and CM_360 of the outer edge of the continental margin are located on the 200 n mile line of Seychelles (Fig. 11.25).

(3) Recommendations

In the northern plateau region, the outer edge of the continental margin beyond the 200 n mile line is based on the 60 n mile formula points, in accordance with Paragraph 4 of Article 76 of the Convention (Fig. 11.25). The commission recommended that these points be used as the basis for delineating the outer limits of the continental shelf in this region, subject to the relevant constraints.

11.4.3.5 The Application of the Constraint Criteria

The fixed points that define the line of the outer limits of the continental shelf are based on the outer edge of the continental margin, taking into consideration the constraints contained in Paragraphs 5 and 6 of Article 76 of the Convention.

Consequently, the fixed points constructing the line of the outer limits of the continental shelf on the seabed, drawn in accordance with Paragraph 4 (a) (ii) of Article 76, either shall not exceed 350 n mile from the baselines from which the breadth of the territorial sea is measured, or shall not exceed 100 n mile from the 2,500 m isobath. For the outer limits of the continental shelf in the northern plateau region, Seychelles provided data and information on both the distance and the depth constraints. In the northern plateau region, the depth constraint is located entirely within the 200 n mile line. Therefore, the applicable constraint is defined by the distance constraint line (Fig. 11.26).

The distance constraint line submitted by Seychelles was constructed by arcs at a distance of 350 n mile from the baseline from which the breadth of the territorial sea of

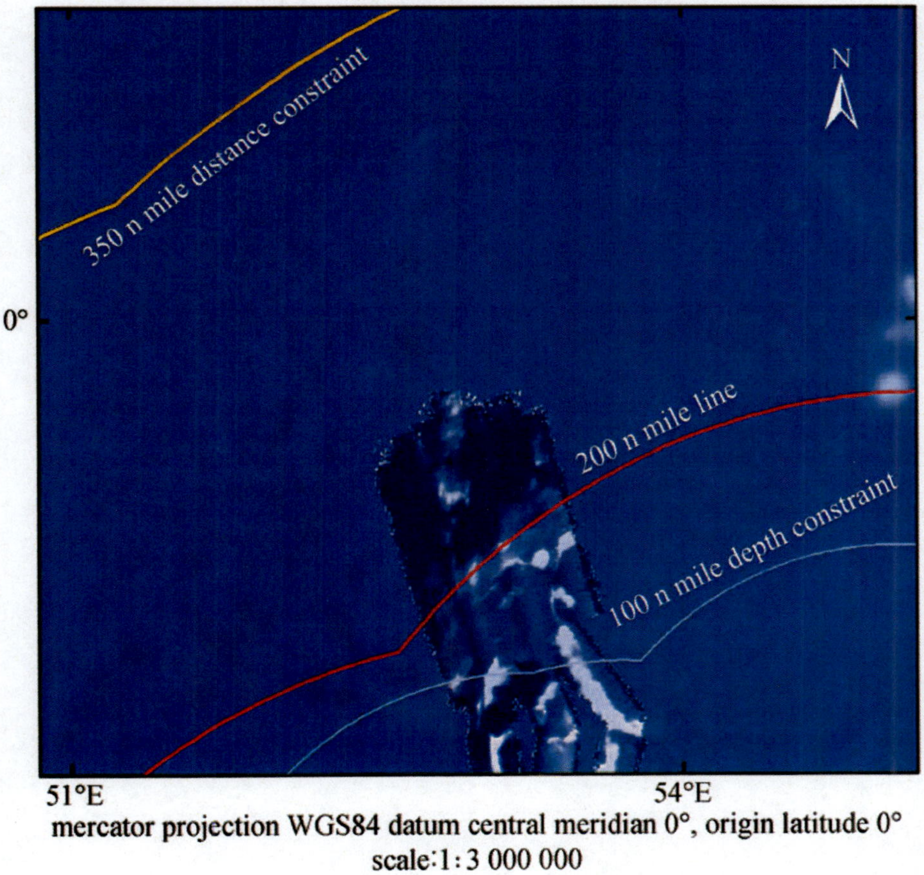

Fig. 11.26 Map illustrating the location of the distance and depth constraint lines

Seychelles is measured (Fig. 11.26). The commission agreed with the procedure applied by Seychelles in the construction of this constraint line and its accuracy.

11.4.3.6 The Outer Limits of the Continental Shelf

The commission recommended that the delineation of the outer limits of the continental shelf in this region be conducted in accordance with Paragraph 7 of Article 76 of the Convention, by straight lines not exceeding 60 n mile in length, connecting fixed points defined by coordinates of latitude and longitude. Further, the commission agreed with the methodology and the accuracy applied in delineating the outer limits of the continental shelf in the northern plateau region, including the determination of the fixed points and the construction of the straight lines connecting those points (Fig. 11.27).

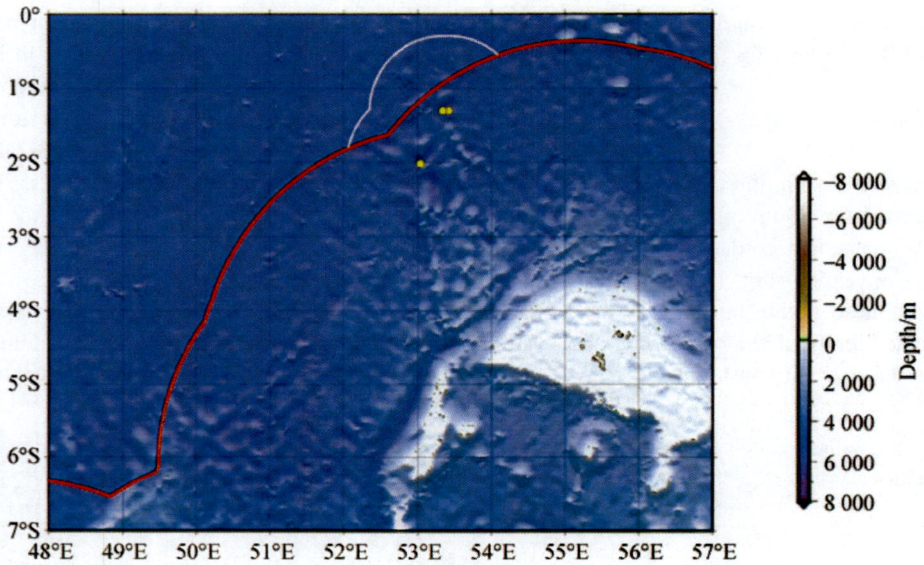

Fig. 11.27 The outer limits of the continental shelf in the northern plateau region, delineated by straight lines not exceeding 60 n mile in length

Appendix

See Figs. 11.28, 11.29, 11.30, 11.31, 11.32 and Tables 11.1, 11.2, 11.3.

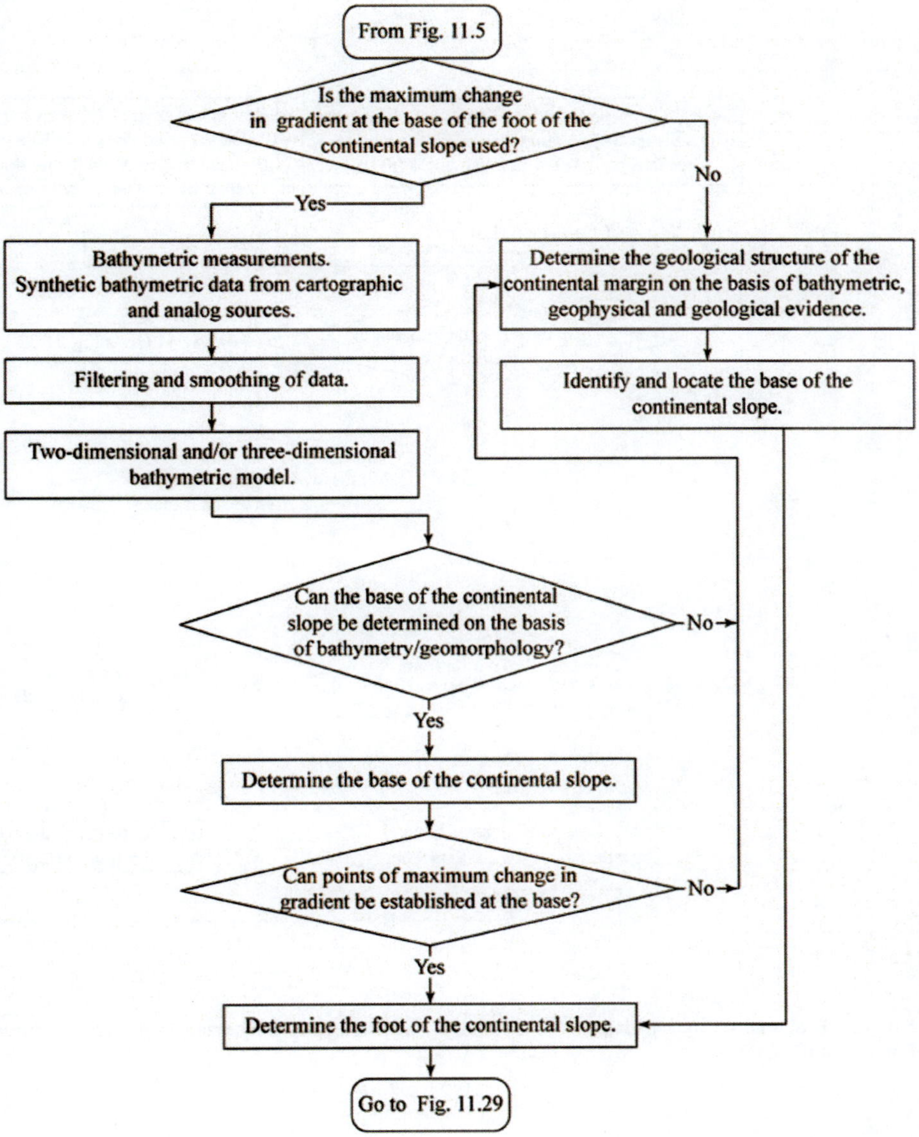

Fig. 11.28 Sub-flowchart showing the determination of the foot of the continental slope—indicated on master flowchart (Fig. 11.5) as Procedure II

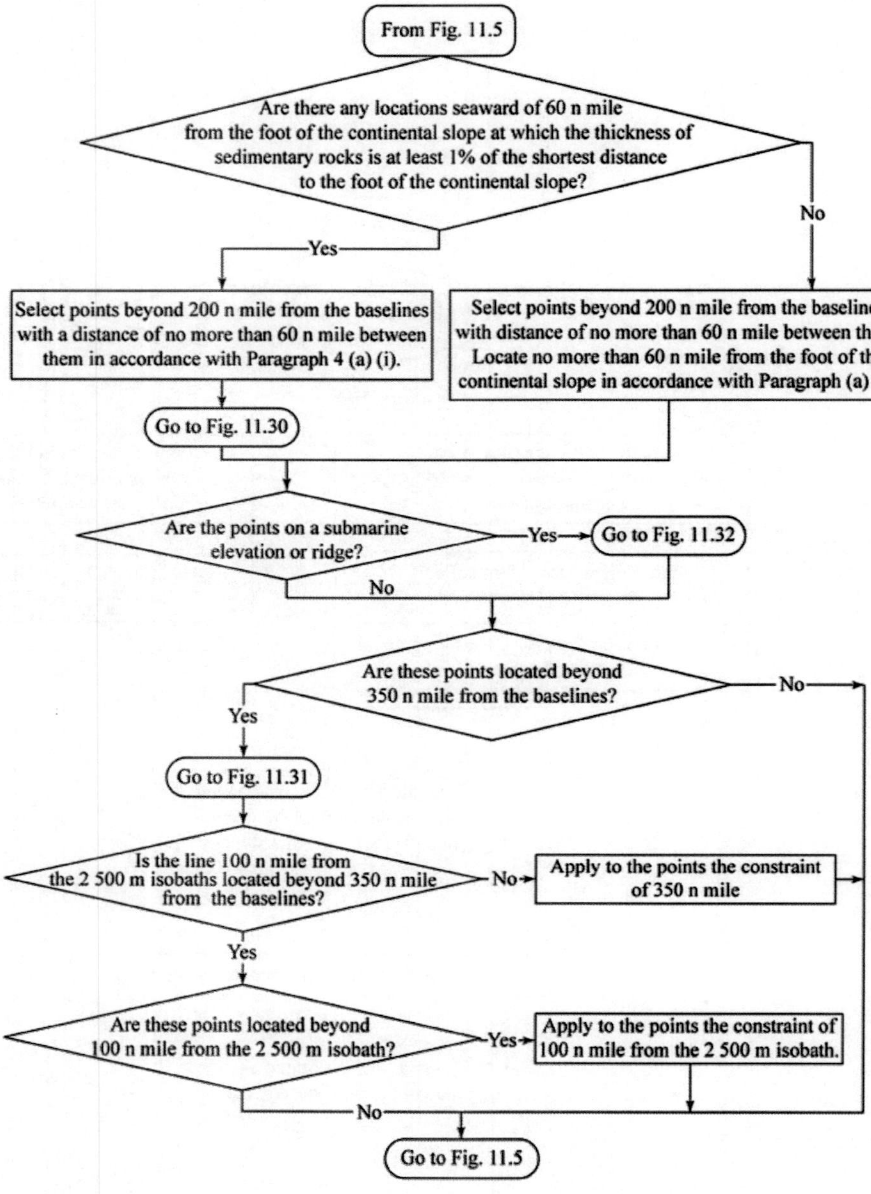

Fig. 11.29 Sub-flowchart showing the application of the distance formulae and constraints—indicated in master flowchart (Fig. 11.5) as Procedures III and V

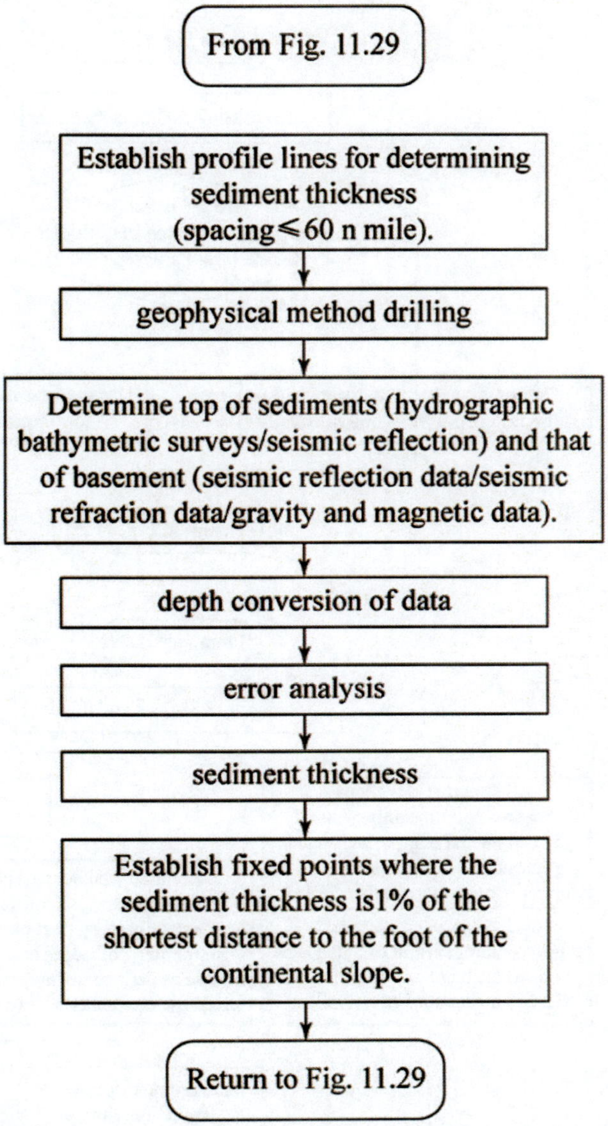

Fig. 11.30 Sub-flowchart showing the establishment of fixed points where the sediment thickness is $\geq$ 1% of the shortest distance to the foot of the continental slope—indicated on master flowchart (Fig. 11.5) as Procedure III-application of formulae

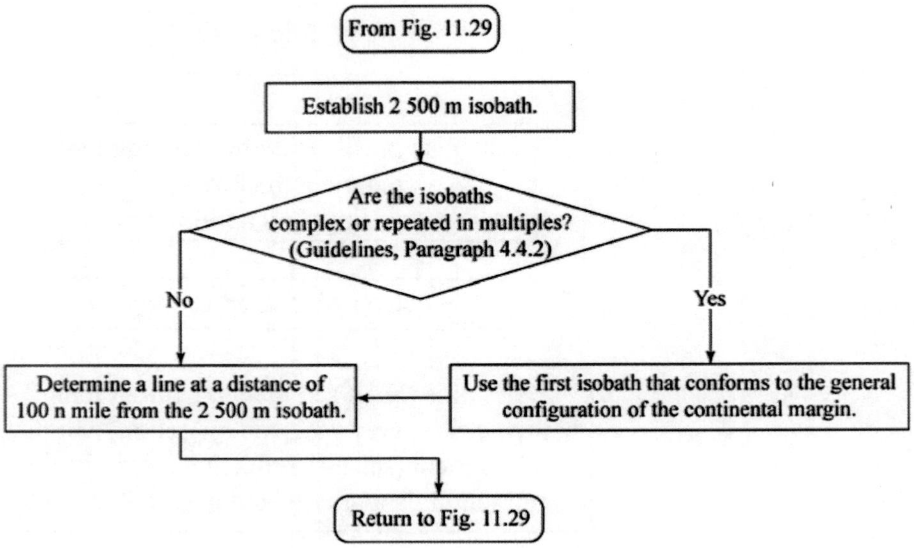

Fig. 11.31 Sub-flowchart showing the establishment of the 2,500 m isobath and determination of the 100 n mile line from that isobath—indicated on the master flowchart (Fig. 11.5) as Procedure V—application of constraints

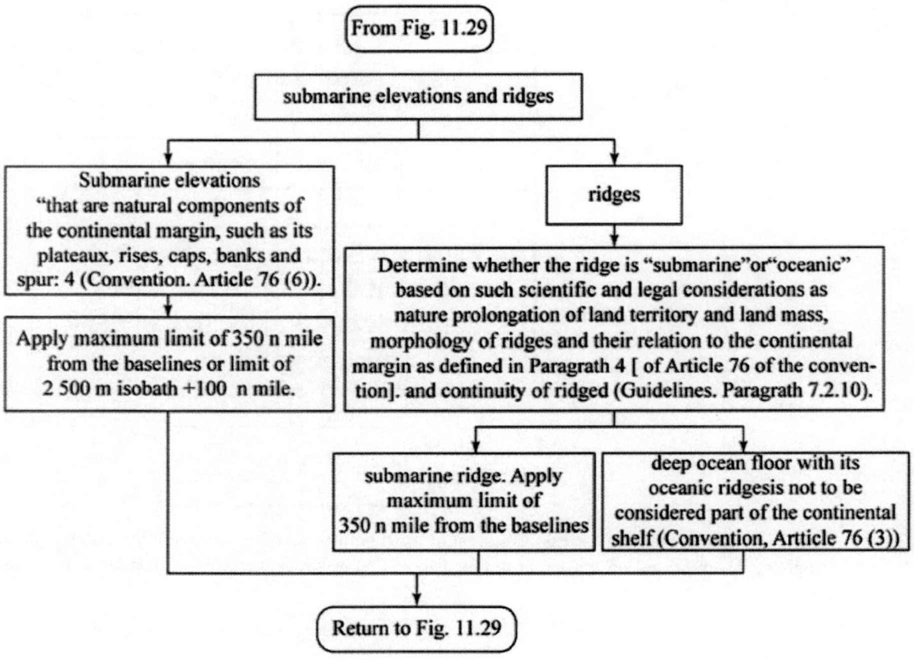

Fig. 11.32 Sub-flowchart showing the formulation of the solution to the problem arising from Articles 76 (3) and (6) and relating to sea floor highs, whether submarine elevations or ridges, indicated on master flowchart (Fig. 11.5) as Procedure IV—determination of the limits in the case of submarine elevations and/or ridges

Table 11.1 Submissions by coastal States

Submission by [State]	Presentation to the CLCS*	Date of submission
1	Russian Federation	December 20, 2001
1a	Russian Federation, partial revised submission in respect of the Okhotsk Sea	February 28, 2013
1b	Russian Federation, partial revised submission in respect of the Arctic Ocean	August 3, 2015
2	Brazil	May 17, 2004
2a	Brazil, partial revised submission, in respect of the Brazilian southern region	April 10, 2015
2b	Brazil, partial revised submission, in respect of the Brazilian equatorial margin	September 8, 2017
2c	Brazil, partial revised submission, related to the Brazilian Oriental and meridional margin	December 7, 2018
3	Australia	November 15, 2004
4	Ireland, Porcupine Abyssal Plain	May 25, 2005
5	New Zealand	April 19, 2006
6	Joint submission by France, Ireland, Spain and the United Kingdom of Great Britain and Northern Ireland—in the area of the Celtic Sea and the Bay of Biscay	May 19, 2006
7	Norway, in the northeast Atlantic and the Arctic	November 27, 2006
8	France, in respect of the areas of French Guiana and New Caledonia	May 22, 2007
9	Mexico, in respect of the western polygon in the Gulf of Mexico	December 13, 2007
10	Barbados	May 8, 2008
10a	Barbados, revised	July 25, 2011
11	United Kingdom of Great Britain and North Ireland, Ascension Island	May 9, 2008
12	Indonesia, northwest of Sumatra Island	June 16, 2008
13	Japan	November 12, 2008
14	Joint submission by the Republic of Mauritius and the Republic of Seychelles, in the region of the Mascarene Plateau	December 1, 2008
15	Suriname	December 5, 2008
16	Myanmar	December 16, 2008
17	France, areas of the French Antilles and the Kerguelen Islands	February 5, 2009
18	Yemen, in respect of southeast of Socotra Island	March 20, 2009
19	United Kingdom of Great Britain and North Ireland, in respect of Hatton Rockall area	March 31, 2009
20	Ireland, in respect of Hatton-Rockall area	March 31, 2009
21	Uruguay	Aprilch 7, 2009
22	Philippines, in the Benham Rise region	April 8, 2009
23	The Cook Islands, concerning the Manihiki Plateau	April 16, 2009
24	Fiji	April 20, 2009

(continued)

Table 11.1 (continued)

Submission by [State]	Presentation to the CLCS*	Date of submission
25	Argentina	April 21, 09
25a	Argentina, partial revised submission	October 28, 2016
26	Ghana	April 28, 2009
27	Iceland, in the Ægir Basin area and in the western and southern parts of Reykjanes Ridge	April 29, 2009
28	Denmark, in the area north of the Faroe Islands	April 29, 2009
29	Pakistan	April 30, 2009
30	Norway, in respect of Bouvetøya and Dronning Maud Land	May 4, 2009
31	South Africa, in respect of the mainland of the territory of the Republic of South Africa	May 5, 2009
32	Joint submission by the Federated States of Micronesia, Papua New Guinea and Solomon Islands, concerning the Ontong Java Plateau	May 5, 2009
33	Joint submission by Malaysia and Vietnam, in the southern part of the South China Sea	May 6, 2009
34	Joint submission by France and South Africa, in the area of the Crozet Archipelago and the Prince Edward Islands	May 6, 2009
35	Kenya	May 6, 2009
36	Mauritius, in the region of Rodrigues Island	May 6, 2009
37	Vietnam, in the north area (VNM-N)	May 7, 2009
38	Nigeria	May 7, 2009
39	Seychelles, concerning the northern, plateau region	May 7, 2009
40	France, in respect of La Réunion Island and Saint-Paul and Amsterdam Islands	May 8, 2009
41	Palau	May 8, 2009
42	Côte d'Ivoire	May 8, 2009
43	Sri Lanka	May 8, 2009
44	Portugal	May 11, 2009
45	United Kingdom of Great Britain and Northern Ireland, in respect of the Falkland Islands, and of South Georgia and the South Sandwich Islands	May 11, 2009
46	Tonga	May 11, 2009
47	Spain, in respect of the area of Galicia	May 11, 2009
48	India	May 11, 2009
49	Trinidad and Tobago	May 12, 2009
50	Namibia	May 12, 2009
51	Cuba	June 1, 2009
52	Mozambique	July 7, 2010
53	Maldives	July 26, 2010
54	Denmark, Faroe-Rockall Plateau Region	December 2, 2010
55	Bangladesh	December 25, 2011
56	Madagascar	April 29, 2011
57	Guyana	September 6, 2011
58	Mexico, in respect of the eastern polygon in the Gulf of Mexico	December 19, 2011
59	United Republic of Tanzania	January 18, 2012

(continued)

Table 11.1 (continued)

Submission by [State]	Presentation to the CLCS*	Date of submission
60	Gabon	April 10, 2012
61	Denmark, in respect of the southern continental shelf of Greenland	June 14, 2012
62	Joint Submission by Tuvalu, France and New Zealand (Tokelau), in respect of the area of the Robbie Ridge	December 7, 2012
63	China, in part of the East China Sea	December 17, 2012
64	Kiribati	December 24, 2012
65	Republic of Korea	December 26, 2012
66	Nicaragua, in the southwestern part of the Caribbean Sea	June 24, 2013
67	Federated States of Micronesia, in respect of the Eauripik Rise	Augest 30, 2013
68	Denmark, in respect of the north-eastern continental shelf of Greenland	November 26, 2013
69	Angola	December 6, 2013
70	Canada, in respect of the Atlantic Ocean	December 6,2013
71	Bahamas	February 6, 2014
72	France, in respect of Saint-Pierre-et-Miquelon	April 16, 2014
73	Tonga, in the western part of the Lau-Colville Ridge	April 23, 2014
74	Somalia	July 21, 2014
75	Joint Submission by Cabo Verde, The Gambia, Guinea, Guinea-Bissau, Mauritania, Senegal and Sierra Leone, in respect of areas in the Atlantic Ocean adjacent to the coast of West Africa	September 25, 2014
76	Denmark, in respect of the Northern Continental Shelf of Greenland	December 15, 2014
77	Spain, in respect of the area west of the Canary Islands	December 17, 2014
78	Oman	October 26, 2017
79	France, in respect of French Polynesia	April 6, 2018
80	Joint Submission of the Republic of Benin and the Togolese Republic	September 21, 2018
81	Liberia	October 23, 2018
82	Mauritius, concerning the Southern Chagos Archipelago region	March 26, 2019
83	Indonesia, in the area of North of Papua (Eauripik Rise)	April 11, 2019
84	Canada, in respect of the Arctic Ocean	May 23, 2019

Table 11.2 Preliminary information

Submission by [State]	Preliminary information submitted by [State]	Received date
1	Angola	May 12, 2009
2	Bahamas	May 12, 2009
3a	Bénin	May 12, 2009
3b	Bénin et Togo	April 2, 2009
4	Brunei Darussalam	May 12, 2009
5	Canada	December 6, 2013
6	Cameroon	May 11, 2009
7	Cabo Verde	May 7, 2009
8	Chile	May 8, 2009
9	China	May 11, 2009
10	Comoros	June 2, 2009
11	Congo	May 12, 2009
12	Costa Rica	May 11, 2009
13	Cuba	May 12, 2009
14	Democratic Republic of the Congo	May 11, 2009
15	Equatorial Guinea	May 14, 2009
16	Fiji in respect of the south east region of the North Fiji Basin, Lau-Colville, Tonga-Kermadec Complex	April 21, 2009
16a	Fiji and Solomon Islands on the Charlotte Bank Region	April 21, 2009
16b	Fiji, Solomon Islands and Vanuatu on the North Fiji Basin	April 21, 2009
17	France, Polynésie française et Wallis et Futuna	May 8, 2009
18	France, Saint-Pierre-et-Miquelon	May 8, 2009
19	Gabon	May 12, 2009
20	Gambia	May 4, 2009
21	Guinea	May 11, 2009
22	Guinea-Bissau	May 8, 2009
22a	Guyana	May 12, 2009
23	Mauritania	May 11, 2009
23a	Mauritius in the Chagos Archipelago region	May 6, 2009
24	Mexico en el polígono oriental del Golfo de México	May 6, 2009

(continued)

Table 11.2 (continued)

Submission by [State]	Preliminary information submitted by [State]	Received date
25	Micronesia (Federated States) for the Eauripik Rise and Mussau Ridge areas	May 5, 2009
25a	Morocco	August 3, 2015
26	Mozambique	May 11, 2009
27	New Zealand—Tokelau	May 11, 2009
28	Nicaragua	April 7, 2010
29	Oman in respect of the area that includes the narrow bathymetric shelf adjacent to the land mass of Oman, the Owen basin, and the Owen Ridge, and that abuts the deep ocean floor of the Arabian Indian Sea (Indian fan)	April 15, 2009
30	Papua New Guinea for the Mussau Ridge and Eauripik Rise Areas	May 5, 2009
31	Republic of Korea	May 11, 2009
32	Sao Tome and Principe	May 13, 2009
33	Senegal	May 12, 2009
34	Seychelles in the Aldabra Island Region	May 8, 2009
35	Sierra Leone	May 12, 2009
36	Solomon Islands	May 5, 2009
37	Somalia	April 14, 2009
38	Spain en el área al Oeste de las Islas Canarias	May 11, 2009
39	Togo	May 8, 2009
40	United Republic of Tanzania	May 7, 2009
41	Vanuatu	August 10, 2009

Table 11.3 Recommendations adopted by CLCS

Submission by [State]	Presentation to the CLCS*	Date of submission	Date of recommendations adopted
1	Russian Federation	December 20, 2001	June 27, 2002
1a.	Russian Federation, partial revised Submission in respect of the Okhotsk Sea	February 28, 2013	March 11, 2014
2	Brazil	May 17, 2004	April 4, 2007
2a	Brazil, partial revised Submission, in respect of the Brazilian Southern Region	April 10, 2015	March 8, 2019
3	Australia	November 15, 2004	April 9, 2008
4	Ireland, Porcupine Abyssal Plain	May 25, 2005	April 5, 2007
5	New Zealand	April 19, 2006	August 22, 2008

(continued)

Table 11.3 (continued)

Submission by [State]	Presentation to the CLCS*	Date of submission	Date of recommendations adopted
6	Joint submission by France, Ireland, Spain and the United Kingdom of Great Britain and Northern Ireland, in the area of the Celtic Sea and the Bay of Biscay	May 19, 2006	March 24, 2009
7	Norway, in the Northeast Atlantic and the Arctic	November 27, 2006	March 27, 2009
8	France, in respect of the areas of French Guiana and New Caledonia	May 22, 2007	September 2, 2009
9	Mexico, in respect of the western polygon in the Gulf of Mexico	December 13, 2007	March 31, 2009
10	Barbados	May 8, 2008	April 15, 2010
10a	Barbados, revised	July 25, 2011	April 15, 2012
11	United Kingdom of Great Britain and Northern Ireland, Ascension Island	May 9, 2008	April 15, 2010
12	Indonesia—northwest of the Sumatra Island	June 16, 2008	March 28, 2011
13	Japan	November 12, 2008	April 19, 2012
14	Joint submission by the Republic of Mauritius and the Republic of Seychelles, in the region of the Mascarene Plateau	December 1, 2008	March 30, 2011
15	Suriname	December 5, 2008	March 30, 2011
17	France, areas of the French Antilles and the Kerguelen Islands	February 5, 2009	April 19, 2012
21	Uruguay	April 7, 2009	Augest 19, 2016
22	Philippines, in the Benham Rise region	April 8, 2009	April 12, 2012
23	The Cook Islands, concerning the Manihiki Plateau	April 16, 2009	August 19, 2016
24	Fiji	April 20, 2009	
25	Argentina	April 21, 2009	March 11, 2016
25a	Argentina, partial revised submission	October 28, 2016	March 17, 2017
26	Ghana	April 28, 2009	September 5, 2014
27	Iceland, in the Ægir Basin area and in the western and southern parts of Reykjanes Ridge	April 29, 2009	March 10, 2016
28	Denmark, in the area north of the Faroe Islands	April 29, 2009	March 11, 2014
29	Pakistan	April 30, 2009	March 13, 2015
30	Norway, in respect of Bouvetøya and Dronning Maud Land	May 4, 2009	April 30, 2019
31	South Africa, in respect of the mainland of the territory of the Republic of South Africa	May 5, 2009	March 17, 2017
32	Joint submission by the Federated States of Micronesia, Papua New Guinea and Solomon Islands, concerning the Ontong Java Plateau	May 5, 2009	March 17, 2017
39	Seychelles, concerning the Northern Plateau region	May 7, 2009	August 27, 2018

References

Douglas DH, Peucker TK (1973) Algorithms for the reduction of the number of points required to represent a digitized line or its caricature. Cartographica: Int J Geograp Inform Geovisualization 10 (2):112–122

Dyment J (1991) Structure et evolution de la lithosphere oceanique dans I'ocean indien reppoert des anomalies magnetiques. Dissertation, University Lius Pasteur de Strsbourg

Feng SZ, Li FQ, Li SQ (1999) Introduction to Marine Science (in Chinese). Higher Education Press, Beijing

Hedberg HD (1976) Ocean boundaries and petroleum resources. Science 191:1009–1018

Huang MT, Zhai GJ, Ouyang YZ et al (2001a) Fusion processing technology for multi-beam and single-beam sounding data (in Chinese with English abstracts). Acta Geod Cartogr Sin 30(4):299–303

Huang XY, Zhai GJ, Sui LF et al (2001b) The effect of optimizing training samples on bathymetric outliers in the LS-SVM algorithm (in Chinese with English abstracts). Acta Geod Cartogr Sin 40 (1):22–27

Liu ZJ, Liu J, Jin JY (2007) Study on automatic drawing of the FOS line of the outer continental shelf (in Chinese with English abstracts). Comput Eng Appl 43(31):240–241

Lyu WZ, Zhang HW, Fang YX et al (2012) Continental shelf limits: the scientific and legal interface (in Chinese). China Ocean Press, Beijing

Ou ZQ, Vanicek P (1996) Automatic tracing of the foot of the continental slope. Mar Geod 19(2):181–195. https://doi.org/10.1080/01490419609388078

Peng RC, Dong J, Zheng YD et al (2010) Comparison of the efficiency of the offset method and the Douglas-Peucker algorithm in removing redundant vertices (in Chinese with English abstracts). Bull Surv Mapp 3:66–71

Peng RC, Wang JY (2002) Research on buffer construction technique based on earth ellipsoid (in Chinese with English abstracts). Acta Geod Cartogr Sin 31(3):270–273

Peng RC, Wang JY, Tian Z et al (2005) Research on the selection of the base point of the territorial sea based on convex hull construction technique (in Chinese with English abstracts). Acta Geod Cartogr Sin 34(1):53–57

Peter JC, Carleton C (eds) (2000) Continental shelf limits:the scientific and legal interface. Oxford University Press, London

Schlich R, Dyment J, Munschy M (1990) Structure and age of the Masacrene and Madagascar Basins. Abstract In: Volcanisme intraplaque le point chaud de la Reunion, Conference Proceedings. lle de la Reunion: University De la Reunion et l'l.P.G.P

Sun CB, Li YK, Zhang ZH (2012) Image segmentation based on Douglas-Peucker algorithm (in Chinese with English abstracts). Geomat Spat Inf Technol 35(5):33–38

United Nations (1993) Article 76 of the Convention on the Law of the Sea. Available via DIALOG. http://www.un.org/Depts/los/convention_agreements/texts/unclos/part6.htm. Accessed 12 July 2019

Wang HD, Chai HZ, Wang M (2011) Robust Kriging fitting of multi-beam bathymetric data (in Chinese with English abstracts). Acta Geod Cartogr Sin 40(2):238–248

Yang FL, Li JB, Wu ZY et al (2008) Fine processing method for shallow water multi-beam survey data (in Chinese with English abstracts). Acta Geod Cartogr Sin 37(4):444–457

Yang FL, Li JB, Wu ZY et al (2009) Multi-beam bathymetric instantaneous attitude error correction method (in Chinese with English abstracts). Acta Geod Cartogr Sin 38(5):450–456

Zhang LH, Jia SD, Wu C et al (2011) Digital water depth model interpolation method considering uncertainty (in Chinese with English abstracts). Acta Geod Cartogr Sin 40(3):359–365

Zhang LH, Song GD et al (2012) Adaptive grid water depth modeling method based on regional average vertical uncertainty (in Chinese with English abstracts). Acta Geod Cartogr Sin 41(3):454–460

Zhao YQ, Xie CJ, Qiao YL et al (2008) Douglas-Peucker compression algorithm based on the extremum point (in Chinese). Software Guide 11(7):60–62

12 Mineral Resources Assessment of the International Seabed

The global seafloor contains a variety of potentially useful mineral deposits whose distribution closely corresponds with seafloor topography. Such deposits include polymetallic nodules in ocean basins, cobalt crusts on seamounts, and polymetallic sulfides along mid-ocean ridges. Motivated by the potential importance of these types of deposits as strategic mineral resources, China has carried out scientific investigations of seafloor mineral resources for more than 30 years, and has applied for several exploration contract areas from the International Seabed Authority (ISA) to lay the groundwork for possible commercial exploitation. This chapter is based on relevant research carried out by Chinese scientific research cruises and related progress in international research and mineral exploration.

As early as the first half of the 19th century, scientists had begun to recognize the ocean's potential as a source for mineral resources. During the 1980s, commercial mining of polymetallic nodules was attempted, although it was eventually shelved because of fluctuations in world metal prices, technical reasons, and environmental problems. The United Nations Convention on the Law of the Sea, which came into force in 1994 and was ratified by China in 1996, establishes a legal framework for conducting exploration of deep-sea mineral resources under the auspices of the ISA. Accordingly, the international seabed as defined by the jurisdiction of the ISA (i.e., the area) comprises: the regime for the seabed and ocean floor and subsoil thereof beyond the limits of national jurisdiction, the exclusive economic zone and continental shelf, and is not subject to any national jurisdiction. Thus defined, the international seabed covers an area of 251.7 million km^2 or 49% of the earth's total surface area.

In recent years, marine mining companies have concentrated on the integration of advanced technologies from around the world, and subsequently deep-sea mining technology has greatly improved. With continued technological improvements, it is likely that viable commercial exploitation of deep-sea mineral resources will be achieved. With the gradual depletion of terrestrial mineral resources, exploration of the international seabed for mineral resources is being pursued with increasing vigor and competitiveness.

By 2019, there have been totally 30 contracts for polymetallic nodules, polymetallic sulfides and cobalt crusts exploration in the area approved by the ISA (Table 12.1), which nearly covered the international seabed area with more than 1.5 million km^2. The number of the contract areas for polymetallic nodules, polymetallic sulfide and cobalt crust resources exploration are 18, 7 and 5 respectively. As for polymetallic nodules, sixteen contract areas lie in the Clarion–Clipperton Zone (CCZ) of the East Pacific Ocean, and the other two are located on the central Indian Ocean Basin, and western Pacific Ocean, respectively. The contract area for polymetallic sulfides exploration are located on the mid-ocean ridges of the Atlantic Ocean and the Indian Ocean, respectively. Most of the contract areas for cobalt crusts lie on the seamount area of the western Pacific Ocean, and only one of them is located on the Rio Grande Rise, South Atlantic Ocean.

Z. Wu et al., *High-resolution Seafloor Survey and Applications*,
https://doi.org/10.1007/978-981-15-9750-3_12

Table 12.1 Exploration contracts for polymetallic nodules, polymetallic sulfides, and cobalt crusts in the area

Contractor		Contract commencement date	Guarantor	Approximate location of contract area	Termination date of contract
Polymetallic nodules					
1	Interocean metal Joint Organization	March 29, 2001	Bulgaria, Cuba, Czech Republic, Poland, Russian Federation and Slovakia	Clarion-Clipperton Fracture Zone	March 28, 2021
2	Southern Production Association for Marine Geological Operations	March 29, 2001	Russian Federation	Clarion-Clipperton Fracture Zone	March 28, 2021
3	Government of the Republic of Korea	April 27, 2001	Republic of Korea	Clarion-Clipperton Fracture Zone	April 26, 2021
4	China Ocean Mineral Resources R & D Association	May 22, 2001	China	Clarion-Clipperton Fracture Zone	May 21, 2021
5	Deep Sea Resources Development Co., Ltd.	June 20, 2001	Japan	Clarion-Clipperton Fracture Zone	June 19, 2021
6	IFREMER	June 20, 2001	France	Clarion-Clipperton Fracture Zone	June 19, 2021
7	Government of India	March 25, 2002	India	Central Indian Ocean Basin	March 24, 2022
8	Federal Institute for Geosciences and Natural Research	July 19, 2006	Germany	Clarion-Clipperton Fracture Zone	July 18, 2021
9	Nauru Ocean Resources	July 22, 2011	Nauru	Clarion-Clipperton Fracture Zone (reservation)	July 21, 2026
10	Tonga Offshore Mining	January 11, 2012	Tonga	Clarion-Clipperton Fracture Zone (reservation)	January 10, 2027
11	UK Seabed Resources Ltd.	February 8, 2013	Britain	Clarion-Clipperton Fracture Zone	February 7, 2028
12	G-TEC Sea Minerals NV	January 14, 2013	Belgium	Clarion-Clipperton Fracture Zone	January 13, 2028
13	Marine Minerals Singapore Ltd.	January 15, 2015	Singapore	Clarion-Clipperton Fracture Zone	January 14, 2030
14	Marawa	January 19, 2015	Kiribati	Clarion-Clipperton Fracture Zone (reservation)	January 18, 2030
15	UK Seabed Resources Ltd.	March 29, 2016	Britain	Clarion-Clipperton Fracture Zone	March 28, 2031
16	Cook Islands Investment Corporation	July 15, 2016	Cook Islands	Clarion-Clipperton Fracture Zone (reservation)	July 14, 2031
17	China Minmetals Corporation	May 12, 2017	China	Clarion-Clipperton Fracture Zone (reservation)	May 11, 2032
18	Beijing Pioneer High-Tech Development Corporation	October 18, 2019	China	Western Pacific Ocean	October 17, 2034
Polymetallic sulfides					
1	Government of the Republic of Poland	February 12, 2018	Republic of Poland	Mid-Atlantic ridge	February 11, 2033
2	Government of Indian	September 26, 2016	India	Central Indian Ocean	September 25, 2031
3	Federal Institute for Geosciences and Natural Research	May 6, 2015	Germany	Central Indian Ocean	May 5, 2030

(continued)

Table 12.1 (continued)

Contractor		Contract commencement date	Guarantor	Approximate location of contract area	Termination date of contract
4	IFREMER	November 18, 2014	France	Mid-Atlantic ridge	November 17, 2029
5	Government of the Republic of Korea	June 24, 2014	Republic of Korea	Central Indian Ocean	June 23, 2029
6	Government of the Russian Federation	October 29, 2012	Russian Federation	Mid-Atlantic ridge	October 28, 2027
7	China Ocean Mineral Resources R & D Association	November 29, 2011	China	Southwest Indian ridge	November 28, 2026
Cobalt crusts					
1	Government of the Republic of Korea	March 27, 2018	Republic of Korea	Western Pacific Ocean	March 26, 2033
2	Companhia De Pesquisa de Recursos Minerais	November 9, 2015	Brazil	South Atlantic Ocean	November 8, 2030
3	Ministry of Natural Resource and Environment of the Russia	March 10, 2015	Russian Federation	Western Pacific Ocean	March 9. 2030
4	Japan Oil, Gas, and Metals National Corporation	January 27, 2014	Japan	Western Pacific Ocean	January 26, 2029
5	China Ocean Mineral Resources R & D Association	April 29, 2014	China	Western Pacific Ocean	April 28, 2029

12.1 Polymetallic Nodules

12.1.1 Introduction

It has been about 150 years since the first discovery of the polymetallic nodules in deep-sea. On February 18, 1873, the British *Challenger* expedition led by HMS *Challenger* collected polymetallic nodules for the first time from the seafloor—approximately 300 km southwest of Faroe Island in the Canary Islands west of Morocco—during its first circumnavigational survey of the three oceans from 1872 to 1876. In 1891, Murray and Renard (1891) published the first reports on the occurrence, shape, and composition of polymetallic nodules based on the *Challenger* samples. Extensive sampling of polymetallic nodules from the Pacific Ocean was conducted by the American research vessel *Albatross* on cruises extending from 1899 to 1900 and from 1904 to 1905. In 1902, Agassiz confirmed the widespread distribution of nodules in the southeastern Pacific Ocean based on the *Albatross* survey data and drew a nodule distribution map. Importantly, this map identified a high-density, east-west-trending ore belt distributed between 6°30′N and 20°N.

Prior to the end of World War II, research into polymetallic nodules remained relatively limited. After World War II, from 1947 to 1948 the Swedish Deep Sea Exploration Team obtained a large amount of data on polymetallic nodules and cobalt crusts in the equatorial region of the central and western Pacific Ocean (Landergren et al. 1964; Landergren and Joensuu 1965). Following this, intermittent investigations of polymetallic nodules were conducted with investigations being primarily exploratory and academic. At the time, inadequate analytical techniques and technical limitations in marine surveying prohibited accurate analyses of the composition and potential resource value of polymetallic nodules.

Based on an analysis of polymetallic nodule samples from 110 stations, Mero (1962) was the first to identify the potential economic value of polymetallic nodules. Consequently, the United States, France, the Soviet Union, Japan and other countries began large-scale investigations and studies of marine polymetallic nodule resources. The scope of surveys for polymetallic seabed resources started in the Atlantic Ocean, Indian Ocean and Pacific Ocean, and gradually shifted to focus on the South Pacific, Central Pacific, Northeast Pacific and equatorial Pacific regions. Investigations have since tended to focus on the Clarion-Clipperton Fracture Zone (CCFZ). The most abundant nodule resources have been found in the contract exploration areas obtained by countries and the relevant international consortia (Fig. 12.1). India has also acquired exploration areas for polymetallic nodules in the Indian Ocean (Fig. 12.2).

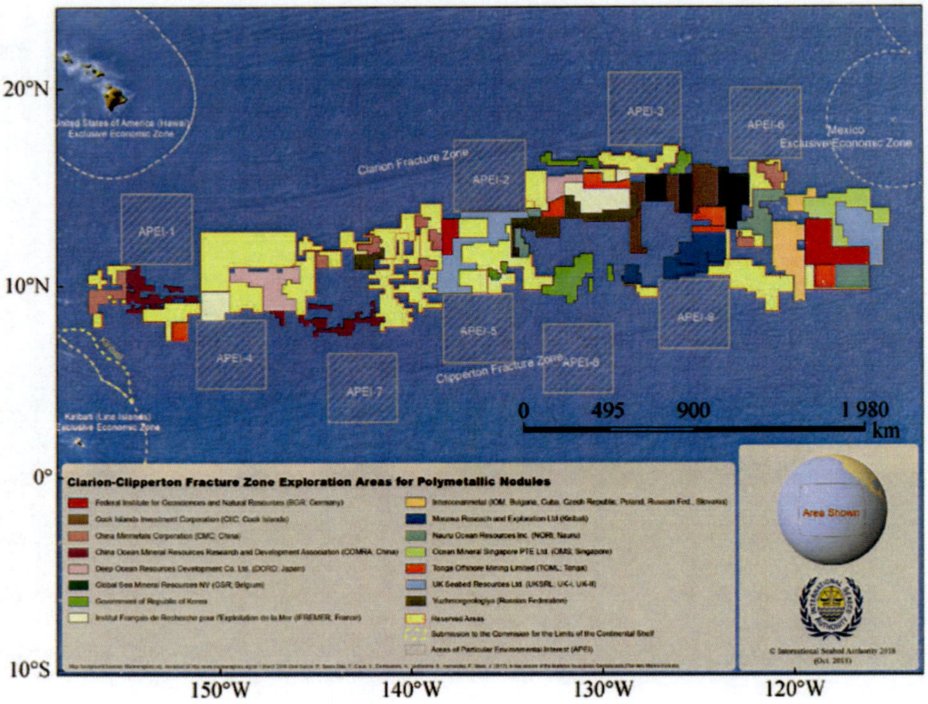

Fig. 12.1 Locations of exploration contract areas for polymetallic nodules and the international seabed authority's reserve areas in the Pacific Ocean (https://www.isa.org.jm/)

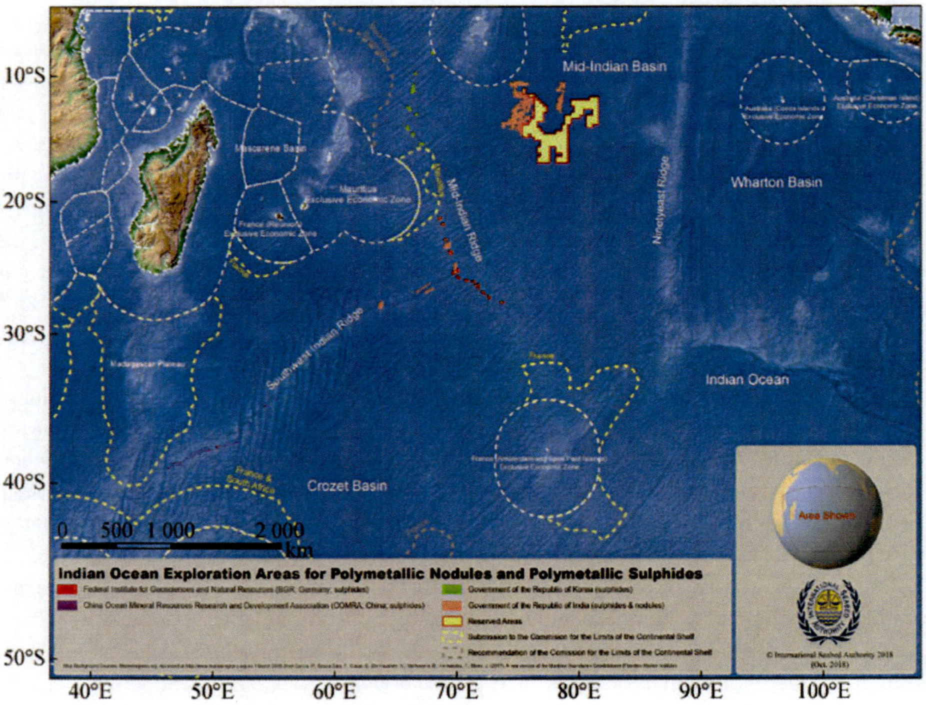

Fig. 12.2 Locations of exploration contract areas for polymetallic nodules and the international seabed authority's reserve areas in the Indian Ocean (https://ww.isa.org.jm/)

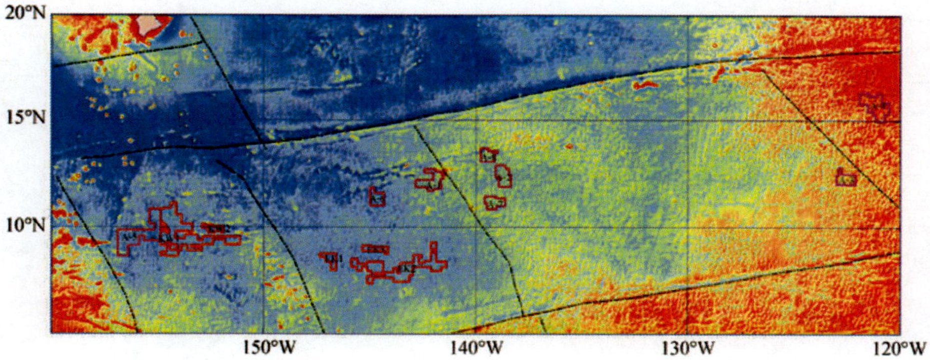

Fig. 12.3 Distribution of Chinese contract areas for exploration of polymetallic nodules. The western (KW1, KW2) and eastern (EK1–EK3) areas with red blocks are exploration contract areas signed by the China Ocean Mineral Resources Research and Development Association and the International Seabed Authority. The A-1-A-8 areas with magenta blocks are exploration contract areas approved for the China Minmetals Group Co., Ltd. Black lines are the fault zones (https://www.isa.org.jm/)

On August 22, 1990, through the China Ocean Mineral Resources Research and Development Association (COMRA), China submitted to the Preparatory Committee of the United Nations Seabed Authority the Application of the Government of the People's Republic of China for Registration of China Ocean Mineral Resources Research and Development Association as a Pioneer Investor. Approval was granted on March 5, 1991. After India, France, the Soviet Union and Japan, China has become the fifth "pioneer investor in deep-sea mining" registered with the United Nations and has obtained exploration rights for 150,000 km^2 of polymetallic nodules in the CCFZ of the eastern Pacific Ocean. Following two regional abandonments in 10 years, China obtained rights polymetallic nodule exploration areas covering 75,000 km^2 (Fig. 12.3). In 2001, the COMRA signed an exploration contract with the ISA to implement a three-stage exploration work plan to be carried out from 2001 to 2016. The exploration contract expired on May 21, 2016. Because the technical, economic and environmental conditions were not yet suitable for commercial mining, COMRA submitted an application to the ISA in November 2015 for a five-year extension of the exploration contract. With its approval, the exploration contract has been extended to 2021.

In addition to COMRA's polymetallic nodule exploration contract area, China Minmetals Group Co., Ltd. submitted an application to the Preparatory Committee of the United Nations Seabed Authority on August 8, 2014, for a work plan for polymetallic nodule exploration, which was approved on July 20, 2015.

12.1.2 Basic Patterns in the Distribution of Polymetallic Nodules

Geographically, polymetallic nodules are widely distributed across the global seafloor. Figure 12.4 displays data from 3,651 polymetallic nodule stations around the world. Polymetallic nodules are distributed most densely in the Pacific Ocean, followed by the Indian Ocean and the Atlantic Ocean. Globally, promising areas for polymetallic nodules are primarily located in the central and eastern Pacific Ocean and in the Indian Ocean—an area covering more than 1.6×10^8 km^2. Especially promising areas include the seafloor area of the CCFZ, the Central Pacific Basin, the Western Pacific Basin, the Peru Basin and the Central Indian Ocean Basin. The most promising area is the CCFZ (Hein et al. 2013). Previous research (Petersen et al. 2016) has indicated that the sedimentation rate of the polymetallic nodule-rich CCFZ in the East Pacific Ocean is less than 10 mm/ka (Fig. 12.5). The manganese, copper, cobalt and nickel contents of polymetallic nodules in the CCFZ reach 30%, 1.5%, 1.0% and 2.0%, respectively (Hein et al. 2013). The combined metal contents of copper, cobalt and nickel can reach 3.5%. The average abundance of polymetallic nodules in this region is 10 kg/m^2, and the total amount of dry nodules is approximately 2.1×10^{11} t (Hein et al. 2013). The Central Indian Ocean Basin is another potential mining area for polymetallic nodules. There, the combined metal content of copper, cobalt and nickel in nodules can reach 2%, but the abundance of nodules is lower than that of the Pacific CCFZ (Lv et al. 2008).

The spatial distribution of nodule abundance and percent cover are closely related to seafloor topography and water depth (Hein et al. 2013). Sedimentary nodules are primarily distributed in the basins surrounding seamounts, whereas sedimentary-diagenetic nodules are primarily distributed on deep-sea hills, plains and plateaus (Hein et al. 2013). Near seamounts, nodule morphology is typically smooth, conjoined and clastic. In contrast, cauliflower nodules are recovered in areas dominated by low gentle hills. Almost all types of nodules are found in intermountain basins. Myrica nodules are found only in deep-water basins and low-gentle hills.

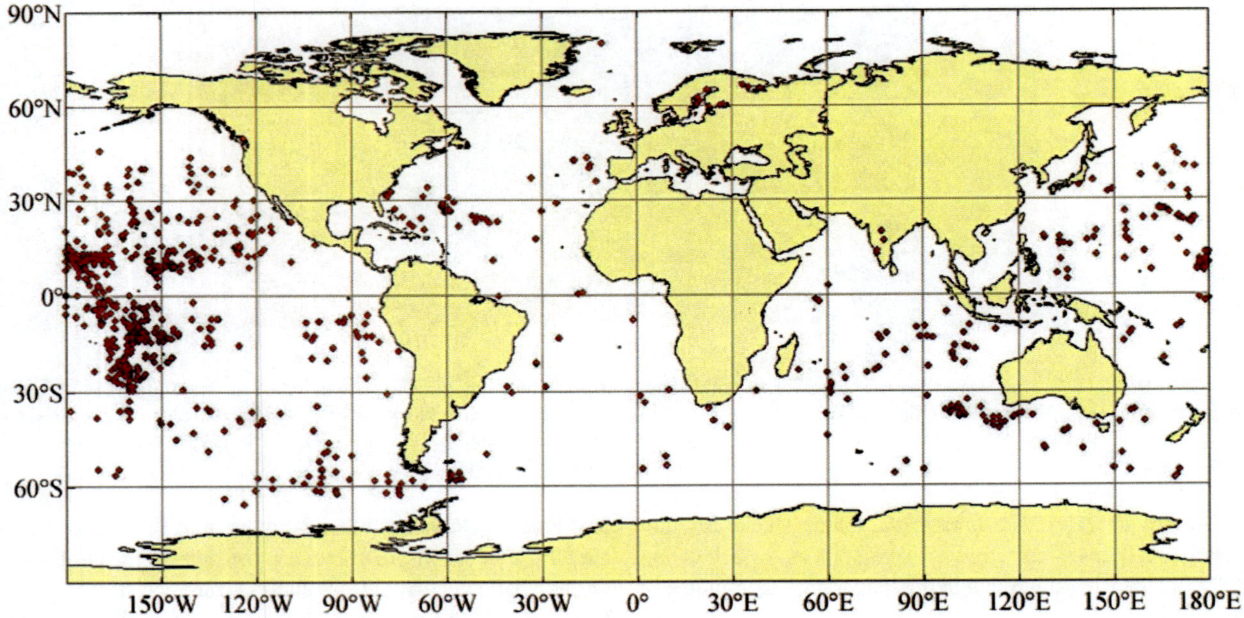

Fig. 12.4 Global distribution of polymetallic nodules (https://www.isa.org.jm/)

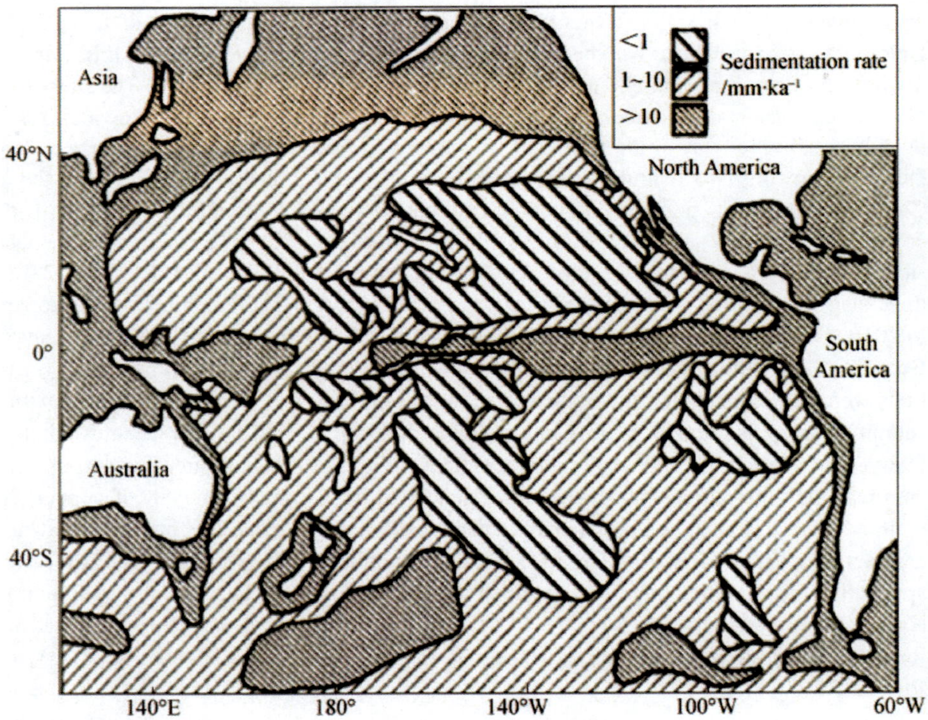

Fig. 12.5 Distribution map of seafloor sedimentation rates in the Pacific Ocean (Piper and Williamson 1977)

The distribution of nodule abundance varies with seafloor topography and differences in geomorphology. Generally, nodule abundance decreases gradually from seamounts or chain seamounts to deep plains or hilly areas. Nodule abundance also varies in different parts of the same topographic unit. Nodule abundance is the lowest on steep slopes >10°, followed by gentle slopes <5° and is the highest on intermediate slopes of 5°–10°. Nodule abundance also varies in different regions of the same terrain, generally increasing from north to south. Nodule size varies with different topographic units. Generally, large- and medium-sized nodules are dominant in low-gentle hilly areas, medium and small-sized nodules are dominant in seamount or chain-seamount areas, and large-, medium- and small-sized nodules are distributed in intermountain basins (Zhang 2001).

12.1.3 Method for Evaluating Polymetallic Nodule Resources

The main variables involved in the evaluation of polymetallic nodule resources and the delineation of mining areas are nodule abundance, nodule Mn, Cu, Co and Ni contents, nodule water content, water depth, seafloor topography, hydrometeorology, sediment type and the engineering and mechanical properties of subsoils (Zhang 1995). The selected methods for evaluating polymetallic nodule resources in mining areas directly affect the accuracy of calculations. Polymetallic nodules are deep-sea mineral resources with a two-dimensional distribution. Estimation methods for these resources include the arithmetic average method and the Kriging method (Zhang 1995; Zhang et al. 2001b).

12.1.3.1 Arithmetic Average Method

The arithmetic average method has been widely applied in the calculation of polymetallic nodule resources resulting in relatively satisfactory estimates. The main variables involved in estimating polymetallic nodule resources are total area, polymetallic nodule abundance, grade (e.g., total Mn, Cu, Co and Ni contents) and water content. The arithmetic average method utilizes information on nodule abundance, metal grade and water content to estimate the arithmetic mean value of resource quantities. The quantities of wet nodules, dry nodules and metals (Mn, Cu, Co and Ni) in the mining area are calculated via the average arithmetic method according to the following formulas:

$$P = S \times \bar{F} \tag{12.1}$$

$$Q = P \times (1 - \bar{W}) \tag{12.2}$$

$$M_i = Q \times \bar{C}_i \, (i = \text{Mn, Cu, Co, Ni}) \tag{12.3}$$

where P is the wet nodule weight in the mining area; S is the area of the mining area; $\bar{F}$ is the average nodule abundance within the mining area; Q is the dry nodule weight in the mining area; $\bar{W}$ is the average water content of nodules in mining area; and M_i is the total metal contents consisting of Mn, Cu, Co and Ni.

12.1.3.2 Kriging Method

The Kriging method makes evaluations based on the theory of regionalized variables using a variogram to analyze natural phenomena including geological phenomena that are spatially distributed with specified randomness and structure. The method places weights on sample data in or near the field being evaluated. Steps of the Kriging method are described below.

1. Block segmentation

With information from the exploration stage and the distributional characteristics of exploration engineering, the exploration area is divided into blocks such that the geological bodies are maximally distributed to all blocks.

2. Determination of adjacent areas of block segment

Adjacent areas are specified according to the range of the semivariogram and shielding effect.

3. Estimator algorithms

$$C_k = \sum \lambda_i C_i \tag{12.4}$$

$$C = \frac{1}{n} \sum C_k \tag{12.5}$$

where C_k is the factor estimator of a block (e.g., grade); λ_i is a Kriging optimal weight coefficient, obtained by solving Kriging equations with variograms; C_i is a known factor value of a block and its adjacent area; C is a total factor estimator, and n is the number of blocks in a mining area.

4. Resource quantity algorithms

$$Q_i = S_i \times C_i \times (1 - \overline{W}) \tag{12.6}$$

where Q_i is the nodule resource quantity (kg), S_i is the nodule area (m^2); C_i is the nodule abundance (kg/m^2), and $\overline{W}$ is the average water content of nodules in a mining area.

$$P_i = Q_i \times C_i / 100 \tag{12.7}$$

where P_i is the mass of the metal in a block (kg); and C_i is the average percentage grade of the metal in the nodules in that block.

12.2 Cobalt Crusts

12.2.1 Introduction

From 1872 to 1876, the cruise of the HMS *Challenger* discovered a variety of cobalt crusts and encrustations collected from shallow water depths (<370 m) during its expedition around the world. Analysis by Buchanan, a chemist on board the HMS *Challenger*, and his colleagues demonstrated that the samples contained important trace elements such as Cu, Ni and Co, and also Th and Tl, which were discovered later. Following these analyses, no further investigation of cobalt crust resources was carried out for the next 60 years.

In the 1970s, the potential economic value of cobalt crusts was again recognized, attracting the attention of governments and academics around the world. Professor Halbach of the Clausthal-Zellerfeld University of Technology, Germany, presided over the first voyage of the R/V *Sonne* during 1981, and carried out the first investigation of the distribution of cobalt crusts in the Line Islands south of Hawaii in the central Pacific Ocean (Halbach et al. 1982). Since then, the former Soviet Union, the US, France, Australia, South Korea, Japan and China have all invested a great deal of human, material and financial resources in the investigation and study of cobalt crust resources.

The cobalt crusts are generally considered to be enrichment of many metal elements such as Co, Ni, Cu, Mn and Pt (He et al. 2005). There have been five contract areas for cobalt crust exploration in the area since 2014, which were assigned between the ISA and COMRA, Japan Oil, Gas and Metals National Corporation, Ministry of Natural Resource and Environment of the Russia, Companhia De Pesquisa de Recursos Minerais (Brazil) and the Government of the Republic of Korea. Four of them are located on the Marcus-Wake and Magellan Seamounts of the western Pacific Ocean, and one of them lies on the Rio Grande Rise of the South Atlantic Ocean.

12.2.2 Distribution of Cobalt Crusts

Cobalt crusts are concentrated on the sediment-free tops and slopes of seamounts, ridges and platforms. Seamounts are the primary hosts of cobalt crust growth. There are more than 8,000 seamounts with >1,000 m in the Pacific Ocean and relatively fewer in the Atlantic and Indian Oceans. Among these, only a small proportion (<15%) of seamounts have been investigated with detailed submarine topographic surveys, mapping and geological sampling (Hein 2004) (Fig. 12.6). Cobalt crusts in the Pacific Ocean, from the Aleutian Trench in the north to the area surrounding the Antarctic Ridge in the south, exhibit a slight decrease in Mn, Co, Ni and a slight increase in Fe, Cu from the mid-Pacific to the western continental margin (Hein 2004). From the mid-ocean ridge to the western continental margin of the Atlantic Ocean, Mn exhibits a slight downward trend, while Fe, Co, Ni, Cu exhibit a slight upward trend. From the mid-ocean ridge to the eastern continental margin of the Indian Ocean, Mn, Fe, Co exhibit an upward trend, while Cu exhibits a slight downward trend, and Ni exhibits a parabolic distribution. Generally, the Fe contents of cobalt crusts in all three ocean basins exhibit an upward trend towards the continental margin, whereas variations in the

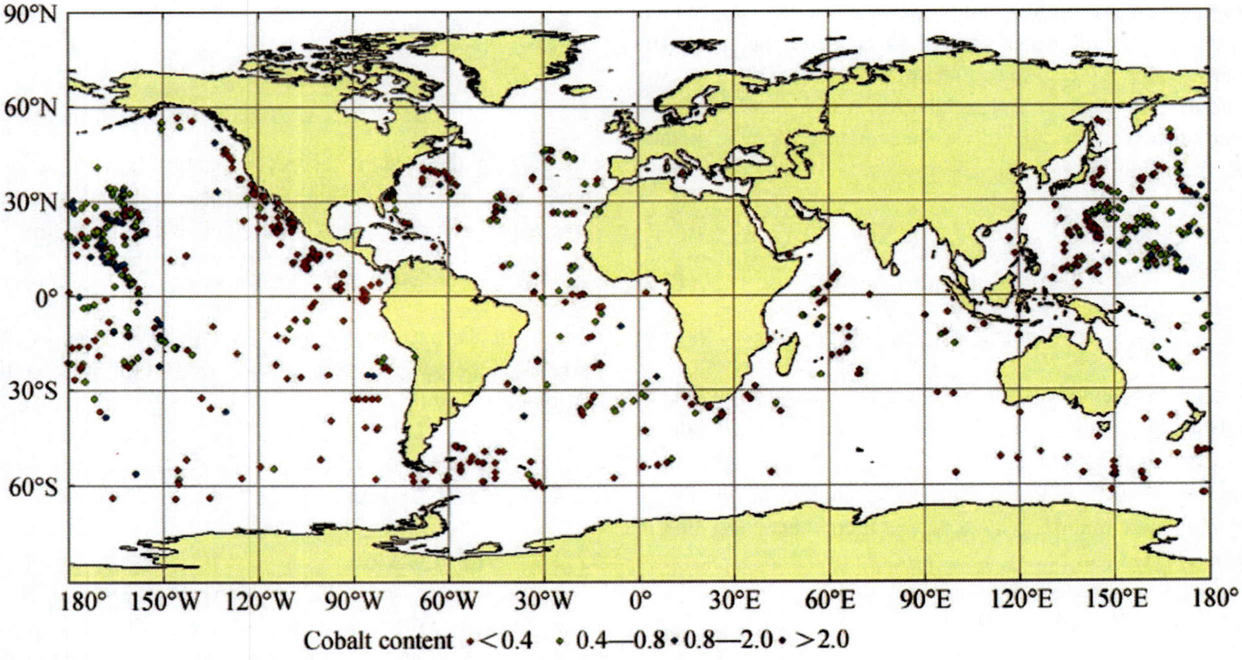

Fig. 12.6 Global distribution of Co content in cobalt crusts

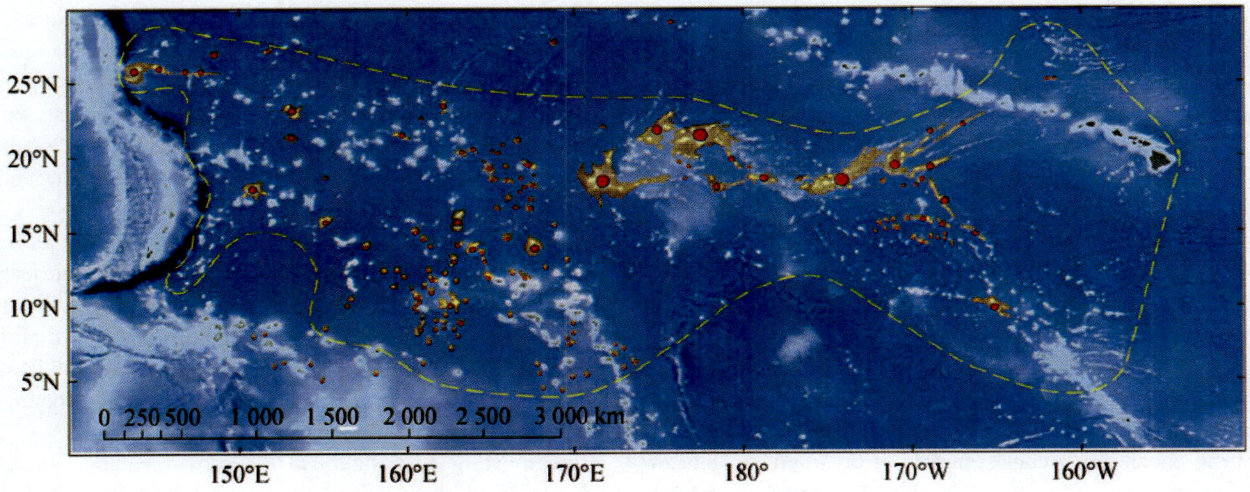

Fig. 12.7 Map of most permissive area for cobalt crusts in the western Pacific Ocean (Hein et al. 2009)

concentrations of other elements are more complex, perhaps related to the complexity of the topography and geomorphology of the seafloor. The western Pacific Ocean is an important area for cobalt crust resources. Hein et al. (2009) have presented an initial delineation of prospective seamounts for cobalt crust resources in the western Pacific Ocean (Fig. 12.7).

The complex morphology of a seamount is caused by long-term geological tectonic movements, as well as ongoing erosion and sedimentation. The two main topographical units at the summits and bases of seamounts are the primary zones of sedimentation as the relatively gentle slope can maintain long-term accumulations of sediments. Because of the relatively steep slope, the main body of the seamount is not conducive to the formation of large-scale sedimentation, and only locally flat or weakly concave areas have the conditions necessary for sufficient sedimentary accumulation. Thus, seafloor topography plays a fundamental role in restricting the spatial distribution of sediments, including the local spatial distribution of cobalt crusts.

As early as 1984, Cronan (1984) proposed that ferromanganese crusts usually occur in water depth of 400–4,000 m with most occurring between the depth of 1,000–3,000 m. Cobalt crusts typically develop in the vicinity of the Oxygen-Minimum Zone (OMZ, 800–2,200 m). However, cobalt crusts have been found in water depths as shallow as 25 m and as deep as 7,117 m (http://www.ngdc.noaa.gov). Thus, while water depth controls the growth and development of cobalt crusts to a certain extent, there is no absolute limit on the water depth suitable for growing cobalt crusts. The statistical results of cobalt crust stations with water depth <4,500 m (Fig. 12.8) show that most crusts are distributed at a depth of 1,000–2,500 m (Hein 2004). The thickest crust in the Pacific Ocean (1,500–2,500 m) corresponds to the outer zone of the top and upper slope of a Cretaceous-era seamount (Cronan 1984). The water depth corresponding to the thick cobalt crusts exhibits regional variations, with cobalt crusts in the South Pacific generally occurring at shallow depths because the Oxygen-Minimum Zone is not well developed; and the cobalt-rich and thickest crusts are located in the water depth 1,000–1,500 m (Cronan 1984). A bathymetric analysis of cobalt crust resources shows that the cobalt crusts in the western Pacific Ocean are mainly distributed at water depth <2,800 m, and that cobalt crusts in the central Pacific Ocean are mainly distributed in the water depth <2,500 m (Zhang et al. 2008a). Cobalt crusts of the Hawaiian

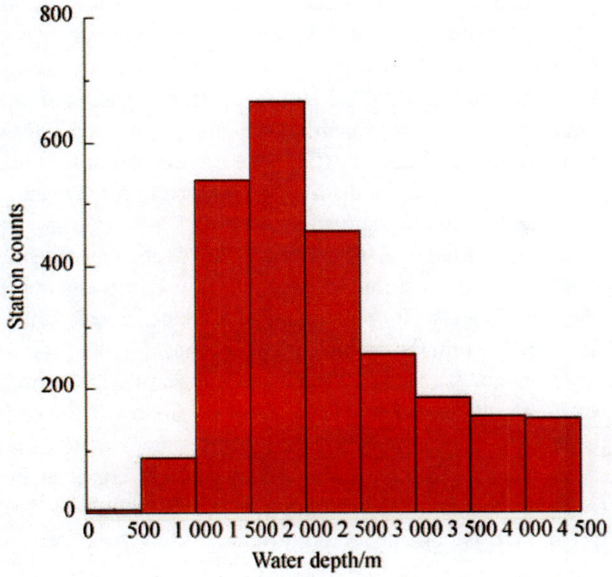

Fig. 12.8 Distribution of cobalt crusts according to water depth

seamount chain in the eastern Pacific Ocean are mainly distributed at the depth of 800–2,400 m (Craig et al. 1982). Across the Pacific Ocean, the general trend is that the depth of maximum distribution of cobalt crusts becomes deeper from east to west.

Continuous photography of the 1,700–3,500 m section of Lamont Guyot (Fig. 12.9) reveals distinct variations in seafloor sedimentation according to depth. For example, the depth interval 1,700–2,200 m is rich in sediment, and cobalt crusts are intermittently exposed. These cobalt crusts consist mainly of board crusts, whereas granule crusts and nodules are rare. The 2,200–2,600 m depth interval exhibits relatively light sedimentation, and cobalt crusts are more exposed, though some are covered with thin layers of sediment; cobalt crusts consist mainly of board crusts with few granule crusts. The 2,600–3,000 m depth interval exhibits abundant sediment, bedrock is exposed, and nodules appear on the surfaces of some cobalt crusts. The 3,000–3,500 m depth interval is rich in sedimentation and cobalt crusts consist of a small amount of board crusts and granule crusts. In general, the cobalt crusts of Lamont Guyot are mainly distributed in the 2,000–2,600 m depth interval, and the cobalt crusts consist primarily of board crusts, with fewer granule crusts and nodules (Fig. 12.9). Water depth thus has a great influence on the distribution of cobalt crusts, even at the scale of a single seamount.

In 1986, Manheim (1986) used single-channel seismic data to study the distribution of unnamed cobalt crusts in the Marshall Islands. It was found that thick cobalt crusts developed continuously on the upper slopes of the investigated seamounts, that moderate thicknesses of cobalt crusts were present on the middle and upper slopes of the seamounts, that cobalt crusts on the lower slopes of the seamounts were relatively thin, and that cobalt crusts at the foots of the seamounts' slopes are rare (Manheim 1986). Cobalt crusts in the middle and lower parts of the seamounts' slopes are primarily composed of encrustations, which are different from those that occur on rock outcrops. Typical cobalt crusts present on rock outcrops have thicker layers. Many seamounts or ridges covered by sediments do not grow cobalt crusts on the tops of seamounts. Cobalt crusts are usually thinner on slopes below 3,000 m water depth on islands and atolls (Hein et al. 1985).

Studies of cobalt crusts on the Magilanov Seamounts in the Pacific Ocean have shown these to be primarily present on bedrock surfaces in the 2,000–2,500 m depth interval where the slope of the seamount surface is less than 20° (Jin 1991). Seamount surfaces with slopes greater than 20° lack cobalt crusts and are instead covered with a very thin coating of iron-manganese oxide.

Investigating a seamount in the central Pacific Ocean, Yamazaki and Sharma (2000) found that the distribution of cobalt crusts is closely related to the slope of the seafloor. When the slope is 0°–3°, nodules are the predominant form of cobalt-rich encrustation. Slopes of 3°–7° are generally characterized by sedimentary cover, slopes of 7°–15° are typically characterized by nodule and cobalt crust transition areas, with the proportion of cobalt crusts increasing gradually with increasing of slope gradients and slopes greater than 15° typically lack substantial sedimentation such that cobalt crusts predominate (Yamazaki et al. 1994, 1995).

Zhang et al. (2006) analyzed the seafloor slope and cobalt crust resources of the Il'ichev Guyot of the Magellan Seamounts and Lamont Guyot as well as three satellite seamounts of the Wake-Marcus Seamounts. Regions of the Il'ichev Guyot where slope <15° are primarily located on the top platforms (shallower than 1,700 m) and lower slopes (deeper than 3,500 m) of the guyot. Cobalt crusts are primarily distributed on the top platforms, consisting of board crusts with some granule crusts and nodules. The 15°–23° slope region is typically zonally distributed along the middle and lower slopes of the guyot. Cobalt crusts in this region

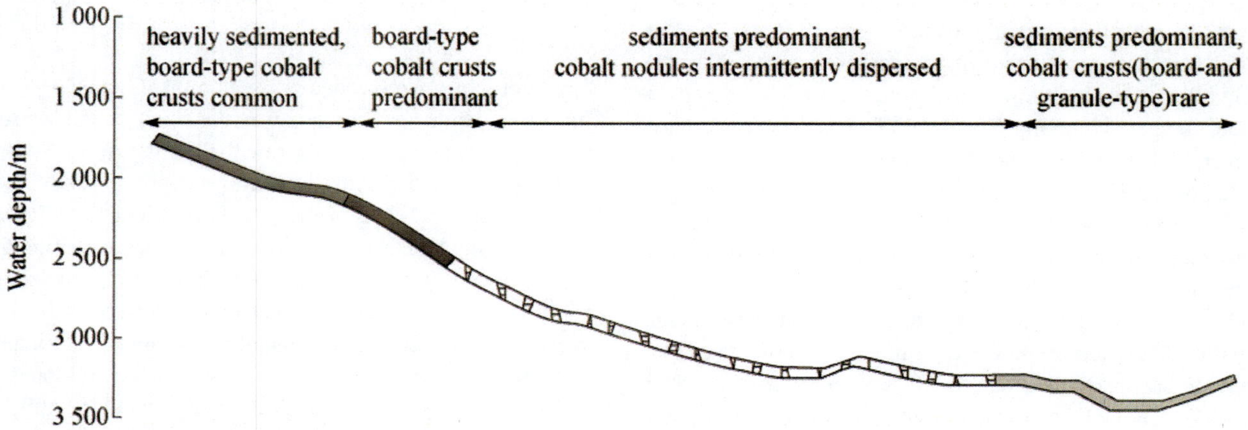

Fig. 12.9 Schematic diagram of cobalt crust distribution along a profile of Lamont Guyot

consist primarily of board crusts, where granule crusts are rare and nodules are not typically present. Steep slopes 23° are mainly distributed in the middle and upper parts of the guyot slopes outside the summit platform. Cobalt crusts in this region are primarily boarded crusts with fewer granule crusts and no nodules.

Regions of Lamont Guyot and its surrounding 3 satellite seamounts where slope <8° are located on the top platforms and the lower portions of the seamounts' slopes (deeper than 3,500 m). Cobalt crust types are more complex. Crust types are most commonly board crusts followed by granule crusts and nodules. The 8°–18° slope region is primarily distributed zonally along the middle and lower slopes of the seamounts. All three types of crusts are present in this region, with board crusts being the most common, followed by granule crusts and nodules. Zones of steep slope $\geq 18°$ are mainly distributed in the middle and upper slopes outside the guyot platforms and the tops of the seamounts. The slope is very steep, and crusts types consist mainly of board crusts, granule crusts, and nodules. Cobalt crust at the top of the seamounts is relatively well developed.

From these investigations, it can be seen that the development of cobalt crusts varies among different seamounts and slope segments and that the morphology of cobalt crusts can vary, even along different slope segments of the same seamount. Based on the investigation of more than 20 seamount sampling stations in the western and central Pacific Ocean where cobalt crusts have been found, the range of slopes on which cobalt crusts occur is 0.06°–33.85° and the average slope is 11.38° (Zhang et al. 2011a). From the data shown in Fig. 12.10, regions of seafloor slope can be divided into 0°–10°, 10°–15° and 15°–28° intervals. The 0°–10° slope region represents 45.22% of the total stations. The range of water depth in this region is 1,713–2,249 m with an average water depth of 2,061 m, and the range in the abundance of cobalt crusts is 57.13–94.67 kg/m^2 with an average abundance of 77.97 kg/m^2. The range of water depths for the 10°–15° slope region is 2,239–2,590 m with an average water depth of 2,408 m, and the range in the abundance of cobalt crusts in this region is 64.45–81.34 kg/m^2 with an average abundance of 73.67 kg/m^2. The range in water depth for the 15°–28° slope region is 2,205–2,583 m with an average depth of 2,392 m; and the range in the abundance of cobalt crusts is 24.33–83.70 kg/m^2 with an average abundance of 60.35 kg/m^2 (Zhang et al. 2011a).

In addition to the relationships between the distribution of cobalt crust resources, water depth and seafloor slope, some scholars have also researched the relationship between the distribution of cobalt crust resources and seamount type. For example, Russian scholars have carried out extensive investigations of cobalt crusts in the Magellan Seamounts and have concluded that, because there are many flat-topped seamounts and fewer pointed seamounts in the Magellan Seamounts, the Magellan Seamounts are more advantageous to the development of cobalt crusts than central Pacific seamounts (Schkoriniko et al. 2001). Because of the relatively small area at the top of the seamounts, there is no prospect for extensive development of cobalt crusts, and the cobalt crust of the flat top is mostly produced in the favorable water depth (the depth of economic value in the Magellan Seamounts is 1,000–3,000 m, that is, the top and sides of the seamounts, but less than 2,500 m in the central Pacific seamounts). The thickness of cobalt crusts in the Magellan Seamounts is twice that of seamounts in the central Pacific Ocean, and exhibit high level of coverage and abundance (Shen et al. 1999; Schkoriniko et al. 2001). In 2001, the Russian Southern Marine Geological Exploration Consortium used the R/V *Gelendzik* to drill 7 shallow boreholes on the IOAN platform in the Magellan Seamounts,

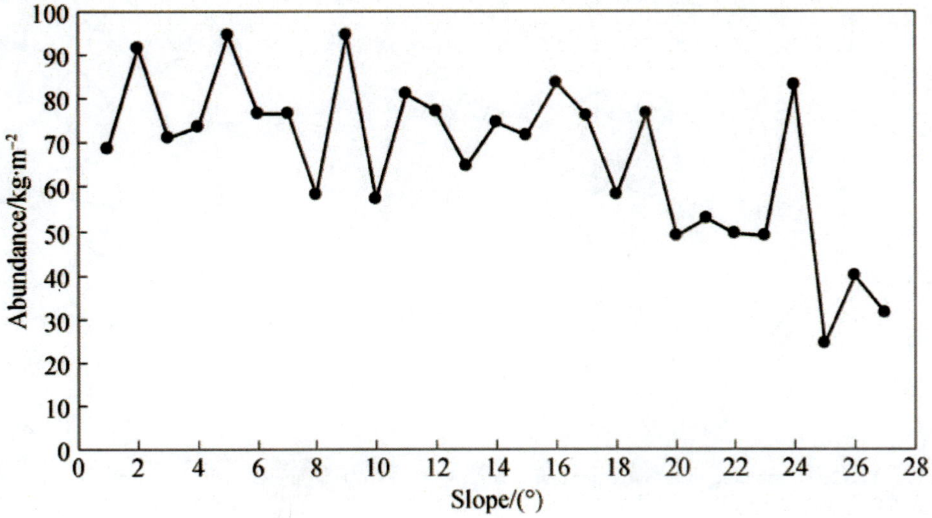

Fig. 12.10 Distribution of seafloor slope and the abundance of cobalt crusts on seamounts in the western and central Pacific Ocean (Zhang et al. 2011a)

Table 12.2 Distribution and metal grade of cobalt crusts categorized by geomorphological unit

Geomorphological unit	Thickness/cm	Mn/%	Fe/%	Co/%	Ni/%	Cu/%	Sample number
Flat tops of guyots	5.81	22.08	14.09	0.57	0.51	0.12	86
Tops of seamounts	5.07	19.58	13.07	0.45	0.38	0.09	4
Steep slope	5.66	21.89	14.46	0.60	0.48	0.12	72
Valley	6.06	22.91	14.35	0.60	0.50	0.12	12
Ridge	6.67	22.42	13.43	0.55	0.52	0.12	70
Foothills	5.52	21.91	14.33	0.58	0.48	0.12	38
Average	5.97	22.09	14.06	0.57	0.50	0.12	282

demonstrating that the crust thickness and the available metal content in the flat area at the top of the IOAN platform are respectively larger and higher than those of cobalt crusts occurring on the slope (Xu 2002). Chinese scholars have studied the types and distribution of cobalt crusts on seamounts in the central Pacific Ocean and have demonstrated that resources occurring on seamounts are superior to those of guyots (Chu et al. 2005; Ma et al. 2013). Based on the topography and geomorphography of seamounts in the western and central Pacific Ocean, the distribution of sampling stations located on the flat tops of guyots and the tops of seamounts, as well as slopes, ridges, valleys, and foothills can be tabulated (Table 12.2). From these data, it can be seen that there are many survey stations on slopes, ridges and flat-topped seamounts, while the tops of seamounts, valleys and foothills contain relatively few survey stations. The distribution of survey stations is based on an understanding of the distribution of cobalt crusts. Long-term investigations show that cobalt crusts are well distributed in geomorphologic units such as slopes, ridges and flat-topped seamounts. Grade and thickness data follow similar distribution rules. It should be noted that there are fewer seamounts in the Wake-Marcus Seamounts and fewer sampling stations located on the top of the seamounts. Thus, the distribution characteristics of cobalt crusts at the tops of seamounts need to be further studied.

In addition to seamount type, water depth and slope, the distribution of cobalt crusts on seamounts is affected by the orientation and slope directions of the seamounts (Zhang et al. 2008b, 2011b). To understand the distribution of resources based on the orientation of seamounts, different types of typical seamounts in the western and central Pacific Ocean were selected. Dividing seamounts into 11.25° tolerance angular sections taken from the center of the seamount centered on azimuthal directions with 0° being due north and degrees increasing clockwise, we collect data from 16 directions (0°, 22.5°, 45°, 67.5°, 90°, 112.5°, 135°, 157.5°, 180°, 202.5°, 225°, 247.5°, 270°, 292.5°, 315°, 337.5°). Unit area resources for each section (T/S) and the difference of the unit area resources of each section (δT/S) are calculated using the unit grid data of the adjacent area method, respectively (Zhang et al. 2006). Results are shown in Fig. 12.11. Based on the analysis of the types and directions of seamounts in the western and central Pacific Ocean and the distribution of cobalt crust

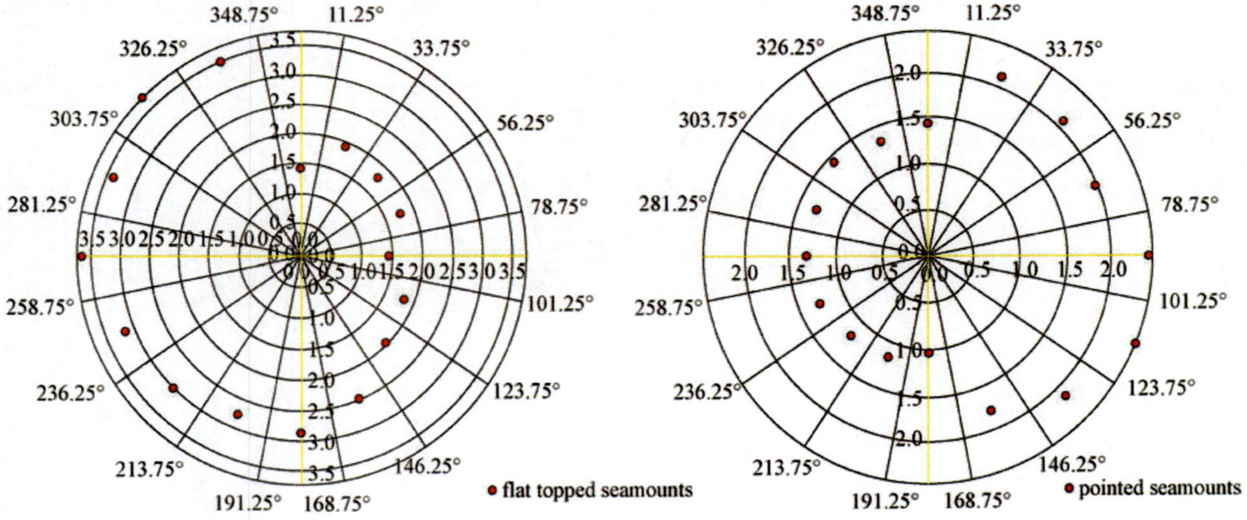

Fig. 12.11 Distribution of resource per unit area (T/S, 10 kt/km^2) in different azimuthal directions for flat topped (left) and pointed seamounts (right) (Zhang et al. 2008b)

resources, it is shown that the tectonic positions of seamounts are different and that the resources per unit area of sections facing different azimuthal directions are additionally highly variable. The relationship between azimuthal directions and cobalt crust resources also varies depending on the type of seamount, whether flat-topped or pointed (Zhang et al. 2008b).

12.2.3 Methods for Evaluating Cobalt Crust Resources

The evaluation, prediction, and calculation of solid mineral resources on land are mature practices that encompass numerous well-developed estimation methods. A brief, non-exhaustive list of some more common methods includes geostatistics (e.g., Hou and Huang 2001), the GIS-based cell cluster method (e.g., Zhang et al. 2001a), the box variable coordinate system and multicomponent reserve calculation integral method (e.g., Cao and Chen 2003), the reserve calculation integral method (e.g., Cao and Chen 1998), the Jackknife method for grade estimation (e.g., Edisoma and Hearst 1996), the analytic triangle truncation method (e.g., Shi 1995), the Grey modeling method for gold resources, and the Kriging method (e.g., Wang and Hou 1994; Kang 1997; Xing et al. 1997). The calculation and evaluation of resource volumes have gradually developed from a single calculation method to a combination of several or even multiple methods. Additionally, mapping systems or GIS software play an increasingly important role in resource evaluation (e.g., Chi et al. 1997; Zhang et al. 1997; Chen et al. 1999; Song et al. 1999; Wang and Fang 2001). Cobalt crusts are thin, layered deposits distributed along the seafloor with generally few sampling points at the prospecting stage. If the thickness and density of cobalt crusts on the seamounts can be approximated as uniform and simplified as a two-dimensional deposit distribution, the arithmetic average method can be adopted. The number of resources can then be estimated using the geological block and adjacent area methods. In the stage of detailed exploration stage or mining stage, there are many sample stations and abundant data are available including cobalt crust thickness and density, among others. Thus, three-dimensional volume and Kriging methods can be used to estimate the total amount of resources. Detailed descriptions of the arithmetic averaging method and the Kriging method can be found in the preceding section on polymetallic nodules and are not detailed further here.

12.2.3.1 Geological Block Method

The geological block method is an improved method for calculating resource quantity based on the arithmetic average method. The principle of the method is to project an orebody onto a plane, according to different types and grades of ore, to identify geological characteristics, such as different resource levels, to divide an orebody into several ideal blocks of different thickness, that is, block segments, and then to use the principle of arithmetic averaging (i.e., the grade weighted averaging method) in each block segment to calculate the amount of resources in each block segment. The total resources are then estimated by summing the resources of the individual grid units (Department of Reserves, Ministry of Land and Resources 2000).

The geological block method encompasses the vertical projection geological block method, the horizontal projection geological block method, and the inclined projection geological block method, which are differentiated by choice of different projection directions. The vertical projection geological block method is suitable for a deposit with a steep dip angle, whereas the horizontal projection geological block method is applicable to deposits with a relatively flat dip angle. The inclined projection geological block method is not commonly used because the calculation is more complicated (Department of Reserves, Ministry of Land and Resources 2000). Investigations of cobalt crusts show that micro-landforms of the seamounts have an effect on the distribution and the thickness of cobalt crusts, with crusts being thickest near the transition from the top of the seamount to the slope, but becoming very thin if the slope angle >45°. Even considering the distribution of the iron-manganese oxide films present on steep slopes, cobalt crust resources are primarily distributed in the areas of slope <45°. Thus, the resource estimation of cobalt crusts is primarily accomplished using the horizontal projection geological block method.

1. Block division

Block segments are shown in Fig. 12.12. Cobalt crusts are generally distributed in horizontal strips along the isobaths on the seamount slope with the grade of cobalt crust varying according to the water depth. Thus, cobalt crusts can be effectively divided according to water depth.

2. Estimation method

Estimation of the total cobalt crust resources may be accomplished using the following formulas:

$$M_{Wi} = S_i' F_i \tag{12.8}$$

$$M_{Di} = M_{Wi}'(1 - W) \tag{12.9}$$

$$M_{Ei} = M_{Di}' C_i \tag{12.10}$$

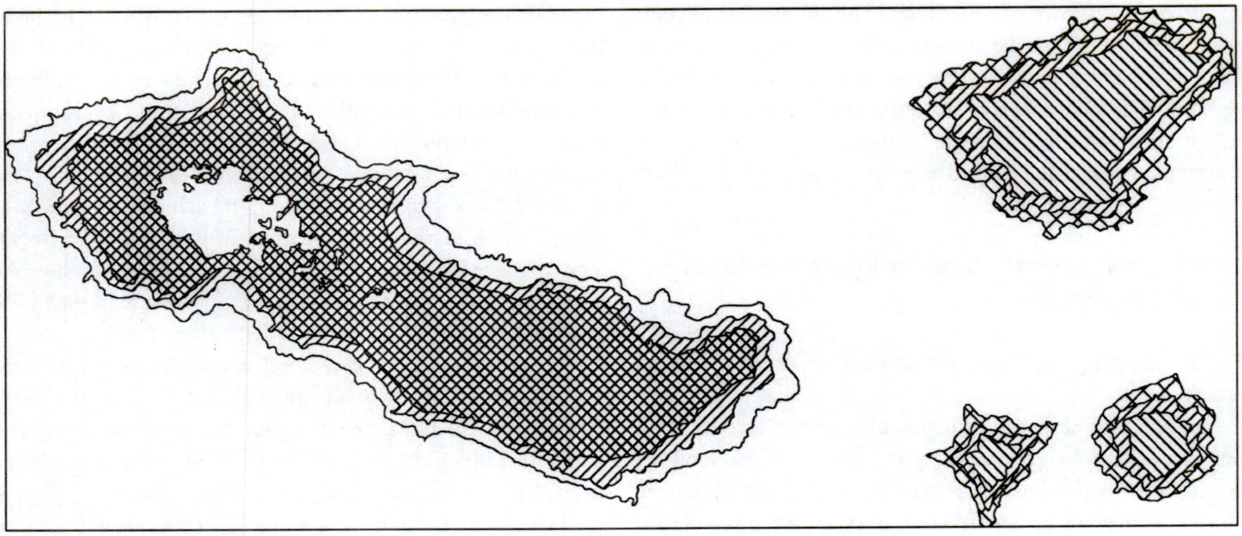

Fig. 12.12 Diagram showing the application of block division used in the geological block method (Zhang et al. 2006)

In the above formulas, M_{Wi} is the wet cobalt crust resources of a particular block i, M_{Di} is the dry cobalt crust resources of the block, M_{Ei} is the metal resources of the block, C_i is the average grade of metal in the block, F_i is the average abundance of the cobalt crust in the block, W is the moisture content of the crust in the block, and S_i is the area of the block. C_i and F_i are calculated as follows:

$$F_i = \frac{1}{n}\sum_{j=1}^{n} F_j \tag{12.11}$$

$$C_i = \frac{1}{n}\sum_{j=1}^{n} C_j \tag{12.12}$$

where F_j is the abundance of cobalt crust at a particular sampling Station j in block i, C_j is the metal grade at a particular station within the block, and n is the total number of stations in block i.

3. Estimation of resource tonnage

The total resource tonnage can be estimated according to the following formulas.

Wet cobalt crust resource tonnage:

$$M_W = \sum_{i=1}^{m} M_W i \tag{12.13}$$

Dry cobalt crust resource tonnage:

$$M_D = \sum_{i=1}^{m} M_D i \tag{12.14}$$

Metal resource tonnage:

$$M_E = \sum_{i=1}^{m} M_E i \tag{12.15}$$

In these formulas, M_W is mineral resources of wet cobalt crust in the mining area, M_{Wi} is mineral resources of wet cobalt crust within a certain block, M_D is mineral resources of dry cobalt crust in the mining area, and M_{Di} is mineral resources of dry cobalt crust within a certain block. M_E is metal resources of the mining area; M_{Ei} is the block- metal resource, and m is the total number of blocks.

12.2.3.2 Neighbor Area Method

In essence, the neighbor area method is used to simplify the irregularly shaped orebody into a series of polygonal cylinders whose volume is easy to calculate. The orebody is divided into a series of closely connected polygonal areas according to the 1/2 distance between each exploration project and adjacent projects in the orebody area delineated on the plan of resource volume calculation. In practice, block segments are divided based on investigations carried out at the exploration stage, with each exploration project defining the center of a polygon. Lines connecting different projects intersect and define the locations of apexes, halfway between two adjacent projects. Thus, the mining area is divided into a series of polygons whose interior points are closer to the exploration project located at the center of the polygon than to any other project location (see Fig. 12.13). Following block division, total mineral resources are calculated according to engineering data collected from stations at the center of each polygonal area. Following block division, calculation methods for estimating the total mineral resources are identical to those of the geological block method (see Sect. 12.2.3.1).

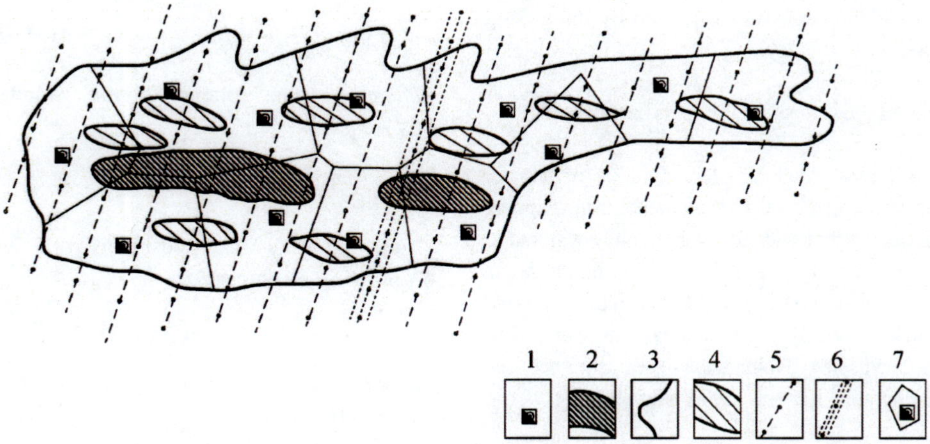

Fig. 12.13 A schematic diagram of block division using the neighbor area method (Wu et al. 2000). 1. sampling station; 2. ore-free or low production benefit window delineated by conditional parameters; 3. orebody outer boundary; 4. ore-free or low production benefit window for statistical calculation; 5. underwater video route and corresponding points; 6. estimation of mining boundary; 7. adjacent block

12.2.3.3 Three-Dimensional Volume Method

In the three-dimensional volume method, the mining area is divided into orthogonal grid units, and the mineral resources of each grid unit are calculated separately. The total resource quantity of all the units is then summed to estimate the total resources of the mining area.

1. Block division

Following grid division, mining elements are divided into regular rectangular elements within the main body of the mining area and non-rectangular irregular elements located along the mining area boundary. Resource estimation for regular rectangular elements is related to grid generation, whereas that for irregular elements becomes a type of a trapezoidal integration problem (Fig. 12.14b).

Block division is accomplished by first dividing the mining area into rectangular prisms, which are then subdivided triangular prisms along diagonal planes between vertices (Fig. 12.14a). The volume of each of these prisms is then calculated (Fig. 12.14b). In practice, isobaths are typically used to define one of the sets of planes dividing the blocks. If the angle between the isobath line and the isobath line is small, the dividing line should be chosen so as not to be orthogonal to the large angle of the isobath line. In addition to triangular prisms formed from regular rectangular blocks, irregular elements along the boundary of the mining area consist of trapezoidal truncated blocks whereby the boundary of the mining area constitutes one side of the block (Fig. 12.14b). As with regular rectangular blocks, irregular blocks are subdivided into triangular prisms, for which the volume can then be easily calculated. The surface area and

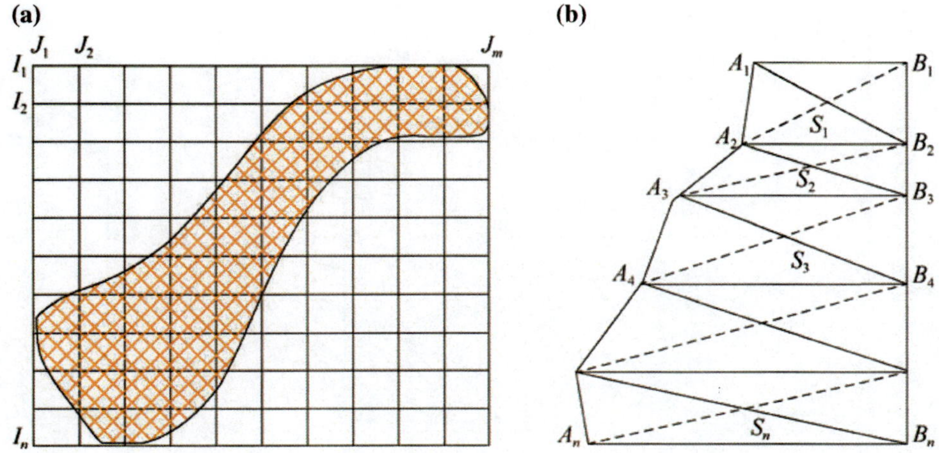

Fig. 12.14 Gridding and sectioning of the model

volume of irregular elements can be obtained by integrating along the mining area boundary.

2. Evaluating the effectiveness of the grinding process

Mining areas appear as collections of continuous or annular regions, presented and stored as vector elements in a mapping system or GIS software, while gridded mining units are stored as grids. Therefore, it is necessary to establish a direct link between the two. There is an inclusion relationship between the rasterized mining area unit and the boundary of the area mining area. The angling method can be used to evaluate the relationship between the discretized mining area unit and the mining area.

3. Method for calculating the resource quantity of a mining area

A schematic diagram of a grid unit used in the three-dimensional volume method is shown in Fig. 12.15a. In this diagram, the Triangle ABC represents a portion of the ocean floor; Triangle $A_1B_1C_1$ represents the base of the cobalt crust, and Triangle $A_2B_2C_2$ represents the base of the grid unit. The distances between corresponding vertices of Triangles ABC and $A_1B_1C_1$, which represent the thicknesses of the cobalt crust, are d_1, d_2 and d_3, respectively. The distances between corresponding vertices of Triangles ABC and $A_2B_2C_2$, which represent the thicknesses of the grid unit, are h_1, h_2 and h_3, respectively. Based on these definitions, the volumes of the triangular prisms $ABC\text{-}A_2B_2C_2$ (V_1) and $A_1B_1C_1\text{-}A_2B_2C_2$ (V_2) are

$$V_1 = \frac{1}{3}S_2(h_1 + h_2 + h_3) \tag{12.16}$$

$$V_2 = \frac{1}{3}S_2(h_1 + h_2 + h_3 - d_1 - d_2 - d_3) \tag{12.17}$$

Therefore, the volume of the cobalt crust element $ABC\text{-}A_1B_1C_1$ is

$$V = V_1 - V_2 = \frac{1}{3}S_2(d_1 + d_2 + d_3) \tag{12.18}$$

If the unit has a cobalt crust density of Φ, then it contains the following resources (q):

$$q = V\Phi \tag{12.19}$$

Integrating all units allows one to obtain the total amount of mining resources:

$$Q = \int_{i=1}^{M}\int_{j=1}^{N} q_{i,j}d_xd_y = \int_{i=1}^{M}\int_{j=1}^{N} v_{i,j}\Phi_{i,j}d_xd_y \tag{12.20}$$

where $V_{i,j}$ and $\Phi_{i,j}$ are the volume and density values for each cell.

If the thickness and density of seamount crusts are approximately uniform, the calculation formula of mining resources can be simplified and approximated as follows:

$$Q = S \times d \times \Phi = d\int S_{i,j}\mathrm{d}s \tag{12.21}$$

where $S_{i,j}$ is the surface area of each unit, S is the total surface area of the mining area, and d is the crust thickness. The unit surface area $S_{i,j}$ can be calculated by Heron's formula according to the value of water depth and the length and width of the unit.

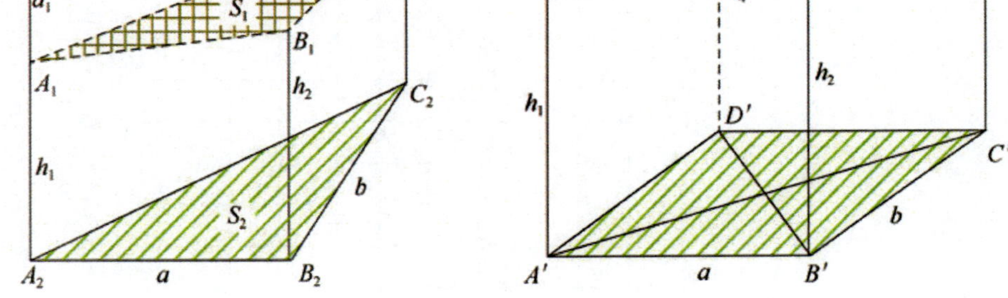

Fig. 12.15 A schematic diagram of a grid unit used in the three-dimensional volume method

12.3 Polymetallic Sulfides

12.3.1 Introduction

Polymetallic sulfides are important products of seabed hydrothermal activity. In 1972, the National Oceanic and Atmospheric Administration conducted the Trans-Atlantic Geotraverse (TAG) program, discovering a hydrothermal field on the Mid-Atlantic Ridge (Scott et al. 1974). In 1979, the first high-temperature, focused hydrothermal vent (sometimes called a black smoker) was discovered at the Galapagos Spreading Center in the eastern Pacific Ocean (Corliss et al. 1979). Polymetallic sulfides are primarily composed of sulfide minerals including pyrite, pyrrhotite, chalcopyrite and sphalerite. Under certain conditions, millions of tons of polymetallic sulfide minerals can be deposited (Hannington et al. 2011). Seafloor polymetallic sulfide deposits are typically characterized by shallow water depths, rapid growth rates, spatially concentrated ore mineral distributions, and high concentrations of various useful metal elements including Cu, Zn, Pb, Ag and Au (Heirzig et al. 2000; Hannington et al. 2011).

Since 2011, the International Seabed Authority (ISA) has signed exploration contracts with the China Ocean Mineral Resources Research and Development Association (COMRA), the Russian Federation, the Government of the Republic of Korea, the Institut français de recherche pour exploitation dealer (IFRMER), the German Federal Institute for Geosciences and Natural Resources, the Government of India and the Government of Poland. Maps of the exploration contract areas are shown in Figs. 12.2 and 12.16. These exploration areas are primarily located on slow- to ultra-slow-spreading ridges of the Indian Ocean and the Atlantic Ocean. In 2005, COMRA began to organize cruises aimed at exploring hydrothermal activity and polymetallic sulfide resources. So far, 42 confirmed and inferred hydrothermal fields based on hydrothermal anomalies had been found on the East Pacific Rise, the Southwest Indian Ridge, the Carlsberg Ridge and the southern Mid-Atlantic Ridge (data from http://www.comra.org/).

12.3.2 Distribution of Seafloor Polymetallic Sulfides

At the end of 2013, the number of known active vent fields was 521 (http://www.interridge.org), including 245 visually confirmed vent fields and 276 vent fields inferred to exist based on physical and chemical evidence in the water column. Vent fields at mid-ocean ridges (MORs) comprise half (52%) of the total is known active vent fields, with about a quarter each located at volcanic arcs (25%) and back-arc spreading centers (BASCs, 21%). A large proportion of known active vent fields along MORs are located along with slow and superfast-spreading centers, while most of those known at BASCs are at intermediate spreading centers (Fig. 12.17). In the past 10 years, the number of hydrothermal fields discovered to exist on slow to ultra-slow-spreading ridges has exceeded the number found along back-arc spreading centers.

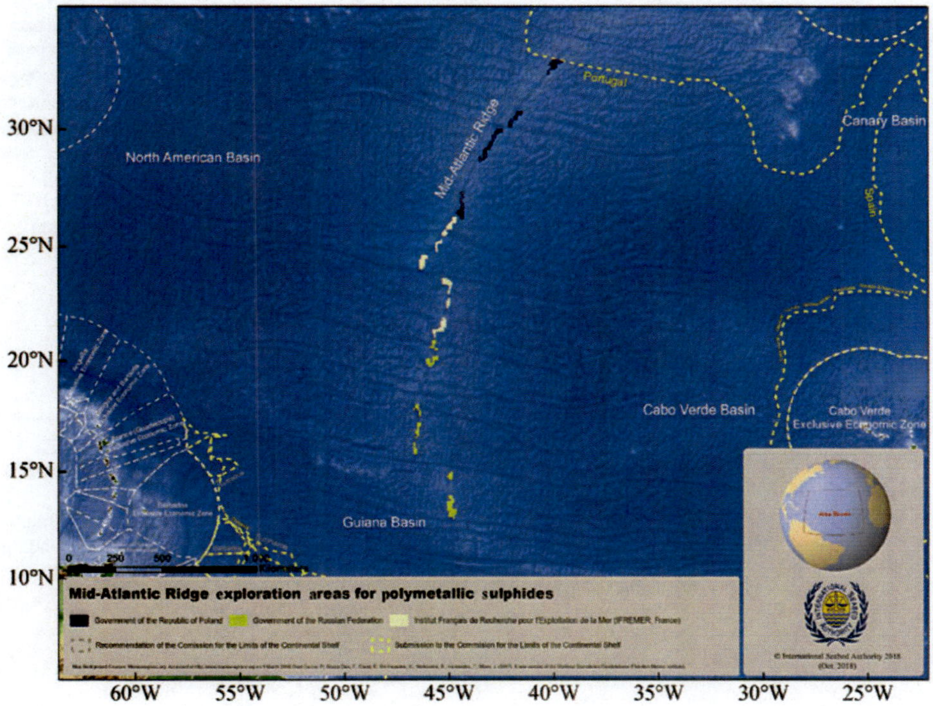

Fig. 12.16 Maps of polymetallic sulfide exploration areas (https://www.isa.org.jm/)

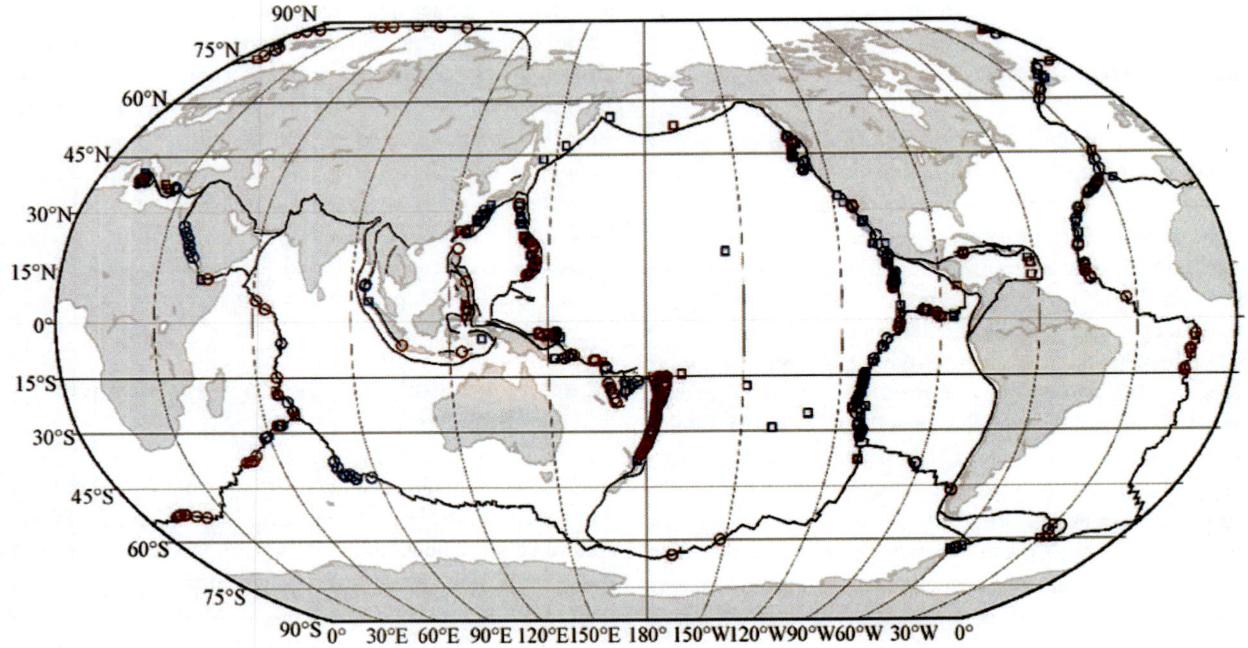

Fig. 12.17 Locations of all known active vent fields, distinguishing confirmed (square) and inferred (circle) and whether discovered/confirmed prior to (blue) or after the (red) year 2000 (Beaulieu et al. 2013)

The depth distribution of known vent fields is strongly bimodal, with a peak in shallow waters (0–200 m depth bin) corresponding to vents along volcanic arcs, and a peek at bathyal depths (2,400–2,600 m depth bin) corresponding to vents at fast-spreading MORs (Fig. 12.18). Vents at arc volcanoes dominate the upper 1,500 m, back-arc vents are prominent at 1,500–2,500 m and MOR vents dominate at depths greater than 2,200 m. Moreover, 90% of known active vent fields are at depths >200 m. In general, shallower vents are of lower temperature, bounded by the temperature-pressure relationship of the two-phase boundary of seafloor hydrothermal fluids. Thus, measured maximum temperatures at vents hosted by arc volcanos tend to be less than those of back-arc- and MOR-hosted vents. In addition, some of the shallow arc vents may include meteoric water as a source of water in combination with seawater (e.g., vents along the submarine flanks of subaerial island arc strato-volcanos). Accounting for differences in depth, vents along arcs, back-arc-, and intraplate settings tend to be lower in temperature than MOR vents (Beaulieu et al. 2013). The highest stable temperature measured so far at a black smoker was 407 °C at Turtle Pits vent on the southern Mid-Atlantic Ridge, which is equivalent to the critical point of seawater at 3,000 m water depth (Koschinsky et al. 2008).

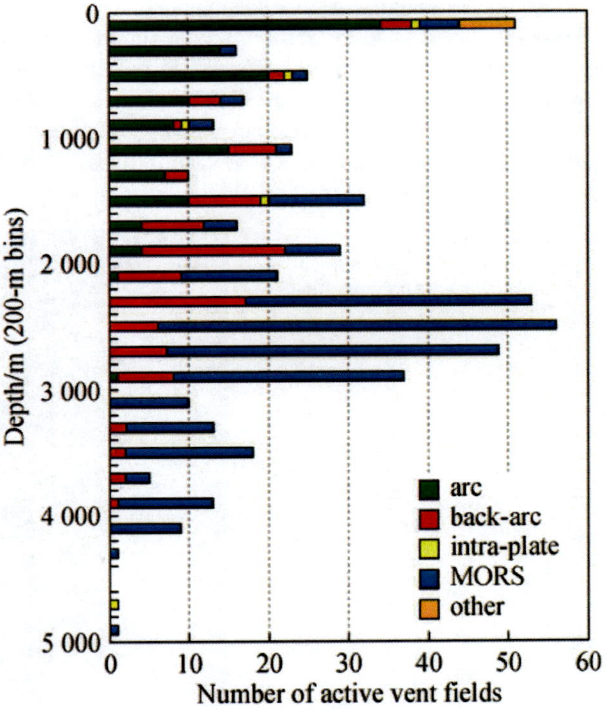

Fig. 12.18 Depth distribution of active submarine hydrothermal vent fields in different tectonic settings (Beaulieu et al. 2013)

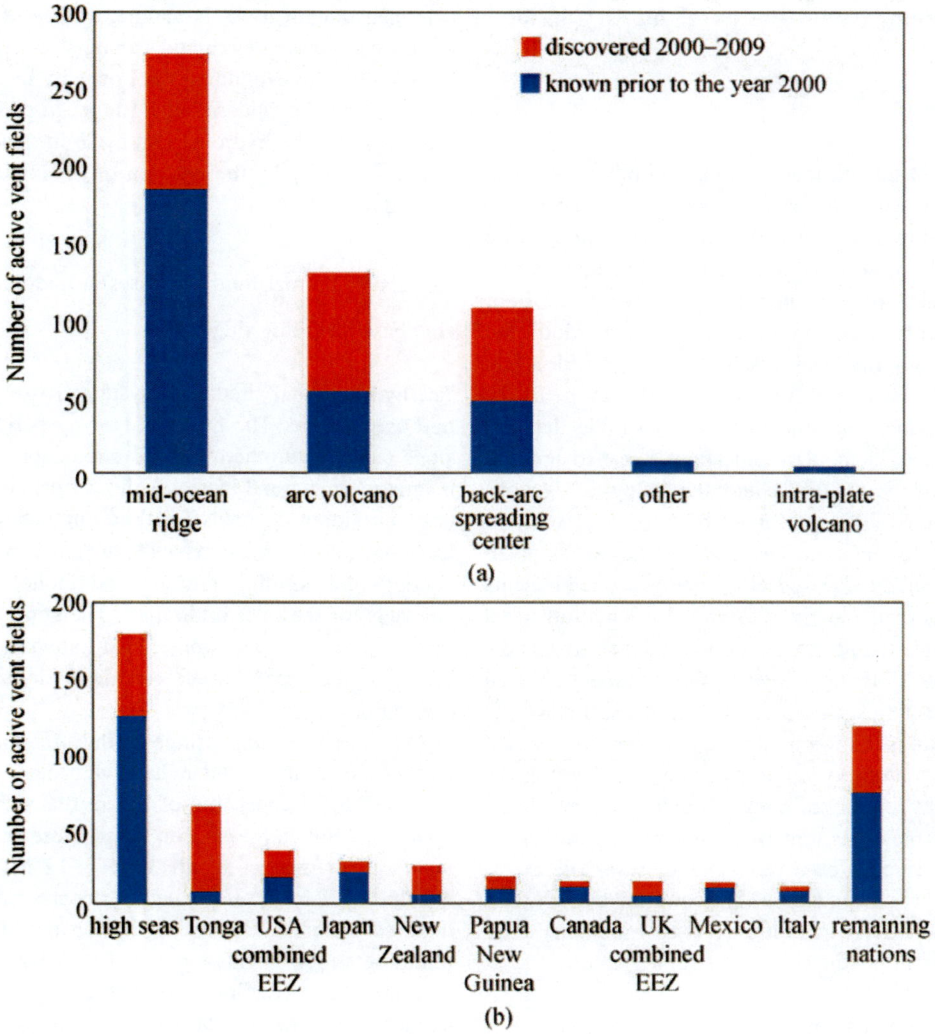

Fig. 12.19 Depth distribution of active submarine hydrothermal vent fields in different tectonic settings. **a** Tectonic setting category; and **b** National jurisdiction (Beaulieu et al. 2013)

Over the last two decades, the number of known active submarine hydrothermal vent fields has almost doubled (Beaulieu et al. 2013). Many discovered hydrothermal fields have changed the understanding of hydrothermal activity distribution. For example, half of the total known active vent fields are located along MORs and roughly a quarter are located along volcanic arcs and along BASCs.

The increased proportion of known active vent fields along volcanic arcs and BASCs is a direct consequence of systematic surveys for hydrothermal activity and has served to increase awareness of the relevance of exclusive economic zones in evaluating seafloor hydrothermal mineral deposits. Additionally, countries have begun to pay more attention to the possible economic benefits of seafloor hydrothermal sulfide minerals. So far, Tonga, New Zealand, Japan and the USA are the countries that have discovered the most hydrothermal vents within their respective exclusive economic zones (Fig. 12.19).

12.3.3 Polymetallic Sulfide Mineralization

Submarine hydrothermal activity is a complex geological process involving interactions between the hydrosphere, lithosphere, and biosphere that connects the ocean and lithosphere. The lithosphere and the hydrosphere are closely linked through transfers of materials and exchanges of energy (Zeng et al. 2002; Zhai et al. 2007; Elderfield and Schultz 1996; Stein and Stein 1994; Alt 2003). The formation of seafloor polymetallic sulfide deposits is controlled by large-scale and local-scale factors. Large-scale factors include tectonic settings (e.g., mid-ocean ridges and back-arc basins) and volcanic activity (e.g., neo-volcanoes ridges and rift valleys). Local-scale controls include mixing between hydrothermal fluid and seawater, permeability, hydrothermal system stability, depth, and types of seafloor cover, among others (Hou 2003).

12.3.3.1 Seafloor Hydrothermal Fluid Circulation

1. Material source

It is generally considered that seafloor hydrothermal fluids are primarily derived from leaching reactions between seawater and crustal rocks under high temperature and pressure conditions. Early water-rock reaction experiments carried out in the wake of the initial discovery of seafloor hydrothermal vents largely confirmed this view (Mottl and Seyfried 1977; Seyfried and Bischoff 1977; Seyfried and Dibble 1980; Hajash and Archer 1980; Dickson 1977). However, the spatial relationship between sulfide deposits and magma bodies in modern and ancient seabed igneous rocks (Yang and Scott 1996) and the study of igneous inclusions collected from presumed hydrothermal reaction zones indicates that magma-sourced fluids rich in dissolved metal ions (especially Cu) and other non-gaseous elements represent additional early forms of hydrothermal ore-forming fluids (Lowenstern et al. 1991; Kamenetsky et al. 2001, 2002). Moreover, at sediment-covered hydrothermal vent fields such as the Guaymas hydrothermal field, reactions between high-temperature fluids and seabed sediments are also important factors influencing the composition of hydrothermal fluids (Giambalvo et al. 2002). Therefore, the materials entering seafloor hydrothermal fluids may be derived from a variety of sources. The extent to which different sources ultimately contribute to the extent of hydrothermal activity and the factors that influence mineral deposit formation still need to be further studied.

2. Heat source mechanisms

Magma intrusion is considered to be the main heat source driving hydrothermal circulation along fast-spreading ridges. The US "US RIDGE" and "RIDGE 2000" programs to conduct long-term continuous observation in EPR 9°–10° N made that ridge the most highly researched area among fast-spreading ridges. Continuous observation of a hydrothermal vent at the EPR 9°50' vent field indicates that the fluid chemical composition of the vent is closely related to dike intrusion in April 1991 (Von Damm et al. 1995, 1997). German and Lin (2004) combined measurements of crust heat flux and numerical calculations to constructing a "Quantum Event" geological model reflecting the relationship between rapid magma intrusions and hydrothermal activity.

In contrast, slow-spreading ridges are characterized by stable tectonic environments and relatively insufficient magma supply. Heat provided by continued cooling of the ocean crust is insufficient to support long-term stable hydrothermal circulation (German and Lin 2004). Heat provided by exothermic serpentinization reactions involving exhumed mantle rocks is another source of heat that may drive hydrothermal circulation along slow-spreading ridges. For example, in examining the Lost City hydrothermal field at 30° N on the Mid-Atlantic Ridge, Ludwig et al. (2006) proposed that the hydrothermal cycle in that field was controlled primarily by the reaction of the mantle peridotite to serpentine.

3. Hydrothermal fluid processes at mid-ocean ridges

(1) Fast-spreading ridges

The hydrothermal circulation system crosses the axis of the mid-ocean ridge. The heat released by periodic dike intrusions drives hydrothermal fluid circulation above the axial magma chamber and increases the permeability of the upper crust providing channels for fluid circulation (German and Lin 2004). As the plate expands, the crust spreads, gradually cooling and forming fissures and faults, which become channels for seawater infiltration. The upper boundary layer of the axial magma chamber, also known as the melt lens, provides the heat source to drive hydrothermal fluid circulation.

On fast-spreading ridges, the interaction between hydrothermal fluids and a layered oceanic crust (i.e., the Penrose Model consisting of basalt, diabase and gabbro) is mainly controlled by reaction temperature and water-to-rock mass ratio (Bach et al. 2013) (Fig. 12.20). The Mg-rich, alkaline (pH = 8.0) seawater permeates into the several hundred meters basalt crusts causing the alteration of surrounding rock. This process mainly occurs under saponite-phase conditions of regional metamorphism, the temperature is low (<150–200°C), and water-to-rock mass ratios are relatively high (up to several tens of kg water per kg rock). A simple Mg–Ca ion-exchange reaction between seawater and basalt leads to the formation of Ca-rich, and Mg-poor hydrothermal fluid with a correspondingly large amount of magnesium-rich silicate precipitated in the surrounding rock. The concentrations of other major elements very little between hydrothermal fluids and seawater (Alt and Teagle 1998; Bach et al. 2013). The more violent and complex hydrothermal alterations, including chlorination, idolization, sericitization, and silicification mainly occur in the concentrated upwelling zone during hydrothermal circulation (Bach et al. 2013). Epidotization is common in volcanic arc-hosted systems and ophiolites, and chlorination is common in mid-ocean ridges. Teagle and Alt (2004) proposed that basalt at the Bent Hill hydrothermal field on the Juan de Fuca Ridge is altered by a continuous series of reactions with hydrothermal fluids. First, plagioclase is converted into albite ± quartz ± chlorite and then forms quartz-chlorite. Orthopyroxene is first altered into chlorite ± quartz ± actinolite and then altered to

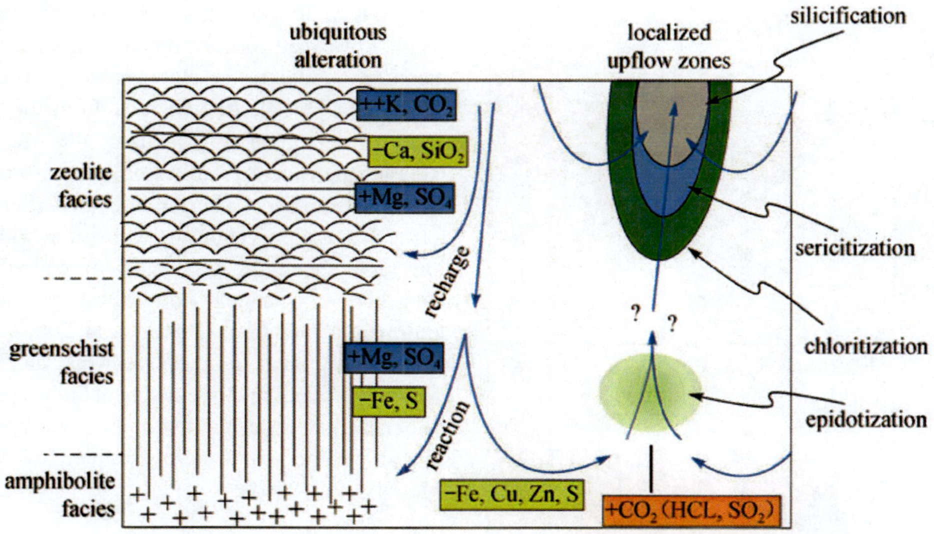

Fig. 12.20 Sketch of hydrothermal circulation, associated mass transfers, and metasomatic zones (Bach et al. 2013)

quartz-chlorite. Chlorite is the most common type of alteration mineral. The relative Fe and Mg content of chlorite reflects the characteristics of hydrothermal fluid and the degree of mixing with seawater (Saccocia and Gillis 1995).

(2) Slow-spreading ridges

Hydrothermal fluid circulation processes on slow-spreading ridges are typically more complex than those along fast-spreading ridges. With the exception of ridges affected by volcanic hotspots, axial magma chambers beneath slow-expanding ridges are difficult to detect, and the magma supply is mostly concentrated at the center of the ridge (Lin et al. 1990). The interaction and restriction of magmatic activity and tectonic expansion process during hyperplasia of the ocean crust and slow expansion of the oceanic ridge results in the coexistence of symmetrically and asymmetrically dilated ridges, leading to greater variability in the thermal structure of lithosphere and the types of tectonic faults and basement rocks compared with fast-spreading ridges (Escartín et al. 2008) (Fig. 12.21).

At asymmetrical slow-spreading ridges, the lack of magma supply causes asymmetric plate spreading. When tectonism dominates the process of crustal accretion, a large number of low-angle strike-slip normal faults (oceanic detachment faults) form ocean core complexes, which are considered to be the primary factor controlling hydrothermal circulation (Escartín and Canales 2011; Tucholke et al. 2008). Compared with asymmetric ridges, symmetric ridges have sufficient magma supply that hydrothermal activity is

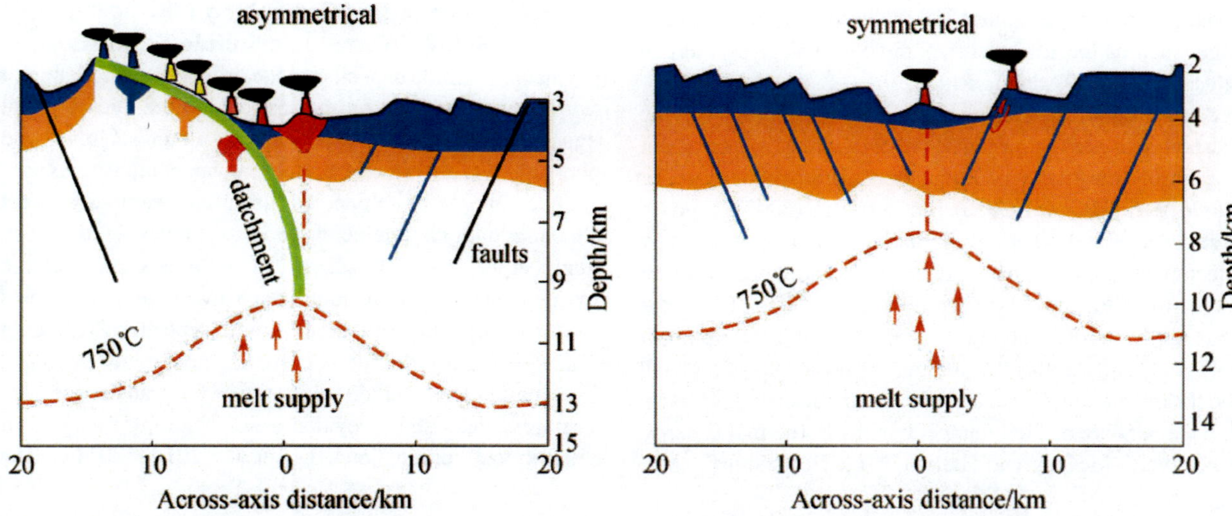

Fig. 12.21 Axial cross-section corresponding to symmetrical and asymmetrical accretion and associated processes (Escartín et al. 2008)

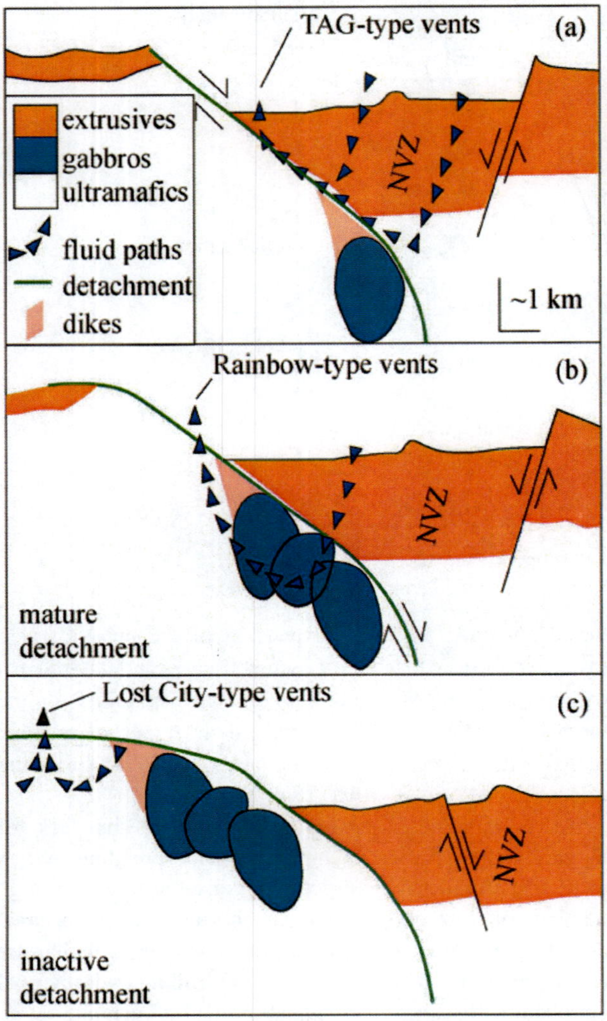

Fig. 12.22 Models for fluid circulation and types of hydrothermal venting in and around detachment faults (McCaig et al. 2007)

developed along the central rift valley of the neo-volcanic ridge, such as the Lucky Strike and Snakepit hydrothermal fields on the Mid-Atlantic Ridge.

At asymmetric spreading ridges, the heat source driving hydrothermal circulation can be crustal cooling, emplacement of gabbroic magma, serpentinization of ultramafic rock, or mantle heating (Lowell 2010). McCaig et al. (2007) proposed three styles of hydrothermal circulation systems controlled by detachment faults, named after type hydrothermal fields: TAG, Rainbow and Lost City (Fig. 12.22). The TAG-type vents are hosted in hanging-wall basalts. Both the Rainbow and Lost City hydrothermal systems are hosted in the footwall of detachment faults hosted in ultramafic rocks and represent high temperature and low-temperature hydrothermal systems, respectively (McCaig and Harris. 2010). Oceanic drill cores (ODP and IODP) and extensively exposed altered rocks reveal mantle flow, melt formation and migration, local strain, and alteration processes in slow-spreading ridges (Bach and Früh-Green. 2010). Altered rock from the Atlantis Massif detachment fault at 30° N along the Mid-Atlantic Ridge exhibits extensive crystallization, plastic-brittle deformation and talc-hornblende- chlorite alteration (Schroder and John 2004). These observations are consistent with complex reactions (at temperatures <500 °C) between hydrothermal fluids and gabbro and mantle peridotite host rocks under high-stress deformation conditions (Boschi et al. 2006). The presence of mechanically weak minerals such as talc, serpentine, and chlorite surrounding the detachment fault is proposed to result in localized strain of the mylonitic shear zone that supports several million-year-scale hydrothermal fluid circulations (McCaig et al. 2007).

(3) Ultra-slow spreading ridges

Ultra-slow spreading ridges are a special type of ridge (Dick et al. 2003; Zhou and Dick 2013). Traditional magma budgets cannot explain the distribution of hydrothermal activity on ultra-slow spreading ridges (Baker and German 2004). So far, few studies have investigated the hydrothermal system and mineralization of ultra-slow spreading ridges. Ultra-slow spreading ridges have poor magma supply and consist of volcanic and non-volcanic ridge segments (Dick et al. 2003). However, mantle heterogeneity may result in sufficient local magma supply and thicken crust (Standish et al. 2008). Additionally, cooling of mantle rocks and exothermic serpentinization reactions provide sufficient heat to drive hydrothermal systems, which makes the frequency of hydrothermal activity on ultra-slow spreading ridges deviate from that predicted by the theoretical "magma budget" hypothesis (Baker and German 2004).

12.3.3.2 Mechanisms Controlling Mid-Ocean Ridge Polymetallic Sulfide Deposits

Polymetallic sulfide deposits are the primary accumulation of hydrothermal products on the seafloor. These hydrothermal deposits mainly consist of chimneys, mounds, gravel and sphericities, with mineral and chemical compositions primarily composed of sulfide, sulfide-sulfate, carbonate, silica sediment, and clay mineral deposits (Tivey et al. 1995; Herzig et al. 1998a, b; Zeng 2011). In a stable tectonic environment, long-term, multi-stage hydrothermal systems can form large hydrothermal sulfide deposits (Zierenberg et al. 1998). Large sulfide deposits are not simply the result of accumulations sulfide over time, but rather complex structures that show evidence of multiple stages of hydrothermal production (Li et al. 2010, 2011). The formation of seafloor polymetallic sulfide deposits is influenced by different geologic factors occurring at a variety of spatial and temporal scales including fluid mixing, crustal permeability, stability of the hydrothermal system, boiling, geological cover, fluid-rock/sediment interactions, regional tectonism, crustal structure, host rock lithology, topography, fluid flow and transformations (Fouquet et al. 1996; Zeng 2011).

1. Effects of spreading rate on seafloor polymetallic sulfide deposits

On fast-spreading ridges, which typically exhibit relatively thin crust along the axis, a widespread continuous magma lens provides sufficient heat supply to drive hydrothermal circulation. It is generally believed that the spatial distribution of hydrothermal activity along fast-spreading ridges is controlled by the thermal structure of the lithosphere. Dike intrusions open fluid channels and provide the thermodynamic energy necessary for the development of hydrothermal circulation distributed along the ridge axis. On the East Pacific Rise, shallow magma chambers and continuous magmatism leads to the relatively shallow circulation of high-temperature hydrothermal fluids (1–2 km). However, because the eruption rates are so high, new polymetallic sulfide deposit is frequently covered by volcanic eruptions (German and Parson 1998). As a result, the deposits at fast-spreading ridges tend to be abundant but small (<1 Mt), and exhibit large chimneys and simple morphologic features (Hannington et al. 2005).

Intermediate spreading ridges (full spreading rates of 4–6 cm/a) typically exhibit thick crust and lack long-term magmatism. During the tectonic and/or intrusive period, ridges exhibit intermittent volcanism, as exemplified by the Juan de Fuca Ridge, Gorda Ridge and Galapagos Ridge, which have low magma supply and deep rift valleys that can host large chimneys. For example, the Endeavour segment on the Juan de Fuca Ridge hosts 50–100 chimneys divided into 6 groups within a 2–3 km by 15 km area in the valley (Delaney et al. 1992; Kelley et al. 2001). These chimneys are often concentrated on rift walls or axial fissures.

Compared with fast-spreading ridges, slow-spreading ridges (full spreading rates of 2–5.5 cm/a), are characterized by thicker crust and lower magma supply. Localized eruptions, long periods of hydrothermal activity, and more stable tectonic environments favor the formation of large hydrothermal sulfide deposits. For example, the Mid-Atlantic Ridge between 13°–15° N has strong tectonic expansion characteristics, and a large number of transform faults are formed by detachment faults (Escartín et al. 2008; Smith et al. 2008). On slow-spreading ridges, the inner corners of non-transform discontinuous and transform faults become favorable environments for seafloor hydrothermal activity (Blackman et al. 2009; McCaig and Harris 2010). Typical hydrothermal sulfide deposits exhibit a characteristic vertical zonation. The base of the ore body is composed of a disseminated and vein-mass mineralized layer that is associated or stacked together with the overlying layers of massive sulfide mineral deposition.

Relatively few studies have investigated sulfide mineralization on ultra-slow spreading ridges. One example is the investigation of Mount Jourdanne, located on the eastern flank of the Southwest Indian Ridge, near the Rodriguez Triple Junction. Mineralogical and geochemical studies of sulfide deposits show that Zn-rich sulfides have high concentrations of trace elements such as Pb, As, Ag and Au, similar to the trace element profiles of felsic-hosted hydrothermal systems in back-arc or arc volcanic regions.

2. Morphology and structure of sulfide deposits

Basement rock lithology, hydrothermal fluid properties, and composition are the main factors determining the morphology, mineral and geochemical composition of seafloor hydrothermal sulfide deposits. Mafic and ultramafic rocks are the two main types of host rocks for seafloor hydrothermal sulfide deposits on mid-ocean ridges. The effects of water depth, phase separation, hydraulic fracturing and permeability in ore body structure lead to notable differences between sulfide ore bodies in each of these two types of host rocks (Fouquet et al. 1996). Mafic-hosted hydrothermal systems are more "brittle", and tend to develop black smoker chimneys. Mafic-hosted hydrothermal systems are additionally characterized by high topography and a domed deposit shape. In contrast, ultramafic-hosted hydrothermal systems have strong rock plasticity, and the hydrothermal fluid is mainly composed of near-bottom high-temperature circulation. As a result, the sulfide deposit is relatively smooth, characterized by a flat mound and larger area of mineralization.

Fluids venting from ultramafic-hosted hydrothermal systems are characterized by high pH, low SiO_2 (aq), low H_2 (aq), and high concentrations of metallic elements which precipitate as metal sulfide minerals including pyrite, isometric cubanite, and chalcopyrite. In contrast, sulfide minerals in deposits developing from mafic-hosted hydrothermal systems are mainly sphalerite and pyrite. Additionally, marcasite is an important minor-mineral in mafic-hosted hydrothermal systems, but it is seldom found in deposits formed by ultramafic-hosted hydrothermal systems. Sulfide deposits developed from ultramafic-hosted hydrothermal fields are often rich in Cu and Ag. Sulfide deposits developed from ultramafic-hosted hydrothermal systems are often Cu- and Zn-rich, with concentrations reaching 20%–30% higher than that of mafic-hosted hydrothermal sulfide deposits. Fouquet et al. (2010) proposed that serpentinization of ultramafic rock may be the main source of Cu and Zn in ultramafic-hosted sulfide deposits.

The internal physical structure and chemical processes of hydrothermal sulfides are largely controlled by local factors occurring at the hydrothermal vent field scale, such as mixing between hydrothermal fluids and seawater, zone purification and mineral precipitation/dissolution reactions (Tivey et al. 1998; Li et al. 2010, 2011). However, given the limited empirical investigations of in situ fluid chemistry at the site of mineral deposition, most studies of deposit formation processes are based on numerical modeling. For example, Dickson et al. (1995) simulated the fluid migration state of sulfide under different permeability and Rayleigh numbers through a two-dimensional model of a homogeneous porous medium. Pascoe and Cann (1995) used a simple tubular model to simulate the formation and evolution of hydrothermal fluids in the black smoker chimney injection zone. Based on the data from oceanic drill cores of the TAG hydrothermal field, Li et al. (2010, 2011) constructed a three-layer sulfide model, simulating the flow field and temperature field inside the hydrothermal body, and proposed that heterogeneity of the deposit and bathymetric relief have an important influence on the evolution of the hydrothermal sulfide deposit. Li et al. (2009) established an internal fluid mixing model of the sulfide deposit to simulate the mixing process of seawater and hydrothermal fluids and explained the relationship between black smoker and white smoker chimney fluids at the TAG hydrothermal field and the mineral precipitation mechanisms of anhydrite.

3. Mineralization period of seafloor hydrothermal sulfide deposits

Seafloor sulfide deposits are generated during specific chronological cycles coinciding with intervals of volcanic eruptions. A radiological study of the TAG hydrothermal field shows that the hydrothermal chimney is active periodically within the same region, and that period of activity and inactivity may alternate over a timespan of thousands of years (Lalou et al. 1995). The slow-spreading ridge is characterized by a lower magma supply, larger tectonic control, and fewer eruptions roughly once every 100–1,000 years. At ultra-slow-spreading ridges, the eruption interval may be on the order of thousands of years. The character, depth, and scale of heat sources, host rocks, and the permeability of faults interact to control periodic releases of fluids and sulfide deposition (Schardt and Large 2009). Previous studies have indicated that the formation of a large (>1.7 Mt Zn or 2 Mt Cu) and an oversized deposit (>12 Mt Zn or 24 Mt Cu) requires a fluid solution containing $>10 \times 10^{-6}$ metals, a deposition efficiency >10% and a sufficiently long time to accumulate useful metals (Schardt and Large 2009).

12.3.4 Estimating the Global Abundance of Seafloor Polymetallic Sulfide Deposits

Because the ocean floor covers more than 70% of the Earth's surface, many expect the oceans to contain a proportionate amount of the world's mineral resources, comparing the resource potential of seafloor massive sulfide deposits to that of deposits now mined on land for Cu, Zn, Pb, Ag and Au (Rona 2003; Cathles 2011). However, smaller, higher-grade massive sulfide deposits are likely to be the first targets for deep-sea mining. For example, Nautilus Mining has acquired more than 4.9×10^4 km^2 of sulfide exploration area in the exclusive economic zones of South Pacific Island countries including Tonga, Fiji, Solomon Islands, New Zealand and Vanuatu, consisting of approximately 20×10^4 km^2 of authorized area and 29×10^4 km^2 of application area.

To estimate the abundance of seafloor massive sulfide deposits in the new volcanic zones of the global oceans, Hannington et al. (2011) identified seafloor sulfide accumulation regions based on the output, distribution and scale of massive sulfide deposits and evaluated the potential of global seafloor sulfide mineral resources through examination of highly studied sulfide deposits. Hannington et al. (2011) examined deposit densities in 32 control areas of roughly equal size (5° of latitude by 5° of longitude) containing 129 vents. The areas were chosen to represent permissive geology throughout the oceans and include roughly 7,200 km of the ridge and 2,800 km of arc and back-arc spreading centers (Hannington et al. 2011). An average of four deposits was included in each of the 5° × 5° control areas; 90% of the control areas have densities of two or more deposits per 100,000 km^2, 50% have densities of six or more deposits, and 10% have densities of ten or more deposits. In Hannington et al. (2011)'s study, they consider only those deposits that are located within the new volcanic zone. Thus, deposit densities can also be expressed in terms of the linear dimensions of ridge or arc segments between vent field occurrences (Fig. 12.23a). The deposits are farther apart on the slow-spreading ridges (average of 174 km) and closer together on the fast-spreading ridges (average of 54 km). Deposit densities on back-arc spreading centers are similar to those on mid-ocean ridges with similar spreading rates;

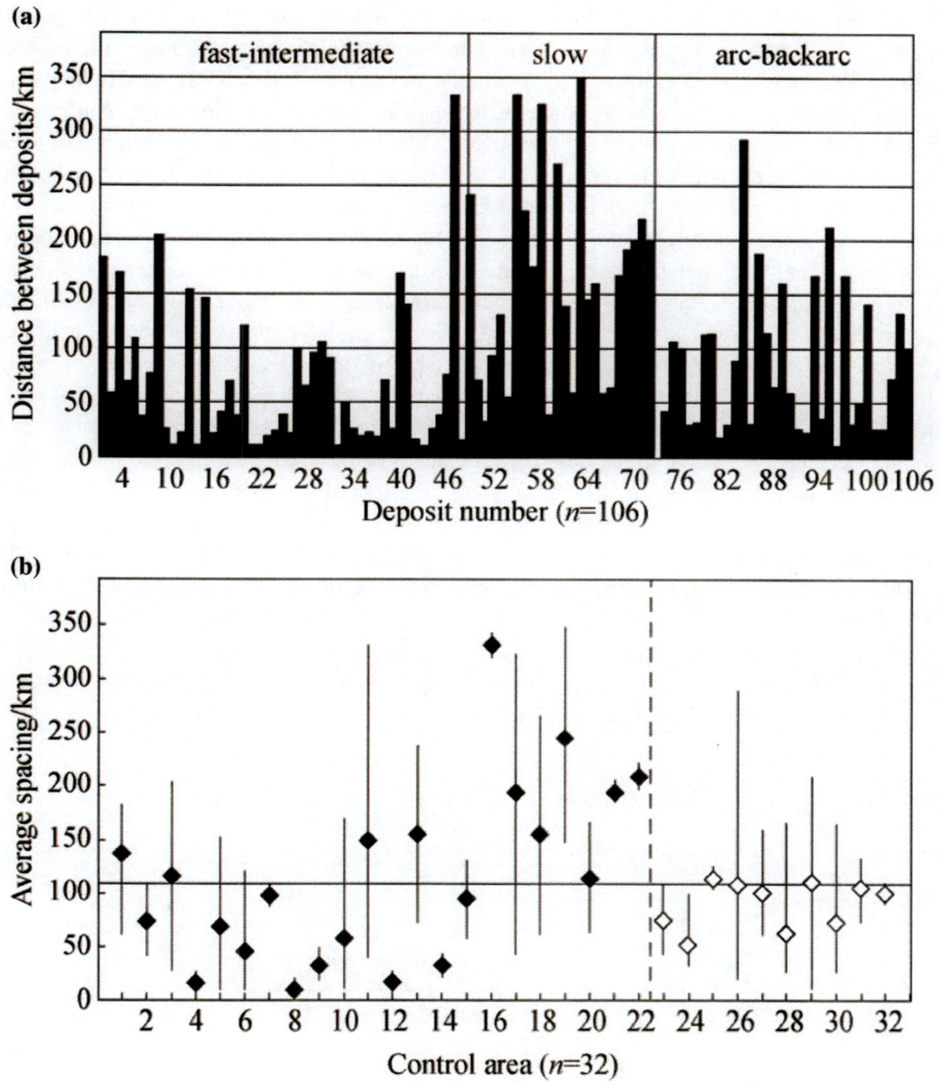

Fig. 12.23 Density distributions of seafloor massive sulfide deposits. **a** Distances between adjacent deposits along 10,000 km of ridge, arc and back-arc spreading centers. Data are included only for those deposits or clusters of deposits more than 10 km apart (n = 106). **b** Mean deposit densities (average spacing) in each of the 5° by 5° control areas; ranges are the minimum and maximum distances between adjacent deposits in each control area. Closed symbols—mid-ocean ridges; open symbols—arcs and back-arc spreading centers. The horizontal line indicates the combined average value for all 32 control areas (107 km) (Hannington et al. 2011)

along volcanic arcs, the spacing is more closely linked to the spacing of individual volcanoes (de Ronde et al. 2003). The combined average spacing for all 32 control areas is 107 km (Fig. 12.23b). Based on the cumulative strike length of the oceanic plate boundaries (89,000 km) and an average density of one deposit every ~100 km, the total number of deposits expected in the global no volcanic zones is ~900. Taking into account the range of deposit densities, the expected number of deposits is 500–5,000.

Because of limitations in surveying the deep ocean, it is difficult to estimate the total sulfide accumulation accounted for by seafloor massive sulfide deposits despite the fact that the total number of deposits in the global no volcanic zones has been reasonably estimated (Hannington et al. 2010). In several cases, drilling information is available to evaluate the continuity and thickness of sulfide outcrops. Four deep drill holes that penetrated the Bent Hill deposit and Ocean Drilling Program (ODP) coring of the mound at Middle Valley (ODP Legs 139 and 169) indicate a combined tonnage of 1–1.5 × 10^7 t (Zierenberg et al. 1998). At the large TAG mound on the Mid-Atlantic Ridge, 17 holes drilled to a maximum depth of 125 m indicate 2.7 × 10^6 t of massive sulfide in this deposit (Herzig et al. 1998a, b; Hannington et al. 1998). Data from extensive drilling of the Solwara 1 deposit in the Eastern Manus Basin (146 holes to a maximum depth of 20 m) have been used to estimate a resource of 2.1 × 10^6 t in a chimney zone covering 90,000 m^2 (Lipton 2008).

Hannington et al. (2010) estimated the tonnages of the 62 best-mapped deposits from the outcrop areas in the global database. Of the 62 deposits used in the tonnage model, only eight have dimensions indicating sizes larger than 2 × 10^6 t. Three of these have been confirmed by drilling (i.e., the

TAG, Middle Valley and Solwara 1 deposits); the others have sulfide outcrop dimensions that are consistent with similarly large tonnages. However, the median deposit size is only 70,000 t, and more than a third of the deposits are considered to be smaller than 3,000 t. Roughly 70% of the potential sulfide resources are contained in the largest 10% of deposits. Using these data, an estimate of the cumulative tonnage of massive seafloor sulfides in the global no volcanic zones can be made by populating the tonnage model with the estimated numbers of deposits discussed above (Hannington et al. 2011).

If there are ~1,000 deposits with a minimum size of 100 t, a maximum size of 1×10^7 t and based on the size distribution is given by Hannington et al. (2010), the total amount of massive sulfide in the new volcanic zones of the global oceans is on the order of 6×10^8 t, and the largest 10% of deposits would contain $\sim 4 \times 10^8$ t. Because slow-spreading centers account for ~60% of the total ridge length and host the largest deposits, they are expected to account for more than 85% of the total tonnage of massive seafloor sulfides at ridges (Fig. 12.24). The contained metal is more difficult to estimate, owing to the highly variable metal grades of surficial seafloor samples (Hannington et al. 2010). For a median grade of 5wt% Cu + Zn, the total amount of metal (Cu + Zn) along the global no volcanic zones is expected to be $\sim 3 \times 10^7$ t (Hannington et al. 2011).

Studies have confirmed that modern seafloor sulfide deposits are similar to ancient massive sulfide deposits. However, the average resource of modern seafloor sulfide deposits is about two orders of magnitude lower than that of ancient massive sulfide deposits (Hannington et al. 2011). Few large massive sulfide deposits have been found on the seafloor, most are small hydrothermal vents, isolated chimneys, and mounds, with deposits only comprising a few thousand tons. For example, more than 60 hydrothermal sulfide deposits were found in the southern section of the Explorer ridge (about 8 km long), most of which were distributed in two smaller areas of about 250 m × 200 m. At present, giant deposits like the Kidd Creek (135×10^6 t), the Neves Corvo in Portugal (262×10^6 t), and the Iberian pyrite belt ($1{,}765 \times 10^6$ t) have not been discovered on the seafloor (Herzig 1999). However, because large hydrothermal sulfide deposits may exist in the as yet unexplored sections of mid-ocean ridges, we also need to consider the scope of the model when making comparisons between ancient and modern deposits (Hannington et al. 2011).

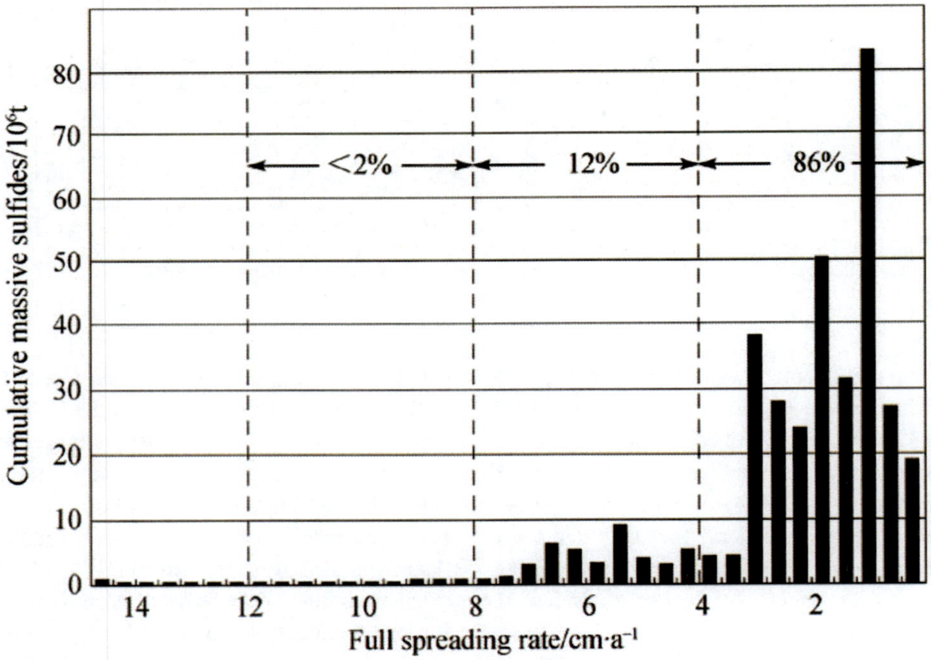

Fig. 12.24 Expected distribution of massive seafloor sulfide on mid-ocean ridges as a function of spreading rate. The proportion of massive sulfide deposition corresponding to different spreading rates is estimated from the length of the ridge segments (Carbotte and Scheirer. 2004), the deposit density as a function of spreading rate, and the expected sizes of the deposits (see the Data Repository [see footnote 1]). The total length of ridges represented in this plot is 53,000 km; the expected number of deposits (n = 537) have a total estimated tonnage of $\sim 3.6 \times 10^8$ t. The large proportion of massive sulfide expected on the slow-spreading ridges reflects the cumulative length of the slow ridges plus the large size of the known deposits (Hannington et al. 2011)

References

Alt JC (2003) Hydrothermal fluxes at mid-ocean ridges and on ridge flanks. Geochemistry 335:85–864. https://doi.org/10.1016/j.crte.2003.02.001

Alt JC, Teagle DAH (1998) Probing the TAG hydrothermal mound and stockwork: oxygen-isotopic profiles from deep-ocean drilling. ODP Scientific Results 158:285–295. https://doi.org/10.2973/odp.proc.sr.158.222.1998

Bach W, Früh-Green GL (2010) Alteration of the oceanic lithosphere and implications for seafloor processes. Element 6(3):173–178. https://doi.org/10.2113/gselements.6.3.173

Bach W, Jöns N, Klein F (2013) Metasomatism within the Ocean Crust. In: Harlov D, Austrheim H (eds) Metasomatism and the chemical transformation of rock: the role of fluids in terrestrial and extraterrestrial processes. Lecture Notes in Earth System Sciences Springer, pp 253–288. https://doi.org/10.1007/978-3-642-28394-9_8

Baker ET, German CR (2004) On the global distribution of hydrothermal vent fields. Mid-Ocean Ridges: Hydrothermal Interactions Between the Lithosphere and Oceans. Geophys Monogr Ser 148:245–266. https://doi.org/10.1029/148GM10

Beaulieu SE, Baker ET, German CR et al (2013) An authoritative global database for active submarine hydrothermal vent fields. Geochem Geophys Geosyst 14(11):4892–4905. https://doi.org/10.1002/2013GC004998

Blackman DK, Canales JP, Harding A (2009) Geophysical signatures of oceanic core complexes. Geophys J Int 178(2):593–613. https://doi.org/10.1111/j.1365-246X.2009.04184.x

Boschi C, Früh-Green GL, Delacour A et al (2006) Mass transfer and fluid flow during detachment faulting and development of an oceanic core complex, Atlantis Massif (MAR 30° N). Geochem Geophys Geosyst 7(1):Q01004. https://doi.org/10.1029/2005GC001074

Cao YP, Chen ZJ (1998) Summary of the integral method for the calculation of reserves (in Chinese with English abstracts). Henan Geol 16(2):130–136

Cao YP, Chen ZJ (2003) Brief introduction of box al tarnation coordinate system and multiple reserve calculation integration (in Chinese with English abstracts). China Mining 12(9):58–63. https://doi.org/10.3969/j.issn.1004-4051.2003.09.018

Carbotte SM, Scheirer DS (2004) The global mid-ocean ridge. Hydrogeol Oceanic Lithos CD-ROM 1:59

Cathles LM (2011) What processes at mid-ocean ridges tell us about volcanogenic massive sulfide deposits. Miner Deposita 46(5–6):639–657. https://doi.org/10.1007/s00126-010-0292-9

Chen YL, Wu W, Liu DY (1999) An automatic integrating method of statistical results from several predicting models in GIS (in Chinese with English abstracts). Comput Techniq Geophys Geochem Explor 21(4):319–322. https://doi.org/10.3969/j.issn.1001-1749.1999.04.006

Chi SD, Zhou SP, Wu XL (1997) Geological Anomaly analysis and experience prognosis of metallic mineral under GIS (in Chinese with English abstracts). Earth Science—Journal of China University of Geosciences 22(1):99–103

Chu FY, Sun GS, Li XM et al (2005) The growth habit and controlling factors of the cobalt-rich crusts in seamount of the Central Pacific (in Chinese with English abstracts). J Jilin Univers (Earth Sci Edn) 35(3):320–325. http://doi.org/10.3969/j.issn.1671-5888.2005.03.008

Corliss JB, Dymond J, Gordon LI et al (1979) Submarine Thermal Sprirngs on the Galápagos Rift. Science 203:1073–1083. https://doi.org/10.1126/science.203.4385.1073

Craig JD, Andrews JM, Meylan AM (1982) Ferromanganese deposits in the Hawaiian Archipelago. Mar Geol 45:127–157. https://doi.org/10.1016/0025-3227(82)90183-9

Cronan DS (1984) Criteria for the recognition of areas of potentially economic manganese nodules and encrustations in the CCOP/SOPAC region of the central and southwestern tropical Pacific. South Pacific Marine Geological Notes 3(1):17. https://doi.org/10.1016/0198-0254(84)93436-8

de Ronde CEJ, Massoth GJ, Baker ET et al (2003) Submarine Hydrothermal Venting Related to Volcanic Arcs. In: Simmons SF, Graham I (eds) Volcanic, Geothermal, and Ore-Forming Fluids: Rulers and Witnesses of Processes within the Earth. Soc Econ Geol, Incorporated, pp 91–110. https://doi.org/10.5382/SP.10.06

Delaney JR, Robigou V, McDuff RE et al (1992) Geology of a vigorous hydrothermal system on the Endeavour segment, Juan de Fuca Ridge. J Geophys Res 97(19):663–682. https://doi.org/10.1029/92JB00174

Department of Reserves, Ministry of Land and Resources (2000) Compilation of calculation methods for mineral resources reserves (in Chinese). Geological Publishing House, Beijing

Dick HJB, Lin J, Schouten H (2003) An ultraslow-spreading class of ocean ridge. Nature 426(6965):405–412. https://doi.org/10.1038/nature02128

Dickson FW (1977) The role of rhyolite-seawater reaction in the genesis of Kuroko ore deposits in Newfoundland. J Geophys Res 76:3179–3206

Dickson P, Schultz A, Woods A (1995) Preliminary modeling of hydrothermal circulation within mid-ocean ridge sulphide structures. Hydrotherm Vents Process, Spec Pub, Geol Soc Lond 87 (1):145–157. https://doi.org/10.1144/GSL.SP.1995.087.01.13

Edisoma GS, Hearst MG (1996) Grade estimation and precision of mineral resources-Jackknife method (trans: Li X). Min Eng 11:14–20

Elderfield H, Schultz A (1996) Mid-ocean ridge hydrothermal fluxes and the chemical composition of the ocean. Annu Rev Earth Planet Sci 24(1):191–224. https://doi.org/10.1146/annurev.earth.24.1.191

Escartín J, Canales JP (2011) Detachments in oceanic lithosphere: Deformation, magmatism, fluid flow, and ecosystems. Eos Trans Am Geophys Union 92(4):31. https://doi.org/10.1029/2011EO040003

Escartín J, Smith DK, Cann J et al (2008) Central role of detachment faults in accretion of slow-spreading oceanic lithosphere. Nature 455:790–795. https://doi.org/10.1038/nature07333

Fouquet Y, Cambon P, Etoubleau J et al (2010) Geodiversity of hydrothermal processes along the Mid-Atlantic Ridge and ultramafic-hosted mineralization: A new type of oceanic Cu-Zn-Co-Au volcanogenic massive sulfide deposit. In: Rona PA, Devey CW, Dyment J et al (eds) Diversity of hydrothermal systems on slow spreading ocean ridges. Washington DC, USA, American Geophysical Union, pp 297–320. (AGU Geophysical Monograph, 188)

Fouquet Y, Knott R, Cambon P et al (1996) Formation of large sulfide mineral deposits along fast-spreading ridges. Example from off-axial deposits at 12°3'N on the East Pacific Rise. Earth Planet Sci Lett 144(1–2):147–162. https://doi.org/10.1016/0012-821X(96)00142-2

German CR, Lin J (2004) The thermal structure of the oceanic crust, ridge-spreading and hydrothermal circulation: How well do we understand their inter-connections? In: German C, Lin J, Parson L (eds) Mid-ocean ridges: hydrothermal interactions between the lithosphere and oceans. Washington DC, USA, American Geophysical Union, pp 1–18. (AGU Geophysical Monograph, 148)

German CR, Parson LM (1998) Distributions of hydrothermal activity along the Mid-Atlantic Ridge: interplay of magmatic and tectonic controls. Earth Planet Sci Lett 160(3):327–341. https://doi.org/10.1016/S0012-821X(98)00093-4

Giambalvo ER, Steefel CI, Fisher AT et al (2002) Effect of fluid-sediment reaction on hydrothermal fluxes of major elements, eastern flank of the Juan de Fuca Ridge. Geochim Cosmochim Acta 66 (10):1739–1757. https://doi.org/10.1016/S0016-7037(01)00878-X

Hajash A, Archer P (1980) Experimental seawater/basalt interactions: effects of cooling. Contrib Miner Petrol 75:1–13. https://doi.org/10.1007/BF00371884

Halbach P, Manheim FT, Otten P (1982) Co-rich ferromanganese deposits in the marginal seamount regions of the central Pacific basin–results of the Midpac '81. Erzmetall 35:447–453

Hannington MD, Jamieson J, Monecke T et al (2010) Modern seafloor massive sulfides and base metal resources: Towards an estimate of global seafloor massive sulfide potential. In: Goldfarb RJ, Marsh EE, Monecke T (eds) The Challenge of Finding New Mineral Resources: Global Metallogeny, Innovative Exploration, and New Discoveries. Soc Econ Geol, Incorporated, pp 317–338. https://doi.org/10.5382/SP.15.2.001

Hannington MD, de Ronde CD, Petersen S (2005) Sea-floor tectonics and submarine hydrothermal systems. In: Hedenquist JW, Thompson JFH, Goldfarb RJ et al (eds) One Hundredth Anniversary Volume. Soc Econ Geol, Incorporated, pp 111–141. https://doi.org/10.5382/AV100.06

Hannington MD, Galley AG, Herzig P et al (1998) Comparison of the TAG mound and stockwork complex with Cyprus-type massive sulfide deposits. In: Herzig PM, Humphris SE, Miller DJ et al (eds) Proceedings of the ocean drilling program, scientific results, Vol. 158, College Station, TX, pp 389-415. https://doi.org/10.2973/odp.proc.sr.158.217.1998

Hannington MD, Jamieson J, Monecke T et al (2011) The abundance of seafloor massive sulfide deposits. Geology 39(12):1155–1158. https://doi.org/10.1016/0198-0254(84)93436-8

He QH, Li AQ, Zou XF (2005) Investigation progress and exploitation technology of ocean cobalt -rich crusts (in Chinese with English abstracts). Metal Mine 347(5):4–7. https://doi.org/10.3321/j.issn:1001-1250.2005.05.002

Hein JR, Conrad TA, Dunham RE (2009) Seamount characteristics and mine-site model applied to exploration- and mining-lease-block selection for cobalt-rich ferromanganese crusts. Marine Geores Geotechnol 27(2):160–176. https://doi.org/10.1080/10641190902852485

Hein JR, Mizell K, Koschinsky A et al (2013) Deep-ocean mineral deposits as a source of critical metals for high-and green-technology applications: Comparison with land-based resources. Ore Geol Rev 51:1–14. https://doi.org/10.1016/j.oregeorev.2012.12.001

Hein JR (2004) Cobalt-rich ferromanganese crusts: global distribution, composition,. origin and research activities. In: International Seabed Authority (ed) Minerals other than polymetallic nodules of the international seabed area. International Seabed Authority, Kingston, p 188–256

Hein JR, Manheim FT, Schwab WC et al (1985) Ferromanganese crusts from Necker ridge, Horizon guyot and SP Lee guyot: geological considerations. Mar Geol 69(1–2):25–54. https://doi.org/10.1016/0025-3227(85)90132-X

Heirzig PM, Petersen S, Hannington MD (2000) Polymetallic massive sulfide deposits at the modern seafloor and their resource potential. In: International Seabed Authority Technical Study 2, Report on thè UN Workshop on Seafloor Mineral Resources. Kingston, June 2000

Herzig PM (1999) Economic potential of sea-floor massive sulphide deposits: ancient and modern. Philosophical Trans R Soc Lond A: Math, Phys Eng Sci 357(1753):861–875. https://doi.org/10.1098/rsta.1999.0355

Herzig PM, Hannington MD, Arribas JA (1998a) Sulfur isotopic composition of hydrothermal precipitates from the Lau back-arc: implications for magmatic contributions to seafloor hydrothermal systems. Miner Deposita 33(3):226–237. https://doi.org/10.1007/s001260050143

Herzig PM, Humphris SE, Miller DJ et al (1998) Proceedings of the ocean drilling program, scientific results, Vol. 158, College Station, TX

Hou JR, Huang JX (2001) Application of geostatistics in classification for resources/ reserves of solid fuels and mineral commodities (in Chinese with English abstracts). Geol Prospect 37(6):61–66. https://doi.org/10.3969/j.issn.0495-5331.2001.06.017

Hou Z (2003) Seafloor hydrothermal mineralization. Geological Publishing House, Beijing

Jin Q (1991) Newly-found solid mineral resources in the ocean-floor and structural aspects in the northwestern Pacific (in Chinese with English abstracts). Gresearch Ecol South China Sea 3:1–8

Kamenetsky VS, Binns RA, Gemmell JB et al (2001) Parental basaltic melts and fluids in eastern Manus backarc Basin: implications for hydrothermal mineralization. Earth Planet Sci Lett 184:685–702. https://doi.org/10.1016/S0012-821X(00)00352-6

Kamenetsky VS, Dacidson P, Mernagh TP et al (2002) Fluid bubbles in melt inclusions and pillow-rim glasses: high-temperature precursors to hydrothermal fluids? Chem Geol 183(1–4):349–364. https://doi.org/10.1016/S0009-2541(01)00383-7

Kang Z (1997) Calculation of mineral reserves by use of computer (in Chinese with English abstracts). Miner Res Geol 11(61):353–360

Kelley DS, Delaney JR, Yoerger DA (2001) Geology and venting characteristics of the Mothra hydrothermal field, Endeavour segment, Juan de Fuca Ridge. Geology 29(10):385–491. https://doi.org/10.1130/0091-7613(2001)029<0959:GAVCOT>2.0.CO;2

Koschinsky A, Garbe-Schönberg D, Sander S et al (2008) Hydrothermal venting at pressure-temperature conditions above the critical point of seawater, 5° S on the Mid-Atlantic Ridge. Geology 36 (8):615–618. https://doi.org/10.1130/G24726A.1

Lalou C, Reyss JL, Brichet E et al (1995) Hydrothermal activity on a 105-year scale at a slow-spreading ridge, TAG hydrothermal field, Mid-Atlantic Ridge 26° N. J Geophys Res: Solid Earth 100 (B9):17855–17862. https://doi.org/10.1029/95JB01858

Landergren S, Joensuu O (1965) Studies on trace element distribution in a sediment core from the Pacific Ocean. Prog Oceanogr 3:179–189. https://doi.org/10.1016/0079-6611(65)90017-0

Landergren S, Muld W, Rajandi B (1964) On the geochemistry of deep-sea sediments. Reports Swedish Deep-sea Exped 10(5):57–178

Li HM, Zhai SK, Tao CH et al (2010) Numerical simulation for fluids mixing within seafloor hydrothermal sulfide deposit: taking the Trans-Atlantic Geo-Traverse (TAG) field for example. Acta Oceanol Sin 29(1):33–41. https://doi.org/10.1007/s13131-010-0005-3

Li HM, Zhai SK, Tao CH et al (2011) A numerical study of hydrothermal circulation patterns within the large hydrothermal sulfide deposit (in Chinese with English abstracts). Acta Oceanol Sinica 33(4):111–120

Li HM, Zhai SK, Yu ZH et al (2009) Evolving patterns of the fluids within the TAG hydrothermal field. Sci China (Ser D): Earth Sci 52 (1):29. https://doi.org/10.1007/s11430-009-0008-0

Lin J, Purdy GM, Schouten H et al (1990) Evidence from gravity data for focused magmatic accretion along the Mid-Atlantic Ridge. Nature 95:4909–4923. https://doi.org/10.1038/344627a0

Lipton I (2008) Mineral resource estimate, Solwara 1 project, Bismarck Sea, Papua New Guinea: NI43–101. In: Technical Report for Nautilus Minerals Inc. Available via DIALOG. http://www.nautilusminerals.com/i/pdf/2008-02-01_solwara_43-101.pdf. Accessed 4 July 2019

Lowell RP (2010) Hydrothermal circulation at slow spreading ridges: analysis of heat sources and heat transfer processes. In: Rona PA, Devey CW, Dyment J et al (eds) Diversity Of hydrothermal systems on slow spreading ocean ridges. Washington DC, USA, American Geophysical Union, pp 11–26. (AGU Geophysical Monograph, 188)

Lowenstern JB, Mathood GA, Rivers ML et al (1991) Evidence for extreme partitioning of copper into a magmatic vapor phase. Science 252:1405–1409. https://doi.org/10.1126/science.252.5011.1405

Ludwig KA, Kelley DS, Butterfield DA et al (2006) Formation and evolution of carbonate chimneys at the Lost City Hydrothermal Field. Geochim Cosmochim Acta 70(14):3625–3645. https://doi.org/10.1016/j.gca.2006.04.016

Lv WZ, Huang YX, Zhang GZ et al (2008) Geology of the China pioneer area of polymetallic nodule deposit in the Pacific Ocean (in Chinese). China Ocean Press, Beijing

Ma WL, Jin XL, Chu FY et al (2013) Comparing research on the cobalt-rich crusts mineralization and distribution of conical and flat-topped seamounts from the Central Pacific Seamounts (in Chinese with English abstracts). Acta Oceanol Sinica 35(2):90–112. https://doi.org/10.3969/j.issn.0253-4193.2013.02.011

Manheim FT (1986) Mar Cobalt Res. Science 232:600–608. https://doi.org/10.1126/science.232.4750.600

McCaig AM, Cliff RA, Escartin J et al (2007) Oceanic detachment faults focus very large volumes of black smoker fluids. Geology 35 (10):935–938. https://doi.org/10.1130/G23657A.1

McCaig AM, Harris M (2010) Hydrothermal circulation and the dike-gabbro transition in the detachment mode of slow seafloor spreading. Geology 40(4):367–370. https://doi.org/10.1130/G32789.1

Mero JL (1962) Ocean-floor manganese nodules. Econ Geol 57 (5):747–767. https://doi.org/10.2113/gsecongeo.57.5.747

Mottl MJ, Seyfried WE (1977) Experimental basalt/seawater interaction: rock- vs seawater-dominated systems and the origin of submarine hydrothermal deposits. Geol Soc Abstracts Program:1104

Murray J, Renard AF (1891) Deep-sea deposits. Rep Scientific Results HMS Challeng, pp 1873–76

Pascoe A, Cann JR (1995) Modelling diffuse hydrothermal flow in black smoker vent fields. Hydrotherm Vents Process, Spec Pub, Geol Soc Lond 87(1):159–173. https://doi.org/10.1144/GSL.SP.1995.087.01.14

Petersen S, Krätschell A, Augustin N et al (2016) News from the seabed-Geological characteristics and resource potential of deep-sea mineral resources. Mar Policy 70:175–187. https://doi.org/10.1016/j.marpol.2016.03.012

Piper DZ, Williamson ME (1977) Composition of Pacific Ocean ferromanganese nodules. Mar Geol 23(4):285–303. https://doi.org/10.1016/0025-3227(77)90036-6

Rona PA (2003) Resources of the sea floor. Science 299:673–674. https://doi.org/10.1126/science.1080679

Saccocia PJ, Gillis KM (1995) Hydrothermal upflow zones in the oceanic crust. Earth Planet Sci Let 136(1–2):1–16. https://doi.org/10.1016/0012-821X(95)00155-5

Schardt C, Large RR (2009) New insights into the genesis of volcanic-hosted massive sulfide deposits on the seafloor from numerical modeling studies. Ore Geol Rev 35(3):333–351. https://doi.org/10.1016/j.oregeorev.2008.11.008

Schkoriniko et al (2001) The occurrence conditions, distribution laws and formation features of cobalt-rich ferromanganese on guy-ots from the Magellan Seamounts. Тихоокеанска 20(2):76–86

Schroeder T, John BE (2004) Strain localization on an oceanic detachment fault system, Atlantis Massif, 30° N. Mid-Atlantic Ridge. Geochem Geophy Geosy 5:Q11007. https://doi.org/10.1029/2004GC000728

Scott RB, Rona PA, Mcgregor BA et al (1974) The TAG Hydrothermal Field. Nature 251:301–330. https://doi.org/10.1038/251301a0

Seyfried WE, Bischoff JL (1977) Hydrothermal transport of heavy metals by seawater: the role of seawater/basalt ratio. Earth Planet Sci Let 34:71–77. https://doi.org/10.1016/0012-821X(77)90107-8

Seyfried WE, Dibble WE (1980) Seawater-peridotite interaction at 300 °C and 500 bars: implications for the origin of oceanic serpentinites. Geochim Cosmochim Acta 44:309–321. https://doi.org/10.1016/0016-7037(80)90139-8

Shen YJ, Zhong X, He ZQ et al (1999) Present status of investigation and development of ocean cobalt crust resources (in Chinese with English abstracts). Min Metallurgical Eng 19(2):11–13

Shi R (1995) Analytic triangular wedge method for calculation of reserves (in Chinese with English abstracts). Technol Chem Mines 24(2):29–33

Smith DK, Escartin J, Schouten H et al (2008) Fault rotation and core complex formation- Significant processes in seafloor formation at slow-spreading mid-ocean ridges (Mid-Atlantic Ridge, 13–15° N). Geol Geophys Geosyst 9:Q03003. https://doi.org/10.1029/2007GC001699

Song GY, Xiao KY, Zhu YS (1999) The mineral resources evaluation in USA (in Chinese with English abstracts). Comput Techniq Geophys Geochem Explor 21(4):301–306. https://doi.org/10.3969/j.issn.1001-1749.1999.04.003

Standish JJ, Dick HJB, Michael PJ et al (2008) MORB generation beneath the ultraslow spreading Southwest Indian Ridge (9–25° E): Major element chemistry and the importance of process versus source. Geochem Geophys Geosyst 9(5):Q05004. https://doi.org/10.1029/2008GC001959

Stein CA, Stein S (1994) Constrains on hydrothermal heat flux through the oceanic lithosphere from the global heat flow. J Geophys Res 99:3081–3196. https://doi.org/10.1029/93JB02222

Teagle DAH, Alt JC (2004) Hydrothermal alteration of basalts beneath the Bent Hill massive sulfide deposit, Middle Valley. Juan de Fuca Ridge. Economic Geol 99(3):561–584. https://doi.org/10.2113/99.3.561

Tivey MA, Johnson HP, Bradley A et al (1998) Thickness of a submarine lava flow determined from near-bottom magnetic field mapping by autonomous underwater vehicle. Geophys Res Lett 25 (6):805–808. https://doi.org/10.1029/98GL00442

Tivey MK, Humphris SE, Thompson G et al (1995) Deducing patterns of fluid flow and mixing within the TAG active hydrothermal mound using mineralogical and geochemical data. J Geophys Res Solid Earth 100(B7):12527–12555. https://doi.org/10.1029/95JB00610

Tucholke BE, Behn MD, Buck WR et al (2008) Role of melt supply in oceanic detachment faulting and formation of megamullions. Geology 36(6):455–458. https://doi.org/10.1130/G24639A.1

Von Damm KL, Buttermore LG, Oosting SE et al (1997) Direct observation of the evolution of a seafloor "black smoker" from vapor to brine. Earth Planet Sci Lett 149:101–112. https://doi.org/10.1016/S0012-821X(97)00059-9

Von Damm KL, Oosting SE, Kozlowski LG et al (1995) Evolution of East Pacific Rise hydrothermal vent fluids following a volcanic eruption. Nature 375:47–50. https://doi.org/10.1038/375047a0

Wang QM, Fang YP (2001) GIS in investigation and evaluation of mineral resources (in Chinese with English abstracts). Chinese Geology 28(4):39–44. https://doi.org/10.3969/j.issn.1000-3657.2001.04.007

Wang ZM, Hou JR (1994) Cokriging and its Application in Estimation of Mineral Reserves (in Chinese with English abstracts). Geol Explor 30(3):39–48

Wu GH, Zhou HY, Yang SF (2000) Combined application of the methods of the nearest domain and geological block to resource evaluation of Cobalt-rich crust on a seamount in the Pacific (in Chinese with English abstracts). Mar Geol Quatern Geol 20(4):87–92

Xin HX, Ju TZ, Lin JY (1997) The ordinary kriging's application to mineral calculation of reserves (in Chinese with English abstracts). Geol Explor 33(4):39–48

Xu DY (2002) Research development of Russian oceanic sold mineral resource (in Chinese with English abstracts). Mar Geol Let 18 (10):21–22

Yamazaki T, Sharma R (2000) Morphological features of Co-rich manganese deposits and their relation to seabed slopes. Mar Georesour Geotechnol 18(1):43–76. https://doi.org/10.1080/10641190009353782

Yamazaki T, Sharma R, Tsurusaki K (1995) Influence of distribution characteristics and associated seabed features on exploration of cobalt-rich manganese deposits. Paper presented at the First ISOPE Ocean Mining Symposium, Tsukuba, Japan, 21–22 November 1995

Yamazaki T, Sharma R, Tsurusaki K (1994) Microtopographic analysis of cobalt-rich manganese deposits on a mid-Pacific Seamount. Mar Georesour Geotechnol 12(1):33–52. https://doi.org/10.1080/10641199409388252

Yang KH, Scott SD (1996) Possible contribution of a metal-rich magmatic fluid to a sea-floor hydrothermal system. Nature 383:420–423. https://doi.org/10.1038/383420a0

Zeng ZG (2011) Submarine hydrothermal geology (in Chinese). Science Press, Beijing

Zeng ZG, Li J, Jiang FQ et al (2002) Sulfur isotopic composition of modern seafloor hydrothermal sediment and its geological significance. Acta Oceanol Sin 21(4):519–528

Zhai SK, Li HM, Yu ZH et al (2007) Advances in the Investigation Technology of Modern Seafloor Hydrothermal Activities (in Chinese with English abstracts). Adv Earth Sci 22:769–776. https://doi.org/10.3321/j.issn:1001-8166.2007.08.001

Zhang FY (1995) Calculation of resource evaluation method of polymetallic nodules (in Chinese with English abstracts). Acta Oceanol Sin 17(4):85–93

Zhang FY (2001) Theory and method of polymetallic nodule resources evaluation and delineation of ore regions (in Chinese). China Ocean Press, Beijing

Zhang ZF, Gao FY, Ma ZM et al (2001a) A mineral resource prediction method using in areas with only one model cell: GIS-supported eigenfaces (in Chinese with English abstracts). Geol Explor 37 (6):51–54. https://doi.org/10.3969/j.issn.0495-5331.2001.06.013

Zhang J, You L, Qin J (1997) Method of mineral resources evaluation by function of GIS spatial analysis (in Chinese with English abstracts). Comput Techniq Geophys Geochem Explor 19(3):283–286

Zhang WY, Zhang FY, He GW et al (2001b) Two-dimensional geostatistics model of polymetallic nodule deposit (in Chinese with English abstracts). Mar Sci Bull 20(2):66–77

Zhang WY, Zhang FY, Hu GW et al (2008a) Relationship between shape classification of Pacific seamount morphology and distribution of cobalt-rich crust resources (in Chinese with English abstracts). Acta Oceanol Sin 30(6):76–84. https://doi.org/10.3321/j.issn:0253-4193.2008.06.009

Zhang WY, Zhang FY, Yang KH et al (2006) Fractal research on quantity of cobalt-rich crust resources distributed on seamount topography in the pacific (in Chinese with English abstracts). Acta Sedimentol Sin 24(5):705–713. https://doi.org/10.3969/j.issn.1000-0550.2006.05.012

Zhang FY, Zhang WY, Zhu KC et al (2008b) Distribution characteristics of cobalt-rich ferromanganese crust resources on submarine seamounts in the western pacific. Acta Geol Sin 82(4):796–803. https://doi.org/10.3321/j.issn:1000-9515.2008.04.009

Zhang FY, Zhang WY, Zhu KC (2011a) Evaluation of cobalt crust resources in pacific seamounts (in Chinese). China Ocean Press, Beijing

Zhang WY, Zhang FY, Zhu KC et al (2011b) Fractal research on seamount topography in the pacific ocean and its geological significance (in Chinese). Paper presented at the International conference on Fuzzy System and Neural Computing (FSNC 2011), Hong Kong, 20–21 February 2011

Zhou HY, Dick HJB (2013) Thin crust as evidence for depleted mantle supporting the Marion rise. Nature 494:195–201. https://doi.org/10.1038/nature11842

Zierenberg RA, Fouquet Y, Miller DJ et al (1998) The deep structure of a sea-floor hydrothermal deposit. Nature 392:485–488. https://doi.org/10.1038/33126